Origins and Destinies

IMMIGRATION, RACE, AND ETHNICITY IN AMERICA

Origins and Destinies
IMMIGRATION, RACE, AND ETHNICITY IN AMERICA

SILVIA PEDRAZA
University of Michigan

RUBÉN G. RUMBAUT
Michigan State University

Wadsworth Publishing Company
I(T)P® An International Thomson Publishing Company
Belmont • Albany • Bonn • Boston • Cincinnati • Detroit •
London • Madrid • Melbourne • Mexico City • New York • Paris •
San Francisco • Singapore • Tokyo • Toronto • Washington

Editor: Eve Howard
Assistant Editor: Susan Shook
Editorial Assistants: Julie McDonald, Carrie Kahn
Production: Professional Book Center
Print Buyer: Karen Hunt

Permissions Editor: Jeanne Bosschart
Cover Design: Vargas/Williams/Design
Composition: Professional Book Center
Printer: Malloy Lithographing, Inc.

For more information, contact Wadsworth Publishing Company:

Wadsworth Publishing Company
10 Davis Drive
Belmont, California 94002, USA

International Thomson Publishing Europe
Berkshire House 168–173
High Holborn
London, WC1V 7AA, England

Thomas Nelson Australia
102 Dodds Street
South Melbourne 3205
Victoria, Australia

Nelson Canada
1120 Birchmount Road
Scarborough, Ontario
Canada M1K 5G4

International Thomson Editores
Campos Eliseos 385, Piso 7
Col. Polanco
11560 México D.F. México

International Thomson Publishing GmbH
Königswinterer Strasse 418
53227 Bonn, Germany

International Thomson Publishing Asia
221 Henderson Road
#05–10 Henderson Building
Singapore 0315

International Thomson Publishing Japan
Hirakawacho Kyowa Building, 3F
2–2–1 Hirakawacho
Chiyoda-ku, Tokyo 102, Japan

Library of Congress Cataloging-in-Publication Data is available for this title.
ISBN 0-534-21444-4

CREDITS: Cover art: "Land of Bliss and Wonder, California," Hans Hofmann, 1960, oil on canvas. Courtesy of Jan and Frederick Mayer, Denver, Colorado. Hofmann was born in Germany in 1880 and immigrated to the United States in 1932; p. 60 from George P. Rawick, *From Sundown to Sunup: The Making of the Black Community.* Reprinted with permission of Greenwood Publishing Group, Inc., Westport, CT. Copyright © 1973; p. 107 from "*La Carreta Made a U-Turn:* Puerto Rican Language and Culture in the United States," *Daedalus,* Spring 1981, 110:2. Reprinted by permission; p. 450 from the *Dear Abby* column by Abigail Van Buren. © Universal Press Syndicate. Reprinted with permission. All rights reserved. Several chapters were adapted from previously published articles: Ch. 2 from R. Rumbaut, "Origins and Destinies: Immigration to the United States Since World War II," *Sociological Forum, 9*(4), Dec. 1994, pp. 583–621. With permission from Plenum Publishing. Ch. 24 from R. Rumbaut, "Vietnamese, Laotian, and Cambodian Americans." In P. G. Min (ed.), *Asian Americans: Contemporary Trends and Issues.* Thousand Oaks, Sage. With permission of Sage Publications, Inc. Ch. 32 from A. Portes and R. Schauffler, "Language and the Second Generation: Bilingualism Yesterday and Today," *International Migration Review, 28*(4), Winter 1994, pp. 640–661. With permission of the Center for Migration Studies.

 This book is printed on acid-free recycled paper.

To all my family—both those who immigrated to America and those who remained in Cuba. And to my students—past, present, and future.
Silvia Pedraza

For Rubén Darío, father and son, origin and destiny.
Rubén G. Rumbaut

Acknowledgments

This book owes a great deal to many people who took its promise to heart and helped it come to fruition. Our largest thanks are for all the people at Wadsworth Publishing Company in California, who, despite some difficult moments along the way, stood by us and helped the book succeed. Gary Carlson, editor in chief; Serena Beauparlant, our original sociology editor; Eve Howard, our current sociology managing editor; Julie McDonald, editorial assistant; Susan Shook, assistant editor; and Hal Humphrey, production manager, all did their best to help us preserve the intellectual and professional integrity of this complex undertaking. We also thank Jennifer Ballentine and the staff of Professional Book Center in Colorado, who were responsible for many aspects of the production process.

Our heartfelt thanks are also due to many others in our two universities who helped our endeavor. At the University of Michigan, Lisa Neidert, research associate at the Population Studies Center, helped us obtain the U.S. 1990 census data samples of the many racial and ethnic groups in American society that were mined to illustrate many of the essays. In addition, Sabrina Williams, Sandra Stevens, Barbara Polcyn, and Patricia Preston expertly typed various drafts of the many essays. Cassandra Munguia, Silvia's student in the Summer Research Opportunity Program, gladly assisted various aspects of the effort, as did Christine Reinhard, a student in the Undergraduate Research Opportunity Program. At Michigan State University, Melissa Gilroy, Debi Hogle, and Irene Unkefer were extraordinarily helpful throughout.

The photographs that illustrate this book were chosen as the result of an extensive research effort that involved looking at hundreds of photographs in various photo archives all over the country. Many fine people, as enthusiastic about photographs-as-social-documents as she, assisted Silvia Pedraza in the research. At the Library of Congress, Barbara Natanson was particularly helpful, assisting in the selection of photographs that portrayed the history of immigration, race, and ethnicity in America, and solving problems along the way; Eveline Overmiller and Michael Cooper also deserve thanks for the production of the photographs. At *The Miami Herald,* Lisette Nabut provided access to their rich collection of photographs on the contemporary Cuban and Haitian immigration. At the Historical Museum of Southern Florida, Rebecca Smith made the collection of photographs on the old Cuban immigration accessible, and Dawn Hugh was helpful in their production. At the National Parks Service, Terry Adams and Rosa Wilson were also most helpful, providing access to their collection of photographs on Washington, D.C. and its monuments, America's symbols. At AP/Wide World Photos, Joan Carroll helped choose a photo of the 1992 Los Angeles riots. Billy Cypress, of the Seminole reservation in Florida, helped identify the historical moment depicted in another photograph with precision. At the Filipino American National Historical Society, Dorothy Cordova also helped in the selection process. At the Walter P. Reuther Library of Wayne State University, Thomas Featherstone assisted in the selection of photographs of Cesar Chavez and the Mexican farmworkers, a labor movement. At the Bentley Historical Library of the University of Michigan, Nancy Barlett and Anne Frantilla helped with the selection of photographs on the immigration of southern and eastern European immigrants at the turn of the century, as well as from papers donated by the Allmendinger family of Ann Arbor, whose descendant, Harriett Birch, also participated in the selection process. A very special thanks goes to Michael Villareal, Silvia's former student in the Summer Research Opportunity Program at the University of Michigan. After graduating from college, he also assisted in this research effort at the Library of Congress while working as an intern for the Federal Reserve Board in Washington, D.C., for neither wages nor a grade, but to help with the book.

Several of the contributors to this book helped to illustrate it by contributing their own photographs. We thank Russell Thornton, Carole Marks, Ricardo Romo, Héctor Carrasquillo, Walter Kamphoefner, Hasia Diner, Dominic Pacyga, Leo Chavez, Robert Manning, Matthew Snipp, David Williams, and Charlotte Steeh for providing us with many from which to choose. No one, however, deserves as large a thanks as Steve Gold, who is not only a fine sociologist, but also a fine photographer. Over and over again, he made his fine collection of photographs available to us. The effort to illustrate this book would hardly be the same without his contributions.

We owe our deepest debt to our colleagues, the authors of the three-dozen essays this volume comprises, whose diligence and steadfast commitment to the project were demonstrated at every turn. Scholars in full command of their subjects, each of them made creative efforts to write engagingly for a general public, combining fresh information with broad synthesis and deep insight to bring to life some central themes of the American experience. We also thank the following external reviewers for their incisive and thoughtful comments: Rodney Coates, Miami University; Howard Ehrlich, Towson University, Prejudice Institute; John Lie, University of Illinois, Urbana; Maura Toro Morn, Illinois State University; and Morrison Wong, Texas Christian University.

Finally, we are particularly grateful to everyone at Wadsworth for undertaking a new venture in supporting our efforts to deliver the finely textured messages of the book, first as a collection of original essays by social scientists—based on their established research record—and, second, through visual images—in the best tradition of American historical and social photography.

Contents

Preface

SILVIA PEDRAZA

In many ways, I began working on *Origins and Destinies* when I first began teaching. After completing my Ph.D. at the University of Chicago, in the early eighties, I began to teach at Washington University in St. Louis. Like most beginners, I focused my first courses on what the textbooks usually called race and ethnic relations, with texts that usually portrayed the experience of only a few groups of color in American society, such as Black Americans, Mexican Americans, Native Americans, Chinese Americans. My commitment to issues that pertain to race and ethnicity is woven through every aspect of my career—teaching, research, and administration, both in my own university and in my discipline. Nonetheless, if a good teacher is one that leads one to oneself, and one that is capable of articulating our experience as human beings within an analytical frame of questions and issues that can help us to make sense of it, such an approach, particularly at that time, allowed me to communicate with remarkably few students.

That in itself came as a surprise to me. As a graduate student at the University of Chicago, like most of my fellow students, I had begun to acquire teaching experience by teaching in other city schools, such as Roosevelt College, Loop Junior College, and Elmhurst College Hispanic Extension Program downtown. In those settings, a large part of my student body had been working class, minority, ethnic, immigrant—all of which had made my teaching experience into a form of field work from which I learned a great deal about my very subject. Real difficulties accompanied that teaching experience—the schools' facilities were often poor, the students' prior academic preparation was not always adequate, adults who worked all day were often tired during evening classes, leaving little time for homework, and so on. But never did I have to persuade anyone of the importance of the subject matter of immigration, race, and ethnicity in America because their daily lives had already taught it to them.

By contrast, when I began teaching at Washington University, and afterwards at the University of Michigan, I encountered a student body whose prior academic preparation was solid, in schools with excellent resources for teaching and learning, and who could be counted upon to study everyday. But my students now lacked social experience because their middle- and upper middle-class backgrounds had given them plenty of fine opportunities but failed to expose them to many aspects of the world in which we live.

Thus, I began to search for a way to include both my students and myself in something that had the qualities of critical engagement, both personal and intellectual, that I associate with good teaching and learning. In those early years of teaching, I made several decisions, which today are reflected in the organization of *Origins and Destinies: Immigration, Race, and Ethnicity in America.*

The first decision was never again to teach from a textbook because textbooks are not often based on the contentious struggle between ideas and facts that we call research. In addition, while supposedly covering all the major perspectives, textbooks have a tendency to homogenize issues, which also dismayed me. To my mind, problems are best understood when we research them from a variety of methodological and ideological perspectives; hence, multiple points of view need to be preserved. All of the contributors to *Origins and Destinies* were invited to join us on the basis of the impressive research record they had already established, and were asked to write a new piece based on their research that would present the material in a manner accessible to our students and the general public, while preserving their own perspectives and points of view. *Origins and Destinies* is a research-based collective effort whose intent is pedagogical.

The second decision was to find an analytical framework that would be broadly inclusive, so that I would not be teaching only to and about the minority students in my midst, glad as we both were to be in a course that incorporated their own experience in what was otherwise a remarkably Eurocentric curriculum. That larger analytical umbrella that could cover all of us—Americans old and new—developed, progressively, over years of teaching and research, first at Washington University in St. Louis, and later at the University of Michigan. I have made an effort to distill it in the introduction and conclusion to *Origins and Destinies.* The notion that every American is an immigrant (even, in some ways, the Native Americans)—people whose origins are various but whose destinies became American—is what opens the analytical umbrella that remains open over the course of four major waves of American history.

More inclusive, that framework also encompasses the Americans of color, the issue of race. In fact, it was at the meeting of the American Sociological Association, a few years ago, in Atlanta that Rubén G. Rumbaut and I decided to work on this book together, as an expression of our friendship. We also walked together, like on a pilgrimage, to visit the area a few blocks away where Martin Luther King, Jr., lived most of his life and preached from the pulpit of his church underneath a stained-glass image of Christ alone and suffering, wanting guidance, as King must often have been. While there I remembered that, walking along in a protest march, Martin Luther King, Jr., had once explained to a reporter that America was a closed society and that the goal of the Civil Rights Movement was to make it open. Thanks to the movement itself, our society today is far changed, though it still remains necessary to strive to open it to all. *Origins and Destinies* is about both immigration and race in American society.

An important part of the large analytical umbrella that I progressively developed over years of teaching and research is the research paper that students write on the history of their families (gathered through interviewing family members) and the history of their own immigrant or ethnic group (gathered through library research). A copy of this assignment is included in the appendix, as over the years I developed a set of questions as guidelines for the students to pay attention to, both in the interviews and the library research. As soon as I began using this assignment, the class changed in a number of ways. Above all, the quality of the research papers suddenly improved remarkably—some even became beautiful as they were driven by a deep-seated engagement that sought to make sense of the world, that is always, first of all, *our* world and thereafter, others' world. In addition, suddenly all experiences were represented, all were valid, all were worth examining. A classroom experience that until then had always been exclusive (of some) became inclusive (of all).

Even more, I discovered, as Oscar Hammerstein once put it in *The King and I,* that "by your pupils you are taught." As I listened to my students who wanted to catch me after class to share a discovery, as I read paper after paper that brought to life the cinnamon peelers of Sri Lanka coming to America, the Germans celebrating in the communal beer halls in St. Louis, the Black sharecroppers of Mississippi in Chicago hoping for a better future, the young Jewish men singing as cantors on the boat over from Europe, I myself was enthralled, gazing at the fully open fan of the history of immigration, race, and ethnicity in America. And once again my teaching became a form of field work. That is to say, that the social world in which we live—perplexing as it is—once again walked into my classroom. Everything that I learned from my students was plowed back into the class—and eventually, into *Origins and Destinies.*

Origins and Destinies brings many of the best of our colleagues—sociologists, historians, anthropologists—together, colleagues whose research in this interdisciplinary area has already left a mark in their disciplines and who have now directed their writing not only to the profession but also to those who should also be informed by the best that our discipline can give: our students, the larger public, policy makers.

In "The Origins of Sociology as a Discipline," Arthur Stinchcombe (1984, p. 60) pointed out that "As a discipline, sociology as a social structure tends toward scholasticism" because its internal prestige system leads us away from studying the real people involved in a social system (living, sweating, and crying) to study, instead, abstractions of people (theory and methods). Thus, while theory and methods should serve to understand and explain human behavior, they instead tend to become closed off to behavior by reaching for ever higher levels of abstraction.

Stinchcombe further argued, however, that while this basic dynamic of the discipline leads us away from the human reality "in the direction of Talmudic elaboration of theory and method " (p. 57), a number of sources of commitment produce an impulse in the opposite direction of empirical engagement, of commitment to the human reality of living, sweating, and crying human beings we always try to abstract. Of the five sources of empirical commitment that Stinchcombe highlighted (applied interests, general intellectualism, reflections on one's own life, sectarian commitments, and teaching), two seem particularly germane to *Origins and Destinies:* teaching and reflections on one's own life.

I have already detailed the ways in which my students' interests intruded into the making of *Origins and Destinies.* I know it was also true for all our contributors, because part of what students do is to force all their teachers to think about their students' experiences—in this case, as immigrants, ethnics, and minorities in America. As our student body grows increasingly more diverse, so will the faculty of the future, so will our discipline. Moreover, one only needs to look at the list of contributors to *Origins and Destinies,* to also realize how often our very lives, what shaped our own experiences, sustain our interest in the reality at hand and are a source of commitment to understanding this empirical area. Like Stinchcombe, I also believe that, ultimately, our discipline is the better for it—more scholarly, less scholastic.

Prologue

RUBÉN G. RUMBAUT

At the entrance to the National Archives in Washington, D.C., there is a massive rock bearing this inscription: "What's Past is Prologue." The words, from Shakespeare's play *The Tempest,* may serve for openers to goad if not guide reflection on the formation of ethnic America and the myriad ways in which the unequal destinies of American racial and ethnic groups reflect their diverse origins. The development of social and economic inequalities based on race and ethnicity has been a central theme—and a central dilemma—of the history of the United States, shaped over many generations by the European conquest of indigenous peoples and by massive waves of both coerced and uncoerced immigration from all over the world. Indeed immigration and conquest—by hook or by crook—have been the originating processes by which American ethnic groups have been formed and through which, over time, the United States itself has been transformed into aguably the world's most ethnically diverse society, a global sponge remarkable in its capacity to absorb tens of millions of people from all over the world. The national self-image forged by that history reflects the experience of a country that has time and again been revitalized and renewed by immigration. But our chimerical conceptions of "race"—pigments, as always, of the imagination—also derive from those fateful encounters, those social relations formed between strangers: physical and cultural differences came to be associated with steep gradients of privilege and power, and became hardened into invidious, indelible, outward markers of social status and identity in caste and class hierarchies. It took a bloody civil war in one century, and a civil rights revolution in the next, to end slavery and the legal underpinnings of racial exclusion, but not their bitter legacy (cf. Staples, 1993; Woodward, 1974; Massey and Denton, 1993). As a nation, we reap what we sow. The past is prologue.

The dynamics of immigration, racialization, and ethnic stratification are not just the stuff of bygone days; they continue today, full of irony and surprise. After a lull of a few decades—during which the flow of immigrants to America went from its historical zenith to its nadir in the wake of national-origins quota laws passed in the 1920s, the Great Depression, and World War II—another massive "new" immigration to the United States, accelerating rapidly since the 1960s, has been transforming anew the American mosaic. Not since the peak years of the pre–World

War I era have so many millions of strangers sought to make their way in America, and never before has the United States received such diverse groups. To be sure, while the old may be a prologue to the new, the United States in the vertiginously changing world of the late twentieth-century is not the same society that processed the "huddled masses" through Castle Garden and Ellis Island; the structure of its economy and the role of government in public life are far different than was the case in those days; and the vast majority of today's immigrants and refugees hail not from Europe but largely from the developing countries of Asia and Latin America.

In this context ethnicity itself is being redefined, often in surprising ways. There are foreign-born mayors of large American cities, first-generation millionaires who speak broken English, newcomers who include world-class professionals and preliterate peasants, a proliferation of sweatshops exploiting immigrant labor in an expanding informal economy, and new myths that purport to "explain" the success or failure of different ethnic groups. Once again, questions are raised about the assimilability of the newcomers—because of their race, language, culture, or alleged unwillingness to speak English. One-size-fits-all panethnic labels—"Asian," "Hispanic," "Black," "White"—are imposed by the society at large to pigeonhole people who hail from the Philippines, Vietnam, South Korea, India, Cambodia, China, Mexico, Cuba, El Salvador, the Dominican Republic, Haiti, Jamaica, Nigeria, Egypt, Irag, Iran, Israel, and scores of other nations, and who differ widely in class origins, phenotypes, languages, cultures, generations, migration histories, and modes of incorporation in the United States (on the politics, the absurdities, and the debate over racial classifications in the U.S. 2000 census, see Wright, 1994; cf. Hollinger, 1995). At the same time, intermarriage and other interminglings are producing creative and extraordinary hybridities—perhaps most notably among immigrants who came as children and have come of age in the United States (see Pérez-Firmat, 1994, for one superb rendering of such "lives on the hyphen").

Yet for all the fascinating, bewildering newness of the current diversity, there is a faintly familiar ring (echoes of an old script, if with a new cast of characters) to news that the enchilada is vying with the hamburger in every fastfood outlet; that a Korean restaurant in Los Angeles serves Kosher burritos in a mostly Black neighborhood; that in Houston the consecration as bishop of a Cuban-born Jesuit who spoke fluent Vietnamese was attended by an overflow of 6,000 faithful, whom he addressed in three languages; or that, as *The New York Times* reported not long ago, "At the annual Lower East Side Jewish Festival yesterday, a Chinese woman ate a pizza slice in front of Ty Thuan Duc's Vietnamese grocery store. Beside her a Spanish-speaking family patronized a cart with two signs: 'Italian Ices' and 'Kosher by Rabbi Alper.' And after the pastrami ran out, everybody ate knishes" (Rumbaut, 1991a).

This book is a collection of new interpretive essays, written by 46 scholars, which aims to connect these themes of immigration, race, and ethnicity in America in historical and contemporary contexts. The contributions reflect the different disciplines, approaches, methods, and perspectives—indeed, as their biographies suggest, the different *origins*—of their authors, who themselves form a microcosm of the larger society that they seek to comprehend. Clearly, the vastness and complexity of the topic cannot begin to be exhausted or even grasped in a single volume, nor is it our aim to provide an encyclopedic treatment of every identifiable ethnic group in the United States. The portrait of ethnic America that emerges from this collection of essays is necessarily selective, but we hope it will convey, with fresh insight and both breadth and depth of scope, something of the vast human drama that it encompasses, as well as of the paradoxical and permanently unfinished character of American ethnicity.

The book, which begins with two introductory essays on the historical and contemporary aspects of the topic, is organized in five main sections. The first two sections, "Color and Caste" and "Pre–World War I Waves of European Migration," focus on the historical origins of major American racial-ethnic groups through the early twentieth century. The middle section, "The Watershed," focuses on the politics of protest in the 1950s and 1960s, particularly the modern Civil Rights Movement and the Chicano and American Indian Movements, and their aftermath. The last two sections, "Contemporary Waves of Latin American and Asian Migration," and "Contemporary Issues" of color, class, consciousness and community, move the analytical focus to recent developments and current social, cultural, and economic patterns into the 1990s.

The essays in the first parts of the book go back to 1492 and the calamitous "demography of contact" between Europeans and the indigenous peoples of North America; examine the institution of slavery, the making of the Black community, and the Great Black Migration from the rural South to the urban North during World War I; and analyze the original modes of incorporation of Mexican, Puerto Rican, Chinese, and Japanese communities, the latter focusing specifically on the experience of immigrant women. These are followed by a series of contrasting essays on the transatlantic crossings and patterns of social mobility of "old" and "new" European immigrants. A detailed discussion of the rise and changing fortunes of a White Anglo-Saxon Protestant Establishment precedes essays on nineteenth-century German and Irish immigration and early twentieth-century Italian, Jewish, and Polish newcomers.

The essays in the last parts of the book examine major types of contemporary immigration: undocumented immigrants from Mexico and Central America, Cold War refugees from Cuba, Vietnam, Laos, and Cambodia, and the socioeconomic origins and adaptations of Dominicans, Filipinos, and Koreans. Four essays on "Urban Destinies" then shift the focus from intragroup profiles to intergroup relations and the transformative impact of the new immigration on major metropolitan areas: New York, Los Angeles, Miami, and Washington, D.C. Other essays follow on "Color and Class," ranging from analyses of the present socioeconomic status of American Indian groups; poverty, race, and health; and the effects of race on wages and employment opportunities in Chicago's inner city. Finally, the issue of bilingualism versus linguistic assimilation among the new second generation of children of immigrants is considered systematically, as are the questions of optional ethnic identities, changes in American racial attitudes as tracked in national polls since World War II, and the contested meaning of race itself in the post–Civil Rights Movement era. A concluding statement seeks to highlight some enduring contradictions of immigration, race, and ethnicity in America.

In the context of today's debates about multiculturalism and the "disuniting of America," the contested meaning of race, the rise of ethnic consciousness, and the politics of identity, it might help to gain some distance from the objects of contention and listen for a moment to a different voice, less ethnocentric,

more cosmopolitan. In *The Buried Mirror* (1992), his quincentennial reflections on Spain and the New World, the Mexican writer Carlos Fuentes, himself a progeny of that original encounter between the Old World and the New, put the matter this way:

> History begs the question, How to live with the Other? How to understand that I am what I am only because another person sees me and completes me? This question, which arises every time that white and black, East and West, predecessor and immigrant, meet in our times . . . became the central question of conquest and colonization in the Americas. (p. 89)

Today he sees "the universal question of the coming century" posed most forcefully in California, especially in Los Angeles, the world's premier immigrant metropolis and a gateway to both Asia and Latin America: "How do we deal with the Other?" He seeks his answer in his hybrid origins: "We [Hispanics] are Indian, Black, European, but above all mixed, *mestizo*. We are Iberian and Greek, Roman and Jewish, Arab, Gothic, and Gypsy." Thus, for instance, after nearly 800 years of Arab rule in Spain, lasting until the triumph of the *reconquista* with the fall of the last Moorish kingdom in 1492, the Arab cultural influence was pervasive, so that today fully a quarter of all Spanish words are of Arab origin (even in the bullfight, the *olé* salute to the matador comes from an Arab word) (p. 52). For Fuentes the answer lies in forging

> centers of incorporation, not of exclusion. When we exclude, we betray ourselves. When we include, we find ourselves. . . . People and their cultures perish in isolation, but they are born or reborn in contact with other men and women, with men and women of another culture, another creed, another race. If we do not recognize our humanity in others, we shall not recognize it in ourselves. Often we have failed to meet this challenge. But we have finally seen ourselves whole in the unburied mirror of identity, only when accompanied—ourselves with others. (pp. 348, 353)

So long as racial and economic inequalities remain deeply entrenched, however, such an abstract cosmopolitan vision is unlikely to be fulfilled (cf., Hollinger, 1995). But the United States today is undergoing its most profound demographic transformation in a century, and inexorable processes of globalization, especially international migrations from Asia, Africa, and the Americas, are bound to diversify still further the polyethnic composition of its population, and make more exigent the question of incorporation posed by Fuentes. At century's end, at a time when issues of immigration, race, and ethnicity have risen to the top of the policy agenda—and become the subject of vituperative, acrimonious, and xeno-phobic public debates on issues from Proposition 187 to Affirmative Action to Multiculturalism—and when the heat of self-interested polemic and controversy is once again dimming the light of truthful inquiry and understanding, this book invites you to reflect on the nature of the tumultuous histories and social processes that have made "Americans" out of a motley crew, and to see "past as prologue": to place present patterns in historical context, connecting endings to beginnings, destinies to origins.

1

Origins and Destinies: Immigration, Race, and Ethnicity in American History

SILVIA PEDRAZA

Once I thought to write a history of the immigrants in America. Then I discovered that the immigrants were American history.
— *Oscar Handlin,* The Uprooted *(1951)*

The United States is presently experiencing another major wave of immigration that is again transforming the very nature of American society. Immigration is at the very core of American society: with the exception of the Native American, every American is an immigrant. And even the Native Americans, as the result of the policy of removal that displaced them physically from their original land to the reservations, also suffered the profound uprooting, alienation, and cultural clash of those whose roots in their land, tradition, and culture were severed by migration. Americans are immigrants—people whose origins were various but whose destinies became American.

Immigration—voluntary or involuntary—is what created all multiracial and multicultural nations, of which the United States is a prime example. The essays in this volume point to how important this distinction in the nature of the original encounter among peoples is. From the area of origin to the area of destination, sometimes the migrants moved freely, as the result of their own choice—by and large, the experience of the European immigrants to the United States. Sometimes their movement was coerced, the result of processes not of their own making—a central aspect of the experience of African Americans, who were enslaved, as well as of Mexican Americans, Native Americans, and Puerto Ricans, whose experience began with conquest and annexation. Sometimes their movement was semicoerced, semifree—the experience of indentured servants, whether Japanese, Chinese, Irish, Mexicans, or Germans, people whose bondage depended on their unrelenting work. And sometimes people chose to move, but under such adverse conditions that their flight could be said to be coerced. Refugees who flee due to the deliberate and unrelenting persecution they are subjected to are a case in point, as are victims of genocide (state-sponsored mass murder)—the conditions that lay behind the flight of Jewish victims from the Nazi holocaust in Germany in midcentury or the recent flight of Cambodians from the terror the Khmer Rouge created. Without doubt, the origins of these relationships—voluntary, involuntary, or in between—influenced their very nature and the course of their development over time—their destinies.

Rodolfo Alvarez (1973) underscored the difference these various origins make in his analysis of the development of the Mexican American community in the United States over the course of distinct political generations. The second generation, "The Migrant Generation," of Mexicans came to the United States looking for work in agriculture, ranching, railroads, and mining—migration that resembled the voluntary migration of the Europeans to the United States. But they arrived into the situation that had already been created by the original encounter of "The Creation Generation"—shaped by violence and war between the United States and Mexico that ended in conquest and annexation, as a result of which the United States acquired Texas, New Mexico, California, Arizona, and parts of Colorado, Nevada, and Utah. All of this created a situation of extreme racial prejudice, a castelike subordination for the Mexican, especially in Texas, that resembled the situation Black Americans faced in the Deep South (Davis, Gardner, and Gardner [1941] 1965). Thus origins must be taken into account if we are to understand the destinies of various groups of immigrants and racial and ethnic minorities in America today.

Transforming America: Waves of Migration

In their analysis of California's newest immigrants, Thomas Muller and Thomas Espenshade (1985) argued that, over the course of American history, immigration can be broadly understood as consisting of four major waves. The first wave was that of northwest Europeans who immigrated from the settlement of the colonies early in the seventeenth century to the middle of the nineteenth century; the second, that of southern and eastern Europeans at the end of the nineteenth and beginning of the twentieth centuries; the third, that of African Americans, Mexicans, and Puerto Ricans from the South to the North, precipitated by two world wars; and the fourth, that of immigrants mostly from Latin America and Asia from 1965 into the present. The essays in this volume are organized along this distinction of historical timing. For not only has each wave been characterized by a different racial or ethnic composition, but also by the profound changes in the very nature of American society that have been taking place over this time span. The immigrants of the first wave—those from northwest Europe—came to an essentially colonial, agrarian society; the immigrants of the second and third

waves—those from southern and eastern Europe as well as African Americans, Mexican Americans, Puerto Ricans, and Native Americans who migrated from the South to the North—came to an urban society where they provided the cheap labor force essential to industrialization and expansion in America; the immigrants of the fourth wave—those from Latin America and Asia—are coming to an increasingly postindustrial, service-oriented society.

Because immigration *is* American history, yesterday as well as today, immigration is central to the identity of its people—even if only as hyphenated Americans, after the passage of so many generations (see Waters, this volume). It is also central to America's identity as a nation (see Pedraza, conclusion to this volume). Both its identity and the identity of its people are woven from the threads of race and ethnicity that sundry waves of migration joined.

Figure 1, based on the number of legal immigrants for which statistics exist, shows nearly two centuries of immigration.

In 1910, at the peak of immigration from southern and eastern Europe, nearly 15 percent of the total population of the United States was foreign-born. By contrast, in 1990, immigrants, over 80 percent of which now come from Latin America and Asia, constitute only around 8 percent of the population. Given the steady decline in the birth rate for quite some time, however, at present, net immigration constitutes a much larger share of population growth. In the decade of 1981 to 1990, net immigration constituted about 39 percent of population growth, roughly the same proportion as in the decade of 1901 to 1910 (U.S. Bureau of the Census 1992a; U. S. Immigration and Naturalization Service (1990–1994).

This large presence of immigrants in our midst can easily give rise to strong expressions of nativism, especially when people are faced with deteriorating economic conditions, reflected in high unemployment rates. John Higham once aptly described this when, in *Strangers in the Land* (1978), he examined the patterns of American nativism between 1860 and 1925. Yesterday as well as today, nativism surfaces most sharply in the regions and cities in which immigrants became overwhelmingly concentrated. At the end of the nineteenth and beginning of the twentieth century, nativism was expressed in the cartoons of the day in leading magazines, such as *Life*, showing the prejudices of the time by depicting Irishmen as

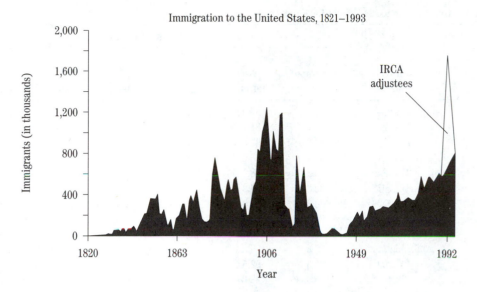

Immigration to the United States, 1821–1993

Figure 1

Immigration to the United States, 1821–1993

Note: Most IRCA adjustees were already living in the United States when they were counted as immigrants; most entered prior to January 1982.

Source: U.S. Immigration and Naturalization Service *Statistical Yearbook of the Immigration and Naturalization Service 1993.* In Philip Martin and Elizabeth Midgley, "Immigration to the United States: Journey to an Uncertain Destination." Population Bulletin 49 (September 1994), Figure 6.

brutes and drunkards, Jews as vulgar social climbers, and Catholics as unthinking Papists (see Higham 1978). In a cartoon that appeared in *Life* in 1906, one year before the Japanese Gentlemen's Agreement that curtailed their immigration, the popular sentiment was that we should "draw the line at the Japanese." Today nativism is expressed in referendums such as Proposition 187—which seeks to deny schooling and the use of social services to undocumented workers and their children—that arose out of the contention that the people of the state of California were being hurt by the large presence of illegal aliens; it is also expressed in the detention of thousands of Cubans at Guantánamo Naval Base, which arose out of the contention that Florida could no longer cope with the excessive burden of refugees and illegal aliens. Whether immigrants, indeed, constitute such an excessive burden or, rather, contribute more than they cost continues to be an unresolved debate that too often makes immigrants into the scapegoats for America's problems (cf. Fallows

1983). Certain, however, is that yesterday as well as today they transformed America into a very diverse nation whose resilience and problems *both* often stem from that very diversity.

The high rates of immigration, coupled with the high birth rates of many minority groups, such as African Americans and Hispanic Americans, also mean that White Americans—the descendants of the older European migrations—are now a declining proportion of the American population. While White Americans still constitute the majority of the population, between 1980 and 1990 their proportion of the U.S. population grew at an extremely low rate in comparison to other groups. In the decade of 1980 to 1990, when the total U.S. population grew by 9.8 percent, Whites increased by only 6 percent in comparison to an increase of 108 percent for Asians (the fastest growing group), 53 percent for Hispanics, 13 percent for Blacks, and 38 percent for Native Americans. Of the 250 million people that in 1990 constituted the nation's total population count, Whites (not of Hispanic

Table 1   Legal Immigration to the U.S., by Region of the World, 1820–1990.

Intercensal Decade	Total Number	Percent of Total					
		Europe Total[a]	North and West Europe[b]	South and East Europe[b]	Asia[c]	Africa	Latin America[d]
1820–1830	151,824	70.1	67.9	2.2	—[e]	—	6.2
1831–1840	599,125	82.7	81.7	1.0	—	—	3.3
1841–1850	1,713,251	93.2	92.9	0.3	—	—	1.2
1851–1860	2,598,214	94.4	93.6	0.8	1.6	—	0.6
1861–1870	2,314,824	89.2	88.1	1.1	2.8	—	0.5
1871–1880	2,812,191	80.8	76.2	4.5	4.4	—	0.7
1881–1890	5,246,613	90.3	78.8	11.5	1.3	—	0.6
1891–1900	3,687,564	96.4	60.6	35.8	2.0	—	1.0
1901–1910	8,795,386	91.6	46.1	45.5	3.7	0.1	2.1
1911–1920	5,735,811	75.3	25.3	49.9	4.3	0.1	7.0
1921–1930	4,107,209	60.0	32.5	27.0	2.7	0.2	14.4
1931–1940	528,431	65.8	38.6	26.7	3.0	0.3	9.7
1941–1950	1,035,039	60.0	49.9	9.8	3.1	0.7	14.9
1951–1960	2,515,479	52.7	39.7	12.7	6.1	0.6	22.2
1961–1970	3,321,677	33.8	18.3	15.4	12.9	0.9	38.6
1971–1980	4,493,314	17.8	6.7	11.0	35.3	1.8	40.3
1981–1990	7,338,000	12.5	7.2	5.3	37.3	3.1	47.1

[a]Includes all of USSR except 1931–1950, when USSR is divided into European USSR and Asian USSR.

[b]Through 1901–1910, north and west Europe includes Austria-Hungary. After 1901–1910, Austria included in north and west Europe, Hungary in south and east Europe. Immigrants recorded as Other Europe (57,182 altogether through fiscal year 1984) are omitted from these two regions as shown in this table.

[c]Asia according to INS definition includes southwest Asia (e.g., Iraq, Israel, Syria, Turkey).

[d]Includes Mexico, Caribbean, Central America, South America.

[e]—Indicates less than 0.05 percent.

Source: Leon F. Bouvier and Robert Gardner, "Immigration to the U.S.: The Unfinished Story," Population Bulletin 41 (1986), Table 1; and U.S. Immigration and Naturalization Service, *Statistical Yearbook 1992*, Tables 1–2.

origin) constituted 76 percent; Blacks, the largest minority, were 12 percent; Hispanic people about 9 percent; Asians are about 3 percent; and American Indians are slightly less than 1 percent (U.S. Bureau of the Census 1993c, Table J).

If current trends in immigration and birth rates persist, by the year 2050 the Latino population is projected to increase more than any other group, to 23 percent; Asians to 9 percent, Black Americans to 14 percent; but White Americans (not including Hispanics) may decline to 53 percent (U.S. Bureau of the Census 1993c, Table J). Hence, immigrants and ethnics are once again transforming America.

The First Wave

The first formal efforts to count immigrants began with an act of 1819 that required the captain of every ship arriving from abroad to deliver to the local collector of customs a list or manifest of all the passengers on board. Between the end of the Revolutionary War and 1819, an estimated 250,000 foreign-born persons had arrived by ship. As Table 1 shows, roughly corresponding to the latter part of the first major wave of immigration, from 1820 to 1880 over 10 million immigrants arrived, predominantly from northwestern European nations such as England, Scotland, Germany, Norway, Sweden, and Ireland. At the turn of the century, they were referred to as the "old immigrants."

When the U.S. Census in 1990 asked people to identify their ancestry or ethnic origin, the top three ancestries most often mentioned came from the descendants of this first wave of migration: German (about 23 percent of the population), Irish (about 16 percent), and English (about 13 percent). Next in line came Italian (about 6 percent) and Mexican (5 percent).

As Everett Lee's theory of migration (1966) underlined, all migration flows correspond both to "push" factors at the place of origin and "pull" factors at the place of destination, although at times one may be stronger than the other. Many of colonial America's first settlers, for example, came fleeing the religious intolerance and persecution then reigning in England and elsewhere in Europe. Quakers, Puritans, Presbyterians, Catholics, Jews (first Sephardic and then German) all concluded that emigration to America would be the road to religious freedom.

Other migrations were driven both by political and economic factors. The Irish, for example, left Ireland for political reasons, especially in the eighteenth century—the English subjugation of Ireland and the prejudice against Catholics—and also, especially in the nineteenth century, increasingly due to economic reasons—the displacement of the peasantry that took place everywhere as part of the transition from feudalism to capitalism. That transition was particularly acute in Ireland because of its delayed industrialization and the catastrophe of the Famine in midcentury (cf. Bodnar 1985).

The actual voyage was often financed by becoming an indentured servant in exchange for the passage. Approximately half of the almost 1 million immigrants who came to America during the colonial period came as indentured servants who agreed to work for a specified period (say, from four to seven years) for a person who paid the passage to America (often a ship captain or a merchant). Most were young men who were unskilled farm laborers.

Eugene Boe (1981, p. 51) told the story of his Norwegian ancestors, the brothers Osten and Henrik Boe, who came to Minnesota looking for land on which to farm:

In the early decades of the nineteenth century reports began drifting in from across the Atlantic that painted the New World in the colors of the Promised Land. A trickle of migrations from Norway to America started in the 1830s. Most of these emigrants set out for the frontiers of Illinois and Wisconsin, where land was available. Their letters home and the eyewitness accounts of visiting missionaries fanned the flames of discontent back in the Old Country and provoked new waves of emigration.

In colonial America the immigrants acquired land and farmed. On a farm, the family is an economic unit of both production and consumption. Women's work in the home, then, was a form of household production critical to the family's survival:

Surely she was busy, indeed. Besides feeding five children and a husband three meals a day and morning and afternoon "lunches," she spun yarn, wove cloth, sewed and darned and knitted and patched the family's wardrobe, made rugs and bedding, washed and ironed, kept the home sparkling clean, tended the orchard, put up preserves and dried fruits, took care of the chickens, often did the milking, churned the butter, baked, and made the candles (by dipping a wick of twisted cord into melted tallow) and soap (from ashes and grease and lye). Then, too, vast reservoirs of energy—and time—always had to be allocated to treating her children's ailments and superintending their religious training. (Boe 1981, p. 62)

Boe particularly stressed the importance of the opportunity created by President Lincoln's 1862 Homestead Act that made land available to all citizens or those who had applied for citizenship who could file a claim for a quarter section (160 acres) of land and could come into ownership of that land after five years' occupation of it. Needing to have only $14 dollars for the filing fee, it was a remarkable opportunity and a remarkable "pull" to further attract immigration from Europe. In Norway, as in many other sending countries, "the government land offices and transportation packagers spread the word throughout the Scandinavian countries and the urge to emigrate became epidemic" (Boe 1981, p. 52).

The vast social transformation and upheaval going on in Europe—the transition from feudalism to capitalism—served to "push" the European emigrants from the Old World. That process was particularly severe in some countries, such as Ireland, where the potato famine of the mid-1840s "pushed" the Irish out of their country in inordinate numbers. It is estimated that the potato famine resulted in the death of 1 million persons due to starvation and related diseases and that one-third of the population of Ireland emigrated—an exodus of 4 million people. In 1914, the population of Ireland was half what it had been in the 1840s (see Diner, this volume). William Alfred (1981, p. 20) recounted the immigrant story of his great grandmother, Anna Maria:

> Of Ireland, she rarely spoke, save to recall that she was often hungry there and for her main meal she often ate cress out of the brooks on oaten bread with a bit of lard. Although she always used to say she had no desire to return to Ireland to live, she lived out of a trunk to her dying day.

The Second Wave

From 1881 to 1930, roughly corresponding to the second major wave of immigration, over 27 million immigrants arrived in the United States, with the southern and eastern European nations, such as Italy, Austria-Hungary, Poland, Russia, playing a larger and larger role. At the turn of the century they were called the "new immigrants." Immigrants from China and Japan arrived in not insignificant numbers to the West Coast. In the decade of 1871 to 1880, 123,201 immigrants arrived from China, numbers that dwindled after the passage of the Chinese Exclusion Act in 1882 barred the immigration of laborers (but not teachers, diplomats, students, merchants, or tourists). The act was rescinded in 1943, when the United States and China were wartime allies (see Chow, this volume). In the decade of 1901 to 1910, 129,797 immigrants arrived from Japan, until the Gentlemen's Agreement of 1907 barred entry to laborers from Japan and a series of Alien Land Acts in 1911 and 1913 in California, together with the barring of immigration from the "Asiatic Barred Zones," sharply curtailed the numbers of Japanese and other Asian immigrants (see Glenn and Parreñas, this volume).

Most of the immigrants from southern and eastern Europe were unskilled labor (pick-and-shovel men), rural peoples who were mostly illiterate. As peasants, their identification was, first and foremost, with their village and their parish. For example, when Italians came to the United States they thought of themselves as Neapolitans, Sicilians, Bolognese,

Romans, or Maltese. Mario Puzo (1981, p. 35) began his autobiography, "Italians in Hell's Kitchen" (in New York), by noting that "As a child and in my adolescence, living in the heart of New York's Neapolitan ghetto," contrary to the stereotypes about Italians as warm, loving, and happy-go-lucky, "I never heard an Italian singing," tired as they were by the hard labor involved in their earning their daily bread. "Italian" was an identity—a pan-identity, in the language of our times—that emerged on American soil (see Alba, this volume). Yancey, Ericksen, and Juliani (1976, 1977) have argued that such emergent identities resulted from the social conditions of the immigrants' lives in American cities—that they shared neighborhoods, social class, and occupations—the shared structural conditions that prompted the immigrants to interact with one another, leading to intermarriage and the development of a common consciousness in America.

The major exception to the peasant origins of the European immigrants were the Jews, who fled Europe in families because of the anti-Semitism and violence they encountered in the particularly virulent *pogroms*—anti-Jewish riots—at the end of the nineteenth century (Howe 1976). They were already urban, very often literate, already skilled, particularly in the needle trades (see Gold and Phillips, this volume). And in Europe they had already played the role of "middleman minority" (Bonacich 1973)—minorities that specialized in trade and commerce and mediated between the poor and the rich, the masses and the elite, often filling a status gap in the society.

It is estimated that approximately one-third of all Jews in eastern Europe (from Russia, Poland, Austria-Hungary, and Rumania) migrated to America—about 2 million people. "Pushed" by circumstances beyond their control, they fled seeking safety, to escape violence and persecution, and came searching for personal and religious liberty. Clearly they were refugees. But immigration law at this time did not legally recognize refugees as a separate type of immigrant. The legal category of refugee did not begin to develop until after World War II, with the 1948 Displaced Persons Act and a series of Refugee Relief Acts and Refugee Assistance Acts in the 1950s and 1960s that authorized the admission of refugees outside of the quota 'limits and provided special assistance for them. Under these latter acts, contemporary refugees were admitted: Hungarians, Cubans (see Pedraza, this volume), and Vietnamese and other Indochinese refugees (see Rumbaut, this volume).

Further immigration was barred from Asia at the turn of the century and from Europe in the 1920s. Prior to this restriction that culminated in the passage of the first Immigration and Nationality Act, immigration law had barred the entry of only those *persons* (not nationalities) considered unfit—physically, morally, or socially: prostitutes, convicts, lunatics, idiots, those likely to become a public charge, those with physical or mental defects, those with tuberculosis, and children unaccompanied by their parents.

In 1892, Ellis Island opened as an immigrant station, through which passed most of the immigrants of this second major wave from southern and eastern Europe, and others who came later until it closed in 1954. It is estimated that today over 100 million Americans can trace their ancestry to the immigrants who passed through the gates of Ellis Island, now restored as a museum of American immigration.

Some of the immigrants who passed through Ellis Island went on to achieve fame in America. Examples of those who achieved illustrious careers were: in music, Irving Berlin, who came from Russia in 1903; in film, Samuel Goldwyn, who came from Poland in 1896; in film and theater, Elia Kazan, who came from Turkey in 1913, Frank Capra, who came from Italy in 1903, and Edward G. Robinson, who came from Rumania in 1903; in law, Felix Frankfurter, who came from Austria in 1894; in religion, Father Edward Flanagan, who came from Ireland in 1904; in labor, David Dubinsky who came from Russia in 1911, and Philip Murray, who came from Scotland in 1902; and in physics, Albert Einstein, who came from Germany in 1933 (National Park Service 1991).

From 1892 to 1931, from Italy alone, over 2.5 million immigrants passed through Ellis Island; from Russia alone, close to 2 million; over 2 million from Hungary, Austria, and Austria-Hungary; over half a million from Germany, England, and Ireland each; a third of a million from Sweden; close to a quarter of a million from Greece, Norway, and the Ottoman Empire each; around a tenth of a million from Scotland, the West Indies, Poland, Portugal, and France each (National Park Service 1991). Upon their arrival, filled with gratitude and relief, the immigrants saw the welcoming torch of the Statue of Liberty, inside of which was inscribed Emma Lazarus' famous poem "The New Colossus" (1883):

Give me your tired, your poor,
Your huddled masses yearning to breathe free,
The wretched refuse of your teeming shore;
Send these, the homeless, tempest-tost to me,
I lift my lamp beside the golden door!

Yet as this wave of migration crested, nativism—the fear of the foreign—took hold in the land. With the passage of a series of laws that culminated in the 1924 Immigration and Nationality Act, immigration was restricted on the basis of national origins, and quotas were set that favored the immigrants from northern and western European nations while curtailing the immigration from southern and eastern European nations. At the same time, the Oriental Exclusion Act banned all immigration from Asia. President Kennedy sent a message to Congress in 1963 that called for a complete revision of the nation's immigration laws and the abolition of the national origins quotas that the McCarran-Walter Immigration Act of 1952 had left in place. In *A Nation of Immigrants* (1964; posthumously published) he expressed the moral shortcomings of the immigration laws then in place:

The famous words of Emma Lazarus on the pedestal of the Statue of Liberty read: "Give me your tired, your poor, your huddled masses yearning to breathe free." Until 1921 this was an accurate picture of our society. Under present law it would be appropriate to add: "as long as they come from Northern Europe, are not too tired or too poor or slightly ill, never stole a loaf of bread, never joined any questionable organization, and can document their activities for the past two years." Furthermore, the national origins quota system has strong overtones of an indefensible racial preference. It is strongly weighted toward so-called Anglo-Saxons, a phrase which one writer calls 'a term of art' encompassing almost anyone from Northern and Western Europe. (Kennedy 1964, p. 77)

That preference for Anglo-Saxons reflected both the Protestant establishment then in place (see Schneiderman, this volume) and the prevailing racism and social Darwinism of the time that saw the Anglo-Saxons as "superior stock" and the real Americans, and the southern and eastern European immigrants then coming to America as "inferior breeds"

(Gossett 1965). In 1911, the Dillingham Joint Commission on Immigration issued a forty-two-volume report that—backed by a social science that was neither scientific nor impartial—had concluded that the new immigrants were racially inferior, inclined toward violent crime, resisted assimilation, and drove "old-stock citizens" out of some lines of work (Archdeacon 1982, p. 163). As Handlin ([1951] 1973, pp. 262–263) put it:

> Restriction involved a rejection of the foreign-born who aspired to come to the United States. It also involved a condemnatory judgment of the foreign-born already long established in the country. . . .
>
> The newcomers could not but feel estranged. In forty-two volumes, under the guise of science, the government had published the record of their shortcomings. Learned men had told them they were hardly human at all; their head shapes were different, their bodily structure faulty, the weight of their brains deficient. If they were Italians, they were not really like the Italians who had a claim to the mantle of Rome; if they were Greeks, they were not genuine Greeks descended from the Hellenes. The Armenians could not comprehend the jargon of the anthropologist who told them they were related physically to the Turks although the shortness and height and flatness of their heads were comparable only to the occiput of the Malays and Filipinos. Yet the derogatory intent of these inflexible classifications was plain enough.

As finally implemented in 1929, of the 156,700 quota immigrants permitted to enter the United States each year, the national origins quotas, basing themselves on the last census of 1920, allotted 82 percent of the Eastern Hemisphere total to countries of northwestern Europe, 16 percent to those of southeastern Europe, and 2 percent to all others. All Asians were totally barred from entry. Prior to the quotas' revision in 1965, this meant that, for example, Great Britain had an annual quota of 65,361 visas and used only 28,291 of them. Ireland's quota was 17,756 and only 6,054 Irish used them. But Poland was permitted only 6,488 and had a backlog of 61,293 Poles wanting to come to the United States. Italy had a quota of 5,666 with a backlog of 132,435. "In short," Kennedy (1964, p. 75) explained, "a qualified person born in England or Ireland who wants to emigrate to the United States can do so at any time. A person born in Italy, Hungary, Poland, or the Baltic States may have to wait many years before his turn is reached." These quotas remained in place until the 1965 amendments to the 1952 Immigration and Nationality Act abolished them.

As Milton Gordon (1964) pointed out, at the turn of the century, three competing ideologies of assimilation took root in America: Anglo-conformity, the melting pot, and cultural pluralism. All three have since been present in American society, although varying in strength over time. Anglo-Saxon conformity demanded that immigrants shed their own cultures (food, dress, religion, customs, values) and assimilate along Anglo-Saxon Protestant lines. The "melting pot" ideal imagined that all immigrant groups would come together and melt (through intermarriage) in a cauldron out of which would come something distinct and new: the American. Cultural pluralism assumed that all ethnic communities could retain their own autonomous cultures with lives of their own, like mosaics of different colors and hues, each of which is distinct and equal, yet all of which together paint a larger picture. In our day, this last, minor ideology has gained importance and has come to be known as *multiculturalism*.

At the turn of the century, of the three, clearly Anglo-Saxon conformity was the dominant one. Social pressure demanded it to achieve success. Conformity to this pressure can be seen in the name changes of many an American movie star who worked for the new American movie industry at the beginning of the twentieth century. That industry that portrayed America as essentially an Anglo-Saxon nation hired many of its talented stars from among the "new immigrants." For example, Doris Day's real name was Doris Kapplehoff; Lawrence Harvey's was Larry Skikne; Tony Curtis's was Bernie Schwarts; Karl Malden's was Mladen Sejulovish; Judy Garland's was Frances Gumm; Ginger Rogers's was Virginia McMath; Mitzi Gaynor's was Mitzi Gerber; Ethel Merman's was Ethel Zimmerman; Vic Damone's was Vito Farinola; Dianne Foster's was Dianne Laruska; Rita Hayworth's was Margarita Carmen Cansino; Aldo Ray's was Aldo Da Re; Donna Reed's was Donna Mullenger; Jane Wyman's was Sarah Fulks; June Allyson's was Alla Geisman; Kirk Douglas's was Issur Anielovitch; Danny Kaye's was Daniel Kaminsky; Dean Martin's was Dino Crocetti; and Jerry Lewis's was Joseph Levitch (in Baltzell 1964a).

Even at the height of this racism towards those that, with the passage of time and the achievement of substantial social mobility, eventually came to be seen as "White," no ceiling was set on immigration from the Western Hemisphere. In effect, an exception was made for the Mexican because the growers successfully argued to Congress that Mexican labor was needed as the main source of agricultural labor. With World War I came a demand for agricultural production and a labor shortage that constituted an enormous "pull" to Mexican immigration. The national origins quotas caused the large growers to turn to unorganized Filipino and Mexican labor. When the Filipinos became organized in the Filipino Labor Union and struck for higher wages, they also ceased to be desirable (see Cariño, this volume). Thus the growers became enchanted with the Mexicans and fought to keep them from being placed under a quota restriction, using the argument that the Mexican would not become a public charge because, like a homing pigeon, "he goes back to roost" (McWilliams 1939; Pedraza-Bailey 1985).

The Third Wave

After the closing of the door to European immigration in 1924, the third major wave of migration was mostly internal—the migration of African Americans, Mexicans, Native Americans, and Puerto Ricans from south to north, that entailed the transformation of a rural people into urban dwellers. Such a migration cannot be captured by immigration statistics, but the social transformation it entailed was as vast as that of the immigrants who crossed the oceans. For example, more than 90 percent of Blacks lived in the agricultural South at the beginning of the twentieth century; by 1960, about half lived in the industrial North and only 40 percent of the Blacks that still remained in the South still lived in rural areas. "Pulled" up by the employment opportunities generated by the industrialization of the North and by World War I, between 1910 and 1920 about half a million Blacks moved to northern and western states (see Marks, this volume). With the further decline of agriculture in the South, when World War II shifted the economy into the high gear of economic expansion, particularly in the Northeast and the Midwest, it is estimated that during the decade of 1940 to 1950 over 1.5 million African Americans left the South, and another 1.5 million left during the following decade

of 1950 to 1960 (Henderson 1967, p. 83). Thus, what had been a mostly rural people became mostly urban. That urbanization constituted a form of economic modernization that Piven and Cloward (1979) argued was necessary for the development of political modernization—of the Civil Rights movement (see Morris and Herring, this volume). The internal migration of these groups (from south to north and rural to urban places and ways of life) was essential for their development of the social movements that were able to break through the systematic exclusion and segregation imposed by Jim Crow segregation, beginning the process of inclusion that to this day remains only partly achieved.

The Fourth Wave

The fourth wave of migration, which Rubén Rumbaut details in the next essay, began in the immediate post–World War II period. It gained particular impetus from the 1965 amendments to the McCarran-Walter Immigration and Nationality Act of 1952, amendments that, in abolishing the national origins quotas of 1924, opened the door to renewed immigration to the United States. The major criteria for legal admission to the United States in this contemporary period became twofold: occupational certification—giving preference to immigrants whose occupations were in a relative shortage in the United States; and family reunification—giving preference to immigrants whose immediate family had been rent apart by migration. David Reimers (1985) has argued that an unintended consequence of this legislative reform was that increasingly the origins of the immigrants shifted to the Third World. As Table 1 also shows, from 1981 to 1990, when the proportion of European immigration had declined to the very low figure of 12 percent, 85 percent of the new immigrants came from Latin America and Asia. From 1961 to 1990, roughly corresponding to the fourth major wave of American immigration, the total number of legal immigrants to the United States was about 18 million people.

Immigrants as a Social Type

As a result of the fourth wave of American immigration, which we are still living through, a veritable boom in immigration research has taken place in the

From the end of the 19th century to the Civil Rights movement, during the Jim Crow period in the Deep South, African Americans endured a caste-like segregation. Here in 1932 in Belzoni, the Delta area of the Mississippi, Black Americans enter a movie theater through the "colored" entrance. (Photo by Marion Post Wolcott. U.S. Farm Security Administration Collection; Library of Congress)

last twenty years. In sociology, the pattern of immigration research is quite clear. Alejandro Portes (1978) has repeatedly stressed that the study of immigrants was closely wedded with the beginnings of social science in America, when immigrants and their plight were the focus of vivid studies. For example, Robert Park ([1926] 1950), one of the founding fathers of American sociology, evolved his famous theory of the race relations cycle. He argued that the race relations cycle consisted of stages of interaction through which racial or ethnic groups were thought to progress irreversibly: contact, competition, accommodation, and the final stage of eventual assimilation, of becoming like the dominant, majority population. Park was also responsible for creating the concept of the "marginal man." Park stressed that marginal human beings—those who, as a result of migration, ended up by living simultaneously in two separate worlds—were not only marginal, never fully

belonging to one world or the other, but also enormously creative and intelligent, because experiencing more than one social world had sharpened their vision and sensibilities. Also part of "the Chicago school," Thomas and Znaniecki (1927), in *The Polish Peasant in Europe and America*, analyzed the social psychological impact of immigration on the immigrants themselves.

Sociologists at the turn of the century were concerned with what the experience of immigration had done to the immigrants' lives and with the outcomes to the process of integrating those who arrived at its shores. These outcomes were usually conceptualized as acculturation and assimilation—as needing to become like the dominant population, which at the turn of the century clearly meant conformity to Anglo-Saxon ways (cf. Gordon 1964).

To understand what the assimilation process was actually like, Gordon (1964) argued that it was crucial to distinguish between cultural behavior and social structure. Thus, Gordon set apart cultural and structural assimilation. Cultural assimilation entailed a process of acculturation on the part of the immigrants, of becoming like the dominant population in cultural patterns, such as language, behavior, and values. Structural assimilation, as Gordon defined it, resulted only when the immigrants had been "taken up and incorporated." Specifically, structural assimilation entailed the full integration of the immigrants and their descendants into the major institutions of the society (educational, occupational, political) and into the social cliques, clubs, and institutions of the core society that lead to intimate primary relationships, including intermarriage. This distinction aimed to provide a more exact conceptual tool to gauge the reality of the assimilation of immigrants and racial minorities in America. To this day, the distinction typifies the contrasting emphases of social science research on race and ethnic relations: the emphasis on ethnic identification (its absence or strength) and the emphasis on tangible outcomes (such as occupation, education, and income).

Research on immigrants and the eventual outcomes of processes of immigration, therefore, was at the very foundations of American sociology. That emphasis began to wane, and in the 1960s, it all but disappeared. Several different trends promoted its disappearance. The Immigration and Nationality Act of 1924 cut the massive waves of European immigration to the United States. Moreover, under the pressures of Anglo conformity, the children of those European immigrants went on to assimilate in American society at a time when the price of success was often one's ethnicity and identity. Paul Cowan's (1982) *An Orphan in History* is the autobiography of a talented New York writer for *The Village Voice* that exemplifies these costs well. Paul Cowan's very name signifies his loss, his feeling of being an orphan in history, for his real name—lost as a result of his parents' great American success—in truth should have been the frankly Jewish and rabbinical name of Saul Cohen. Even more, the confrontations—personal and social—the Civil Rights movement entailed also caused the analytical focus to shift to that of racial and ethnic relations; and in the process, what is really distinctive about immigrants was lost.

What is distinctive about immigrants? First, that they have experienced another whole life in another country, another culture, yet they will live out a whole new set of choices and experiences (depending on their age and circumstances) in the new society to which they migrated. And very often during the immigrant generation they remain extremely involved with the country, village, and family they left behind.

Moreover, immigrants bring a whole host of social resources with them—such as their social class, race, education, gender, family, institutional knowledge, political attitudes, and values—from another society. They arrive in the new society, such as the United States, at times when the social context that greets them presents them with vastly different amounts of opportunity—such as economic growth or recession, government policy programs, a warm welcome or cold reception, sunrise or sunset industries, particular regions or cities of settlement, and acceptance or discrimination. As a result, immigrants from various migrations (or from various waves of the same migration) may well undergo rather different processes of incorporation into American society, as reflected in their contrasting social outcomes (Pedraza-Bailey 1990). That is to say, that their origins are related to their destinies.

Social scientists aim to understand what are the causes and the consequences of particular social patterns and forms of social behavior. Because migration entails change from one society (origin) to another (destiny), the major questions in immigration research can be conceptualized as follows:

What caused the immigrants to make the decision to move? What "push" and "pull" factors impelled them to leave their homelands, resettle, and become uprooted themselves?

What was the nature of the crossing—the literal crossing of the journey? Was it arduous and fraught with danger? What was the nature of the more abstract crossing produced by the policies of the two governments that regulated the migration?

What consequences did the migration hold for them afterwards? How can we best describe that process—as assimilation, adaptation, colonization, or incorporation?

Origins of Migrants

In sociology, the traditional, individual micro approach was best developed by Everett Lee's (1966) theory that made explicit the "push" and "pull" factors involved in the decision people make to move—factors that "hold and attract or repel people"—as well as the intervening obstacles that prove more of an impediment to some than to others. The decision to migrate was the focus of his analysis, although, as he stressed, "indeed, not all persons who migrate reach that decision themselves. Children are carried along by their parents, willy-nilly, and wives accompany their husbands though it tears them away from environments they love" (Lee 1966, p. 51).

In history, the traditional approach concentrated not only on the nature of the personal or familial decision to migrate, but particularly on the impact that the experience of immigration had on the lives of the immigrants themselves. In *The Uprooted*, Oscar Handlin ([1951] 1973) underscored the profound clash of cultures that left the immigrants uprooted, bewildered, and in pain, forever having left the best of themselves and their lives behind them. In this passage, for example, Handlin ([1951] 1973, p. 98) caught the sadness, despair, and nostalgia of everyone who has ever been uprooted:

> Yesterday, by its distance, acquires a happy glow. The peasants look back . . . and their fancy rejoices in the better days that have passed, when they were on the land and the land was fertile, and they were young and strong, and virtues were fresh. And it was better yet in their fathers days, who were wiser and stronger than they. And it was best of all in the golden past of their distant progenitors who were everyone a king and did great deeds. Alas, those days are gone, that they believed existed, and now there is only the bitter present.

In contrast, in the more recent *The Transplanted*, John Bodnar (1985) painted the immigrants less in pain because he sees them as more resilient, as people who sought to gain some measure of control over their lives by the effective use of the institutions they brought with them, particularly the core institution of "the family-household." Together with those "whom experience made their brothers," as Handlin ([1951] 1973, p. 152) put it, the immigrants developed new institutions—unions, churches, mutual aid associations, schools, newspapers, theaters—that responded to their needs as members of a social class (working class or middle class) and as ethnics (Germans, Italians, Mexicans, Poles, Jews). Through their development of and participation in these institutions, whose form were the products of American conditions (such as the Sons of Italy, the United Hebrew Trades, the Benito Juarez Society, the Shrine of St. Jude), the immigrants sought to protect themselves. Immigration entails both bewilderment, alienation, and pain as well as resiliency and institution-building. As Handlin ([1951] 1973, p. 142) underscored, "A possession of infinite value had disappeared in the course of the migration—the inner meaning of their own existence in the universe." That had to be reconstructed along with their new lives.

Selectivity of Migration

The most popular interpretation—both among academics and nonacademics—of the reasons for the enormous waves of migration that peopled the United States is the "pull" of American economic activity. But, as Bodnar (1985) pointed out, if that were true, immigrants would be equally selected from all nations and all regions at all times. Instead, we find that immigration was (and still is) highly selective: selecting immigrants from different countries (for example, over 5 million came from Italy to the United States, but less than half a million from Greece); selecting immigrants from different countries at different times (for example, from Ireland mostly before 1880, from Italy mostly after 1900); selecting immigrants from the same country at different times (for example, in Ireland the level of emigration was relatively low prior to the 1840s, in comparison to the peak reached during the years of the famine in the 1840s, and declining afterwards); selecting immigrants from the same country but from different regions (for example, certain provinces of Great Britain and Germany saw many of its people emigrate, others saw relatively few); selecting immigrants from different occupations and social classes (for example, from Poland came mostly peasants, while from eastern Europe came mostly Jewish traders and merchants); and selecting immigrants from different occupations and social classes at different times (for example, many of the European migrations that were economically motivated were characterized by a pattern where artisans and craftsmen left first, followed by masses of illiterate peasants, to be joined by literate urban dwellers last).

To understand the selectivity of migration, we need to consider the "push" and "pull" factors at the origins and destinations that structured the migration. The European immigrant struggle, Bodnar (1985) emphasized, was shaped by the transition from feudalism to capitalism in Europe—first in the form of the arrival of commercial agriculture in the Old World, which was accompanied by the population explosion; second, in the form of the arrival of industrial manufacturing and urbanization in both the Old World and the New. Bodnar (1985, p. 216) underscored that "Transplanted by forces beyond their control," the immigrants were "indeed children of capitalism." But the intrusion of capitalism was an uneven process that hit certain countries first, then others; certain regions first, others later; certain social classes first, followed by others. Artisans, craftsmen, and illiterate, unskilled peasants emigrated from the feudal societies of Germany, Ireland, Italy, Mexico, and Japan. They immigrated to America and became its working class that supplied the extremely cheap labor that made possible the enormous economic growth that this developing, industrial, capitalist society experienced (see Pacyga, this volume).

The thesis has often been advanced (for example, Rosenblum 1973) that because the immigrants thought of themselves and lived their everyday lives, first and foremost, as ethnics (Catholics, Jews, Poles, Italians, Greeks, Slovenians), they prevented the development of labor solidarity. Hence, that by comparison with the western European experience, the American labor movement remained weak and failed to develop a strong socialist tradition—an American exceptionalism. Bodnar (1985), however, emphasized that the spread of socialism was not only impeded by forces within immigrant communities but also by the actions of labor leaders and organizers who did not fully understand the differential receptivity of different ethnic groups to labor unionism—given their past experiences and political affiliations in their homeland—as well as their strong bias towards skilled, rather than unskilled, workers.

Yesterday, as well as today, the migrants moved through kin chain migrations—migrations in which the most daring and adventurous went first, established themselves in certain places (such as Chicago or New York) and in certain occupations (such as in the steel or garment industry). Once they became established, they offered their help (through immigrant letters then, letters and phone calls later, faxes today) to their relatives and closest friends, offering to support them until they could stand on their own feet, and offering to help them get jobs, often where they themselves worked. Those who remained back at home would signal their intention and plans. In the past, when literacy was a skill not often found among peasants, villagers often went to the village letter writer who, for a fee, wrote the letter the villager dictated to relatives and friends overseas. Today, when villagers cannot depend on letters or phone calls for reliable communication, they often travel to the nearest city, to a central place where, for a fee, they can send a fax to their relatives and friends overseas. So migration results in the geographical concentration and occupational concentration of particular ethnic groups. As Massey et al. (1987) underscored in their analysis of Mexican migration to the United States, once begun, migrants' social networks of family and friends grow and develop, supporting and channeling the migration on a continuously widening scale. Thus, the migration that was initially propelled by an external, structural dynamic and logic increasingly acquires an internal dynamic and logic of its own. "In this sense, migration comes to fuel itself" (Massey et al. 1987, p. 252).

Women, Family, and Migration

As Boyd (1989) pointed out, "the division of labor—the structured activities in a society—is gendered." Thus, gender is also central to the decision to migrate. Sherri Grasmuck and Patricia Pessar's (this volume) analysis of Dominican migration to New York City shows that migration is often a family strategy to meet the challenges that accompany underdevelopment and economic and political transformation in the Third World. As they emphasized, the household is the social unit that makes decisions as to whether migration will take place, who in the family will migrate, what resources will be allocated to the migration, what remittances will be sent back to those left behind, which household members can be expected to return, and whether the migration will be temporary or permanent.

Moreover, the meaning of migration may well be different for women than for men. Difficult as the experience of immigration is, it is often far more positive for women than for men because the enormity of the changes produced by the migration allow women to break with traditional roles and patterns of dependence and assert a newfound (if meager) freedom (Foner 1978). In their analysis of the Dominican migration, Grasmuck and Pessar found that very often the goal of the family strategy is to sponsor the wife's emigration to the United States, where there is less resistance to women working outside the home for wages and where the sons can obtain jobs no longer found in the rural areas of the Third World.

For the women themselves, the act of emigrating also became a way of escaping total dependence on their husbands. Gender became part not only of the decision to migrate but also of the decision to return. While immigrant men often lived in an extremely frugal way, saving all they could to return, women often depleted the family savings by buying large items, such as refrigerators or sofas, that would help prevent the family's return.

To understand the impact of gender on migration, we need to pay attention to the causes and consequences of the demographic composition of a migration flow. A predominantly male migration, such as the Italian or Mexican, often resulted in temporary settlement, large amounts of remittances sent back home, and a high level of return migration. A predominantly family migration, such as the Jewish or Cuban, often resulted in a large concentration of the immigrants in family businesses. A predominantly female migration, such as that of the Irish or Swedish, aided in the occupational concentration of women in domestic service (see Diner, this volume). As the immigrants had no family obligations, they were able to concentrate in domestic service, particularly as "live in" help, and to accumulate savings at an impressive rate. These savings went to bring over other relatives, especially other women, to pay for the mortgaged lands back home, to support their devotions to favorite saints in the Catholic Church, to finance their upward mobility, and to provide a "nest egg" for marriage.

Moreover, women lived the daily experience of immigration as women whose deeply felt needs found expression in their popular, religious tradition. In his analysis of the devotion Italian immigrants poured onto the Madonna of 115th Street in New York, Robert Orsi (1985, pp. 204–205) underscored that while the Madonna came from Italy with the immigrants, and as such was a symbol to all Italian immigrants of nation, history, and tradition, above all she was a woman's devotion—both because women

were its main participants and because "it emerged out of and reflected the special role and position of women in Italian culture." In Italian Harlem, the Madonna also became an expression of the lives of immigrant women as these women turned to the Madonna with petitions for help with the hardship and powerlessness of their lives—as women bound by a strong, patriarchal tradition, and as immigrants mired in poverty, toil, and trouble. That private relation became public at the annual *festa* when both men and women participated as a community. This served to regenerate their culture as Italians and to console them for the physical and spiritual trials of immigration.

Destinies of Migrants

Because migrants have always moved through the social networks of kin and friends who are like kin, migration flows also lead to a substantial amount of concentration in certain places of settlement and in certain occupations. For example, New York was (and is) the quintessential city of immigrants (see Binder and Reimers, this volume). Looking back at American history we can see that from 1850 to 1920, the American economy was expanding and becoming more diversified and industrial. In the first part of this period, the expansion in the economy took place in the older port cities, such as those of Boston, Philadelphia, New York, St. Louis, and Chicago. The older European immigrants—Germans, Irish, Canadian, British—concentrated in these cities. In the second part of this period, during the age of steam and steel, with the new changes in technology the most rapid expansion took place in the midwestern cities of Detroit, Milwaukee, Chicago, and Pittsburgh. The newer immigrants—from Poland, Italy, and Russia—became concentrated there.

Thus, distinct ethnic groups became concentrated in particular cities. Moreover, hand in hand with it went occupational concentration. The Italians, for example, became concentrated in construction; Polish, in steel; Jews in the garment industry. That the immigrants shared the same occupations meant that they also shared the same economic status and similar lifestyles, had common social and economic interests which could result in a common class consciousness, and associated a great deal with each other, leading to marriage (Yancey, Erickson, and Juliani 1976).

Work concentration also gave rise to residential concentration. Since the immigrants needed to live close to work, they settled in immigrant ghettos or ethnic enclaves such as "Little Italy" or "Old Polonia." City, work, and residential concentration in turn gave rise to the development of ethnic institutions in neighborhoods: grocery stores, bars, restaurants, churches, mutual aid societies and fraternal organizations such as The Sons of Italy, and newspapers such as the Jewish daily *The Forward*. These ethnic institutions, in turn, reinforced the immigrants' ethnic solidarity.

Today's immigrants have likewise become concentrated in certain places of settlement and in certain occupations (see Chávez; Bozorgmehr, Sabagh, and Light; and Manning, this volume). Among yesterday's immigrants (Jewish, Italian, Puerto Rican) as well as today's (Guatemalan, Mexican, Chinese), women have become incorporated in the garment industry. Above all, because it relied on a traditional skill that throughout much of the world defined womanhood—the ability to sew—and also because it relied on home work and subcontracting, allowing women to stay at home with their children to care for them. This advantage led women to accept low wages and exploitative conditions, as remains true today. At the turn of the century New York's garment industry mostly hired Jewish, Italian, and Puerto Rican women. Today immigrant women newly arrived from Latin America and Asia continue to supply the labor for the garment industry (Pedraza 1991).

Analytical Models

From the beginnings of social science in America, the question arose of how to best describe and understand the outcomes of the process of immigrant integration in America. For a long time the assimilation model dominated social science research as well as the popular understanding. Best expressed in the work of Nathan Glazer (1971), the assimilation model expected that, as the result of a natural, evolutionary process in due time immigrants and minorities would become like the dominant majority Americans. In essence, the model held out the promise that as immigrants and ethnics became acculturated—took on the values, customs, language, manner, and dress of the majority White population—entry into the major institutions and mainstream of the society would be achieved. Hence, the assimilation model held out the expectation that cultural assimilation would lead to structural assimilation (see Kamphoefner; Portes and Schlauffer, this volume).

The major challenge to assimilation theory came from the proponents of the internal colonialism model, the effort to delineate in what ways the experiences of the racial minorities (Blacks, Puerto Ricans, Mexicans, Native Americans—some of its oldest immigrants and most indigenous native sons and daughters) differed significantly from the experiences and eventual assimilation of the White European immigrants at the turn of the century. Best expressed in the work of Blauner (1969), Barrera (1979), and Acuña (1988), the internal colonialism model underscored that the experience of the racial minorities differed from that of the southern and eastern European immigrants in fundamental ways. First, the racial minorities entered this country involuntarily—the fate of African slaves (see Rawick, this volume)—or were conquered on their own soil—the fate of Native Americans, Mexicans, Puerto Ricans, and Filipinos (see in this volume Thornton; Romo; Carrasquillo and Sánchez-Korrol; and Cariño). By contrast, the European immigrants came voluntarily in search of a better life and transformed their culture at their own pace. Second, the immigrant ghetto was a one- and two-generation phenomenon for the southern and eastern Europeans. By contrast, for Blacks and other racial and ethnic minorities it was a more permanent phenomenon that only a few have escaped. Last, the southern and eastern Europeans generally experienced only a brief period during which their institutions were owned or controlled by outsiders. By contrast, the racial minorities were unable to control their own communities, owned and administered, as these were, by outsiders.

Just as the internal colonialism model posed a sharp critique to the assimilation model, the internal colonialism model itself underwent sharp criticism that essentially pointed to the significant differences in history and social, political, and economic conditions between the domestic patterns of race relations in the United States and what actually took place in Africa and Asia. Thus, critics insisted that the notion of internal colonialism might be of value only as a metaphor.

Despite the shortcomings of both the assimilation and internal colonialism models, they remain fully developed theoretical models that guide empirical research on the integration of immigrants in American society. The shortcomings of both the assimilation and internal colonialism model can be transcended by focusing on the varying ways in which different ethnic groups were incorporated in American society. As Joe Feagin (1978, p. 47) underscored, we need to pay attention to the *manner of incorporation* of immigrant groups in America—to "the initial and continuing placement and access of various groups" within the economic, political, and educational institutions of the society over the course of American history.

The concept of assimilation assumed that, irrespective of the initial placement of an ethnic group within the larger social structure, with the passage of time in a natural and evolutionary way, all immigrant and racial or ethnic groups would become like the dominant group—adopting the core culture, and eventually arriving at institutional parity in income, education, occupation, and political participation, as well as at intimate primary relations, social life, and marriage. The concept of incorporation does not assume eventual parity but, rather, seeks to explain the persistent inequality of certain immigrant, racial, or ethnic groups over time as a function of the difference in their initial and continuing placement within the larger social structure; historical placement that left them with differential access to the economic, political, and educational institutions of their society.

Broadly speaking, two very different patterns of ethnic incorporation existed in America in the historical past that still hold consequences for the present—patterns that we have chosen to distinguish as *discrimination* versus *exclusion* (Pedraza and Neidert 1990). A vast difference exists between the experience of those groups that suffered from discrimination and those that suffered from exclusion. Among those who suffered from discrimination were the Irish, Italians, Greeks, Jews, and Poles—the European immigrants, mostly from southern and eastern Europe, that came to America voluntarily in the mid-nineteenth to early twentieth centuries. Among those who suffered from social exclusion were Blacks, Native Americans, Mexican Americans, and Puerto Ricans—those whose history began as the product of involuntary conquest, annexation, and colonialism, as a result of which they were not al-

lowed to become integrated into the major institutions of the society. Initially also excluded from full social participation, at the turn of the century, Chinese and Japanese immigrants found themselves physically excluded from this society, their numbers remaining rather small. Lieberson (1980) has argued that the small size of these communities was a factor in their relative success, as they were not constantly disrupted by further immigration and were able to find certain "niches" in the economy, such as truck farming and small businesses for Japanese Americans (cf. Bonacich and Modell 1980).

Discrimination barred the southern and eastern European immigrants from equal access to the social rights of citizenship; exclusion kept the racial minorities from access to them altogether. The color line that developed in American society as a result of the initial patterns of conquest and colonization kept the Americans that were socially defined as racial minorities from the realization of the rights to good jobs, quality educations, decent neighborhoods, and political representation. Instead, it condemned them to a society segregated on the basis of color that became, as Booker T. Washington ([1899] 1969) once aptly portrayed it, a nation within a nation. As we know, in America segregation was practiced in all the core institutions of the society—from birth (in a segregated hospital), throughout the course of one's life (in segregated neighborhoods, public schools, colleges, sports, swimming pools, railroad cars, beauty parlors, drinking fountains, restaurants), to death (in segregated funeral parlors, cemeteries) (see Du Bois [1899] 1967; Frazier [1957] 1971; Montejano 1987; Myrdal 1964; Patterson 1982; Spear 1969; Thomas 1967; Woodward 1955; and in this volume see Marks; Morris and Herring; Romo; Snipp; and Thornton). The pervasive segregation the racial minorities suffered in America was more than a form of *discrimination*, it was a form of *exclusion*; it was not just a class subordination but a form of racial caste subordination held in place by law, custom, and violence that was particularly evident in the Deep South (see Davis, Gardner, and Gardner [1941] 1965), but pervaded all of American society. The essays in this volume emphasize the difference in the causes and consequences of the two broad patterns of historical experiences. Prejudice, discrimination, hardship, and poverty was the lot of the first-generation immigrants, mostly from southern and eastern Europe. But the experience of the racial minorities was different—different, as Thurgood Marshall (1978) repeatedly pointed out—not only in degree but also in kind.

Feagin and Eckberg (1980) distinguished different types of discrimination, from the individual discrimination practiced by a bigot or small group to the institutionalized discrimination that may be unintentional but is just as real in its consequences. Lieberson (1980), in his comparison of southern and eastern European immigrants in the North with Blacks who migrated from the South to the North at the turn of the century concluded that, of all the groups, Blacks faced the worst prejudice and the greatest discrimination, which had also served to put them at the bottom of the hiring queue of racial and ethnic groups in the United States at a time when jobs in the skilled sector were already in decline. Color mattered—as it still does (see Tienda and Stier, this volume).

When John Williams (1981) wrote his autobiography, he recalled that his mother, Ola, had migrated from Mississippi to Syracuse, joining after World War I that restless wave of migration from old home to new:

> Descended from a family five generations in America, she traveled within the boundaries of her country and was untouched by the immigration laws of 1882, 1917, and 1924. But she was a stranger who was quickly recognized by and restricted because of her color. . . .
>
> In fact, my mother had come north to work for a White family. She thought the North was salvation; while the streets were not paved with gold, she sensed opportunity. In Syracuse White people did not ride down upon you at night, and they did not lynch you. But she did not know that in Syracuse the White population simply left the Black population to moulder in the narrow alleys along E. Washington Street, where the Negro section was. (Williams 1981, pp. 133–134)

Because patterns of discrimination—individual and institutional—served to exclude some groups from full participation in the society, their access to the historical moments of opportunity that presented themselves over time also varied. For example, Harry Roskolenko (1981) remembers, in his autobiography called "America, the Thief: A Jewish Search for Freedom," the incredible poverty of the Jewish tenements where he spent his childhood in the Lower East Side of New York. His father's life was nothing less than brutal, working as a presser in the garment industry, "for it was the wrong time for a man of half skills who preferred God to the making of money to be in the United States of America." Yet later he adds:

> Who was not a wage slave on the East Side? But with World War I everything changed, in value, price, and wages. . . . There was money in the streets, at last. . . . America was no longer a thief. (Roskolenko 1981, p. 172)

Another example comes from the autobiography of Mario Puzo (1981), author of *The Godfather*, who was an adolescent when World War II broke out. In "Choosing a Dream: Italians in Hell's Kitchen," Mario Puzo tells us that:

> Not even my gift for retrospective falsification can make my 18th to 21st years seem like a happy time. I hated my life. I was being dragged into the trap I feared and had foreseen even as a child. It was all there, the steady job, the nice girl who would eventually get knocked up, and then the marriage and fighting over counting pennies to make ends meet. I noticed myself acting more unheroic all the time. I had to tell lies in pure self-defense, I did not forgive so easily.
>
> But I was delivered. When World War II broke out I was delighted. . . . My country called. I was delivered from my mother, my family, the girl I was loving passionately but did not love. And delivered WITHOUT GUILT. Heroically. . . . The war made all my dreams come true. I drove a jeep, toured Europe, had love affairs, found a wife, and lived the material for my first novel. (Puzo 1981, p. 46)

By contrast, those groups who suffered from exclusion were unable to benefit from these historical moments of opportunity to the same extent. World War II generated the rapid urbanization of Mexicans and Native Americans from the South and Southwest to the North and Midwest, the mass exodus of Puerto Ricans to the cities of the Northeast, and the huge waves of migration of Black Americans from the South to the North of the United States. But when the racial minorities arrived in "the promised land" (cf. Brown 1965), in truth they were not "the

At the turn of the century, the major wave of immigration that arrived then, until the 1924 Immigration and Nationality Act curtailed it, was mostly composed of immigrants from southern and eastern Europe. Carrying their belongings in straw suitcases, here Italian immigrants are just arriving at Ellis Island, around 1918-1920, where they were processed. (Theodore Wesley Koch Collection; Bently Historical Library, The University of Michigan)

newcomers" (Handlin 1959) but the latecomers, as they remained victims of a racial segregation and exclusion that barred them from the best economic and social opportunities (Agueros 1981; Blumer 1982; Montejano 1987; Rodriguez 1989; Wilson 1978). In "Halfway to Dick and Jane: A Puerto Rican Pilgrimage," Jack Agueros (1981, pp. 97–98) recalls his mother describing the reality of his childhood days:

When you were born we had been living as boarders. It was hard to find an apartment, even in Harlem. You saw signs that said "No Renting to Colored or Spanish." That meant Puerto Ricans. We used to say, "This is supposed to be a great country." But with a new baby we were determined not to be boarders and we took an apartment on 111th Street. Soon after we moved, I lost my job because my factory closed down. Your father was making seven or eight dollars a week in a terrible job in a carpet factory. They used to clean rugs, and your father's hands were always in strong chemicals. You know how funny some of his fin-

gernails are? It was from that factory. . . . With me out of work, a new apartment and therefore higher rent, we couldn't manage. Your father was furious when I mentioned home relief. He said he would rather starve than go on relief. But I went and filled out the papers and answered all the questions and swallowed my pride when they treated me like an intruder. I used to say to them, "Find me a job—get my husband a better job—we don't want home relief." But we had to take it.

The Civil Rights movement—the pioneering Black American civil rights movement, followed by the Chicano movement, and the Mexican American striving for *La Causa*, as well as the American Indian movement—was a major watershed in American society. Despite its success in delivering political and civic rights to America's racial minorities, the Civil Rights movement was not able to deliver social and economic rights. That would have entailed solving the problems of poverty and segregation that are still in our midst (see in this volume Snipp; Williams).

But the Civil Rights movement, together with the other movements it spawned—the American Indian movement, the Chicano movement—did break the pattern of racial exclusion, the social yoke under which Blacks, Mexicans, Native Americans, and Puerto Ricans had labored and lived (see in this volume Morris and Herring; Romo; Vigil; and Cornell). The internal migration of these minorities from south to north and rural to urban places and ways of life was an essential precursor to the development of their social movements that were able to break through the systematic exclusion and segregation imposed by Jim Crow legal segregation, beginning the process of inclusion that, while real, to this day remains only partly achieved. In "Time and Tide: Roots of Black Awareness," John Williams (1981) explained that after his experiences in the navy, where Jim Crow walked the planks, on ship and shore, for him to attend Syracuse University while working part time and raising a family was not the same as for the other students:

> It did seem to me that the White students were a great deal more casual about the college experience than I was, and this went for the ex-GIs as well. Perhaps they understood that, being White, a college education was merely frosting on the cake they already had. I looked at college as though it were the hand-truck with which to secure work. My father had to take his with him when he was looking for jobs; a man who didn't own his own hand-truck was seldom hired. College, I thought, was to be my tool. There was a big difference between its being the frosting and the hand-truck. (Williams 1981, pp. 143–144)

While the Civil Rights movement did succeed in opening up what had been a closed society, since that time we have progressively witnessed the development of what has come to be called postindustrial society, a service-oriented society that many argue cannot provide the economic opportunities the European ethnics had in the past (Bluestone 1982; Noyelle and Stanback 1984). For the racial minorities today, the reality is still too often one of sustained impoverishment and dashed hopes (Wilson 1978, 1987).

Those different experiences reside side by side in American cities today. In the United States, early in this century, ethnic enterprise was an important avenue of immigrant social mobility for first-generation Jews, Italians, Greeks, Chinese, and Japanese who, as a result, were able to escape urban poverty. At present, this "middleman minority" role is being played by Koreans, Asian Indians, Arabs, Cubans (especially in Puerto Rico), and Colombians, all of whom have quite directly replaced the old Jewish, Italian, Greek, and Chinese merchants, often by taking over their old businesses. These same groups whose success in America came via their being a "middleman minority" were also the brunt of much conflict. Historically that conflict often erupted between Blacks and Jews, despite their also being allies in the struggle for greater civil rights in America. That conflict between Blacks and Cubans surfaced in Miami in the mid 1980s (see Grenier and Pérez, this volume), and most recently in Los Angeles between Blacks and Koreans (see Min, this volume). As Robert Blauner (1969) pointed out so long ago, American minorities trapped in the hopelessness of the ghetto resent having their communities owned by outsiders, whose empowerment visibly reminds them of their own lack of efficacy.

Conclusion

Without doubt, the racial attitudes of Americans have undergone dramatic change in recent decades (see Schuman and Steeh, this volume). But America is still a racialized society where the very meaning of race and ethnicity continues to be contested (see Omi and Winant, this volume). Many of the new immigrants come to America with substantial social resources (social class, capital, education, institutional knowledge) typical of the "brain drain" migration of skilled professionals and often also found among refugees. As a result, these new immigrants can quickly insert themselves at rather high levels in this society—what Ivan Light (1983, p. 322) has called "leapfrog migration." At the same time, the prospects for working-class Americans and the domestic poor, many of whom are the nation's oldest sons and daughters—African Americans, Mexican Americans, Native Americans, Puerto Ricans—grow increasingly dim, as they bear the brunt of the deindustrialization of America with its persistent and growing poverty (Wilson 1994). The 1992 riots in Los Angeles have served, once again, to bring the old and persistent fact of poverty among minorities in America to our national consciousness. As Martin Luther King, Jr., once said, "A riot is the language of the unheard."

As sociologists, historians, anthropologists, and as Americans—new and old—the authors of this volume collectively hope that its essays will help shed light on the causes and consequences of the social patterning of immigration, race, and ethnicity in America over the course of its history. As Robert Merton (1949) underscored so long ago, to the extent to which we succeed in this aim, we will contribute to clarifying the alternatives for effective social programs aimed to reduce conflict and inequality among groups. And that in itself should contribute to the realization—someday—of America's ideals of equality of opportunity and justice for all.

2

Origins and Destinies: Immigration, Race, and Ethnicity in Contemporary America

RUBÉN G. RUMBAUT

How curious a land is this—how full of untold story, of tragedy and laughter, and the rich legacy of human life; shadowed with a tragic past, and big with future promise! . . . It is a land of rapid contrasts and of curiously mingled hope and pain.

— *W. E. B. DuBois*, The Souls of Black Folk *(1903)*

Those apt and emotive words may serve as an epigraph for America at the end of the twentieth century, a land of rapid contrasts being again transformed by new waves of immigration that rival those of the tumultuous era before World War I. But DuBois was describing the "Black Belt" at the beginning of the century as he traveled by train across Georgia in the "Jim Crow Car," passing by the land of the Cherokees and the Creek Indians—"and a hard time the Georgians had to seize it"—to "the centre of the Negro problem . . . of America's dark heritage from slavery and the slave trade" (DuBois [1903] 1989, 78–79; see also Rawick, this volume). In *The Souls of Black Folk*, written soon after Africa was partitioned by European colonial powers and Puerto Rico, the Philippines, Guam, and the Hawaiian islands were seized by the United States in the aftermath of the Spanish-American-Cuban War, DuBois had prophesied that "the problem of the twentieth century is the problem of the color line—the relation of the darker to the lighter races of men in Asia and Africa, in America and the islands of the sea" (DuBois 1989, 10).

Much has changed since then, mainly in the years after World War II, including the decolonization of Africa and Asia and the English-speaking Caribbean, and the dismantling of Jim Crow in the United States; but much has not, so that another preeminent African American scholar, John Hope Franklin, could "venture to state categorically that the problem of the twenty-first century will be the problem of the color line" (Franklin 1993, 5). Class, not color, had shaped the fates of the "white ethnics"—southern and eastern Europeans—whose arrival by the millions around the turn of the century culminated in the restrictionist national-origins laws of the 1920s (see Alba; Gold and Phillips; and Pacyga, this volume). The Great Depression and World War II combined to cut the flow of immigrants to America to its lowest point since the 1820s. But the new and rapidly accelerating immigration to the United States—from Asia, Latin America and the Caribbean, and increasingly (though still relatively small) from Africa and the Middle East—is unprecedented in its diversity of color, class, and national origins; it is changing fundamentally the racial and ethnic composition and stratification of the American population, and perhaps also the social meanings of race and ethnicity, and of American identity (see Omi and Winant, and Waters, this volume). As were their predecessors, the new immigrants and their children are themselves being transformed in the process, and their destinies, as yet full of untold stories, will bear the stamp of their diverse origins.

The stories that *are* being told in the news media of the day—particularly in the "Immigrant Belt" of global cities like Los Angeles, New York, and Miami—are full of the dramatic contrasts and the "curiously mingled hope and pain" of contemporary immigrants from all over the world who enter, with or without permission, in search of future promise or to escape a tragic past, and of the variety of ways in which the natives respond, often with alarm, to their presence. As in the past, American nativism is exacerbated during periods of economic recession. Nowhere has that been more palpable than in California in the early 1990s, where the deepest and most prolonged recession since the Great Depression, in part a product of post–Cold War cutbacks in the defense industry and military spending, coincided with the largest concentration of legal and illegal immigrants and refugees in the country. Governor Pete Wilson led other politicians of both parties in singling out the cost of providing public services to immigrants, especially in Los Angeles, as the cause of the state's budget deficit, and called for the denial of citizenship to U.S.–born children of undocumented immigrants. This anti-immigrant sentiment culminated in Proposition 187, a California measure that requires schools, hospitals, and other public agencies to report to the Immigration and Naturalization Service (INS) any persons suspected of being in the United States illegally, so that they may be detained, deported, and denied basic public services. Although Proposition 187 appears to violate constitutional laws and is pending review in the courts, it passed by a landslide margin with 59 percent of the vote in November 1994.

Yet in sharp contrast to the poverty and vulnerable status of many newcomers, such as laborers from Mexico and Central America and refugees from Southeast Asia, many immigrant communities throughout Southern California have thrived despite the recession. Iranians make up about a quarter of the high school enrollments in upscale Beverly Hills. In 1992 nearly one in five (19 percent) home buyers in Los Angeles County had a Chinese surname—although the Chinese make up only 2.7 percent of the county's huge population—propping up the moribund housing market and revitalizing entire neighborhoods. Some of these homes are so-called monster houses in rich suburbs for the wives and children of Taiwanese businessmen (called "astronauts" locally) who spend most of their time abroad making money. In 1993 in nearby Monterey Park—known as "Little Taipei" and as "the first suburban Chinatown"—a businessman from mainland China walked into a Taiwanese community bank, was assured by the manager that his transactions would be kept confidential from Chinese authorities, and deposited $10,000 on the spot, returning the next day to deposit another $900,000 by wire transfer. Business is booming in China's coastal provinces, and private fortunes are winding their way through Hong Kong to the United States and Canada; as a report in the *Los Angeles Times* put it, "The Gold Rush, this time, is from west to east" (Schoenberger 1993). And the 1994 lifting of the U.S. trade embargo against Vietnam, whose rapidly growing economy is widely expected to make it the next "Asian tiger," will stimulate Vietnamese businesses throughout California, and Orange County's "Little Saigon" may come to rival the economic dynamism of "Koreatown" just to the north; already Vietnamese developers and realtors in California report a growing flow of capital from Vietnam as entrepreneurs there seek safe havens for their excess cash (Kotkin 1994). No one had imagined such an outcome when a million mostly indigent Indochinese refugees came in the years after the fall of Saigon as part of the largest refugee resettlement program in U.S. history.

Such reports, however, are less common than those that focus on undocumented immigrants, such as nannies and housecleaners for native-born affluent families (and government officials who since 1993 have discovered they had a "Zöe Baird problem"), and those who wait at busy corners throughout Southern California to be hired for day jobs—an estimated forty sites in Los Angeles County alone: "Each year they probably complete close to a million workdays of dirt moving, tree trimming, hauling, moving, painting, yard and pool cleaning, dry-walling, plastering and tiling for the region's homeowners and small contractors. It is labor that might not get done if it weren't for migrants willing to work for $5 or less an hour" (Kelley 1990). Several of the top economic sectors in California—especially agriculture, apparel, and construction—depend almost exclusively on immigrant labor. Still, after the massive earthquake in Los Angeles in 1994, legislation was introduced to deny assistance to foreign-born victims who could not prove their legal status.

Meanwhile, on the other side of the country, the *Miami Herald* reported that just before Christmas 1993, seven Cubans and ten Haitians pooled their resources and sailed from the Bahamas—literally in the same boat—to Florida, where upon arrival the Cubans were duly processed by the INS while the Haitians were sent to detention facilities (García and Cavanaugh 1993). In 1992, the INS approved over 96 percent of Cubans who applied for refugee status, but less than 11 percent of Haitians (INS 1993). Color, Cold War foreign policies, and the political clout of Cuban ethnic communities in the United States have helped to shape the contradictory reception accorded to escapees from Cuba and Haiti who make it to the United States: the Cubans, fleeing an economic crisis deepened by the 1989 collapse of the Soviet Union and a tightened U.S.–imposed trade embargo, were generally guaranteed political asylum and, under the Cuban Adjustment Act of 1966, became eligible for permanent residency a year after their arrival; the Haitians, fleeing the political terrorism of a military regime that deposed the democratically elected president in 1991 (until he was restored after the 1994 U.S. invasion of Haiti), were detained and subject to deportation as economic migrants. Some made it spectacularly, as did a twenty-one-year-old Cuban who in 1994 crossed 110 miles of shark-infested waters from Cuba to the Florida keys riding a windsurfer (the second such crossing in recent years). Many did not make it at all, and thousands of *balseros*, or "raft people," drowned at sea. Dramatic events on the high seas in August 1994 finally led the Clinton Administration to reverse the three-decade-old U.S. preferential policy toward Cubans.[1]

Such examples make for sensational copy, and they humanize and illustrate memorably the extraordinary diversity of the new immigrants, their motives and modes of passage to America, and their contexts of exit and reception. This essay is an effort to make sense of that diversity through a broad-brush sketch of the social and historical origins of immigration to the United States since the end of World War II. The focus will be on the types of immigrants that comprise contemporary flows, their patterns of destination and settlement, their distinctive social and economic characteristics compared to major native-born racial-ethnic groups, and their different modes of incorporation in—and consequences for—American society.

A twofold classification of contemporary immigrants will help to organize our analysis and render their diverse origins and destinies more comprehensible. First, depending on their socioeconomic status, we focus on three salient types of immigrants: professionals, entrepreneurs, and manual laborers. Second, depending on their legal-political status, we distinguish among three basic types: regular immigrants (who enter under provisions of U.S. law that favor family reunification as well as highly skilled professionals), undocumented immigrants (who enter illegally by crossing the border or overstaying a temporary visa), and refugees (or asylees, in the case of those who claim political asylum once in the United States). Each of these types is represented by several nationalities; and conversely, within a single nationality may be found individuals who represent different types (Portes and Rumbaut 1990). In addition, these socioeconomic and legal-political statuses crisscross and combine in ways that magnify the relative social advantages or disadvantages of particular immigrant groups as they make their way in America. For example, the undocumented tend to consist disproportionately of manual laborers, whose legal vulnerability makes them in turn more economically exploitable; professionals are typically found among the first waves of refugee flows, and among "pioneer" immigrants who come under the occupational preferences of U.S. law (whereupon, once established as permanent residents or U.S. citizens, they can sponsor the immigration of their immediate family members). These and other factors—such as the extent of racial discrimination and nativist prejudice, the state of the economy, the structure and cohesiveness of families, and the presence or absence of strong coethnic communities in areas of immigrant settlement—form the contexts within which immigrant groups adapt and shape their diverse fates.

Changes in U.S. immigration laws—in particular the amendments passed in 1965, which abolished the national origins quota system and changed the preference system to give priority to family reunification over occupational skills—have often been singled out as the principal reason for the "new immigration" and the change in its composition. But the causal effects of the Hart-Celler Act of 1965 have been exaggerated; and its most important consequences, such as the removal of barriers to immigrants from Asian and African countries, were largely unintended. The law does matter, of course: it influences migration decisions and constitutes a key context of reception shaping the destinies of newcomers, especially their right to full membership and future citizenship. But it cannot control historical forces or determine the size or source of migration flows.

Migration patterns are rooted in historical relationships established between the United States and the principal sending countries; that is, the size and source of new immigrant communities in the United States today is directly if variously related to the history of American military, political, economic, and cultural involvement and intervention in the sending countries, and to the linkages that are formed in the process which (often unintentionally) open a surprising variety of legal and illegal migration pathways. A remarkable recent example involves the thousands of former Iraqi soldiers and prisoners of war from the 1991 Persian Gulf War who resettled in the United States as refugees. Thus, immigration to the United States and the formation of new American ethnic groups may be seen as the complex and deeply ironic social consequences of the expansion of the nation to its post–World War II position of global hegemony. As the United States has become more deeply involved in the world, the world has become more deeply involved in America—indeed, in diverse ways, it has come to America.

A World on the Move: Immigration to the U.S. Since World War II

International migration has been a major feature of the new world order that emerged from the ashes of World War II, for a complex set of reasons (Massey et al. 1993). Societies have become increasingly linked in numerous ways—economically, politically, culturally—and modern consumption standards (especially American lifestyles and popular culture) have been diffused worldwide at an accelerating rate. Transnational population movements of workers, refugees, and their families are but one of many other global exchanges of capital, commodities, and

information across state borders, all facilitated by a postwar revolution in transportation and communication technologies that have reduced the costs of travel and raised expectations about opportunities abroad. In general, the patterns reflect the nature of contemporary global inequality: capital flows from rich to poor countries and labor flows from less developed to more developed regions. Also, the era since World War II—the era of the Cold War and global superpower confrontation, decolonization and the formation of new states, revolutions and counterrevolutions—has been characterized by continuing flows of refugees primarily from one Third World country to another (Zolberg et al. 1989). Since the disintegration of the Second World at the end of the 1980s and another round of new state formation, new refugee flows have been added in Europe, Africa, and elsewhere.

In absolute numbers, the United States remains by far the principal receiving country: the 19.8 million foreigners counted in the 1990 U.S. census formed the largest immigrant population in the world—indeed, in world history. In *relative* terms, however, only 7.9 percent of the 1990 U.S. population was foreign-born, a percentage exceeded by many countries and a much lower proportion than had been the case earlier in this century, as Table 1 makes clear. Still, net immigration accounted for 39 percent of total U.S. population growth during the 1980s—a proportion larger by far than any other decade in the twentieth century except for 1901–1910.[2] The so-called foreign stock population—the two generations of immigrants plus their children born in the United States—is near an all-time high, some 45 million in 1990, and poised to increase much more rapidly in the coming decades. Although little is known about their adaptation patterns to date, the new second generation—the children of the new immigrants—are bound to represent a crucial component of American society in the years to come.

Table 1 shows decennial trends in the changing size and composition of immigrant flows to the United States for the century from 1890 to 1990, using data from the two main national-level sources: the U.S. census (counts of all foreign-born persons, regardless of their legal status) and the INS (counts of persons "admitted to lawful permanent residence," who either entered the United States with an immigrant visa from abroad or were already in the United States but adjusted their status from temporary to

Table 1 Decennial Trends, 1890–1990, in the U.S. Foreign-born Population (Census Data), in Legal Immigration by Region of Origin (INS Data on Admissions to Permanent Residence), and in Net Immigration Proportion of Total U.S. Population Growth

| Census Year | Census Data: Foreign-born Population | | | Immigration and Naturalization Service (INS) Data: Immigration by Decade and Region of Last Residence | | | | | | |
	N (1000s)	Foreign-born of Total U.S. Population (%)	Decade	N (1000s)	North/West Europe and Canada (%)	South/East Europe (%)	Latin America (%)	Asia (%)	Population Growth Due to Net Immigration (%)
1900	10,445	13.6	1891–00	3,688	44.7	51.8	1.0	2.0	20.3
1910	13,360	14.7	1901–10	8,795	23.8	69.9	2.1	3.7	39.6
1920	14,020	13.2	1911–20	5,736	30.3	58.0	7.0	4.3	17.7
1930	14,283	11.6	1921–30	4,107	53.8	28.7	14.4	2.7	15.0
1940	11,657	8.8	1931–40	528	58.0	28.3	9.7	3.1	1.6
1950	10,431	6.9	1941–50	1,035	63.8	12.8	14.9	3.6	8.8
1960	9,738	5.5	1951–60	2,515	51.8	16.0	22.2	6.1	10.6
1970	9,619	4.7	1961–70	3,322	30.0	16.3	38.6	12.9	16.1
1980	14,080	6.2	1971–80	4,493	10.2	11.4	40.3	35.3	17.9
1990	19,767	7.9	1981–90	7,338 [a]	7.2	5.3	47.1	37.3	39.1

[a]Data include 1,359,186 formerly undocumented immigrants who had resided in the United States since 1982 and whose status was legalized in fiscal years 1989 and 1990 under the provisions of the Immigration Reform and Control Act (IRCA) of 1986. An additional 1.7 million eligible legalization applicants, already qualified under the IRCA, had not yet adjusted their status to permanent resident as of 1990 and are thus not included in this table; they are reflected in INS statistics for fiscal 1991 and subsequent years. Indeed, in 1991 a record total of 1,827,167 immigrants were legally admitted into the United States; of these, 1,123,162 were IRCA legalizees. In 1992 and 1993 respectively, admissions totaled 973,977 and 904,292, of which IRCA legalizations numbered 163,342 and 24,278.
Sources: U.S. Bureau of the Census, *Statistical Abstracts of the United States* (112th ed.), 1992, Tables 1, 5–6, 45; U.S. Immigration and Naturalization Service, *Statistical Yearbooks*, 1990–1993, Tables 1–2.

permanent resident). The INS data do not reflect the entry of unauthorized immigrants into the United States, deaths, or emigration from the United States, which accounts for many of the discrepancies between the two data sets in Table 1. For example, the census data show an increase of some 3 million in the foreign-born population between 1900 and 1910; but INS data show 8.8 million immigrants were admitted during that decade, the largest number on record. Much of this flow was initiated by active recruitment on the part of American employers, and many immigrants returned home after a few years in the United States—"birds of passage," predominantly young single men, whose movements tended to follow the ups and downs of the American business cycle (Piore 1979).

After World War I broke out, immigration began an uneven and then precipitous decline until the trend reversed itself immediately after World War II, and has been increasing rapidly since. In the post–World War II period, legal immigration flows are more apt to be sustained by family reunification preferences in the allocation of immigrant visas and by kinship networks developed over time (Jasso and Rosenzweig 1990) than by economic cycles and deliberate recruitment. While in the first decades of the century 67 percent of all immigrants were men, since 1941 the majority—55 percent—have been women (INS 1991). By 1940, only 13 percent of immigrants were coming from Asia and Latin America, while 86 percent came from Europe and Canada. The pattern had been completely reversed by 1990, with 84 per-

Like yesterday's immigrants, today's also prefer to live and work in the comfort of their own ethnic en-claves. Here a Vietnamese woman entrepreneur is in her shop, filled with Oriental paintings and porce-lains, in Little Saigon, Westminster, California, 1992. (Photo by Steve Gold)

cent coming from Asia and Latin America and less than 13 percent from Europe and Canada (see Table 1).

Contemporary flows include sizable and increasing proportions of (1) *political refugees and asylees*, beginning with the 1948 Displaced Persons Act, which first recognized refugees in U.S. law; (2) *highly skilled professionals*, who have entered variously under employment-based visa preferences, in the first waves of refugee exoduses, or under student or other temporary statuses; (3) *undocumented laborers*, whose numbers began to swell after the termination in 1964 of the Bracero Program (begun in 1942 to meet labor shortages in the southwest during World War II, then maintained during the postwar years of rapid expansion of the U.S. economy), and also ironically after passage of the 1965 law (and after another in 1976) which placed for the first time numerical limits on legal immigration from the Western Hemisphere even as it abolished the national-origins quota system that had governed admissions from the Eastern Hemisphere; and (4) *persons entering on "nonimmigrant" visas* (university students, exchange visitors, temporary workers, and many others). These characteristics provide in part the basis for the typology of immigrants proposed above.

For all of the attention it has received, the 1924 Johnson-Reed Act that set preferential quotas for immigrants from northwest Europe had virtually no effect on immigration from Asia, which had already been barred in 1917, except for the U.S. colony in the Philippines; much earlier, in 1882, the Chinese had been excluded (a ban not repealed until 1943, when the United States and China were wartime allies), and the Japanese were restricted after 1907 (see Chow, and Glenn and Parreñas, this volume). The 1924 law also had no effect on immigration from the Americas. Largely at the urging of American growers and ranchers, no limits were set on Western Hemisphere countries by the 1924 law: it was understood that cheap, unskilled Mexican labor could be recruited when needed, as happened during World War I and the 1920s (Vargas 1993), and again during the Bracero Program (see Romo, this volume), and that those laborers could be deported *en masse* when they were no longer needed, as happened during the 1930s and again during "Operation Wetback" in the mid-1950s. Similarly, the McCarran-Walter Act of 1952 (passed over President Truman's veto, who opposed it because it hampered U.S. foreign policy) basically kept the Eastern Hemisphere quota system and exempted the independent nations of the Western Hemisphere from any limits. Again at the urging of growers and ranchers, the act included a "Texas Proviso" exempting employers from sanctions for hiring illegal aliens that in fact encouraged undocumented immigration, all the more after the Bracero Program was ended in 1964. [This loophole was formally closed by the Immigration Reform and Control Act (IRCA) in 1986.]

To this general picture of post–World War II immigration flows and immigration policies need to be added the internal migration of Blacks from the south to the north and west (which dwarfed the earlier Great Migration of the World War I era—see Marks, this volume), and the island-to-mainland journey of Puerto Ricans (who are also not "immigrants" but travel freely as U.S. citizens by birth), since both form a key part of the labor history of the period. Labor recruitment of Puerto Ricans became widespread among employers in the northeast during and after World War II, when cheap air travel was instituted between San Juan and New York (a one-way ticket cost less than $50). Mass immigration to New York reached its peak in the 1950s and made Puerto Ricans the first "airborne" migration in U.S. history. Net migration to the mainland during the peak decade of the 1950s (about 470,000) was higher than the

immigration totals of any country, including Mexico (Falcón 1991; Rumbaut 1992). Puerto Ricans have since become the nation's third largest ethnic minority, after African Americans and Mexican Americans (see Carrasquillo and Sánchez-Korrol, this volume).

Still, in its size and above all in its extraordinary diversity, the 1980s has rivaled any other decade in U.S. immigration history: nearly half (44 percent) of the 19.8 million immigrants counted in the 1990 census—8.7 million people—arrived during the 1980s. Given current trends, the size of the immigrant population may well be eclipsed and its composition further diversified during the 1990s. Already 1990 and 1991 set annual records for legal admissions (over 1.5 and 1.8 million, respectively) as a result of legalizing formerly undocumented immigrants under the amnesty provisions of IRCA. The Immigration Act of 1990 (implemented in 1992) increased worldwide legal immigration limits by about 40 percent, to 700,000 per year through 1994 and 675,000 thereafter; of these, employment-based visas (reserved largely for professionals) nearly tripled to 140,000, and family-sponsored admissions also expanded to about half a million annually. The law also set aside 55,000 visas annually through 1994 for the spouses and children of immigrants whose status was legalized under the IRCA, and 120,000 "diversity visas" for immigrants from thirty-four countries "adversely affected" by the 1965 law, with the lion's share going to natives of Ireland (tens of thousands of whom had overstayed their temporary visas and remained in the United States illegally in recent years). The number of temporary "nonimmigrants" admitted set a record in 1992: nearly 21 million persons, mostly tourists and businesspeople but also over half a million foreign students and exchange visitors and their families (INS 1993). In addition, since passage of the 1980 Refugee Act, refugees and asylees are admitted under *separate* ceilings determined each year by the administration and Congress. Since the end of the Cold War, refugee admissions have actually increased: in 1992 the refugee ceiling was 142,000, up from 125,000 in 1990; and the asylee ceiling has been raised to 10,000 per year, even though the number of asylum applications doubled to 104,000 in 1992.

The undocumented immigrant population has not only grown but diversified (see Chávez, this volume). The IRCA did not stop the flow of unauthorized migrants, despite a big drop immediately after 1986; the number of apprehensions along the Mexican border increased abruptly after 1989, reaching 1.3 million in 1992, and the number of deportations grew to 1.1 million. Excluding the nearly 3 million formerly undocumented immigrants whose status was legalized under the IRCA (over 2 million were Mexican nationals, followed by Salvadorans, Guatemalans, and Haitians), the INS estimated that the illegal resident population—consisting of "entries without inspection" and "visa overstayers"—totaled about 3.2 million by 1992, and was growing at a rate of perhaps 300,000 annually. Of that total, about 1 million were estimated to be from Mexico, 298,000 from El Salvador, 121,000 from Guatemala, and roughly 100,000 each from Canada, Poland, the Philippines, and Haiti (U.S. INS 1992b).

Passages to America: The National Origins of the New Immigration

Census data on the size, year of immigration, citizenship, and states of principal settlement of the 1990 foreign-born population are presented in Table 2, broken down by the major regions and countries of birth. For the first time in U.S. history, Latin American and Caribbean peoples replaced Europeans as the largest immigrant population in the country. Fully half of those 8.4 million immigrants came during the 1980s. Mexicans alone accounted for 22 percent (4.3 million) of the total foreign-born population, and 26 percent of all immigrants arriving since 1970; Mexico is by far the largest source of both legal and illegal immigration. The Philippines ranked second, with close to one million immigrants and 5 percent of the total. Indeed, Mexicans and Filipinos comprise, respectively, the largest Hispanic and Asian immigrant groups in the U.S. today. Thus, while today's immigrants come from over 140 different countries, some regions and nations clearly send many more than others, despite the equitable numerical quotas provided to each country by U.S. law since 1965. Regionally, the Asian and African immigrant flows grew fastest of all, with most arriving in the 1980s; only 12 percent of immigrants coming in the 1980s hailed from Europe and Canada.

One pattern, a continuation of trends already under way in the 1950s, is quite clear: immigration from the more developed countries has declined over time, while from less developed countries it has grown sharply. Among the more developed countries, this pattern is clearest for Canada and most European countries, with the sharpest reductions occurring since 1960. Although traditional countries of immigration to the United States in the past, by the 1960s their prosperous postwar economies dampened the relative attraction of America. About half of all Europeans and Canadians came before 1960; many British, German, and other European scientists and professionals journeyed to America in the aftermath of the war to pursue opportunities not available in their countries—Britain alone lost 16 percent of its Ph.D.s between 1952 and 1961, half of them coming to the United States (Weisberger 1994, 88). Among the less developed countries, the major countries of immigration are located either in the Caribbean basin—in the immediate periphery of the United States—or are a handful of Asian nations also characterized by significant historical, economic, political, and military ties to the United States. These historical relationships, and the particular social networks to which they give rise, are crucial to an understanding of the new immigration, both legal and illegal—and help explain why most less developed countries are not similarly represented in contemporary flows, as might be predicted by neoclassical economic theories of transnational labor movements.

In fact, just a baker's dozen of countries has accounted for two-thirds of all immigrants since 1970: Mexico, Cuba, El Salvador, Guatemala, Nicaragua, the Dominican Republic, and Haiti in the Caribbean Basin; and the Philippines, Vietnam, South Korea, China, Taiwan, and India in Asia. Not surprisingly, the two largest source countries, Mexico and the Philippines, share the deepest structural linkages with the United States, including a long history of dependency relationships, external intervention, and (in the case of the Philippines) colonization. In both countries, decades of active agricultural labor recruitment by the United States—of Mexicans to the Southwest, and Filipinos to plantations in Hawaii and California—preceded the establishment of chain migrations of family members and eventually of large and self-sustaining migratory social networks.

In the case of Mexico, the process has evolved over several generations (see Romo, this volume). From California to Texas, the largest Mexican-origin communities in the United States are still located in former Mexican territories that were annexed in the

Table 2 Size, Year of Immigration, U.S. Citizenship, and Patterns of Concentration of Principal Foreign-born Groups in the United States in 1990, by Region and Country of Birth, in Rank Order by Recency of Arrival

Region/Country of Birth	Persons (N)	(%)	Year of Immigration to the U.S.				Naturalized U.S. Citizen		States of Principal Settlement		
			1980s (%)	1970s (%)	1960s (%)	Pre-1960 (%)	Yes (%)	No (%)	Calif. (%)	NY/NJ (%)	Florida (%)
Africa	363,819	1.8	61	28	7	4	34	66	18.1	22.2	4.1
Asia	4,979,03	25.2	57	29	9	5	41	59	40.2	15.7	2.3
Lat. Amer./Carib.	8,416,924	42.6	50	28	15	7	27	73	38.7	17.9	12.8
Eur. & Canada	5,095,233	25.8	20	13	19	48	63	37	16.1	24.6	7.5
Cambodia *	118,833	0.6	86	14	0	0	20	80	47.5	3.0	1.1
Laos *	171,577	0.9	73	27	0	0	17	83	42.0	1.8	1.3
Vietnam *	543,262	2.7	64	35	1	0	43	57	49.9	3.9	2.4
El Salvador	485,433	2.4	76	19	4	1	15	85	60.3	10.5	2.1
Guatemala	225,739	1.1	69	22	7	2	17	83	60.2	10.7	5.1
Nicaragua	168,659	0.9	75	16	5	4	15	85	34.6	7.1	42.7
Korea	568,397	2.9	55	37	6	2	41	59	35.2	17.5	1.5
Taiwan	244,102	1.2	65	27	8	1	39	61	42.9	16.8	1.9
Iran	210,941	1.1	50	41	6	3	27	73	54.7	9.1	2.3
Haiti	225,393	1.1	61	26	11	2	27	73	1.2	45.7	36.9
India	450,406	2.3	58	30	10	2	35	65	18.6	26.4	2.7
Philippines	912,674	4.6	51	31	13	5	54	46	52.8	9.8	2.4
Mexico	4,298,014	21.6	50	31	11	8	23	77	57.6	1.3	1.3
Dominican Rep.	347,858	1.8	53	27	17	3	28	72	1.0	79.9	6.7
Colombia	286,124	1.4	52	27	18	3	29	71	10.7	43.0	23.3
Jamaica	334,140	1.7	47	33	15	5	38	62	3.4	50.2	22.1
Hong Kong	147,131	0.7	44	33	19	4	55	45	43.9	24.5	1.8
China	529,837	2.7	55	21	13	11	44	56	39.9	27.5	1.5
Japan	290,128	1.5	53	16	14	17	28	72	33.6	14.2	2.3
Cuba *	736,971	3.7	26	19	46	9	51	49	6.7	15.6	67.5
Soviet Union *	333,725	1.7	34	17	3	46	59	41	23.2	35.3	5.3
Poland *	388,328	2.0	33	11	13	43	62	38	7.5	32.8	6.8
Portugal	210,122	1.1	25	36	29	10	44	56	17.1	23.4	1.8
Greece	177,398	0.9	13	28	28	31	71	29	9.5	32.6	5.1
United Kingdom	640,145	3.2	25	15	20	40	50	50	21.2	16.0	9.5
Canada	744,830	3.8	17	12	20	51	54	46	21.0	9.6	10.4
Ireland	169,827	0.9	19	8	16	57	68	32	11.1	40.0	4.5
Germany	711,929	3.6	11	8	22	59	72	28	14.6	18.9	7.8
Italy	580,592	2.9	5	14	23	58	76	24	8.3	44.9	4.9
Totals:											
Foreign-born:	19,767,316	100.0	44	25	14	17	41	59	32.7	19.3	8.4
Native-born:	228,942,557	100.0	—	—	—	—	—	—	10.2	9.6	4.9

* Denotes country from which most recent migrants to the United States have been officially admitted as refugees.
Source: U.S. Bureau of the Census, *1990 Ethnic Profiles for States*, CPH-L-136, 1993; *The Foreign Born Population in the United States*, 1990 CP-3–1, July 1993, Tables 1, 3; and *The Foreign Born Population in the United States: 1990*, CPH-L-98, 1993, Table 13. Data on year of immigration are drawn from a 5 percent public use microdata sample (PUMS) of the 1990 census, and subject to sample variability.

last century—although other large communities stretch to Chicago and the Midwest—and they are today linked to entire communities on the other side of the border (Massey et al. 1987; Portes and Bach 1985; Rumbaut 1992). In the context of Mexico's internal economic crises in the 1970s, 1980s, and mid-1990s, big wage differentials and strong demand for Mexican labor across the 2,000-mile-long U.S.-Mexico border have acted as a magnet to attract immigrants to America. The 1993 passage of the North American Free Trade Agreement (NAFTA) may stimulate further Mexican immigration to the United States, at least in the short-term.

Unlike Puerto Rico, which also came under U.S. hegemony as a result of the 1898 Spanish-American War, the Philippines secured its formal independence from the United States after World War II (symbolically, on July 4, 1946). This has since led to different patterns of immigration (see Cariño, this volume). During the half-century of U.S. colonization, the Americanization of Filipino culture was pervasive, especially in the development of a U.S.-styled educational system and the adoption of English as an official language. Today the United States is not only the Philippines' major trading partner, but also accounts for more than half of total foreign investment there and for the second largest share of income of the country's gross national product. Since the 1960s, the Philippines have sent the largest number of immigrant professionals to the United States, particularly nurses, and a high proportion of the many international students enrolled in American universities. The extensive U.S. military presence in the Philippines—including until recently the largest American bases in the Asian-Pacific region—has also fueled immigration through marriages with U.S. citizens stationed there, through unique arrangements granting U.S. citizenship to Filipinos who served in the armed forces during World War II, and through direct recruitment of Filipinos into the U.S. Navy. Remarkably, by 1970 there were more Filipinos in the U.S. Navy (14,000) than in the entire Filipino navy (Reimers 1985); and in San Diego, site of the third largest Filipino community in the United States, about half of the Filipino labor force is employed by the U.S. Navy (Rumbaut 1991c).

American foreign policy, particularly the post–World War II doctrine of communist containment, is of key importance in explaining many of the most recent sizeable migrations from different world re-

gions. As of 1990, as Table 2 shows, the most recently arrived Asian groups were the Cambodians, Laotians, and Vietnamese, admitted as political refugees—a dialectical legacy of the U.S. role in the Indochina War (see Karnow 1991; and Rumbaut, this volume). The most recent European arrivals were Soviet Jews and Poles, who also have been admitted mainly as refugees, like other groups from communist countries. From Latin America the most recent arrivals were the Salvadorans, Guatemalans, and Nicaraguans, who fled civil wars and deteriorating economic conditions in Central America in the 1980s in a context long shaped by American foreign policy (LaFeber 1983; Niess 1990). As with Haitians, Salvadorans and Guatemalans have been denied refugee status and entered mostly without documents (although the Immigration Act of 1990 granted "temporary protected status" to many Salvadorans). Cubans, who comprise the oldest and (after Mexico) largest of Latin American immigrant groups, entered primarily in the 1960s, despite the chaotic flotilla of 125,000 *Marielitos* that began the decade of the 1980s (see the chapter on Cubans in this volume). With the exception of the Mariel "entrants," they have been the classic example of the use by the United States of refugee policy as foreign policy (Domínguez 1992). In fact, it has been to "prevent another Cuba" that a variety of U.S. interventions throughout the Caribbean basin have been justified, such as in the Dominican Republic after the assassination of Trujillo in 1961 and the U.S. military occupation in 1965, which opened key immigration pathways that over time have led to the large Dominican population in the United States (see Grasmuck and Pessar 1991; and Mitchell 1992).

Among the other leading countries of recent immigration, linkages unwittingly structured by American foreign policy and military intervention are most salient in the exodus of the Koreans in the aftermath of the Korean War and the subsequent U.S. economic and political involvement and permanent military presence in South Korea (Dudden 1992; Kim 1987a). There, as in India and Taiwan, large-scale U.S. foreign aid, technical assistance, trade, and direct investment (which in India surpassed that of the United Kingdom after decolonization in 1947) helped to forge the channels for many professionals and university students to come to America. Emigration connections forged by U.S. intervention and foreign and immigration policies are also a common denominator in the exodus of the Chinese after the 1949 revolution, and Iranians after the 1978 revolution. Very few of the tens of thousands of students from China, Taiwan, and Hong Kong who came to the United States after World War II on nonimmigrant visas ever returned home, including an elite Chinese cohort stranded as refugees in the United States after the events of 1949 in China (Tsai 1986). Many adjusted their status and gained U.S. citizenship via marriages or occupational connections with American industry and business, thus becoming eligible to send for family members later on, consolidating and expanding social networks that over time give the process of immigration its cumulative and seemingly spontaneous character.

Urban Destinies: The Geography of Immigrant America Today

The impact of the new immigration on American communities is much more significant than might appear at first glance because, as in the past, immigrants tend to concentrate in urban areas where coethnic communities have been established by past immigration. Such spatial concentrations serve to provide newcomers with manifold sources of moral, social, cultural, and economic support that are unavailable to immigrants who are more dispersed. This gravitational pull to places where family and friends of immigrants are concentrated was clearly demonstrated by a recent analysis using 1980 census data for 411 counties with populations above 100,000, covering more than 90 percent of the national foreign-born population: it found that for every 100 immigrants in 1975, 35 new immigrants were added by 1980, whereas only 12 persons were added for every 100 natives (Enchautegui 1992).

In general, patterns of concentration or dispersal vary for different social classes of immigrants (professionals, entrepreneurs, manual laborers) with different types of legal status (regular immigrants, refugees, the undocumented). The likelihood of dispersal is greatest among immigrant professionals, who tend to rely more on their qualifications and job offers than on preexisting ethnic communities, and, at least initially, among recent refugees who are

sponsored and resettled through official government programs that have sought deliberately to minimize their numbers in particular localities. However, refugee groups too have shown a tendency to gravitate as "secondary migrants" to areas where their compatriots have clustered (e.g., Cubans to South Florida, Southeast Asians to California). The likelihood of concentration is greatest among undocumented (e.g., over 25 percent of the 3 million IRCA applicants nationally were concentrated in the Los Angeles metropolitan area alone) and working-class immigrants, who tend to rely more on the assistance offered by preexisting kinship networks, and among business-oriented groups who tend to settle in large cities. Dense ethnic enclaves provide immigrant entrepreneurs with access to sources of cheap labor, working capital and credit, and dependable markets. Social networks are thus crucial not only for an understanding of migration processes, but also of adaptation processes and settlement patterns in areas of final immigrant destination (Portes and Rumbaut 1990).

Over time, as the immigrants become naturalized U.S. citizens, local strength in numbers also provides opportunities for political advancement and representation of ethnic minority group interests at the ballot box (see Table 2 for the proportion of U.S. citizens by national origin). The median time from arrival to naturalization is eight years, though some groups such as Asian and African professionals naturalize more rapidly, controlling for length of residence in the United States. Others, such as Canadians, Mexicans, and the British have low rates of naturalization (INS 1993). In part because of the recency of their arrival, the majority (59 percent) of the 1990 immigrants had not yet acquired U.S. citizenship or begun to vote in elections. Time in the United States alone does not explain why different groups become U.S. citizens at different rates, but this is an important question since, along with higher numbers and greater concentration, citizenship acquisition and effective political participation go to the heart of ethnic politics and to the ability of these groups to make themselves heard in the larger society. The research literature has shown that, among legal immigrants and refugees, the motivation and propensity to naturalize is higher among upwardly mobile younger persons with higher levels of education, occupational status, English proficiency, income and property, and those whose spouses or children are U.S. citizens (Portes and Rumbaut 1990). Undocumented immigrants, by definition, remain disenfranchised and politically powerless.

Table 2 also lists the states of principal immigrant settlement in the United States. While there are immigrants today in each of the fifty states, just six states (California on the west coast; New York, New Jersey, and Florida on the east coast; and Texas and Illinois between the coasts) accounted for three-fourths of the total 1990 U.S. foreign-born population. A decade earlier, those same six states had accounted for two-thirds of the national foreign-born total, a fact that underscores the pattern of increasing concentration in just a few states and localities. California dominates the national figures, accounting for fully one-third of all immigrants in the United States (up from one-fourth in 1980), including approximately half of the undocumented; in fact, by 1990, 22 percent of all Californians were foreign-born, and their U.S.-born children nearly doubled that proportion. New York, New Jersey, and Florida combined for another 28 percent of the foreign-born in 1990, although only 15 percent of the native-born lived in those states.

Patterns of immigrant concentration are even more pronounced within particular metropolitan areas (see, in this volume, the chapters on New York, Los Angeles, Miami, and Washington, D.C.). By 1990, Los Angeles had become the premier immigrant capital of the world, with 2.9 million foreign-born residents (a third of the county's huge population). The New York metropolitan area followed with 2.1 million immigrants (more than a quarter of New York City's population), concentrated mainly in Queens and Brooklyn, then Manhattan and the Bronx. This does *not* include Puerto Ricans, who are not "foreign-born," though 40 percent of their total mainland population of 2.7 million was concentrated in New York City. Remarkably, those two global cities together contained 5 million immigrants, one in four of the national total. Orange and San Diego counties, next to Los Angeles, added 1 million immigrants, while the San Francisco–Oakland–San Jose metropolitan area in northern California added 825,000 more. The Miami metropolitan area contained 875,000 immigrants, including the two cities with the greatest proportion of foreign-born residents in the

United States: Hialeah (70 percent) and Miami (60 percent).

Moreover, as Table 2 shows, different immigrant groups concentrate in different states and metropolitan areas and create distinct communities within each of these cities. Asians, Mexicans, and Central Americans are heavily concentrated in California; Caribbean newcomers join European oldtimers who are still densely overrepresented in New York; and Cubans and Nicaraguans are concentrated in South Florida. Among the largest contingents of recent immigrants, Miami remains the premier destination of Cubans, where they are already a majority of the city's total population, as is New York for Dominicans, Jamaicans, and Soviet Jews. Colombians and Haitians are also most concentrated in New York and Miami. Los Angeles is the main destination for Mexicans, Salvadorans, Guatemalans, Filipinos, Koreans, Iranians, and Cambodians—their communities there are already the largest in the world outside their respective countries—and it is the third choice of Chinese and Indians. After Los Angeles, recent Mexican immigrants have settled in largest numbers in San Diego and El Paso, Filipinos in San Diego and San Francisco, Koreans in New York and Washington, D.C. The Vietnamese are concentrated in Santa Ana, Los Angeles, San Jose, and San Diego, with another major enclave in Houston. Most immigrants from China, Taiwan, and Hong Kong settle in San Francisco, Los Angeles, and New York; more Indians also settle in New York (although among all major immigrant groups Indians tend to be the most dispersed, reflecting their significantly greater proportion of professionals).

Less obvious is the fact that these concentrations often consist of entire community segments from their places of origin—of extended families and former neighbors, and not just compatriots. In some cities in California and elsewhere the links with particular towns or villages in Mexico go back generations and can be traced to the Bracero Program or earlier migration chains. In other cases, the process happens very quickly: for example, 20 percent of the small Salvadoran town of Intipucá was already living in the Adams-Morgan section of Washington, D.C. in 1985, with an organized Club of Intipucá City to assist new arrivals (Schoultz 1992, 189). In the Detroit metropolitan area, home to the largest Arab community in North America, most Chaldeans are from the village of Tel Kaif in northern Iraq and live in South-field; Lebanese Shiites are from Tibnin or Bent Jbail near the Israeli border and prefer Dearborn; most Palestinians are from within ten miles of Jerusalem—Christians from Ramallah, Muslims from El Bireh or Beit Hanina—and live in Livonia; Egyptian Coptic Christians are in Troy, and Lebanese Maronite Catholics on the east side (Stockton 1994, 12A).

Color, Class, and Context

Most new immigrants self-report as nonwhite: the proportion of white immigrants declined from 88 percent of those arriving before 1960 to 64 percent in the 1960s, 41 percent in the 1970s, and 38 percent in the 1980s. Black immigrants increased from 2 percent of pre-1960 arrivals to over 8 percent in the 1980s; Asians from 5 percent pre-1960 to 31 percent in the 1980s; and "other race" from 5 percent pre-1960 to 23 percent in the 1980s. This changing racial-ethnic makeup, which will change in still more complex ways due to rapidly increasing rates of ethnic intermarriage, affects primarily the leading metropolitan areas of immigrant concentration: significantly, half of all black immigrants are concentrated in the New York metropolitan area and another 16 percent in Miami; half of the mestizo and Asian populations are concentrated in California.

Immigrants from the Americas are the most racially mixed, with less than 45 percent self-reporting as white (disproportionately from Argentina and Cuba, then Colombia and Nicaragua), 13 percent black (most from Haiti, Jamaica, and the English-speaking Caribbean, then the Dominican Republic), 1 percent Asian (mostly Indians from former British colonies in Guyana and Trinidad-Tobago), and over 41 percent identified as "other" (predominantly mixed populations of mestizos from Mexico and Central and South America, and mulattoes from the Spanish-speaking Caribbean). Among African immigrants, 52 percent self-classify as black (mostly Nigerians and Ethiopians), 41 percent as white (the Egyptians and many South Africans), and 7 percent as Asian (also remnants of the Indian diaspora during British colonial rule, especially from East Africa). Of immigrants from Asia, 15 percent self-classify as white (mainly the Iranians, Israelis, and Arabs throughout the Middle East). Virtually all Europeans and Canadians indicate they are white (98.5 percent).

Tables 3 and 4 provide a general sociodemographic and socioeconomic portrait of the largest immigrant nationalities in the United States in 1990, compared to principal native-born racial-ethnic groups. While the effort is necessarily constrained by the nature of the available aggregate census data, the selection of variables drawn for this portrait is aimed to draw attention not simply to individual characteristics, but to a variety of larger contexts that shape the incorporation of these groups as they make their way through the American economy and society. Table 3 presents basic information about the age/sex structure of the groups, fertility and selected indicators of family structure, and level of English language proficiency. Table 4 focuses on educational, occupational, and economic characteristics and resources of all of these immigrant groups, ranked in order of their proportion of college graduates, which may serve as a proxy for their social class origins. The overall picture drawn in Tables 3 and 4 is contrasted against the norms for the total native-born population as well as for specific groups of natives—non-Hispanic Whites and Blacks, Puerto Ricans on the mainland, Asian and Mexican Americans, and indigenous groups (American Indians, Alaska Natives, Pacific Islanders). Both tables dramatize the extraordinary diversity and complexity of the new immigration.

By region of birth, very significant contrasts are apparent in the demography of the foreign-born. Europeans and Canadians, most of whom immigrated decades ago, are now a much older and rapidly aging population (median age is fifty-three), disproportionately female (57 percent), with low fertility rates that match those of non-Hispanic white natives (1.8 children ever born per woman aged 35 to 44, which is an approximation of completed fertility), and they generally have, along with immigrants from Asia and Africa, a notably lower proportion of female-headed households and a higher proportion of children under eighteen residing at home with both parents than is the case for native-born Americans. Yet immigrants from Asia include the groups with both the lowest fertility (Japanese) and the highest fertility (Cambodians and Laotians, the latter including the Hmong, the ethnic minority with the highest fertility rate in the United States). In this context, the very young median ages of the U.S.-born Asian and

Table 3 Age, Gender, English Proficiency, Fertility, and Family Contexts of Principal Immigrant Groups in the United States in 1990, in Rank Order by Age, by Region and Country of Birth, Compared to Native U.S. Racial-Ethnic Groups (*continued next page*)

		Age		Gender	Speaks English:[a]		Fertility[b]	Family Contexts[c]	
Country of Birth	Persons (N)	Median Age (yrs)	60 yrs. or Older	Female (%)	English Only (%)	Not Well or at All (%)	Children Born per Woman 35 to 44	Female H'holder (%)	Children <18 with 2 Parents (%)
Region									
Europe & Canada	5,095,233	53	40.0	57.0	45	9	1.8	10.9	86
Asia	4,979,03	35	10.6	51.0	8	22	2.0	10.7	83
Africa	363,819	34	6.2	40.7	25	5	2.2	11.0	75
Latin Amer./Carib.	8,416,924	33	9.6	48.4	13	40	2.7	19.1	69
Above U.S. average									
Italy	580,592	59	48.3	51.8	22	16	2.1	9.8	85
Poland *	388,328	57	46.9	52.8	20	20	1.6	11.1	83
Ireland	169,827	56	43.6	60.0	90	0	2.4	17.1	88
Soviet Union *	333,725	55	45.8	54.8	20	24	1.7	10.8	88
Canada	744,830	53	40.7	58.7	80	1	1.8	12.3	86
Germany	711,929	53	37.0	64.6	41	2	1.8	16.4	75
United Kingdom	640,145	50	33.9	59.8	93	0	1.8	13.9	85
Cuba *	736,971	49	30.1	51.6	5	40	1.8	16.2	72
Greece	177,398	49	26.6	46.2	11	20	2.1	7.4	89
China	529,837	45	25.1	50.5	3	44	1.8	8.2	87
Near U.S. average									
Portugal	210,122	40	17.3	50.0	6	34	2.1	8.0	89
Philippines	912,674	39	14.7	56.7	11	7	1.9	15.1	78
Japan	290,128	38	12.5	62.6	16	25	1.6	14.7	95
Jamaica	334,140	36	12.0	55.2	94	0	2.2	34.6	53
Colombia	286,124	35	8.2	53.6	5	34	1.8	21.5	65
India	450,406	36	6.4	45.1	12	9	2.0	3.3	92
Iran	210,941	35	9.1	41.9	8	12	1.8	7.6	86
Korea	568,397	35	8.1	57.0	7	30	1.8	11.1	87
Haiti	225,393	35	7.3	50.2	6	23	2.4	27.6	56
Dominican Repub.	347,858	34	7.9	54.5	4	45	2.5	41.3	47
Taiwan	244,102	33	3.6	53.0	5	17	1.7	10.2	81

Mexican-origin populations (fifteen and eighteen years, respectively), shown on the last two rows of Table 3, suggest that they consist largely of U.S.-born children of recent immigrants, and show how the higher overall levels of fertility of the foreign-born combine with high levels of current immigration to produce the rapid growth of new ethnic groups in the United States. The low fertility and low immigration of Europeans, by contrast, augur the inexorable diminution of European ethnic communities and their rapid assimilation into the "twilight of ethnicity" (Alba 1985, 1990).

With the exception of Cubans and Jamaicans, the majority of immigrants from Latin America and the Caribbean arrived during the 1980s. As a whole, they appear to be younger, disproportionately male, with

Table 3   (cont.)

Country of Birth	Persons (N)	Age		Gender	Speaks English:[a]		Fertility[b]	Family Contexts[c]	
		Median Age (yrs)	60 yrs. or Older	Female (%)	English Only (%)	Not Well or at All (%)	Children Born per Woman 35 to 44	Female H'holder (%)	Children <18 with 2 Parents (%)
Below U.S. average									
Nicaragua	168,659	30	7.1	51.8	4	41	2.5	21.0	66
Mexico	4,298,014	30	7.0	44.9	4	49	3.3	14.1	73
Vietnam *	543,262	30	5.3	47.4	4	31	2.5	15.3	73
Guatemala	225,739	30	4.2	48.7	3	45	2.6	19.5	66
Hong Kong	147,131	30	3.2	49.9	7	15	1.7	9.7	84
El Salvador	485,433	29	3.9	46.3	3	49	2.7	21.4	61
Cambodia *	118,833	29	5.1	52.4	2	43	3.3	24.3	71
Laos *	171,577	27	4.9	48.3	2	43	4.2	11.9	81
Totals									
Foreign-born:	19,767,316	37	18.0	51.1	21	26	2.3	14.8	74
Native-born:	228,942,557	33	16.7	51.3	92	1	1.9	16.1	73
Native racial-ethnic groups									
White (Non-Hisp.)	188,128,296	35	19.1	51.3	94	1	1.8	11.8	80
Black (Non-Hisp.)	29,216,293	28	11.6	52.8	94	1	2.2	43.2	37
Native American	1,959,234	27	8.5	50.4	76	6	2.5	26.2	56
Puerto Rican	2,727,754	26	6.7	51.0	19	17	2.5	36.6	47
Pacific Islanders	365,024	25	6.3	49.6	69	6	2.6	18.4	68
Mexican (US-born)	8,933,371	18	5.6	49.8	35	10	2.5	21.6	69
Asian (US-born)	2,363,047	15	7.5	49.5	63	6	1.5	12.4	86

* Denotes country from which most recent migrants to the United States have been officially admitted as refugees.

[a] English proficiency of persons aged 5 years or older.

[b] Children ever born per woman aged 35 to 44 years regardless of marital status (an approximate measure of completed fertility).

[c] Percent of family households headed by a female householder with no husband present; and of children under 18 living with two parents.

Source: U.S. Bureau of the Census, *The Foreign Born Population in the United States*, CP-3–1, July 1993, Tables 1, 2; *Persons of Hispanic Origin in the United States*, CP-3–3, August 1993, Tables 1, 2; *Asian and Pacific Islanders in the United States*, CP-3–5, August 1993, Tables 1, 2; and data drawn from a 5 percent public use microdata sample (PUMS) of the 1990 U.S. census, and subject to sample variability.

higher fertility and a higher proportion of female-headed households than non-Hispanic white natives, and are significantly handicapped in their knowledge of English (40 percent do not speak it well or at all); but, as Table 3 makes clear, wide differences between nationalities are concealed at this level of aggregation. For example, the Cubans are much older and have much lower fertility; the Jamaicans and Dominicans have predominantly female populations. The Dominicans have by far the highest proportion

of female-headed households (41 percent) and are the only immigrant nationality with less than half its children residing at home with both parents (see Grasmuck and Pessar, this volume); this pattern is matched only by native Blacks and Puerto Ricans, and approximated to a lesser extent by Jamaicans and Haitians. Salvadorans, Guatemalans, and Mexicans—the groups composed primarily of undocumented immigrants—are disproportionately male, much younger, with fertility rates that are the high-

Table 4 Socioeconomic Stratification of Principal Immigrant Groups in the United States in 1990, in Rank Order of College Graduates, by Region and Country of Birth, Compared to Native U.S. Racial-Ethnic Groups (*continued next page*)

Country of Birth	Persons (N)	College Grads (%)	In Labor Force (%)	Self-Employed (%)	Upper White-collar (%)	Lower Blue-collar (%)	Poverty Rate (%)	Public Assist. (%)	Own Home (%)
		Educat.[a]	Labor Force and Occupation[b]				Income[c]		
Region									
Africa	363,819	47.1	75.1	7.1	37	12	15.7	4.7	34
Asia	4,979,03	38.4	66.4	7.8	32	13	16.2	10.7	50
Europe and Canada	5,095,233	18.6	52.2	9.5	32	12	9.3	5.7	68
Latin Amer./Carib.	8,416,924	9.1	70.7	5.0	12	26	24.3	11.3	37
Above U.S. average									
India	450,406	64.9	74.6	6.3	48	8	8.1	3.4	54
Taiwan	244,102	62.2	64.9	7.5	47	4	16.7	3.7	66
Iran	210,941	50.6	67.9	12.0	42	6	15.7	7.7	55
Hong Kong	147,131	46.8	75.1	5.5	41	7	12.7	3.5	62
Philippines	912,674	43.0	76.3	3.3	28	11	5.9	10.4	61
Japan	290,128	35.0	54.2	7.9	39	7	12.8	2.2	46
Korea	568,397	34.4	63.9	18.0	25	13	15.6	7.9	48
China	529,837	30.9	62.3	7.8	29	16	15.7	10.6	56
Near U.S. average									
Soviet Union *	333,725	27.1	39.7	10.1	31	11	25.0	16.7	47
United Kingdom	640,145	23.1	57.3	8.3	40	6	6.6	3.7	69
Canada	744,830	22.1	52.1	9.5	38	8	7.8	4.8	71
Germany	711,929	19.1	54.7	9.1	33	9	7.7	4.3	73
Poland *	388,328	16.3	50.4	7.9	21	20	9.7	5.4	64
Vietnam *	543,262	15.9	64.4	5.8	17	21	25.5	26.2	47
Cuba *	736,971	15.6	64.1	7.3	23	18	14.7	16.2	56
Colombia	286,124	15.5	73.7	6.6	17	22	15.3	7.5	38
Jamaica	334,140	14.9	77.4	4.0	22	11	12.1	7.8	44
Greece	177,398	14.8	60.9	14.7	29	12	9.1	5.3	67
Nicaragua	168,659	14.6	73.1	4.7	11	24	24.4	8.4	26
Ireland	169,827	14.6	51.5	7.3	29	9	8.4	4.1	60

est among all immigrant groups except for Indochinese refugees, and they also exhibit the greatest degree of English language handicaps. Yet among Latin Americans, the Mexicans, who number 4.3 million immigrants and dominate the aggregate statistics, have the highest fertility (3.3 children ever born among women aged 35 to 44) but also the lowest proportion of female-headed households and the highest percent of children living at home with both par-

ents—indicators of family cohesion that belie their low socioeconomic status.

As Table 3 also shows, for all the alarm about the demise of the English language and the alleged lack of assimilation of today's immigrants, fully one in five (21 percent) speak English *only*, and another 53 percent speak it well or very well, although 44 percent arrived during the 1980s. Only one in four does not yet speak it well or at all, and these are dispropor-

Table 4   (cont.)

Country of Birth	Persons (N)	Educat.[a] College Grads (%)	Labor Force and Occupation[b] In Labor Force (%)	Self-Employed (%)	Upper White-collar (%)	Lower Blue-collar (%)	Income[c] Poverty Rate (%)	Public Assist. (%)	Own Home (%)
Below U.S. average									
Haiti	225,393	11.8	77.7	3.5	14	21	21.7	9.3	37
Italy	580,592	8.6	46.4	10.1	20	18	8.0	5.5	81
Dominican Repub.	347,858	7.5	63.8	5.1	11	31	30.0	27.8	16
Guatemala	225,739	5.8	75.7	5.2	7	28	25.8	8.3	20
Cambodia *	118,833	5.5	48.4	5.2	9	23	38.4	49.5	23
Laos *	171,577	5.1	49.7	2.2	7	41	40.3	45.5	26
Portugal	210,122	4.6	71.6	5.1	9	36	7.0	8.4	62
El Salvador	485,433	4.6	76.3	4.7	6	27	24.9	7.1	19
Mexico	4,298,014	3.5	69.7	4.5	6	32	29.7	11.3	36
Totals									
Foreign-born:	19,767,316	20.4	64.3	6.9	22	19	18.2	9.1	49
Native-born:	228,942,557	20.3	65.4	7.0	27	14	12.7	7.4	65
Native racial-ethnic groups									
Asian (US-born)	2,363,047	35.9	68.8	5.5	34	8	9.8	4.5	63
White (Non-Hisp.)	188,128,296	22.0	65.3	7.7	29	13	9.2	5.3	68
Black (Non-Hisp.)	29,216,293	11.4	62.7	2.8	18	21	29.5	19.7	43
Pacific Islanders	365,024	10.8	70.1	4.1	18	16	17.1	11.8	44
Puerto Rican	2,727,754	9.5	60.4	2.8	17	21	31.7	26.9	26
Native American	1,959,234	9.3	62.1	5.8	18	19	30.9	18.6	54
Mexican (US-born)	8,933,371	8.6	67.2	4.4	16	19	24.5	13.5	54

*Denotes country from which most recent migrants to the United States have been officially admitted as refugees.

[a] Educational attainment for persons aged 25 years or older.

[b] Labor force participation and occupation for employed persons 16 years or older; Upper White-collar = professionals, executives, and managers; Lower Blue-collar = operators, fabricators, and laborers.

[c] Percent of persons below the federal poverty line, and of households receiving public assistance income.

Source: U.S. Bureau of the Census, *The Foreign Born Population in the United States*, CP-3–1, July 1993, Tables 3–5; *Persons of Hispanic Origin in the United States*, CP-3–3, August 1993, Tables 3–5; *Asian and Pacific Islanders in the United States*, CP-3–5, August 1993, Tables 3–5; and data drawn from a 5 percent public use microdata sample (PUMS) of the 1990 U.S. census, and subject to sample variability.

tionately the elderly (especially those in dense ethnic enclaves, such as Cubans in Miami), the most recently arrived, the undocumented, and the least educated. English proficiency increases significantly over time, and English becomes by far the preferred language of use by the second generation (see Portes and Schauffler, this volume); this is implied by the data in Table 3 for the U.S.-born Asian and Mexican-origin populations. In fact, in recent years immigrant children have been consistently not only among the winners of national science talent search contests, but of the U.S. National Spelling Bee as well (Rumbaut 1991c). It is the immigrants' mother tongue that atrophies over time, and quickly; this has been the pattern throughout American history—the third generation typically grows up speaking English only—and explains why the United States has been called a "language graveyard."

In the contemporary period, professional immigrants constitute a major type of immigration. Here a Soviet Jewish cardiologist in her lab coat is surrounded by her diplomas at her office on Wilshire Boulevard, Los Angeles. (Photo by Steven Gold)

The Africans, who are mostly recently arrived young males, have the least difficulty with English (only 5 percent do not speak it well or at all, compared to 9 percent among the Europeans and Canadians). This is not surprising since Egypt and Nigeria alone account for 33 percent of all African immigration; Ethiopia, South Africa, and Ghana add another 25 percent; and all but Ethiopia are English-speaking countries. In fact, three-fifths (61 percent) of all African immigrants come from countries where English is an official language. In addition, economic and political ties have shaped their migration pathways to the United States (e.g., the United States is a principal trading partner of and direct investor in the main sending countries; Ethiopians have entered as Cold War refugees; Egypt since the Camp David Accords receives massive U.S. foreign aid and economic and military assistance, exceeded only by Israel; and while there were virtually no Somalis counted in the 1990 census, U.S. involvement in Somalia in the early 1990s has opened various migration pathways). But there are more complex reasons. The oil-rich Nigerian government in the 1970s tried to build a national university system and sent thousands of Nigerians abroad for advanced studies, but an economic crisis in the 1980s left the students abroad without funds in the middle of their studies and many failed to return. A similar situation in Ghana caused it to experience the highest "brain drain" rate of any country in Africa (Kritz and Caces 1992, 226). One irony is that Nigerians now rank with Indian and Taiwanese immigrants as the most highly educated group in the United States: nearly two-thirds of Nigerian adults over twenty-five are college graduates. However, many are "visa overstayers" (as I learned in 1993 from a Nigerian taxi driver with an M.A. in economic development who was finishing work on his Ph.D.). The 1990 census counted a foreign-born Nigerian population of 55,350; in 1992 the Nigerian illegal immigrant population was estimated at between 24,000 and 30,000 (U.S. INS 1992b).

Table 4 stratifies all of the principal immigrant groups in the United States by their proportion of college graduates, and compares them to native racial-ethnic groups on various indicators of socioeconomic status. The foreign-born as a whole have the same proportion of college graduates (20 percent) as does the native-born population, as well as an equivalent rate of labor force participation and self-employment. They are, however, more likely than natives to be poor and to work in low-status jobs. But again, decontextualized data at this level of analysis conceal far more than they reveal, although that is often the level at which arguments about the supposedly "declining stock" of new immigrants are made (i.e., that relative to earlier decades today's immigrants are less educated and more welfare dependent). By far, the most educated *and* the least educated groups in the United States today are immigrants, a reflection of polar-opposite types of migrations embedded in very different historical contexts. Disaggregated by region and country of birth, the huge differences among them are made clear, underscoring the fact that these groups cannot sensibly be subsumed under supranational categories like "Asians" or "Latinos." The fact that most come from poorer nations does not mean that the immigrants themselves are poor and uneducated. On the contrary, one point that stands out in Table 4 is the extremely high degree of educational attainment among immigrants, especially from the developing countries of Africa and Asia—47 and 38 percent are college graduates, respectively—and they are well above the U.S. average in their proportion of professionals, executives, and managers.

Table 4 lists an upper stratum composed of sizeable foreign-born groups whose educational and occupational attainments significantly exceed the average for the native-born American population. Without exception, all of them are of Asian origin—from India, Taiwan, Iran, Hong Kong, the Philippines, Japan, Korea, and China—with recently immigrated groups reflecting the highest levels of attainment. Also in this upper stratum are several other smaller immigrant groups (not shown in Table 4), notably those from Nigeria, Egypt, South Africa, Kenya, Israel, Lebanon, Ghana, and Argentina. By the mid-1970s, one-fifth of all U.S. physicians were immigrants, and there were already more foreign medical graduates from India and the Philippines in the United States than American Black physicians. By the mid-1980s, over half of all doctoral degrees in engineering awarded by U.S. universities were earned by foreign-born students, with one-fifth of all engineering doctorates going to students from Taiwan, India, and South Korea; and one third of all engineers with a doctorate working in U.S. industry were immigrants. These "brain drain" immigrants are perhaps the most skilled ever to come to the United States. Their class origins help explain the popularization of Asians as a "model minority" and debunk nativist calls for restricting immigrants to those perceived to be more "assimilable" on the basis of color, language, and culture.

By contrast, as Table 4 shows, the lower socioeconomic stratum includes recent immigrants from Mexico, El Salvador, Guatemala, the Dominican Republic, and to a lesser extent Haiti—many or most of whom are undocumented. They have higher rates of labor force participation but much lower levels of educational attainment, are concentrated in low-wage unskilled jobs, and have poverty rates as high as those of native minority groups but much lower proportions of households on welfare. Here also are less educated but less visible European immigrants from Italy and Portugal (34 percent of Portuguese adult immigrants have less than a fifth-grade education, compared to 1.8 percent of the total U.S.-born population, yet only 7 percent of the Portuguese are below the poverty line). And two Asian-origin groups, Laotian and Cambodian refugees, exhibited by far the highest rates of poverty and welfare dependency in the United States. Southeast Asians and to a lesser extent Chinese and Korean workers are much in evidence, along with undocumented Mexican and Central American immigrants, in a vast underground sweatshop economy that expanded during the 1980s and 1990s in Southern California. These data too debunk the stereotypes that have been propounded in the mass media as explanations of Asian success, and point instead to the contextual diversity of recent immigration and to the class advantages and disadvantages of particular groups.

A middle stratum evident in Table 4, composed of groups whose educational and occupational characteristics are close to the U.S. average, is more heterogeneous in terms of their national origins. It includes older immigrants from the Soviet Union, Britain, Canada, and Germany, and more recent immigrants from Vietnam, Cuba, Colombia, and Jamaica. However, not at all evident in Table 4 is the fact that *within* particular nationalities there are often also many class differences that reflect different "waves" and immigration histories. For example, while 31 percent of adult immigrants from China have college degrees, 16 percent have less than a fifth-grade education; this bimodal distribution in part reflects different patterns of Chinese immigration and enterprise between the pre-1965 *Lo Wa Kiu* ("old overseas Chinese") and the post-1965 *San Yi Man* ("new immigrants") (Wong 1987). Desperate Haitian boat people arriving by the thousands in the 1980s and 1990s mask an upper-middle-class flow of escapees from the Duvalier regime in the early 1960s; by 1972 the number of Haitian physicians in the United States represented an incredible 95 percent of Haiti's stock (Rumbaut 1991c). Similarly, the post-1980 waves of Cuban Mariel refugees and Vietnamese "boat people" from modest social class backgrounds differed sharply from the elite "first waves" of the 1959 to 1962 Cubans and the 1975 Vietnamese, underscoring the internal diversification of particular national flows over time.

Among the employed, the percentage of longer-established Canadian and certain European immigrants in professional specialties exceeds the respective proportion of their groups who are college graduates, but the percentage of recently arrived Asian immigrants who are employed in the professions is generally far below their respective proportions of college graduates. These discrepancies between educational and occupational attainment point to barriers, such as English proficiency and strict licensing requirements that regulate entry into the professions, that recent immigrants—most of them non-white, non-European, and non-English speaking—must confront as they seek to make their way in America. In response, some immigrants shift instead to entrepreneurship as an avenue of economic advancement and as an alternative to employment in segmented labor markets. As Table 4 shows, Korean immigrants are the leading example of this entrepreneurial mode of incorporation, with self-employment rates that are higher by far than any other native-born or foreign-born groups (see Min, this volume).

Are Immigrants a Cost or a Benefit to American Society?

Entrepreneurial immigrants create jobs. For example, among Koreans in Los Angeles in 1980, a study found that 22 percent were self-employed, and they in turn employed another 40 percent of Korean workers in their businesses; Korean-owned firms thus accounted for almost two-thirds of all employed Koreans in the Los Angeles metropolitan area (Light and Bonacich 1988). Similar proportions have been reported for immigrants from China, Hong Kong, and South Korea in New York City (Kim 1987b). A panel study of Cuban refugees who arrived in Miami in 1973 showed that by 1979, 21 percent were self-employed and another 36 percent were employed in businesses owned by Cubans. A subsequent survey of Mariel Cubans who arrived in Miami in 1980 found that by 1986, 28 percent were self-employed and another 45 percent were employed by their conationals (Portes and Bach 1985; Portes and Stepick 1993). Indeed, the national study cited earlier of all U.S. counties with populations above 100,000 found that the larger the foreign-born population, the larger the gains in employment: immigrants added twice as many jobs to the county as natives (Enchautegui 1992). Overall, immigrants create more jobs than they themselves fill.

There is little evidence that unemployment is caused by immigrants either in the United States as a whole or in areas of high immigrant concentration, or that immigration adversely affects the earnings of either domestic majority or minority groups (Bean et al. 1988; Borjas 1990a; Butcher and Card 1991; Simon 1989; but see Tienda and Stier, this volume, for an analysis of displacement effects in Chicago's inner city). To the contrary, research studies of both legal *and* undocumented immigration point to significant net economic benefits accruing to U.S. natives. To be sure, some newcomers in areas of immigrant concentration—especially the undocumented and unskilled immigrant women—are exploited as sources of cheap labor in a growing informal sector that is fueled by foreign competition and the demand for low-cost goods and services in the larger economy. In this context, the presence of a large supply of cheap labor does keep wages down: the low wages are paid to immigrants who, under their precarious circumstances, are willing to accept whatever work is offered.

As a rule, the entry of immigrants into urban labor markets helps to *increase* native wages as well as productivity and investment, sustain the pace of economic growth, and revive declining sectors such as light manufacturing, construction, and apparel (New York City, Los Angeles, and Miami provide examples). An influx of new immigrant labor also has the effect of pushing up domestic workers to better supervisory or administrative jobs that may otherwise disappear or go abroad in the absence of a supply of immigrant manual labor. Less skilled immigrants typically move into manual labor markets deserted by native-born workers, who shift into preferred non-manual jobs. And immigrant professionals, such as engineers and physicians, fill significant national needs for skilled talent, and in some respects also serve as a strategic reserve of scarce expertise. For example, given declining enrollments in advanced engineering training among the native-born, the proportion of the foreign-born in these fields has grown rapidly. Foreign-born physicians have similarly performed key functions in American medical care: they have been concentrated in the less prestigious, non-university-affiliated hospitals in underserved areas that do not attract native-born physicians, especially in rural and inner-city hospitals serving Medicaid patients and the uninsured working poor, thereby giving native-born medical graduates more options in choosing jobs.

As Table 4 shows, the highest proportions of households receiving public assistance income are found primarily among groups admitted as *refugees* (mainly from Laos, Cambodia, Vietnam, the former Soviet Union, and Cuba). Refugee assistance programs evolved in tandem with the expansion of the welfare state in the 1960s and early 1970s (Rumbaut 1989b), and indigent persons received as refugees have the right to access public assistance programs on the same means-tested basis as U.S. citizens (legal immigrants do not, at least during their first three to five years in the country). The Dominicans, whose high reliance on welfare in the New York City area is tied to a very high proportion of poor female-headed households, are the main exception. In addition, the increase in the number of elderly parents entering the United States to reunite with their children has also increased the rolls of legal immigrants eligible to receive Supplemental Security Income (the SSI program was founded in 1974 to aid indigent persons who are blind or disabled, as well as elderly Americans who do not qualify for Social Security benefits because they have not worked in the United States). Among the 10 percent of immigrants from China and the Philippines who receive public assistance (see Table 4), many are elderly SSI recipients. Because of their vulnerable legal status, the undocumented typically avoid involvement in welfare services altogether (U.S. INS 1992b).

The costs of providing public services to immigrants—including welfare, education, and health care—has again become the focus of heated debates, particularly in states and localities of immigrant concentration such as California and Florida. Research on the fiscal impacts of different types of immigrants—legal, undocumented, refugees, IRCA legalizees—has sought to estimate both specific taxes paid by immigrants to all levels of government, as well as the specific service costs incurred by local, state, and federal governments. Such estimates are necessarily based on many assumptions. The most cautiously reasoned national cost-benefit assessment to date estimated that immigrants who came to the United States between 1970 and 1992 were paying a total of over $70 billion in taxes of all kinds (including FICA and a dozen other types of taxes) from aggregate incomes in excess of $300 billion, more than 9 percent of all U.S. personal income. Subtracting from those taxes the estimated costs of social services used by immigrants and their children—including costs borne disproportionately by state and local governments, such as primary and secondary education—indicated that immigrants entering between 1970 and 1992 generated a *surplus* of $25 to $30 billion (Passel 1994).

But thinking about immigrants ⟨...⟩ ducible to a fiscal calculus—"perhaps ⟨...⟩ U.S. history where there was an effort ⟨...⟩ dollar value to a group's participation in civil ⟨...⟩ (Suro 1994)—misses entirely their significance ⟨...⟩ American society. The long-term effects of contemporary immigration will ultimately not be measured by the achievements of the immigrants themselves, but of their children and grandchildren. Yet just as no one could have predicted as European immigrants disembarked at Ellis Island in 1910 what the world would be like for their children in the 1930s or their grandchildren in the 1960s, so we cannot foresee what the world of the twenty-first century will hold in store for the diverse progeny of today's non-European newcomers. The fate of the new second generation now coming of age in American cities remains a puzzle only now beginning to be unraveled by new research (see Portes 1994; Rumbaut and Cornelius 1995).

Conclusion

Today's immigrants are extraordinarily diverse, and so is the American society that receives them. Their origins reflect the global reach of U.S. hegemony. Unlike the expanding industrial economy that absorbed the huge flows from Europe at the beginning of this century and then the smaller flows of the first two decades after World War II, since the 1970s new and much larger waves of immigrants have entered an hourglass-shaped economy that has undergone industrial restructuring and downsizing, characterized by the loss of high-wage manufacturing jobs and the growth of low-wage service jobs; and new and larger waves of refugees have entered a welfare state with expanded opportunities for public assistance. The process has reduced opportunities for social mobility, particularly among the less educated, and widened the gap between rich and poor. It has also produced heated nativist reactions, most notably in California, where a third of the nation's foreign-born population is located, and where a prolonged recession exacerbated a deep public discontent that, as in the past, became ripe for political manipulation. Within this changing historical and structural context, the destinies of the new immigrant groups, and of the rapidly growing generation of their children born in the United States, will likely be as divergent

, their origins. Internal characteristics, including their demographic structure, family organization, and social class resources, interact in complex but patterned ways with external contexts of reception—such as government policies and programs, the strength of existing ethnic communities, employer preferences in local labor markets, and the color line—to mold their diverse fates in American society. In the process, as yet full of untold stories, they are becoming, creatively, unevenly, its newest members.

Notes

1. In August 1994, over 30,000 Cubans attempted to cross the Florida Straits in flimsy rafts and inner tubes after the Castro government indicated it would not prevent anyone who wanted to leave Cuba from doing so. The drama on the high seas climaxed several years of increasingly desperate crossings. The Clinton administration, fearing "another Mariel," responded by interdicting the *balseros* (rafters) on the high seas and taking them to the U.S. Naval Base at Guantánamo Bay, Cuba, where they were kept in indefinite detention along with over 15,000 Haitian boat people. In the process, the U.S. government reversed three decades of preferential admission and reception policies toward Cuban refugees. In a pact reached in September 1994, the Cuban government agreed to curtail the dangerous boat crossings conditioned upon the U.S. government's decision to admit 20,000 Cubans annually via regular immigration channels. Later that month, U.S. troops occupied Haiti to ensure the return of the Haitian president who had been ousted in the 1991 coup. The military intervention, ironically, may yet open up new pathways of Haitian immigration to America.

2. Of course, many people also permanently leave the United States, although official data are lacking. The collection of statistics on *emigration* from the United States was discontinued in 1957, but indirect measures suggest that it averaged over 100,000 per year during the 1970s (a substantial if lower proportion than had been the case in the early decades of the century) and roughly 200,000 per year in the 1990s. Of the roughly 30 million immigrants admitted to the United States between 1900 and 1980, an estimated 10 million emigrated (Warren and Kraly 1985).

3

North American Indians and the Demography of Contact

RUSSELL THORNTON

The year 1992 commemorated the 500-year anniversary of the arrival of Columbus in the Western Hemisphere in 1492. This anniversary—the Columbian Quincentennary—was the cause of much celebration for many of the peoples of this hemisphere as well as of the other. They and the countries they formed ultimately trace their origins to the arrival of Columbus in the Caribbean. Certainly *most* of the hundreds of millions of people now populating the countries of North, Central, and South America had much cause to celebrate. They are the products of a remarkable population growth in this hemisphere which has occurred during the 500-plus years since Columbus's arrival.

Another population history exists, however. It is the population history of the native peoples who first populated the Western Hemisphere, thousands and thousands of years before Columbus arrived. The year 1492 also represents a very significant date in their history. The date commemorates the beginning of a long demographic disaster. For following Columbus's discovery of his "New World," Native Americans, who include American Indians, Eskimos, and Aleutian Islanders, experienced a virtual decimation of their population. Their numbers declined sharply; many of their societies and cultures were pushed to the edge of extinction and beyond.

As Paul Harrison (1981, p. 41) underlined, the overseas expansion of European colonialism "was a vast upheaval of races, a reshuffling of genetic pieces on the chessboard of the globe. The arrival of the Iberian conquistadors depopulated large parts of Central and South America. The Indians who escaped their guns succumbed to their diseases, and only in the high Andes and the most inaccessible reaches of the jungle did they survive intact. Africans were uprooted from their native soil and shipped over to cope with the resulting shortage of labor. Asians were transported to East and South Africa and the Caribbean. Even the huge markets of the colonies were not big enough to employ Europe's expanding population in industry: her surplus spread out into North America, Australia, southern Africa and South America." Along with this went a global transfer of plants that provided new staple foods and created the basis for plantation economies. "From South America maize and cassava spread worldwide. Cocoa from Mexico became the principal livelihood for millions of West Africans. Rubber from the Amazon took root in South-East Asia. In its turn, the New

World received the crops of the Old: sugar, a native of Bengal, became the chief produce of the Caribbean and the north-east coast of South America. Coffee from the Middle East came to dominate farming in Colombia, Brazil, and several countries in West and East Africa. Tea was transplanted from China to India and Ceylon, and onwards to East Africa" (Harrison 1981, p. 41). That this population expansion was detrimental to the indigenous populations of the Western Hemisphere is underscored by the fact that most of the world's 5.5 billion people today are descendants of the populations of the Eastern Hemisphere in 1492.

This was not, however, the only demography of contact. Native Americans also experienced fundamental changes in their societies and cultures as a result of the demography of contact, and responded and adapted to its legacy. Even today, Native Americans must cope with the demographic legacy of contact.

Epidemic Disease and Population Collapse

Considerable scholarly debate exists as to the size of the total native population of the Western Hemisphere at the time of Columbus's arrival. In my recent book, *American Indian Holocaust and Survival* (Thornton 1987), I estimated that there were approximately 75 million Native Americans in the Western Hemisphere in 1492. At that time most of the native population was located south of the Rio Grande River, in what is today central Mexico and some of the countries that form Central and South America. In the rest of the world, there were about 500 million people at this time. Of the initial total Native American population, in the following centuries the population declined to only a few million, after which some population recovery occurred. It is well documented that epidemic diseases were the primary reason for this decline, and that soon after European arrival epidemic diseases devastated American Indian populations in areas of present-day Mexico, the Caribbean, and Central and South America. Nevertheless, other important reasons for this population decline included enslavement and genocide, as well as the destruction of native ways of life and subsistence patterns.

Considerable scholarly debate also exists as to the size of the native population of aboriginal America north of Mexico (i.e., the present-day United States and Canada, including Greenland). Estimates have ranged from scarcely one million inhabitants to many millions. Early in the twentieth century, James Mooney (1910, 1928) of the Smithsonian Institution estimated individual tribal populations north of the Rio Grande, summed them by regions, and arrived at a total of only about 1,150,000 Native Americans at the time of, as he expressed it, the "first extensive European contact." According to Mooney, the dates of extensive European contact with the regions varied from 1600 to 1845 across the different regions. Thus his estimate did not necessarily refer to aboriginal populations, since the dates were generally after some European contact in the regions, although Mooney did assert that Native Americans experienced little, if any, population reduction prior to their first extensive contact with Europeans. In subsequent decades other scholars generally reaffirmed Mooney's low population estimate; Alfred Kroeber even considered Mooney's estimate excessive, and reduced it to 900,000. In 1966, however, Henry Dobyns (1966) used ratios of depopulation following European contact to assert an aboriginal population for this area. He asserted that Native Americans reached population nadirs or lows of only one-twentieth to one-twenty-fifth their aboriginal sizes. Using a Native American nadir population of 490,000 for the United States and Canada, Dobyns then multiplied it by 20 and 25 and arrived at an estimated aboriginal population for North America of between 9,800,000 and 12,150,000.[1] In 1983 Dobyns (1983) used mortality rates from epidemics as well as environmental carrying capacities—how many could have been supported in an area—to assert there were some 18 million native inhabitants north of Mesoamerica (hence, the area included northern Mexico as well as the present-day United States, Canada, and Greenland).

Other scholars have arrived at estimates falling between the extreme of Mooney and Dobyns. In the mid-1970s, for example, Douglas Ubelaker (1976) of the Smithsonian Institution summed tribal estimates from the *Handbook of North American Indians* (Sturtevant 1978) to arrive at an aboriginal population north of Mexico of 1,850,011; in the early 1980s I, using a linear depopulation trend, assessed the population at 1,845,183 (see Thornton and Marsh-Thornton 1981). In *American Indian Holocaust and Survival* (Thornton 1987), I analyzed these and other population estimates and concluded that the aboriginal population north of present-day Mexico numbered over 7 million people in 1492—somewhat over 5 million for the United States mainland and somewhat over 2 million for present-day Alaska, Canada, and Greenland combined. Another more recent population estimate for the area north of Mexico is 3.79 million (Denevan 1992). Subsequent scholarly work will likely generate more consensus regarding the estimates.[2]

Consensus does exist that native populations north of the Rio Grande experienced drastic declines following contact with European and African populations. Following 1492, the native peoples north of Mexico experienced a full 400 years of population decline. Not until about 1900 did their population reach its nadir and some population recovery begin. At the dawn of the twentieth century, when the population of the United States and Canada had grown to over 80 million, there were only about 250,000 Native Americans in the land area of the United States and another 100,000 to 150,000 to the north (Thornton 1987; see Ubelaker 1988, for a somewhat higher nadir population figure).

What accounted for such a population decline in the face of the remarkable population growth of the other peoples who colonized this hemisphere after 1492? As mentioned earlier, the primary reason for the Native American holocaust was the diseases the Europeans, and also the Africans, brought with them to the Western Hemisphere; other reasons were important as well: war, genocide, and the destruction of native ways of life. Populations change—grow or decline—over time according to increases or decreases in the three components of population growth: the total death rate, the total birth rate, and net migration (immigration into the area minus emigration out of the area). At this time, migration was not important in the overall picture of Native American population decline. Rather, diseases, war, genocide, and the destruction of ways of life produced population decline as ever numerically smaller generations of Native Americans were born, lived, and died. Each subsequent generation of the Native American population contained fewer people than the one preceding it due to an increase in the death rates and/or a decrease in the birth rates. Hence, Native American

lives were either short-lived or unlived, and the total population declined accordingly.

Epidemic Disease

Smallpox, measles, the bubonic plague, cholera, typhoid, pleurisy, diphtheria, scarlet fever, whooping cough, mumps, colds, pneumonia, malaria, yellow fever, and various venereal diseases such as gonorrhea, chancroid, and possibly syphilis came, sooner or later, with the newcomers.[3]

Native people here had few serious diseases prior to 1492. The reasons for this are not yet fully understood, but surely include the existence of fewer domesticated animals, such as horses, cows, or sheep, from which many human diseases may be traced, as well as the relative lack of large centers of population concentration, such as cities, which foster many diseases. The lower overall population density, a condition that hindered the survival of many diseases, may also possibly be a reason.[4] Europeans and Africans had acquired a degree of immunity to the diseases that developed there, and some diseases had even become childhood diseases for them since many adults were immune. However, Native Americans lacked the acquired immunity to the diseases the newcomers brought from Europe and Africa. Thus, the diseases often resulted in what are called "virgin soil epidemics," whereby a new disease in a population becomes particularly virulent and spreads to virtually all members of a population (see Crosby 1976). Moreover, the newly encountered diseases did not arrive merely once and then disappear; rather, they infected native peoples over and over again. Smallpox (*Variola major*), for example, was an especially deadly killer of American Indians, and epidemics of it among the native peoples of North America occurred from the early 1500s to the latter 1800s, although the latter epidemics were generally of a milder form (*Variola minor*). The overall effects were devastating. The newly encountered diseases caused increased death rates and, without doubt, decreased birth rates as well.

The Pattern in the Southeast

The first century or so after initial European contact was possibly especially destructive to native populations (see Thornton, Warren, and Miller 1992). Scholars argue, for example, that early European ex-

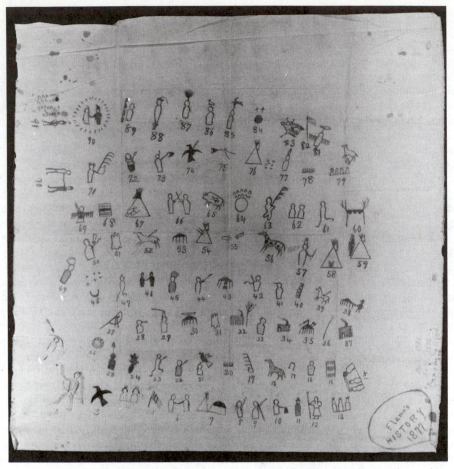

As a memory device, the Winter Count is a pictograph done by the Plains Indians tribes, where every year the most significant event of that year was recorded. They were used by some Plains tribes, such as the Kiowa and the Sioux, to record tribal history. Here a Sioux Winter Count depicts their history from 1786 (year 1) to 1876 (year 90) (numbers were added later by non-Indians for identification purposes). It is very likely that some of the events recorded here may well be someone's success in catching an eagle (21), a meteor shower (48), and a treaty between the Indians and the U.S. government (90). Some years the most significant event was an epidemic disease, such as smallpox (33), whooping cough (28), measles (16), and dropsy (41). (National Anthropological Archives; Smithsonian Institution)

plorations in the Southeast and also the Southwest, as well as other contacts, such as slave raids and shipwrecks on the East Coast, brought "Old World" diseases to North America and spread them among native peoples here (see, for example, Dobyns 1983; Thornton, Warren, and Miller 1992; Upham 1986, 1987).

Documented explorations of the Southeast began with Juan Ponce de León's arrival in *La Florida*[5] in 1513. In 1519, Alonso Álvarez de Pineda journeyed to *La Florida*, and in 1521 Pedro de Quexo and Francisco Gordillo sailed the Santee River of South Carolina looking for slaves. In 1526, a colonizing effort by Lucas Vásquez de Ayllón occurred on either the Georgia or South Carolina coast, from where the colonizers withdrew in a few months but left a number of African slaves. In 1528, Pánfilo de Narváez landed with an army of 400 men and 40 horses on the coast of present-day Florida, near Tampa Bay, to explore the area. Twelve years later, only four men had survived the ordeal, when they made it to Mexico City. The most famous of the four was Alvar Núñez Cabeza de Vaca. From 1539 to 1541, Hernando de Soto explored the Southeast.[6]

Accounts of the explorations of Cabeza de Vaca and de Soto provide some *possible* descriptions of epidemic disease, possibly brought to the area by earlier explorations. In 1542, Cabeza de Vaca wrote an account of his experiences, where he reported, for example, that on Galveston Island, "half the natives died from a disease of the bowels" (Cabeza de Vaca 1961, p. 60), probably dysentery. Cabeza de Vaca also reported encountering a group of Indians in Texas in which the "majority had one clouded eye and others were completely blind" (Cabeza de Vaca 1961, p. 103), which may indicate smallpox.[7]

One of the chroniclers of the Hernando de Soto expedition indicates possible epidemic disease that occurred in one of the chiefdoms prior to their arrival. This chronicler, the Gentleman of Elvas, described the arrival of the expedition at the chiefdom of Cofitachiqui: "About the place, from half a league to a league off, were large vacant towns, grown up in grass, that appeared as if no people lived in them for a long time. The Indians said that, two years before, there had been a pest in the land, and the inhabitants had moved away to other towns" (Bourne 1922, I; 66).[8] Another chronicler, *El Inca* Garcilaso de la Vega wrote later: "a great pestilence with many consequent deaths had ravaged their province during the past year, a pestilence from which their town [Cofitachiqui] alone had been free" (Garcilaso de la Vega 1951, p. 298).[9]

According to Henry Dobyns (1983) and others (see Ramenofsky 1987; Upham 1986), these diseases infected native populations in both the Southeast and the Southwest during the initial decades of the sixteenth century, and, Dobyns argues, frequently culminated in epidemics and larger pandemics that devastated Native American populations of not only these regions but other regions of North America as well.[10] Consequently, Dobyns argues, the aboriginal population of North America was exceedingly large (his estimate is of 18 million north of Mesoamerica), and had already been reduced greatly by epidemic disease prior to significant historical documentation.

Scholarly research has generally refuted Dobyns arguments regarding continentwide pandemics or even pan-regional epidemics during the sixteenth century (see Thornton, Warren, and Miller 1992). Marvin Smith (1984), for example, examined mass and multiple burials, population curves, site sizes, and population movements, yet found little archaeological evidence to support the thesis of population collapse as a result of European disease. He did, however, find considerable evidence for the disintegration of the many Southeastern Indians by the early seventeenth century, and argued this was a result of epidemic disease. Similarly, Dean Snow (Snow and Lanphear 1988), who has been working on a detailed analysis of Iroquois population history following European contact, concluded that the Iroquois did not experience rapid depopulation until the seventeenth century. Furthermore, Jerald Milanich (1990) points out that the Calusa of Florida survived for more than 150 years after European contact. In this regard he notes: "We have more than 600 printed pages of documents about Jesuit mission efforts in the 1560s in La Florida, and there are no references to epidemics among the Calusa" (Milanich 1990, p. 8).

Some research, however, does support the notion of very early epidemic disease in the Southeast. For example, Ann Ramenofsky (1987) focuses primarily on the analysis of settlements and finds archaeological evidence of population collapse during the sixteenth century in the lower Mississippi valley. She does not, however, find evidence for this in the middle Missouri valley or central New York until the seventeenth century, supporting Snow's findings.[11] Still debated is whether sixteenth-century diseases in the Southeast (and by implication, the Southwest) occurred as regionwide epidemics or as more isolated disease episodes: the diseases may very well have "burned themselves out" prior to spreading very far (see Thornton, Warren, and Miller 1992).

Other Reasons for Depopulation

Although it was the primary reason for reducing the total population, not only disease decimated Native Americans. Warfare with Europeans and Americans and, even more, outright genocide—the deliberate, state-sponsored mass murder of a people—at the hands of Europeans and Americans also reduced the native population, even to the brink of extinction for many Native American peoples. For example, tribe after tribe of California Indians was dramatically reduced in size following the arrival of miners and, particularly, of settlers in northern California during the gold rush of the mid-1800s (see Thornton 1987, pp. 107–109). The fate of the Yahi Yana of California and Ishi was but one instance of the racial genocide that occurred: Ishi was the last known survivor of the Yahi Yana, following a series of massacres of his people at the hands of the Californians (see Heizer and Kroeber 1979; Thornton 1987, pp. 109–113). Many other tribes suffered similar fates in the decades following the gold rush, as the California Indians were disposed of by the Euro-Americans. The massive population relocations and removals, as well as population concentrations on reservations, were also destructive demographically. In addition, they broke tribes into different reservations, or threw tribes hostile to one another on the same reservation. Typically, reservations were land areas not desirable for farming by Euro-American methods, but Native Americans were generally encouraged to become farmers on them, even if they were not an agricultural people. At the hands of the Euro-Americans, Native Americans in the southeastern United States experienced particularly large-scale removals to Indian Territory (an area designated specifically for Indian relocation, now the state of Oklahoma) during the first half of the 1800s. The American Indian peoples involved named these removals "Trails of Tears" for good reason (see Thornton 1984a). The removals caused suffering and death; hence, population decline (see the essay by Snipp, this volume). Also, the overall destruction of native resources and ways of life must be added to any list of reasons, though the particular demographic effects are harder to ascertain. For example, how much American Indian population loss can be traced to the deliberate destruction of the North American buffalo during the nineteenth century? We do not know, but American Indian population losses were likely substantial. Euro-Americans slaughtered buffaloes for the economic value of their hides and because they were in the way of westward expansion. The destruction was so great that the estimated 40 million buffalo in 1800 (down from an aboriginal population estimated at 60 million) declined to 20 million by 1850 and to a mere 1,000 by 1895.[12] Such great American Indian peoples as the Arapaho, Cheyenne, Comanche, Crow, Kiowa, Sioux, Pawnee, and Caddo were dependent on the buffalo for food, clothing, housing, tools, bedding, and other material necessities of life, and their societies and cultures—

their ways of life—all but collapsed with its demise (see Thornton 1987, pp. 51–53, 123–125).

The Legacy of Depopulation

Even today, epidemics of disease, geographical removals, war, or genocide often stand as the most significant historical event in the collective memory of a particular American Indian people. What do the Cherokee of Oklahoma remember over and over again? The "Trail of Tears"—the thousands and thousands of Cherokees who died as a result. The Kiowa remember the cholera epidemic of 1849, and the Mandan remember when smallpox quickly struck them in 1837 and reduced their population from several thousand to a few hundred. Similarly, the Cheyenne remember the Sand Creek Massacre of 1864 and the Sappa River Massacre of 1875. Only now, is the Smithsonian Institution returning to the Cheyenne the skeletal remains of some of their ancestors who were killed at Sand Creek! And will any American Indian ever forget the massacre of Sioux men, women, and children at Wounded Knee Creek in 1890? Also significant and also mourned today are the destructions of many native resources and the lifestyles that accompanied them. Dispersed on reservations or concentrated in urban ghettos, many, but certainly not all, Native Americans have not yet achieved integration into American society, economically or otherwise; they live in poverty, with little hope that their children will have better lives.

Beyond American Indians as "Victims"

The issues regarding American Indian population decline continue to dominate many scholars in the field, particularly the issue of aboriginal population size. Certainly, this is an important topic and one in need of further attention. The scholarly research of the Middle Ages was trivialized by asserting scholars sought to ascertain "how many angels could dance on the head of a pin." I certainly do not wish to trivialize Native American historical demography and the closely related topic of Native American historical epidemiology, but scholars should not limit themselves to a preoccupation with numbers and their decline. They particularly should not limit themselves to a consideration, to paraphrase the above, "of how many Indians could dance in North America in 1492." Perhaps a more important topic is the examination of how specific Native American societies and cultures responded to and *adapted* to the demographic legacy of their medical and military conquest by Europeans. Such a focus would not only depart from seeing Native Americans only as victims—though they were, of course, victims of European and African diseases and European colonization—but also is more congruent with demographic thought as to the maintenance of demographic regimes over time by any group of humans—Native Americans or other.

The Importance of "Demographic Regimes"

When discussing "demographic regime" demographers—scholars who study population—refer to the determinants of fertility (for example, age at marriage and patterns of breast-feeding), mortality (for example, occupations and health care practices), and migration (for example, movement to urban areas from rural areas) in a population, which, as they interact together, produce growth, decline, or stability over a particular period of time. These patterns are typically more or less stable, and influence the population's ability to respond to disturbances such as those caused by episodes of a disease. Demographic regimes have been discussed in various ways. E. A. Wrigley in his *Population and History* (1969) noted:

> A year of demographic crisis had the effect of weeding out many of those who were most vulnerable in a population, both in the economic and the psychological sense. Years of crisis were, therefore, normally followed by a period of much lower mortality. If there were untenanted holdings or unworked looms made available by death during the crisis, or if marriages had been postponed because of the hard times, a rush of new marriages followed and a spurt of births would coincide with the reduction in the numbers of deaths to give a few years in which the population could rise rapidly. (Wrigley 1969, pp. 68–69)[13]

Hence, disturbances such as episodes of a disease may result in only short-term population decline as

populations may also return to their previous levels of population growth, decline, or stability. And, as Wrigley indicated, even the effects of depopulations can produce spurts of population growth thereafter (see also Thornton, Miller, and Warren 1991).

Similarly, considering the European case, the historian William H. McNeill (1976, p. 150) noted: "the period required for medieval European populations to absorb the shock of renewed exposure to plague seems to have been between 100 and 133 years, i.e., about five to six human generations." Societies in Western Europe during the sixteenth, seventeenth, and eighteenth centuries were able to withstand severe demographic shocks, in large part because of the particular nature of their demographic regimes. By contrast, other populations have surely been driven to extinction by a series of relatively mild shocks because their demographic regimes were not capable of coping with the exogenous disturbances. All Native American populations in 1492 had previously experienced exogenous biological shocks of one form or another, to one degree or another. Be they the shocks from disease, war, or natural disaster, the Native Americans who were here in 1492 had dealt with these shocks and had survived them. Such shocks were so commonplace elsewhere in the world that it is hard to conceive of the Western Hemisphere being free of them. The arrival of the Europeans, with their weapons and their diseases, was yet but another shock to the Native American demographic regimes, albeit one of a magnitude until then not experienced. We know from history that some Native American demographic regimes survived European contact, although in changed form, while others did not. McNeill (1976, p. 150) also noted that the 100- to 133-year period it took European populations to recover from the plague was about the same time period "Amerindian and Pacific island populations later needed to make an even more drastic adjustment to altered epidemiological conditions." Similarly, the anthropologist Henry F. Dobyns (1966) argued that it took American Indian populations about 130 years to begin to recover their population losses.

Thus, depopulation may have been followed by population recoveries, though not necessarily total ones, and epidemics may not reduce a population permanently or even for an extended period of time. A zigzag pattern of population growth, loss, and growth again seems more realistic for many groups than a downward stairstep pattern of population loss, loss, and further loss (see Thornton, Miller, and Warren 1991).

All Native American groups were subjected to similar forces of depopulation following European contact; however, similar tribal experiences did not necessarily have similar results. As I have stated elsewhere, the *"differential survival of American Indian tribes does not seem to have been a straightforward result of depopulation experience.* Consequently, tribal social and cultural factors may have influenced survival" (Thornton 1984b, p. 188). For example, I did a comparative study of the differential survival of the Tolowa and the Yuki Indians of California after both were depopulated during the gold-rush period, and found that the Tolowa were able to survive as a distinct tribal people to a greater extent than were the Yuki. The study suggested three factors which might account for their differential survival: (1) the relative magnitude of the depopulation (the Yuki were larger and experienced greater relative loss, though both were reduced to small groups); (2) the different reservation experiences (the Yuki were placed on a reservation with other tribes, and intermarried with them); and, (3) patterns of social organization (Tolowa kinship patterns allowed for the easy incorporation of outsiders into the tribe) (Thornton 1986a, p. 129). Hence, the depopulation experience per se was not the only factor determining tribal survival. Similarly, I tried to illustrate in *American Indian Holocaust and Survival* (1987) that Native American populations actually exhibited an exceedingly wide range of responses to the shock and aftermath of European contact. Many of these responses may be seen as attempts to establish ways of surviving and adapting to the European presence and its implications. In this sense, then, Native Americans can be seen not only as victims but also, at least in part, as actively involved in attempting to shape (if not actually shaping) their own destinies.

Responses to Disease and Depopulation

Native American responses to the demographic legacy of conquest were various: directly attempting to deal with European diseases; trying to restore demographic regimes; adapting their regimes to patterns of more or less permanent survival; amalgamation with other groups (Native American or non–Native American); "ethnogenesis;" and revitalization movements.

Most scholars of American Indians and many laymen are familiar with the various descriptions of how disease created havoc among American Indian peoples. The reactions to smallpox, for example, are typically described as attempted cures by bathing in cold water, or sessions in a sweat house, or panic and self-destruction in the face of almost certain death. Certainly these occurred. Native Americans also attempted to deal with diseases in other ways, however. John Moore has argued that the Cheyenne dispersed themselves upon the plains in response to epidemic disease. Such a response may have spread the disease to other tribes, but it also operated to reduce the spread of the disease among the Cheyenne themselves. Moore (n.d., p. 1) stated: "when properly organized, the dispersion tactic can remove the population from proximity to disease vectors, and can mathematically decrease the effects of the secondary attack rate." The Apache had a similar response:

> Their Medicine Men could do little or nothing for the victim of smallpox. But they realized that if the groups remained together there probably would be more deaths than if they separated. Magoosh held a council and it was decided that the Lipans should divide into small bands, each to be under the command of a leader or headman and composed of relatives. . . . Arrangements were made that, when and if the sickness ceased, we were to reassemble when and where the chief should select a time and place.[14] (Ball 1980, p. 269)

Ubelaker and Wiley (1978) have observed changes in Arikara mortuary practices that *may* have resulted from exposure to European disease. The Arikara, they argued, historically buried their dead soon after death, but certain historical, archaeological, and entomological evidence points to a "prehistoric" pattern of scaffolding before burial. It is possible that the Arikara changed their burial practices in response to European disease. "The Arikaras of the plains may have changed from a scaffolding style of burial to direct interment to avoid the hazard of increased exposure to corpses infected with epidemic disease" (Meyer and Thornton 1988, p. 11). Along similar lines, scholars have argued that various American Indian healing societies and curing rites either arose from or were modified in an attempt to combat European diseases. The case could even be made that the well-known Ojibwa *Midewiwin* (Medicine Lodge) curing society was modified, if not actually created, to attempt to deal with European disease.

Depopulations from disease or other reasons also resulted in Native American responses to attempt to maintain—to the extent possible—their demographic regimes. For example, the Tolowa Indians of northern California were able to recover from severe demographic losses due in large part to a series of massacres by Euro-Americans in the mid-1800s. They did so by having the few surviving males each take several wives from neighboring tribes. A number of children were born to these unions in a relatively short time. Since Tolowa society was both patrilocal and patrilineal, the children were considered Tolowa. Thus, "Tolowa marriage and kinship patterns allowed the incorporation of non-Tolowa females into the tribal group, and the consequent propagation by a few males of numbers of children defined as Tolowa. Thus, demographic re-creation of Tolowa society occurred. Had these customs not prevailed, it is difficult to understand how the Tolowa would have continued to today" (Thornton 1984b, p. 194).

By contrast, some Native American populations chose to subsume their populations under the demographic regimes of other Native Americans. Many remnant American Indian groups in the eastern United States joined with the Iroquois and were adopted by them; for example, the Tuscarora who fled northward from the Carolinas to escape the slave trade. Similarly, the migration of various tribes into the Mississippi River valley and their amalgamation there has been documented. Jeffery P. Brain (1971) noted that the Natchez Indians of Mississippi changed their marriage rules in order to adopt other Indians as relatives. It may also be noted that the Yuki Indians of northern California merged with other tribes of the Covelo Indian Community of Confederated Tribes of the Round Valley Indian Reservation, and, through intermarriage and a loss of identity, remnants of some California Indian tribes simply lost themselves in the general California population (see Thornton 1986a).

Conversely, various new Native American groups have been created in response to the demographic events of Euro-American contact. The most well known of the groups is surely the *Metis* of Canada and the U.S.–Canadian border. This Indian-White racially mixed group was created, they say, "nine months after the first white man set foot in Canada." Other new Native American groups include the Lumbee of North Carolina and many, varied triracial isolates scattered throughout the Atlantic, southeastern, and southern states that are composed of populations of varying degrees of American Indian, White, and Black mixture. New groups also include such historically prominent American Indian tribes as the Catawba of the Carolinas.

Finally, other responses to American Indian depopulation and the social and cultural collapse often associated with depopulation include a variety of revitalization and other social movements. These include the Ghost Dance of 1870, centered in California, and the Ghost Dance of 1890, centered in the plains (Thornton 1981, 1986b). The Ghost Dances were new religions created by American Indians, and were deliberate attempts to confront the changing circumstances in which western American Indian peoples found themselves in the late 1800s. The American Indian peoples involved sought divine intervention to revitalize their societies and cultures by restoring them to what they had been before the tremendous depopulation at the hands of the Americans. They believed that they could recapture their past ways of life, that game such as the buffalo would be restored, and even that American Indian dead would return to life if they performed certain ceremonies involving a "ghost" dance.

To fully understand American Indian population dynamics and adaptations scholars need to cease considering Native Americans as mere victims of European and African diseases and European and American military technology. Scholars instead need to begin to examine the ways specific Native American groups attempted to respond to the demographic implications of contact. American Indians were not just victims, they were also people attempting to deal with the circumstances in which they found themselves after 1492.

Twentieth Century Patterns and Adaptations

The Native American population north of Mexico began to increase about the turn of this century. In part, this population recovery was a result of lower mortality rates and increases in life expectancy as the effects of "Old World" diseases and other reasons for population decline lessened. The population recovery also resulted from adaptation through intermarriage with nonnative peoples and changing fertility patterns during the twentieth century, whereby Native American birth rates have remained higher than those of the average North American population. The census enumerations indicate a Native American population growth for the United States that has been nearly continuous since 1900 (except for an influenza epidemic in 1918 that caused serious losses) to 1.42 million by 1980 and to over 1.9 million by 1990. To this may be added some 740,000 Native Americans in Canada in 1986 [575,000 American Indians, 35,000 Eskimos (Inuit), and 130,000 *Metis*] plus some increase to 1991, and perhaps 30,000 Native Americans in Greenland. The total then becomes around 2.75 million in North America north of Mexico—obviously a significant increase from the perhaps fewer than 400,000 around the turn of the century.[15] However, it remains far less than the over 7 million circa 1492. It is also but a fraction of the total populations of the United States (250 million in 1990) and Canada (over 25 million in 1990) (see Thornton 1994a, 1994b).

Native Americans today are distributed unevenly throughout North America, a reflection more of events following European arrival than of aboriginal patterns. The 1990 census enumerated the largest number of American Indians in the states of Oklahoma, California, Arizona, and New Mexico. The 1990 census also indicated that slightly over one-half of Native Americans live in urban areas; cities with the largest Native American populations are New York City, Oklahoma City, Phoenix, Tulsa, Los Angeles, Minneapolis–St. Paul, Anchorage, and Albuquerque. About one-fourth of American Indians in the United States live on 278 reservations (or pueblos or rancherias) or associated "tribal trust lands," according to the census (see Maps 1 and 2). The largest of these is the Navajo Reservation, with 143,405 Native Americans and 5,046 non-Indians living there in 1990. About 60 percent of the Native American population of Alaska live in "Alaska Native Villages" (Thornton in press a).

The increase in the twentieth century in the Native American population reflected in successive censuses of the United States was also due to changes in the identification of individuals as "Native American." Since 1960 the United States census has relied on self-identification to ascertain an individual's race. Much of the increase in the American Indian population from 523,591 in 1960 to 792,730 in 1970 to 1,366,676 in 1980 to over 1.8 million in 1990 resulted from individuals not identifying as American Indian in an earlier censuses but identifying as such in a later censuses. It has been estimated, for example, that as much as 60 percent of the population "growth" of American Indians from 1970 to 1980 may be accounted for by these changing identifications (see Thornton 1987, pp. 220–221)![16] The political mobilization of American Indians in the 1960s and 1970s along with other ethnic pride movements may have lifted some of the stigma attached to American Indian racial identity. This would be especially true for persons of mixed ancestry who formerly may have declined to disclose their American Indian background. Conversely, however, individuals with only minimal American Indian background may have identified as American Indian out of a desire to affirm a "romanticized" notion of being American Indian. Today over 300 American Indian tribes in the United States are legally "recognized" by the federal government and receive services from the U.S. Bureau of Indian Affairs. In addition, there are some 125 to 150 tribes seeking federal recognition and dozens of others who may do so in the future. The Bureau of Indian Affairs uses a "blood quantum" definition—generally a one-fourth degree of American Indian "blood"—and/or tribal membership to recognize an individual as American Indian. However, each tribe has a particular set of requirements—typically including a "blood quantum"—for membership (enrollment) of individuals in the tribe. Requirements vary widely from tribe to tribe. For example, some tribes, such as the Navajo, require at least a one-half Indian (or tribal) blood quantum, many tribes require a one-fourth blood quantum; some tribes, generally in California and Oklahoma, require a one-eighth, one-sixteenth, or one-thirty-second blood quantum; and many tribes have no minimum blood quantum requirement but only require some degree of Ameri-

Table 1 Blood Quantum Requirement of American Indian Tribes

Blood Quantum	Number of Tribes
More than $\frac{1}{4}$	21
$\frac{1}{4}$ or less	183
No minimum requirement	98

Note: Information not available on 17 tribes.

Table 2 Blood Quantum Requirement by Reservation Status

Blood Quantum	Percent Reservation Based
More than $\frac{1}{4}$	85.7
$\frac{1}{4}$ or less	83.1
No minimum requirement	63.9

Note: Information not available on 17 tribes.

can Indian lineage (see Thornton 1987; Meyer and Thornton 1990) (see Table 1). Typically, those tribes located on reservations have higher blood quantum requirements for membership than those not located on reservations (see Table 2). This pattern of requiring low percentages of Indian "blood" for tribal membership and dealing with the federal government to certify it also may be seen as a result of the demographic legacy of contact. As the numbers of American Indians declined and American Indians came into increased contact with Whites, Blacks, and others, American Indian peoples increasingly married with non-Indians. As a result, they have had to increasingly rely on "formal" certification as proof of their "Indianness."

In the early 1980s the total membership of these approximately 300 tribes was about 900,000. Therefore, many of the 1.37 million individuals identifying themselves as American Indian in the 1980 census were not actually enrolled members of federally recognized tribes. In fact, only about two-thirds were. In the late 1980s the total membership was about 1 million; hence, only about 55 percent of the 1.9 million people identifying themselves as Native American in the 1990 census were actually enrolled. Such discrepancies varied considerably from tribe to tribe. Most of the 158,633 Navajos enumerated in the 1980 census and the 219,198 enumerated in the 1990 census were enrolled in the Navajo Nation; however, only about one-third of the 232,000 Cherokees enumerated in the 1980 census and the 308,132 enumerated in the 1990 census were actually enrolled in one of the three Cherokee tribes [The Cherokee Nation of Oklahoma, the Eastern Band of Cherokee Indians (of North Carolina), or the United Keetoowah Band of Cherokee Indians of Oklahoma] (see Thornton 1990). Thus the Navajo Nation is the American Indian tribe with the largest number of enrolled members, but more individuals identifying as Native

American identified as "Cherokee" in the 1980 and 1990 censuses than as any other tribe.

Similarities and differences occur with the situation in Canada. Officially, to be an Indian in Canada one must be registered under the Indian Act of Canada. A person with Indian ancestry may or may not be registered, and categories of Canadian Indians include "status" or registered Indians, individuals registered under the act; and "nonstatus" or nonregistered Indians, individuals who either never registered or who gave up their registration and became enfranchised. Status Indians may be further divided into treaty and nontreaty Indians, depending on whether their group ever entered into a treaty relationship with the Canadian government. Of the 575,000 American Indians in Canada in the mid-1980s, some 75,000 were nonregistered and some 500,000 were registered (see Thornton 1987, p. 224; 1994a). The *Metis* are a group of individuals of Indian and White ancestry who are not legally recognized as Indian.

Canadian provinces with the largest number of registered Indians are Ontario, British Columbia, Saskatchewan, and Manitoba. In 1986, some 40 percent of the 740,000 Native Americans in Canada lived in cities; the largest American Indian populations were in Vancouver, Edmonton, Regina, Winnipeg, Toronto, and Montreal.

Many Canadian Native Americans live on what are called "reserves." However, about 70 percent of the registered Indians live on one of 2,272 reserves (Thornton 1994a). In the early 1980s, there were 578 organized bands of Canadian Indians. Most bands contained fewer than 500 members; only eight bands contained more than 3,000 members: the Six Nations of the Grand River (11,172), the Blood (6,083), the Kahnawake (5,226), the Iroquois of St. Regis (4,098), the Saddle Lake (4,020), the Wikwemikong (3,493), the Blackfoot (3,216), and the Lac La Ronge (3,086) (Thornton 1994a).[17]

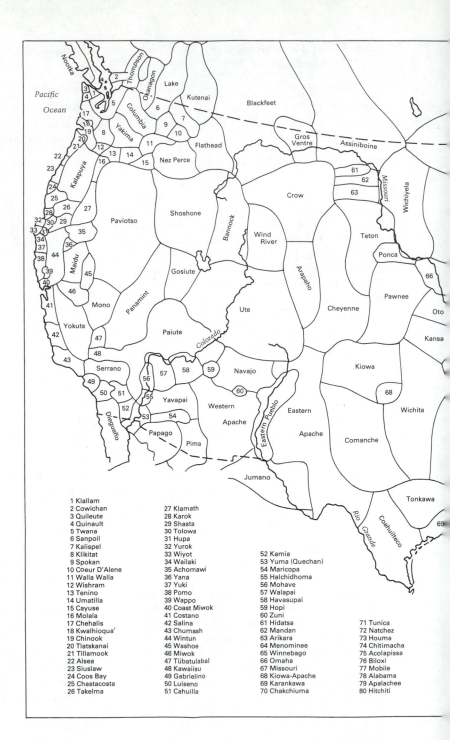

Map 1

American Indian Tribes, Circa 1600

Source: From C. Matthew Snipp, *American Indians: The First of this Land,* New York: Russell Sage, 1989; adapted from Murdock, in Stephen Thernstrom et al., *Harvard Encyclopedia of American Ethnic Groups,* Cambridge, MA: Harvard University Press, 1980.

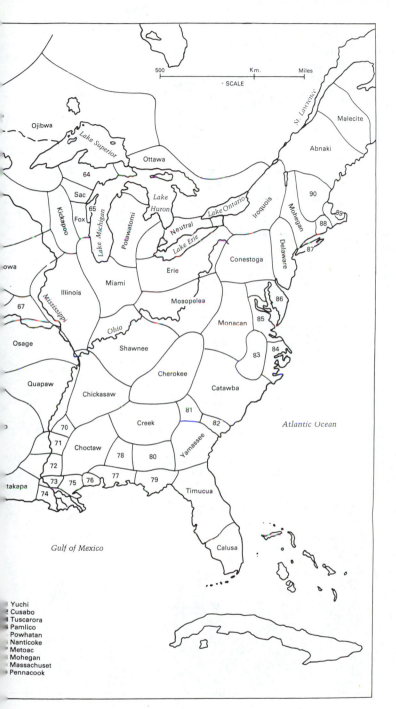

Ojibwa

Lake Superior

Ottawa

Malecite

Abnaki

64

Sac

65

Kickapoo

Fox

Lake Michigan

Potawatomi

Lake Huron

Neutral

Lake Erie

Lake Ontario

Iroquois

Mohegan

90

88

89

87

owa

Illinois

Miami

Erie

Mosopelea

Conestoga

Delaware

86

85

67

Mississippi

Osage

Ohio

Shawnee

Monacan

83

84

Quapaw

Cherokee

Catawba

Chickasaw

81

82

Atlantic Ocean

70

Creek

Yamassee

71

Choctaw

78

80

79

72

77

73   75   76

Timucua

takapa

74

Gulf of Mexico

Calusa

St. Lawrence

500   Km.   Miles

· SCALE

1 Yuchi
2 Cusabo
3 Tuscarora
4 Pamlico
5 Powhatan
6 Nanticoke
7 Metoac
8 Mohegan
9 Massachuset
10 Pennacook

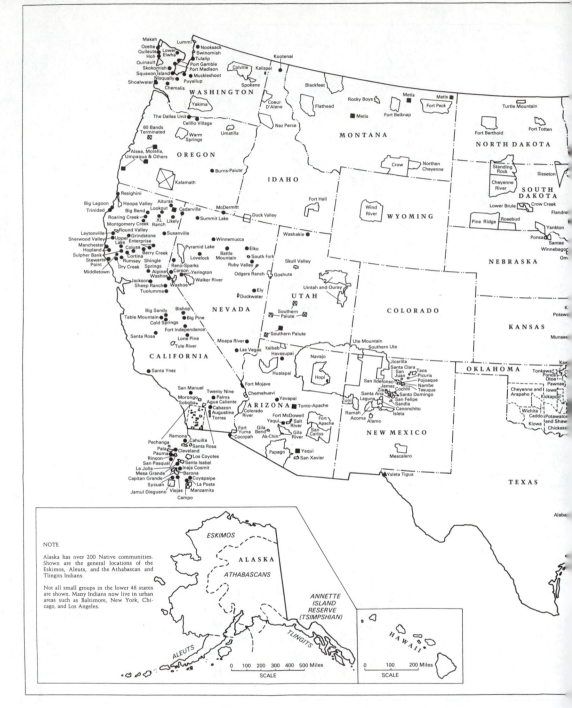

Map 2

Indian Lands and Communities

Source: Council of Energy Resource Tribes (1985).

Federal Indian Reservation

Former Reservations in Oklahoma

State Reservations

Indian Groups Without Trust Land

Federally Terminated Tribes and Groups

50 0 50 100 150 200 Miles
SCALE

Conclusion

The five centuries from 1492 to 1992 represent two vastly different population histories that took place in the Western Hemisphere, especially in what is now the United States and Canada. One history is that of those populations represented by Christopher Columbus on his arrival to his "New World;" the other history is that of those populations already here when Columbus arrived. Little wonder that the date 1492 carries opposite symbolic meanings for the two populations. Yet these histories are intertwined, and Native Americans have had to respond to the racial prejudice and colonialism inflicted upon them. Native Americans adapted and survived, both as physical populations and tribal peoples as they encountered and confronted the demographic implications of prejudice and colonialism; however, they were forever changed.

Notes

1. Scholars have noted that Dobyns nadir population is too high; a more correct nadir population is 350,000 (see Thornton and Marsh-Thornton 1981).

2. Ubelaker (1988) recently revised his population estimate slightly upward to 1,894,350.

3. Tuberculosis was seemingly present in both hemispheres by this time, as were other diseases, including treponemal infections, but probably not syphilis which is caused by *Treponema pallidum* (see Crosby 1972).

4. It has also been argued that the "migration" of *Homo sapiens* across cold, inhospitable Beringa "filtered out" certain pathogens causing human diseases (see Thornton 1987, 40).

5. The Southeast and a much larger area extending all the way to present-day Canada was called *La Florida* at one time; then *La Florida* became more restricted, more or less coinciding with the Southeast today, and, of course, eventually "Florida" became only the present-day state by that name.

6. Hernando de Soto's explorations are well known, although his probable route is being revised by Charles Hudson and his colleagues (see Hudson, Smith, and DePratter 1984; Hudson et al. 1985).

7. The narrative also includes many other reports of sickness, pains, and headaches among Indian peoples encountered by Cabeza de Vaca.

8. Hence, the "epidemic" would have been in 1538. If it happened, possibly it came from the Ayllon "colony."

9. Garcilaso de la Vega (1951, p. 315) also wrote that there were "four large houses" in the town "filled with the bodies of people who had died of the pestilence." However, other accounts of the de Soto expedition do not mention this "disease episode" (see Bourne 1922), and the accuracy of Garcilaso de la Vega has been questioned as well as has the interpretation of the word "pestilence" (see Henige 1986, p. 301).

10. Dobyns (1983, pp. 8 254–262, 270, Table 25) also argues for other, earlier epidemics in what is now Florida in 1513 to 1514 (malaria?), 1519 to 1524 (smallpox), and 1528 (measles or typhoid fever).

11. Similarly, it is possible that smallpox was present early in the sixteenth century in the Southwest, as Upham (1986, 1987) has argued using simulations of smallpox epidemics.

12. Today, there are well over 50,000 buffalo (Thornton 1987, pp. 52–53).

13. Various other scholars have also discussed the demographic regimes of historic European and other populations, and I have discussed those of American Indian populations (see Thornton, Miller, and Warren, 1991).

14. The Lakota Sioux of the Dakotas actually increased in size, according to Richard White (1978), during the approximate period that their neighbors the Mandan were being decimated by smallpox. The village Mandan could not disperse during an epidemic in the 1830s, since they were prevented from doing so by surrounding enemies, but the Lakota Sioux were scattered across the plains; hence, the effects of smallpox may have been lessened.

15. Douglas Ubelaker (1988) suggests the Native American population of North America exceeded 2.5 million in 1985.

16. The 1980 U.S. census obtained information that some 7 million Americans had some degree of Native American ancestry. Native American ancestry ranked tenth in the total U.S. population in 1980. In descending order, the ten leading ancestries were

English, German, Irish, Afro-American, French, Italian, Scottish, Polish, Mexican, and Native American.

17. The largest "group" of Canadian Indians in terms of language and culture are the Chippewa-Ojibwa. Adding the more than 100,000 Ojibwa enumerated in the 1990 U.S. census to them would make for a total population (registered and nonregistered) that rivals the 220,000 self-declared Navajos in 1990, if not the more than 300,000 self-declared Cherokees.

4

From Sundown to Sunup: Slavery and the Making of the Black Community

GEORGE P. RAWICK
(EXCERPTED BY SILVIA PEDRAZA)

Black history in the United States must be viewed as an integral, if usually antagonistic, part of the history of the American people. Without understanding the historical development of Black society, culture, and community, comprehension of the totality of America's development is impossible. Slavery was a fundamental part of the history of the whole American people, just as its aftermath continues to pose a fundamental question for our national life.

Master and Slave

American Black slavery was a human institution, albeit an exceedingly inhumane one. Yet rarely has the discussion of slavery in North America proceeded from this premise. Rather, almost all historians have presented the Black slaves as dehumanized victims, without culture, history, community, change, or development. . . .

But if the slave had a history, then his behavior changed over time as he learned from the past and met new experiences. Men, however, do not move on their own behalf or make revolutions for light and transient reasons. Only when they no longer can stand the contradictions in their own personalities do they move in a sharp and decisive fashion. The victim

For this synthetic description and sociological analysis of slavery in the South of the United States, the late George P. Rawick based himself on the Federal Writers' Project slave narrative collection of the Works Project Administration compiled during the years 1936 to 1938 to preserve the life stories of ex-slaves. It consisted of over 10,000 pages of typescript containing over 2,000 interviews with ex-slaves. Rawick's *From Sundown to Sunup: The Making of the Black Community* is volume 1 of the nineteen volumes which constitute the oral history collection of slave narratives, and that is organized by states of the Confederacy. Quotations from the slave narratives are cited as FWPSN, by state and volume.

At the time *From Sundown to Sunup* was written and published, the late George P. Rawick was Associate Professor of Sociology at Washington University in St. Louis. Years later, as an Assistant Professor of Sociology at Washington University, Silvia Pedraza served on numerous dissertation committees with him. In his memory, with pleasure she excerpted this essay from his book.

is always in the process of becoming the rebel, because the contradictions demand this resolution.

As the German philosopher Hegel understood in the famous passage on master and slave in *The Phenomenology of Mind* (1970, I:183ff), the slave fights against the master by wrestling with his own internal conflicts. The will of the master and the will of the slave both appear as a contradiction within the slave.

We can see this process working in many accounts left by ex-slaves telling of their flight from slavery. Even when writing their stories as abolitionist propaganda, they indicate their fears, self-doubts, even guilt, about running away. The greatest of all abolitionist leaders, the ex-slave Frederick Douglass, indicates this ambivalence in his autobiography:

> I have been frequently asked, when a slave, if I had a kind master, and do not remember ever to have given a negative answer; nor did I, in pursuing this course, consider myself as uttering what was absolutely false; for I always measured the kindness of my master by the standard of kindness set up among slave holders around us. ([1845] 1963, 20)

Most of Douglass' book deals with the years of preparation that he went through in order to be ready to run away. At all times he was unsure of himself. Even after making his way to freedom in New York City he reports that shortly after arriving there, "I was again seized with a feeling of great insecurity and loneliness. . . . But the loneliness overcame me." ([1845] 1963, 106).

Douglass felt all of this probably even more keenly because in fact he never had a home. His father had been a plantation owner with whom he had no relationship and he had seen his mother four or five times since infancy. Nevertheless, the ex-slave Frederick Douglass . . . became a powerful, influential leader in the struggle against slavery.

But Douglass' achievement must be set into the context that while tens of thousands of other slaves, in tribute to human courage, did indeed follow the North Star to New England, the upper Midwest, and Canada, many thousands more either failed in their attempts to escape or never made the effort. . . .

Running away from slavery was objectively difficult. It entailed a journey of hundreds, even a thousand miles from the Deep South, on foot, with only the food one managed to get from the land, or an occasional slave, under circumstances in which virtually every White man, and even an occasional Black, was a potential captor. The patrol system operated throughout the South and the patrollers and their bloodhounds were in fact in pursuit of runaways. Harriet Beecher Stowe's portrayal of a slave runaway in *Uncle Tom's Cabin* was not particularly exaggerated, although not as interesting as the stories of the hundreds of real runaways recorded in the slave autobiographies and narratives.

We get a picture of the magnitude of the task involved in running away in the autobiographical account of Solomon Northrup. Northrup, born a free man in New York State, was kidnapped and spent twelve years as a slave, almost all of the time in a relatively primitive, inaccessible area of the frontier territory of western Louisiana. In his volume *Twelve Years a Slave* ([1857] 1968, 101–107) he tells us that he made only one effort to escape, and that effort demonstrated to him that such an attempt was fruitless.

The obstacles that the slaves had to overcome in making their way across the many miles to freedom were too great for most, particularly since these obstacles were both subjective and objective. Convinced of their desire to be free, they were nevertheless afraid of the consequence of reaching freedom. . . . A slave must inevitably make some adjustment to his situation, an adjustment which must include some view that the condition of being a slave is the normal one for people such as himself. The mark that necessary adjustment left on Frederick Douglass was not the sense of being childlike but the much more poignant sense of being "lonely" when finding himself finally free.

Culture and personality are not like old clothes that can be taken off and thrown away. . . . "New" cultures emerge out of older cultures gradually and never completely lose all the traces of the old and the past. Human society is a cumulative process in which the past is never totally obliterated. . . . Culture is a historical reality, not an ahistorical, static abstraction. Thus, the process whereby the African in the New World changed in order to meet his new environment was dependent upon his African culture. While it is certainly true that the African under American slavery changed, he did so in ways that were recognizably African.

Coming for the most part from West Africa, these people, who had been stolen or captured and taken from their homes, *brought virtually nothing with them except themselves.* Coming from a large area of West Africa which was subdivided among dozens of distinct peoples, . . . they were jumbled together on board the slave ships, "seasoned" by the "middle passage" across the Atlantic and then "seasoned" again in their first years in the New World. As slaves, not only were they prevented from bringing material objects with them, but they also could not even bring over their older social relations or institutions.

Such was not the case with the Europeans. They brought their churches with them. They brought their own foods with them and could continue to get supplies of specific items from the old country if need be. They brought their own dress with them and could choose to wear it or abandon it as they saw fit. They brought their own marriage customs, their own rites of passage, their own kinship system. The Europeans preserved their old customs for as long as they were needed and gradually modified them as they moved into the main society. While some met opposition for being foreigners, they were not stripped of their foreignness overnight.

But the African slaves could do none of this. Overnight they were transformed from merchants, or Arabic scholars, or craftsmen, or peasant farmers, or cattle-tenders into American slaves. They ate what they were given, not what they wanted. They dressed in the clothes that were given them, not those they had known in the past. African women were removed from a stable social order which gave them a specific place and function, which protected them with a traditional morality—and indeed exploited them in a traditional way—and made them commodities, unprotected by a traditional morality, without specific places and functions, sexually exploited by the master and even deprived of a full relationship with their children.

The Africans had to give up their own languages and learn to express themselves through other media of communication. They had to give up their old kinship systems and create new ones. They had to give up almost all of their culture and become American slaves.

That was demanded of them. But on the other side, the slave could not really do any of these things, for he brought with him his past. He brought with him the content of his mind, his memory; he recognized as socially significant that which he had been taught from childhood to see and comprehend as sig-

nificant; he gestured, laughed, cried, and used his facial muscles in ways that he had learned as a child. He valued that which his previous life had taught him to value; he feared that which he had feared in Africa; his very motions were those of his people and he passed all of this on to his children.

He faced this contradictory situation in a context which was complicated by the fact that while there were many similarities among West African peoples, there were also many differences. Each people had its own language, history, technology, and so forth. . . .

Slaves in the United States had come from many different African cultures. They were thus faced with the difficult task of adjusting not only to their new environment and their new social relationships, but also to each other; they had as well to build a culture out of the interactions of Africans with other Africans. Therefore, while all Africans were slaves and slaves were supposed to act in a specific way, none knew what this way was. There was no model to follow, only one to build. . . .

Out of the interaction of the men and women who carried a varied African heritage in their minds and memories along with the environment of a harsh plantation system based on slave labor, there emerged over time an independent Black community. It showed the marks of the African experience while at the same time being insistently American.

This community took its form in the slave quarters of the plantations, among the dispersed but large number of slaves who lived on small farms and plantations, and in the back alleys of the cities. It developed its own church, one designed to meet the needs of slaves and Afro-American freedmen in the New World. It had its own value system, reflective of the attitudes of African countrymen, but modified by contact with and the need to live with those of European origin. It employed modes of communication which, whatever their particular origins and forms, were neither African languages nor the same languages spoken by Europeans in their own countries—but new and different (see Mintz 1971). It had its own class system based on the division of labor on the plantation. It even had its own subordinate economic structure and activities and its own political system. . . .

We shall view slavery as an ongoing social institution in a developing and relatively open society in which a rough democracy and egalitarianism prevailed for the Whites. In such a society there was room for maneuver, for tactics and strategies, for Blacks as well as for Whites. The conditions under which Blacks lived were, of course, different from those of Whites, but this did not mean that Blacks were totally denied social space within which to struggle to meet the natural and social environment.

To reject Elkins' (1958) thesis of the slave as absolute victim, . . . slave society as totalitarian, . . . and to assert that there was room for maneuver is to argue that the slaves were not totally isolated. They were, in fact, in constant communication with free men, Black and White. . . . The slave community was daily influenced by the White community. It was actively confronting the White world. . . .

Above all the slave community was making itself. . . . The slave community was ultimately more important to the slaves than the nuclear family, because while the family could be—and was—regularly broken up, the individual slave could be taken to another plantation hundreds of miles away and become part of another community where he knew how to behave and where he would be accepted. This reality of community was the major adaptive process for the Black man in America.

But more than that, the White community was influenced by the slaves through both imitation and reaction to the slaves' tactics and strategies. Not only did the slave community make itself, it also directly and indirectly helped make the White world. The South was one single social system, one single society, with two communities, one White and one Black. . . .

What were the possible strategies of the American slaves?

They could have worked as hard as required, disciplined themselves to accept the conditions of slavery, and lived out their lives, finding as much satisfaction as they could in the routines of daily life.

They could have accepted their lot as slaves in this world and, by having hope in another world to come and acting this out in religious practices, given themselves the necessary inner strength to meet the problems each day brought.

They could have presented a collectively contented face to the master class while individuals could and would get out of line and behave violently towards their oppressors.

They could have thought of themselves as inferior beings and accepted their lot, gaining as much

satisfaction as possible from emulating and serving the master class.

They could have done their work but at the same time struggled through strikes and sabotage, actions which required some common community and some prior discussion among groups of slaves.

They could have supported individual acts of rebellion by protecting those of their number who acted violently toward a member of the master class or his agent, or who ran away and hid, or who tried to make it to freedom.

They could have supported the actions of those who ran off into the woods or swamps and created communities or joined Indian groups.

They could have spent a great deal of time and social energy in adapting the West African cultures that they brought with them to meeting the new conditions of the United States.

They could have become revolutionaries, while waiting for the right time to strike.

And above and through all of these possible approaches was the ever-present, ever self-creating and -renewing strategy of building the slave community. The individual slave was never alone except when he ran away, and even then he often went from one community of slaves to the next, aided in his flight by his fellow slaves united into communities by the processes of slave production.

The slaves could have chosen any of these strategies. In fact they chose all of them and they all were interrelated. If any of these approaches had been abandoned, major transformation in the other strategies would have become necessary. This was the set of choices that faced an entire community and it was this total situation that dictated the context in which the community functioned, although certain individuals were socially specialized in particular approaches.

In order to understand this we have to view Black slaves in America as people whose ancestors had come from somewhere, who in the American South lived in a particular kind of environment and social system, who out of the past and present—a present in which the majority of the population were of European ancestry—wove a future. Black slaves were descendants of West Africans; they were field hands or house slaves or urban slaves in the American South's slave social system, one which was embedded in the matrix of world capitalism, despite the fact that slavery as a mode of production was itself different from capitalism (see Genovese 1966, 1969). And the slaves were to become free men as a result of struggles in which they themselves played important roles.

The slaves labored from sunup to sundown and sometimes beyond. This labor dominated part of their existence—but only part. Under slavery, as under any other social system, those at the bottom of the society were not totally dominated by the master class. They found ways of alleviating the worst of the system and at times of dominating the masters. They built their own community out of materials taken from the African past and the American present, with the values and memories of Africa giving meaning and direction to the new creation. They lived and loved from sundown to sunup.

The Religion of the Slaves

In many concrete ways the African past and the behavior of Afro-Americans under slavery were linked (see Bascom 1941, 1969). . . . Africans migrated involuntarily to North America in order to be worked as chattel slaves on plantations, in mines, in forests, on roads, on small farms of relatively diversified crops, and in town and cities. They labored generally from sunup to sundown and often into the evening. Their work was usually difficult and tedious, they were driven to it by threats, fear, the lash, and the accustomed sense of work routines common to all settled agricultural and urban societies. While their life was dominated by the social relations of work, they did have a life outside of work in the time that was, or that they made, their own. . . .

Only if we understand this side of the slaves' lives can we understand how their personalities were kept from destruction, how they developed and built their communities. . . . Close to the center of the slaves' lives from sundown to sunup was religion. The African slaves in the New World had come from societies in which there was no distinction between sacred and secular activities. The holy and the sacred were experienced as part of all activities. Divisions between this world and the next, between flesh and spirit, between the living body and the spirits of the dead were not conceived of as absolute. Men were thought to be able to slip across these boundaries with comparative ease.

For people from such a world, religious activities were areas of considerable potential creativity and social strength. The slaves in the New World used religion as the central area for the creation and recreation of community. . . .

In the nineteenth century there was a strong attempt by Whites to use religion as a form of social control. If we read the thousands of interviews with ex-slaves taken in the 1920s and 1930s, as well as the accounts of slavery from contemporary sources of the nineteenth century, we can see an attempt on the part of the masters to superimpose a formal religion on the slaves. But that religion never seemed to gain the total adherence of the slaves, who continued to carry on prayer meetings at night. . . . While religion certainly may at times be an opiate, the religion of the oppressed usually gives them the sustenance necessary for developing a resistance to their own oppression.

The religion of the slaves kept alive in them the desire and basis for a struggle for freedom. On a more immediate level, it made their daily lives bearable. . . . In prayer meetings and night sings Africans became American slaves while American-born slaves renewed their contact with the African experience through exchanges of ideas with newly arrived migrants from Africa. . . . Through these prayer meetings the bonds among people were tightened. . . .

The Black slaves in North America utilized West African concepts in a new and totally different context. In so doing they transformed those West African forms into something that was neither African nor European-American but a syncretic blend of the two. . . .

The intense relationship in the Black church between the preacher and the congregation is dependent upon the congregation being a community, a sacred family, in which the preacher is the leader and the head of the community. This relationship is similar to that of the elder in a West African village extended family compound. . . .

The Black minister functioned in a context controlled and limited by a shadowy but powerful group of elders or deacons and by older "sanctified" women. . . . These roles have West African analogues. . . . Their power is not derived from some legal or constitutional authority but from the traditional respect afforded elders because they are believed to act not out of selfish self-seeking motives, but out of their deep contact with the soul of the congregation and the community. The cries of "Amen," "Halleluja," "Tell it to them, Preacher," and the like that punctuate the sermon of the Black preacher are in effect affirmations that he is in tune with the soul of the community.

While the roots of his behavior are African, the form is not, nor are all of the meanings. Additional meanings have been added on so that the affirmations of the preacher's sermons are also in part a reflection of the urban, American Black ghetto scene with its rough democracy in which each man and woman is urged to "tell it like it is." . . .

The Black church came out of three distinct experiences: the slaves' own religion, both in its pre-Christian and in its Christian form; the Christian churches of the freedmen in the cities; and the White church. We must remember that until the end of the eighteenth century, few slaves had been converted to Christianity. Yet they practiced some sort of religion, although we have virtually no description of it. It was, however, clearly out of this West African set of practices that the slaves' own religion—the one they practiced on weekdays and in the evenings, in the hollows and the holes in the fields, in the cabins protected by the iron pot, at the crossroads where a little man appeared before the sinner and announced salvation—came. It was this religion which fertilized the freedman's churches, although these were always Christian in doctrine. Upon the slaves' own religion was superimposed the official religion of the masters. . . .

It was out of the religion of the slaves, the religion of the oppressed, the damned of this earth, that came the daily resistance to slavery, the significant slave strikes, and the Underground Railroad, all of which constantly wore away at the ability of the slave masters to establish their own preeminent society. Even the few but significant slave revolts . . . came out of Black religion.

Master and Slave: Treatment

While there has been a great deal of discussion about the life of the American slave at work, there has been relatively little focus and even less consensus about the quality of slave life "from sundown to sunup." How common was whipping? Were slaves decently fed, housed, and clothed? What was the structure and strength of the slave family? Did slaves rebel? . . .

Partially as a result of the lack of direct evidence, the debate over these questions has usually revolved around several extreme theses which clearly serve ideological needs. The first of these ought to be immediately dismissed: the gamine and magnolia tale about happy, well-fed, pampered, banjo-plucking "darkies," wearing simple but clean going-to-meeting clothes, eating chickens stolen from "massa," smiling watermelon-slice-sized grins. . . .

Two opposite views of slavery, derived from the view that the slaves in North America were indeed poorly treated, are deserving of more serious attention. One of these argues that the treatment of the slaves was so bad in North America that their condition was analogous to that of the inmates of German concentration camps in the Hitler era. . . . The very severity of North American slavery made the slave into Sambo—the shuffling man-child, the absolute victim (Elkins 1958). . . .

Others . . . have reached an opposite conclusion. Slavery was so bad that the slaves were almost always either plotting insurrection or actually at the barricades. There were hundreds of slave revolts and rebellions (Aptheker 1943). . . .

The first has become the historical justification for the theory that Blacks in America are psychological victims (cf. Moynihan 1965). . . . The second is the grounding for all sorts of views that transform Blacks into instant revolutionaries (cf. Aptheker 1943). . . .

The picture of slave treatment and of slave response that we get from the slave narratives and interviews . . . enable us to see that there were certain areas of autonomy carved out by the slaves in a situation which usually produced neither absolute victims nor instant revolutionaries [and] . . . enable us to see the conflicts between master and slave that took place day by day, and how these developed over time. . . . Resistance flowed from the network of informal organization of that community. . . .

The ex-slaves' accounts of their treatment make clear that most slaves suffered beatings and whippings; that they were often poorly fed, clothed, and housed; that they were often overworked; and that slave women were regularly used as sexual objects by Whites, while slave men were often used as breeding bulls, and slave children were frequently abused. They also indicate that the slaves had many ways of adding to the food and clothing supplied

Because slaves were the property of their masters, and could be bought and sold without regard for their family bonds, the impact of slavery on the African American family has been the subject of much intellectual debate. Here an African American family, five generations under slavery, is in front of their home in J. J. Smith's plantation, Beaufort district, South Carolina, in 1862. (Photo by Timothy O'Sullivan, stereograph. Library of Congress)

them by their masters; that they were given and found ways of having nonregimented social relations with their fellow slaves; that they were often able to break through the legal boundaries of slavery in such matters as learning to read despite prohibitions against teaching them this skill; and that the social structure of the slave community with its divisions of house slaves, field slaves, slaves who hired out their own time, and freedmen provided a circulation of needed news and information in that community. . . .

Physical coercion was necessary to slave society, particularly when the slaves often greatly outnumbered the masters and nonslaveholding Whites (see Fitzhugh [1857] 1960). . . . Almost all accounts of punishment indicate that whippings very often were the consequence of individual acts of resistance. . . .

Eli Coleman, born a slave in Kentucky in 1846, recalled:

Massa whooped a slave if he got stubborn or lazy. He whopped one so hard that the slave said he'd kill him. So Massa done put a chain round his legs, so he jes' hardly walk, and he has to work in the field that way. At night he put 'nother chain round his neck and fastened it to a tree. After three weeks massa turn him loose and he the prodes' nigger in the world, and the hardes' workin' nigger massa had after that. (FWPSN, Texas, vol. 1, p. 237)

The patrol system was a very important part of the system of social control over the slaves. Poor Whites were used to chase runaway slaves, to punish them, and generally to frighten and intimidate the slave population. They acted in a very brutal fashion; they would often simply go into the slave quarters at night and with the slightest provocation whip slaves and generally torment them. Elige Davison, born a slave in Virginia, graphically and perceptively described the patrol system:

Us couldn't go nowhere without a pass. The patterrollers would git us and they do plenty for nigger slave. I's went to my quarters and be so tired I jus' fall in the door, on the ground, and a patterroller come by and hit me several licks with a cat-o-nine-tails, to see if I's tired 'nough to not run 'way. Sometimes them patterrollers hit us jus' to hear us holler. (FWPSN, Texas, vol. 1, p. 299)

Slavery, as is the case with any social system, created reciprocity between slaves and masters. Since they did not live in worlds hermetically sealed from each other but had intimate, face-to-face contact, such matters as status, rank, and masculine prowess became involved in the relationship. . . .

One former slave, Tom Woods of Alderson, Oklahoma, born in the early 1850s, understood how important was the relationship of slaves to poor Whites:

Lady, if de nigger hadn't been set free dis country wouldn't ever been what it is now! Poor White folks wouldn't never had a chance. De slave holders had most of de money and de land and dey wouldn't let de poor White folks have a chance to own any land or anything else to speak of. Dese White folks wasn't much better off dan we was. (FWPSN, Oklahoma, p. 354)

There is little doubt that there were many masters who were decent, some even with abolitionist sentiments. But individual sentiments only slightly alter social behavior because the pressures of the social system on any individual, even though he is part of the dominant class, are very great. Not only was it impossible for individual slave owners with abolitionist sentiments to abolish slavery; despite their ideology, all they could be was particularly paternalistic and kindly masters, good patrons. Even those who treated their slaves well, perpetuated the basic social relations of slavery. . . .

While the average slave did not starve, he was not fed lavishly by the master. The usual diet provided by the masters consisted of cornmeal, sidebacon, and molasses. . . . George Kye, born in Virginia in the 1820s, described the typical diet: "We had stew made out of pork and potatoes, and sometimes greens and pot liquor, and we had ash cake mostly, but biscuits about once a month" (FWPSN, Oklahoma, p. 172). . .

House slaves were often better fed than field slaves and they often would take food to give to those, particularly children, who were not allowed in the "Big House." The slave economy was a make-do one in which "taking" (which was different from "stealing"—one *took* from the master, but *stole* from a fellow slave) was a crucial part. Mary Raines, born a slave in South Carolina in 1836 or 1837, tells it this way:

I was a strong gal, went to de field when I's twelve years old, hoe my acre of cotton, 'long wid de grown ones, and pick my 150 pounds of cotton. As I wasn't scared of de cows, they set me to milkin' and churnin'. Bless God! Dat took me out of de field. House servants 'bove de field servants, them days. If you didn't git better rations and things to eat in de house, it was your own fault, I tells you! You just have to help de chillun to take things and while you doin' dat for them, you take things for yourself. I never call it stealin'. I just call it takin' de jams, de jellies, de biscuits, de butter and de 'lasses dat I have to reach up and steal for chillun to hide 'way in deir little stomachs and me, in my big belly. (FWPSN, South Carolina, pt. 4, p. 2)

Slave houses in the nineteenth century were generally rude, one-room boxlike affairs. . . . Daniel Dowdy, born in 1856 in Georgia as a slave, described how they lived. His mother was the cook in the Big House:

> We lived in weatherboard houses. Our parents had corded-up beds with ropes and we chillun slept on the floor for the most part or in a hole bored in a log. Our house had one window jest big enough to stick your head out of, and one door, and this one door faced the Big House which was your master's house. This was so that you couldn't git out 'less somebody seen you. (FWPSN, Oklahoma, pp. 111–112)

As a rule, however, clothing was barely adequate by minimal standards for all but privileged house servants. . . .
. . . Slavery need not have been universally harmful to the slave in order to have been far less than a minimally satisfactory form of human existence.

The Black Family Under Slavery

The myth that the slaves had no normal, significant family life, that for the most part, they lived promiscuously, jumbled all together, with no male having a regular relationship with his children (see Stampp 1956; Franklin 1967) dies hard. . . . The picture of the slave family that emerges from a study of the slave narratives and interviews indicates that . . . the Afro-American family under slavery was part of a distinct, viable Black culture, adapted to slavery and deprivation (see Frazier 1957).

While it is true that slaves were not allowed to make legal marriage contracts, it is also true that men and women under slavery did not simply breed promiscuously. There were a variety of socially approved and culturally sanctioned relationships. . . . In certain of these relationships, men had acknowledged kinship relationships with specific slave women, but did not live with them. There were also socially recognized marriages in which the father

and mother lived together under the same roof with their children. The entire living unit functioned socially and economically as a single family unit, not dissimilar to the European-American kinship pattern, although sometimes broken up by the sale of members of the family to different masters, a situation which often happened upon the death of a master.

In some situations, certain male slaves were encouraged to have sexual relationships with more than one slave woman, without any permanent alliance being required. Masters, their male offspring, White overseers, and other non-slave-owning Whites were usually permitted to make sexual alliances on either a casual or more formal basis with one or more slave women.

Selections taken at random from the interviews from one state with ex-slaves give us a picture of the range of possible family arrangements and marriage customs under slavery. . . .

Betty Foreman Chessier, born a slave in Raleigh, North Carolina, in 1843, reported:

> My mother was named Melinda Manley, the slave of Governor Manley of North Carolina, and my father was named Arnold Foreman, slave of Bob and John Foreman, two young masters. They come over from Arkansas to visit my master and my pappy and mammy met and get married, 'though my pappy only seen my mammy in the summer when his masters come to visit our master and dey took him right back. I had three sisters and two brothers and none of dem was my whole brothers and sisters. I stayed in the Big House all the time, but my sisters and brothers were given to the master's sons and daughters when dey got married and dey was told to send back for some more when dem died. I didn't never stay with my mammy doing of slavery. I stayed in the Big House. I slept under the dining room table with three other darkies.
>
> After the War, I went to mammy and my step-daddy. She done married again, so I left and went to Warrington and Halifax, North Carolina, jest for a little while nursing some White chillun. I stayed in Raleigh, where I was born till 7 years ago, when I come to Oklahoma to live with my only living child. I am the mother of 4 chillun and 11 grandchillun.
>
> When I got married I jumped a broomstick. To git unmarried, all you had to do was to jump back-

wards over the same broomstick. (FWPSN, Oklahoma, pp. 27–28)

John White claimed to have been born in 1816, which would have made him 121 years old when interviewed. He told an interesting story about his parents:

Of all my mammy's children I am the first born and the longest living. The others all gone to join Mammy. She was named Mary White, the same name as her Mistress, the wife of my first master, James White.

About my pappy. I never hear his name and I never see him, not even when I was the least child around the old Master's place 'way back there in Georgia more'n one-hundred-twenty years ago!

Mammy try to make it clear to me about my daddy. She married like the most of the slaves in them days.

He was a slave on another plantation. One day he come for to borrow something from Master White. He sees a likely looking gal, and the way it work out that gal was to be my Mammy. After that he got a paper saying it was all right for him to be off his own plantation. He come a'courting over to Master White's. After a while he talks with the Master. Says he wants to marry the gal, Mary. The Master says it's all right with Mary and the other White folks. He finds out it is and they makes ready for the wedding. Mary says a preacher wedding is the best but Master say he can marry them just as good. There wasn't no Bible, just an old Almanac. Master White read something out of that. That's all and they was married. The wedding was over!

Every night he gets a leave paper from his Master and come over to be with his wife, Mary. The next morning he leaves her to work in the fields. Then one night Mammy says he don't come home. The next night is the same, and the next. From then on Mammy don't see him no more—never find out what happen to my pappy. (FWPSN, Oklahoma, p. 325)

White commented on the sexual relations between Whites and slave women:

Sometimes the White folks go around the slave quarters for the night. Not on the Davenport plantation, but some others close around. The slaves talked about it amongst themselves.

After a while they'd be a new baby. Yellow. When the child got old enough for chore work the master would sell him (or her). No difference was it his own flesh and blood—if the price was right! (FWPSN, Oklahoma, p. 325)

. . . If the slave family was so weak (cf. Bernard 1966), how can we then account for the following facts which emerge with incessant urgency from the interviews?

As soon as the Civil War was over, and even in the few years before the end of the war when the discipline on the plantations was virtually destroyed, thousands of slaves went looking for and found their mothers, fathers, sisters, brothers, sons, and daughters from whom they had been separated.

Hundreds of the ex-slaves interviewed knew something of their family histories, including stories about parents and grandparents who had come from Africa.

Most ex-slaves knew their precise relationship with brothers and sisters and half-brothers and half-sisters and obviously valued these relationships.

The language in which the ex-slaves describe the fairly ubiquitous marriage ceremony of jumping the broom indicates that these and similar ceremonies were not there because the masters required them but because the slaves wanted them. . . .

The slave community acted like a generalized extended kinship system in which all adults looked after all children and there was little division between "my children for whom I'm responsible" and "your children for whom you're responsible." . . . There was always some older person who would, with relative ease, take over the role of absent parents—as is usually the case today in the Black community.

A kind of family relationship in which older children have great responsibility for caring for younger siblings is obviously more functionally integrative and useful for slaves. . . .

Indeed, the activity of the slaves in creating patterns of family life that were functionally integrative did more than merely prevent the destruction of personality that often occurs when individuals struggle unsuccessfully to attain the unattainable. It was part and parcel . . . of the social process out of which came Black pride, Black identity, Black culture, the Black community, and Black rebellion in America.

Master and Slave: Resistance

Under slavery, as under any other social system, those at the bottom were not totally dominated by the master class. . . . While Blacks were oppressed and exploited, they fought back in a constant struggle by all available means. . . .

We must conceive of the slave personality as an ambivalent one. On the one hand are submissiveness and a sense that one deserves to be a slave; on the other hand is a great deal of anger in ways that protect the personality and have objective results in the improvement of the slave's situation and eventual liberation, at least from chattel slavery.

Unless the slave has had a tendency to be Sambo he can never become Nat Turner. One who has never feared becoming Sambo, never *need* rebel to maintain his humanity. . . . The slaves went from being frightened human beings, thrown among strange men, including fellow slaves who were not their kinsmen and who did not speak their language or understand their customs and habits, to what W. E. B. DuBois (1964, pp. 55–83, 112) once described as the general strike whereby hundreds of thousands of slaves deserted the plantations, destroying the South's ability to supply its army. The over 200,000 Blacks who joined the Northern army maintained the ability of the North to mount the military initiative after the battle of Gettysburg and win the war. How did this happen? . . .

. . . The slave narratives are the richest source we have ever had for a description of the slave personality. In them the same individuals describe themselves as simultaneously Objects and Subjects and tell us of the consequences of their self-transformation from one to the other. On the one hand is the Object: the man who does not receive enough food, clothing, or shelter to keep alive, and does not work well because of incompetence; the man who is whipped and humiliated; the man who calls upon those who listen to him to have pity and be merciful. On the other hand is the Subject: the man with needs and wants of his own. . . . He may demand better and more food, clothing, and shelter. He may demand higher status, dignity, and the time and opportunity to carry on flirtations, to laugh, dance, sing, make love, loaf, play with his children, and raise them as he sees fit; he may demand the end of being the whipped Object and become the one who chooses not to work well as an act of rebellion. The Subject wants liberty and freedom and the opportunity to appropriate for himself and his family the best that is available in his time and place. . . .

Let us look at some descriptions of the development of the slave's social personality in the slave narratives, . . . (for example) Mingo White's description of the beating of Ned White in which White was made "to pull off ever-thang but his pants an' lay on his stomach 'tween de pegs whilst . . . dey whupped him 'twell de blood run from him lack he was a hog." Mingo White went on to describe the aftermath of this beating:

Afer ol' Ned got sech a terrible beatin' fer prayin' for freedom he slipped off an' went to de North to jine de Union Army. (FWPSN, Alabama, pp. 416–418)

Slaves ran away after being whipped and then got caught. Sometimes they would commit suicide rather than go back to slavery:

Ever once in a while slaves would run away to de North. Most times dey was caught an' brought back. Sometimes dey would git desp'rit an' would kill demse'ves 'fore dey would stand to be brought back. One time dat I hear of a slave that had 'scaped and when dey tried to ketch him he jumped in de creek an' drown hisse'f. (FWPSN, Alabama, p. 390)

The slaves would consciously plot against the masters and their agents, the patrollers. . . . In order for large numbers to resist with any degree of success, slaves had to know that other slaves resisted, and how this was accomplished.

At the center of any community is a network of communications and social relations. . . . There were many ways that news traveled. Slaves were allowed passes to travel from one plantation to another; some slaves regularly worked on one plantation and slept on another if they were married to a woman on the second plantation; slaves would go off on their own to all-night prayer meetings and "frolics;" one slave commented that the church service on Sunday was the big time for gossip; the arrival of new slaves from far away places would bring in news of the outside. . . .

Not all news passed by word of mouth. On virtually every plantation there were a few slaves who knew how to read and who would read newspapers that had been left lying about. While it was illegal to teach slaves to read and write, many, both as children and as adults, learned how to read, and a few learned how to write. . . . On some plantations slaves were taught how to read by the master so that they could read the Bible and keep records. . . . Solomon Northrup, who had been born free and kidnapped into slavery, could write, but was unable for many years to acquire a pen and ink or paper (see Nichols 1963, p. 45). Slaves who were caught with books or writing material were usually severely punished.

That most slaves were illiterate is not particularly surprising in an era in which many masters and overseers were themselves illiterate or nearly so. What is noteworthy is the efforts made to prevent the slaves from learning how to read. Either the master class was suffering from extreme paranoia or it had good reason to suspect that slaves who were literate would spread sedition. . . .

It is equally noteworthy that many of the slaves did learn to read. Frederick Douglass writes in his autobiography that when his master forbade his mistress to continue teaching him to read because it would make him "unfit to be a slave," he discovered that "the pathway from slavery to freedom" was literacy. He therefore went out and found another way of learning to read:

The plan which I adopted, and the one by which I was most successful, was that of making friends of all the little white boys whom I met in the street. As many of these as I could, I converted into teachers. With their kindly aid, obtained at different times and in different places, I finally succeeded in learning to read. (Douglass [1845] 1963, pp. 40–41)

The several hundred thousand Black freedmen, anomalies of the slave system, also provided sources of information for the slaves with whom they had regular contact. Many of the freedmen were literate and could get access to newspapers and could even read to the slaves. It was through this route, most likely, that slaves heard of David Walker's appeal to them to rise and of Nat Turner's revolt. Moreover, through such sources the slaves heard of the activities of the abolitionists and were encouraged. . . . Time after time runaway slaves were sheltered by freedmen, many of whom had previously been run-

aways. Freedmen made the most courageous conductors on the Underground Railway. . . .

The slave revolts themselves had grown out of the total Black community, which included both slaves and freedmen. The slave revolts came out of that independent Black religion. . . . This was so both in terms of impetus and of organization. . . . Nat Turner was an exhorter, a field hand preacher. . . .

In the Civil War, the Blacks played the decisive role (see McPhearson 1965, pp. 191, 234–235). . . . It was the fact that Blacks joined the Northern army and navy, the fact that Blacks had taken guns in their own hands in their own behalf, that settled the matter of slavery. . . . The Emancipation Proclamation had freed all slaves behind the Southern lines and was aimed at gaining the support of the slaves. . . . Lincoln, who had never believed that Blacks would be able to become part of American society, and who was hard pushed by events to support abolition, by the end of the war had come to understand that the war was also fought for a "new birth of freedom." . . . Thus, in his Second Inaugural Address he carried the idea forward when he declared:

Fondly do we hope—fervently do we pray—that this mighty scourge of war may speedily pass away. Yet, if God will that it continue until all the wealth piled by the bondman's two hundred and fifty years of unrequited toil shall be sunk, and until every drop of blood drawn with the lash, shall be paid by another drawn with the sword, as was said three thousand years ago, still it must be said, "the judgments of the Lord are true and righteous altogether." (in Basler 1953, p. 333)

But the last word must not remain with the White president but with a Black soldier who understood what the war and the Blacks' role in it was all about. Corporal Thomas Long of Thomas Wentworth Higginson's regiment, acting as chaplain one Sunday, told the troops of his all-Black regiment:

If we hadn't become sojers, all might have gone back as it was before; our freedom might have slipped through de two houses of Congress and President Linkum's four years might have passed by & notin been done for we. But now tings can never go back, because we have showed our energy & our courage & our naturally manhood. (McPhearson 1965, p. 213)

Racism and the Making of American Society

America was born nearly free and racist. Class division among Whites and the sense of class were much less sharp than in Europe. There was no extensive feudal aristocracy, although there was a degree of class privilege. There was a seemingly endless supply of land. In such a society, men could contract one with another voluntarily to construct a new society.

But almost from the beginning American Indians and Blacks were permanently excluded from the social contract. Race and ethnic consciousness was more evident than class consciousness. As long as that has been true, the promise of American life, the full promise of the Declaration of Independence of "life, liberty, and the pursuit of happiness," has been denied for both White and non-White. . . .

5

Farewell—We're Good and Gone: Black Migration from the PostBellum South

CAROLE MARKS

In the two years from 1916 to 1918, more Blacks fled the southern United States in search of a new life in the North than had ever migrated before in the history of North America. The land they left behind was undeveloped and isolated. It had less industry, fewer schools, and fewer basic services than almost any other region of the country. The absence of radio and limited mail delivery meant that many Southerners did not hear about major current events until long after they happened (Baker 1908, p. 100). Yet between 1916 and 1918, more than 400,000 Blacks left the South for northern industrial centers. In two years, 5 percent of the Black population of the United States was redistributed (Marks 1989). This paper addresses four questions: Why did so many Blacks leave the South between 1916 and 1918?, Who left?, How was information about opportunities for Blacks in the North communicated to those living in remote areas of the South?, and Was the migration a success?

Origins of the Black Migration

World War I and the resultant interruption in the flow of European immigration to the United States, the boll weevil infestation that devastated the Southern cotton industry (Ransom and Sutch 1977, p. 196), and nascent industrialization in other sectors were all factors in the migration of Blacks northward. The rise of Jim Crow, the legalized system of racial apartheid in the South, further exacerbated the situation as a segregation from cradle to grave gave Blacks little social or political standing. The exodus brought vast changes in the distribution of the population (Myrdal 1944). By the late 1920's, there had been a net migration of one million from the South—one-tenth of the Black population of the United States (Florant 1942). Approximately 400,000 moved between 1916 and 1918 (Scott 1920; Tuttle 1974). What is perhaps more surprising than the scale and rapidity of the migration is the fact that Blacks waited so long after President Lincoln's Emancipation Proclamation went into effect in 1863 to leave the land of their oppression in significant numbers. As late as 1900, for example, over 7 million Blacks lived in the South, while just over 900,000 resided in all other regions of the country combined (Katznelson 1930).

By 1915, however, the labor needs of the North were changing. World War I was under way in Europe, and northern U.S. industry was expanding in response to it. At the same time, the supply of immigrant labor from Europe was diminishing (Florant 1942). From 1900 to 1914 over 12 million immigrants had found their way to the United States. More than 1 million came in 1914 alone (Hutchinson 1953). The next year this figure was cut to about one-third; in 1916, to about one-fourth; and, by 1918, only 110,000 immigrants landed on the shores of the United States, while 95,000 immigrants left (Palmer 1967). Many who left were called back to their countries of origin to fight. Other sources of labor were needed, and southern Blacks were the largest available and willing substitute.

Employment in the North came at an opportune time. Northern industrialists had gained control over the economic development of the South in the late 1880s and continued to control that development into the twentieth century (Fligstein 1981). As a result, northern business exploited southern natural resources without establishing any indigenous industry. The North also controlled southern railroads. With this control, they charged cheap rates for shipment of raw materials and high rates for shipment of finished products (Woodward 1951). It was less expensive for Southerners to buy their finished goods from the North than to produce these goods themselves. The South's lack of ownership and control of its own industries created a dependent relationship throughout its economic system. In addition, the boll weevil was destroying southern cotton, the only cash crop of many Black Belt sharecroppers, further depressing the agricultural areas. More Blacks than Whites were displaced by the infestation (Fligstein 1981).

Southern Blacks moved to northern industrial centers, where they were readily incorporated into the industrial workforce. Blacks had not been recruited before because industry had access to cheap foreign labor, heavily recruited since the 1880s, and no reason to raid the labor supply of the South, especially when the North's economic well-being depended on an abundant supply of cheap southern cotton. But the war changed that.

Southern migrants moving north were not uniformly distributed throughout the northern economy. They were overrepresented in industries such as meat packing and steel and incorporated into a ra-

cially split labor market (Bonacich 1972)—that is, a labor market divided along racial lines where the price of labor differs for the same work. The poverty of the southern Blacks forced them to accept long hours, low wages, and hazardous working conditions, terms nonmigrant workers in the North considered unacceptable (Bonacich 1976).

Who Left

Compilation of perfectly timed, historically unique factors underlying the migration, none of which are disputed here, tell us little about those who left. Who were these 400,000 people who abandoned the land of their birth?

The assessment that a majority of the Great Migration migrants were from rural areas has been frequently assumed but never proved. While numerous investigations document a large exodus from rural areas to southern cities, that the same population entered northern communities is assumed but not proved. Lyonel Florant highlights the problem in pointing out that

> True, large numbers of Southern Negroes moved (from rural areas) to nearby towns and cities, but whether or not these same individuals comprised the majority of those who later arrived in Northern cities has not been subject to verification. (Florant 1942)

Migrants are thought to represent an undifferentiated mass. Of the Great Migration it is said, "People, particularly Blacks, left rural areas to take jobs in towns and cities of the South and North" (Fligstein 1981, p. 104). It has also been stated, "the areas from which the majority of the migrants came were predominately rural" (Johnson and Campbell 1981, p. 79). Little documentation follows such assertions beyond observations such as, "To the poorly paid Southern farm hand the wages paid by Northern industries—paid in cold cash by the week instead of in store credit once a year—seemed fabulous sums" (Kennedy 1930, p. 44). Such observations obscure more than they reveal.

The image of the rural sharecropper, leaving his plough in the field to make the journey north is pervasive in the literature of the Great Migration. The high percentages of the Black population employed in agriculture make such assessments both routine and unquestioned. Yet while the South is often depicted as a region steeped in premodern agrarian values, its economic picture is more complex. As Gavin Wright (1986) has pointed out, from 1869 to 1909, "the pace of Southern industrial growth was not slow, with the value of production growing at a rate of about 7 percent per year." This is not to suggest that the South's industry thrived. Rather, "Southern industry was concentrated in low wage, low skill sectors, using imported technology and usually imported machinery" (Wright 1986, p. 62). The South's problem at this time "was not backwardness but isolation." As Wright points out, "much of the actual flow of information in long-distance industrial labor markets operates through informal channels, such as letters from relatives and word-of-mouth talk within ethnic groups" (Wright 1986, p. 74). But the South, through a combination of distinctive factors, was not directly linked to these informal channels and hence not connected to northern labor markets. Southern isolation meant that in such things as wages, separate systems developed. In 1890, Southern 'common laborer' earnings were 30 to 50 percent below North Central levels. With such wage differentials, it was logical that southern workers would gravitate to the higher market in the North. But, concludes Wright, they did not know about them. Creating links to labor markets elsewhere was an important first step in the migration process.

Migration and the Nonagricultural Worker

Blacks toiling in the South before the exodus performed a number of tasks. In 1910, over 60 percent were engaged in a variety of agricultural pursuits from small farm ownership to sharecropping, 18 percent were in domestic and personal service, and the remaining 20 percent were employed in all other nonagricultural pursuits from teaching to street cleaning (U.S. Bureau of the Census 1978).

Many of those employed in nonagricultural labor lived in urban areas. By 1910, the Black population was 22 percent urbanized. Nearly 2 million Blacks lived in southern cities at least six years before the start of the Great Migration. Southern cities gained even more Black population after 1910. The proximity to migration networks of these workers was obvi-

ous. Pointed out investigator Leavell (1983), "It was the town Negroes who went first."

But other nonagricultural workers lived in rural areas. In the rural saw and planing mills, for example, Blacks represented nearly 63 percent of the common and 40 percent of the semiskilled laborers by 1910. Gottlieb argues that a significant proportion of these workers also migrated North (Gottlieb 1987, pp. 12–62).

The combination of urban residence and industrial work experience among a significant sector of the migrating population does more than belie the common image of the "ignorant" and "ill-equiped" rural migrant (Jones 1992, p. 210). For development theorists, such work experience is often cited as the very basis of modernization. More importantly, it reflects the selectivity of labor migrations, a selectivity that explains why migrants often advance more quickly in a host society than comparable groups of nonmigrants.

Those engaged in manufacturing and mechanical pursuits, a census category roughly comparable to the industrial sector, represented 9 percent of the workforce. This relatively small percentage did not represent a negligible number of people. In 1910, over 90 percent of the entire Black population of the United States resided in the fourteen states of the South. According to the census, nearly 600,000 Blacks worked in manufacturing and transportation in that year. The migration, which was made up of both workers and their families, easily could have been populated entirely by this group. While obviously the migration was not composed simply of nonagricultural workers, the overemphasis on the rural farmer neglects not only a significant segment but the basic economic foundation of a labor migration.

It is perhaps surprising to argue that a population receiving a higher wage would be more drawn into moving than the more numerous, low wage farmer. But migrations are frequently selective; that is, those drawn into moving often have more skills, more education, and sometimes more income than those staying behind. In 1910, approximately 35 percent of Black nonagricultural workers were engaged in skilled and semiskilled trades. Some of these were from the old artisan class of slavery—blacksmiths, masons, and carpenters—a small class who once held a monopoly within certain trades, but gradually were getting pushed out by competition and obsolescence (Woodward 1951, pp. 360–361). Many lived in the urban areas of the South.

Many of the remaining 65 percent were common laborers, working in the newly developed factories of the South in tobacco, lumber, coal, cotton oil refinement, and iron and steel production (Rabinowitz 1978, p. 66). For example, one-third of the railroad firemen and brakemen and over half the trackmen in the South were Black (Worthman and Green, p. 52). Two-thirds of the Virginia shipbuilders and the New Orleans dockworkers also were Black (Worthman and Green, pp. 52–53). This nonagricultural employment was found in both urban and rural locations.

Wages in the South were low. For those in nonagricultural work, they ranged from about $1.25 a day for laborers to $3.00 a day for artisans. While those in agriculture working for wages earned about $0.75 per day, a majority received no wages at all, working instead for payments in kind. On the average, wages in the South were only about three-fourths of those in the North (Scott 1920, p. 17). Migrants could make more in the worst jobs in the North than they had in the best jobs in the South. This was constantly pointed out to them and was a large part of the North's appeal (Tuttle 1974, pp. 81–84).

Moreover, the boll weevil infestation threw hundreds of thousands of agricultural laborers off the land, out of rural areas, and into Southern cities. There developed an even greater competition for jobs. Unemployment among Blacks, which was about 13 percent in 1910, reached nearly 20 percent in 1920. Writes Emmett Scott:

> A host of idle persons thrown suddenly on the labor market could have no other effect than to create an excess in the cities to which they flocked, make laborers easily replaceable, and consequently reduce wages. (Scott 1920, p. 15)

A Southern newspaper, in commenting on this situation declared, "There is nothing for this excess population to do.

These people must live on the workers, making the workers poorer" (quoted in Scott 1920, p. 15).

Southern economic development did not mitigate high unemployment. Explain Worthman and Green, "Living in the primitive huts of upland company towns and segregated in the disease-infested 'slave quarters' of Southern cities, Black workers had good

cause to wonder why they had left the farms" (Worthman and Green, p. 53).

As is typical of industrialization at its early stages, there were insufficient jobs for the number of willing workers and a fierce competition resulted. The result was a decline in real earnings. Observed one migrant, "Everything had gone up but the poor man's wages" (Scott 1919b, p. 426). Race also became an issue of some importance in this competition. Jobs had always been defined along the color line with the most "dirty work" jobs reserved for Blacks at the bottom.

With unemployment high and jobs scarce, White workers were not willing to write off jobs as "Negro work" as they formerly had done. Commented one migrant:

Houston, Texas: April 29, 1917

Dear Sir: I am a Negro, age 37, and am a core maker by trade having had about 10 years experience at the business and hold good references from several shops, in which I have been employed. It is hard for a Black man to hold a job here, as prejudice is very strong. I have been discharged on account of dissatisfaction with my work, but I have been 'let out on account of my color.' (Scott 1919b, p. 425)

Competition was also a political phenomenon. Beginning at the turn of the century, laws were established that prohibited Blacks from numerous occupations, from holding any skilled positions on the railroads, and from a variety of service jobs where, it was argued, they should be replaced by Whites who were out of work. The result was that many Blacks who had formerly worked found their positions threatened. Observes Simkins, "If a new garment or shoe factory came to a Southern town, only Whites were employed. 'Give jobs to White men and women,' and 'Blood is thicker than water' were the cries" (Simkins 1947, p. 292).

All of these factors served to make Black nonagricultural workers, particularly those who had spent some time employed in urban areas, disadvantaged. They faced competition from both Black and White rural workers willing to undercut the wage levels they had enjoyed previously. They faced displacement from trades by White workers who used political disenfranchisement to legislate them out of positions formerly held. And they had no legal recourse or protection.

In the face of deteriorating conditions, a significant portion left; a selective resource the South could ill afford to lose. Moving north in search of better opportunity, these workers rivaled other labor migrants who abandoned a lower priced market for a higher one. But as suggested previously, their ability to do this hinged on the creation of informal migration channels.

Lines of Communication

Migrations begin, it is suggested, not with well-established migration streams, but with the development of information links regarding the possibility of movement. Advanced economic centers to which migrants flock represent communications systems, to borrow a phrase from Webber (1944, pp. 23–54), "vast complex switchboards through which messages (about migration) are routed." Among the most important of these are recruitment messages: messages about job opportunities and transportation. Recruitment activities, according to Piore (1979), explain both the timing of particular population movements, and the particular areas between which migrant flows develop.

Lines of communication set up two-way channels of information between sending and receiving areas on employment, housing, education, climate, and general conditions of social life. These links may be set up by a variety of sources from labor agents paid to recruit workers to family and friends who have tested the waters and encouraged relocation.

It is unlikely that any single line of communication would represent a sufficient link for the mobilization of migrating populations. Successful agents, to the extent that they remove available labor from sending areas, are often restricted by local governments in their activities and, in some cases, are prevented from carrying out their jobs. Family and friends are rarely systematic in providing information on housing and employment to potential migrants. It is in combination that these lines of communication form the necessary links that make large numbers of individuals aware of opportunities in receiving areas.

Agents

In the Great Migration, the initial line of communication was established by agents of northern companies, acting as intermediaries between employers and potential labor migrants. Agents were paid a fee for each worker they were able to produce. Most "active in large cities where their presence [was less] conspicuous," some were paid to merely "walk briskly down the street through a group of Black workers [and], without turning, say in a low tone, 'Anybody want to go to Chicago, see me'" (Scott 1920, p. 38). According to Scott, "that was sufficient." Scott (1920) notes that "they were reported, at one time or another, in every section from which migrants went." Agents were surprisingly selective of their recruits, considering they received a flat rate for each, favoring men over women, young over old, and healthy over infirm. Frequently, stipulations as to age and health were written directly into the contracts. Misrepresentation was grounds for instant dismissal. Larger companies, in particular, made use of labor agents. In the summer of 1916, for example, the Pennsylvania Railroad used them to import "12,000 Blacks to do unskilled labor" (Thomas 1973, p. 333; Tuttle 1974, p. 88). The Illinois Central Railroad "also issued passes," through labor agents, "on which hundreds of Blacks traveled to Chicago" (Scott 1920, p. 61).

The federal government was a less frequently mentioned though nonetheless active source of recruitment. In 1916, the "Immigration Bureau" of the Department of Labor, a bureau whose function was to place all Americans in jobs wherever they might be, as a matter of patriotic duty, helped hundreds of Black workers find jobs in the booming war industries of the North. An agency of the state was operating as an employment bureau in the interests of those who wanted to transport cheap labor from one region to another.

Because of their constant activity, researchers assume that the trip north was paid for by labor agents (Scott 1920, p. 37). Yet the volume of migration directly attributable to them seems to be minimal. After initially providing "passes," the practice was soon abandoned in favor of one where transportation costs were advanced and then deducted from wages. Neither system, however, proved to be effective.

During the years of Jim Crow "separate but equal" segregation, many public facilities were segregated by race throughout the South. Here a man is drinking at a separate water fountain in the streetcar terminal of Oklahoma City, in 1939. (Photo by Russell Lee. U.S. Farm Security Administration; Library of Congress)

Northern employers began to complain about the workforce gathered by agents. Wanting to exercise more control, they demanded greater selectivity of labor agents and lent money to migrants to move their families only after they had been on the job "believing that you will get a better type of man when he is willing to pay his own way."

Passes and prepaid tickets distributed by labor agents were only necessary during the initial stages of the migration. After that, the movement became self-generating and those desirous to leave the South had to pay their own fares (Henri 1975, p. 67; Kennedy 1930, p. 53; Scott 1920, pp. 47–48; Tuttle 1974, p. 88; Spear 1969, p. 1330). Nonetheless, the early activities of labor agents not only served to transport numbers of workers from south to north, but also to plant the seeds of awareness of opportunities in those who remained behind.

Family and Friends

Information provided by a trusted informant, such as a family member or friend, is the keystone of communication. According to Scott (1920, p. 34), "Personal appeals in the form of letters have a recognized weight in influencing action." Perceived as having nothing to gain by providing false impressions, their accounts of life and opportunity in receiving areas are persuasive. Tienda (1980, p. 384) has observed, "family members who migrated at an earlier time may be influential in attracting new migrants to an area and might also be available to cushion the potentially disorganizing consequences of immigrant adjustment."

In the Great Migration this is found to be true. Once begun by agents, migration streams developed on their own. "The arrival of each new migrant in the

North created a new contact with potential migrants, and personal communication made the labor agents superfluous" (Spear 1969, p. 133). Letters from migrants to relatives and friends were said to generate the "moving fever," particularly when the letter "contained money" and "offered concrete evidence of success in the North" (Spear 1969, p. 134; Tuttle 1974, p. 86). Some suggested that nine out of every ten migrants came because of letters from friends or relatives.

Some communities would send a small, advance group to test the waters. For a week or more there would follow a tense period of waiting. Finally a card would arrive bearing the report, "Home ain't nothing like this."

Letters from the North were often read in churches and, according to Scott (1920, p. 26), "fresh news on the exodus was given out." Out of these discussions, clubs were formed to facilitate the migration. Leaders were "to notify Chicago industries, newspapers, and placement agencies that a certain number of Black people were available for employment and to request train tickets" and arrange financing. The groups "ranged in size from two or three families to more than 1,000 persons" (Henri 1975, p. 123).

Letters expressed all the hopes and fears of those unfamiliar with "the promised land." Commented Tilly (1968, p. 157), "For people moving without a guarantee of a job the presence of friends and relatives matters a great deal more than such things as the housing supply or availability of public assistance."

Service Organizations

Service organizations represent the third line of communication. Voluntary associations, it has been noted, "operate as an adaptive mechanism in situations of rapid change undergone by urban migrants" and "to strengthen the consciousness of a groups' culture of origin" (Sassen-Koob 1979, pp. 319–320). The Urban League may be viewed as one such organization for the Great Migration. The Chicago Urban League was established in 1917 during the "period of heaviest migration of [Blacks] to the city" (Chicago Commission on Race Relations 1920, p. 146). It is estimated that between 1917 and 1919, as many as 55,000 sought jobs and dwellings from

the organization (Tuttle 1974, p. 99). Its Industrial Department alone, during 1920, placed more than 15,000 migrants in positions, made industrial investigations in sixteen plants, and provided lectures for working men in plants and for foremen over Black workers. It also investigated worker complaints, selected and fitted men for positions, secured positions for Blacks where Blacks had never worked before, and assisted in other ways the adjustment of Blacks in industry (Chicago Commission on Race Relations 1920, p. 147). That migrants were aware of league activities and consulted them before journeying north is evident from many letters.

In general, however, the league was more important as a resource in northern than in southern cities. The excitement that was generated by the other lines of communication caused some, paying their own railroad fares, to leave for the North without a promise of a job or, indeed, without any concrete plans. Many of these would head for the offices of the *Chicago Defender*, the largest Black newspaper in Chicago, upon their arrival. The *Defender* would, then, send them on to the offices of the Urban League.

The league collected information on jobs and housing, provided clothing for those inadequately prepared for the northern climate, and granted any other social services as could be handled, as has been suggested. In these years before the New Deal, when social welfare activities were left in the hands of state and local agencies and when the private assumption of responsibility was in fact much greater than the public, it provided a needed service for incoming migrants. Its participation in the mobilization process as a refuge for those who could not turn to family and friends, made it an important link in the migration chain. It is likely that it had a "multiplier effect" on the migration, helping to establish newcomers without personal networks, who could in turn, after their settlement, encourage family and friends to follow.

The Ethnic Press

Ethnic presses in particular attempt to satisfy a new and pressing need for information generated by migration. An explosion of reading and writing even among populations illiterate in their language of origin has been frequently noted (Thomas and Znaniecki 1920).

Although a Black press existed even before the end of the Civil War, it became a useful weapon for the Black community at the turn of the century. An important step was the founding of the *Boston Guardian* by William Monroe Trotter in 1901. Trotter "rejected the conventional etiquette of cloaking protest in the niceties of circumspection and genteel language" (Kornweibel 1975). In 1905, Robert Abbott's *Chicago Defender* was born. Abbott, who began with $0.25, a kitchen table, and a half-dozen pencils, developed what eventually became a nationally distributed newspaper. The *Defender* ushered in the "role as an organ of racial propaganda" featuring bold, provocative headlines and, at the same time, an emphasis on corruption and scandal.

The paper also attempted to report on every lynching in the South and every case of burning and pillage. In one issue, it described in graphic detail the severed head of a lynched Black man that had been "thrown into a crowd of Negroes on the principal Negro street" (Leavell 1983, p. 29). According to federal investigator Leavell, "a photograph of what purports to be the head as it lies on the deserted street is published under the telling caption, 'Not Belgium—America.'" Leavell characterized the *Defender* as "extreme" and suggested it was "all the more effective, because there is a natural tendency to minimize such justification as may exist" (Leavell 1983, p. 29).

Several southern cities, in retaliation, attempted to prevent its circulation by confiscation. There are reports that in some communities the *Defender* had to be dropped off in rural sections and brought into the cities in the middle of the night. Other reports suggest that when circulation was prohibited, Blacks had to rely on subscription copies delivered through the mails (Tuttle 1974, 89). Despite numerous attempts to halt its distribution, the *Defender* was circulated. Stated federal investigator T. J.Woofter (1919, p. 103), the Chicago paper "makes its lurid appeal to the lowly class of Negroes. In some sections it has probably been more effective in carrying off Negroes than all the labor agents put together."

Yet the *Defender* did more than simply editorialize about the benefits of leaving. They published advertisements of employers and those seeking work were instructed to apply to those companies; information not available in southern presses White or Black:

Wanted: Men for laborers and semi-skilled occupations. Address or apply to the employment de-

partment, Westinghouse Electric & Manufacturing Co. (Quoted in Scott 1920, p. 17)

These ads stressed opportunities for young, able-bodied males, "able to take charge of their positions." It was these announcements to which migrants responded:

Baton Rouge, La. April 26, 1917

Dear Sir: I saw your advertisement in the Chicago Defender. I am planning to move North this summer. I am one of the R.F.D. Mail Carriers of Baton Rouge. As you are in the business of securing jobs for the newcomers, I thought possibly you could give some information concerning a transfer or a vacancy, in the government service, such as city carrier, janitor, or something similar that requires an ordinary common school education. (Scott 1919, p. 302)

The paper filled its pages "with photographs, cartoons and even poems" and emphasized "in the most convincing ways the great advantages which were awaiting [those] who would go North." It also, in the words of investigator Leavell, "promoted discontent with treatment received locally at the hands of Whites in the courts, in the schools, in political life, and in the distribution of public improvements" (Leavell 1983, p. 30).

Clearly the *Defender* was instrumental in prompting "thousands of Blacks to venture North." The 1919 Chicago Race Commission which investigated the riots stated, "Many migrants in Chicago attributed their presence in the North to the *Defender* encouraging pictures of relief from home with which they became increasingly dissatisfied as they read" (Chicago Race Commission 1920, p. 147).

The Black press represented an important vehicle for the articulation of the goals of the migration. Ethnic presses in general may be said to be the final, reinforcing link in the migration chain.

Lines of communication are crucial to understanding how information about opportunities in advanced economic centers is conveyed to those in outlying, remote areas. Not only does this communication make migrants aware of opportunities and provide a vision of a better life, it also encourages them in new modes of thought. Black migrants from the South deliberately and consciously cast off south-

Over the course of two major waves of migration, around World War I and World War II, Black Americans migrated, in large numbers, from South to North, looking for better job opportunities than those provided by agriculture in the South and for fairer treatment as Americans than the deep-seated racism they had endured. Here college-trained men work in the Packard Motor Car Company, in Detroit, in 1918.

ern caste traditions in journeying north. They had, through migration, learned to want the same accommodations.

A Successful Migration?

More than anything, migrants dream of "making it" in their new land. Black migrants were hopeful that leaving the South would signal an end to oppression and the beginning of a new life. What they discovered in the North was not the stuff of which dreams are made. Observed James Comer's mother in his book *Maggie's American Dream:* "While still down South, I thought to myself, I'll wait and see what it's like. And, sure enough, East Chicago wasn't what I had been told. It was quite a letdown. You know when people in the South long years ago said Chicago or East Chicago, Indiana, they claimed; this is heaven. But I never did believe it" (Comer, 1988, p. 34).

The more lucrative North Central region was the obvious choice of migrants. Chicago, home of the largest Black newspaper, the *Chicago Defender*, and transportation nexus for the United States, was a logical destination for many. Jobs were plentiful and well advertised, and wages were comparatively high, especially in such industries as slaughtering and meat packing (Grossman 1989; Tuttle 1974). Said one, "I could not rest in my bed at night for thought of Chicago." Before the onset of the migration, the Black community had been small, somewhat dispersed, and mainly invisible. The sudden influx of thousands of southern Blacks looking for jobs in industries previously closed to the Black population was met both with puzzlement and, at times, hostility. Black natives found themselves "confused" with the newcomers and went to great lengths to create as much distance as possible. Migrants were discouraged from joining churches, isolated in the schools, and excluded from social clubs. "While filling positions and jobs that negroes have never been known to hold before," Northerners were "somewhat ashamed of being identified with the Southerners and have not fully accepted the responsibility of his initiation" (p. 17).

Resentment and fear of the newcomers led to their segregation into the worst areas of the city. Wrote Richard Wright of Chicago, "We remain to live in the clinging soot just beyond the factory areas, behind the railroad tracks, near the river banks, under the viaducts, by the steel and iron mills, on the edges of the coal and lumber yards (Wright 1969, p. 103). The impoverishment of many southern migrants created a host of social problems in their communities leading to the labels of lawlessness and licentiousness. And even when innocent, migrants were blamed for everything.

Regardless of their previous work experience, Blacks moving north, like many labor migrants crossing international borders today, were given jobs at the bottom; they were low paying, hazardous, and dead end. Observed one migrant working for Graham Paige in Detroit:

Many workers would pass out. The boys would say, 'The bear has got you.' When we got real hot, we'd see little dots in front of us. We worked on a swing shift. We'd get through after a continual half running pace all day, fifteen minutes before the whistle. If we sat down we often caught the cramps in our legs and all over. We couldn't move, sometimes we had to wait fifteen or thirty minutes before we could get up and go home.

Often these inequities were formalized by labor agents who would bind migrants to low wage contracts for a specified period before they even left the South. Migrants had to work through the contracts before they could improve their wages.

White workers resented the competition, particularly fearing a decline in their wages. When these fears were translated into strikes, as they often were, more migrants were brought in as strikebreakers, further adding to the problem. While all immigrant groups had faced some of these patterns of discrimination, Blacks faced greater disadvantage by virtue of the unique combination of their ethnic exclusion and economic vulnerability, a combination that employers had unsuccessfully attempted to impose on other workers in the past. Employers discovered that the fostered antagonism of Irish against Pole that had failed to stick found its measure in appeals of White against Black. According to Piven and Cloward (p. 194), "One measure of the success of this strategy was the outbreak of White mob violence against Blacks in the post–World War I period," for "some twenty-five race riots were touched off in American cites during the last six months of 1919" as an upsurge of unemployment followed the slowing of war production and the demobilizing of the armed forces.

As a result, "Blacks were slowly forced out of many of the industries where they had obtained wartime employment, and they were excluded from unions as well" (Piven and Cloward, p. 194). A substantial gap developed between Black workers and the rest of society. In his report of the situation in Minneapolis, investigator Abram Harris points out, "in 1919 the Bureau of Labor Statistics considered a weekly wage of $43.51 essential to maintain an acceptable standard of living for a family of five." In contrast to this, Harris found that of the 222 married Negro men with families whom he interviewed in Minneapolis, only 12 were receiving more than $40.00 a week, 200 were earning between $15.00 and $30.00 a week, and the median weekly wage was $22.55.

The exploitation precluded advantages that otherwise may have accrued to migrants. Skilled artisans and common laborers alike were recruited for the lowest level jobs at the lowest pay. Reported one organization from New York City, of the 2,000 skilled workers surveyed only "one was employed at his calling." The rest were "rendering menial service as porters, elevator operators, chauffeurs, waiters, common laborers and so on." Some of these workers were graduates of Hampton, Tuskegee, and other industrial schools of the South who had been attracted north by promises of better wages and better conditions. It is clear that their actual wages as unskilled laborers were not commensurate with their skills, and advancement was almost impossible.

Conclusion

Despite their exploitation, the incorporation of southern migrants into the more economically advanced North, however, did provide a degree of security unknown in the South. The increased size of the Black community "brought a new disposition to fight and defend themselves, to fight White workers for the right to work, to confront the police in eviction actions, and to join the struggles of the unemployed against the relief system" (Piven and Cloward, p. 204). Perhaps the greatest change came in the area of politics. Moving from an environment of nonparticipation to citizenship, "Blacks evinced for the first time since the demise of Reconstruction a growing leverage on state and national affairs" (Marable, 1984, p. 14; McAdam 1982, p. 79). By 1930, they joined with other groups to kill President Hoover's nomination of a White supremacist to the Supreme Court and were instrumental in the defeat of several senators who had supported the nominee (p. 80).

The patterns of development and change experienced in the South during this period would be replicated in peripheral societies throughout the world brought into the orbit of core society expansion. Migrants are afforded the opportunity to begin at the bottom. The North by contrast, offered them a job, the ballot, and educational opportunity for their children. In the past, such contrasts have been sufficient to support the argument that the North, while not a promised land, would eventually provide advancement and progress. But it would take a "Second Reconstruction," as Manning Marable (1984) has called it—a protracted struggle for civil rights—before the migration dream could be pronounced a success.

6

Mexican Americans: Their Civic and Political Incorporation

RICARDO ROMO

People of Mexican origin have lived in the United States for nearly four centuries; the first mestizos, the name for those of mixed Mexican Indian and Spanish heritage, settled the upper Rio Grande valley in 1598, and from that date they have occupied a broad region of territory extending from San Francisco Bay in northern California to Nacogdoches in east Texas (Bannon 1970; Castaneda 1936–1958; Chipman 1992). During the first 200 years of their settlement in the United States, Mexicans founded pueblos and developed ranching, farming, and mining industries. Over time, they evolved principally into classes of peasants, ranchers, and merchants, while a small number gained elite status through extensive land ownership, trade with Anglos (English-speaking Americans), and cattle and sheep holdings (De Leon 1982; Perrigo 1971). Today, 83 percent of the Mexican American population resides in the Southwest, although they dwell in every state. Their status in American society is unique because they can claim indigenous origins, being among the first of this land, yet are recent immigrants.

Milton Yinger (1985, p. 32) defined civic and political integration as a critical aspect of assimilation, underscoring that a group is integrated "to the degree that its members are distributed across the full range of associations, institutions, and regions of society in a pattern similar to that of the population as a whole." To understand how Mexican Americans fit into the boundaries of the national community of the United States, an understanding of their experiences with citizenship and equality—their process of civic and political integration, particularly over the last 150 years—is essential. Kenneth Karst (1989, p. 210) reminds us that "many of the most important changes in the American civic culture have accompanied the recognition—sometimes gradual, sometimes abrupt —of new groups as fully participating members of the community of equal citizens." While Mexican

I have greatly benefited from and appreciate the critical reading and editing assistance of the following people: Robert Crunden, Mark Smith, Silvia Pedraza, Rubén G. Rumbaut, R. Rolando Hinojosa, and my wife, Harriett D. Romo. I am also grateful to Maria Velez of the Center for Mexican American Studies and Dorothy Harris of the Office of the Executive Vice President and Provost with the University of Texas at Austin for their assistance in preparing this manuscript.

Because the Mexicans have consistently supplied the cheap labor for American agriculture, when the immigration of southern and eastern Europeans was curtailed by the Immigration and Nationality Act in 1924, an exception was made for the Mexican migration. Here a Mexican farmworker is growing corn and beans in Chamisal, New Mexico, in 1940. (Photo by Russell Lee. U.S. Farm Security Administration Collection; Library of Congress)

Americans gained legal citizenship in 1848 as a result of the Treaty of Guadalupe Hidalgo, socially they remained second-class citizens. A prolonged historical struggle toward achieving civil rights in American society resulted in some measure of civic and political integration.

Historical Incorporation

The historical forces concerned with the evolution of Mexican American incorporation into American society fall into four distinct periods. During the first phase, Mexicans settled parts of the Southwest and introduced Mexican culture and civic government to the region. Due to the geographic isolation that existed between the initial New England colonies and the Spanish frontier, Mexicans had little contact with English-speaking Anglos. In the second phase, Anglos succeeded in gaining control of the Southwest territories. The first sign of declining Mexican influence occurred in 1835 when an Anglo militia defeated the Mexican commander of San Antonio. By 1848, a series of American military triumphs in New Mexico, California, and Texas, as well as in the interior of Mexico, culminated in Mexico's loss of the Southwest territories. Mexican Americans acquired American citizenship, but much like Native Americans and Black Americans, they were denied full rights as citizens. In the third phase, 1930 to 1964, Mexican Americans reasserted their civic and political rights, challenging their second-class status in American society. This new assertion began in 1930 when Mexican American leaders successfully disputed school segregation designs in Del Rio, Texas, and Lemon Grove, California. Fourth, three decades later, with the passage of the 1964 Civil Rights Act, Mexican Americans gained some measure of equality. The Civil Rights Act offered Mexican Americans broad legal protection for the first time. It also prompted Mexican Americans to begin to press for full equality and for the opportunities to improve their social and economic status.

Mexican Hegemony

By the mid-1500s, colonial Mexico's outer provinces provided the bulk of the new colonists for their communities north of the Rio Grande. From the frontier states of Nuevo León, Sinaloa, Sonora, and Coahuila, the citizens of Mexico traveled relatively short distances to settle the northern frontier. The movement north frequently resulted in the establishment of frontier pueblos with saint names: San Francisco, Los Angeles, Santa Fe, San Antonio, and Corpus Christi (Bolton 1963; Cruz 1988). Because Spaniards virtually ceased migrating from the motherland after 1600, we can characterize most of this population as mestizo (Vigil 1980). Over a period of 300 years, colonial Mexico and its people transformed, culturally and economically, a vast region of what is today the United States but was then a territory contiguous to and belonging to Mexico (Gutierrez 1991).

With the support of the Spanish Royal Crown and the Catholic Church, the frontier settlements of California, Arizona, New Mexico, Colorado, and Texas took hold. By the eighteenth century, Spanish-speaking communities, including missions and presidios, dotted the landscape of the Southwest (Almaraz 1989; Cruz 1988). For nearly 300 years Spain maintained its colonies in North America. Spanish influence in America, however, declined with Thomas Jefferson's bold purchase of the Louisiana Territory in 1803, where the first Mexicans to acquire American citizenship lived. The purchase served to bring Mexicans and American colonists closer together geographically. From Santa Fe, Mexicans drove their wagons, filled with furs and silver, east to St. Louis. Likewise throughout the early nineteenth century, Mexicans in San Antonio traded cattle and horses with Americans in New Orleans (Meinig 1969).

The aftermath of Mexican independence from Spain in 1821 brought renewed governmental pressures on Mexicans to migrate north. Officials in Mexico City feared the growing immigration of Anglo-Americans into Texas throughout the region west of the Sabine River and south of the Red River. In an attempt to curtail American migration, Mexico sought to promote European settlement by offering generous land grants to settlers. But Europeans came in very small numbers. By contrast, Americans, especially Southerners, came in droves, lured by both the prospect of cheap land and the possibility to expand slavery. The Mexican government's immigration policy was a miscalculation that resulted in a disastrous set of events.

Anglo Assertions

The incorporation of Mexicans in Texas took on a new meaning when Texas became an unofficial American colony in the 1830s. In the early 1820s, Mexican officials had given Stephen F. Austin a generous grant of land in the Brazos region of Texas. Austin encouraged American immigration, and in just ten years his "American" colony grew to over 30,000. By 1830, Americans outnumbered the Mexicans in the territory by more than five to one. In 1835, Anglo-Texans successfully pushed for independence and a year later established the Republic of Texas. Many Mexicans fled south to Mexico to escape the violence that followed the war for an Anglo Texas (Horsman 1981; Weber 1973). The 1836 Texas Constitution gave Mexicans full rights as citizens, although it maintained the enslavement of Blacks.

With the Texas Republic established, the political and social standing of Texas Mexicans changed dramatically. Stephen Austin, who had once been friendly to the Mexicans and had benefited tremendously from the generous Mexican land grants, now changed his attitude. He described Mexicans as a "mongrel Spanish-Indian and Negro race, [posed] against civilization and the Anglo-American race" (De Leon 1983, p. 12). Since the Texans were also involved in exterminating the Indians, referring to Mexicans as Indians entailed serious consequences. Not surprisingly, some members of the upper-class Mexican Tejanos, such as Lorenzo de Zavala, a former Mexican Senator; Jose Navarro, a signer of the Texas Constitution; and Juan Seguin, mayor of San Antonio, tried to maintain ties to the Anglo leadership, hoping that they would not be victimized by the new rulers of their homeland. But they were mistaken. Seguin, for example, fled to Mexico in fear for his life during the turbulent Republic of Texas years (Montejano 1987). Not until Texas joined the Union in 1845 did Texas Mexicans officially become Americans citizens.

Literary and historical studies illustrate the feelings of Americans toward Mexicans in other regions of the Southwest besides Texas during the late 1800s. In their writings, authors such as Lansford Hastings (1840), Alfred Robinson (1846), and idealist philosopher Josiah Royce (1887) introduced the Mexican people to an American society in the East that was anxious to learn about the Southwest. These authors described the Mexicans as fatalistic, superstitious, and prone to violence. Royce (1887, p. 28), for example, noted that "in politics, as in morals, and in material wealth (the California Mexican) was unprogressive." No doubt, these attitudes made incorporation more difficult for Mexican Americans.

One consequence of the war fought between Mexico and the United States on Texas soil was that for the Mexican people of Texas, as David McComb (1989, p. 46) pointed out, "the war cast seeds of prejudice that still bear bitter fruit—Mexicans became a despised and distrusted minority." McComb goes on to emphasize that "the term *spik* and *greaser* became a part of the Texas vocabulary in the 1850s" as Mexicans became second-class citizens. "To be sure, the highest wall of prejudice in Texas, one buttressed with segregation laws, was that between Blacks and Whites." Mexicans, however, "lived in barrios, remained impotent in politics, and became almost faceless in Texas society until the middle of the twentieth century" (McComb 1989, p. 46).

When the U.S. Senate ratified the Treaty of Guadalupe Hidalgo that ended the war between Mexico and the United States in 1848, they granted citizenship to Mexicans living in the conquered territory. While the Indian Treaties of that period did not allow Native Americans to vote or to participate in the judicial process, the 1848 treaty did specify that Mexicans would enjoy "all rights as citizens." The treaty also stipulated that Mexicans would be "protected in the free enjoyment of their liberty and property, and secured in the free exercise of their religion without restriction" (Miller 1937). However, that equality was only a promise. The first conflict over incorporation followed the discovery of gold in northern California. Mexican American miners from southern California were among the first to arrive to the gold-rush region. But their elation over reaching

the goldfields early was short-lived, as Anglo-Americans used violence to drive the Mexicans away (Chavez 1984, p. 43; Limerick 1987). Indeed, Californians treated the region's Mexican American population as if they were a foreign group. Anglos who were racially prejudiced toward Native Americans and African Americans extended their prejudices to Mexican Americans. The philosophy of Manifest Destiny was based on these sentiments.

Segregation

Manifest Destiny as an expression of White superiority is but one explanation for what became a clear rise of anti-Mexican sentiments in the 1850s. European Americans had long denigrated Mexicans, and their actions against "brown" people were consistent with their treatment of other racial groups. Determined to keep Mexicans and Asians from being a part of the greater American dream, they passed the "Greaser Law" and a "foreigner miners tax" in 1851 that restricted Mexican economic participation (Camarillo 1979, p. 108; Pitt 1966, p. 60). A politics of exclusion soon followed. In the late 1850s, for example, Manuel Dominguez, a country supervisor in Los Angeles, was prohibited from testifying in a California court because of his Mexican background (Pitt 1966, p. 202). These negative experiences in the years following the Mexican War curtailed the assimilation of Mexican migrants into their southwestern communities.

The problem was compounded by the number of Anglo Southerners who had migrated west bringing their racially prejudiced attitudes with them. They feared that people of color—Blacks and Mexicans—would unite to challenge their political and economic hegemony. Southern segregation penetrated as far west as California. For example, the *San Francisco Herald* in 1859 referred to Mexicans as "a degraded race; a part of them are so Black that one needs much work to distinguish them from Indians; there is little difference between them and the Negro race" (Pitt 1966, pp. 204–205). The same editorial also warned that plans were under way to create a territorial government in southern California and that Mexican Californios "would establish friendship with the Negro slaves, (and eventually) would be united with one another, until all would be amalgamated and all would be slaves" (Pitt 1966, pp. 204–205). Na-

Because of the deep-seated prejudice and discrimination they encountered, Mexican Americans had to engage in social protest to achieve a measure of civic and political incorporation. Here farmworkers march to Sacramento, California, in 1966, for the National Farm Workers Association under the banner of the Mexican patron saint, the Virgen de Guadalupe, followed by the flags of the United States and Mexico. (Cesar Chavez Collection, Archives of Labor and Urban Affairs; Walter Reuther Library, Wayne State University)

tivists and racists who desired to keep Mexicans in their place frequently reminded others of how undesirable dark-skinned people were and how dangerous the potential for unity between people of color was.

Uneven economic development and political instability largely account for the vast differences in the unification efforts of Mexicans and Anglos. Rodman Paul (1988, p. 141) writes that when it became apparent to the Mexican *ricos* (the wealthy class) that their future lay with the United States, many "sent their sons to St. Louis, New York, or other American cities to be educated, or apprenticed them to one of the mercantile houses that handled supplies and freighting across the plains, so that they could learn English and American business methods." In some New

Mexican counties, however, Mexican Americans were able to retain their land and managed by their sheer numbers to influence the political process. In the upper Rio Grande valley region of Texas, the elite Mexicans often formed commercial and political alliances with the Anglos. And, in south Texas, Anglo political bosses won local elections by courting local Mexican bosses who controlled large blocs of votes. These arrangements lasted as long as they were mutually beneficial (Campa 1979; Rosenbaum 1981).

During the 1870s and 1880s, Mexican Americans relied extensively on ranching and farming to support their families (Camarillo 1979; Rosenbaum 1981). As the men joined the work gangs of the large ranches, they faced long periods of isolation and boredom. Women worked alongside men doing subsistence agriculture, but they too found the frontier life harsh. Cattle ranching and small family farms dominated the Southwest region. These industries grew slowly, however, due to a lack of irrigation, unskilled labor, and poor transportation networks. The small Mexican American communities depended on limited water supplies; they often barely survived by using canal irrigation, ditches, and flooding techniques long applied by Indians and Mexicans.

The population of the U.S.–Mexico border states grew slowly at the turn of the century. The area attracted only small numbers of European and Mexican immigrants. With the exception of Los Angeles and San Francisco, this region had no cities with a population of over 100,000 in 1900. The population of the entire Southwest area at this time comprised less than 10 percent of the total national population, and urban growth in the area reached merely 50 percent of the level found in the East and Midwest. The region's largest metropolitan areas—Los Angeles, San Francisco, San Antonio, El Paso, and San Diego—did not experience rapid growth until after 1900 (Romo 1977, p. 184). No doubt, Mexican immigration contributed to the growth as well as to the cultural revitalization of Los Angeles, San Antonio, Tucson, and El Paso, the oldest Mexicans communities in the United States. The Mexican-born population of the United States in 1870, estimated at 42,435, increased to over 100,000 by 1900, and was mostly concentrated in three states: Texas, Arizona, and California. By 1900, Texas had the largest Mexican-origin population, nearly 69 percent of the total, while only 2 percent lived in California (Romo 1977, p. 194; U.S. Bureau of the Census 1932). In many instances, proximity to Mexico meant that the new migrants were able to encourage other family members to emigrate. In many of the barrios, one could find neighborhood clusters of extended family and friends from the same homeland town.

Between the years 1900 and 1930, when the first major wave of massive Mexican immigration to the United States occurred, the most striking influence on the Mexican American communities in the Southwest was the constant movement of people north across the border. Subsequently, this migration gave rise to many new Mexican American barrios. More than 1 million Mexicans, mainly but not exclusively unskilled workers, moved into the Southwest in the first three decades of the twentieth century. As Alejandro Portes and Ruben Rumbaut (1990, p. 226) un-

derscored, "contrary to the conventional portrait of Mexican immigration as a movement initiated by the individualistic calculations of gain of the migrants themselves, the process had its historical origins in North American geopolitical and economic expansion that first restructured the neighboring nation and then proceeded to organize dependable labor flows out of it." Sociologist Rodolfo Alvarez (1973) adds that the new wave of Mexicans entering the United States in the post-1900 era encountered a social situation very different from that of immigrants from other lands. Alvarez (1973, pp. 927–928) asserts that the experience of Mexicanos "upon entering the United States was predefined by the well-established social position of pre-1900 Mexican Americans as a conquered people—politically, socially, culturally, economically, and in every other respect."

Many rural and isolated communities grew out of the labor camps created in the Southwest during this period. The communities began when the large industrial companies sent their *enganchistas* (labor recruiters) into the interior of Mexico to locate workers. The *Los Angeles Times* (Sept. 18, 1916) reported, "Every week five or six trains are run from Laredo, carrying Mexicans who have been employed by labor agents, and similar shipments are being made from other border points." The *Los Angeles Times* concluded: "The demand for these laborers is so great that they are employed as fast as they cross the Rio Grande." Men, women, and children arrived at the U.S.–Mexico border and contracted for work in mining and railroad construction camps between California and Texas. In time these labor camps, or colonias, became some of the small towns of California, Arizona, New Mexico, Colorado, and Texas.

Between the years of the Mexican Revolution and the end of the Great Depression (1910 to 1940) nativistic attitudes directed against Mexicans became more pronounced. Rarely did Americans distinguish between the foreign-born immigrants and native-born Mexican Americans. Too many Americans assumed that Mexicans were illiterate and indifferent to public and civic affairs. Although some in the Southwest defended Mexican laborers as an economic asset to the region, social attitudes toward the Mexican American community reflected hostility toward their incorporation as well as negative stereotypes about their contributions to the United States. University of California Professor Samuel J. Holmes addressed these attitudes during a 1926 speech on the political consequences of making Mexicans citizens of the United States. Holmes favored the restriction of Mexican immigration, because Mexicans could be a "menace to a democracy." Doubting that Mexican residents could be an "addition to our political or our social life," Holmes offered that "the Mexicans are a race almost as distinct as the Negro, especially the Indians who form a very large component of that race. We are inviting another race problem for solution" (Holmes 1926, pp. 24–27). Indeed, many Anglo Southwesterners came to think of Mexicans as a new race problem and certainly not as a group that could be rapidly assimilated into U.S. communities.

In 1921, the U.S. Congress passed an emergency immigration act that restricted immigration from Europe to 3 percent of the number of nationals from each country living in the United States in 1910. Despite these restrictions, more than 500,000 Europeans entered the United States. In 1924, however, Congress adopted the National Origins Quota Act, which in effect cut immigration permits to 150,000 a year and allocated most of the permits to western Europeans (Divine 1972, p. 30). As a consequence, Italian, Russian, and Polish immigration decreased significantly, while German, English, and French immigration continued unabated. Southwestern farmers and growers successfully lobbied for the exemption of Mexicans from the 1924 Quota Act, thus assuring that Mexicans would become the preferred labor for this region.

In the mid-1920s, Paul S. Taylor began conducting field work in the agricultural areas of the Southwest and Midwest that culminated in his collection of oral histories on the Mexican immigrant's life story. He found many Mexican Americans living in relative isolation and maintaining little contact with mainstream social, political, and economic institutions. This isolation was largely a consequence of the limited employment and housing opportunities available to foreign-born workers in rural communities. The interviews show that Mexicans were viewed as seasonal intruders; most White farmers held negative stereotypes and prejudices about Mexicans, which stifled their opportunities for upward mobility.

Both in California and Texas, Paul Taylor found many Anglo farmers convinced that Mexicans workers were lazy, passive, and dull. Labor historians have checked these stereotypic notions by pointing to the evidence that every year farmers rushed to

find Mexican farmhands, contradicting the "lazy Mexican" notion. These farmers found few other workers willing to labor 12 to 14 hour days under a burning sun for a piffling wage. Still, the contradiction escaped these employers. Southwestern farmers told Paul Taylor that Mexicans remained farmworkers because they were incapable of doing work that required skill or intelligence. A ranch executive added that "White labor (was) necessary in the packing sheds, as the Mexicans generally are not particularly good at this kind of work." While employers claimed that "White packers are used because a higher degree of intelligence is required for this work," racism also played a part in the decision. One manager reported that "race prejudice is a factor (because) the introduction of Mexicans in work commonly performed by whites would lead to serious disorders between the groups" (Taylor [1928] 1971, p. 41).

Mexican American Reassertion

As the Southwestern Spanish-speaking communities grew in the two decades after World War I, the social problems faced by Mexican Americans did not escape notice. Writing in 1926, Sociologist Max Handman, a Southwestern scholar observed that "the Mexican has been keeping to himself, but as he begins to share more and more in the occupation and activities of his American neighbor—even in a subordinate capacity—the situation is becoming strained. The theory that the Mexican is a White man is receiving its acid test" (Handman 1926, p. 36). Three years later, political scientist O. Douglas Weeks' research in south Texas also supported the notion that Mexican American and Anglo relations badly needed reform. Weeks (1929, pp. 257–258) asserted that Mexican Americans want "to eliminate as much as possible race prejudice on both sides of the dividing line and to gain for the Mexican American equality before the law, equal facilities for educational and other forms of improvements, and a reasonable share of political representation in the affairs of the community, state and nation."

To combat problems of discrimination, Mexican Americans created their own community organizations. These institutions, particularly the ethnic newspapers and mutual aid societies, were common in every region of the Southwest by the early 1930s.

Spanish-language schools and labor associations were regional in nature, but they too were created as a response to the racial barriers to incorporation into the mainstream society. These barrio institutions survived well beyond the traditional second- and third-generation patterns commonly found among the European ethnic communities. Their longevity was largely due to the persistence of class and racial discrimination in the host society when Mexicans attempted to enroll their children in Anglo schools or sought to trade in all-White retail establishments.

When Mexican Americans filed their first desegregation suits in the early 1930s, their principal goal was incorporation. Segregation lacked dignity and Mexican Americans viewed the separation of schoolchildren on the basis of race as particularly offensive (Garcia 1991; Gonzalez 1985; San Miguel 1987). In 1930, Mexican American parents filed an injunction against the Del Rio, Texas, school board to protest the attempt by officials to increase school taxes to pay for a new two-room building for the education of Mexican children. Mexican American lawyers established in court that the two schools existed for the purpose of separating Mexican and Anglo children. The court decreed that Mexican American children could not be *racially* segregated, but allowed the separation of Mexican and Anglo children on the pretext that Mexican children needed special language instructions due to their migrant status *(Del Rio Independent School Dist. v. Salvatierra,* 1930, 33 S.W. 2d 790 Tex. Civ. App.). In its analysis, however, the court failed to ascertain *how many* of the children came from migrant families and assumed that *all* the children had inadequate language skills (Romo 1986, p. 344). In a "Black and White" world, Mexican American children ended up classified as "White." Nonetheless, Texas schools remained defacto segregated for the next two decades.

This *Salvatierra* case also raised the issue of Spanish language usage in Mexican American communities. No doubt, Mexican-origin residents living in rural and urban areas of the Southwest sought to maintain the language and other cultural traditions of their home country. Cultural persistence can be attributed to several factors, however. First, the Southwest had been solely Spanish-Mexican for three centuries and, as such, native customs and language remained long after annexation by the United States. Second, the Southwest region had been absorbing new Mexican immigrants in substantial numbers for

several generations. These new immigrants constantly reinforced all aspects of Mexican culture, enabling the Spanish language and Mexican customs to thrive. Third, the proximity to the homeland enabled the Mexican-origin population to visit Mexico frequently and inexpensively. This constant interaction gave Mexican Americans added reason to maintain their Spanish language skills. However, the desire to maintain homeland customs and language took on new meaning during the depression era. The economic crisis would once again test American attitudes toward the Mexican community.

The advent of the Great Depression forestalled the incorporation of Mexican Americans as the economic crisis gripped the nation. During the two previous decades, most Mexican communities had grown substantially and, attendant to the economic opportunities created by World War I, Mexican Americans had made inroads into better-paying urban jobs in construction, manufacturing, and service industries. With the stock market crash of 1929, followed by massive bank closures, every community in America saw large-scale shutdowns of factories and businesses and a subsequent loss of jobs. Thousands of "Okies" and "Arkies" became dislocated, providing a new source of cheap labor and lessening the need for unskilled Mexican labor in California. Faced with a large unemployed Mexican population and dwindling welfare resources, public officials in California, Arizona, and Texas communities decided to repatriate, or deport, Mexican residents. In Los Angeles, county authorities staged dramatic neighborhood raids designed to uncover those Mexicans thought to be in the United States illegally. The apprehended were given only a short notice to pack and leave. Officials provided free railroad transportation to the border, so as to clear Mexicanos out of Los Angeles County. This deportation of "aliens" heightened the hostility between the Anglo and Mexican American communities.

Many civil rights violations occurred during these forced deportations. The officials who participated in repatriation raids were often oblivious to the citizenship of those they deported. Thousands of children born in the United States were also taken to Mexico along with their parents who lacked citizenship. This persisted even though those born in the United States qualified to remain in the United States (Balderrama 1982; McKay 1982, p. 560). In Texas, where almost half of the deportees resided, Reynolds

McKay (1982) found that the civil rights violations included "not permitting returnees to dispose of their property or to collect their wages, deporting many not legally subject to deportation because of their length of Texas residency, separating family, and deporting the infirm." Communities that only a few years earlier had praised the diligent work of Mexican laborers now turned their backs on Mexican residents at a time of dire economic need (Hoffman 1974; McWilliams 1933).

America's entry into World War II forced President Franklin D. Roosevelt to abandon many of his popular domestic economic policies and respond instead to the urgency of the war. Funding for federal employment initiatives, such as the Work Progress Administration (WPA) and Civilian Conservation Corps (CCC), collapsed as German tanks threatened to overrun western Europe and Japanese soldiers captured large regions of Indochina. In late 1941, FDR called up the civilian reserves and launched an aggressive armed service recruitment program. Over the next four years, more than 15 million U.S. men and women went into uniform to join the war effort. Over the same period, more than 5 million women entered the workforce in defense-related factory and farm work, boosting the number of women in the war effort to nearly 20 million.

The war effort created a domestic labor shortage, and farmers and industrialists were quick to plead their case for cheap labor replacements. At the urging of U.S. employers, the governments of the United States and Mexico signed an international agreement to allow Mexican contract laborers (braceros) to enter the United States for a six-month period to work in agriculture, railroad construction, and maintenance. The conditions under which the braceros were to work and live were specifically outlined in the agreement: the braceros were to receive free transportation and food, guaranteed wages, safe working conditions, and sanitary living quarters. With these protections, Mexican workers flocked to participate in the program (Cornelius 1978; Samora 1971).

Bracero laborers proved extremely valuable to the western American industries that required large numbers of unskilled workers. Peak years for the recruitment of labor from Mexico occurred in 1944 and 1945. Labor economist Ernesto Galarza (1964, p. 53) estimated that during the peak months of 1945, some 58,000 contract laborers worked in agriculture and nearly 62,000 worked on the railroads. Throughout the war, the railroad industries employed the largest share of the braceros. In the last two years of the war, 135,283 workers came under contract to the United States. When the bracero program was scheduled to expire at the end of the war, agriculture and railroad interest groups successfully lobbied to continue it. Congress approved its extension several times during the next twenty years. Overall, more than 5 million braceros worked in the United States during the program's existence (*Commonweal*, pp. 275–278).

Opposition to the bracero program came from many sectors. Organized labor opposed the importation of foreign laborers, arguing that the use of such workers threatened unionization efforts and wage gains. Mexican American workers of the borderland states, where much of this contract labor was utilized, also opposed braceros largely because foreign laborers competed for their jobs. Although the bracero agreement clearly prohibited the displacement of native-born workers, such restrictions did not prevent the wholesale denial of jobs to domestic labor with bracero labor in the West. According to the National Farm Labor Union, the Imperial and San Joaquin valleys of California could count on the availability of 60,000 to 80,000 native workers for harvesting the crops; but braceros were preferable, according to *Fortune* magazine, because they worked for 60 cents an hour (20 cents less that the prevailing rate), which included board [*Fortune* 43 (April 1951): 60–61].

The issue of discrimination also surfaced during the bracero years. Mexican officials who traveled or studied in the United States resented the second-class treatment extended to persons of Mexican descent. During the war years, the Mexican government went so far as to exclude the state of Texas from the bracero program because of the state's record of allowing overt discrimination against Mexicans to go unchecked. Eventually, Texas was allowed to participate in the program, but Mexican officials remained disappointed with the state's lack of progress toward the extension of equality to people of color.

With increasing opposition from Mexican American workers, unions, and religious groups, the Lyndon B. Johnson administration allowed the bracero program to lapse in 1964. With the termination of the bracero program, agricultural labor in California was finally able to organize effectively for higher wages and better working conditions. Mexican American community leaders came out in support of the war effort, and nearly 500,000 Mexicans joined the armed forces and served overseas. Two cases involving Mexican American civil rights highlighted the campaign to end overt discrimination and segregation in southern California during the war years. The arrest of hundreds of Mexican American youngsters linked with a rising crime wave in Los Angeles instigated the first case. *People v. Zammora* demonstrated the extraordinary extent and persistence of prejudice in a western community thought to be exemplary in race relations (R. Romo 1990). Popularly referred to as the "Sleepy Lagoon" case, it concerned the prosecution of twenty-four (twenty-three Mexicans and one Anglo) young men for the death (alleged murder) of one Mexican youth in the summer of 1941. Indeed, it was one of the largest murder trials ever held in the United States.

The trial lasted thirteen weeks and was given full coverage by the Los Angeles press. An all-White jury found three of the defendants guilty of first-degree murder and gave them life sentences; they also found nine guilty of second-degree murder and gave them from five years to life (R. Romo 1990, p. 390). When *People v. Zammora* went before a state's appeals court in October 1944, the judges found that "the evidence was insufficient to show that the defendants had conspired to commit the crimes charged" (1944, p. 167). The court also found the trial judge "guilty of judicial misconduct in making undignified and intemperate remarks tending to disparage or cast reflection on defendants' counsel" (1944, p. 152). When the young men won their release from jail as a result of the appellate decision, the community celebrated their victory.

The second case, *Mendez v. Westminster*, was a successful school-desegregation suit in Orange County, California, initiated in 1946 by the Mexican American community. Judge William Denman wrote in his concurring opinion that it was not only in Orange County that "public officers (were) guilty of such perversions of the privileges long recognized as common law (sic) as essential to the orderly pursuit of happiness by free men" (*Mendez v. Westminster* 783). He saw no reason to defend segregation in California, adding, "Were the vicious principle sought to be established in Orange and San Bernardino Counties followed elsewhere, in scores of school districts the adolescent minds of American children would become infected" (R. Romo 1990, p.

404). The decision struck down school segregation for children of Mexican, Native American, and Asian descent, and was a precursor of the *Brown v. Board of Education* decision in 1954, the case that succeeded in striking down the "separate but equal" doctrine that was the legal underpinning of Jim Crow segregation in the United States since 1896 (Wollenberg 1974, 1978).

America's involvement in World War II overshadowed the nation's internal race problem. The war served as an occasion for Mexican Americans to once again demonstrate their loyalty to America. Men and women from the barrios also contributed to the combat effort by entering the industrial workforce. Like other Americans, many joined because they believed their government's assertion that the actions of the Axis forces threatened freedom everywhere. To their credit, some government authorities also expressed hope that those "patriotic sentiments would serve as a unifying agent among races, ethnic groups, and classes" (Polenberg 1980, p. 46). As one official noted, "by making this a people's war for freedom, we can help clear up the alien problem, the Negro problem, the anti-Semitic problem" (Polenberg 1980, p. 47). Consequently, thousands of Spanish-speaking men and women gave their lives to secure an Allied victory in the battlefields of Asia and Europe.

In some instances, those who returned home safely found new opportunities that raised their chances for incorporation into the middle class. Thousands of Mexican Americans entered high schools, colleges, and universities with support from the "G.I. Bill," the Veteran's Entitlement Act. Many others bought their first homes with the assistance of G.I. Bill veteran's benefits. In the post–World War I years, a number of Texas-Mexican veterans had organized the League of United Latin American Citizens (LULAC) for the purpose of improving the social and political standing of the Mexican American people. Likewise, one of the most significant changes following the end of World War II was the establishment by these veterans of new political organizations, such as the Community Service Organizations in California, and the American G.I. Forum, founded in Texas by Hector Garcia.

In other instances, veterans returned home from the war to find many unsolved social and economic problems in their communities. Throughout the Southwest, many Mexican Americans still lived in segregated communities, attended segregated schools, and were denied the use of many public facilities (Allsup 1982). In Three Rivers, a small south Texas ranching community, Jim Crow segregation practices prohibited the use of the only funeral home in town for the grief-stricken family of Felix Longoria, despite the fact that Longoria, a veteran, had died serving his country in the war effort. The case of Felix Longoria became a rallying cry for Dr. Garcia and the newly formed American G.I. Forum organization.

The G.I. Forum joined the Longoria case as symbolic of the racism that still plagued Mexican Americans and acted aggressively. Unable to work out arrangements for the use of the funeral home by the Longoria family, Dr. Garcia and the veterans' group arranged with the young Texas Senator Lyndon B. Johnson to have Longoria buried at Arlington National Cemetery near the nation's capitol. The Longoria incident represented one of the first successful efforts of the Mexican American community in its battle against segregation. Indeed, as a consequence of more organized efforts, Mexican Americans experienced success after success in challenging segregation.

From the 1940s on, Mexican American organizations formed coalitions in their drive to improve their community's social and political status. Throughout the late 1940s and 1950s, LULAC and the American G.I. Forum challenged segregation policies in public schools, residential neighborhoods, and public accommodations (San Miguel 1987). However, LULAC and the American G.I. Forum bylaws prohibited their involvement in politics. Recognizing that political involvement would hasten their incorporation, Mexican American political leaders established PASSO—the Political Association of Spanish-Speaking Organizations—in Texas, and MAPA—the Mexican American Political Association—in California (Guzman 1976).

Presumptive Equality

Jim Crow segregation laws generated a discontent that became the foundation for the modern Mexican American civil rights movement. The 1964 Civil Rights Act and 1965 Voting Rights Act helped to usher in unprecedented advances, including the destruction of an "entrenched system of segregation

and disfranchisement" which had long humilated Blacks and Mexican Americans (Nieman 1991, p. 188). However, while there was a "dramatic expansion of federal power to protect civil rights and to achieve genuine equality," problems remained (Nieman 1991, p. 188). It quickly became apparent to a new generation of Mexican American leaders that the end of Jim Crow did not remove all of the barriers to equal opportunity (U.S. Commission on Civil Rights 1971). High school and college students became especially impatient with the slow pace of political change. In Crystal City, Texas, the drive for better schools resulted in a large-scale social and political referendum (Foley 1988; Garcia 1989). For the first time, Mexican Americans, many of them in their twenties, won elective positions to the school board and city council in several south Texas communities. In Los Angeles and other large urban areas, Mexican American students boycotted schools that they felt were indifferent to the rising dropout rate and lack of college preparatory classes (Munoz 1989). In college campuses across the Southwest, Mexican American activists formed student organizations that successfully lobbied for the inclusion of ethnic studies classes in the curriculum and helped to open the universities to more minority students and faculty (Castro 1974).

These new civil rights groups also benefited from the Mexican American farmworkers' vigorous labor campaigns in the early 1960s. When the United Farmworkers' Union (UFW) was formed, most farmworkers had only seasonal or part-time jobs with giant agricultural farms. The farmworkers often lived in company housing in rural isolation, earning as little as 50 cents an hour (Taylor 1975). In 1962, Cesar Chavez, Dolores Huerta, and other UFW leaders began organizing thousands of farmworkers in an effort to improve the lives of the lowest paid and most exploited workers in America. To inform Americans of the terrible living and economic situation facing farmworkers in the Central Valley of California, Chavez took his rural movement into the Mexican American urban community. The grape boycott brought students and industrial workers together in the battle for economic justice.

By the early 1970s, the UFW had achieved some successes in the areas of collective bargaining, minimum wages, and workman's compensation. The UFW's extended battle for these basic labor rights captured the heart of the Mexican American community and included many sympathetic White community and included many sympathetic White community activists. Chavez used the unique strategies of social movements founded on the principle of non-violence, such as long marches of more than 100 miles and extensive hunger strikes to call attention to the cause.

Many of the early UFW volunteers followed Chavez's example of organizing communities. Two UFW volunteers in Texas, Willie Velasquez and Ernest Cortes, applied UFW tactics in building new political organizations in the Southwest (Rodgers 1990). In the early 1970s, they formed new organizations dedicated to the social, political, and economic well-being of the Mexican American community. Under the dedicated leadership of political organizers Willie Velasquez and Andy Hernandez, the Southwest Voter Registration and Education Project (SVREP) extended voting power to many previously disenfranchised Mexican Americans. By the 1970s, these organizations often worked together to challenge discrimination in the labor force, violations of voting rights, and school segregation. To this day, the extension of the franchise via voter registration drives remains a major goal.

From its foundings in the late 1960s, the Mexican American Legal Defense and Education Fund (MALDEF) has also been at the forefront of major litigation involving voting rights, employment discrimination, and school desegregation. In 1975, MALDEF attorneys challenged a newly passed Texas law that prohibited undocumented Mexican children from receiving an education by attending Texas public schools. State legislators had passed this new law on the premise that both higher education costs and higher taxes were related to increased Mexican immigration. The state of Texas went so far as to argue that barring immigrants would improve "the quality of education of others." The U.S. Supreme Court heard the immigrant children case in 1982 and ruled in, *Plyler v. Doe*, that these children had been "denied their constitutional rights under the Fourteenth Amendment" and also found no evidence that illegal entrants imposed a significant burden on the Texas economy. [Under the Fourteenth Amendment, the parents and their children were entitled to the protection of "life, liberty, or property," as well the "equal protection of the laws."] Indeed, the Court concluded that, to the contrary, "the available evidence suggest that illegal aliens underutilized public services, while contributing their labor to the local economy and tax money to the state" (*Plyler v. Doe* 1982, p. 25), particularly since most of the immigrant children in Texas would one day go on to gain citizenship.

Significantly, while the Mexican American community was not directly affected by the immigrant education law, it supported MALDEF's concern for the rights of immigrant children. If the undocumented population was to have any chance to enter mainstream society, it too had to have the right to a public education.

In most instances, political incorporation follows economic integration. But for many in the Mexican American community, the prospects for economic integration in the immediate future remain poor. A recent *New York Times* article reported: "Their numbers swollen by a flood of recent immigrants, people of Mexican descent in the United States are, as a group, much worse off than non-Hispanic Americans by many economic and educational measures. They also lag behind the average for Hispanic Americans" (*New York Times* 19 January 1992). Indeed, the median earnings of Mexican Americans in 1992 was $8,874, far off from the majority whites (non-Hispanic) earnings of $11,885. That same year (1992) the income level of Mexican Americans stood at around 75 cents for every dollar earned by a non-Hispanic (*New York Times* 19 January 1992).

Continued immigration has had significant consequences for the incorporation of Mexican Americans into U.S. communities, especially in the Southwest. Most recent Mexican immigrants have less education than Mexican Americans and other non-Hispanic groups. Indeed, while only 1.7 percent of non-Hispanics had less than five years of schooling, 15.5 percent of the Mexican-origin adult population (over twenty-five years of age) has not gone beyond the fifth year of school. The low schooling experience of Mexican newcomers may account for some of the high poverty (28.4 percent) and high participation in low-skill labor and manufacturing jobs (*New York Times* 19 January 1992).

Today, much of the Mexican immigrant population remains outside the American economic and social mainstream. This has especially been true of Mexican immigrants living in the U.S.–Mexico border communities. In these communities, the Mexican-origin population earned substantially less than other Americans. For example, while the annual median income of a non-Hispanic male in 1990 was

$22,207, a Mexican American male in the United States earned an income of only $12,894 (U.S. Bureau of the Census 1991b, p. 12).

Moreover, the average Mexican American woman who worked earned less than $10,000 a year. While one-quarter of Americans had completed college, less than 10 percent of Mexican Americans had a college diploma. Indeed, the dropout rates in many communities continue to be staggering. Southwest Voter Registration and Education Project Director, Andy Hernandez, warned: "A large and growing population that remains on the margins of society cannot contribute to the economic and social progress of that community. Furthermore, it is inconceivable that democracy can remain vital and effective without Hispanic participation" (Southwest Voter Research Institute 1991, pp. 88–89).

As always, the recent political and economic incorporation of Mexican Americans is closely linked to grassroots organizations such as the Industrial Areas Foundation (IAF) and Southwest Voters Registration Project (SVREP). In communities that traditionally have been poor and economically impoverished, the drive for political self-determination has proved challenging, though not insurmountable. In the last decade, the IAF has enlisted more than 100,000 volunteer families dedicated to improving their communities. Their reform platforms have brought about vast improvements in public schools, urban infrastructures, and economic well-being. Also, the SVREP has had extraordinary successes in the political arena; during the past decade they conducted more than 1,000 voter registration education campaigns in over 200 cities and Native American reservations in the Southwest. The result has been increased voter registration in the Southwest. In Texas alone, for example, the number of Mexican American elected officials rose from 500 in 1973 to nearly 2,000 in 1992. More recently, the battle over the lack of political representation for barrio residents has intensified, leading to redistricting. For example, in 1990, Mexican Americans won the right to secure an ethnic district in Los Angeles County where they constituted one-third of the population but had been denied political representation for more than a century in county elected offices.

Mexican Americans continue to play active roles in contesting civil rights violations. The community is no longer intimidated by the legacy of the notorious segregation statutes that previously governed interracial contacts in public places and public facilities to exclude Mexican Americans, Blacks, Indians, and Asians from most accommodations available to Whites. For as long as Mexican Americans were denied basic constitutional rights, they could not view themselves as equal before the eyes of the law or of society. For Mexican Americans, Jim Crow legislation and social custom encompassed school segregation, denial of access to restaurants, hotels, and theaters, and exclusion from many well-paying jobs. Thus, the post–World War II generation of Mexican American community leaders believed that their work for social justice and equal opportunity had just begun.

Conclusion

Nearly 150 years ago, the war between Mexico and the United States forever changed the process of political integration for people of the Southwest. In the aftermath of political change, a new group of U.S. citizens of Mexican origin sought to develop cultural and economic relations with their fellow Americans. But full incorporation into American society was delayed for most Spanish-speaking residents of the Southwest. Nearly a century after the Mexican citizens joined the U.S. community as citizens, they were still faced with virtual exclusion. Nonetheless, exclusion and inequality could be contested in the courts. While the Constitution protected racial and ethnic minorities from discrimination, prior to the 1964 Civil Rights Act and the 1965 Voting Rights Act—the major achievements of the Civil Rights movement—the Supreme Court's decisions still allowed unequal social and political treatment. Only in the last four decades has the Mexican American community been able to successfully challenge that inequality.

The battles for incorporation, however, were won not only in the courts but also on the streets and in the political arena. Grassroots organizations, political groups, and the youth movement helped to keep the Mexican American engaged in a series of important confrontations over inclusion. Over the last thirty years, the gains have been impressive, and subsequent elections may well show a new surge of political activity among the Mexican population of the United States. As the courts and political institutions respond to the concerns of the Mexican American community, improved socioeconomic conditions should follow.

Moreover, the changing demographics of American society may also influence future civic and political incorporation (Hayes-Bautista, Schink, Chapa 1988). For example, as a result of the contemporary wave of migration in the last decade, from 1980 to 1990, the old barrios of the border states, particularly those in California and Texas, gained more than 5 million new Spanish-speaking residents who came not only from Mexico but also from a number of Central American countries, especially El Salvador and Guatemala. Likewise, the number of Hispanic residents more than doubled in sections of California's Orange County, the San Francisco Bay area, San Diego, and Dallas, Texas. These population shifts have already begun to translate into increased political incorporation as more Spanish-speaking public officials are voted into public office. As one prominent political organizer underlined: "As America becomes more Hispanic, much of its future will be determined by the extent to which its Hispanic population is integrated into the nation's economic, social, and civic life" (Southwest Voter Research Institute 1991).

7

Migration, Community, and Culture: The United States–Puerto Rican Experience

HÉCTOR A. CARRASQUILLO
VIRGINIA SÁNCHEZ-KORROL

Strolling down the streets of Sunset Park, in the borough of Brooklyn in New York City, one not only experiences sights, sounds, smells, and tastes associated with the culture of the community's Puerto Rican population, but one can also travel over the remnants of an often-neglected part of the history of the relationship between Puerto Rico and the United States. Looking down at the paving blocks on Thirty-seventh Street, between Fourth and Fifth avenues, one might observe that they are made of a molded iron blue-slag material that a trained eye would identify as *adoquines*.[1] These stones were used as ballast, as weights to provide ships with stability, in the early days of travel between Europe and the New World. Those particular examples can be traced back to a ship that went from Spain to Puerto Rico around the turn of the seventeenth century. Subsequently, Brooklyn became a major port of entry for Puerto Rican produce and it was on a nineteenth-century vessel laden with sugar that the *adoquines* traveled, this time becoming personal possessions of a local builder who purchased them for use as paving material. Little did he realize that decades later many Puerto Rican people would follow the *adoquines* to reside on these very same streets.

The journey of these stones from Spain to Puerto Rico, and then to the United States, was much simpler than the migration of the people. The stones changed their function over time, yet they retained their basic character. If the need were to arise they could once again be placed in the holds of a sailing ship to stabilize its journey across rough waters. But over the course of the 350 years since the stones' initial voyage, Puerto Ricans and their culture have been molded and reshaped several times over, adapting to sociocultural and economic circumstances well beyond their control. Moreover, whereas the paving stones lie passive on silent streets, the Puerto Ricans who live on them, and who reside throughout the United States, have played an active role in reconstructing communities and transforming their mainstream culture. The paths of the stones and the people are connected, nevertheless, and this chapter endeavors to initiate the story of this association.

The authors thank Professors Antonio Nadal and María Pérez González from the Department of Puerto Rican Studies at Brooklyn College for their support and suggestions.

The Island Background to Migration

From a European perspective, the history of Puerto Rico is said to begin with Columbus's claim of the island in 1493. Yet, as is well known, Columbus's "discoveries" were such only for the Europeans. The native Tainos were living on the island, which they named *Boriquen*, for hundreds of years before Columbus and other Spanish explorers incorporated it into the imperial crown of Spain.

To the Europeans, the wealth of the "New World" was spectacular. The silver and gold of parts of Mexico and Peru astounded the conquerors and made those areas prime targets for imperialist expansion. Puerto Rico did not have the same quantities of wealth, but it was rich nonetheless in its climate, soil, and people. Spain's primary interest in the island was based on the exploitation of its mineral resources, but that was soon replaced by exploitation based on the island's strategic location as a major link in the chain of military fortresses protecting the Spanish mainland.

An agricultural economy rooted in the cultivation of sugar, coffee, and tobacco emerged during the nineteenth century. This period was characterized by a plantation economy that depended on labor intensive methods to produce agricultural products for commercial export. Once harvested, those then-luxurious items—tobacco, coffee, and sugar and sugar by-products such as rum and molasses—would be traded abroad.

As was the case elsewhere in the Americas, plantation economies were driven by cheap labor provided by enslaved people imported from Africa, and a Creole landed and landless peasantry. Over time, the original inhabitants of the island blended into the multiracial mix that was becoming characteristic of Puerto Rican society and that to this day provides the island with its cultural diversity. As with its history, the culture of Puerto Rico today is thus heavily impacted by a heritage of colonialism and combines indigenous, Spanish, and African roots.

The economic poverty of the island was a direct result of its colonial heritage. The interest of the Spanish metropole was less on forging an economically viable colony that would be settled by emigrants from the colonizing country, as was the case with the British colonies in the Americas, than on ex-ploiting the land for profits that would largely be returned to the metropolis. As such, developmental priorities were not based on the evolution of a balanced economy that could in time become self-sustaining, but on the cultivation of those agricultural products that would increase the wealth of the ruling elite. By the nineteenth century a pattern was set into place in which, despite significant agricultural expansion, Puerto Rico became dependent for its basic foodstuffs on imports from abroad.

As its plantation economy flourished, however, trade patterns changed. Whereas the sugar grown on Puerto Rican soil was initially shipped to Spain, by the last third of the nineteenth century the United States became one of the island's major trading partners (López 1980). With this, and the growing interest of the United States in establishing a sphere of influence throughout the Caribbean and Latin America, the ties between the United States and the Spanish colony of Puerto Rico were set for future expansion.

After the Spanish-Cuban-American war of 1898, Puerto Rico found itself free of Spanish domination, only to become a colony of the United States. Ironically, the island became the property of the very country that a mere three decades before had provided safe havens for a multitude of eastern and south central European immigrants. Similarly, since the 1868 aborted cry for independence known as *el Grito de Lares*, political exiles from Puerto Rico who were dedicated to the cause of Antillean liberation also waged their battles from the concrete fortresses of New York. As a result of the Treaty of Paris, which ended the war of 1898, the island was ceded to the United States as a booty of war. The basic structural reforms that could have positioned the island's economy on the road to stability became less likely, and Puerto Rican interests continued to be subordinate to that of its colonial master.

In 1899, the San Ciriaco hurricane virtually wiped out the island's agricultural sectors, and given the U.S. investment in sugar production to the exclusion of all other crops, coffee, and tobacco, failed to survive the colonial transition. As poorly paid plantation work became increasingly unavailable, thousands of Puerto Rican men, women, and children were forced to emigrate to other areas under U.S. hegemony, even as far as Hawaii. In ships chartered by U.S. plantation owners in need of cheap labor, organized by intermediary agents whose business it was to sup-

Celebrations that express a group's culture are important markers of identity. Here the Puerto Rican Merchant's Association designed a float for the Puerto Rican Day Parade in New York City, that shows the island's map, as well as the beauty of its landscape, women, and children behind the flags of both the United States and Puerto Rico. (Justo A. Martí Collection; Centro de Estudios Puertorriqueños, Hunter College)

ply such labor, Puerto Ricans traveled to remote destinations that included San Francisco, New Orleans, and Honolulu. Such destinations became the sites of the earliest Puerto Rican communities in the nation.

As early as 1903, 539 Puerto Rican children were already enrolled in Hawaiian schools. In California, the creation of associations like the Puerto Rican Club of San Francisco (1911) and the *Liga Puertorriqueña de California* (1923) signaled the initial stages of community formation. At least one history dissertation now exists (Norma Carr, *The History of Puerto Ricans in Hawaii*) that documents this phase of the U.S.–Puerto Rican experience.

Often contract workers were treated more like indentured servants than free laborers, and tales of abuse and attempted escapes abound in oral histories and newspaper accounts of the period. What is important to note is that even though the laborers were not virtual slaves of the agents with whom they signed contracts, their decisions to leave the island were complex and not always voluntary. If structural economic factors had been so arranged for the island to support its own population, many of those who left Puerto Rico might have elected to stay. The domination of the island by the United States accelerated various factors, including the move towards a capitalist, monocultural economy, and in the process perpetuated a legacy of dependency. Moreover, how and where Puerto Rican workers would make a living would be determined, in great measure, by the U.S. control of the means of production and exchange, wages, taxation, education, and foreign relations. Until the island evolved to the position to determine its own future, migration and colonialism would henceforth be linked together.

A critical event in the history of the relationship between Puerto Rico and the United States was the imposition of American citizenship on Puerto Ricans under the Jones Act of 1917. This further encouraged and facilitated the population movements between the two countries. The prosperity of the American economy from the period of World War I and into the 1920s, led to demands for low-paid workers in such areas as manufacturing, restaurants and hotels, domestic services, and laundries. Puerto Ricans were among the groups that entered the United States as part of this workforce. Between 1920 and 1930 about 53,000 Puerto Ricans came to reside in the United States and by 1930, the census reveals that Puerto Ricans lived in every state of the nation (Fitzpatrick 1987; Sánchez-Korrol 1981).

However, it would be a gross oversimplification to consider that such dislocations could be explained by implying that the newcomers were merely "seeking a better life." Despite the fact that a massive program for the industrialization of Puerto Rico—Operation Bootstrap—was implemented on the island in the 1950s, its limitations made it clear that the large-scale chronic unemployment evident for so long would not disappear. This became more significant given the rapid increases in the island's population and the manifestation of such expansions in the rise of migration.

The movement of a significant number of Puerto Rican men and women to the United States over the past half-century thus cannot be separated from the island's political and economic history. Poverty in Puerto Rico is not limited to the island's supposed lack of resources or a nonindustrious workforce, but rather is driven by a historically distorted process of growth without development. If Puerto Ricans developed a propensity toward emigrating in search of economic betterment, it is not because they undervalued their culture and island nation, but because the island simply became incapable of supporting them within its economic structure.

In this, it should be noted, Puerto Rican emigration had much in common with many of the human waves in the nineteenth and twentieth centuries, including European immigration into the United States. Very often the immigrant experience is summed up by such easy phrases as "seeking a better life in the New World," or looking for "streets paved with gold," as if this were a voluntary choice. Whether emigration became the means of escaping starvation in the Ireland of the potato famine in the mid–nineteenth century or the brutal poverty and virtual starvation characteristic of life in southern Italy, there was little choice other than to starve there or risk the chance of living a life marked by marginality and scarce resources for oneself and one's family. Similarly, for Puerto Ricans the difficult journey to a new land was made from such necessity.

Reconstructing Community

Along with the paving stones that wandered from Puerto Rico to New York in the nineteenth century came a relatively small band of Puerto Rican traders, professionals, and political exiles, joined in less than three decades by an increasing number of contracted and noncontracted laborers. As the century came to a close, New York City became the nexus for political exiles from Cuba and Puerto Rico who sought to expunge the yoke of Spanish domination. The movements for Antillean independence centered the political activities of numerous groups and organizations, including the Puerto Rican branch of the Cuban Revolutionary Party founded in 1897. Among other exiles, the writer, Lola Rodríguez de Tío, supporter of Puerto Rican independence and author of the island's fiery revolutionary anthem, *La Borinqueña*, lived and worked in the United States during this period. Another case in point is the Puerto Rican Afro-Caribbeanist Arturo Schomberg who founded and headed associations like *Las Dos Antillas*. As a young Puerto Rican man in search of his Afro-Antillean roots, Schomberg resided among his compatriots in the New York Puerto Rican community and participated in shaping the direction of its organizational structure. Nonetheless, his greatest contribution would be the collection of African and African American documentation relating to the Black experience in the Americas. It would ultimately form the core of the prestigious Schomberg Center in New York City. In time, many kinds of organizations would structure expatriate Puerto Rican communities, particularly in Manhattan and Brooklyn, the port of call for ships traveling from the island.

Despite the widespread geographic dispersal of the diaspora evident by the decade of the 1930s, it was New York that received the greatest number of Puerto Rican migrants, and it was this city that became the center of Puerto Rican culture in the United States. Structured by the institutions and organizations that evolved to nurture growth and survival under harsh socioeconomic conditions, the New York community would also provide the locus for the great period of immigration that followed World War II and the island's industrialization.

In 1940, scarcely 70,000 Puerto Ricans resided in the United States, over 80 percent of whom lived in the New York metropolitan area, but by 1950 these figures rose to over 300,000. Two decades later nearly 1.5 million, excluding a growing number of Puerto Ricans who were third generation, lived in the United States. According to the 1990 census figures, some 2.7 million Puerto Ricans reside in the States today, with another 3.5 million living in Puerto Rico. Population estimates through the year 2000 project the total increase among both sectors to be more than 7 million individuals (Falcon 1993, p. 5).

For Puerto Ricans in New York City and other urban areas, life in the cold barrios or ghettos of the North could be quite harsh, but at least there were prospects for work and for making a living that consisted of more than mere survival. The 1950s, like the 1920s, was a time of prosperity for the United States, and in such times there is always a demand for cheap labor. The cost of working in the United States, however, was quite high. Not only were migrants displaced from family and friends who provided them with emotional and practical support, but they were faced with living in substandard housing, often with limited opportunities available for advancement in spite of their hard work or education. Fleeing from marginal economic conditions on the island, Puerto Rican workers often found themselves equally marginalized in the United States. As urban dwellers relegated to live in the poorest sections of the cities they inhabited, Puerto Ricans sought jobs in blue-collar industries that were themselves in the process of decline and geographic relocation. A predominately poor community, by the decade of the 1970s, Puerto Rican families were disproportionately headed by women who, in turn, labored in low-paying, precarious employment sectors and received inadequate social services. And while education might have provided a vehicle for mobility, it too, failed to meet the needs for many of the community's children. This is not to say that there did not emerge a significant middle class in the Puerto Rican communities, but that on the whole Puerto Ricans continued to battle the poor economic conditions that seem to have always plagued their existence.

Yet as economic indicators continue to paint the canvas of U.S.–Puerto Rican diaspora communities in dismal, borderline hues, less is appreciated about the positive and proactive contributions of the group, particularly in confrontational arenas like community empowerment, education, civil rights, language enrichment, and the expressive and performing arts. Following in the historical paths of earlier community activism and reflecting the struggles of the African American and Chicano communities, Puerto Ricans also claimed their cultural and human rights. By the late 1950s a grassroots leadership, particularly in the New York community that continued to dominate the U.S.–Puerto Rican scene, began to confront social and economic issues through coalition building and the creation of representative organizations. Composed of island-born and second-generation Puerto Ricans, this leadership understood the roots of their oppression and united in attempts to bring about solutions. Unlike the period before World War II, many of these groups projected interests on a broader state and national scale while others continued to work at a regional level. Among the organizations created were the Migration Division (1947), which attempted to ease the devastating effects of the migration experience, and the Puerto Rican Association for Community Affairs (1956), instrumental in the creation of numerous other groups, including the Puerto Rican Forum (1957). The forum was patterned after the NAACP and served as a vehicle for strategic planning in the development of community resources. These were followed by the formation of critical groups that spearheaded programs for the education of Puerto Rican youth (ASPIRA in 1961, and the Puerto Rican Family Institute in 1963). Along with the Puerto Rican Legal Defense and Education Fund, ASPIRA launched a class action suit on behalf of the non-English speaking children in New York City schools against the city's board of education. The result was the Aspira Consent Decree (1974), which guaranteed bilingual education to all who needed it.

Radical politics, and other community-based organizations such as the proliferation of home-town clubs organized under the umbrella of the *Congreso del Pueblo*, fleshed out the proactive structure of the community. Finally, Puerto Rican educators reflected in their own internal organizations the broad parameters of the community's struggles. The teachers who brought about bilingual education and the licensing of bilingual teachers in the public schools, and the activists who demanded and achieved entry into the universities through the creation of departments and programs exemplify but two of the major contributions to U.S. communities. These facets of Puerto Rican history in the United States, including the building of diaspora communities, cannot be underestimated. As we move toward the new millennium, the multiculture of the Puerto Rican people has achieved greater visibility and continues to exert influence on the wider society.

The Transformation of Puerto Rican Culture on the Mainland

As was noted previously, Puerto Rican culture synthesizes several strands into a dynamic whole. The genius of the Puerto Rican people is evident in being able to take different patterns of life and to bind them into one that is simultaneously Taino and non-Taino, Spanish and non-Spanish, African and non-African, American yet something more than American. Adaptation has been the hallmark of the culture, and it has continued in the United States, with enormous impact on the Puerto Rican in the United States and on the mainstream culture itself. Unlike other groups, Puerto Ricans did not consider the United States to be a melting pot into which they might assimilate, but rather as a place where ethnic identity, multiple allegiances, and multilingualism could coexist.

As Puerto Rican barrios—neighborhoods or *colonias*—emerged in New York City and other urban centers, commerce and other formal and informal institutions associated with Puerto Rican culture also developed. Soon after Puerto Ricans established themselves on the mainland, barrios became filled with *botánicas*, stores that sell herbs and religious articles, *bodegas*, which sell not only goods typical of the United States but also tropical produce and ingredients of Puerto Rican/Caribbean cuisine, music shops, and other commercial enterprises that served the needs of the community. Walking down the streets of the barrio one is confronted with sights, smells, and sounds that would not be out of place in San Juan or in a Puerto Rican village. While groups of men sit around improvised tables playing dominoes, a game very popular in Latin America, women tend to the affairs of making *bacalaítos* (cod fish fritters) or *tejidos* (crocheting). The sounds of all these activities mingle with the *salsa* rhythms that blare from store windows and boom-boxes. Music of a different sort, but just as rhythmic, can often be heard coming out of the many storefront Pentecostal churches, a testament to the growing influence of this Protestant sect both on the island and among Puerto Ricans living in the United States. Although Puerto Rican communities are for the most part poor, they vibrate with a rich and diverse culture. Even when they leave the island, Puerto Ricans do not leave their culture behind.

At the same time, it would be impossible for Puerto Rican culture to be untouched by a movement to the States, with its different traditions, socioeconomic conditions, and climate. Thus one finds that some of the basic values of Puerto Rican society have been modified or shifted as migrants adapted to their new situation. The ability to adapt has been the hallmark of Puerto Rican culture, and such adaptational powers continue to transform the lives of members of the Puerto Rican diaspora, consistently presenting a challenge to the mainstream culture of the United States.

For all its valuing of individual rights, the mainstream culture of the United States has often not lived up to its ideal of tolerance. Ethnophobia has, in fact, been a source of difficulty for virtually every immigrant group that has come into the country. The Statue of Liberty may provide a hearty welcome to those entering New York Harbor in search of a better life, but that welcome is soon replaced by ethnic name-calling and discrimination. Because Puerto Ricans are racially mixed, they have, however, been subject to the most persistent and malignant types of American intolerance. This, too, has been an important factor in the development of Puerto Rican culture.[2]

The culture of the Puerto Rican diaspora, and that of the island, is also embedded in long-standing and persistent poverty. The rich sights of the barrio, real as they are, cannot and should not mask the often desperate poverty that is all too common in Puerto Rican barrios; the joyful sounds all too often cannot drown out the sounds of gunfire in communities already ravaged by crime and drugs; the astuteness of the Puerto Rican people in adapting their culture to a new environment should not lead us to ignore patterns of inequity that would, in any event, prevent most individuals from penetrating the mainstream of mainland society. With nearly one-third of Puerto Rican families living below the poverty line, life in Puerto Rican communities remains difficult (Rodriquez 1991, p. 33).

The early Puerto Rican migrants often held an extremely romanticized view of the United States and the opportunities that it offered them. Coming from an island whose economy had been distorted by the forces of colonialism, they saw in the United States a land where all were accepted and where hard work would inevitably be rewarded. Given this spirit of optimism, the pioneers of Puerto Rican life in America hoped to establish stable communities that would allow them to thrive in ways that the poverty of island life made impossible. These pioneers often found hard work, but not the expected reward; what they found was discrimination. Consequently, their dreams of a better future generally foundered and their struggle to form stable communities often failed before their eyes. This loss is vividly recalled in the following passage from a work by Jack Agüeros, in which he recalls the shift between the early days of Spanish Harlem in New York City (albeit seen through the naive eyes of a child who did not understand his parents' struggle) and the more difficult era that followed:

What caused the clean and open world to end? Many things. Into an ancient neighborhood came pouring four to five times more people than it had been designed to hold. Men who came running at the promise of jobs were jobless as [World War II] ended. They were confused. They could not see the economic forces that ruled their lives as they drank beer on the corners, reassuring themselves of good times to come while they were hell-bent toward alcoholism. The sudden surge in numbers caused new resentments, and prejudice was intensified. Some were forced to live in cellars, and were characterized as cave dwellers. Kids came who were confused by their new surroundings; their Puerto Ricanness forced us against a mirror asking, "If they are Puerto Ricans, what are we?" and thus they confused us. In our confusion we were sometimes pathetically reaching out, sometimes pathologically striking out. Gangs. Drugs. Wine. Smoking. Girls. Dances and slow-drag mu-

sic. Mambo. Spics, Spooks, and Wops. Territories, brother gangs, and war councils establishing rules for right of way on blocks and avenues and for seating in the local theater. Pegged pants and zip guns. Slang. (Agüeros 1981, p. 93)

Both the richness and the dangers of Puerto Rican life in the United States are inextricably linked to the difficulties of that life.

The traditions and the adaptations, the strengths and the problems of Puerto Rican life can especially be found in that most central of social institutions, the family. This should not be surprising. The family is among the most central of human institutions and it is very often the institution which people are most apt to fall back on when making difficult transitions. Such is the case in making a move to a foreign land, where the languages and customs are very different from what one is used to. This may, moreover, be especially true in the case of the Puerto Rican culture, where the centrality of the family within the social setting is stressed.

For those growing up in traditional Puerto Rican society, the family stands at the very root of their existence. The Greek philosopher Aristotle observed that were persons without citizenship, without political identity, were in a sense monstrous, lacking something that was essential to their humanity. For the Puerto Rican, much the same can be said of those who are not connected to their family. Such persons are seen as drifting about without a central core to their life, nor the basic identity that can provide security in an often trying world. Because a Puerto Rican tends to relate to someone in this culture not as an individual but as a member of a family, a person without a family may seem to be lacking in something essential. Therefore, others may not know how to relate to or act toward such a person.

For a Puerto Rican, "family" indicates not only a unit consisting of mother, father, and children, as is typical in contemporary U.S. society, but an extended family that embraces grandparents, aunts, uncles, cousins, and others. Extended family members often live close to each other throughout their lives. Very often several generations of family members can be found living together in the same home. In cities, where people may need to come and go because of work or educational opportunities, one can often find that the residents of a particular household may change frequently as circumstances require family members to move about and stay for periods of time with different members of the extended family.

Social relationships tend to be rooted more in the friendships that develop among families than in pure individual relationships. This linkage of family groups into the larger society is made possible in part by an extension of the extended family beyond biological relationships, particularly through the *compadrazgo* (godparent) relationship. Although this relationship has its roots in Catholic, particularly Mediterranean, religious practices, it has received extensive elaboration in Puerto Rican culture. A godparent and his or her godchild are expected to nurture deep emotional ties to one another and the godparent is obliged to help the godchild spiritually and materially when the need arises. This indicates how strong the importance of the family is in traditional Puerto Rican culture. Even when people look outside the family for social resources they integrate those resources within a familial structure.

In the United States, Puerto Ricans found that the family was less central and was often strictly defined in terms of its nuclear components. One student remarked that when he came to the United States the very names of familial relationships, terms such as "uncle" and "aunt," seemed to not quite fit with their Spanish translations. It was not that Spanish lacked precise words for the biological relationships denoted by these terms, but as used in everyday life the *social* reality of the family was more important than the *biological* reality. Rather, terms like *Tío* or *Tía* connoted a complex relationship that had both emotional and practical aspects that went well beyond the English usage. What had been personal in Puerto Rico was perceived as formal in the United States.

As Puerto Ricans moved to the United States in large numbers, many of them attempted to retain at least several aspects of the familial orientation (Rogler and Santana-Cooney 1984, p. 74). Often several members of an extended family would come to the United States and settle in the same city or neighborhood. As in Puerto Rico, they would be available for mutual support in time of need and provide a source of personal security in what were often difficult times. Even today, the extended family remains a central institution of Puerto Rican life both in the States and on the island.

As in other areas of life, however, the poverty of Puerto Ricans in the United States has played a ma-

jor role in the development of family patterns. On the one hand, this poverty has to some extent helped maintain the extended family by forcing community members to pool their resources for survival. But on the other hand, it has placed the family system under great pressure and in some cases brought it to the brink of dissolution. This has been especially manifested by a great increase in female-headed households. By the time of the 1980 census, one-third of Puerto Rican families were headed by women. Currently the rate is approximately one-half (Rodriquez 1991, pp. 33–34). This is a significant change for a culture that is family oriented. Although there is no definitive proof that such a change in familial pattern is causally associated with the increased level of drug use and crime in the Puerto Rican community, it is quite likely that the loss of the male-headed family is a symptom of social disintegration resulting from the poverty and discrimination of the Puerto Rican in the United States.

As Puerto Ricans in the United States went out into the work world, they were faced with social situations quite different from what they had been used to in their homeland. The poverty, social, and economic difficulties that many faced after migration often outstripped the resources available through the extended family. Many Puerto Ricans needed social services that provided health care and help with housing and employment; this necessitated interaction with agencies whose formal bureaucratic nature did not conform to the social assumptions of traditional Puerto Rican life. The formal nature of these agencies, and of much of the culture of the United States, conflicted with the cultural value of personalism, a belief that one should treat others not in terms of their social role but in terms of their individuality. Relationships formed under such a value system are informal, although at times highly ritualized, and are based on knowing who a person is—in essence, what family he or she belongs to. In speaking to someone about a business matter it would be considered quite rude to immediately start in on the business at hand; to do so would be to imply that the other person was of interest merely as an instrument of that particular matter, rather than as a human being. Thus, this type of interaction would not convey appropriate respect toward the other person.

In their interactions with other North Americans, particularly in institutional settings like hospitals and social service agencies, Puerto Ricans were exposed to a very different way of defining social relationships. Rather than persons, they were reduced to "cases." What was important was not so much who they were as human beings, rooted in a family, but their abstract status as citizens enmeshed in a particular plight. Rather than beginning an interaction with the requisite small talk that indicated the personalistic nature of the relationship and mutual respect, the encounter was structured to move straight to the business at hand in a way that often seemed dehumanizing to the client. Faced with precarious living conditions, Puerto Ricans in the United States could seldom avoid such encounters, and having been raised in a very different social climate they often could not help but feel degraded by them. The coldness that was experienced by many Puerto Ricans in the United States was often as much emotional as meteorological.

Here we see how a basic value of Puerto Rican life, one which enriches even those lives that are poor in material goods, can make adaptation more difficult. Social services, for all of their obvious limitations and faults, are designed to provide a safety net for the needy within society. Ideally, in fulfilling that role they should help and encourage their clients to become self-sufficient. Because visits to the social service agencies are so depersonalized and degrading, however, they tend to alienate clients from full participation in society by indicating to them that they do not belong. Although the intent of these institutions may not be to maintain the needy in a culture of poverty, all too often that is exactly what they do.

However, it was not only in their relations with the world outside the family and the barrio that Puerto Ricans had to adapt. This is perhaps best seen in the shifting definition of relationships between men and women. In traditional Puerto Rican society there is a clear distinction between *trabajo de hombre* and *trabajo de mujer*, essentially the domain and activities of the man and the domain and activities of the woman (Rogler and Santana-Cooney 1984). In the modern American family such distinctions also exist in practice, but in theory at least there has emerged an ideal of equality and lack of differentiation. In traditional Puerto Rican society, on the other hand, the male is defined as the head of the family and the woman is asked to accept a more subordinate role that allows her a lesser degree of freedom of action.

Thus, males are expected to embody *machismo* and the female to embody *marianismo*. The cultural roles not only indicate an ideal of behavior to be practiced by men and women but also particular psychological traits that are said to be characteristic of the two sexes. Thus, it would not be enough for a female to be subservient to her husband. The ideal required that she regard him as her superior. This is not, however, to say that marital affection and mutual respect played no role in the family; to treat the male role as formally superior would be a violation of the personalist orientation of the culture. Rather, it is felt that each of the gender roles deserves respect. A husband who brutalized his wife, for example, would be perceived not as acting out the ideal of *machismo*, but violating it. Although *machismo* attributes to males a natural aggressiveness and Don Juanism, it also demands that he provide for and protect his family. His life in the community goes beyond the family but remains rooted to it.

Women, on the other hand, are ideally considered to live almost completely within the confines of the family. Their role is to be a homemaker, bear children, and care for the family. *Marianismo* is the female equivalent of *machismo*. It relates the woman's role to the Virgin Mary, considered to be the model of chasteness, fruitfulness, and faithfulness (Stevens 1973). Thus, women are expected to remain virgins until marriage and to comport themselves with modesty. Moreover, contact with the world outside of the family is supposed to be limited. Even such a household-related duty as shopping for food is often done by the man, rather than having the woman enter into such a public sphere.

This view of gender relations not only conflicted deeply with the assumptions of mainstream American society but also with the realities of life in the United States. Faced with expensive living conditions and poor-paying jobs, it was often impossible for migrant women to avoid working outside the home. Moreover, the traditional skills women had acquired —such as sewing—were often more marketable than those acquired by men, a situation that made it easier for a wife to get a job than her husband. From the point of view of the traditional Puerto Rican culture this led to the awkward situation of the woman becoming the breadwinner for the family. As women gained more economic control over the resources of the family and gained more experience in the outside world, traditional ideals and expectations changed, leading to inevitable social conflict. That women had their horizons expanded to some extent was a positive impact of the new social situation; however, the inability of men to fulfill their culturally appropriate role may at times encourage them to reject their responsibilities, thereby increasing the social problems of the community.

In the case of younger or second-generation women, the changes might be even more radical. Growing up in a new environment where they had greater freedom, many of the old constraints were lost, leading to behavior that in the island setting would have been construed as beyond the pale of respectability. Dating, wearing what would be considered to be immodest dress, and engaging in premarital sexual activity are violations of the code of *marianismo*, and although these behaviors certainly were not unknown on the island, their prevalence on the mainland provides clear evidence of shifting social values and conditions.

These shifts are rarely, however, without hesitation. Even members of the third generation in the United States may feel ambivalent about their transformed life. In talking to a young Puerto Rican middle-class couple who were born and, except for short periods, lived their entire lives in the United States, one could sense their uneasiness about their new life and all of the opportunities it had brought them. Both members of this couple had well-paying jobs and they had deferred having children until the woman could reach a certain level of advancement in her career. She felt that if she took time off to have children too early it would sidetrack her advancement so severely that she would not ever be able to make up for the lost time. This decision was, however, made with great anxiety. Both members of the couple, but most particularly the husband, noted that there was a feeling that a mistake was being made, that they were foregoing what was essential to their lives for false values and a more comfortable life. Moreover, both of their families expressed dissatisfaction with the decision and they found themselves caught between the expectations of those they most loved and their own ambitions. They were, as the husband noted, living somewhere in the ocean between New York and San Juan.

The adjustment of Puerto Rican culture to the mainland is thus fraught with anxiety and hesitancy, but it also has been a process that has, as in the Puerto Rican past, elicited great cultural creativity in any number of arenas. Even the language spoken by Puerto Ricans within their families and communities

has come to embody the middle ground occupied by the migrants and their ingenuity in finding ways to occupy that ground. Rather than simply retaining Spanish in their home life, or adopting English to it, Puerto Ricans, and to some extent other Latinos, have developed a mode of communication (called code switching) that shifts back and forth between the languages, often in a single sentence, as if the speakers were trying to express two very different views of the world in a coherent manner. Moreover, even particular words in "Spanglish," as the new language has come to be called, may combine English and Spanish forms, as if the very objects and emotions that were being referred to had acquired fused identities that cannot be expressed in a word in a single language. While some may regard these forms as degraded versions of two "pure" languages, Spanglish may also be interpreted as a creative and logical response of a group that has come to live within American society but maintains a unique identity and a strong connection to a homeland culture.

The positive nature of this and other cultural adaptations can be found, for example, in their artistic expression. Just as much modern literary prose and poetry seeks to go beyond the expressive resources of ordinary language to express changes of identity, the syncretic Puerto Rican ethos is well expressed by poets, other artists, and writers who are exploring the multiple expressive resources of their new life. Yet even here we can feel the ambivalence of their situation, as in the following poem by the Nuyorican poet Tato LaViera (1992):

> i want to go back to puerto rico
> but i wonder if my kink could live
> in ponce, mayaguez and carolina
> *tengo las venas aculturadas*
> *escribo* in spanglish
> abraham in *español*
> abraham in english
> *tato* in spanish
> "taro" in english
> *tonto* in both languages

The world of Puerto Rico, with its unique ways and physical and emotional warmth, seems to many to be very inviting. Yet it seems to Laviera, as it appears to many Puerto Ricans, that they would be as truly at home on the island as they are in the United States. Just as the native Tainos, the Spanish settlers, and the enslaved Africans merged to form a new identity, so it seems the Puerto Ricans in the United States are adapting to deal with their new circumstances.

The Future

For those that may still hold a vision of the United States as a melting pot that takes in people of many cultures and eventually transforms them into members of a homogenous mainstream culture, the duality of Puerto Rican lives may seem to be an aspect of their recent arrival in the United States. As the generations moved forward, it was expected that Puerto Ricans would shed their identity and become just "Americans." That this has not yet happened, it might be thought, is a result of the discrimination to which Puerto Ricans have been subjected. It has also been stated that perhaps once their socioeconomic condition improves they will move quickly to the mainstream. As is claimed to be the case with other ethnic groups, ethnic identity may become something that is assumed, if at all, merely on special occasions.

It is uncertain whether this model of assimilation is accurate even in relation to those immigrant groups of the late nineteenth and early twentieth centuries; it is even more questionable to expect it to apply to more recent immigrant groups. Immigration in the contemporary world is different from that which occurred in the pre–World War I era. To try to understand today's immigrants in light of earlier experiences may be a mistake. Times have changed, and the nature and meaning of patterns of movement from one country to another have changed with them. The availability of relatively cheap and rapid air transportation allows immigrants to travel more freely to and from their original homelands, making both the physical and cultural breaks with the past less radical and less severe. Moreover, changes in communications technology have made contact with the Old World and its culture easier to maintain. Recent immigrants maintain contact with those left behind; in the city they often have radio and television programs available in their own language. When the earlier immigrants came to the United States, contact with the country of origin was infrequent. This may

have encouraged a relatively rapid adoption of mainstream attitudes and values. With the ongoing contact that recent immigrants have with their countries of origin, the patterns of becoming "American" will greatly differ.

Moreover, in the case of Puerto Ricans, both geographical proximity and their status as citizens of the United States makes residency in both countries much more practical than for earlier groups. Members of the pioneer generation often expressed a belief that their period of residence in the United States would be merely temporary; they would spend some time making money in the United States, but would eventually return to their homeland. Very few, however, actually did return. Particularly since the 1980s, there has been considerable return migration to the island. Even more important is the fact that a pattern of fluid back and forth movement between Puerto Rico and the United States has emerged; no longer, it seems, is it necessary to choose a single country of residence or a single ethnosocial identity.

It is not, however, only the situation of the immigrants that has changed. Economic conditions in the United States and the world have changed rapidly, and this could not fail to affect the immigrant experience. At the turn of the century, the United States was still changing from a nation in which vast agricultural resources played the most important role in economic development to an industrial nation. With a rapidly expanding economy there was considerable room for energetic immigrants to begin working at menial jobs, and through hard work and luck, they gradually climbed higher on the economic and social scale. For these immigrants, the United States may have really seemed like the land of unlimited opportunity that publicists claimed it to be. Today, however, the United States is a powerful, but also mature, economic power that is facing increased economic competition from other nations of the world. Economic growth has been relatively stagnant in recent years and the opportunities for economic advancement somewhat limited, even for longtime residents. While many recent immigrants have achieved success in this difficult economic climate, confidence in one's ability to begin at the bottom and work one's way upward might be questionable. This, too, can be expected to make the current immigrant experience different from its Ellis Island counterpart.

Conclusion

Puerto Rican culture is itself a syncretic product that was created in response to social conditions generated by centuries of imperialism; as such, Puerto Ricans have long participated in the New World phenomena of forging a national identity out of diverse elements. Seen in this light, the Puerto Rican adaptation to life in the United States might be seen as one more manifestation of the creative response of the Puerto Rican people to conditions of adversity; that is, it may be merely a stage in the further development of a culture. Nevertheless, whether strolling down the streets of Sunset Park or Old San Juan, the *adoquines* remain a symbol of the strength and continuity of Puerto Rican culture and community.

Notes

1. The *adoquines* were removed in the summer/fall of 1992 and all that is left is about twelve stones of the sidewalk in memory. The unofficial story tells that they were sold for $50,000 and taken back to be placed in Old San Juan. See The Brooklyn Historical Society, *Brooklyn's Hispanic Communities* (Brooklyn, N.Y.: The Brooklyn Historical Society, 1989), p. 6.

2. For a particularly good analysis of the impact of race on Puerto Rican life see Rodriguez 1991, pp. 49–84.

8

Family, Economy, and the State: A Legacy of Struggle for Chinese American Women

ESTHER NGAN-LING CHOW

Migration studies have traditionally focused on male migrants and emphasized racial and ethnic factors in migration adaptation and experience. Despite the prominent presence of women migrants in the United States since the colonial years, their experiences have been largely neglected (Houstoun, Kramer, and Barrett 1984; Pedraza 1991). Particularly for women of color, this omission has led to an incomplete and distorted understanding of the causes, processes, and consequences of migration. Unlike their middle- and upper-class White counterparts, immigrant women of color have always supported their families not only as unpaid workers in the household, but also as helpers in ethnic enterprises and as paid laborers in the market economy. For these women, the public sphere of work and the private one of family life have been closely interconnected.

Structural forces of race, class, and gender combine in an overarching system of domination embedded in the social institutions of the state, the economy, and the family. This chapter examines how structural elements have shaped the immigration process and experience of Chinese women and men over time. First, I review a prevalent model for studying the relations between women's work and their family lives at different historical junctures. I then propose a reformulation of this model that is more consistent with the experiences of Chinese American women and men. I explore how social institutions have influenced their work and family, as Chinese Americans have struggled to overcome the challenges of racism, classism, and sexism in the United States.

Women, the Economy, the State, and the Family

The modern family that emerged at the end of the eighteenth and beginning of the nineteenth century in the United States was qualitatively different from the household of colonial times (Hareven 1976, Wells 1991). The arrival of industrialization changed the

The author wishes to thank Silvia Pedraza, Rubén G. Rumbaut, Morrison Wong, Elaine Stahl Leo, and Patricia Masters for their constructive comments and suggestions. Susan Barlow McMillan's research assistance is also gratefully acknowledged.

mode and location of production and altered family patterns over time. A new ideology accompanied this economic transformation: the "cult of true womanhood" glorified the woman's domestic role in the home, where she performed her "sacred calling," service to her family (Andersen 1993). Emphasizing demographic and economic factors, historians Tilly and Scott (1978) describe how women's labor has always been related to family economy and how women's work in England and France from 1700 to 1960 passed through three general phases: family-based economy, family-wage economy, and family-consumer economy. Their work has been extended to apply to women's work and family life in the United States (Andersen 1993, Baca Zinn and Eitzen 1993).

In the family-based economy of colonial America (roughly from 1700 to the onset of industrialization), the household was the basic economic unit. Households were relatively small, consisting of husband, wife, children, and servants (slaves or free laborers). Little distinction was made between economic and domestic life because household members produced the necessities that the family consumed. Families were not private, but rather domains of work and living closely linked to the larger society. In a division of labor based on age and sex, the work of women, men, and children was interdependent in both urban and rural settings. In addition to directly participating in agricultural production and domestic labor, women supervised children and laborers. Despite their substantial contributions, however, women remained subordinate to men, their social status based on their roles as daughters and then wives.

The family-wage economy evolved as goods and services were increasingly produced outside the household in centralized factories and shops, which generated problems of coordinating women's productive and domestic activities. As industrialization and urbanization proceeded, more and more workers earned a living outside the home, and families were supported primarily by the wages of male breadwinners. The individual's worth was measured in terms of wages earned. Families turned inward, specializing in providing domestic services and offering intimacy and emotional support. The functions of the family narrowed to procreation, social class maintenance, socialization, and consumption. Some women assumed dual roles, working as paid laborers and unpaid homemakers. But male control of

the wage-labor system coupled with the capitalist drive to maximize profits resulted in women and children working for low wages and holding insecure positions as a reserve labor pool. Responsible for child-rearing and household tasks with few other options, women became dependent financially on men, and their domestic labor with little exchange value in the economy was devalued.

The family-consumer economy, a late nineteenth and twentieth century phenomenon, is the latest phase in which women have steadily increased their labor force participation (except for a brief interruption following World War II). Many of them have been drawn into tertiary employment and specialized more in child-rearing and managing the family's consumer activities. The majority of these women have remained in occupations highly segregated by gender and race. Public institutions have assumed many family functions—from early childhood socialization in the schools to care of the elderly in publicly subsidized housing. Technological advances, increasing occupational differentiation, mass production, and globalization of the economy have led to a growing availability of consumer goods. Increasingly, women have come to be defined primarily as consumers, even though many women contribute significantly to the financial support of their families, with large numbers of divorced women and single mothers heading households. All women continue to do the "lioness's share" of domestic labor.

Drawing from rich historical documents, Tilly and Scott's (1978) model is useful in four major respects. First, it represents one of the few early efforts to focus on women's lives by analyzing how demographic and economic changes interact with the family and the labor force to produce change in women's positions. Second, it provides an analytical framework that is dynamic rather static, linking the worlds of work and family in three main stages that have shaped women's roles and experiences historically. Third, it broadens the meanings of work to include both wage labor outside of the family and unpaid labor in the household. Finally, it documents that families hold shared values, including a collective commitment to economic survival, and demonstrates how family needs determine when women work.

Tilly and Scott's model is relevant to understanding the work and family lives of early, predominantly working-class Chinese Americans as well as other immigrant groups (Baca Zinn and Eitzen 1993, Jones 1985), and the separate public and private spheres model is inconsistent with the experiences of Chinese American women (Chow and Berheide 1988). Economic necessity required some to engage in production outside the home to support their family. These women also worked as unpaid laborers in the home and in small family enterprises, exemplifying how social reproduction and production, family and work lives, intertwine.

Using Tilly and Scott's (1978) model to study the work and family life of Chinese American women is, however, problematic; the model requires some modifications before it can be directly applied. First, while broadly applicable, the model is primarily based on the experiences of women in working- and middle-class families of European background. Industrial processes affected different groups of people in different ways at different historical periods. Second, the interlocking of race, ethnicity, gender, and class affects the ways in which resources, prestige, and power are allocated, the kinds of work performed, and the types of families organized. Chinese American women historically faced multiple forms of oppression, including not only economic exploitation and sexism, but also racial discrimination. The dominant Western "cult of domesticity and true womanhood" buttressed traditional Chinese gender-role ideology. Both ideology and culture interact with gendered social institutions and perpetuate Chinese women's subordination and gender inequality in the United States.

Third, determinants of work and family linkages go beyond the demographic and economic factors identified by Tilly and Scott (1978). The political economy within the United States and its international politics affect its legal system and immigration policies. Initially, Chinese women were almost entirely excluded from immigration, a policy that lasted for six decades. Through its immigration policies, the state regulated Chinese women's as well as men's lives to discourage family formation, to separate families, and to control Chinese population growth in this country. The prolonged effects of the state as a public patriarchy need to be taken into account in the historical analysis.

Fourth, various family forms emerged, some of which are race specific and class relevant. The work and family lives of early Chinese Americans were quite different from those of White nuclear families.

For more than half a century, their marriage and family lives were severely affected by split households divided by the Pacific Ocean. The result was the formation of bachelor communities offering mutual aid critical for survival. For Chinese immigrant men, and later women, families, and extended kin or quasi-kin, communities provided much-needed support in coping with hardships, discrimination, and exclusion from the larger society (Li 1977).

Fifth, rigid periodization of the three stages of family economies is not directly applicable to the work and family transformation of Chinese immigrants in the United States. Reexamination of the stages of family economy grounded in historical evidence reveals a more appropriate periodization for studying the work and family lives of Chinese American women. Four historical periods instead of three more accurately characterize their experience: community-based economy, family-based economy, family-wage economy, and wage work and consumption economy. While these periods are conceptualized as having followed each other linearly, in actuality some characteristics of an earlier period often continued as a new one emerged. Nevertheless, a distinct household and work pattern typified each stage of the Chinese immigration experience.

The Community-based Economy, 1848–1882

In the mid-1850s, responding to expansion of the nation's capitalist economy and rapid growth of the frontier, Chinese were recruited as low-wage laborers by employers in the Western states and in Hawaii. These recruitment efforts coincided with economic crises, political upheavals, natural disasters, and social turmoil in China that compelled many to emigrate seeking a better life.[1]

Attracted by the 1848 California gold rush and the thriving economy in the West, Chinese immigrant males initially came as sojourners to earn money for themselves and their family members who remained in China (Siu 1952). During the open immigration period between 1850 and 1882, over 300,000 left southern China to work in California and the West (Lyman 1974). Many were imported here as coolies (means "bitter labor" in Chinese), cheap contract laborers, and strikebreakers to replace White workers. For fi-

Table 1   Chinese Population in the United States by Sex and Sex Ratio, 1860–1990

Year	Total	Male	Female	Males per 100 Females
1860	34,933	33,149	1,784	1,858.1
1870	63,199	58,633	4,566	1,284.1
1880	105,465	100,686	4,779	2,106.8
1890	107,488	103,620	3,868	2,678.9
1900	89,863	85,341	4,522	1,887.2
1910	71,531	66,856	4,675	1,430.1
1920	61,639	53,891	7,748	695.5
1930	74,954	59,802	15,152	394.7
1940	77,504	57,389	20,115	285.3
1950	117,629	77,008	40,621	189.6
1960	237,292	135,549	101,743	133.2
1970	431,583	226,733	204,850	110.7
1980	812,178	410,936	401,242	102.4
1990	1,645,472	821,124	824,348	99.6

Source: U.S. Bureau of the Census. 1960. *Historical Statistics of the United States: Colonial Times to 1957*; and U.S. Bureau of the Census, Population Reports for 1963, 1973, 1983, 1991, and 1992.

nancial reasons, many Chinese workers did not bring their families along. Since the workers did not support families here and were not expected to stay, their presumed temporary status gave White employers and landowners added incentive to extract as much profit as they could from this inexpensive labor. Many of these early laborers did eventually return home; there were never more than 108,000 Chinese recorded by the U.S. census at any one time until 1950 (see Table 1). Structural racism made it difficult for the Chinese migrants to form stable families in the United States, forcing them to live in ethnic enclaves and creating a situation in which the young Chinese women who managed to immigrate were sexually exploited.

Exclusion and Absence of Chinese Women

Because the Chinese were viewed as cheap, temporary labor, federal and state government policies tended to favor young, able-bodied men and to exclude young Chinese women. Under patriarchal control in China, Chinese women were also deterred from leaving by family responsibilities, cultural restrictions, costly or risky travel, and fear of physical or sexual assault. Restrictive immigration policies, cultural barriers, hardships in the American West,

and anti-Chinese sentiment all resulted in an extreme imbalance in the sex ratio, a situation that persisted into the twentieth century. In 1852, of the 11,787 Chinese that had come to the United States, only seven were women (Bonacich 1984). The U.S. census records no more than 5,000 Chinese women on the U.S. mainland during the three decades of unrestricted Chinese immigration. Table 1 shows how the skewed sex ratio increased from eighteen men to every woman in 1860 to twenty-one to one in 1890.

Despite the barriers erected to discourage Chinese women from immigrating, small numbers managed to enter the United States during the nineteenth century. Mostly concentrated in the western United States, these women worked as indentured servants, unskilled laborers, slaves, and prostitutes; a few of the more fortunate married Chinese merchants. In the 1860 and 1870 California censuses, most Chinese women were listed as prostitutes; others were listed as housekeepers, laundresses, seamstresses, miners, and railroad workers (Yung 1986).

The lack of marriageable Chinese women created a lucrative market for prostitution that drew many women into its net. Chinese prostitutes were generally portrayed by men, Chinese and non-Chinese, as degraded and cheap sex commodities. They worked under slavelike conditions, received no regular wages, and were subjected to physical and sexual abuses so severe that many were unable to outlive their contracts. Most lacked both the individual and collective means to cope with their miserable circumstances; thus suicide, madness, and violent death were prevalent.

In 1874, the Women's Occidental Board established the Presbyterian Mission Home as a shelter for Chinese girls, *mui jai* (poor girl servants), and prostitutes in San Francisco. With the help of the police, Donaldina Cameron and Margaret Culbertson successfully rescued 3,000 of them over a period of forty years (Yung 1986). The story of Wong Ah So typifies the lives of many rescued prostitutes (Yung 1986, p. 20). Born into a poor Chinese family, Wong Ah So married a Chinese laundryman at the age of nineteen and was deceived into immigrating to America, where she was forced into prostitution. Her father's friend found her and enlisted help from the Presbyterian Mission to rescue her. She was subsequently placed in the mission home and eventually learned both English and weaving to enable her to earn a living.

Public opinion turned against Chinese prostitution (while sparing White prostitution), leading to the enactment of federal and state laws to prevent a vice that was said to spread disease and corrupt the young. As early as the 1850s, San Francisco barred Chinese women intended for prostitution from entering the country. As prostitution declined in the 1870s, an increasing number of Chinese women married merchants and laborers. The 1880 census reports listed 46 percent of Chinese women in California as housekeepers (Hirata 1979). Relatively protected from harassment, physical harm, and possible kidnapping, most Chinese women labored primarily at home, living a sheltered life; some helped in family businesses. The wives of wealthy merchants were more visible in social life within the Chinese community.

The Experience of Early Immigrants

Due to the white majority's low receptivity and even outright hostility toward Chinese Americans, a community-based economy sprang up among the early immigrants. The virtual absence of women and limited family life spawned what were known as "bachelor communities." Survival strategies relied on individual struggle and mutual aid through the informal social networks of kin and friends and, later on, through clan associations. Chinese bachelors were considered desirable workers in farming, mining, fisheries, railroad construction, and manufacturing industries; their White employers could easily move them around and house them cheaply in dormitorylike accommodations. Due to a shortage of White women workers in the West, Chinese men were also recruited to work as domestic, laundry, and food service workers, traditionally women's jobs. A few Chinese immigrants became self-employed, working in mining, tenant farming, truck gardening, fishing, and laundries (Amott and Matthaei 1991).

Eventually, the Chinese congregated in urban Chinatowns. These ethnic enclaves provided daily amenities, job opportunities, and mutual support. Small businesses, such as laundries, import-export enterprises, restaurants, apothecaries, and brothels began to flourish (Mei 1984). Chinese merchants hired Chinese workers at meager wages, contracting them out to White employers. Many single Chinese males lived alone in rented rooms or in crowded lodges with other "bachelors," spending their limited leisure time with fellow countrymen and participating in Chinese festival celebrations. Some expected to earn enough to return to China, while others, particularly those who were married, regularly sent remittances back home to support their families and villages in China.

Formed as a self-defense mechanism, Chinatowns enabled the immigrants to insulate themselves from the racial conflicts and hostility they experienced in the larger society and also kept the White majority from closely scrutinizing Chinese community life. Led by Chinese merchants, ethnic associations, trade guilds, and political parties were established by the Chinese laborers based on common interests. Among the most important associations, tongs (from tang, meaning "hall"), tight-knit fraternal organizations, functioned as powerful ruling bodies and judicial groups, representing and defending the Chinese community.[2] As centers of work and communal life for early Chinese Americans, tongs settled disputes; dispensed credit and loans; provided support to the needy; cared for the young, the sick, and the old; and buried the dead (Chan 1991, Chow 1995).

During the economic recession of the 1870s, industrialists, labor unions, and White workers used Chinese laborers as scapegoats, resulting in the exclusion of Chinese from jobs and social life in the

West. To escape anti-Chinese hostility, many immigrants fled to other parts of the country. Some settled in remote rural areas, engaging in farming and other agricultural production. In some cases, they replaced freed blacks on the Southern plantations (Loewen 1988). Economic discrimination against the Chinese was reinforced by the state at different levels, as immigration policies and other related laws aimed to exploit labor and discourage family formation. At the federal level, the Naturalization Act of 1870 denied Chinese immigrants the rights to citizenship that had been accorded to European immigrants by a 1790 naturalization law.[3] The alien status of the Chinese enabled passage of numerous laws that further limited their legal rights, imposed special taxes on Chinese businesses, and restricted Chinese from entering certain occupations such as teaching, government, and manufacturing jobs.

Ultimately, this sinophobia (anti-Chinese sentiment) culminated in violence and massacre, incited by political and labor leaders who attacked the Chinese in San Francisco, Los Angeles, Denver, Rock Springs, Tacoma, and Seattle (Lyman 1971). This led to what came to be called the "Great Driving Out," a campaign to expel the Chinese from small towns and rural areas. The anti-Chinese movement succeeded in passing the Chinese Exclusion Act in 1882. This federal law prohibited any further immigration by Chinese laborers and their alien-born wives with a few exceptions.[4] These economic and legal exclusions severely restricted Chinese population growth in the United States. They also set the pattern for the social invisibility that Chinese women have experienced and resisted against throughout their settlement history.

Two Tracks: Family-based and Family-Wage Economy, 1882–1960

In U.S. society, one of the most detrimental effects impinging on the lives of Chinese American was the gender bias in the immigration laws and policies that were racially motivated and class instigated, targeting Chinese first and then other Asian Americans. The Chinese Exclusion Act of 1882 prohibited Chinese laborers and their alien-born wives from immigrating, but exempted officials, students, merchants, and their relatives. The act was originally enacted to last for only ten years; however, its provisions were

extended indefinitely in 1892 and were not rescinded until the Magnuson Act of 1943. The Exclusion Act of 1924 extended legal restrictions to other Asians. The insidious Cable Act of 1922 prohibited U.S.-born women of Asian ancestry from marrying aliens ineligible for citizenship (such as men from China); if such women violated the law, they lost their own citizenship.[5] Other miscegenation laws also forbade Chinese laborers as noncitizens in most western states from marrying Whites. These acts had a profound impact on the subsequent settlement history of Chinese Americans, making it distinctly different from that of European immigrants. Political and legal exclusion truncated development of Chinese families at a critical point when their working-age, able-bodied men might have married or sent for their wives and children.

These legal exclusions resulted in five major coping strategies for Chinese male laborers. The first was to return home permanently, which at least half of the immigrants did once they had earned enough money and fulfilled their contracts. Otherwise, if single, Chinese men could stay and live with other bachelors. Some of them became isolated and disintegrated personally, indulging in gambling, drugs, and other vice. Others with U.S. citizenship returned to China to marry and came back to the United States as wage laborers. If married before arrival in the United States, they maintained separate households and returned to China for occasional visits if possible. Finally, a few Chinese laborers were able to marry before the law enactments and worked as wage earners to support their families. This led to the development of a series of long-term sojourners' strategies that reinforced the Chinese man's position as the wage laborer to support a family, whether the household was split or not.

The Chinese economy in the United States had coalesced by 1900 and developed two major occupational patterns (Light 1972). One group of Chinese emerged as self-employed small producers, while another was employed in the labor market. By 1920, the former group had become dominant, because until 1940 employment discrimination greatly diminished job opportunities for Chinese in the general labor market. The Chinese who were self-employed had established family lives and had developed a family-based economy in which production by family members was targeted to a limited market and for household consumption.[6] The Chinese workers who depended on wages to support themselves and their families, be the latter in the United States or China, developed the family-wage economy pattern. The family-based economy overlapped with the emerging family-wage economy from 1882 to the 1960s. Beginning in the 1940s, Chinese men were increasingly incorporated into the paid labor market, making the family-wage economy pattern increasingly predominant. However, the family-based economy remained strong, with many Chinese relying on small businesses for household survival even after World War II.

Family-based Economy

After the Chinese Exclusion Act became effective, the number of Chinese females registered by the census dropped to just 3,868 in 1890, yielding a skewed sex ratio of twenty-seven to one, an all-time extreme (see Table 1). The Chinese population dwindled from 107,000 in 1890 to 61,000 in 1920 and grew only slightly thereafter until the repressive immigration laws were repealed in 1943. Wives of merchants were allowed to immigrate because of their husbands' status as members of an "exempted class."

It is worth noting that the experience of Chinese in Hawaii differed from those on the mainland. In Hawaii, Chinese laborers had a better chance of establishing families because employers encouraged immigrant workers to bring wives and children, who were seen as a potential labor source as well as a stabilizing element. After Hawaii was annexed to the United States in 1898, the Chinese Exclusion Act applied to the Chinese who were there (Bonacich 1984; Yung 1986), but its effects were less deleterious to family formation because Hawaiian law permitted interracial marriage. Despite myriad attempts to restrict the entry of Chinese women into the United States and thus curtail family formation, settlement history from the late nineteenth to the early twentieth century reflects the growing presence of Chinese American families. During this time, several significant demographic changes occurred, as Table 2 shows.

First, between 1900 and 1940, the number of foreign-born Chinese in the United States gradually declined. After nine decades of Chinese settlement, by 1940, native-born Chinese outnumbered those born in China who had immigrated here. Second, the number of children increased slowly but steadily

over time. In 1890, only 500 children were recorded (constituting less than 1 percent of the population), but by 1900, 9,000 children (comprising 3.4 percent of the total Chinese population) were included in census reports (Chan 1991, p. 107). From 1920 to 1930, the proportion of children ages fourteen or younger nearly doubled, increasing from 12 percent to 20 percent. Third, the proportion of single Chinese men diminished from 69 percent in 1890 to about 37 percent in 1990, while the percentage of those who married steadily grew from 26.1 percent in 1890 to 58.5 percent in 1990. Fourth, Chinese women consistently married in greater numbers (ranging from 59 percent to 69 percent) than their male counterparts at all times. Finally, the number of Chinese families grew steadily and the Chinese population increased substantially beginning in 1950 due to changes in the immigration laws, reflecting a strengthening of family life as the second and third generations of Chinese Americans were born.

The family-based economy of first-generation Chinese immigrant households was similar to that of colonial America. Within the small, nuclear household, with help sometimes from boarders, kin, or fictive kin, work and family lives intermingled. Economic production and domestic reproduction activities were carried out at or near the home. The division of labor was based on gender and age, with task responsibilities assigned consistent with role prescriptions, capacity, and experience within a family that was headed by males.

Unfamiliar with the language and culture, Chinese women encountered hardships and discrimination as they struggled to build communities and integrate their work and family lives. Their rural origins prepared many to adapt readily to economic adversities by working hard to bring in outside income, as well as to carry out their domestic responsibilities. The "cult of true womanhood" as a cultural ideal applied only to a privileged few Chinese families. Wives of wealthy merchants, for example, were cared for by *mui jai*, enabling these wives to socialize with others of their own class and enjoy a comfortable life (Mei 1984, Yung 1986).

Distinctions existed between urban and rural Chinese immigrant families. In urban areas, family life was highly privatized. Most Chinese women labored side by side with their men, doing sewing and needlework, cooking, washing, rolling cigars, cleaning, and making slippers and brooms while simulta-

Table 2   Chinese Population in the United States By Nativity, Young Children age 14 and under, Marital Status, and Labor Force Participation

Year	Foreign-born (%)	Children Age 14 or Less (%)	Marital Status				Labor Force Participation Rate of Chinese Women
			Single		Married		
			Females	Males	Females	Males	
1870	99.8						
1880	99.0						
1890	99.3		32.3	69.0	63.5	26.1	
1900	90.0	3.4	24.3	58.6	67.3	38.0	
1910	79.1	7.0[1]	23.0	53.3	68.2	41.1	
1920	69.9	12.0	21.8	46.4	69.1	49.7	14.1
1930	58.8	20.4	23.3	48.7	68.2	46.3	16.0
1940	48.1	21.2	33.4	43.9	57.4	53.0	25.0
1950	47.0	23.3	28.2	35.3	64.6	58.1	30.7
1960	39.5	33.0	22.0	35.1	68.9	57.8	44.2
1970	47.1	26.6	27.7	36.7	61.5	58.0	49.5
1980	63.3	21.1	28.4	37.2	59.3	57.7	58.3
1990	69.3	19.4	28.5	36.7	59.2	58.5	59.2

[1]Based on figures for three states only—California, Oregon, and Washington. No comparable figures for the U.S. total.

Source: U.S. Bureau of the Census, Population Reports for the 1872, 1883, 1895, 1902, 1913, 1922, 1933, 1943, 1953, 1960, 1963, 1973, 1983, 1992, 1993c, and 1993d.

neously caring for their children and families (Yung 1986). Their lives were isolated; they rarely saw other women, and their closest bonds were to their children (Chan 1991). The birth of children created burdens for Chinese women, adding child-rearing responsibilities to existing household chores and production activities. Within the family, mothers were the disciplinarians who emphasized the importance of education for their children's future success.

In comparison to their urban sisters, Chinese women living in rural areas enjoyed greater freedom of movement, closer connections to the larger society, and less social constraint. Christian missionaries interested in introducing religion and American ways of life made house visits and taught Chinese women and children English and the Bible. Nevertheless, many rural women were subjected to economic hardship and had few comforts. They lived in segregated Chinese communities where they observed their own customs and maintained their cultural heritage, partly in response to racial prejudice (Yung 1986). In addition to shouldering their family responsibilities and working in the family business, some Chinese women also supplemented family in-

come by cooking for others, taking in laundry and sewing, gardening and canning produce, tending livestock, and taking in boarders. Some women exchanged food with their white neighbors and helped them out in emergencies. Other Chinese were successful in truck gardening and tenant farming and were able to buy enough land to establish ranches and farms (Chan 1984).

Job discrimination, limited capital, and scarce opportunities in the general labor market led the Chinese to work as small producers in ethnic enterprises, alone or in partnership (Amott and Matthaei 1991; Glenn 1983). The Great Driving Out forced many Chinese to lose their land and businesses and to move from rural areas to the Chinatowns. The alien land laws of 1913 made it illegal for the Chinese to purchase land for farming on their own. At the same time, the shortage of domestic help created an occupational niche for the Chinese, who went on to establish laundries and later grocery stores, restaurants, import-export outlets, and other retail stores.

The growth of ethnic enterprises reflected the Chinese Americans' strong belief in the Horatio Alger tradition that business ownership is a "symbol of opportunity," a classic avenue of upward mobility for the disadvantaged (Light 1972). By 1920, more than 50 percent of Chinese Americans were employed or self-employed in restaurants, laundries, and other ethnic enterprises. Because Chinese Americans retained Chinese cultural preferences in economic consumption (for example, for Chinese vegatables, teas, and noodles), their ethnic enterprises were small and highly labor intensive, creating jobs and supplying ethnic commodities and services unavailable in the larger society. Chinese restaurants attracted the patronage of the White working class because they served inexpensive food and appetizing meals. In the Deep South, Chinese catered to the needs of a Black clientele whom White merchants refused to serve (Loewen 1988).

By consolidating their limited economic resources, using available labor sources, and integrating household members and kin, small producers and ethnic enterprises firmly established the family-based economy among Chinese Americans. Once the Chinese became merchants, they could also bring wives and children over, and their newly arrived families were the main labor source for home-based economic production.

Family-Wage Economy

Patriarchy, embedded in public institutions such as the state and economy and in private institutions such as the family, was the cornerstone of the family-wage economy. Under patriarchal rule, men were socially defined as the household heads and the major breadwinners, expected to earn a "family wage" to support the family. The patriarchal system thus relegated women to a subordinate status, a social location further reinforced by a gender-role ideology that devalued women's contributions. Their gendered division of labor inside and outside the home generated hidden value, paid or unpaid, which benefited the economy and the family. This situation, applying to women in general also formed the material and ideological conditions for the development of a family-wage economy for Chinese American women. However, the latter lived under a "double patriarchy," as the Chinese and the American modes of patriarchy combined historically to produce a social reality that was further compounded by the racism and classism that existed in U.S. society. The simultaneity of race, class, and gender created structural circum-

stances (that is, commuter marriages and split households) that were not common among women in general.

First of all, the Chinese Exclusion Act of 1882 institutionalized what is called *Gum-Shan-Poo,* or "Gold Mountain Lady," referring to those women in China who maintained transoceanic commuter marriages and split households while their husbands worked in the United States. During the open immigration period from 1850 to 1882, a vast majority of young Chinese laborers came to the United States alone on a credit ticket system that obligated them to work for a fixed term, generally seven years, to pay for their passage to the United States (Glenn 1983; Mei 1984). To ensure family continuity and loyalty to kin, some young men (who sometimes borrowed money from their relatives for the U.S. trips) were married and fathered a child before leaving. It was estimated that over half of Chinese "bird of passage" laborers were married and left their wives and families behind in China (Coolidge 1909). Others emigrated as single men, returned to their homeland for an arranged marriage and conception of a (preferably male) child before they came back to "the Gold Mountain," as the United States was called.

By the 1920s and 1930s, the split household family was firmly established as an interdependent and cooperative unit, performing the major functions of the family in separate geographic locations. The family's production function was primarily carried out by the male laborers in the United States and other procreation, socialization, and consumption functions were performed by the split household in China. Thus the household division of labor was globalized. Chinese men were expected to send money to support not only their families but also extended kin in the village. With the exception of a few who accumulated enough capital to establish small enterprises as either sole owners or partners, Chinese male sojourners were employed as wage workers in ethnic enterprises or the general labor market. Those who saved enough money eventually could afford to return to China to stay. In the absence of their husbands, Chinese women in the homeland were controlled by the patriarchal family, which required that they practice the "Three Obediences" (first to her father, then to her husband, and, if widowed, to her eldest son) and that they acquire the "Four Virtues" (chastity and obedience, shyness, a pleasing manner, and domestic skills). Wives usually lived with and under the supervision of their in-laws, raised children, did domestic work, and sometimes even engaged in subsistence production to support their husbands' family. The mother-child relationship was particularly strong, while that between father and child was formal and remote.

Limited family unification was possible because loopholes in the Chinese Exclusion Act permitted the relatives of citizens to immigrate to the United States. Chinese men with U.S. citizenship could return to China, marry, and father children, after which their sons, though not their daughters, were eligible for entry here. It became a common practice for these sojourners to report the birth of sons after they visited China, thus creating a slot for them to come later.[7] Even if these men did not have children of their own, they could claim fictitious sons and sell the immigration rights to "paper sons" of other Chinese at high prices. The "paper son" then assumed the name and identity of the alleged son. When coming of age, boys often joined their fathers or even grandfathers at work in the United States, forwarding remittances home generation after generation to wives and mothers who never left China.

In spite of the anti-Chinese sentiment and discriminatory laws, however, some *Gum-Shan-Poo* did manage to circumvent the laws (for example, by falsifying marriage) and come to "the Gold Mountain." Most who arrived between 1910 and 1940 first went through the painful experience of a long detention at the Angel Island Immigration Station in San Francisco Bay before they were released to join their husbands. During the depression, for family survival, most Chinese women worked extra-long hours at menial jobs while also raising children. Nevertheless, they enjoyed their newfound freedom from the social constraints of the traditional family in China, being able to live with their husbands and investing hopes in their children.

Most second-generation Chinese Americans who grew up in the United States faced the contradictions of following Chinese cultural traditions at home while adapting to American ways of life. According to Yung (1986, p. 40), Chinese American women of this period, influenced by Western ideas of gender equality, showed a greater degree of independence and social consciousness than their predecessors and participated more fully in the larger society. Rejecting the traditional practice of foot-binding, they began to unbind their feet, appear often in public places, work

outside the home to subsidize family income, educate their children, join ethnic associations, and form women's groups.

Until World War II, intergenerational mobility of Chinese Americans was blocked by prejudice and discrimination based on race, class, gender, and nationality in the schools, the economy, the legal system, and the state. Both the 1943 repeal of the Immigration and Nationality Act that had set a quota of 105 immigrants per year and the 1945 War Brides Act allowing foreign wives of U.S. servicemen to immigrate permitted family reunification to begin. In the late 1940s and early 1950s, almost all immigrants from China were women and children (Yung 1986). The new arrivals brought fresh vitality to their families and communities.

Meanwhile, the war economy and the postindustrialization in the United States created new opportunities for incorporating Chinese Americans into the larger wage labor market. This market had been highly segregated by gender, race, ethnicity, and class. Prior to World War II, the Chinese male labor force was primarily composed of service and operative workers and small businessmen, while Chinese women engaged in domestic service and manufacturing jobs. During the war, increasing numbers of Chinese men and women served in either the military or the paid labor market in response to the labor shortage created by the war economy. By 1950, Chinese men were employed as service workers, managers, officials, proprietors (except farm), operatives, clerks, and sales workers (Wu 1980; Yuan 1969). The Chinese female labor force grew slowly, from 10.4 percent in 1900 to 16.3 percent in 1930, and then increased markedly to 22.3 percent in 1940 and 30.8 percent in 1950 (Amott and Mathaei 1991; Yung 1986). By 1950, for the first time, Chinese women's labor force participation rate, 30.8 percent, was higher than that of their White female counterparts, 28.1 percent.

Chinese American women's occupations, though still highly segregated by gender, shifted to clerical, sales, and professional and technical jobs, but one fifth of these women were still employed as operative workers in the 1940s and 1950s (Wu 1980, pp. 23–25). The most common manufacturing jobs held by Chinese women were in garment work and canning, where unions had made breakthroughs in the all-White labor movement. A few Chinese women entered the professions as teachers, journalists, lawyers, social workers, nurses, and physicians. Their occupational distribution started to resemble that for all U.S. females. However, except in Hawaii, Chinese women faced discrimination that slowed their access to white-collar jobs in comparison to White women (Amott and Mathaei 1991). The new occupational opportunities set the stage for the improvement and challenges of Chinese American women's emerging economic roles.

Wage Work and the Family Consumption Economy, 1960 to Present

The fourth stage is characterized by demographic changes leading to new household formation, an increase in Chinese American women's participation in the labor force, and the rise of the family as a consumer unit which had acceded many family functions to other social institutions. These changes came about gradually with the repeal of repressive immigration policies, the improved political relations between China and the United States, the labor demand created by U.S. economic globalization, social movements, and the sociopolitical climate of the postwar era.

With the easing of immigration restrictions following the war, the Chinese American population grew nearly fourfold, from 236,080 in 1960 to 812,178 in 1980, and then doubled from 1980 to 1990 to approximately 1.6 million. It also achieved a balanced sex ratio (102.4 in 1980 and 99.6 in 1990) and became more diverse in terms of social class, generation, age distribution, and national origin (U.S. Bureau of the Census 1992c and 1993d). In 1990, 69 percent of the Chinese population was foreign-born, originating primarily from different parts of mainland China, Taiwan, and Hong Kong (U.S. Bureau of the Census 1993d). Chinese students and refugees from communism were allowed to immigrate under several anticommunist legislative acts (for example, The Refugee Relief Act of 1953). More Chinese entered as a result of the 1965 amendment to the Immigration and Nationality Act of 1952, which abolished national origin quotas and permitted relatives of citizens and permanent residents to immigrate. In addition, occupational preference was given to scholar-professionals to meet the labor demands of the expanding technological and service economy of a postindustrial society.

As a result of the Civil Rights and women's movements in the 1960s, discrimination eased somewhat, affirmative action policies benefited racial minorities and women including Chinese American, and a more favorable social climate led to improved opportunities and conditions. Meanwhile, Chinese American women were becoming aware of their rights both as members of a racial minority and as women. As they developed a social consciousness, they became more actively involved in balancing their work, family, and community lives (Chow 1987a).

Since 1960, the extent and nature of Chinese American women's incorporation into the wage economy has changed phenomenally. Chinese American women experienced a marked increase in their labor force participation rate, leaping from 30.7 percent in 1950 to 44.2 percent in 1960 and 59.2 percent in 1990 (U.S. Bureau of the Census 1953, 1993d), a rate which surpassed that of White women and all U.S. women. Economic necessity, often due to the social and economic losses sustained as a result of immigration, prompted many of these women to engage in income generating activities to support their families. Their occupations diversified as they moved into industries outside of Chinatown and began doing white-collar and service work. Once employers recognized that Chinese American women were skillful, reliable, and hardworking, they were subjected to a new stereotypical portrayal as obedient "office wives" (Amott and Mathaei 1991, p. 209). By 1990, 38 percent of Chinese American women did technical, sale, administrative, and clerical work, while 17 percent held professional jobs and 15 percent managerial positions. Nearly one-third of them were employed as service and operative workers (U.S. Bureau of the Census 1993d). However, the labor incorporation of Chinese American women into the wage economy modified rather than undermined patriarchal control over them. Men still reign as the major breadwinners in households with multiple wage earners including women (Chow 1987b).

The household has become a consumer unit rather than an economic one, with production and other functions principally carried on outside the home. Since immigrants came in family groups, typically a couple with their unmarried children, most households remain small and nuclear, though some have in-laws and other kin living with them. According to the 1990 census, the average family size for Chinese Americans was 3.62 persons; 84.2 percent were married-couple families (U.S. Bureau of the Census 1993d). About half of the families had two wage earners, and one-fifth of them had three or more workers who contributed financially to the family. Chinese women often work a double shift of paid labor outside and unpaid labor inside the home. Children or relatives may also work, full or part time, contributing financially to the household.

The work and family life of the Chinese American household varies with social class. The legal criteria for immigration created an influx of two distinct Chinese immigrant groups, characterized by those of the "downtown" and the "uptown" Chinese suggested by Kwong (1987). Most downtown Chinese are immigrant families who support themselves as wage earners in either ethnic enterprises or the paid labor market. Some Chinese who once had split household families were finally able to send for their families and relatives. Many who came from rural areas settled in or near Chinatowns, where they could easily find jobs in ethnic businesses without knowing much English, locate less expensive housing, and live in a familiar cultural milieu. Some who came from urban areas, such as Hong Kong or Taiwan, were accustomed to the Western lifestyle and language, and their adjustments were less dramatic. The arrival of the new immigrants revitalized the Chinatowns that had declined somewhat as second-generation Chinese had entered the general labor market and eventually moved out. The diplomatic normalization of relations between the People's Republic of China and the United States in 1973 also created another spurt of immigrants to replace the first and second waves of Chinese immigrants in Chinatowns. Later, Southeast Asians, especially Vietnamese of Chinese descent, joined in and built ethnic enterprises near Chinatowns.

Chinatown has thus become a way station for some and a final destination for others. It gives a sense of place to impoverished immigrants who work for Chinese American businessmen and other American employers in the secondary labor market. Several studies of the Chinatowns in San Francisco and New York City document that the majority of women residents were employed in garment and other manufacturing industries; in clerical and sales jobs; and in restaurants, along with their spouses, as cooks, waitresses, janitors, and store helpers (Loo

and Ong 1982; Nee and Nee 1974; Zhou 1992). In some cases, women do home-based production, subcontracting as pieceworkers for factories, with children and sometimes kin or fictive kin helping out (Specter 1991). Due to lack of education or skills, low proficiency in English, and limited job experience, many work long hours in dead-end jobs for little pay (Chow 1994; Yung 1986). Family relationships occasionally suffer as some of these women struggle to combine work and family living with meager economic resources available, little emotional support from their spouse, lack of support networks, and too little time and energy for meaningful family communication. Rebellion, juvenile delinquency, and gang violence among youths have sometimes been the results.[8]

The uptown Chinese, on the other hand, consist of better-educated professionals, businessmen, top financial managers, and former government officials who were anticommunist refugees admitted under the McCarran-Walter Act of 1952. This group, mostly Mandarin speaking, is socioeconomically better off than the Cantonese-speaking Chinese immigrants of earlier generations who came primarily from southern China. While "brain drain" has perhaps depleted China, Hong Kong, and Taiwan, it has contributed a high quality labor source to the United States. Other Chinese initially arrived as college students, eventually married, and started their families here afterward. Experiencing better occupational opportunities and less hostile residential segregation, some of them, including women, are employed as professional, technical, managerial, and clerical workers. They live in suburban middle-class neighborhoods with scant connection to the Chinatowns. The American-born Chinese who speak fluent English and are assimilated to a greater extent than their parents' immigrant generation have joined the ranks of this group.

It is the uptown Chinese group that bolsters the generally held stereotypes that Chinese and Asian Americans are a "model minority." Those who have succeeded simply transplanted their educational and occupational resources and achievements from overseas to the United States and strive for upward mobility here. But many others still experience language and cultural barriers, mismatch of skills, licensing hurdles, undervalued education credentials from overseas, and low entry-level jobs. With similar levels of education, and performing jobs comparable to those of whites, many receive lower income returns (Woo 1995). A substantial number of Chinese immigrant women, many of their painful stories untold, have experienced downward mobility, underemployment, and discrimination (Chow 1994; Glenn 1983). For a considerable time after immigrating, they are either unemployed or employed in the secondary market sector doing traditionally female jobs at low pay.[9] The false impression given by the "model minority" image masks the virulent sinophobia, the prevalence of discrimination, and the persistence of poverty downtown Chinese Americans often experience. The 1990 census documents that Chinese have a poverty rate slightly higher than that among all Americans, despite a higher median family income.

Whether Chinese American women belong to the uptown or downtown groups, the main family survival strategy for both husband and wife is to find jobs as wage workers and pool their income to meet necessary family expenditures. As their women's wages become an important source of family income, they not only acquire skills, labor experience, and economic independence but also elevate their social status and increase their influence within the household. Though still subordinated to men, they challenge the traditional gender ideology that a Chinese woman's place is in the home (Chow 1987b). As their families integrate into the larger society, they seek social services, schooling for children, legal advice regarding immigration, and medical help from mainstream social institutions. District and clan associations exert less influence on the new immigrants. Sometimes without immediate kin to give help readily, the families of the uptown and downtown Chinese build their support networks and communities, inside and outside of Chinatown, to preserve their cultural heritage and identity, to develop social ties, and to aid one another when needed.

However, Chinese American women are still subjected to patriarchal control, occupational segregation, underemployment, discrimination, and lower earnings than their male counterparts. Some of them confront sexual and racial stereotypes (for example, that they are exotic sex objects, prostitutes, mail-order brides, obedient wives, and cheap laborers) and ill treatment in the workplace and society, experiences also encountered by other women of color and immigrant women workers (Chow 1994; Dill 1988; Grasmuck and Pessar 1991; Pessar 1984; Sassen-Koob 1984; Segura 1989). With limited access to so-

cioeconomic resources and legal-political power, they often have had little choice but to tolerate sexist and racist attitudes, behavior, and even violence against them (U.S. Commission on Civil Rights 1992).

Conclusion

The historical experience of immigration has transformed the sense of time and place for Chinese Americans as structural conditions have shaped their life experiences and ethnic communities socially, economically, and politically over time. The interdependence between work and family life, circumscribed by an intersection of race, class, and gender, forms a system of domination. This domination system is embedded in the social institutions—the family, the economy, and the state—that constitute the main structural sources of multiple oppression for Chinese American women in particular and for other racial and ethnic minority women in general. In the case of Chinese American women, the state, under different sociopolitical economies and racial circumstances at various times, has played a critical role, excluding Chinese immigrants legally for more than half a century and preventing them from realizing fulfilling work and family lives.

Contextualizing the gender experience of Chinese Americans, the present study demonstrates the usefulness of Tilly and Scott's (1978) model in providing an analytical framework to examine how demographic and economic conditions have historically transformed work and family life and the interconnection of these private and public spheres. Compensating for the limitations of their model, this study seeks to contribute to its theoretical utilization and reformulation in eight main respects. First, the study departs from Eurocentric analysis to ground analysis in the experience of Chinese Americans. Second, the study broadens the sociopolitical contexts by identifying not only the demographic and economic, but also the social and political-legal factors determining how work links to family life and how this linkage has changed over time. Third, the study shows how the interlocking of race, class, and gender is embedded structurally in the family, economic, and political institutions that have produced sets of changing circumstances, affecting how the worlds of work and family interwine. The legacy of

Chinese women's struggle in this country provides a concrete case illustrating how racism, classism, and sexism simultaneously impinge on their work and family life in particular historical times and places. Fourth, the study uncovers the community as an integral, significant component in the immigration history of Chinese Americans. Like other women, Chinese American women have engaged in balancing work, family, and community life differently according to sociocultural context and historical periods.

Fifth, the study reformulates Tilly and Scott's (1978) model by offering a more flexible periodization that reflects the actual experience of Chinese Americans. Instead of three periods as originally suggested, four historical periods—the community-based economy, the family-based economy, the family-wage economy, and the wage work and family consumption economy—provide a better fit to the immigration history and experience of Chinese Americans, particularly women. Each period is characterized by differences in economic strategies for subsistence, family patterns, household division of labor, relationships between production and social reproduction, and family relationships. Sixth, the model reformulation is based on historical evidence that shows that the transformation of work, family, and community living for Chinese Americans has not been strictly linear. Seventh, the study clearly reveals the human agency of Chinese American women as actors, struggling in everyday life, practicing a culture of resistance, and shaping their history as well as destiny. Finally, the reformulated model may be applicable to other racial and ethnic minority women depending on their immigration patterns and history; the extent to which other women's immigration experiences may vary needs further exploration.

Throughout their history, Chinese American women have struggled for what they believed was good for their families, their communities, and their own survival. Some have experienced upward mobility into the mainstream, while others have had to wrestle with economic hardship, discrimination, and problems of assimilation. Although the political and legal barriers have supposedly been removed and immigration to the United States now offers new opportunities for Chinese Americans, their families and communities still bear the long-term scars of past cultural assault, socioeconomic exploitation, and political-legal discrimination. Presently, Chinese American women are more socially conscious of their civil rights as a racial minority and as women, are more able to question social inequality and injustice, and are politically more active than ever before. Having experienced this social awakening and gained political efficacy, they are facing new challenges with tenacity, determination, and resourcefulness as their predecessors did.

Notes

1. The Chinese diaspora spread not only to the United States, but also to Japan, Korea, the Philippines, Southeast Asia, India, Australia, India, Canada, Europe, Africa, and Latin America.

2. The fraternal organizations known as *tongs* (*tangs*) originated in China as "Triad" secret societies (Chan 1991, p. 67). Some tongs were found to engage in illegal activities.

3. The 1790 naturalization law restricted the privilege of citizenship to "free white persons," which was later modified to include "aliens of African nativity and persons of African descent," but the law continued to prohibit Chinese immigrants from becoming naturalized.

4. The immigration exclusion of Chinese females started as early as 1875, when the federal government passed an act, commonly known as the "Page Law," that forbade most Chinese women whom the government suspected of being prostitutes from entering the United States. The successive Chinese exclusion laws passed between 1882 and 1904 primarily targeted Chinese male laborers, but the courts decided to treat the wives of laborers under the same exclusionary regulations (Chan 1991, pp. 105–106).

5. A 1931 amendment to the Cable Act permitted them to regain their citizenship through naturalization, and the entire act was repealed in 1936.

6. The family-based economy is primarily characteristic of agricultural colonial America. The Great Driving Out compelled most of the early Chinese immigrants to seek refuge in Chinatowns as urban villagers. Here the main charateristics of the family-based economy were developed and maintained while the society at large had become industrialized.

7. Taking advantage of the 1906 San Francisco earthquake and fire that destroyed most municipal records, Chinese residents could claim to be of

American birth and officials could not disprove these claims.

8. In my interviews with Chinatown community workers and mental health professionals in Washington, D.C., New York, and Philadelphia, I learned that numerous Chinese women reported hidden histories of family violence, especially wife abuse, as well as desertion and divorce, which had often resulted in their disillusionment, despair, and even mental breakdown.

9. During a career spanning nineteen years, a Chinese woman whom I interviewed was first a "barefoot" doctor in the countryside and later a physician in a major hospital in mainland China. In the United States, she was not able to enter the medical field even as a nursing aide due to educational and licensing requirements. After employment as a waitress for seven years, she eventually found a job working in a nursing home. The labor experience of other Chinese and Asian immigrant women is documented in Chow (1994).

9

The Other Issei: Japanese Immigrant Women in the Pre–World War II Period

EVELYN NAKANO GLENN
RHACEL SALAZAR PARREÑAS

In the 1960s, Japanese Americans organized a grass-roots campaign to fund a sociological study of their ethnic group. The funds established the Japanese American Project at UCLA, which conducted a survey of 2,000 respondents representing three generations of Japanese Americans to assess intergenerational mobility and assimilation. In the case of the Issei, the first generation, the research plan called for community volunteers to interview male heads of household. Women were interviewed only if their husbands had died, in which case they were asked about their husbands' immigration and occupational histories. This research approach reflected the prevalent attitude that the relevant social unit was the household and that the male head represented all its members. As a result, the opportunity to record Issei women's experience of immigration, settlement, and occupational mobility was lost.

This omission continued a long tradition in the scholarship on Japanese Americans. Studies of Japanese immigrants assumed that Issei men were the sole economic actors and overlooked women's work and contributions to the community. Women were mentioned only peripherally as dependents of male immigrants—anonymous wives and mothers. Yet Issei women constituted a substantial portion of Japanese immigrants. Their presence ensured a different course of development for prewar Japanese American communities than for other Asian American communities which lacked women.

The focus of this essay is on Issei women's immigration, labor, and family life. Seeking to recover the women's own perspectives, we highlight their strategies for survival in the face of cultural assaults, restrictions on their rights, and a stratified labor market. We hone in on variations in these strategies depending on the constraints and opportunities presented by local economies. Issei women emerge as and agents, rather than as passive adjuncts of men or blind followers of tradition.

A larger question underlying this effort is how does centering on women's experiences and activities alter our understanding of Japanese American immigration and labor in the pre–World War II period? Focusing on women highlights important issues that are overlooked when only men's labor is considered. One critical issue is how labor systems affect the family and cultural systems of immigrants. Women's entry into the labor market is often the first major change that occurs in the family economy as immigrants adapt to the urban economy. The family economy of most rural immigrants is based on household production, with women contributing unpaid labor or bringing in income through informal market work. When women become wage workers outside the household, their work is no longer under the direct control of their husbands and their contributions become more visible. We need to explore the implications of these changes for women's position in the family.

Focusing on women's lives exposes the inadequacy of concepts derived from the study of immigrant men's labor. The work of women requires separate analysis because their relationship to work is distinct. Their position in the labor market is shaped simultaneously by racial-ethnic and gender hierarchies, which interact to create a distinct labor market for immigrant women. Additionally, women's work encompasses unwaged reproductive activities: labor that maintains the current labor force (feeding, clothing, and cleaning for male workers) and creates the next generation of workers (bearing, nurturing, and socializing children). Women also are the ones who sustain ties with extended kin and transmit cultural traditions, for example, organizing family celebrations and preparing traditional foods. These tasks are especially arduous for immigrant women. Attending to these varied activities expands the definition of what constitutes labor.

Japanese Women and Labor Migration

The case of Issei women is a microcosm of a worldwide phenomenon—the movement of people from less developed regions to fill labor demands in economically advanced centers. Though primarily seen as a post–World War II phenomenon, labor migration from less developed regions has been a critical element in the development of United States capitalism since at least the mid-nineteenth century. The source of immigrants, the regions from which they are drawn, and the economic sectors into which they are recruited have changed to meet the shifting demands of capitalist development.

The influx of Asian immigrants to California and other parts of the American West in the second half of the nineteenth century was tied to the demand for "cheap" labor to develop the region's economic infrastructure. Starting in the 1850s hundreds of thousands of Chinese men were recruited to fill the demand. With the passage of the Chinese Exclusion Act of 1882, labor immigration from China was cut off. Since the demand for labor did not abate, the Japanese were recruited to fill the gap. Over 80,000 Japanese men immigrated to the continental United States from the late 1880s until 1910. They took over many of the jobs formerly filled by the Chinese in agriculture, lumbering, canneries, coal mines, railroads, and domestic service (Ichihashi 1932, pp. 106–177). Initially, the Issei were highly mobile, moving from city to country and back, and up and down the coast and inland in search of employment (Ichihashi 1932, p. 106; Millis 1915, p. 28). Some settled in towns and cities and established small Japanese enclaves usually on the fringes of Chinatowns (Kitano 1960).

Over time they became increasingly concentrated in agriculture, moving from laborers to sharecroppers on leased land and specializing in truck farming. Those who managed to accumulate a little capital started small businesses such as stores, boarding houses, and laundries catering to the immigrant male population. Like the Chinese before them, Japanese male immigrants were denied the right to become naturalized citizens and were slotted into a labor system designed to extract maximum profits from them as individual units of labor. Policies that prevented family formation among the Chinese were applied to the Japanese. Japanese women were prohibited from entering or discouraged by special scrutiny: officials suspected them of being prostitutes or of harboring diseases. As a result, few women immigrated during the peak years of Japanese immigration from 1890 to 1908. Women constituted only 985 of the 24,326 resident Japanese counted in the 1900 census.[1] Of the 985, 855 were fifteen years or older and only 410 (47.9 percent) women were married (See Table 1). By 1910 the number of Japanese women over fourteen years old had risen to 6,648 and 83.9 percent were married. In 1920, when the population of adult women reached 24,242, 91.1 percent were married. By comparing the proportion of adult women who were married in 1900 versus 1910 and 1920, we can deduce that most women who arrived before 1900 came as single women, whereas those who came after 1900 came to join husbands. Given the disproportionate number of men, antimis-

Table 1   Japanese Females in the United States, 1900–1920

Year	All	15 Years Old or Older	Married	Married Women 15 Years or Older (%)
1900	985	855	410	47.9
1910	9,087	6,648	5,581	83.9
1920	38,303	24,242	22,193	91.1
1930	57,063	29,411	23,930	81.4
1940	54,980	39,399	21,461	54.5

Source: Statistics computed from Ichihashi 1932, pp. 71–72; U.S. Bureau of the Census, 1933. Table 7, "Marital Conditions of the Population of Minor Races, Population by Sex, Color, and Nativity, 1930;" U.S. Bureau of the Census, 1943, "Nonwhite Population 15 Years Old and Over by Marital Status, Age, Race."

cegenation laws that prohibited marriage between Japanese men and White women, and the legal status of women as dependents of men, we can speculate that many of the Japanese women who entered the United States before 1900 were prostitutes.

Early Female Immigrants: *Ameyukisan*

The names of Japanese women were recorded in lists of prostitutes kept in the Chinese licensed quarter of Denver's red-light district in the 1860s (Ito 1973, p. 76), and their presence was noted in Butte, Montana, a mining town, in 1884 (Ichioka 1977, p. 3). By the 1880s Japanese prostitutes worked in brothels and "pink hotels" in the red-light districts of Seattle and other West Coast cities and in Hawaii. They did not provide their services solely to immigrant Japanese. In fact, they were divided into groups depending on the race of men they served: *hakujin-cho* served White clients, for example, while *shinajin-cho* served Chinese men (Ito 1973, p. 768). Although their exact numbers cannot be determined, observers have estimated several hundred on the mainland and an equal or larger number in Hawaii (Hayashi 1989; Ichioka 1977). Old-time residents of Seattle reported as many as 200 in that city working in at least five or six Japanese brothels (Ito 1973, p. 768). Japanese prostitutes were observed in a number of smaller towns as well. Bunshiro Tazuma who arrived in 1902 and worked as a cook for migrant Japanese laborers recalled, "To my surprise, I found at least two to six Japanese prostitutes in every town where I went between Seattle and St. Paul, a range of two thousand miles" (Ito 1973, p. 169).

These women were part of a large-scale prostitution trade that flourished in Japan from 1870 to 1920. At the height of the trade, many thousands of women were sent abroad to become prostitutes (*karayukisan*) in Singapore, Hong Kong, and Vladivostok, and even Australia, India, and Africa (Hayashi 1989). The profits from the trade and the millions of yen generated by their work constituted an important source of foreign currency for Japan's modernization. Japanese political leaders vigorously defended the morality of exporting prostitutes by linking it to both filial duty and patriotism. The official view was stated by Hirobumi Ito, the first prime minister of Japan, in an interview: "Some of these prostitutes should be respected for they are helping financially impoverished parents. Those who do good things while doing a painful job can come back as a member of respectable society" (Hayashi 1989, pp. 18–19).

The center of trafficking in prostitution was Nagasaki, a major international port located in a region of poverty-stricken agricultural and fishing villages. Many prostitutes were drawn from the nearby islands of Amakusa, which lacked arable land and had a tradition of sending young men and women to other regions in search of employment. Some of the women were sold into prostitution by impoverished parents. Others were abducted or lured aboard ship by male procurers (*zegen*), who tricked them with promises of jobs abroad as babysitters or maids (Ichioka 1977, pp. 3–4). According to Hayashi (1989) some *karayukisan* already worked as prostitutes in Japan but agreed to be sold abroad because they could not earn enough to pay off their debts. Male procurers bribed crews of freighters to smuggle the women in boxes. The women were kept in the bottom of the hold and given little food or water. Many lost their lives enroute. Some women were shipped straight to America, while others were sent first to Singapore and then to Canada and taken across the border.

One early Issei reported, "On the boats the women received the attentions of the captain, boatswain, and officers. At the ports of Victoria, Vancouver, Bellingham, Olympia and so on, there were no immigration offices, so the girls were unloaded at night. There was a boss also on the American side who took the girls at some hundred dollars each, and sold them to local bosses. Oppressed by a boss, it is said they could do nothing but weep for four or five years." (Ito 1973, p. 771)

In her pioneer study of nineteenth-century Chinese prostitutes, Lucie Cheng (1984) argued that prostitution needs to be understood as, above all, a form of exploited labor. Japanese prostitution in America can be seen in this same light. The foreign currency prostitutes generated was crucial to the Japanese economy at a time when silk exports and other sources of foreign capital were drying up. Those involved in the prostitution trade, procurers, middlemen, ships' crews, crooked immigration officers and police, pimps, and brothel owners all profited.

Prostitution also contributed to the economic development of Japanese communities in America. Profits from prostitution enabled many a contractor or pimp to establish himself in a "legitimate" business in the United States. Ichioka discusses the example of Genji Hasegawa, whose brothels gave him the capital needed to become a labor contractor for the Southern Pacific. Chushichi Tanaka, a pimp, gave his mistress to a Chinese contractor in exchange for labor contracting rights with the Oregon Shortline, and thereby earned the distinction of becoming the first Japanese railroad labor contractor (Ichioka 1977, pp. 11–12).

By the early 1890s Japanese prostitutes had become the focus of stepped up anti-Japanese attacks. Fearing hostility similar to that which led to the Chinese Exclusion Act, leaders of the Japanese American community urged the Japanese government to stem the prostitution trade. The Japanese government cooperated and the United States government also enacted legislation prohibiting entry of "certain classes of aliens, among them prostitutes, contract laborers, and paupers" (Ichihashi 1932, p. 15). These efforts succeeded in severely reducing the entry of prostitutes by the mid-1890s.

Virtually no information exists about the fate of individual women. Many undoubtedly died of venereal and other diseases. The case of Waka Yamada, who later became a prominent activist for women's rights in Japan, suggests that some managed to return to Japan. Still others married and settled in the United States. In 1906 a group of Japanese Christians founded the Humanitarian Association in Seattle to rescue prostitutes. They placed them in the Women's Baptist Home run by Miss Fife, which attempted to place them in jobs and help them to marry (Ito 1973, pp. 772–775). There is no indication of how many women were "rescued" or how many ended up in legitimate marriages. It seems likely, given the shortage of women, that some of them married and helped found new families in America.

Wives and Picture Brides, 1907–1924

After 1908 the proportion of women immigrating increased dramatically; ironically, this was due to a measure intended to reduce Japanese immigration. In 1907 the Japanese government signed a "gentleman's agreement" with the United States. Under the agreement, the Japanese government agreed to stop issuing passports to laborers, but a provision permitted entry to relatives of Japanese already residing in the United States. This provision created a loophole that Issei men could use to marry and bring their brides. Following Japanese custom, the marriages were arranged by parents through a *baishakunin* (go-between). Many Issei men managed to save or borrow money to return to Japan to meet their prospective brides and take part in a wedding ceremony. Others, who could not afford to return, had matches arranged by a go-between through an exchange of photographs: hence the term "picture marriage." A ceremony was conducted by proxy. The union was legalized by registering it in the husband's home prefecture. Between 1909 and 1923, 45,706 Japanese women entered the United States, of whom 32,628 (71.4 percent) were listed as wives (Gulick n.d., p. 29).

The arrival of large numbers of Japanese women inflamed anti-Japanese forces. The practice of picture marriages was condemned as immoral and a ruse to get around the gentleman's agreement. Further, the alleged fecundity of Japanese females was viewed as a particular threat to the "White" race. V. S. McClatchy, the publisher of the *Sacramento Bee* and a director of the Associated Press, projected that the Japanese population in the United States would grow to 100 million by the year 2063. He testified before the U.S. Senate Japanese Immigration hearings, apparently convincing many that

The Japanese are less assimilable and more dangerous as residents of this country than any other of the peoples ineligible under our laws. . . . They come here specifically and professedly for the purpose of colonizing and establishing here permanently the proud Yamato race. They never cease being Japanese. . . . In pursuit of their intent to colonize this country with that race, they seek to secure land and to found large families. (Daniels 1973, p. 99)

To appease anti-Japanese sentiment, the Japanese government voluntarily stopped issuing passports to picture brides in 1921. The anti-Japanese forces were not satisfied, however, until they achieved complete exclusion with the passage of the 1924 Immigration Act, which prohibited all further immigration from the Asian subcontinent. The flow of Japanese women was thus concentrated in the brief sixteen year span between 1908 and 1924. This narrow time frame led to a community made up of distinct generational cohorts. The typical Issei woman was born in the last two decades of the nineteenth century or the opening decade of the twentieth century. She was in her late teens to mid-twenties when she arrived, having recently married a man ten years her senior, who had lived for some years in the United States and was employed as a wage laborer, small entrepreneur, or farmer. Their children (the Nisei) were born between 1910 and 1940 (Strong 1933).

The majority of Issei women came from farming and small entrepreneurial families in southern Japan. As part of the Meiji reforms, universal education had been established in 1872, so most Issei women had received some formal education (Embree 1939). A later study found that they had an average of eight years of schooling, identical to that of Issei men and higher than average for Japanese women during that era (Strong 1933).

The Immigration Experience

The Issei women had a wide range of attitudes and feelings regarding marriage and immigration. For the most part, the women felt that they had little say in selecting a husband. Yet the extent to which the women felt coerced varied, as did their reasons for "accepting" matches that would require them to emigrate.

At one extreme was Mrs. Takagi, who recalls that her father tricked her into going to stay with her adoptive grandfather in another town on the pretext that she would receive training to become a midwife:

Otherwise, I wouldn't have gone, you see. I knew my mother needed help . . . I stayed one week and helped my uncle (a doctor). I was thinking I would stay to help him. Pretty soon, they took me to see this man. I'd never seen or heard of him. He was my second cousin. You don't know the Japanese system: they just pick out your husband and tell you what to do. So I just did it, that's all. . . . I never gave my parents a fight.[2]

Another Issei, Mrs. Nishimura, fell somewhere in the middle of the continuum. Persuaded by her father, she got married when she was only fifteen:

In the Japanese style we used a go-between, and the husband would come to Japan to pick up his bride. My father was rather new in his thinking so he told me that rather than stay in Japan to attend school, I should come to the United States. My mother told me even that I was too young. But it's something that had to be done. . . . I was rather big for my age. . . . But I cried at the time and I'll always remember that. My parents felt a little guilty about it, almost as if they forced me to come, and apparently they kept asking about me, about how I was doing, until they died.

At another extreme we have Mrs. Shinoda, who dreamed of going to the United States even as a child:

I told my father that I wouldn't get married unless I could come to the United States.
Did your parents oppose you?
Yes, they were all against me.
How did you know you wanted to come to the United States?
I don't know. When I was small, in elementary school, we had to write an essay on "What I Wish For." I wrote in that essay that I'd like to go to America. My friends read it and told what I had written. That's funny, huh?

In leaving their families and going to the United States, the Issei women were following usual Japanese practice. Custom dictated that a woman leave her parents' household or village to live in her husband's home. The Issei were simply traveling a much greater geographic and cultural distance. Despite the pain of separation and fear of the unknown, most Issei women came with some positive expectations. Just as men came to better their lot, to avoid military conscription, to seek adventure, Issei women came with their own hopes—to further their education, to help their families economically, and to experience new adventures. They expected to achieve their goals and then return to a better life in Japan.

Arrival and Adjustment

The boat trip to the United States, usually from Yokohama to Seattle or San Francisco, took at least a month. The women felt homesickness and physical illness, although they developed friendships with other women during the voyage as well. Upon arrival, the first ordeal was getting through immigration procedures:

> Most women arrived as third-class steerage passengers for whom an inspection was a grim experience. Inspectors examined them more scrupulously than first- or second-class passengers. . . . Many questions worried women to no end. Were their papers in order? Each wife had to have a valid passport, a certified copy of her husband's family registry, and a health certificate. Would she pass the physical examination? That she had been found free of trachoma and hookworm in Japan was no guarantee that she would. (Ichioka 1980)

A second shock awaited many a picture bride when she encountered her new spouse for the first time. The man meeting her frequently bore little resemblance to the person shown in the photograph: "Men often forwarded photographs taken in their youth or touched up ones that concealed their real age. . . . Some had improved their overall appearance. They had all traces of facial blemishes and baldness removed. . . . Suave, handsome-appearing gentlemen proved to be pockmarked country bumpkins" (Ichioka 1980, p. 347).

Mrs. Yoshida, who traveled with a number of other picture brides, recalled the reactions of some of her companions to their first glimpses of their husbands:

> A lot of people that I came together with said, "I'm going back on this very boat." I told them, "You can't do that; you should go ashore once. If you really don't like him and you feel like going back, then you have to have a meeting and go back." . . . Many times the picture was taken twenty years earlier and they had changed. Many of the husbands had gone to the country to work as farmers, so they had aged and become quite wrinkled. And very young girls came expecting more and it was natural.

Mrs. Yoshida herself was disappointed with her husband's appearance, but she had wanted to go the United States so much that, "I didn't care what the man looked like." Although her own marriage worked out well, Mrs. Yoshida added, many picture marriages did not.

The newly landed Issei women were taken straight from clearing immigration to be completely outfitted, exchanging the accustomed comfort of kimonos and slippers for constricting Western dresses and shoes. Besides the discomfort brought by the new customs, seeing the conditions under which they would live also appalled many Issei women. Though they knew that the streets would not be paved with gold, they were nonetheless unprepared for a total lack of amenities. The majority of Issei women followed their husbands to the rural areas of the Pacific region. Some went to remote labor camps that were built for railroad workers, coal miners, sugar beet fieldhands, or to lumbering camps and sawmills. Mrs. Sugihara recalled:

> We went to Seattle and then to Oregon. . . . After we got to Oregon, we had to clear off the mountainside in order to prepare a field for planting—with shovels. Unlike California, Oregon is mountainous and the trees—pines six feet in diameter—would have to be cleared away before anything could be planted. We'd use long saws like this to cut up the pine trees. . . . During the day my husband would work [in the sawmill], then at night we would light some lanterns—kerosene lanterns—and work outside in our field. It rained a lot there, so on those days, we'd wear

Like many other immigrant groups to the West Coast at the turn of the century, the Japanese initially provided the farm labor for American agriculture. Quite successful, they often established their own small farms, selling fruits and vegetables. Here Japanese immigrants are working on their own berry farm in California. (Theodore Wesley Koch Collection; Bentley Historical Library, The University of Michigan)

raincoats and work outside. That's how we made the land ready for planting . . .

Many women settled in California where their husbands worked in the fields and tilled the soil as tenant farmers.[3] In addition to working alongside their husbands, women in labor camps and farms drew their own water, gathered wood to cook and heat the house, and fought to keep dirt out of houses that were little more than shacks.

Women who went to rural areas got very little help in adjusting and were often isolated. Those who settled in more urban areas, such as Seattle or the San Francisco Bay area, were more fortunate. Sometimes husbands who had worked as domestics taught wives how to shop, cook, and clean. Community agencies such as the YWCA, church missions, and public schools sponsored housekeeping and English classes for newcomers. These women too

worked long hours and kept house in crowded quarters, but their conditions were less primitive and the presence of an ethnic community eased their adjustment. Living quarters were usually secured within the ethnic enclave. Many couples rented rooms in a house and shared kitchen and bathroom facilities with several other families. Help and comfort were close at hand as other women in the community provided assistance and shared vital information with newcomers.

In most cases, they quickly became too busy to venture much outside the confines of the family and the ethnic community, except perhaps to work as domestic servants. Births followed shortly after marriage and immigration. Large families were the norm, particularly in rural areas (Watanabe 1977, p. 24). Harsh conditions drove some wives to desert their husbands, usually in the company of another man. Because of the imbalanced sex ratios, there were many detached men to whom women could turn. Still, most Issei women persevered even in the face of an unhappy marriage.

Economic Activities of Issei Women

The arrival of substantial numbers of women shifted the Japanese American community from a bachelor society to a conjugal family society. Sex ratios continually dropped from 1900 to 1920: in 1900, there were 2,360 Japanese men for every 100 women; in 1910, there were 694 men for every 100 women (Ichihashi 1932, Tables 7 and 8). By 1920, the ratio had dropped to 189 men for every 100 women. Nee and Wong (1985) credit the early formation of conjugal families for the relatively rapid economic mobility of the Japanese in contrast to the delayed mobility of the Chinese. Nee and Wong are rather vague about the nature of the connection between conjugal family formation and economic mobility, however. We suggest that the unpaid labor of wives and children was the ingredient that enabled Japanese men to make the transition.

The marriages that brought Issei women to the United States were instigated by male sojourners at a point when they realized that they were not going to make their fortunes and return home as quickly as they had originally planned. They hoped that a wife would assist them in reaching their goals. Thus, quite apart from any sentimental desire for family life, men had pragmatic reasons to send for wives. A wife would provide much-needed labor in the form of services in the home and income-producing activities outside it. Any Issei woman arriving in the United States could expect to pull her weight economically. Therefore, although they were not recruited directly as laborers, Issei women were quickly absorbed into the workforce. As early as 1915 an observer of the "Japanese problem" in California noted that

> the great majority of farmers, barbers, and small shopkeepers take a more or less regular place in the fields or shops of their husbands, while a smaller number accept places in domestic service, or in laundries or other places of employment. Thus a large percentage of those admitted find a place in the "labor supply." (Millis 1915, p. 20)

According to U.S. census figures, 20.8 percent of all Japanese women ten years of age and over were gainfully employed in 1920. This is similar to the proportion of White women gainfully employed (20.7 percent).[4] Since virtually all Japanese women in the population were married (see Table 1), however, the Issei rate of employment was in fact remarkably high. In the population at large, only 9 percent of married women were in the labor force.[5] Moreover, the 20 percent rate is undoubtedly an underestimate, given the Japanese concentration in agriculture and small enterprises, fields in which wives performed unpaid labor that frequently went unrecorded by census takers.

It is difficult to arrive at an exact occupational distribution for Issei women in the prewar period. They frequently divided their time among housework, unpaid labor for the farm or business, and wage work. Strong (1933), who surveyed 1,716 Issei women in the early 1930s, classified 998 (58 percent) as housewives, 438 (26 percent) as part-time assistants to their husbands, 53 (4 percent) as full-time assistants, and 227 (13 percent) as engaged in independent occupations. He notes, however, that

> undoubtedly the last two figures are too low and the first figure too high. Accuracy in this connection was very difficult to secure because many of these women speak very little English and are unaccustomed to talk to strangers, and in some cases the Japanese men prevented or interfered in the interviewing of their wives. (Strong 1933, pp. 108–109)

The most comprehensive source of detailed occupational data, the U.S. census, suffers from the same problems and, furthermore, overlooks important categories of women's work by systematically underreporting unremunerated family labor. The census nonetheless provides a rough estimate of the proportions of Issei women engaged in various fields from 1900 to 1940 (see Table 2). These data show that agricultural labor, including employment in plant nurseries (an early Japanese specialty), was the largest field of employment through the prewar period, accounting for about one-third of all employed Issei women. Outside of agriculture, domestic service was by far the most common form of employment. In 1900 over half of all wage earning women were so employed; however, with only 266 counted as gainfully employed, their numbers are so small as to make the data inconclusive. Twenty years later, when 5,289 Issei women workers were listed, 1,409 were in domestic service. The number had dropped to 690 by 1940, but it remained the largest single nonagricultural occupation throughout the prewar years.

Japanese women also specialized in personal services and retail trade, which rose from 25 percent in 1920 to 35.7 percent by 1930. The rise reflects the movement of Japanese men out of wage labor and into self-employment in small business, such as "Oriental" gift stores and shops catering to a Japanese clientele. Wives were engaged as paid and unpaid sales and clerical help in these ethnic enterprises. A small and fluctuating percentage of women found work in manufacturing, primarily in food processing and garment making. The proportion dipped from 7.1 percent in 1920 to 5.2 percent in the depression trough of 1930, but rebounded to 12 percent in 1940. The establishment of ethnic community institutions created a small demand for professionals such as teachers in Japanese-language schools and instructors in calligraphy, music, and other Japanese arts. The proportion of Japanese women listed as professionals peaked at 4.9 percent in 1930. Perhaps the most important category of female professions was midwifery. Up to the mid-1930s, Issei women gave birth at home, attended by a Japanese trained midwife (*osamba*). The *osamba* provided both prenatal and postnatal care, starting with the preparations of the layette and the manual manipulation of the fetus to prevent a breech delivery, extending to advice on breast-feeding and postnatal nursing services.

Census statistics do not tell the whole story. Information garnered from community informants revealed that women engaged in a great deal of informal market work. The tourist traffic in San Francisco created a demand for handmade "Oriental" goods that women could manufacture at home. Embroidery, hand sewing, and paper flower making were common home industries. The contractor, usually a store owner or wholesaler, delivered raw materials and picked up finished products. For example, during the 1920s a couple named Isoe sold hand-embroidered dresses to exclusive apparel shops. The husband sketched the designs and the wife taught the stitches to Issei women, who then embroidered in their spare time at home (Glenn 1986, p. 75).

Regional variation in patterns of occupational concentration reveals the importance of *opportunity* in

shaping immigrant women's economic strategies. Local labor markets offered different choices, depending on the mix of industries and the composition of the local labor force. A comparison of the occupational distribution of Issei women in Seattle and San Francisco, two cities with roughly equal Japanese populations is instructive. In his study of prewar Seattle, Frank Miyamoto (1939) found the majority of Japanese there engaged in running small retail and service establishments, such as stores, hotels, restaurants, barbershops, and pool halls. The demand for such services grew out of Seattle's position as a stopover point for transient workers in the lumber and canning industries of the Pacific Northwest. Not surprisingly, in 1940 a majority of wage-earning Issei women (63.8 percent) in Seattle were involved in three occupations related to small enterprises: proprietors and managers, clerical and sales personnel, and service workers outside private households. Opportunities for small enterprises of this kind were more limited in San Francisco, where the proportion of Issei women who worked in the same three types of occupations was only 34.6 percent, about half that of Seattle. In San Francisco, the market was already filled by the Chinese, who had arrived first and specialized in trade and services. Although the Japanese did run hotels, restaurants, and shops in the bay area, they catered primarily to a Japanese clientele. Thus, the number of such businesses was necessarily limited. Lacking the option of working in small business, a high proportion of Issei women in the bay area (50.4 percent) were forced to "choose" domestic service, while in Seattle only a small percentage of Issei women (3.3 percent) worked as domestic servants[6] (U.S. Bureau of the Census 1943a, Table 38).

Regardless of geographic location, whether in the formal or informal market, Issei women's occupations shared several characteristics: the work could be fitted around family responsibilities (for example, it was done at home, or children could be taken to work, the hours were flexible); it involved tasks that were an extension of women's work in the home (for example, food preparation, laundry, and sewing); it was in a low-technology, labor-intensive field where low wages and long hours reduced competition from White women; and it took place in a family-owned or ethnic enterprise where language difficulties and racial discrimination did not constitute barriers to employment.

Table 2   Occupations of Employed Japanese Women, 1900–1940[a]

	1900		1920		1930		1940[b]	
	No.	%	No.	%	No.	%	No.	%
Agriculture workers, including farm and nursery labor	13	4.9	1,797	34	2,041	30.3	2,525	37.7
Servants, including cooks and chambermaids	151[c]	56.8	1,409	26.6	1,195	17.7	690	10.3
Other personal service workers, including barbers, waitresses, lodging housekeepers, and laundry operatives	57	21.4	951	28	1,463	21.7	1,579[d]	23.6
Tradeworkers, including saleswomen and clerks	9	3.4	369	7.0	946	14.0	683[e]	10.2
Dressmakers, including seamstresses and tailors	23	8.6	124	2.3	121	1.8	—[f]	—
Other manufacturing, mechanical workers	8	3.0	378	7.1	348	5.2	801[g]	12.0
Professionals (teachers, nurses)	5	1.9	145	2.7	329	4.9	214	3.2
Clerical workers	—	—	75	1.4	271	4.0	—[h]	—
Other	—	—	41	0.8	27	0.4	201	3.0
Total Employed	266	100	5,289	99.9[i]	6,741	100	6,693	100

[a] Data for 1910 are omitted because occupational figures for Japanese and Chinese were combined in the census report.

[b] Only foreign-born (Issei) women are included in the figures for 1940. The 1940 census for the first time separated native- and foreign-born. The figures for 1930 include some native-born (Nisei) women, but they probably constitute a small proportion of the total. Because of immigration patterns, most Nisei were born after 1910.

[c] Includes some waitresses.

[d] Made up of "proprietors, managers, and officials, except farm" and "service workers, except domestic."

[e] Named "clerical, sales, and kindred workers" in the 1940 census.

[f] No longer separately reported; presumably these occupations are included below under "manufacturing."

[g] Named "operatives and kindred workers" in the 1940 census.

[h] Included under "trade," above.

[i] Due to rounding.

Sources: For 1900: U.S. Bureau of the Census, 1904, Table 35: "Distribution by Specified Occupations of Males and Females in the Chinese, Japanese, and Indian Population Gainfully Employed, 1900." For 1920: U.S. Bureau of the Census, 1923, Table 5: "Total Persons 10 Years of Age and Over Engaged in Each Specified Occupation." For 1930: U.S. Bureau of the Census, 1933, Table 6: "Chinese and Japanese Gainful Workers 10 Years Old and Over by Occupation and Sex, for the United States and Selected States, 1930." For 1940: U.S. Bureau of the Census, 1943, Table 87: "Non-white Employed Persons 14 Years Old and Over, by Major Occupation Group, Race, and Sex, for the United States, by Regions, Urban and Rural, 1940."

Regardless of the specific occupation or region, women's labor was crucial for the survival of the household. The unpaid family labor was essential for the small-scale farming in which the Issei specialized. After the passage of the 1913 Alien Land Act in California, the Issei could not own land or lease it for longer than three years. They had to take short-term leases on small, often undesirable, parcels and work on them intensively. Mrs. Sasaki, a Nisei who grew up in the Sacramento Delta reported:

> My father was a sharecropper. My mother did farming, helping her husband. She worked as hard as my father did in the fields . . .

Women's labor was just as essential in urban areas. For families engaged in small-scale enterprise, women's unpaid labor made it possible for families to eke out a living from laundries and other small businesses that could not have succeeded if they relied on paid help. An example was the case of Mrs. Watanabe's family. Mrs. Watanabe, who grew up in San Francisco in the 1930s, recalled that when her older sister and brother were small, her parents were employed in a commercial laundry. These children were sent to Japan to be raised by relatives so that both parents could work. By the time Mrs. Watanabe was born, the parents had managed to acquire a small laundry, where her mother worked in the shop alongside her father. Among wage-earning households, men's earnings rarely constituted a family wage. Both husbands and wives worked in low-wage jobs, and their pooled earnings were needed to support the family. A common combination in the San Francisco Bay area was a husband employed as a gardener or cook with a wife employed as domestic day worker.

The Issei Household Economy and Gender Relations

In every society the family is the central institution defining women's place and social identity. Its importance was magnified for Issei women because of the encompassing nature of the family system in Japanese society. Its very structure embodied Japanese cultural values and was the means by which traditional patterns of male dominance and privilege were perpetuated. Yet the family, though conservative, was not impervious to change; its structure and relationships were supported by and dependent upon larger economic and political systems. Thus, relations in the immigrant family underwent realignment in response to new external contingencies.

Despite dramatic changes in the external circumstances following immigration, certain factors encouraged continuity in Issei family values and household structure. Because of their sojourner status, the Issei did not completely sever ties with their households of origin. [They kept one foot in each society by maintaining kinship obligations in both countries.] In many cases, migration was not an individual act but a family strategy. The immigrant was "sent abroad" to work and remit money for the support of the kin group. The immigrant took on the role of economic provider while the kin group at home assumed much of the reproductive burden. Even those who formed conjugal families abroad still retained some obligations to provide for parents and other relatives left behind. Mrs. Taniguchi, for example, mentions scrimping in order to send money to her husband's father, who was having trouble hanging on to the family property because of the profligacy of his own father. Several women also sent money to their own parents and siblings. In return for remittances, relatives in Japan could be called upon to perform services such as caring for children who were left behind or sent back by parents working in America. The larger kin network provided extra flexibility and could be mobilized to absorb some of the burden produced by the multiplicity of problems the Issei faced. Domestic cooperation thus spanned international boundaries.

A second factor that fostered continuity of household structure among a large segment of the Issei was their specialization in farming and ethnic enterprise. The household economy resembled that of agrarian and small business families in Japan. Among this segment there was little break with tradition, since the household as the fundamental social unit was easily transplanted to a new setting. As in Japan, the household served as the basic unit of ownership, production, and consumption. "Work" and "family" were integrated. The husband managed the farm or store and oversaw the unpaid labor of wife and children. Under these circumstances wives continued to work under the authority of their husbands. They earned no independent income, so they had little control over economic resources. They were also prevented from meeting new people and forming relationships outside the family. Husbands continued to make major decisions, rarely consulting their wives (Yanagisako 1975).

Despite their economic success in the United States, their color deeply affected the lives of Japanese Americans who, during World War II, were evacuated from the West Coast and placed in camps under a U.S. Army war emergency order. Here Japanese families are waiting for registration in the Santa Anita reception center, Los Angeles, in 1942. (Photo by Russell Lee. Library of Congress)

What about the Issei who were drawn to wage labor? The family economies of this large segment differed considerably from of those of farm and small-business families. Husband, wife, and older children were individually employed, mostly in marginal, low-paying jobs. Each worker's earnings were small, but the pooled income was sufficient to support a household and to generate some surplus for savings, remittances, and consumer goods.

The strategy of the family-wage economy was in some ways consistent with the values of the Japanese ie (household) system. Because several wages were needed, economic interdependence among family members was preserved. Further, the employment of women was consistent with the assumption that women were full economic contributors. In other ways, however, the strategy marked a departure from the tradition. With wage work, the individual, rather than the family, was the unit of production, and work and family life were separated. Husbands no longer controlled their wives' labor.

These contradictions are reflected in the ambivalence of Issei men toward their wives' employment. Some men opposed their wives' employment on the grounds that their services were needed at home. Other men demanded that their wives pull their full weight and take a job, regardless of the women's own inclinations. Thus, whereas Mrs. Amano defied her husband's wishes by going out to work secretly, Mrs. Togasaki indicated that she felt pressure to seek outside work:

My husband didn't bring in enough money, so I went out to work. I didn't even think about it. If I didn't take a job, people would have started to call me "madam" (i.e., would presume that she considered herself too much of a lady to work). It was like a race; we all had to work as hard as possible.

The contradictions extend to the effects of wage work on women's position in the family. To the extent that the traditional division of labor and male privilege persisted, wage work added to be the women's burdens. But to the extent that wage work reduced women's economic dependence and male control over their labor, it helped women transcend traditional gender roles in the family. Evidence of both tendencies emerges from Issei women's accounts, though the increased burdens are more obvious.

Male Privilege and Female Overload

Although women shared responsibility for economic support, they were expected to assume all responsibility for domestic chores such as housework and child care. All but one of the Issei women interviewed claimed that their husbands did little or no work in the home. Mrs. Nishimura was the most explicit in citing the sense of male privilege assumed by husbands:

> No, my husband was like a child. He couldn't even make tea. He couldn't do anything by himself. He was really Japan-style. Sometimes I had too much to do, so although I would always iron his shirts, I might ask him to wait awhile on the underwear, but he'd say no. He'd wait there until I would iron them. People used to say he was spoiled. He was a completely Japanese man. Some people divorced their husbands for not helping around the house, but that never entered my mind. I thought it was natural for a Japanese.

Although Mr. Nishimura might be viewed as an extreme case even by other Issei, the frequency with which Issei women used the term "Japanese" to describe their husbands is striking. They saw their husbands' intransigence as peculiarly Japanese and attributed it to their early upbringing.

The pattern of male privilege was so firmly entrenched that even when the wife was the sole earner, as was the case if the husband was ill or unemployed, she continued to do most of the housework. In a subculture that placed a premium on male superiority, unemployed men may have had a special need to avoid losing status by taking on "women's work." The result of male privilege was that women experienced considerable overload. The men worked long hours, often at physically exhausting jobs, but the women's days were longer. Their work began before other members of the household arose with the preparation of the morning meal, and ended after others were relaxing with the cleanup following the evening meal. In between they had to fit in laundry and cleaning. Some women were endowed with natural vitality and got by on little sleep. Mrs. Nishimura described her schedule during the years she was working as a garment operative:

> Since I had so many children, I asked my mother-in-law to take care of the children. I would get up at five o'clock to do the laundry—in those days we'd do it by hand—and hang up the laundry, then go to Oakland. I would come home, and since my husband didn't have much work then, he'd get drunk and bring the children home. I would cook and eat, and then go to sleep. They all asked me how long I slept at night. But since I was in my twenties, it didn't affect me too much.

Others, like Mrs. Togasaki, who worked as a laborer in a flower nursery, were exhausted at the end of the day and let things slide at home: "My house was a mess. I went to work in the morning, and when I came back from work, I'd cool a little then go to sleep, and that's about all."

While the Issei had extremely stable marriages, one should not romanticize their marriages or exaggerate the degree of harmony within the family. Economic interdependence and common cultural values bound members to each other. However, gender divisions generated serious conflicts: the discrepancy in power and privilege, the unequal division of labor and child care, and the separation of male and female social and emotional worlds made men's and women's interests fundamentally different. Overwork and poverty exacerbated conflict. Far from responding passively, many women actively contended with their husbands. Mrs. Taniguchi, who had to send two children to Japan and work in a laundry to support her sickly husband said:

> My life in the United States was very hard in the beginning because my husband was ill so much and we had such totally different personalities. We were both selfish, so we had many problems. But after I started going to church, I became more gentle. So we had fewer quarrels. I think that is a gift from God.

Mrs. Nishimura also reported that she and her husband quarreled a great deal:

> Well, he was rather short-tempered. . . . There were times when I thought he was stubborn, but we were far apart in age, so I would attribute our differences to that. Being apart in age does create quite a lot of differences . . . but I bore it all.

A plausible interpretation of the conflict is that her husband tried to exert his patriarchal authority and expected her, as a younger female, to acquiesce to his dominance. Both she and Mrs. Taniguchi expressed the traditional Japanese attitude that women must bear up under any hardship. Yet it is evident that they did not do so silently.

World War II Incarceration and Aftermath

By the eve of World War II, the Issei had for the most part achieved some degree of economic security by dint of hard work and sacrifice. They had started out as sojourners, intending to accumulate capital to return to Japan. The establishment of families and the growth of an American-born generation, the Nisei, marked a transition to permanent settlement. Still, some Issei were embittered by the racism that they confronted and became more committed to remaining "Japanese" and not assimilating. Throughout the 1930s many Issei families, at least vaguely, thought they might someday go back. This illusion was shattered by the trauma of World War II; when forced to decide, the overwhelming majority of Issei realized they were too deeply rooted to leave.

On the eve of the war, the average Issei man was in his early sixties and the average Issei woman in her early fifties, and together they had spent most of their adult life in the United States, establishing a home and raising children. In April 1941, 50,000 Issei were among the 120,000 Americans of Japanese ancestry residing in the Pacific states who were ordered to report to assembly centers, bringing only what they could carry. From there they were transported to one of ten internment camps. In addition to losing their freedom, they lost their homes and most of their possessions. The economic status the Issei had achieved so painfully was gone. They had not been allowed to become U.S. citizens, and now they were decried as enemy "aliens." Because of their alien status and lack of fluency in English, the Issei

generation lost its authority. Nisei children became the representatives of the family with the camp authorities.

In addition to losing their life's work, many Issei lost their communities as they resettled in other regions after the war. Farming families who returned to the Pacific Coast found they had been displaced in agriculture and could not regain leases on land, while urban Japanese discovered others had taken over their retail and service establishments and moved into their old neighborhoods. Having invested their youthful energy and dreams in establishing their farms and businesses before the war, the average Issei couple no longer had the vigor to start from scratch (Bonacich and Modell 1980, p. 107). Once again Issei women's labor was critical in reestablishing the Issei family. Instead of returning to unpaid family labor on the farm or in the shop, women moved into wage work. With few possessions and little remaining savings, the Issei rushed to get work as quickly as possible. Mrs. Amano described the situation as "like a rat race." Issei women took whatever jobs were available. The most frequently found first jobs were in domestic service, which offered the possibility of living quarters, or in packing sheds and factories. Such jobs were seen as temporary, but some Issei women found themselves resigned to long-term employment in these occupations. Strikingly, the rate of employment in 1950 was 55 percent higher than in 1940 (Bonacich and Modell 1980, p. 99), and remained high among older Japanese women, many of whom worked into their seventies and eighties (Glenn 1986, p. 81). As their husbands aged and passed away, they became the caretakers, and finally the lone survivors.

Conclusion

To return to the question we posed at the beginning of the paper: How does focusing on women's experiences alter our understanding of Japanese immigration and settlement? First, this account has highlighted the role of women's labor in the development of the Japanese American community. Much of this labor has been unacknowledged or taken for granted and its significance therefore overlooked. Issei women's work encompassed a wide range of activities. The historical evidence on prostitution in Japanese American communities, while sparse, suggests that it was an important economic activity, one that was crucial to the rapid entry of some Issei men into labor contracting and other "legitimate" enterprises. The labor of wives in farming and small entrepreneurial families has often been invisible—hidden by its unpaid and often unreported nature. In addition to contributing unpaid field labor, these women engaged in subsistence activities, such as growing and preserving foodstuffs and making clothes. The manufacturing of goods that are directly consumed by members of the household is often not recognized as an economic activity, but it was essential to the family's standard of living. The economic contribution of wage-earning women is perhaps more visible because it was monetized. However, even when women were employed, they also performed a great deal of domestic work that was "hidden in the household."

More generally, this account highlights the importance of gender as an analytic concept for understanding the immigrant experience. Gender is a central organizing feature of labor systems and the family. The gender division of labor both in and out of the family creates differences in power and resources. Men and women do not contribute equal labor and they do not enjoy equal benefits from joint labor. Unequal power and labor means that men and women may have different standards of living within the same household. For example, some Issei men kept a large portion of their own earnings to spend for themselves, leaving their wives to meet household expenses by scrimping. The immigrant family emerges, not as a monolithic entity with unitary interests, but as a differentiated institution divided along gender and generational lines. Descriptions and personal accounts of immigrant family life reveals conflict between both men and women and between generations.

Immigration may change the distribution of work and resources between men and women in complex ways. Migration from the economic periphery to the advanced center is often thought to have a liberating effect for women, as traditional patterns of family life give way to individualism and as women gain economic independence through outside employment. The experiences of Japanese immigrants in the prewar period supports the argument that women's entry into the labor force, rather than exposure to "modern" ideas, is crucial in the realignment of gender politics. Farming and small-business families who

were engaged in household-based production reproduced traditional relations in the household. In contrast, Issei families whose economies shifted to multiple wage earning displayed greater discontinuity.

Changes accompanying the shift to a family-wage economy were not, however, uniformly liberating for women. The changes had contradictory implications for women's reproductive work and for conjugal relations. The tradition that women contributed economically was easily adapted to the requirements of wage labor; women simply transferred their productive activities from the household to the labor market. The parallel tradition of male privilege was harder to alter.

In many ways, the immigrant situation increased women's overall workload. Women had to compensate for the inability of Issei men to bring in an adequate income not only through employment outside the home, but also through increased labor in it. They had to produce clothing, basic foodstuffs, and other goods at home to stretch the budget. The sojourner strategy of working hard and sacrificing short-term comfort, combined with the externally constrained work rhythms of wage labor, made for a more relentless and constant pace of work, fewer occasions for communal projects that combined work and socializing, and fewer festivities to provide periodic release. Additional problems were created by the shift to wage work, the most notable being the lack of adequate care and supervision of children. Under the family production system, children were surrounded by caretakers, both parents, and sometimes grandparents and other relatives. Issei women were left on their own to make arrangements for having their children minded while they were out working.

The gains that women made with migration and entry into the labor force were more subtle than the costs. There was, of course, the tangible benefits of added income, part of which could be retained for individual savings or spending. A less tangible but perhaps more significant gain was some degree of control over their economic circumstances. Husbands did not or could not always provide adequately for reasons ranging from misfortune to improvidence. Unlike their mothers and sisters in Japan, Issei wives had the option of engaging in wage work to provide for themselves and their children.

What is the relevance of this discussion for the situation of today's Asian immigrant women? Like Issei women in the prewar period, first-generation Asian American women face labor market stratification and the double day. Korean, Chinese, Filipino, and South Asian immigrants generally have the same two options of working in ethnic enterprise using family labor or relying on multiple wage earning. Although a large proportion of post-1965 Asian immigrants are highly educated and from urban backgrounds, many experience downward mobility. Lack of facility in English and nonrecognition of foreign credentials means that they have to take relatively low-paying jobs or go into small enterprises, such as green groceries, restaurants, and cleaners. Husbands and wives put in long hours, often seven days a week. And like Issei wives, Asian immigrant wives bear primary responsibility for domestic tasks. In the San Francisco Bay area, Chinese immigrant women constitute the majority of seamstresses in the garment industry, often working in small Chinese-owned subcontracting firms. Asian immigrant women also make up 24 percent of Silicon Valley's assembly-line workers. These women's job options are constrained by the need to find jobs that do not require English, are close to home, and can be fit around their family responsibilities. The double burden of employment and housework leaves them little or no time to learn English or acquire job training. Like Issei women of the earlier period, working-class Asian immigrant women suffer from a lack of services, such as child care. Poverty and stress exacerbate family conflicts, and wife battering has finally been exposed as a frequent problem. Today's Asian immigrant women need the same services that would have benefited prewar Issei women: child-care services, English language classes, job training, and—above all—fair wages and job benefits.

Notes

1. Interestingly, five women were among the 53 Japanese sent to the United States in the 1870s to learn about Western technology (Hayashi 1989)

2. All quotes that are not cited are from interviews conducted by Evelyn Nakano Glenn for her research on Japanese American women (Glenn 1986). Pseudonyms are used to protect confidentiality.

3. From 1900 to 1930, California had the largest population of Japanese. In 1900, 10,151 Japanese, constituting 41.7 percent of the U.S. Japanese population resided in California; in 1910, 41,356 (57.31 percent); in 1920, 71,952 (64.8 percent); and in 1930, 97,456 (70.2 percent). Computed and compiled from Ichihashi 1932.

4. Statistics of Japanese women taken from *Fourteenth Census of the United States Taken in the Year 1920, vol. 4, Population*, "Occupations" (Washington, D.C.: U.S. Government Printing Office, 1923), Table 5: "Total Persons 10 Years of Age and Over Engaged in Each Specified Occupation, Classified by Sex, Color, or Nativity, and Parentage, for the United States, 1920." Percent of white women working taken from *Historical Statistics of the United States, Colonial Times to 1957*, series D 26–35: "Civilian Labor Force by Color and Sex, and Marital Status of Women, 1890 to 1957."

5. U.S. Bureau of the Census, *Historical Statistics of the United States, Colonial Times to 1970*, Bicentennial ed., pt. 2, series D 49–62: "Marital Status of Women in the Civilian Labor Force, 1890 to 1970."

6. As Glenn (1986, 79) has argued previously, the size of the domestic labor force is highly elastic since the demand always exceeds the supply. Women rarely choose domestic service when offered other choices. Only when other options are unavailable does the proportion of women in domestic service rise.

10

The Protestant Establishment: Its History, Its Legacy— Its Future?

HOWARD G. SCHNEIDERMAN

At its height, the Protestant Establishment, dominated by a group of wealthy and powerful consanguine families of old-stock, white, Anglo-Saxon Protestants, or as they are now generally called WASPs, provided the nation with an authoritative core of important and influential leaders. These old-stock WASP families exercised national authority in the presidency, as well as the cabinet, in diplomatic circles, and in other governmental jobs (Schneiderman 1991, pp. xxi–xxii). They took the lead in legal and juridical roles, and as heads of important corporations, banks, investment houses, and other commercial interests, not to mention churches and colleges throughout America. In short, they set the tone for American political, business, and cultural life. As such, WASPs were less an ethnic group than a high status group. The Protestant Establishment's power and authority, as well as its social class and status, rather than its ethnicity, were driving forces in American history and contemporary society. Yet it must be said that the Protestant Establishment is now all but gone from the American scene; or as Robert C. Christopher (1989, pp. 267–268), says, "relatively few members of the national social aristocracy have made truly major contributions to the political or economic development of the United States in the last four decades."

Taking the lead, and setting the tone, in a democracy is never easy, neither for individuals nor for a social class. Complex and imperfect, democracy has been called, by no less a judge than Winston Churchill, the worst form of government—except for all the rest. The problem of leadership in America is particularly difficult to resolve because our democracy is beset with two special paradoxes not shared by more hierarchically oriented societies. The first paradox—the paradox of leadership itself—involves the ideal of egalitarianism and the reality of inequality of social class, status, power, and authority. These elements of inequality violate the egalitarian ethos at the core of American culture, while being indispensable to its survival—an uncomfortable fact of democratic life with which we must always live. The second paradox—the paradox of social mobility—also involves the ideals of egalitarianism, which stress the principle of an open class system in which everyone has the opportunity to rise to positions of leadership and success based on merit. But this meritocratic principle of achievement often leads to a very different reality in which ascription, as well as achievement, come to bear on success. After all, successful men and women always want to pass on to their children whatever advantages they themselves have earned, in the form of enhanced and unequal opportunities for education and a leg up in their careers, thus violating the basic premises of our egalitarian values. Both of these paradoxes are always at play in American society. As we shall see, they are part of the story of how the Protestant Establishment was able to command deference and to exercise authority in America until about the middle of this century.

The Protestant Establishment

Oddly, it was not when the Protestant Establishment was at the height of its authority, a period referred to by Joseph Alsop (1992) in his memoirs as the "WASP Ascendancy," and somewhat earlier by E. Digby Baltzell and myself (1988, 1991) as the period of "WASP Hegemony," but only when it was declining, that the acronym "WASP" came into existence. So much a part of our contemporary language is the term "WASP" that many readers may find it curious that it was nowhere to be found in the best and most definitive dictionaries before the mid-1960s. This is probably partly due to the fact that like many newly minted words, no standardized usage had been associated with it. The term "WASP," however, has been in use since the 1950s, at least among intellectuals, but did not enter our general usage until it was made popular by Baltzell (1964b) in his widely read book, *The Protestant Establishment: Aristocracy and Caste in America*, which chronicled the decline of WASP authority in American society.

While lacking a standardized definition, the term "WASP" is closely identified with the American upper class, as well as with various elites in the American system of political and social stratification. For instance, *Webster's Third International Dictionary* defines WASP as "an American of northern European and especially British stock and of Protestant background; especially a member of the dominant and most privileged class of people in the U.S." Expanding on this same theme of stratification Christopher Hitchens (1990, pp. 34–35) wrote about WASPs that

. . . somewhere in the subtext of all this is the ticklish question of race and the awkward matter of class. Ethnic hierarchy in America actually con-

fuses the two things in a revealing minor way, since the word WASP, which denotes a racial and religious group, is only ever applied to a certain social layer of it. (George Bush is a WASP. George Wallace may have been a white Protestant of Anglo-Saxon descent, and even rather vocal on all three points, but a WASP he was not.)

To this last point we might add that while presidents such as Theodore Roosevelt, William Howard Taft, Woodrow Wilson, and Franklin Delano Roosevelt were WASPs, Lyndon Johnson, Richard Nixon, Gerald Ford, Jimmy Carter, and William Jefferson Clinton are not; and although after his presidency Ronald Reagan was knighted by Queen Elizabeth, neither was he a WASP despite being wealthy, white, Protestant, and of Irish descent. But if only some Americans of Anglo-Saxon heritage are WASPs, how did they become so? How did a relatively small group of families become a national upper class, and how did a relatively small subgroup among these families come to have disproportionate power and authority in America from just after the Civil War to just after World War II?

A century and a half ago America was very different than it is today. The vast majority of Americans were white, of Anglo-Saxon origin, and Protestant. They lived in small towns, not large cities, or on farms, not in suburbs. Wealth, status, and power were diffused throughout these small communities in ways that are largely foreign to us now. For instance, Alexis de Tocqueville (1945a, p. 52), the most prescient observer of American society, could easily write in 1835 that "in America there are but few wealthy persons; nearly all Americans have to take a profession." And Tocqueville knew that these professions were sources of prestige and authority in the small communities of which America was composed. Not great wealth, but rather eminence as a minister in a local church, or as a lawyer or jurist, or merchant, or doctor commanded deference and authority. Thus, Henry Adams, the eminent author and historian, born in 1838, and scion of one of America's leading establishment families—son of Charles Francis Adams, the American Ambassador to Great Britain, grandson of President John Quincy Adams, and great-grandson of President John Adams—could look back on his own childhood and remember that "down to 1850, and even later, New England society was still directed by the professions. Lawyers, physicians, professors, merchants were classes, and acted not as individuals, but as though they were clergymen, and each profession were a church" (Adams 1931, p. 32). All of this is to say that authority in America until just after the Civil War was highly localized, as was deference, power, and leadership in general.

Emergence

Great changes were in store for America, however, in the aftermath of the Civil War. Throughout the 1870s and 1880s urbanization and industrialization, coupled with the growth of railroads and large corporations, broke down small town ties of community, and in Richard Hofstadter's (1955, p. 136) terms "transformed the old society and revolutionized the distribution of power and prestige." This status revolution was accompanied by a significant shift in population from rural to urban America. In 1850, for instance, only 15 percent of Americans lived in cities, but by 1880 this percentage had just about doubled, and by 1890 about 35 percent lived in urban areas. The tidal waves of immigration that began during this period helped swell the urban population even further, and by 1920 more than half of all Americans lived in urban areas. The trend toward urbanization continues to our day. In 1870 only about one in four Americans lived in cities; since 1970 this percentage has been stood on its head, with only one in four Americans *not* living in cities or their immediate vicinity.

Immigration also changed the ethnic makeup of America. In 1790 over 80 percent of the population was White, and between 60 and 70 percent of the White population was of English ancestry. Of the rest of the White population, about 10 percent was Irish, 8 percent Scotch, and 9 percent German in ancestry. By 1920 over 90 percent of the population was White, but only 44 percent of the White population was of English ancestry (while Americans of Irish ancestry were about 11 percent of the population, the proportion of German Americans had nearly doubled to about 16 percent.) Also, by 1920 other ethnicities— Italian, Polish, Russian, and Spanish, for example— accounted for almost 30 percent of White Americans. By 1990 about 80 percent of the population was White, but only about 25 percent of the White population was of English ancestry; Irish Americans, however, now accounted for about 20 percent of the

White population, and German Americans matched those of English ancestry at about 25 percent.

But the great post–Civil War transformation was especially noticeable with respect to wealth. When Tocqueville published the second part of *Democracy in America* in 1840, there were not even twenty millionaires in the entire United States, and he was entirely correct when he wrote that "though there are rich men, the class of rich men does not exist; for these rich individuals have no feelings or purposes, no traditions or hopes, in common; there are individuals, therefore, but no definite class" (Tocqueville 1945b, p. 160). But the industrialization of America helped change this situation. In the thirty years between 1870 and 1900 the national wealth quadrupled, and the number of millionaires grew enormously. By 1893 it was estimated that there were over 4,000 millionaires in the United States, and at least 120 men who were worth over $10 million each. In fact, in 1892 the U.S. Census Bureau estimated that about 9 percent of American families accounted for over 70 percent of the nation's wealth. This trend toward accumulating wealth has continued, and in this regard it is notable that according to a recent estimate published in *The New York Times* (21 March 1993), there are over 3.2 million households in the United States today worth over $1 million. It is also worth noting that among America's seventy-three billionaires listed in *Forbes'* "Four Hundred Richest People in America" for 1992, there were a few WASPs such as Henry Lea Hillman, but many more self-made men such as Henry Ross Perot. And while David Rockefeller, Laurance Spelman Rockefeller, and Paul Mellon represent a few old-money families, John Werner Kluge, Samuel and Donald Newhouse, Ronald Perelman, Ted Arison, Edgar Bronfman, Leslie H. Wexner, Jay and Robert Pritzker, Samuel LeFrak, Kirk Kerkorian, and Milton Petrie, among others, represent the enormous drive and achievement of talented individuals from religious and ethnic groups that have been excluded from the precincts of the Protestant Establishment, but who have succeeded in leaving most old-stock families trailing in the dust.

Inclusion

While today great wealth is scattered among millions of families of all racial, ethnic, and religious backgrounds, during the period of the status revolution it was concentrated in the hands of relatively few. This newly created wealth formed the basis of a new class of rich men, a largely WASP business aristocracy of the sort that Tocqueville had said was absent from American society as late as the 1840s. Writing in 1955, a famous sociologist, Talcott Parsons, described the process by which this business aristocracy had come into existence as follows:

> There is a continuing tendency for earlier economic developments to leave a 'precipitate' of upper groups, the positions of whose members are founded in the achievements of their ancestors, in this case relatively recent ones. By historical necessity these groups are strongest in the older parts of the country. Hence the cities of the Eastern seaboard have tended to develop groups that are the closest approach we have . . . to an aristocracy. They have generally originated in business interests . . . (Parsons 1955, p. 125)

Following Parsons, another sociologist, E. Digby Baltzell (1957), perhaps our premiere analyst of the American upper class, showed how the newly rich industrialists and their families eventually formed a national upper class, which then drifted apart from local community roots.

This new national upper class was formed in American cities in the latter part of the nineteenth century when successful moneymakers—businessmen, merchants, and bankers—many of whom came from low social class backgrounds, founded families whose individual members through marriage, private education, and club membership associated themselves with old-money families. Through these associations, generation after generation, they came to constitute a business aristocracy in which upper-class values, wealth, and power were preserved. This business aristocracy, as Baltzell (1957, p. 5) put it, "too often placed the desire for material comfort and security above the duties of political and intellectual leadership." Nevertheless, up to World War II, upper-class WASP families, living in fashionable neighborhoods and listed in the *Social Register*, helped to shape the business and cultural life of cities across the country. They educated their children at private prep schools, such as The Taft School, Groton, Hotchkiss, St. Mark's, St. Paul's, Foxcroft, and Choate, and colleges such as Princeton, Harvard, Yale, Vassar, and Smith among other elite colleges. They belonged to exclusive clubs—there are over 250 listed in the *Social Register* today, from the Acorn in Philadelphia and the Algonquin in Boston to the Wilmington Country Club in Delaware and the Yale Club in New York—and helped shape the business and cultural life of cities across the country. At the beginning this WASP upper class was relatively open to new talent and assimilated newly arrived individuals and their families into its ranks, but within a short time upper-class status appeared to be limited to old-stock Protestant families, which led Baltzell (1957, p. 395) to ask, "what is the future function of a predominantly Anglo-Saxon and Protestant upper class in an ethnically and religiously heterogeneous democracy?"

Exclusion

This question invokes the image of declining class authority in America reminiscent of a similar situation described in Tocqueville's *The Old Regime and the French Revolution* (1955), which showed that when new men of talent, wealth, and power were refused membership in the French aristocracy by noblemen who drew a caste line making membership a matter of birth alone, class authority failed and set the stage for the French Revolution. A similar crisis of leadership in America had developed by the mid-twentieth century because the white, Anglo-Saxon, and Protestant national upper class had, by excluding talented individuals and their families from membership because they were from ethnic and racial minority groups, become a privileged caste rather than an authoritative establishment.

The old local upper classes had been, by and large, based on social considerations of family, rather than on business connections and wealth, and they had been open to newer ethnic groups as well as to old-stock Protestant Americans of Anglo-Saxon heritage. Thus, as late as the 1870s, Jews were assimilated into the local upper classes of American cities such as Boston, where Louis D. Brandeis, a future justice of the U.S. Supreme Court, could join exclusive patrician clubs. The son of a wealthy Jewish merchant who had immigrated to Louisville, Kentucky, from Prague in 1848, Brandeis entered Harvard Law School in 1875 and immediately rose to the top of his class. He soon became friends with many Boston brahmins such as Dennis Warren, Jr., the son of a rich manufacturer, with whom he founded a law firm after graduation. Brandeis had no difficulty in joining the Union Club, or the Exchange, nor did anyone object to his joining and eventually becoming one of the directors of the Union Boat Club (Baltzell and Schneiderman 1991). Brandeis's experience was hardly singular, for before the 1880s and the coming of a new national WASP upper class, wealthy Jews, especially German Jews, were able to assimilate into the local upper classes of most American cities. Thus the Morgenthaus, Ochses, Sulzbergers, Strauses, Lehmans, and Guggenheims—families that are still well-known today—were an integral part of the nineteenth century New York establishment.

The new national upper class that formed in the 1880s was different than the old local upper classes. For one thing it was clearly anti-Semitic. Historians and sociologists have suggested that WASP anti-Semitism was a reaction to the enormous number of immigrants from eastern Europe, and some have tried to distinguish this upper-class bigotry from other types by calling it "genteel" or "social" anti-Semitism, although there is hardly a difference. In 1880 there were only about 250,000 Jews in America, but by 1920 there were over 4 million, and over 350,000 were cramped into New York City's Lower East Side alone. About 80 percent of the 4 million American Jews in 1920 were of east European, as opposed to German, extraction. These new immigrants brought with them very different values, languages, and cultures than did the German Jews who were already well established here by the 1880s. Whatever the reasons, the new national upper class now became exclusively white, Anglo-Saxon, and Protestant, and began excluding Jews from the institutions that were associated with it. As Baltzell (1991, p. 48) has shown, "upper class anti-Semitism was perhaps more blatantly displayed in the five decades after 1880 than at any other time in our history."

By the 1890s Jews were being excluded from membership in patrician clubs throughout the country. In 1893, New York's Union League blackballed the son of one of its Civil War–era founders, Jesse Seligman, because he was Jewish. And by 1913 President Woodrow Wilson decided against having Louis Brandeis in his cabinet, after opposition to his appointment was voiced by patrician New Englanders who had once invited Brandeis into their social circle in the 1870s and 1880s. Bankers, lawyers, real estate magnates, railroad tycoons, stockbrokers, and others from families such as the Astors, Baldwins, Cuttings, Depews, Elliotts, Fishes, Griswolds, Hewitts,

Iselins, Jaffrays, Kanes, Livingstons, Martins, Newbolds, Otises, Posts, Roosevelts, Schuylers, Twombleys, Vanderbilts, and Winthrops were part of the national metropolitan WASP upper class, as were the Harrimans, Rockefellers, Tafts, Cabots, and Lodges. There were no Jews or other ethnic minorities among them.

This national WASP upper class paid more attention to wealth than to family, and was more conscious of ethnicity than the earlier local upper classes throughout America. Excluding all but WASPs, this national upper class gained a profound self-consciousness through a network of exclusive clubs, boarding schools, resorts, and Ivy League colleges that promulgated a subculture of common values and common norms of behavior—that is to say common morals and manners. The need for commonalty can be easily understood. The geographic distribution and larger numbers of the new national upper class required criteria beyond the geographical propinquity, face-to-face dealings, and local intermarriages that had prevailed in the preindustrial local upper classes. In spite of its anti-Semitic and antiethnic bigotry, however, this new national upper class served an important purpose in America: it provided a pool of national leaders generally committed to liberal democracy (Schneiderman 1992b, p. 82). This was of no small value, for as a famous political scientist, V. O. Key wrote in 1949,

> the operation of democracy may depend on competition among conflicting sections of the 'better element' for the support of the masses of voters. Hence, the workings of democracy require a considerable degree of disagreement within the upper classes. . . . In the absence of popular leadership from the 'better element,' the breach may be filled by persons and groups with few scruples and often little ability. (Key 1949, p. 181)

Political Leadership

Nothing illustrates Key's point better than Theodore Roosevelt's succession to the presidency in 1901, which marked the beginning of a cycle of presidents of upper-class origins, a cycle that lasted forty-four years, ending with the death of Franklin Delano Roosevelt in 1945. This cycle, which began at the turn of the century, reflected a new national upper-class's contribution to the political leadership of the nation. The patrician reformer Theodore Roosevelt, having inherited considerable wealth, had a certain loathing for the self-made millionaires of his day, whom he said "made the till their fatherland." He was followed in office by two other establishment presidents, William Howard Taft, a Republican as was Roosevelt, and Woodrow Wilson, a Democrat (Schneiderman 1989).

The election of 1912 shows the remarkable range of the new WASP upper-class's interest in politics, and of its capacity to compete for the support of the masses as Key (1949) suggested it must. In this election the three patricians who served, or would serve, as president from 1901 through 1921 were all candidates, and represented different political parties and different political platforms. Theodore Roosevelt running as a Progressive outpolled Taft, the incumbent Republican, but lost the election to the Democrat, Woodrow Wilson. Following Wilson's administration, the 1920s were presided over by presidents of upper-middle-class backgrounds—Harding, Coolidge, and Hoover—thus interrupting, but not ending, the cycle. Franklin Delano Roosevelt and a host of fellow patricians of inherited wealth and privileged backgrounds came to Washington in the 1930s and 1940s, thus keeping the cycle going until at least 1945. Although big government was here to stay by the beginning of the century, the entry of the patrician into national political life infused it with a new sense of noblesse oblige (Baltzell and Schneiderman 1988). Establishment figures such as Theodore Roosevelt, Taft, Wilson, and FDR entered political life to see that the huge machinery of government was administered honestly and efficiently.

Not all members of the new upper class, however, contributed to the leadership of the nation. In fact, relatively few did. Many observers actually felt that there was a shameful lack of responsible leaders in this country at the turn of the century. For example, James Bryce, author of *The American Commonwealth*, one of the best books ever written about our country, concluded that the majority of citizens pay so little attention to public affairs that "they willingly leave all but the most important to be dealt with by the few." Attributing this lack of interest to individualistic tendencies, Bryce (1921, p. 557) suggested that an individual's interests and priorities are ordered as follows:

. . . first, the occupation by which he makes his living . . ., secondly, his domestic concerns, his family and relatives and friends, thirdly, . . . his religious beliefs or observances, fourthly, his amusements or personal tastes, be they for sensual or for intellectual enjoyments, fifthly, his civic duty to the community.

Although Bryce was aware that the order of these interests varies from citizen to citizen, he was certain that civic duty would almost always rank last. As he said in *The American Commonwealth*, "individualism, the love of enterprise, and the pride in personal freedom, have been deemed by Americans not only their choicest, but their peculiar and exclusive possessions" (Bryce 1889, pp. 406–407). That these very American values born of democracy should be an obstacle to fulfilling the democratic spirit was an irony not lost on Bryce any more than it had been on Tocqueville before him (Schneiderman 1993).

In a famous lecture at Yale University in mid-October 1908—just before the election in which Taft was elected to the presidency, replacing Roosevelt, and preserving the WASP establishment's hold on the nation's top office—Bryce blamed the privileged classes for their inattention to civic duties. Such indolence in his eyes "made it a government of the many who don't care by the few who do." Noting the effect of selfishness in a business civilization such as our own, Bryce (1993, pp. 92–93) said that

. . . absorbed in business or pleasures, we think little of what our membership in a free nation means. The eloquent voice of a patriotic reformer sometimes breaks our slumber, but the daily round of business or pleasure soon fills the mind and public duty again fades into the background of life.

But Bryce's chief complaint was that among the educated classes, as he called the privileged upper-middle and upper classes, there was too much reluctance by "the fellows who are too good to run for minor offices," to take the lead in governing their local communities, states, and the nation. This was an old complaint against democracy, one which Tocqueville had thought was basic to the destruction of the *Ancient Regime* in France, and which Bryce had lodged against the "better classes" in America many times before. Basic to this complaint was an understanding that, as Bryce (1921, p. 501) later said, "as a rule, that which the mass of any people desires is not to govern itself, but to be well governed."

Also of concern to Bryce was that the higher-status, better-educated classes from which public leaders were most likely to be drawn were most apt to be affected by the love of money. Bryce, who was the British ambassador to America, and who was well acquainted with many members of the upper classes, observed that they had the duty to, and in practice did, set the moral standard for all the rest. Indeed, he called the leading political stratum drawn from the upper classes "the tone-setting class," because he believed that it is this class that "forms the standard not only for those who conduct public business but also to a great extent for the whole community." A tone-setting class, or what we would today call an establishment should, according to Bryce (1993, p. 119), set a high standard. If it sets a low standard, tolerates base motives and actions, "it depraves the morality of the community . . . and politics are defiled and debased, selfishness and trickery are taken to be natural, and public life becomes the favorite hunting ground of unscrupulous and reckless men." Once the moral or civic standard is lowered it is difficult to raise it again, and Bryce advocated legislation to prevent men from getting rich through public life.

Like his establishment friends, such as Henry Adams and Theodore Roosevelt, Bryce was clearly an advocate of responsible leadership by the "better" classes, and to him this did not mean successful moneyed classes per se, but, rather, solidly educated achieving classes willing to take the lead in civic duty. His ideal in this was his good friend Roosevelt who, he told his audience at Yale (Bryce 1993, p. 32), as a young man "more than twenty years ago did not consider himself above going to the lower house of the New York Legislature and who has now become an eminent statesman."

But Roosevelt, who had served as the police commissioner of New York City from 1895 to 1897—just a few years before becoming President of the United States—was indeed the exception in terms of local politics. And while the WASP upper class contributed men such as Roosevelt to national business and politics, it effectively abandoned local politics to men of other ethnic backgrounds. As America grew, there was plenty of room for talented individuals from outside this class to exercise their talents, especially in the cities of America.

While a few cities, such as Philadelphia, which did not elect an Irish Catholic mayor until 1964, maintained WASP leadership through most of this century, most, such as Milwaukee, which elected its first Irish Catholic mayor in 1863, or Boston which followed suit in 1885, were dominated by ethnic elites since the latter part of the nineteenth century. Thus New York City elected its first Irish Catholic mayor, William Grace, in 1881, its second, Hugh Grant, in 1889, followed by Thomas Gilroy, in 1893. With the exception of two WASP mayors, Seth Low and George McClellan, who were in office from 1902 to 1909, New York had an unbroken string of seven Irish Catholic mayors from 1910, when William Gaynor was elected, through 1934 when an Episcopalian, Fiorello LaGuardia, son of a lapsed middle-class Italian Catholic father and an Austrian Jewish mother, was elected to the office. LaGuardia, who was mayor until 1945, was followed by another Irish Catholic mayor, William O'Dwyer, then by an Italian Catholic, Vincent Impellitteri, an upper-class German Catholic, Robert Wagner, and finally, after a half-century of ethnic mayors, by a WASP, John Lindsay, in 1966. Following Lindsay's departure in 1973, New York elected its first Jewish mayor, Abraham Beame, in 1974, and its second Jewish mayor, Edward I. Koch, in 1978. Koch was followed by New York's first black mayor, David Dinkins.

If many of the great American cities were governed by ethnic minorities during the last hundred or so years, what effect did these new powerholders have on the authority of the Protestant Establishment? For one thing, the old-stock establishment based its authority on deference, in much the same way that Walter Bagehot said the English aristocracy had in the nineteenth century. Ironically, this process was at work in the Democratic Party, which has been, in this century, by and large, the party of immigrants, ethnics, and minorities. In the words of Andrew Hacker (1957, p. 1015), a political scientist, "Ed Flynn might boss the Bronx, but he would defer to Franklin D. Roosevelt (of Harvard); Carmine De Sapio rides behind Averill Harriman (of Yale); and Jake Arvey cleared the way for Adlai Stevenson (of Princeton). The seeming inconsonance of the fact that the party of the immigrant accepted old-stock patricians as its leaders is good evidence of the deference that was paid to the *ancien regime*." The Protestant Establishment might not have governed the cities of America, but for the first half of this century it set the tone and the atmosphere for culture and politics at the national level.

But if the WASP establishment set the tone and atmosphere for the exercise of national authority, how and why had it been able to do so? Asked another way, what is an establishment? Simply put, it is a group of leaders within an elite whose families are part of an upper class. Those few members of any upper class who form an establishment stand in contrast to the majority who are satisfied to have high status but little or no authority, in other words, with those who form a caste. The American establishment has been made up of WASP families such as the Adamses, Bayards, Breckinridges, Frelinghuysens, Harrisons, Lees, Livingstons, Lodges, Roosevelts, Stevensons, Tafts, and Washburns, who have followed the aristocratic ethos that emphasizes the duty to lead, rather than the snobbish caste ethos that emphasizes only the right to privilege. Just four families—Adams, Harrison, Roosevelt, and Taft—for instance, have not only provided the nation with seven Presidents (thus accounting for more than one-sixth of all our presidents), but have also contributed eleven governors, ten ambassadors, ten U.S. senators, twenty-seven members of the U.S. House of Representatives, six cabinet officers, and two justices of the U.S. Supreme Court, not to mention five mayors of major cities, and over fifty state and local legislators and officials.

These four families are representative of the sort of leadership provided by the Protestant Establishment. In fact, the U.S. presidency, the most visible position of leadership in the nation, has been dominated by establishment figures. Of our forty-one presidents, seventeen—Washington, both Adamses, Jefferson, Madison, Monroe, both Harrisons, Polk, Tyler, Taylor, both Roosevelts, Taft, Wilson, Kennedy, and Bush—have come from clearly upper-class backgrounds. And even though Kennedy was our only Catholic president, he was clearly upper class and educated as a WASP. But aside from the presidency, establishment families have provided more than their share of governors, justices and chief justices of the U.S. Supreme Court, diplomats, U.S. senators and congressmen, and cabinet officers.

The Establishment's Future

A moral force within the putatively amoral world of politics and power elites, an establishment of leaders drawn from upper-class families can protect freedom in modern democratic societies. Such an establishment of political, business, cultural, religious, and educational leaders succeeds in its moral function when it sets, follows, and enforces rules of fair play in contests of power and opinion. This point was brought home by the German sociologist Ralf Dahrendorf (1967, p. 277) in his brilliant analysis of the post–World War II political elite in Germany. "This elite is a mere sociological category without social reality," he wrote, and "it is my thesis that the German political class has developed from unity without plurality, to plurality without unity, and that both these states are highly detrimental for the constitution of liberty. . . . Liberal democracy demands an established political class . . . [and] a democratic elite has to be united in status, and divided in politics." Similarly, in one of his most penetrating insights into the sociological functions of an establishment, Tocqueville (1955, p. 142) wrote that "a powerful aristocracy does not merely shape the course of public affairs, it also guides opinion, sets the tone for writers, and lends authority to new ideas." In other words it has hegemony over society. Whereas Marxists see hegemony as a social evil, Tocqueville, like Dahrendorf, saw it as necessary to the well-being of society. Establishments give coherence to society. They do not eliminate conflict, but prevent it from ripping society apart.

One of the ironies of American history is that while power has been accessible to many, its sources remain hidden from the sight of most. This irony did not escape the eye of Reinhold Niebuhr, a leading theologian and social critic. "The knight of old knew about power. He sat on a horse, the symbol of military power," wrote Niebuhr (1952, pp. 12–13), "but the power of the modern commercial community is contained in the 'counters' of stocks and bonds which are stored in the vaults of the bank. Such a community creates a culture in which nothing is known about power, however desperate the power struggles within it." Those who have and control power in America are, as Niebuhr well knew, often protected from sight just below the edge of public scrutiny by egalitarian mores and traditions that create ambiguous feelings about power and authority, and often deny that the powerful should exist.

Decline

It is this blind edge of the power structure of American society that marks the line between the upper-class authority of the establishment and the power elites. The genius of an establishment lies in its capacity to put moral brakes on power by applying an upper-class code of conduct and responsibility to it. This capacity was at work, for instance, when Joseph Welch, senior partner in the old-stock, patrician Boston law firm of Hale and Dorr, serving gratis as the U.S. Army's special counsel, stood up to Senator Joseph McCarthy and finally brought him down, ending the career of one of the most dangerous demagogues in American history. Nevertheless, McCarthyism marked the beginning of the end of establishment rule in American national politics. After all, it was the establishment itself that McCarthy attacked. Alger Hiss, a patrician from an old-stock Baltimore family, a graduate of Harvard Law School, was associated with other establishment figures, such as U.S. Supreme Court Justice Oliver Wendell Holmes, for whom he clerked, President Franklin Roosevelt, for whom he was a special advisor at Yalta, his friend former Secretary of State Dean Acheson, and the new Secretary of State, John Foster Dulles. In 1950 Hiss was convicted of perjury for denying the accusations that he was a communist made by Whittaker Chambers. That same year Senator McCarthy began his four-year campaign to rid the government of communist sympathizers, and his attack on the establishment, especially on patricians such as Dean Acheson, whom McCarthy branded "a pompous diplomat in striped pants, with a phony British accent." It finally took the courageous establishment lawyer from Boston, Joseph Welch, who was unafraid to stand up to McCarthy, to help end one of the most disgraceful, and dangerous, episodes in our history. But the Protestant Establishment was beginning to lose its legitimacy, which it finally lost in the anarchic 1960s, when authority itself was broadly called into question in this country.

During the 1960s legitimacy and authority came into question and began to fail, fostering the growth of cynicism about many hierarchical elements of society, including the establishment. This cynicism has carried through to the present. Thus in *Cultural Literacy*, a well-known book that has been criticized as elitist by some on the academic left, E. D. Hirsch (1987) listed "the Establishment" as one of the ideas literate Americans should know. But for all the charges of elitism against him Hirsch defined "the Establishment" in a cynical manner his critics would applaud. In the *Dictionary of Cultural Literacy* (1988), Hirsch described the establishment as follows:

> Individuals and institutions that exercise social, economic, and political authority over a society. The term has a pejorative connotation because it suggests that political and economic power is in the hands of the few.

Perhaps all sociologists would agree with the first part of this description, but only a few would openly disagree with the pejorative connotation attached to the term.

A deceptive myth in liberal democracies like America is that civil liberties and freedom of expression are valued by most members of society, and most certainly by those at the bottom and middle of the social structure. But as Samuel Stouffer amply demonstrated in his classic study *Communism, Conformity, and Civil Liberties* (1955), and as others have found since, this is not the case. Such freedoms are always most highly valued and protected by the elite few who are better educated and who believe in the liberal democratic tradition, and the last bulwark of freedom may well be, as Baltzell (1964b, p. 293) has suggested,

> a unified Establishment from within which the leaders of at least two parties are chosen, who, in turn, compete for the people's votes of confidence, from differing points of view and differing standards of judgement, yet both assuming the absolute necessity of using fair means in accusing their legitimate opponents of fallibility rather than treason.

The sociological alternatives to rule by establishments have been, historically, rule by functionaries and bureaucrats or rule by demagogues, neither of which has proven satisfactory in protecting freedoms.

During this last century America has undergone a social revolution in leadership. Whereas the WASP upper class dominated America's political, economic, and intellectual elites up to the late 1940s, thus controlling the nation, today WASPs no longer have that control, nor perhaps does any other group. Thus Peter Schrag ended his cogent book *The Decline of the WASP* (1970, p. 255) by suggesting that

> America is not on the verge of becoming two separate societies, one rich and white, the other poor and black. It is becoming, in all its dreams and anxieties, a nation of outsiders for whom no single style or ethic remains possible.

American leadership has for the last half-century been increasingly dominated by atomized elites, like those in Germany described by Dahrendorf (1967), rather than by an establishment. The difference is of some consequence, because an establishment is a real group, a *gemeinschaft*, or community, subject to all the forces of social control, that at least potentially can instill and enforce moral standards in its members, if not society at large. An elite, however, is a *gesellschaft*, not a real group, but merely a convenient sociological category not subject to the forces of social control, and not a moral entity.

The castelike exclusivity and bigotry of the Protestant Establishment served to destroy its class authority. Yet America's best democratic traditions might still be best served by having an establishment of some sort to provide moral authority and leadership to the nation. Of course, those traditions dictate that we should choose the most talented available men and women to represent us in positions of moral and political leadership, regardless of their class, religious, ethnic, or racial backgrounds. But we should not be surprised if these individuals try to pass on to their children all the advantages they can, nor if these individuals and their families intermarry with each other and constitute a new establishment—albeit one more open to talented newcomers, and less exclusive than its predecessor.

A New Establishment?

In this regard it is notable that the most publicized and interesting recent political marriage in America was not between WASPs, and may reflect a new establishment in the making. In June 1990, Andrew Cuomo, the son of New York's Governor Mario Cuomo, married Kerry Kennedy, the daughter of the late Robert Kennedy, who was U.S. senator from New York, U.S. Attorney General, and candidate for the presidency when he was assassinated in 1968.

This very newsworthy marriage brought together two of America's most powerful political families, and Catholic families, no less. The Kennedys, perhaps America's best known Irish Catholics, are by birth, education, and achievement, one of the few non-WASP families in what remains of the Protestant establishment. The bride is the great-granddaughter of John Francis Fitzgerald, a mayor of Boston, and a U.S. congressman, the granddaughter of Joseph Patrick Kennedy, a self-made millionaire, and ambassador to Great Britain, the niece of President John F. Kennedy, and Senator Edward Kennedy, the daughter of Senator Robert Kennedy, and the sister of Congressman Joseph P. Kennedy II. She has a law degree, and is executive director of the Robert F. Kennedy Memorial Center, a human rights organization. The Cuomos, by contrast, are a newcomer family to America's political elite, an Italian American family which produced one of the nation's most powerful governors. The groom, a second-generation politician appointed to a post in the Clinton administration, has prepared himself for politics by running a New York City organization that builds transitional housing for the homeless. That Kennedys and Cuomos now stand shoulder to shoulder with Rockefellers, Tafts, Duponts, and other scions of the old Protestant Establishment hints that the open-class principle may be overcoming the tendency to caste.

One comment by Governor Cuomo on the day of the wedding, however, as reported in *The New York Times*, reveals some of the differences between new men of power, such as the governor, and establishment types such as the Kennedys. After having been urged repeatedly by his son, Andrew, to wear morning coat and ascot, the governor finally refused, saying "penguins look good in tails, birds look good in tails. Tails are the ultimate expression of form over substance. Who likes them? Probably three people in the country, and I would neither vote for those people nor appoint them to any office." Perhaps the governor had forgotten just how splendid the bride's uncle, President Kennedy, had looked in tails at his inaugural ball. But then again, the governor, a man of substantial intellect who graduated summa cum laude from St. Johns University in 1953, did not have the privileged upbringing of President Kennedy, who graduated from two of the Protestant Establishment's favorite schools, Choate and Harvard, where he learned the morals and manners of the upper class into which he became assimilated. Undoubtedly, the most famous and memorable words ever spoken by President Kennedy were those from his inaugural address: "Ask not what your country can do for you, but what you can do for your country." What most people do not know is that in his welcoming address to the boys in JFK's class at Choate, the headmaster, George St. John, had uttered similar words: "Always remember, it is not what Choate can do for you, but what you can do for Choate." Thus did the future president learn lessons about both form and substance, and later felt as much at home in tails as in the Oval Office. One would expect the governor's son, now part of an establishment family, to learn to feel more at home in tails than his father. If the remnants of the old Protestant Establishment can manage to assimilate new families into its precincts, and inculcate ideals of public service and commitment to liberal democracy, it may yet survive and provide the nation with leadership as it did earlier in the century. But will it do so this time as a properly open establishment, accepting achievement, rather than ascription, as its basic principle of inclusion? There are reasons to think that it may, as well as that it may not.

Conclusion

If the Protestant Establishment's values of achievement, conscience, industry, antisensuality, and civic-mindedness have become the backbone of American values, as Richard Brookhiser, in *The Way of the WASP* (1991), makes the case that they have, since the 1960s they have been challenged by a new emphasis on ascriptive values. No establishment can survive without deference, and there can be no deference without a commonly shared culture that maintains at least a modicum of consensus (Schneiderman 1992b). Throughout this century successful individuals and their families of Irish, Jewish, Italian, and Black backgrounds, were excluded from the WASP upper class; therefore they established their own upper classes. And according to Nelson W. Aldrich, Jr. (1988, p. 280), they bought into WASP culture even as they founded their own

. . . schools, summer resorts, clubs—and bred their own patricians and aristocrats. They also established their own suburban enclaves and cultivated their own ethnic consciousnesses. Still, it seemed that these non-WASPs not only wanted to imitate WASP culture but yearned somehow to *be* WASP.

This may no longer be the case.

American society has become increasingly fragmented, and there is little or no consensus to be found regarding our political, business, and cultural life. Deference for the authority of the Protestant Establishment, or for any establishment for that matter, has been replaced in many quarters by an active, conscious hostility and opposition. Whether or not Cuomos marry Kennedys may count for very little if we adhere to no common set of social standards, obey no collective moral injunctions, and subscribe to no single culture. This situation was analyzed with great insight by Francis Fitzgerald in *America Revised* (1980, p. 104), in which she says that the message of most contemporary textbooks is that "Americans have no common history, no common culture, and no common values, and that membership in a racial or cultural group constitutes the most fundamental experience of each individual. The message would be that the center cannot and should not hold." Under these conditions it is difficult to imagine the resuscitation, or reconstitution of the establishment. Then again, there are still the Kennedys and Cuomos, and countless other Irish, Italian, Black, Jewish, Polish, and other ethnic families who have adopted the WASP values of achievement and success, even while American society seems to be increasingly emphasizing ascriptive values in their place. As such, they may represent our brightest prospects for the future. As of now, it is too early to tell whether they will be successful in re-creating an establishment, especially one open to all talented individuals and their families, no matter what their racial, ethnic, or religious backgrounds, or whether the fragmentation of our society now under way will preclude such a development.

11

German Americans: Paradoxes of a "Model Minority"

WALTER D. KAMPHOEFNER

In May 1917, novelist Booth Tarkington and other prominent citizens of Indianapolis brought a petition protesting the fact that, in the bilingual schools of that city, the "Star Spangled Banner" was being sung in German translation (Ellis 1954, p. 372)! Depending on one's point of view, this could be seen either as an outrageous example of the lengths to which presumptuous German Americans would go in promoting their ethnic culture, or as a reassuring sign that patriotism and political loyalty is totally independent of mother tongue. Either way, it is an illustration that the image of rapid and complete German American assimilation often encountered today is at best a serious oversimplification if not entirely off base.

Examples of this image of rapid and complete assimilation abound. A *Harvard Encyclopedia* article found that, "The Germans in the 1970s were among the least visible of American ethnic groups." Similarly, the most recent synthesis of immigration history contends that, "If the cultural disappearance of German Americans has been often overstated . . . German Americans, as an effective ethnic group in contemporary America, just do not exist" (Conzen 1980, p. 406; Daniels 1990, p. 164)

Commentators at the time of heaviest immigration painted a much different picture. Already during the colonial period, Benjamin Franklin had complained about "Palatinate Boors" who were threatening to "Germanize us instead of our Anglifying them, and will never adopt our Language or Customs." Such fears died down as immigration slacked off in the generation following the American Revolution. Use of the German language declined except where it was reinforced by separatist religious communities such as the Mennonites and their Amish offshoots, Moravians, and other Anabaptist groups. But just as predictably, resentments flared up again with the resumption of mass immigration in the 1840s and 1850s. Apprehensive of the growing political power of immigrants, the *Buffalo Commercial Advertiser* in 1857 raised charges that have since become familiar against other groups:

> The fourth, fifth, sixth, and seventh wards are as little American as the duchy of Hesse Cassel; their population speaks a foreign language, reads foreign newspapers, isolates itself from the American element, and steeped in ignorance of American politics, it clings to the bald name of Democracy, and claims the right to subject the sons

of the soil to the despotism of the brute force of numbers.

Buffalo was not an isolated case; similar cultural patterns, and similar complaints, could be found with respect to Cincinnati's "Over the Rhine," Milwaukee's northwest side, or south St. Louis.

With the passage of time, however, Germans came to be looked upon as more desirable, particularly when contrasted with newer arrivals from southern and eastern Europe. The *Philadelphia Enquirer*, editorializing on the record numbers of immigrants in 1891, was probably thinking of Germans when it asked: "What kind of people are these new citizens? Some are honest men seeking a home. They will go West, take up land," and then went on to contrast them with the "hoards of Huns and Poles" and the "scum of Italy and Sicily." But even in this era one occasionally heard a tolerant voice, such as Isaac Hourwich (1912, p. 67) arguing that "every complaint against the immigrants from Eastern and Southern Europe is but an echo of the complaints which were made in an earlier day against the then new immigration from Ireland, Germany, and even from England."

Geographic Concentration

The preference for rural immigrants who would go on to become farmers was based as much on prejudice as on objective criterion. But paradoxically, the public image of German Americans—dominated as it was by rural scenes—was in any case largely undeserved. Despite the fact that emigrants were recruited most heavily from the rural elements of society, Germans in America throughout the nineteenth century were much more urbanized than either the compatriots they left behind or the inhabitants of the United States as a whole. In the newly founded German Reich of 1871, only 12.6 percent of the inhabitants resided in cities with populations of more than 20,000. By contrast, over 39 percent of German Americans in both 1870 and 1880 resided in the fifty largest U.S. cities, which ranged upwards in population from 25,000 and 35,000, respectively. In fact, there was a larger share of German immigrants living in towns of more than 25,000 inhabitants than of people residing in towns of 2,000 or more among the folks back home. Thus German immigration in-

cluded many raw recruits to urban life—unskilled laborers as well as the artisans who have usually dominated their image (Kamphoefner 1987, 80–82).

The urbanization rate of German immigrants is all the more striking when viewed in relation to the American population as a whole. During the entire second half of the nineteenth century, eight large metropolitan centers alone were home to between 20 and 30 percent of all German Americans, compared to only 7 to 9 percent of the total U.S. population. In relative terms, Germans in 1850 made up nearly four times their share of the metropolitan population. A similar pattern emerges when one includes smaller cities: from 1870 to 1890, the percentage of Germans in the population of the forty-four largest American cities was about 2.5 times their share nationwide. Still, although their absolute numbers in U.S. cities continued to climb from decade to decade, as a percentage of all Germans and relative to all inhabitants of the United States, their urbanization rate was highest in 1850 and declined steadily through 1890. Thus Germans were clearly swimming against the cityward tide of Americans in general. This suggests that many had been reluctant urbanites, and moved on to farming areas when they had accumulated enough money to purchase land.

Nevertheless, German Americans throughout the nineteenth century continued to be more urban and less concentrated in the farm population than most of the other "old immigration" groups, even those who arrived later and should have had more difficulty gaining access to land. The Irish and the British were practically the only "old" immigrants who were more heavily urbanized than Germans. The Dutch, Swiss, Norwegians, Danes, and even Czechs were overrepresented in the farming population as late as 1920, while Germans (and to a lesser extent Swedes) remained consistently underrepresented (Kamphoefner 1984; Hutchinson 1956).

German Americans were nevertheless quite visible in farming because they were such a large group—in raw numbers they surpassed any other immigrant nationality in agriculture. But much the same could be said for their presence in American cities. During the whole second half of the nineteenth century, Germans were the leading element in American cities. Since the runner-up was the English-speaking Irish, Germans were by far the largest foreign-language group. As late as 1920, a quarter-century after mass immigration from Germany had

WITTER'S

GERMAN-ENGLISH PRIMER

AND

N E W

FIRST GERMAN READER

FOR

PUBLIC SCHOOLS.

ILLUSTRATED WITH MANY WOODCUTS.

REVISED EDITION.

PREPARED BY

TEACHERS OF THE ST. LOUIS PUBLIC SCHOOLS.

ST. LOUIS, MO.

C. WITTER, PUBLISHER AND BOOKSELLER,

Witter's

Deutsch - Englische

Schreib- und Lese - Fibel

— und —

Neues Erstes Lesebuch

für

Amerikanische Freischulen.

Neue durchaus umgearbeitete und verbesserte Auflage.

Mit vielen Bildern.

Bearbeitet von Lehrern der öffentlichen Schulen in St. Louis.

St. Louis, Mo.

Verlag der C. Witter'schen Buchhandlung.

Bilingual education remains a controversial public policy issue today; yet in American history many immigrant groups practiced it for the benefit of their children. Here is the bilingual title page of an 1881 primer used in the St. Louis public school German program. (Reprint edition, Eberhard Reichmann, ed. Indianapolis: Indiana German Heritage Society, Inc., 1987)

ceased, only Italians made up a larger share of America's urban population than Germans. The largely Jewish immigrants from Russia were the only other group that even came close in numbers.

But Germans in cities did stand out from other immigrant groups in their regional concentrations. Very few settled in New England, which had a very heavy Irish contingent. Germans came in a close second to the Irish in the Mid-Atlantic region. But in nearly every city west of the Appalachians, Germans were the largest immigrant nationality and by far the largest foreign-language group. This held true not only for such leading and well-known German centers such as Milwaukee, Cincinnati, and St. Louis, but also for Chicago, Detroit, Indianapolis, and the Ohio cities of Cleveland, Columbus, Dayton, and Toledo (Ward 1971). This gave Germans major ad-

vantages in their attempts at preserving their native culture and language.

Reality notwithstanding, the popular image of a German immigrant was that of a farmer, and a superior one at that. As Conzen (1980) observed, "The stereotypical 19th century German farmer was regarded by contemporaries as stable, hard-working, dependable, and thrifty—some even said penurious." Others went further, crediting them with greater sagacity in their choice of location and higher productivity than their Anglo-American neighbors. Empirical investigations from Texas to Minnesota, however, have found little evidence of this alleged superiority, or indeed much distinctiveness of any kind in German American agriculture. Their cropping systems were very similar to that of their Anglo-American neighbors, their yields per acre usually no higher, nor did they prove to be particularly good judges of land quality. One aspect in which they did stand out was in landownership: already in 1850, a higher proportion of German immigrants were landowners than their Anglo-American neighbors in both Texas and Missouri (Conzen 1980, 1985; Jordan 1966; Kamphoefner 1987).

Once German Americans gained a foothold in farming, they tended to maintain and expand it. Although they were underrepresented in agriculture throughout the nineteenth century, the 1950 census showed second-generation Germans at 50 percent above their "quota" among farm operators, compared to their share of the total labor force. As the farm population continued to shrink, Germans managed to hold on better than others. Figures on ancestry from the 1980 census show Germans second only to Norwegians in their degree of concentration in the farm population. And in raw numbers, they outnumbered any other group down on the farm. Out of 5.6 million farm dwellers, over 1 million were "single ancestry" and another million "mixed ancestry" Germans—together they made up well over one-third of the 1980 agricultural population (Kamphoefner 1987).

Gains in agriculture, especially in the nineteenth century, were not simply a result of Germans' superior practices; they were more reflective of distinctive cultural values. One element of the German image and self-image which is borne out in fact is their greater locational persistence compared to the "restless" Anglo-Americans. This tenacity in holding onto the land was the vestige of a peasant mentality, the memory of population pressures in Europe, and a tendency to equate land with security. It contrasted sharply with the American pioneer view that land was an unlimited commodity. However, some of the German gains were made at costs that were considered unacceptable to most Anglo-Americans. Land was often accumulated at the cost of children's educations, and by putting wives and daughters as well as sons to work in the fields and milking barns. In this clash of cultures, the Yankee viewpoint was sketched by agrarian novelist Hamlin Garland ([1893] 1956) in his portrayal of a daughter of prosperous but traditional German immigrants:

> She had a little schooling . . . but her life had been one of hard work and mighty little play. Her parents . . . could speak English only very brokenly.

> . . . Her life was lonely and hard. . . . She knew that the Yankee girls did not work in the fields—even the Norwegian girls seldom did so now, they worked out in town—but she had been brought up to hoe and pull weeds from her childhood, and her father and mother considered it good for her.

Immigrant Cultural Preservation

German Americans were a diverse lot. Political unification of the mother country was not achieved until 1871, and considerable numbers of German speakers also originated from areas outside the German Empire. Religion was also a divisive factor. Catholics were the largest single denomination, but they were outnumbered by German Protestants of various competing affiliations. Further complicating the picture was a considerable anticlerical element, and a small but important contingent of German Jews who were sometimes a part of and sometimes apart from the larger German community. Thus language and culture were among the few elements around which most Germans could unite.

Whether rural or urban, and regardless of religious persuasion, German immigrants were perhaps nowhere more at odds with the dominant Anglo-American culture than in their views on alcohol and leisure. Typical is the complaint of an 1873 *Atlantic Monthly* article regarding the German immigrants: "Wherever they have settled in any numbers, they hold . . . the balance of power, and it would be almost impossible to pass a Maine Liquor Law, or a Sunday Law, or if passed, to enforce it. The principle that Christianity is part of the common law is fast disappearing wherever they settle" (cited in Helbich 1988, p. 133). If such conflicts have been absent in the last half of this century, it is not because German Americans adjusted, but rather because the dominant culture came around to the Germans' (and other immigrants') point of view. But a number of bitter battles were fought along the way. For example, an attempt in 1855 to raise Chicago license fees from $50 to $300 annually led to a three-day "Lager Beer War," pitting the nativist mayor against German saloonkeepers and their clientele. Three companies of infantry and one of artillery were called out to defend city hall, and one German was killed and a policeman maimed before order was restored. But the fees were lowered again to $100. A dispute with Know-Nothings over Sunday closings of theaters and other entertainments formed the background of an anti-German riot in 1854 that left eight people dead in St. Louis. Baltimore was the scene of a similar disturbance in the same year. But the worst example of nativist violence was Louisville's "bloody Monday" of August 1855, an election riot with a death toll of twenty-two. As late as 1872, the *Louisville Glaubensbote* was still reminding its readers which politicians had protected the Germans and which had supported the Know-Nothings (Kamphoefner, Helbich, and Sommer 1991).

After the Civil War, such battles were usually fought in the political arena, rather than in the streets. Here too German Americans more than held their own. In both Iowa and Wisconsin, the first Republican defeats of the postbellum era came when their German constituents were alienated by antialcohol legislation. It is also doubtful that nationwide prohibition would have ever been enacted had it not been for the anti-German sentiments aroused by World War I.

The German Press

A staunch defender of "personal liberty" and the most obvious manifestation of cultural preservation was the huge network of German-language publications. While the roots of the German American press reach back well into the eighteenth century, its heyday fell between the years 1848 and 1917, with both dates tied to larger historical events. The well-educated, idealistic political refugees of 1848 were only a tiny minority compared to the masses of economically motivated migrants, but they leavened the whole loaf. They gave the ethnic press an electrifying impulse just at the time that the Germans overtook the Irish to become the largest immigrant nationality. By the Civil War there were German daily newspapers in at least sixteen cities around the country, with two or more dailies competing in many of the largest cities. Except for Brooklyn Germans, who were well supplied from New York, the largest urban group without its own daily were the 9,000 Germans of Cleveland (Arndt and Olson 1965; Wittke 1957).

The immigrant was astoundingly well supplied with reading material, even compared to folks back home. With four German dailies, New York in 1850 was better off than the Prussian capital of Berlin or the publishing center of Leipzig. By 1872, the *New Yorker Staatszeitung*, with its circulation of 55,000, claimed to be the world's largest German newspaper. In 1876, the centennial of American independence was celebrated by seventy-four German-language dailies, printing one copy for every fifth German American. Counting weekly circulation, there was more than one paper printed in German for every two immigrants. Every one of the then forty-eight states except for Maine, Mississippi, and New Mexico had a German-language paper at some point.

Representing the largest foreign language group in the country was not the only advantage held by the German American press. The great bulk of Germans were literate by the late nineteenth century. And the group had achieved at least a moderate level of prosperity, making subscriptions more affordable. A county population of between 1,000 and 2,000 Germans constituted a group large enough to support a weekly newspaper. As the 1885 Iowa census shows, the only Iowa county with more than 2,000 Germans that did not support a newspaper was populated largely by separatist Amana colonists. In fact, the *Ottumwa Journal* survived from 1871 to 1912, despite having only 569 Germans in its home county (Jackson 1885).

At that time, nearly four-fifths of all foreign-language periodicals in the United States were German. As the number of German publications reached a peak in 1894 of nearly 800 (including 97 dailies), they still constituted two-thirds of the foreign-language press. But from this point on, as immigration dwindled to insignificance, the German-language press embarked on a slow but inexorable downhill slide. Even in such a heavily German city as Milwaukee, the combined circulation of German newspapers declined from 92,000 in 1884 to 50,000 in 1910, and more significantly, from double that of the English competition in 1884 to merely one-third their total in 1910. The anti-German hysteria of World War I did not initiate this decline, but it did markedly accelerate it. In 1912, half of all foreign-language periodicals in the United States were still German, more than 600 in all. But by 1920, wartime censorship, intimidation, and loss of advertising revenue had reduced the German total to less than 300, merely one-fourth of the foreign-language press (Park 1922; Wittke 1957).

While contemporary America has seen papers such as the *Miami Herald* branch out into the Spanish market, two major and numerous minor German publishers made the transition to English. Best known is the immigrant Joseph Pulitzer, who got his start as a reporter for the St. Louis *Westliche Post*. But the Knight-Ridder publishing chain also owes the second half of its name to a German Catholic enterprise that achieved notoriety by flying the German flag over its New York headquarters at the outset of World War I.

Bilingual Education

The ethnic press was closely involved in a mutually supportive relationship with German-language elementary schools, public as well as private. Although Catholics had the largest and best-known parochial system, accounting for about one-third of the pupils learning German, there were also 100,000 pupils in 2,100 German Lutheran parochial schools across the United States at the turn of the century. Some, perhaps most of these, were bilingual, although in Wisconsin in 1888 there were 130 German parochial schools that taught no English whatsoever. However, 42 percent of all elementary students who learned German in 1900 did so in public schools at public expense (Conzen 1980).

Beginning with Pennsylvania and Ohio in 1838, at least a dozen states required foreign-language instruction in public grade schools wherever a certain number of parents or school board members requested it. The practice was probably even more liberal than the law, particularly in homogeneous rural settlements. Missouri, for example, made no provision for foreign-language elementary instruction, yet the state school superintendent complained in his 1888 report:

In a large number of districts . . . the German element . . . greatly preponderates and as a consequence the schools are mainly taught in the German language and sometimes entirely so. Hence, if an American family lives in such a district the children must either be deprived of school privileges or else be taught in the German language. . . . Some of the teachers are scarcely able to speak the English language.

In heavily German Gasconade County, Missouri, half of the public grade schools were taught partly in German (Kloss 1977).

At the municipal level, bilingual education extended even further. Many systems—St. Louis, Chicago, and New York, for example—only taught the language as a subject for one hour a day. But others such as Indianapolis, Baltimore, Cleveland, and Cincinnati went to surprising lengths to meet immigrants halfway, establishing fully bilingual systems, with the German language as a means of instruction. The Indianapolis system went so far as to publish its own translations of English texts when suitable German books were not available. At the turn of the century, bilingual instruction cost Cleveland an estimated $100,000 a year, a sum roughly equal to the annual earnings of 200 workers. It divided its school week equally between English and German in its bilingual schools, eleven hours of each. Children in the German track got less than five hours per week of English reading, grammar, and spelling, compared to nearly eight hours in the normal track. But this apparently caused no ill effects, because on the high school admission exams (conducted entirely in English) a higher percentage of children from the bilingual track passed than from the English only. An Anglo-American principal from Cincinnati reported similar results for his school system (Kamphoefner, Helbich, and Sommer 1991).

In early twentieth-century Ohio, the leading state in terms of bilingual education, nearly 8 percent of all grade school students were studying German at a time when German immigrants and their children made up less than 13 percent of the state population, so about half of all German-stock pupils must have been covered by these programs, in which some non-German children also participated. What were the motives for offering foreign-language instruction in grade schools or for participating in these programs? While testimony of school officials and others showed that employment considerations played a role, a commission looking into the question in Cleveland in 1906 concluded that "the reason for the teaching of German in the primary and grammar grades . . . is not educational, but chiefly national and sentimental." Here again, the testimony of the time contradicts the image of eager and early German assimilation.

All the public foreign-language programs in elementary schools were wiped out by the anti-German hysteria of World War I, but the nativist crusade did not stop with public schools or with the end of the war. By early 1919, fifteen states had also passed laws requiring English as the means of instruction in all schools, public or private. Not satisfied with that, seven midwestern and plains states forbade any in-

Though the migration of Germans to the United States, particularly the Midwest, was quite successful overall, around the time of World War I those with German names and accents met with a substantial amount of prejudice, provoked by the war, and often suffered a reversal of their fortunes. Such was the story of the family of David F. Allmendinger, founder of the Allmendinger Piano and Organ Company in Ann Arbor, Michigan, who was born in Germany in 1848. Here he is with his family, four generations, in front of their Ann Arbor home in 1907. (Allmendinger Family Collection; Bentley Historical Library, The University of Michigan)

struction in any language other than English in any elementary schools. A revived Ku Klux Klan in the early 1920s targeted not only blacks, but also Catholics, Jews, and even such Protestants as dared to speak a foreign language. In Oregon, a constitutional amendment backed by the Klan was ratified in 1922, essentially outlawing parochial or private schooling for all children between the ages of eight and sixteen. German Lutherans, also supported by Catholics, challenged the Nebraska language law. In Oregon, Catholics led the fight against the school law, with Lutheran, Jewish, and other backing. In both cases, the Supreme Court decided for religious and linguistic liberty, but by then the damage had largely been done. Most of the German-language programs

in parochial elementary schools disappeared by 1930 (Jorgenson 1987, 205–215; Luebke 1974, 1990.

One interesting but elusive question is the long-term impact of school programs on language preservation. The ironic conclusion, from the limited evidence we have, is that the language hung on most tenaciously in rural areas, despite the concentration of public bilingual programs in cities.

The 1940 census was the first to pose the question of mother tongue to persons beyond the immigrant generation. Even at this late date, there were over 900,000 German speakers who had no immigrants closer than their grandparents. They represented the largest group among the nearly 3 million old-stock Americans who grew up speaking a language other than English. Also among the second generation, Germans were slower to adopt English than most other "old immigration" groups, even those who had arrived later. While the second generation was all born in America, it made a difference whether your parents were just off the boat or had themselves come over as children. Germans were outranked in English ability by the Danes, French, Swiss, Dutch and Swedes, and barely outscored the Rumanians and Norwegians. Once again, the reality of German immigrant life is at odds with the popular image (U.S. Bureau of the Census 1943b).

Another surprising feature is the way German language preservation varied between rural and urban environments. Traditionally both scholars and the American public have had a strong prorural bias. Frederick Jackson Turner considered the frontier to be one of the strongest forces promoting Americanization. Critics of the "new immigration" from southern and eastern Europe in the late nineteenth and early twentieth century, considered rural settlers to be most desirable and had strong reservations about urban ghetto dwellers. The 1940 census data on second-generation Germans show just the opposite: English acquisition took place most quickly in the big cities and was slowest down on the farm. Overall, the second generation was divided almost evenly between English and German speakers. In the central cities of large metropolitan areas, however, over 57 percent grew up speaking English. In suburbs and smaller cities, the figure was 55 percent. But in rural areas, German speakers predominated; they made up nearly 54 percent of second-generation Germans in the rural nonfarm population and a striking 64 percent among farm dwellers. This held true even though cities contained more recent arrivals, and length of residence should have worked in favor of rural residents (Kamphoefner 1992).

The place where the German language held on most tenaciously was in the farming areas of the upper Midwest. Wisconsin, the most heavily German state nationwide, led the way with over two-thirds of its second generation growing up speaking German rather than English. In the neighboring states of Iowa, Nebraska, Minnesota, and North Dakota over 60 percent of second-generation Germans reported German as their mother tongue: the same held true for Texas, another heavily rural state. Indiana, Missouri, Kansas, South Dakota, and Montana are were the only other states with an appreciable German population where over 50 percent of the second generation grew up speaking German. Large urban populations in Illinois, Michigan, and Ohio, all leading immigrant destinations, put these states just below the 50 percent mark.

Several factors help explain the greater language persistence in rural areas. To begin with, rural ethnic communities were truly communities, often formed by chain migration and including a high proportion of people from the same county or village in Europe. Granted, there was some clustering by place of origin in city neighborhoods as well, but seldom could one group dominate as they often did in rural areas. Family migrants were also more prone to settle in rural areas, whereas those who came as young, single adults often preferred the cities. Thus, the ties reinforcing ethnic identity among rural Germans were more emotional and intense than they were in the cities. It was not an abstract sense of nationality, but ties reinforced by a common dialect, local roots, and a shared migration experience that commanded loyalty. An urban environment undermined such provinciality much more quickly, with the German American melting pot preparing the way for the American one (Kamphoefner 1984).

Ironically, the bilingual education offered in the cities could do little to change this. The fears of the urban "huddled masses" that pervaded the 1911 Immigration Commission Report and formed the background of the discriminatory legislation of the 1920s were totally off base. It was practically impossible to shield the youth from American popular culture in the big cities. Hence, it was in the rural ethnic enclaves, not in the urban ghettos where foreign languages persisted the longest.

Contemporary policy makers should be reassured by another finding from the 1940 census data: children of German immigrants who grew up speaking German suffered no great economic disadvantages over those who grew up speaking English. How favorably their occupational profile compares with that of Anglophones depends largely on how highly one rates the occupation "farmer." These people could not have been too marginal a part of the agricultural population, given the persistence of German Americans farmers to the present. In other sectors of the economy, German speakers tended to be found in more traditional occupations. But they actually had higher rates of home ownership and self-employment than English speakers. Among employees the wage rates of Germanophones were only 15 percent below those of Anglophones overall, and actually surpassed them in several occupational categories. Similarly, there was only a 5 percent gap in the value of owned or rented dwellings between the two groups overall. Among farmers, German speakers were actually ahead. All this suggests that the alleged dangers of bilingualism have been somewhat exaggerated (Kamphoefner 1992).

Conclusion

Given this evidence of language and cultural preservation on the part of Germans, what accounts for their image as a group that was quite easily Americanized? Part of the explanation lies in the Anglo-Saxon race ideology at the turn of the century, which caused people to see what they wanted to see. Also, observers such as the U.S. Immigration Commission neglected the element of time, comparing "new immigrants" fresh off the boat with "old immigrants" who had been around at least several decades, if not a couple of generations. But perhaps most importantly, the Germans who were least assimilated were also the most isolated and least visible—those in ru-

ral enclaves. And even they did not appear to pay a heavy economic price for their cultural preservation.

The immigration restriction laws of the 1920s that set up the national origins quotas and curtailed the immigration of southern and eastern Europeans brought one final irony. The emotional excesses of the "100% Americanism" movement and charges of "hyphenism" during World War I were directed above all against German Americans. Yet Germans, like other northwest Europeans, came off rather well under the national quotas that were established. Poles and Czechoslovakians, two groups who for their own nationalistic reasons had most enthusiastically supported the Allied cause in the Great War, were restricted to annual quotas of 6,500 and 2,800, respectively, while that for Germans exceeded 26,000 (Archdeacon 1982). Even this influx of new blood was not sufficient to keep many German cultural institutions alive. Already before the war, second-generation Germans had outnumbered actual immigrants two to one. With them, acculturation was far enough advanced that, when forced to make a choice—unwelcome as this choice may have initially been—they moved easily into the American mainstream. But it was only from this point on that the behavior of German Americans began to match their image as wholesale assimilationists.

12

Erin's Children in America: Three Centuries of Irish Immigration to the United States

HASIA DINER

Like all immigrant groups, the Irish[1] made their way to America in distinct stages. Each stage—in immigration history aptly termed "wave"—reflected the specific conditions of the time in terms of the dual forces of pressures at home ("the push") and the opportunities in the receiving society ("the pull"). Each wave differed from the others in terms of the causes of the exodus, the nature of the migrant pool, and the process of adaptation, adjustment, and reception in the place of destination. Yet earlier and later migrants from the same homeland shared certain experiences and functioned in common institutions—and in America they created an ethnic identity that transcended historic differences.

Irish immigration to the Americas has been continuous, and it is fair to say that since the end of the sixteenth century Irish people have always been present in the Americas. They began to migrate to Britain's American colonies in the seventeenth century, joining up with the first White settlers and shaping the new nation, helping to make it "American" rather than English. They continue to flow across the Atlantic at the end of the twentieth century, making up a significant part of the contemporary immigration. Many are today, indeed, "illegal" or undocumented aliens.

Yet the middle of the nineteenth century saw the heaviest flow of Irish immigration, and in the decades from the 1840s through the 1880s, in terms of sheer numbers, the Irish made their greatest impact on American society. In these years they engaged in an intense process of institution building. The idiosyncratic nature of their economic adaptation, their unparalleled skill in the political arena, and the sharp social and cultural contest that ensued between them and the American Protestant majority made their immigration a key event in American social history.

The Irish immigration to America can be divided into four eras: pre-1840s—that is, prefamine; a famine immigration of the late-1840s and early 1850s that saw men and women fleeing the devastation of the potato blight; a postfamine migration that reflected the revolution in Irish society wrought by the aftershocks of the famine, which stretched from the 1850s through the 1880s (but actually continued until the 1920s); and after a lull between 1920 and 1960, a new immigration of the 1970s and 1980s.[2] Certain characteristics link all of the eras of Irish migration to America. First, economics played the most significant role in propelling the Irish to America. Theirs was an escape from poverty, varying in severity, but political or religious crises at home played little or no part in sending them overseas. Second, the Irish were ardent institution builders in America, quick to perceive the need to provide buffers between themselves and the American majority to protect them from the alien and often hostile environment in their new home. Third, the Catholic Church became the premier institution in their communities, and the Irish transformed a small, insecure Catholic Church in America into an Irish institution which volubly proclaimed the need for the Irish to remain apart from American culture. Finally, the Irish migration was a permanent one. Unlike other immigrants—Italians, Greeks, and Poles—the Irish, no matter how lyrically they spoke about Ireland and its green hills, no matter how intensely they crusaded for the cause of Irish independence, had moved for good. With the exception of a nostalgic trip back home to show off American success, the Irish had come to America forever.

Before the Famine

For much of American history, the terms "Irish" and "Catholic" have been inextricably bound together, synonymous words denoting the near universality of Catholicism among the Irish immigrants, the significance of the Catholic Church in the lives of the Irish newcomers and their children, and the powerful influence that the Irish exerted over American Catholicism. Yet among the first Irish who migrated to America, Protestants—Presbyterians, in particular—also joined in the migration stream, and in the seventeenth and eighteenth centuries, to be Irish did not necessarily mean to be Catholic. In the seventeenth and eighteenth centuries a religiously and economically heterogenous migration from Ireland to the Americas reflected the turbulence of life at home and the opportunities in the colonies.

The centuries of English colonization of America coincided with a revolutionary period in Irish history. In this vast sweep of time, the English successfully consolidated power over Eire, displacing politically and religiously both the traditional Catholic majority and the small, but important, Presbyterian—but Irish—minority. Local Catholic landowners lost possession of their estates to the English, who had been settled there purposely by the Crown to solidify English authority. In the process of this forcible

transfer of land, the Irish people became increasingly pauperized. The Anglican Church, known ironically as the Church of Ireland, pushed legislation through Parliament basically criminalizing Catholicism, as well as suppressing various dissenting Protestant denominations. Catholics, for example, could not purchase land from Protestants, they were forbidden to lease land for more than thirty-one years, and they could not send their children abroad to school, effectively barring them from the ranks of the educated elite. New English legislation destroyed the traditional inheritance system, and now when the head of a family died, land could not be passed to a single heir but had to be equally divided among the sons. The Anglican rulers treated Presbyterian and Quaker dissenters harshly also, although by the beginning of the eighteenth century, they softened their blows against these Protestants who lived primarily in the province of Ulster. This, in turn, exacerbated Catholic-Protestant animosity in Ireland, which had been strong to begin with. In addition, the English administrators of Ireland made tremendous efforts to stamp out the Gaelic language, replacing it with English, and sought to root out vestiges of traditional culture (Hechter 1975; Quinn 1991).

At the same time, economic changes in Ireland contributed to the earliest flow out of the island to America. British policy here could be summarized as enforced modernization in order to make Ireland an effective element in its imperial mercantile system. During the seventeenth century Ireland changed from a primitive woodland economy to one based on commercial agriculture. In addition, the development of a textile industry producing linen and woolen goods fit into the English mercantile scheme, although when its success began to threaten English manufacturers, Parliament imposed restraints on it, which lead to a cycle of constant depression followed by brief economic rallies between 1740 and 1780. Not surprisingly, the majority of agricultural and industrial wealth rested in the hands of Anglicans (Cullen 1981).

The social and economic structure mirrored the British domination over the Irish, the degradation of the Catholics, and the awkward, middle position occupied by the Irish Presbyterians. As of 1755 British landowners controlled almost all of the land, with Catholics owning only 5 percent. Of the tenants who could enter into long-term leases, Presbyterians predominated, while Catholics made up the bulk of the cottiers, the majority who rented land from farmers and paid back their debts by labor. Even lower, Catholic rundale occupiers lived in together on patches of unproductive remote land in a kind of co-operative system. The Irish cities were home to a British-dominated Protestant elite of merchants and industrialists, with a small handful of better-off Catholics (Freeman 1957).

It should not be surprising that such an unstable environment would lead to a slow but steady exodus of people. Irish Catholics, Presbyterians, and Quakers all had reason to leave. Importantly, Irish Catholics had migrated seasonally for centuries, going to England and the European mainland as mercenary soldiers and agricultural laborers whenever their services were needed (Miller 1985, p. 34). The Irish began to appear in America initially as regular paying passengers, although most came as indentured servants whose ship's passage was paid by their labor for a fixed term. Some indeed were brought to America, both to the West Indies as well as North America, as involuntary immigrants—political and religious prisoners and their dependents—who were transported to the colonies in lieu of incarceration in Ireland (Galenson 1981; Smith 1947).

Pennsylvania and Maryland served as the most important destinations for these Irish immigrants, although New York, the Carolinas, Virginia, and New Jersey also housed sizable Irish enclaves of Catholics, Quakers, and Presbyterians. Philadelphia, by 1800, was home to about 6,000 individuals born in Ireland, representing all three of the religious communities, and it was America's premier Irish city. Their increasing numbers can be gleaned from the fact that some of the colonies began to pass legislation restricting the entry of "papists" and South Carolina in 1698 and Maryland in 1699 legislated directly against the importation of Irish Catholic servants. Such efforts proved to be futile and the Irish flow continued. In most of the places they settled they suffered few civil disabilities. Catholic churches began to crop up in most of the colonies, and the majority of the parishioners hailed from Ireland. Concrete figures about the Irish population in general and the relative strength of the faith communities within it remain elusive, but a good estimate emerged from the first U.S. census of 1790. It counted approximately 400,000 individuals who claimed an Irish background, making them the largest non-English group during the colonial period. In general, the earlier the

period under scrutiny, the more Protestant the Irish migration, the later, the more Catholic, and by the time of the American Revolution, Catholics predominated (Doyle 1981).

These Irish Catholic migrants came from across the class spectrum, although that in and of itself represented a fairly limited spread given the severe economic and religious persecution of the Catholics in Ireland. A few, like Charles Carroll of Maryland, a signer of the Declaration of Independence, enjoyed vast economic wealth. Others, like Thomas Fitzsimons, a Philadelphia signatory to the Constitution, and Mathew Carey, an early economist, counted themselves among the comfortable merchants of Philadelphia, New York, and other coastal cities, who founded America's earliest Catholic churches (Shea 1886–1892). Parishes like New York's St. Peter in 1785, St. Patrick in 1809, and St. Mary in 1826 (Dolan 1975, p. 13) were heavily Irish and were dominated politically by the well-off merchants who belonged to them. They were joined in creating these early Irish institutions by the artisans, particularly tailors, who probably represented the majority of the American Irish. Others who had come as indentured servants experienced some degree of economic mobility, and many moved out of poverty into modest landowning and small business ventures. For sure, pockets of Irish poverty existed in the early centuries of settlement in America, and not all of the Irish would have been comfortable merchants or artisans. In addition, in the seventeenth and eighteenth centuries, Irish men outnumbered Irish women as immigrants, and the young, able-bodied men constituted the backbone of the immigration. As such, they resembled most of the immigrants to America at that time, indicating the relatively fluid nature of the American economy, which offered quite impressive opportunities for men willing to work hard on the land and in a variety of urban enterprises (Miller 1985, pp. 137–168).

From the point of view of American society in the prefamine period, the Irish constituted a distinct, although not necessarily despised, element among the British majority. Likely to be Catholic in a Protestant sea, the small enclaves of Irish in American cities drew little attention to themselves as different. Where anti-Catholic sentiment existed, it tended to remain in the realm of rhetoric, rather than spill over into either discrimination or violence. Although they tended to be more urban than rural, some of the

Irish did settle in the countryside. Their economic profile did not put them at conspicuous variance with the American population as a whole, and among their number could be seen wealthy merchants, affluent landowners, and comfortable artisans, with a smattering of the poor as well. Men, rather than women, predominated among the newcomers, and Irish immigrants and their children, modest in means, did not seem to be a "problem" in the eyes of Americans. They were seen as a distinct element in the society whose presence did not disrupt the social fabric (Knobel 1986).

The Great Watershed: The Famine and the Exodus to America

It is difficult to overstate the significance of the famine of the late 1840s to Irish society. It not only halved Ireland's population, chopping the island's number from 8 million to 4 million, but it set in motion a series of processes that would reverberate for decades, utterly transforming Irish social and cultural life. Although the famine's shadow would loom over the Irish people for a half-century or more, its immediate aftermath directly affected the process of immigration and it needs to be viewed as an era in the history of that migration in and of itself.

While it is true that widespread emigration from Ireland had begun before the devastation of the potato blight which first hit in 1845, it is also the case that the migration was vastly spurred on by this ecological nightmare. Between 1845 and 1851, some 4 million people disappeared from Ireland. Half of this number died due to starvation and the attendant diseases that came with weakening from malnutrition. The other half emigrated. In 1845 alone, 77,000 people left Ireland; the following year 106,000 abandoned the country. In 1851 the largest exodus occurred, with over 250,000 leaving. The United States loomed as the single most significant destination for these men and women who were fleeing from starvation (Edwards and Williams 1957; Woodham-Smith 1962). Additionally, the famine migration differed from the migration that preceded it, or those that would follow it, in that it was generally less selective. The famine swept up Irish people, regardless of age, gender, or skill level.

Basically, unlike more deliberative emigrations, the famine exodus involved anyone who could possibly get out, without any kind of strategizing of where to go, when to go, or who should go. The migration of these years came from all of Ireland, although immigrants from the midlands and the south predominated because those areas suffered the most from the blight, and it was here that peasants depended most on the cultivation of potatoes (Miller 1985).

In order to understand the significance of the famine and its importance in the immigration of the Irish to America, it is necessary to turn back to Ireland and briefly examine the nature of the Irish economy in the eighteenth and early nineteenth centuries. Between 1793 and 1815 Ireland enjoyed relative prosperity. Because of the constant warfare in Europe and the expanding British population, Irish farmers grew cereal crops on land owned primarily by English landowners. These cereal crops enjoyed high prices at the market, and landlords tolerated peasants who planted potatoes on small patches of unenclosed land for their own sustenance. The potato had been introduced into Ireland in the early eighteenth century, but it was not until the English had completed their domination of the Irish landholding system that the Irish Catholic majority turned to potato cultivation and potato consumption as their means of physical support. Potato cultivation, which accompanied the impoverishment of the Irish, gradually dominated all of Ireland and triggered a constant process by which the land was broken down into smaller and smaller holdings. At the same time the Irish population was growing as the age of marriage dropped, the birth rate skyrocketed, and the population grew (Seavoy 1986, pp. 297–345).

Despite the fact that in the decades before the famine emigration from Ireland to all sorts of places had been ongoing, the population at home still continued to grow. This, in turn, placed severe pressure on the land. The growth in population and the increasing dependence on the potato particularly characterized the people on the lower end of the economic scale. As a consequence, when the blight hit the potato crop in the summer of 1845 and in several subsequent summers, the poorer classes suffered the most, both in terms of starvation and disease. Hence it was from their ranks that the majority of the emigrants were drawn.

The uniqueness of the famine emigration from Ireland can be demonstrated in a number of ways. First, the earlier Irish movements to America had been timed to occur in the spring or early summer, in part to coincide with the greater availability of jobs on the other side of the Atlantic. In 1846 when the first crop failure hit, Irish emigrants fled in the winter as well as the summer, risking the hazards of a winter journey; most seem to have embarked without any preparation or provision. This precipitous flight from hunger and disease continued in 1847 and 1848. In addition, the Irish who fled in these years were equally divided between men and women, with children and older people—not necessarily the best of immigrants—flooding in as well. The Irish newcomers who streamed into Boston, New York, Philadelphia, Baltimore, and a few other U.S. ports were generally poorer than those who came before them and seemed to possess fewer marketable skills. Most fell into the general category of laborers and servants, with just a thin sliver being farmers or artisans, as earlier. While the poorest of the Irish population did not come to America, because they could not afford the passage, the famine immigrants still represented a drop in the social and economic status of the Irish population (Kennedy 1973).

Owing to their depressed state before immigration, the Irish arrivals of the famine years were more in need of assistance and more dependent on charity than their predecessors had been. In addition, they did not choose their destinations with any degree of care, since they did not have the luxury of studying the relative merits of one place over another. Thus, the cheapest route to the United States took them through the port of Boston, which was a city lingering in economic stagnation. The New England city had no jobs to offer them, and boatloads of Irish immigrants came to a place with a long history of anti-Catholicism and no established Irish Catholic community to render services to these needy newcomers (Handlin 1941).

Wherever they landed, the Irish immigrants of the famine years found themselves relatively trapped. They had no capital with which to venture out in search of greener pastures, perhaps to the expanding American West, although Irish men in particular were highly mobile geographically, and they shifted around the country taking the least skilled, lowest paying, and most dangerous jobs on canal construction crews and in railroad building gangs. Irish women, if single and unencumbered by children, flooded into American homes as domestic servants and carved out a particular niche for themselves in the economy that would persist through the century's end. From the 1840s on, domestic service became *the* occupation of Irish immigrant women (Diner 1984).

The Irish settled predominantly in the cities of disembarkation: New York, Boston, and Philadelphia. Within those cities they clustered together, forming densely packed Irish enclaves in the poorest neighborhoods. In New York, for example, the most destitute Irish during the famine years squatted in a vast shanty town of about 20,000 on the northern fringes of the city. In the expanse of land which in the 1860s would become Central Park, they put up shacks and Irish men and women scavengered for food to eat, hired themselves out as day laborers, while their pigs and goats roamed at will (Ernst 1949).

Against this backdrop, it is not surprising that the era of the famine migration, atypical though it was, set in motion a wave of anti-Irish, anti-Catholic hysteria. Americans, largely Protestant and evangelical, viewed with alarm the floods of Catholic immigrants coming into America and believed that American republican institutions stood vulnerable to "papist" attack. Public fervor against the Irish and their church sparked into violence: angry mobs stormed convents and churches, attacking priests and nuns as the visible symbols of an alien invasion (Billington 1938; Higham 1963).

The anti-Catholicism of the 1840s and 1850s led to the emergence of a variety of nativist organizations that fed on a deeply rooted Protestant antipathy toward Catholics. But in the wake of the famine migration, prejudice and discrimination specifically directed at the Irish, articulated in ethnic or national terms, entered the national discourse. The Know-Nothing (or American) Party proved to be the most potent of the anti-Irish political organizations of the 1850s. Tapping into pervasive fears of America being inundated by undesirable Catholic immigrants, the Know-Nothings captured several governorships, elected a handful of congressmen, and in 1856 ran former President Millard Fillmore as their presidential candidate.

The Irish were held up as the source of America's ills, and in the popular press, in political speeches, on the stage, and in a range of other texts American commentators blamed the Irish for the crime, disease, filth, and disorder of nineteenth-century American cities. Irish men and women were ridiculed or castigated for their fecklessness, drunkenness, and violence. Stereotyped as primitive and stupid, the Irish were caricatured in cartoons with simianlike features, sporting protruding foreheads, prominent jaws, and bulging eyes. Middle-class Americans oriented toward the reform of society and bent on curing social ills launched a series of movements—temperance, public school reform, and penal reform, to name but three—that specifically sought to make over the Irish into good, upstanding Americans (Knobel 1986).

American hostility to the Irish received further ammunition as the new arrivals from Eire jumped into the American political process relatively quickly. The Irish aligned themselves almost immediately with the Democratic Party, and with a zeal probably unparalleled by any other immigrant group—before or since—learned that political participation brought with it concrete rewards: jobs and favors. For example, to become a peddler required a license; so did opening a saloon. For either of these typically Irish occupations of the mid-nineteenth century, knowing someone in the political machine helped.

While historians and political scientists are still debating why the Irish succeeded so dramatically in American urban politics, they agree on the point that Irish timing was superb. The Irish in New York City, for example, arrived just when conflict between a rural-dominated, Whig, upstate-tilted legislature sought to dominate and control the affairs of New York City. As New York City became increasingly Irish—and Democratic—tensions between the legislature and the municipal government became more acerbic, and issues of ethnicity and religion charged the already contentious intergovernmental relationship. They arrived just at that moment when cities began to take on more and more functions—police, fire, garbage disposal—and when municipalities launched massive projects building bridges, tunnels, and roads. All of these brought with them something the impoverished Irish needed: work (Bridges 1984).

Given the escalating rhetoric, the evangelical orientation of many American public institutions, as well as the sheer press of numbers, the Irish, particularly under the aegis of the Catholic Church, began to create a series of internal community institutions: parish churches, schools, and orphanages. Before the famine migration, American Catholics were relatively few in number and supported a skeletal communal infrastructure. With the mammoth new influx, the hierarchy, through the effort of clerics like Bishop John Hughes of New York, sought to quickly respond to the needs of the newcomers. Hughes in 1846, for example, invited the Irish order of nuns, the Sisters of Mercy, to set up operations in America and provide services for the thousands of Irish women (Diner 1984, pp. 120–138; Dolan 1975, pp. 1–26). In addition to church-based community institutions, the Irish began to form benevolent societies, banks, newspapers, labor associations, nationalist societies, and a variety of other formal voluntary associations. The newcomers, poor as they were, received a good deal of assistance from the long-established Irish, who provided the financial support to establish such new institutions.

On a neighborhood level, the Irish famine immigrants who lived crowded together in the most destitute wards of America's cities provided informal support networks for themselves. New arrivals from Ireland came directly to family members who provided lodging and advice. Neighborhood saloons and groceries became social centers, and neighborhoods like New York's Sixth Ward throbbed with Irish street life. While middle-class Americans—native-born and Protestant—saw in these streets poverty, dirt, noise, and disorder, to the Irish residents, they represented family and community (Groneman 1977).

One result of both the extreme poverty and the high levels of American hostility toward the Irish was the emergence of a specifically "Irish" identity in America. In Ireland, loyalty had generally been directed at family, town, and county, rather than toward a national entity called "Ireland." Not coincidentally, the forging of a national identity began to surface in the "old world" as well, as a result of the famine and the perceived inaction and callousness of the British toward the suffering of the peasantry. But the process in America outpaced that in Ireland. Thrown together in shanty towns and neighborhoods, on construction crews, on the docks, and in the kitchens of middle-class "Yankee" homes, men and women from Kerry in the west and Cavan in the midlands to Carlow in the east and Down in the north saw themselves as sharing a common destiny and having *everything* in common as compared to the Protestant American majority from whom they felt distance and alienation (Miller 1985).

The migration of the famine years differed from earlier and later Irish migrations. Men and women—in roughly equal number—fled to America rather

Migration flows that were dominated by women, such as that of the Irish immigrants, resulted in the centrality of work in their lives and their occupational concentration. Here Irish women served as delegates to the 1886 Knights of Labor Convention. Elizabeth Rodgers, an immigrant, is holding the baby. (Library of Congress)

than systematically migrated. They fled without strategizing where to go and when. They came without jobs and without plans. But in the process, they created the structures of community, which in the decades to follow would welcome continuous waves of new arrivals, and the particular circumstances of their arrival shaped American opinion about the men and women of Erin, attitudes that would persist through the end of the nineteenth century.

Seven Decades of Postfamine-Migration, 1850–1920

The Irish migration to America did not end with the return of healthy potato crops in the 1850s. Rather, it continued, but in a very particular kind of way. From

the 1850s on, and becoming more pronounced as the century waned, the Irish migration to the United States became increasingly female, and Irish communities in America took on many of their characteristics from the nature of the migration.

Again, it is necessary to look at developments in Ireland to understand the new Irish immigration. In the decades after the famine, Irish family structure and landholding patterns changed dramatically. By and large, the Irish became a people who, in order to consolidate their landholdings and avoid the disasters of the famine years, refused to subdivide their lands, be they owners or renters. As such, land was passed down from parents to sons intact, meaning that only one son in the family would inherit the land. The emerging pattern also postponed marriage until the parents named the heir, since the eldest son was not necessarily selected. Only a single son in a family would inherit and marry; only one daughter in a family would receive a dowry (brought in by the dowry of her one marrying brother). For Irish women after the 1850s, no marriage would take place without endowment (Connell 1968). By 1870, Ireland led the world as the nation with the latest age and the lowest rate of marriage (Kennedy 1973).

The emergence of this new pattern, which actually had been anticipated in the behavior of the more prosperous classes even before the famine, created legions of nonmarriers who faced perplexing questions about their future. What prospects did a single man or woman have in nineteenth century Ireland? These single men, always referred to as "boys," regardless of age, and the "girls," had no urban opportunities as alternatives. The population of Ireland's cities, Dublin for example, declined rather than grew, because British imperial policy stymied industrialization. Single men did have some work prospects at home, as their newly married brother would need help on the family farm; and indeed the Irish countryside came to be heavily male as the century wore on. Migration abroad became the common life choice. The centripetal forces set loose by the new arrangement cast women out as emigrants more often than men, because single females had no options for either marriage or work at home once the land had been transferred to their one brother who "brought in" his new wife (Arensberg and Kimball 1940). For Irish women, physical movement constituted not just an "emigration" from Ireland—a place where they had no prospects, occupational or matrimonial—but it was also an "immigration" to the United States. As more and more young women left Ireland, it became an increasingly unpleasant place for others to live: sisters, female friends, and cousins had disappeared, defeminizing villages. On the other hand, the presence of these women in the United States, in cities across the country, made America a place where bonds of affection could be reforged. The draw of those women already transplanted in Chicago, San Francisco, Pittsburgh, New Orleans, Rochester, and Cleveland, and male kin as well, went beyond the realm of emotion.

Irish women in America served as valuable sources of information about jobs: the great goal of the immigration. Particularly in the area of domestic service, Irish women brought each other over, helping to pay the passage of sisters and cousins and

friends. They also made arrangements for work and helped orient the newly arrived Irish girl to the ethos of American domestic service, the single most important job for Irish women (Diner 1984).

Because so many Irish women at one time or another worked in domestic service, it is useful to explore briefly the nature of the job and examine why it attracted so many Irish women. First, as work for unskilled, uneducated women—salient characteristics of the Irish newcomers—it paid better, under healthier conditions, than any other possible work. For example, the salaries of a servant exceeded those of a textile mill hand, a factory worker, or a seamstress. Domestic servants experienced none of the dangers so rife in unregulated nineteenth-century American industry. Second, for women migrating without parents or spouse—again, salient characteristics of the Irish newcomers—it provided housing, usually in proximity to other Irish women, who also labored as domestic servants, and constituted a kind of mini-community. Third, a domestic servant not only drew wages, but had food and shelter and sometimes clothes provided by the employer. These women, unlike mill hands, could save money and use it as they chose. Nineteenth-century Irish women in America set aside much of their earnings as their own personal dowries for the future. They also sent vast amounts of money back to Ireland to help out their families. They offered generous contributions to Catholic churches and other Irish communal institutions, and they also seem to have spent money on themselves for clothes and leisure, unheard of possibilities in the "ould sod" (the old country) (Diner 1984).

In addition, domestic service attracted Irish women specifically because it repelled so many others. Americans had long complained about a "servant problem," and most native-born White American women adamantly refused the work, which they found demeaning (Katzman 1981). Most other immigrant women who migrated in family groups, with spouses and parents, also turned away from the idea of working as a servant. Indeed, Black women constituted the only other group to show up in large numbers in these decades as domestic servants, but they tended to live out rather than live in, and by and large worked in Southern cities where the Irish did not settle.

Thus, Irish women had somewhat of a monopoly on the job. They seem to have been able to set some of the terms of their employment and quit their positions when the conditions were not to their liking. American housewives had few other choices in terms of household help, and the labor of Irish women was eagerly sought.

The employment patterns of Irish women differed dramatically from those of Irish men. Both labored in a narrow range of unskilled positions, but Irish men worked on construction crews, in meatpacking plants, on railroad building gangs, as longshoremen on the docks, and as carters on city streets, giving them less of an edge. They competed with other poor men, immigrants and native-born, White and Black. They occupied the lowest rung on the pay scale and had almost no opportunity to save money. Over the course of the last half of the nineteenth century and into the early twentieth century, as the Irish migration lost steam and as fewer Irish newcomers arrived in America, the Irish did experience some economic mobility, although it was slow (Thernstrom 1964, 1973). In part, they were pushed up by the entry into America of millions of newer immigrants from eastern and southern Europe. In the large industries like steel, mining, and textiles, the Irish climbed up to more skilled, supervisory positions (Brody 1960).

Part of the explanation for Irish mobility was also generational: the American-born children of the immigrants found niches in the economy that surpassed those of their parents. The daughters of Irish immigrants actually experienced a more rapid and dramatic level of movement out of the laboring class than did their brothers. The daughters of domestic servants trained to become schoolteachers, nurses, stenographers, and other white-collar workers. For Irish men, mobility came through small entrepreneurship—saloon-keeping, grocery store owning, and the like—and more importantly through police, fire, and other forms of municipal employment (Ernst 1949). Sons often labored in the same occupational fields as their fathers, albeit in better paying slots, although it was not unheard of for American-born sons to replicate their father's jobs.

For some, mobility out of poverty came at a price. In order to maximize economic prospects, Irish women and men married later and less often than other Catholic groups.[3] Irish women seem to have very skeptically weighed the meaning of marriage, knowing that it brought with it all the attendant problems of poverty, abandonment, and widowhood.

They balanced marriage against remaining in domestic service, which offered them salaries they could spend as they pleased, as well as the security of a place to live and food to eat. Irish men likewise deferred or eschewed marriage, and biographies of successful Irish men offer plentiful examples of bachelorhood as the ticket to a more comfortable existence (Greeley 1972).

The tremendous Irish involvement in trade union activities also played a role in the process of economic advancement, especially for improving working conditions for both Irish women and men. Interestingly, the Irish had brought with them from Ireland no tradition of trade union activities. Yet in America individuals like Terrence Powderly of the Knights of Labor, Mary Harris Jones ("Mother Jones"), and "Big Jim" Larkin of the Industrial Workers of the World (IWW), as well as Mary Kenney O'Sullivan, Leonora O'Reilly, Timothy Healy, James O'Connell of the American Federation of Labor, and millions of other Irish American rank-and-file unionists demanded a "living wage" for America's working people (Pelling 1960).

The rise of some Irish women and men out of dire poverty created a range of economic classes within Irish America. For some, the experience of moving to America had, indeed, fulfilled the goals of the emigration. These Irish, known colloquially as the "lace curtain" Irish, became homeowners and served as the backbone of the Catholic parishes, charities, and Irish voluntary associations. Their involvement in local politics and labor unions complemented their steady climb into the middle class. For others, the life of the Irish in America continued to be plagued by a series of social and economic catastrophes. Not just for the destitute of the famine years, but even from among those who migrated after careful planning during the postfamine decades, urban existence in America posed tremendous challenges. The mortality figures of the "shanty Irish" demonstrated their concentration in poverty, their penchant for alcohol, and their poor living conditions in the least sanitary neighborhoods of American cities. The Irish died at alarming rates from alcohol and alcohol-related illnesses, as well as from industrial accidents, and Irish infant mortality statistics were among the highest in nineteenth-century America. Irish women were more likely than any other group of women to be left widowed or abandoned with children to care for, and American prisons, mental institutions, and almshouses were home to many Irish people into the early twentieth century (Diner 1984, pp. 106–124).

That a sizable, visible segment of the Irish population remained quite poor, kept alive profound anti-Irish sentiment in America. The negative imagery of the famine period persisted, and the stage, cartoons, jokes, popular literature, and even early cinema projected a stock portrait of the Irish as somewhat happy-go-lucky, inebriated buffoons, speaking in a lilting brogue. Irish servant girls were invariably called "Bridget," regardless of their real names. Mainstream American hostility toward the Irish and toward Catholics persisted well into the early twentieth century. The defeat of Governor Alfred E. Smith of New York, the first Catholic presidential candidate, in 1928 could be largely attributed to this deep American feeling that Irish Catholics were something other than "real" Americans.

Negative images of the Irish and claims that they constituted a people outside of the American mainstream tended to be bolstered by the long involvement of some Irish Americans with the Irish nationalist movement in Ireland. Though Irish men flocked both to the Union and the Confederate armies during the Civil War, fought in the Spanish-American War, and enlisted in record number during World War I, nativists questioned their loyalty. They worried that Irish Americans continued to be involved with the question of the status of Ireland. An Irish Republican Brotherhood, known as the Fenians, was founded in 1857 in New York, and they made a number of direct efforts in the United States to free Ireland from British rule. In 1867, one faction operating out of New York sparked an aborted uprising in Ireland, while another attacked Canada. From then on, a series of nationalist organizations cropped up, and well into the 1920s with the creation of the Irish Free State, Irish Americans spoke volubly, held rallies, and marched down the streets of American cities in support of freedom for Ireland (Brown 1966).

The discrepancy between the better-off and the struggling Irish in America, the tensions between them, and the question of the relationship between Irish Catholics and a Protestant-dominated America emerged as the central theme in the cultural and institutional life of Irish American communities. Much of Irish American popular culture played on the differences between the "lace curtain" and the "shanty Irish." Novelists like Mary Anne Sadlier and Louise Imongene Guiney, satirists like Peter Finley Dunne

("Mr. Dooley"), and the key figure in the development of serious American drama, Eugene O'Neil, each in their unique ways explored the cleavages within Irish communities and the disjunction between the Irish and the American Protestants around them. Successful Irish men and women used literature and the Irish community press to defend the Irish from criticism from the outside and to prove that not all Irish Americans fit the nativist, anti-Irish stereotypes. The ever-increasing institutional network of the Irish-dominated Catholic Church sought to elevate the poor Irish to the level of their more successful brethren, while the Irish political machine in almost every large American city thrived on the preservation of a relatively large needy class of Irish voters, often opposing reforms that would have benefited the poor but which also would undercut its own power.

Yet by the early decades of the twentieth century, the Irish had gradually demonstrated to native and Protestant Americans their basic worth to America and their ability to fit into American society. A few of them achieved prominence in theater, literature, sports, and public life. Many had moved into the reasonable comfort of the middle class, while a persistent segment stayed in the working class. They created an ethnic culture that professed loyalty at once to America and at the same time to "being" Irish. The arrival of large numbers of new immigrants from Italy, Poland, and elsewhere in eastern and southern Europe both pushed the Irish economically upwards and made the sons and daughters of Erin seem less alien by comparison.

The Irish had been America's first mass migrants. The Irish forced the Protestant evangelical majority to confront the nature of American national identity. They created the apparatus of urban politics and perfected the system of machine politics, by which newer groups came in and used local government as a means for mobility. The Irish provided much of the brute strength that fueled America's industrial revolution. Through their work in domestic service they allowed middle-class Americans to live in a genteel lifestyle. Despite the tremendous poverty, the high rates of disease and accident, despite indeed the whole "tangle of pathology" they endured for more than a generation in American cities, the Irish cast their lot with America and in the process transformed both themselves and America.

Notes

1. For the purposes of this essay, "Irish" will refer to Catholics. Irish Protestants, primarily Presbyterians will be dealt with briefly for the seventeenth and eighteenth centuries. But they seem to have blended into the English Protestant majority by the eighteenth century and ceased to be an identifiable element in the American ethnic mosaic. It is important to note however that little serious scholarship exists about the Protestant Irish—sometimes erroneously called "Scotch Irish"—and all statements about their ethnic identity and integration into America need to be viewed somewhat critically.

2. The post-1970 Irish immigration still awaits serious and detailed study and for the purposes of this essay will be treated in a cursory manner. Since the nineteenth century represents the most significant period, it will be the focus here.

3. This pattern continued well into the late twentieth century. A comparative study of different Catholic ethnic groups in Rhode Island in 1970 found that the Irish married later and had the largest number of nonmarriers of any of the examined groups (Goldscheider and Kobrin 1978, pp. 20, 48).

13

Italian Americans: A Century of Ethnic Change

RICHARD D. ALBA

Italian Americans are a litmus test for ethnic changes taking place among the majority of Americans who trace their ancestry to Europe. Arriving as peasants from the underdeveloped Mezzogiorno, or southern Italy, in the early part of the twentieth century, Italian immigrants seemed to many Americans then to be unassimilable, even racially distinct. For decades thereafter, their descendants were typed in the popular consciousness in the role of the solid, ethnic working class—capable in manual trades, including those involving a high degree of craft; enmeshed in tight-knit urban communities that kept outsiders away; but little motivated to seek social mobility, especially that involving high educational attainment or separation from family and community. Yet not quite a century after the high point of their immigration, Italians have, for the most part, entered the American mainstream. Theirs is a remarkable story, one laden with implications for our understanding of ethnicity in America.

The Immigrant Period

Few observers would have foreseen this story when Italian immigrants arrived in the United States. The Italians were the largest of the "new immigration" groups, which came from southern and eastern Europe and dominated the immigrant stream to the United States after the 1890s. Immigration from Italy has taken place over a long period of time, from the earliest U.S. record-keeping of immigration (1820) until the present, and has included northern Italians as well as southerners. Yet it has been remarkably concentrated in time and in its Italian source. In the short span of the 15 years between 1900 and 1914, more than 3 million Italians arrived on American shores—nearly 60 percent of the total Italian immigration in the history of the American nation. The immigrants who made up this high tide—initially men, followed by women and children—were largely dislocated peasants, uprooted by the demographic and economic transformation of agriculture in the Mezzogiorno. Data published by the U.S. Immigration Commission of 1911 reveal that in the crucial period 1899 to 1910, 32 percent of Italian immigrants with previous work experience described themselves as farm laborers and an additional 43 percent as laborers (Kessner 1977, pp. 33–34). The general category of 'laborer,' or *bracciante* in Italian, included many who had only recently been forced out of agricultural work (Sori 1979).

Given their rural origins, it would have seemed natural for Italian immigrants to have gravitated toward rural areas in the United States. But they did so only to a limited extent. More often than not, they settled in urban places, frequently the most urban. In 1910, at the height of immigration, New York City was home to 340,000 Italians, putting it well ahead of its nearest rivals—Boston, Chicago, and Philadelphia (Nelli 1983, p. 62). This settlement pattern was partly due to limited economic resources and partly to the "pull" of chain migration: many immigrants came with little money and could not afford to purchase agricultural land; or they went to cities where relatives had preceded them and could assist them in finding housing and jobs. More important, however, is that Italians often immigrated in the hope of a change in fortune. Many came intending to earn money as quickly as possible and to return to Italy to purchase their own land. Thus, it made sense to work for others and save as much as possible, and employment was most easily found in cities.

The economic niches that these immigrants found open to them generally ranked near the bottom, for many of the immigrants had few skills that were highly valued in an urban, industrial economy. In 1905 in New York City, nearly 60 percent of Italian household heads did unskilled or semiskilled manual labor, working, for instance, on construction gangs or as longshoremen (Kessner 1977, pp. 52–59; La Sorte 1985). In fact, many who came to America fresh from the peasantry discovered that, in the telling phrase of the immigrant writer Constantine Panunzio, "peek and shuvil" (pick and shovel) work was their only option. Yet it would be wrong to ignore the diversity that existed even in such an economically disadvantaged group. About 20 percent of New York City Italians were skilled blue-collar workers, and an equal number could be classified in the "lower white-collar" category, which included many small shopkeepers but also a fair number of street peddlers. Occasionally, immigrants attained great riches: one example is Amadeo P. Giannini, who in 1904 founded an immigrant bank, the Bank of Italy, that he transformed into the huge Bank of America (Nelli 1983, pp. 84–85). Moreover, it was not at all uncommon for immigrant workers to experience some upward movement in occupational terms: In New York City at the turn of the century, about 60 percent

of Italians who started in unskilled labor improved their occupational position in a decade (Kessner 1977, p. 117).

Women made an essential economic contribution in many immigrant families. Usually, however, this contribution did not stem from employment outside the home. Whatever the reason—whether because work outside the home contradicted southern Italian cultural prohibitions against contact between women and male strangers or, perhaps more simply, because of their family responsibilities—women tended to work inside the home. Many families took in boarders, who were invariably other immigrants and frequently from the same hometowns or even kinsmen; and many women took in other forms of homework, such as laundering or the manufacture of artificial flowers (Bose 1984; Kessner and Caroli 1978; Yans-McLaughlin 1977).

The adjustment of Italian immigrants to American society was complicated by several factors. For one, Italians frequently intended to return to Italy; thus, they were "sojourners," in the language of sociologists of ethnicity. The number who ultimately returned to the Mezzogiorno is uncertain, but clearly it was large; one estimate is that 1.5 million Italians repatriated in the years between 1900 and 1914 (Caroli 1973, p. 41). Obviously, many did not return, but even so, they may have realized only gradually that their stay in the United States was to be permanent. Thus, the sojourner's orientation toward the homeland, rather than to the new one, remained powerful, delaying the immigrants' acquisition of citizenship and learning of English. In 1930, at the end of mass immigration, in a number of major cities, Italians ranked at or near the top among immigrant groups in the percentages who did not speak English and who were not citizens (Alba 1985, p. 54; Lieberson 1963, pp. 206–218). This slow adaptation to American conditions inevitably had a large impact on the group, disadvantaging it even relative to other immigrant groups who arrived about the same time.

The adaptation of Italians was also influenced by the initial weakness of an Italian identity that could serve as the basis for immigrant organization. It is often said that Italians only became Italians in the United States. At the time of immigration, Italians generally thought of themselves in more parochial ways, typically in terms of village or provincial identities, as Neapolitans, Sicilians, Calabrians, Romans, and so forth. This phenomenon was known in Italian

Ellis Island was the major port of entry for immigrants who arrived on ships from countries in southern and eastern Europe. Here immigrants in the main hallway of Ellis Island are being checked by an immigration official, mainly for their health and normalcy. At this moment, many long and difficult names were simplified and changed—an instant "Americanization." (Theodore Wesley Koch Collection; Bentley Historical Library, The University of Michigan)

as *campanilismo*, after the church tower—*il campanile*—which often formed the focal point of a Mezzogiorno village (Vecoli 1978). That many immigrants spoke only in regional dialects, which were not always mutually intelligible, helped to sustain these regional identities. They were also reflected in the group's settlement patterns in American cities. Though Italians did frequently settle in large Italian "colonies"—neighborhoods dominated by the immigrants and by an Italian ambiance, expressed in language, customs, stores, and other institutions—on closer inspection these areas were internally diverse, with the immigrants from one village or province concentrated on a particular street or even in a single tenement (Kessner 1977, p. 16). Chain migration played an important role in generating such regional concentrations, as one immigrant assisted other new arrivals in finding nearby housing. Finally, regional

identities were reflected in the organizational life of the immigrants. In the early years, the chief form of organization was that of the mutual benefit society, which provided a forum for socializing as well as help in times of need. These societies were typically based on common village or regional origins, leading to fragmentation: at the turn of the century, the Italians of Manhattan, for example, supported more than 150 such organizations (Nelli 1980, p. 552). It took a few decades for organizations that were open to Italians from all regions, such as the Order of the Sons of Italy, to grow into prominence. Founded in 1905, by the early 1920s the Sons of Italy had grown to about 300,000 members in more than 1,000 local lodges, making it the largest of the Italian organizations (Nelli 1983, pp. 115–118).

Also important for understanding the adjustment of Italians was their family-centered culture. In pointing to it, there is a danger of exaggerating the causal role of culture—culture is malleable, not a fixed entity rigidly influencing social life; the culture of the group, moreover, was not homogeneous but undoubtedly varied by the regional and social class origin of the immigrants; and the culture of Italian immigrants in the United States may already have been different from that of the Mezzogiorno, since it was affected by the conditions of immigrant life and continued to shift in the American environment (Gabaccia 1984). Nevertheless, certain features of the life of this group were very striking to observers, both outsiders and insiders. Most striking was the primacy accorded the family and its interests in a group with peasant origins. This emphasis was expressed in a sharp distinction between kin and others, with much less expected of and given to the latter; a fictive form of kinship, godparentship (the *compare* and *comare*, the cofather and comother), for incorporating outsiders into the charmed circle of kin; family influence over the marriage choices of young people (and, at the extreme, parental arrangement of marriages); protectiveness toward, even surveillance of, unmarried women of the family to preserve their chastity; and an expectation that children would contribute economically to the family's well-being (Campisi 1948; Glazer and Moynihan 1970; Nelli 1983; Ware 1935). Even though these cultural traits underwent important changes in American society, they also influenced the use Italians made of American opportunities. For instance, the expectation that children would contribute economically to the family led sometimes to their departing from school and entering the labor market as early as the law allowed (Covello 1972).

Finally, the American reaction to the immigrants was a major barrier to their assimilation, for they entered as one of the most despised of the European immigrant groups. Stereotyped by many Americans as "swarthy" foreigners, Italians found that even their membership in the same race as native-born, White Americans was in question. This is highlighted by one of the common epithets for them—"guinea," which probably derives from a term for slaves from the West African coast (Alba 1985, p. 68). They were also stereotyped as having dangerous social tendencies, especially a tendency towards criminality. The belief that the Mafia and other criminal societies came to America in the midst of the southern Italian immigration sprang up almost at the same

time as that immigration in the late nineteenth century (Higham 1970). Italians were believed to be prone also to crimes of passion and vengeance, a tendency symbolized for Americans by the stiletto. An 1890 *Popular Science Monthly* article entitled "What shall we do with the 'Dago'?" vividly depicted the stereotype. The author, a penologist, wrote, "The knife with which he cuts his bread he also uses to lop off another 'dago's' finger or ear. . . . He is quite as familiar with the sight of human blood as with the sight of the food he eats" (quoted by Higham 1970, p. 66).

After Immigration: The Italian Americans Emerge

After large-scale immigration had come to an end during the 1920s, Italians continued to stand out in the perception of other Americans as a group beset by social problems and unlikely to assimilate easily. Yet appearances were deceiving. With the gates of immigration closed, Italians increasingly recognized that their future lay in the United States, and the adaptation of the group continued. In the post–World War II period, a convergence between the opportunities of Italian and other White Americans more and more took hold. Coming too late to affect the life chances of the immigrant generation, this convergence especially impacted upon the second and third generations, the children and grandchildren of the immigrants.

The Italians' educational performance initially seemed to type the group as destined to remain in the working class. Italian children, including those born in America, had high truancy and dropout rates in the decades immediately following immigration. In 1930, only 11 percent of Italians who entered New York City high schools graduated from them, at a time when over 40 percent of all the city's high school students stayed on to receive their diplomas (Covello 1972, p. 285). The root problem was undoubtedly the conflict between the schools and the expectations of working-class Italian families. Families with peasant memories did not usually see great value in academic achievement (unless it was directly linked to a vocational outcome) and may have feared that educated children would be lost to them. Frequently, they expected their children to contribute to the family economy at an early age (Steinberg 1981, chap. 5).

Nevertheless, changes were taking place. Some Italian children did achieve much more than the minimum. In one of the most famous studies of Italian-American youth, William Foote Whyte (1955) found a split among the young men of a 1930s Boston neighborhood—between "college boys," oriented toward mobility into the larger society, and "corner boys," for whom loyalty to their peers apparently counted more than educational and occupational mobility. As a result of such college boys—and girls, too—the educational gap gradually began to close. This was demonstrated by a study of the second generation of Italians and other immigrant groups in national data, conducted by Stanley Lieberson (1980). He found that the gap in years of schooling between the children of immigrants and other White Americans started to close in the late 1930s. It was fairly constant among Italians born before 1915, but began to decline among those born in the 1915 to 1925 period (most of whom would have completed their educations by twenty years later). The educational gap virtually disappeared for the group born between 1925 and 1935.

Occupational shifts were under way, too. To be sure, a concentration of Italians in working-class occupations characterized the group for decades after immigration. In the 1950s and 1960s, urban Italians were still disproportionately represented in modest positions in such industries as construction, sanitation, and food-related businesses (groceries, restaurants, butcher shops, and the like). For example, in the New York City region in 1950, two-thirds of second-generation Italian men worked in blue-collar or service positions (Glazer and Moynihan 1970, p. 206). Nevertheless, movement into middle-class, white-collar jobs, including the professions, was occurring at the same time. By 1960, the rate of entry to the professions of young Italian men had leaped upwards and was almost the same as that of other White Americans (Lieberson 1980, p. 330).

The group remained most visible to other Americans because of the large Italian neighborhoods in many cities, neighborhoods in which an ethnic atmosphere was palpable and seemed to set the group apart from the surrounding society. "Urban villages" was a name often applied to these areas by social scientists, after the title of a famous book by Herbert Gans (1962), to indicate that a transplanted village-centered southern Italian culture could be found in them. Nathan Glazer and Daniel Patrick Moynihan

(1970) charted some of these neighborhoods for the New York City of the 1950s and 1960s. Their portrait was one of stable neighborhoods, well-tended modest homes, and watchful family-based networks, suspicious of outsiders.

Behind the facade of stability, however, residential mobility was taking place for many Italians. Studies of residential segregation (e.g., Guest and Weed 1976) showed a gradual decline in the segregation of Italians, which had been very high in 1930. Equally portentous, if little noticed, was the presence of Italian Americans in the exodus to the suburbs which began after 1950. This is shown, for instance, by the dispersal of the second generation in the New York City region. By 1970, about 40 percent of second-generation Italians were living in the region's suburbs (defined here as places with less than 50,000 residents). This figure was only slightly lower than that for all the region's Whites, 45 percent; but it was well above that for first-generation Italians, 29 percent (Alba 1985, p. 88).

Acculturation was the keynote of some other changes, although the end result was not always the adaptation of American culture in an unaltered state. As other groups did, too, the Italians added something of their own to the culture of the host society. Cultural change was at least in part a two-way process, even if American culture was largely dominant.

In language, American culture clearly was dominant. Even in the immigrant generation, linguistic adaptation took place, as immigrants imported words from English for objects that were unfamiliar in the rural Italian areas where their dialects had developed. Thus, "shoppa" (for "shop") and "frigidaria" (for "refrigerator") made their entry into immigrant speech (Lopreato 1970, p. 57). But the more important assimilation took place across the generations. Italians are no exception to the principle of three-generation language shift that has been found among most American ethnic groups: the immigrant generation maintains fluency in the mother tongue, although many of its members may also learn English; the second generation is bilingual, having been raised in homes where the mother tongue was usually spoken, but its predominant language is English; and the third generation typically speaks only English in everyday life, although some of its members may have fragmentary knowledge of the mother tongue (Veltman 1983). Among the Italians, this process was probably helped along by the absence of linguistic unity among the immigrants, who spoke mostly dialects. Data collected by the U.S. Census Bureau in the late 1970s are suggestive of this process. The data indicate that over 4 million claimed Italian as a mother tongue, spoken in their childhood homes, making it one of the most frequently cited mother tongues. But fewer than 1.4 million then spoke the language in their current homes, testifying to a high rate of language loss. Since the number of immigrant Italians, who would have been very likely to use Italian regularly, was then about 800,000, it is clear that only a small proportion of the second and third generations spoke Italian at home (Alba 1985, p. 90; Stevens 1992).

Changes also took place within the family as traits that had meaning mainly in the Mezzogiorno setting declined; but, at the same time, a distinctive Italian-American ethos emerged. A study conducted by Caroline Ware in the 1930s offers a snapshot of family norms in motion. She divided her respondents by age, roughly corresponding with generation. The contrast between the older Italian-born and the younger American-born respondents in terms of agreement with southern Italian family norms was remarkable. For example, only a third of the older group disputed the norm that a "husband's authority should be supreme," whereas two-thirds of the younger group rejected it. Likewise, more than half of the older group accepted the notion that single young women should not keep company with men unless they are engaged to them; but only one of every six younger Italians agreed with this (Ware 1935, p. 193).

Yet the Italian American family remained distinctive; wholesale acculturation to American family patterns did not occur for the second and even a substantial part of the third generation. To outside observers, the most conspicuous characteristic of Italian American life was its family-centeredness. Second-generation Italian Americans chose to spend much of their social life apart from work in a family circle of relatives and family friends. In a classic study of an Italian neighborhood in Boston in the 1950s, *The Urban Villagers*, Herbert Gans (1962) described this phenomenon as the "peer group society." On a national plane, Andrew Greeley (1971, pp. 77–78) found that Italian Americans were more likely than others to live near relatives, especially parents and siblings, and to visit with them frequently. Infrequent divorce was another characteristic of Italian American families. Family size, however, rapidly began to approach American norms. Although there was little attempt by immigrant families to limit their family size, and large families with five or six children were common, matters were entirely reversed in the second generation. The Catholic Church's rulings on birth control apparently were widely ignored by Italian Americans, and in the second generation, family size dropped, with two to four children being the rule, as for urban Americans in general (Femminella and Quadagno 1976).

Finally, the very identity of the group itself changed as an Italian American ethnic identity emerged. Largely gone, especially for the second and third generations, were the parochial regional identities of the immigrants. So, too, was the identification with Italy and Italian nationalism that replaced them, especially during the 1930s, when Italy's power was ascendant in Europe and North Africa. The Italian American identity was very much an American identity, rooted in the American city experience, an identification with an ethnic group rather than a foreign country; and it was salient among the ethnic identities that blossomed during the 1960s and 1970s, as faith in the power of the melting pot waned (Conzen et al. 1992). It contributed at the time to what seemed a resurgence of ethnic identity among Americans, throwing into question whether assimilation would ever occur. The spirit of the time was captured in the title of a popular book, *The Rise of the Unmeltable Ethnics*, by Michael Novak (1972). But just how unmeltable were Italian Americans?

The Contemporary Situation

Trends favoring the assimilation and social mobility of Italians, reflected in their improved educational and occupational chances as well as their increasing suburbanization and intermarriage, manifested themselves in the second half of the twentieth century. As a consequence of these accumulating and interrelated changes, Italian Americans have entered the American mainstream almost en masse as the century draws to a close. This is not to deny that Italian American communities can be found in many cities where immigrants settled and that ethnicity remains in some sense significant to many members of the group. Nevertheless, the conclusion is an important one, given the large size and the ethnic prominence of the Italian group: the 1990 census counted 14.7 million Americans with some Italian ancestry, nearly 6 percent of the total population, the fourth largest European ancestry in the nation.

The changes are clearly visible in the educational attainment of the group. Since they have unfolded through historical shifts in opportunities, a direct way of viewing these changes is through differences across what demographers call "birth cohorts," or groups of people born in the same period. Different cohorts of the Italian American group have matured under quite different conditions. Older cohorts were substantially disadvantaged in comparison with members of the then core group in the United States—White Americans whose ancestry was from the British Isles or elsewhere in northern and western Europe. For instance, among men born in the United States before 1916, those of British ancestry attended college and graduated at rates 50 percent higher than the rates for Italians. Among women, the discrepancy was even greater: the rates for those of British ancestry were more than 100 percent higher. Yet this disadvantage was erased for those cohorts born after World War II. As of 1980, slightly more than half the men of British and Italian ancestry in these cohorts have attended college, and a quarter of each group have graduated; the figures are just slightly lower for women (Alba 1988).

Of course, such data cannot demonstrate the absence of any ethnic discrepancy in educational opportunity. In all likelihood, some differences remain, such as in the access to elite private schools. Moreover, local variations may exist. Insofar as Italians in some cities are still concentrated in the working class, they may suffer also from educational disadvantages, though probably more due to their social class than to their ethnic backgrounds.

In tandem with improving educational opportunity has come occupational mobility that has dispersed Italians more evenly throughout the American occupational structure. This is demonstrated by the growing parity in the occupational status scores of Italian American and British American men. For instance, as of the late 1970s, Italian American men born in the period 1930 to 1945, who were then at or near the maturity of their careers, had an average occupational prestige that was virtually the same as men of British ancestry of the same ages (Alba 1985, pp. 122–123; Neidert and Farley 1985). Hand in hand with this mobility has come a decline in the strength

of Italian occupational niches. Viewed in historical perspective, Italians have had their share of occupational niches—among them, barber, tailor, shoemaker, and opera singer. But these niches, though they still exist to some degree, have become less typical of the group as a whole (Lieberson and Waters 1988, chap. 5). Thus, Italians are now much less easily stereotyped according to specific occupational callings.

A marked change also has taken place in the residential settings where the group's members can be found. In this respect, a strong contrast exists between the contemporary situations of Italian Americans and their image, even among scholars. There is still great attention given to inner-city ethnic pockets, such as Bensonhurst in New York City, the scene of violent racial strife in the late 1980s. Yet despite the continuing image of the group as settled in stable urban villages, Italian Americans now reside predominantly in suburbs. As of 1980, about two-thirds were in suburbs throughout the nation, the same proportion as found for other White Americans. The suburban proportion was even higher (nearly three-quarters) among persons who are only partly descended from Italians; these individuals are probably a good tracer for the residential patterns of Italians who have intermarried. Further, suburban Italian Americans reside for the most part in non-Italian neighborhoods, even though some suburban neighborhoods do have a decided Italian flavor (Alba 1992).

Of course, Italian neighborhoods continue to exist, especially in the cities of the northeast. But, in many cases, they are far from static. Manhattan's "Little Italy," one of the most famous, is now largely a street-level facade of Italian restaurants and stores, with Chinese and other Asian immigrants living upstairs. Another area that has been venerated as an authentic urban village, Belmont in the Bronx, has seen its Italian American population shrink to about 5,000 from about 30,000 in 1950 (LaRuffa 1988). Social mobility and suburbanization have eroded the stability of these areas. The difficulties such neighborhoods face in maintaining stable populations in cities undergoing fairly dramatic racial and ethnic changes probably contribute to the defensive racism they sometimes exhibit. Nevertheless, even diminished neighborhoods can serve important ethnic functions when they become meccas for the ethnic tourism of both Italians and non-Italians wishing to reexperience the "feel" of an ethnic world.

The ultimate test of assimilation for an ethnic group is intermarriage; by this test, the growing integration of Italians with the American mainstream is beyond question. Intermarriage is, at one and the same time, a sensitive barometer of the acceptability of an ethnic group to others, a reflection of the state of contacts across ethnic lines and the ethnic context in which the next generation will be raised. Intermarriage by Italian Americans has surged in recent decades, a trend not unique to the group, but one that reflects the broader decline in the role of ethnic and religious backgrounds in determining marriage choices among White Americans (Alba and Golden 1986; Kalmijn 1991). Among young Italian Americans in the 1980s, the rate of marriage to non-Italians reached between two-thirds and three-quarters, the precise figure depending on whether one looked at women or men and on how one defined intermarriage (Alba 1995). Consistent also with the broader pattern, these intermarriages were overwhelmingly to other Americans of European ancestry. Marriages by Italians are no longer confined mostly to other Catholics, however. Among members of the group born in the post–World War II era, marriages are almost evenly divided between those with other Catholics and those with non-Catholics, mainly Protestants (Alba 1985, pp. 145–150).

As a result of rising intermarriage, a massive shift is occurring in the composition of the Italian group, as is true also for some other White ethnic groups. In the recent past, most of the adults in these groups were individuals who, by virtue of their undivided ethnic heritage and their families' recency of immigration, had grown up in quite ethnic family and community environments. In the near future, this will no longer be true. Today, the majority of young people entering adulthood come from ethnically mixed families, belong to the third if not the fourth generation, and were raised outside of ethnic communities. Among the Italians, more than three-quarters of those born during the late 1960s and the early 1970s, who are entering their early adult years during the last decade of the twentieth century, come from mixed ethnic backgrounds. By contrast, only about one-quarter of Italian Americans born before World War II have mixed ethnic ancestry (Alba 1985, p. 111). This contrast implies a profound shift of cardinal importance for ethnicity. On average, persons with mixed ancestry are much less exposed in their upbringing to ethnic traits, from language to values,

than are persons of undivided ancestry. Therefore, as the Italian group is increasingly composed of those of mixed ancestry, the group as a whole will become less and less ethnically distinctive.

Ethnic Identity Under Conditions of Assimilation

What is, then, to become of the Italian American identity? Can it survive the growing number who have grown up in ethnically mixed families and outside of very ethnic environments? The evidence is hardly definitive, but it does suggest that Italians—like many other White ethnics—are preserving some subjective attachment to their ethnic origins, at the same time that the significance of ethnic group membership in determining life chances is declining, in some respects dramatically. Why ethnicity nonetheless continues to be subjectively salient, and how this phenomenon should be interpreted, are questions that remain unsettled.

Evidence of the relative strength of Italian American identity comes from a survey on ethnic identity in the Albany, New York, region (Alba 1990; see also Waters 1990), which shows that nearly three-quarters of individuals of Italian ancestry identify, at least to some extent, with their Italian origins. Further, virtually all who do so believe that their ethnic background is important to them, whether because it provides a sense of roots, is a source of their personal values, or for some other reason. In these respects, Italians stand apart from most other White ethnics, such as the Irish or Germans. Italians are also more likely than other White Americans to manifest their ethnic origins by eating ethnic foods, talking with others about their ethnicity, and teaching their children about it.

However, ethnic identity has its own unavoidable dynamics. The greater intensity of Italians' ethnic identity is linked to their demographic distinctiveness as a group, a distinctiveness that is being eroded by rising intermarriage, educational and residential mobility, and generational change. Compared to other White Americans, adult Italian Americans are still concentrated in generations where the immigrant experience is recent and a high proportion come from unmixed ethnic backgrounds. Such characteristics mean that a large number of Italian Americans have had direct contact with immigrants or with an ethnic community, growing up either in an immigrant home or a second-generation one, where the Italian ethnic presence was still very strong and the immigrant generation was visible in the form of grandparents and other relatives. When envisioning an ethnic identity for themselves, such individuals have more vivid materials, especially family experiences and memories, to draw upon. Moreover, the collective aspects of Italian ethnicity that are reflected in neighborhoods, festivals, and businesses can again be traced to the historical recency of Italian immigration. But those characteristics making for Italian ethnic distinctiveness are also now changing, and quite rapidly. Those now entering the adult ranks of the ancestry group are increasingly likely to belong to the third and even fourth generations, to come from mixed ethnic backgrounds, and to have been raised outside of an ethnic milieu. Soon there will be numerous young adults who not only come from partly Italian backgrounds, but whose Italian connection is through the mixed ancestry of one parent, resulting in still greater distance from intense expressions of Italian American ethnicity.

What will the endpoint of the process, then, look like? What will be the forms of Italian American identity under conditions of widespread assimilation? Certain hallmarks of ethnic identities are now found among many White Americans and provide clues to the future of Italian American identities. One hallmark is variability in the salience, or intensity, of identity (Alba 1990; Waters 1990). Variability means not only that the salience of identity varies from one individual to another, but also that, for the same individual, it may vary from one situation to another. Ethnic identities are increasingly malleable and situationally specific. Their malleability is partly due to the complexity of ancestry and to the weak social constraints on how White Americans identify themselves. Accordingly, only a small minority of White Americans hold intensely to ethnic identities that they are likely to manifest in a wide range of situations; at the other extreme, an equally small group believes that ethnic background has little or no personal importance. In between these extremes is a large group, probably a majority in the many parts of the United States where most Italians are found, for whom ethnic background and identity are of middling significance. For these individuals, ethnic identity is typically an occasional feeling, experienced perhaps with certain family members, at weddings

and other family celebrations or on holidays like St. Patrick's Day. This form of identity cannot be regarded as a fixed attribute, but as situationally variable and fluid in form (Alba 1990, chap. 2).

Another hallmark is the weak reflection of ethnic identity in everyday experience for most White Americans. Most who identify with their ethnic backgrounds can point to some experiences they view as ethnic. But few have many such experiences or very regular ones (aside, that is, from eating ethnic foods). Moreover, the common ethnic experiences seem to fit well with the social milieus in which most White Americans find themselves, involving continual contact among persons of varied ethnic backgrounds. The prime example is eating ethnic foods. This is an experience that is mostly in the private, rather than public, realm; it is innocuous, unlikely to give offense or attract negative comment; and it need not be ethnically exclusive (ethnic foods can be shared with people of other backgrounds; certainly, this is true of Italian cuisine, one of the most popular in the United States). Much less common are experiences having aspects of ethnic exclusion, favoritism, or conflict—experiences such as receiving business or professional help from a coethnic, suffering from discrimination, or involving oneself in ethnic political issues. Among Italians, incidentally, these are now fairly uncommon experiences—in the Albany survey, just 4 percent reported, for instance, that they had experienced discrimination in the preceding five years (Alba 1990, p. 149). Italians, however, are more likely than other White Americans to have encountered ethnic stereotypes, such as that connecting them with organized crime. (Not all stereotypes of Italians are so negative, to be sure.)

Contemporary ethnic identities can also be described as a "privatization" of ethnicity, reducing its expression to largely personal and family terms. For many White ethnics, Italians included, ethnicity is hard to separate from their notions about their own families, and a larger ethnic community is only hazily discerned at best. One of the most important ways this privatization is revealed is in the kind of ethnic identity White Americans hope to instill in their children, insofar as they imagine one at all. Frequently this seems little more than a sense of family background, as when intermarried parents, or those of mixed ancestry themselves, list the ethnic variety of their children's ancestries with fairly equal weighting. Moreover, few White Americans care strongly about their children's identities; most accept what Herbert Gans (1988) has called "middle American individualism." That is, they have little concern about whether their children take an interest in ethnic background, apparently because they see this as ultimately a matter for the children to decide. As a consequence, the generational transmission of ethnicity, the indispensable foundation of ethnicity in a new generation, is increasingly problematic.

The final hallmark is the loose connection that exists between ethnic identity and ethnic social structures larger than the family. Ethnic identities surely exist, but they do not lead most White Americans to regular involvement in an ethnic community. This loose linkage underscores the highly personal character of many ethnic identities, which do not depend on social relationships that have an ethnic character. Of course, most White ethnics live and work in highly intermixed social settings (at least as far as European ancestries are concerned): they reside outside of ethnic neighborhoods; their circle of friends is heterogeneous; and they are intermarried or come from mixed ethnic backgrounds. But even when it comes to voluntary memberships that make only modest claims on people's lives, such as membership in ethnic organizations, few White ethnics join. This is true also of the Italians, despite their stronger ethnic identities at present. For instance, the total membership of the Sons of Italy, which is still the largest national Italian American organization, is just 90,000, a tiny fraction of an ancestry group that today counts nearly 15 million members (Rotondaro 1991, p. 120).

Conclusion

Taken together, these features of many contemporary ethnic identities seem to fit the profile of what Herbert Gans (1979) has described as "symbolic ethnicity"—an ethnicity that is confined to symbolic occasions and does not intrude much into everyday life. It, therefore, also does not interfere with the ethnic intermixing that is a necessary part of the social worlds of most White Americans. Gans is agnostic regarding whether symbolic ethnicity will eventually give way to complete assimilation. But, based upon the Italian American experience, it would seem that ethnic identities are likely to survive on a mass scale for some time to come. The combination of extensive integration into the mainstream in occupations, neighborhoods, social networks, and the like, together with persisting ethnic identity suggests that a new stage of ethnicity is emerging among Italians and other Americans of European ancestry. This new stage will undoubtedly coexist for a long time—perhaps permanently—with older forms of ethnic identity, but it seems destined to grow in prominence. Ultimately this symbolic ethnicity may prove to be quite important, but not because it is just the residue of a more traditional ethnicity in decline. Rather, it could become the basis for a new ethnicity, drawing upon the European-American experience and finding its specific meaning in a society still deeply divided along racial and ethnic lines.

14

Mobility and Continuity among Eastern European Jews

STEVEN J. GOLD
BRUCE PHILLIPS

The first Jews came to colonial America from the Dutch West Indies in 1654. They were of Sephardic origins and traced their ancestry to Spain and Portugal. By the time of the first U.S. census in 1790, some 2,000 were spread throughout the colonies (Baltzell 1970, p. 327). In the 1850s, about 50,000 highly urbanized German Jews entered the United States and rapidly moved into mercantile occupations, including commercial banking and department store ownership. Nearly 150 years later, Jews continue to enter the United States in significant numbers from Russia, Israel, South Africa, and Iran. Despite American Jewry's long history and diverse national origins, the vast majority of this group trace their beginnings to the eastern European migration of 1880 to 1924. These eastern Europeans (often simply called Russian Jews) brought with them the language (Yiddish), cultural heritage, diet, and way of life that have come to stand for Jewishness in American society.

Escaping economic dislocation and anti-Semitism, eastern European Jews arrived as a part of the great southern and eastern European exodus that entered the rapidly industrializing American society during the late nineteenth and early twentieth century. These Jews experienced a much faster rate of upward mobility than other nationality groups who arrived during the same time period. Their meteoric rise in social and economic standing has been a defining feature of the American Jewish experience.

At present, according to nearly every available measure, the level of social achievement of American Jews exceeds that of other ethnic groups, including high-status denominations of native-born WASPs (White Anglo-Saxon Protestants) (Lipset 1990). Now constituting about 6 million persons, or 2.5 percent of the U.S. population, Jews have risen to lofty levels of accomplishment in all areas of social, economic, political, intellectual, and cultural endeavor (Mayer 1991). For example, in 1993 there were forty-three Jewish members of the U.S. Congress (thirty-three representatives and ten senators) (Hoffman 1993). Jews are also represented well beyond their proportion of the U.S. population among American Nobel Prize winners in science and economics, among faculty at elite universities, among top positions in media industries as well as Washington and New York law firms, and among leading civil service positions. More than 80 percent of college-age Jews attend institutions of higher learning, a proportion twice that of the population at large, and these students are disproportionately elected to Phi Beta Kappa, which demonstrates their academic excellence (Lipset 1990, p. 3–4).

Despite their rapid movement to the mainstream of American society, Jews continue to maintain a social and demographic profile—in economics, occupations, education, politics, and communal life—that distinguish them from non-Jewish Americans of comparable characteristics. In this regard, the American Jewish experience violates the assumptions of assimilationist theories which predicted that economic mobility would be accompanied by the attenuation of distinguishing social features (Gordon 1964; Phillips 1991).

Instead, the distinct patterns of Jews' behavior are more accurately described by theoretical formulations which see ethnicity as an asset to social mobility (Glazer and Moynihan 1963; Cummings 1980). These include cultural pluralism (a society encouraging the maintenance of diverse social patterns among its constituent groups rather than fostering a single culture among all) and situational ethnicity (the use of shared identity as a vehicle around which groups can organize to build viable communities and accomplish desired goals) (Cohen 1969; Greeley 1974; Lyman 1977; Nielsen 1985; Olzak 1983). American Jews evidence a unique combination of mobility and group persistence. As such, the origins of eastern European Jewish economic mobility are worth exploring without disregarding American Jewry's most pressing current concern—how to maintain a viable ethnic and religious community despite high levels of social and economic achievement.

Immigrants in the Golden Medinah

Humble Beginnings in Immigrant Slums

The current accomplishments of American Jews appear all the more remarkable when one considers the conditions associated with recently arrived eastern European Jewish immigrants less than 100 years ago in New York City's Lower East Side and other immigrant slums throughout the northeast and midwest. The lives of these "greenhorns," or first-generation immigrants, offered little evidence to suggest the eventual accomplishments of their descendents.

Housing conditions in the Lower East Side were abysmal. With nearly 700 persons per acre, the dis-

trict was more crowded than the worst slums of contemporary Bombay (Rischin 1962, pp. 87–88). Tenement fires were common. Because of the outhouses, cooking, and industrial activities, the polluted environment was described as "the eyesore of New York . . . the filthiest place in the Western continent" (Sowell 1981, p. 83). Alcoholism was rare, but Jewish immigrants suffered from other diseases associated with overcrowded urban settings such as tuberculosis, venereal disease, mental health problems, and diabetes (Rischin 1962, pp. 87–88). Despite the low quality of housing, costs were still high and thousands of families found themselves evicted because of their inability to pay the rent (Sowell 1981, p. 83).

Like other poor ethnic groups, Jewish communities were plagued by social pathologies and crime (Goren 1970, pp. 147–148). For example, Brooklyn's heavily Jewish neighborhood of Brownsville produced illustrious gangsters including Mickey Cohen, Bugsy Siegel, Meyer Lanksy, and "Murder Inc." (Landesman 1969, p. 323). Despite the notoriety of the latter, Jewish criminality was generally of the nonviolent type, involving protection rackets, prostitution, commercial violations, and labor racketeering (Howe 1976; Joselit 1983).

Just after the turn of the century, authorities noted that within a heavily Jewish area of the Lower East Side consisting of one square mile "there were approximately two hundred disorderly houses, three hundred and thirty-six gang 'hang-outs' and over two hundred pool hall-cum betting establishments; dance halls, a rendezvous of pimps and procurers, were found every two and a half blocks, while gambling establishments blanketed the neighborhood" (Joselit 1983, p. 24). In 1898, the Yiddish-language[1] newspaper, the *Forward* warned its readers, "It is better to stay away from Allen, Chrystie, and Forsyth streets if you go walking with your wife, daughter, or fiancee. There is an official flesh trade in the Jewish quarter. In the windows, you can see human flesh instead of shoes" (Howe 1976, p. 99).

As a result of the immigrants' criminality, in 1905, over a third of the inmates of two New York correctional institutions were Jewish (Joselit 1983, p. 15). While his estimate is now considered an exaggeration, in 1908, New York City Police Commissioner Theodore A. Bingham asserted that 50 percent of the city's criminals were Jews (Goren 1970; Joselit 1983, p. 23). As a result of these widely published allegations, however, the Jewish community created

For almost a hundred years, the garment industry has employed recent immigrants as workers, whether Jewish immigrants like these, in 1924, or recent Latin American and Asian immigrants. In large garment firms, such as this one, then known as "Casino Dresses," work was specialized by gender and by ethnicity, with the men doing the actual machine sewing and the women the hand finishing. In smaller firms, the division of labor was not necessarily gendered. Rather, everyone did whatever was necessary. (United States Collection; Yivo Institute)

several community-building organizations (Goren 1970).

Finally, while many scholars attribute the unprecedented mobility of Jewish immigrants to their love of education, ample evidence suggests that the school performance of turn-of-the-century Jewish immigrants was far from outstanding. For one, immigrants were poorly educated. Twenty-six percent of those who arrived during the first decade of the century were illiterate, including 40 percent of the women. This contrasts with a 7.7 percent rate of illiteracy for the entire U.S. population and a 12.7 percent rate of illiteracy for all foreign-born Whites in 1910 (U.S. Bureau of the Census 1975a, p. 382). Yet in comparison to other southern and eastern European immigrants of the time, these were rather high levels of education. For example, the illiteracy rates of south Italians, Lithuanians, and Poles entering the United States were 54 percent, 49 percent, and 35 percent, respectively. However, immigrants from western Europe were far more literate than Jews. Only 2.6 percent of the Irish and 0.7 percent of the Scots who arrived during the same decade could not read or write (Joseph 1914, pp. 193–194) (see Table 1).

Furthermore, like other working-class and poor children, Russian Jews attended inferior, overcrowded schools and quit when they finished, if not before (Gorelick 1981, p. 121). A 1910 study of twelve American cities found two-thirds of the children of Polish Jews to be below the normal grade for their age; a survey of over 5,000 Russian-Jewish students conducted a year later found 41 percent to be behind their grade level (Lieberson 1980; Sowell 1981, p. 87). Mental ability evaluations conducted during World War I determined that soldiers of Russian (mostly Jewish) origin had the lowest scores of any group tested (Sowell 1981, p. 88).

Table 1 Social Characteristics of European Immigrants to the United States, 1899–1910

Nationality	Less than Age 14 (%)	Female (%)	Returns per 100[a]	Illiterate[b] (%)	Skilled Prior to Migration[a] (%)
Jewish	24.8	43.4	8	26	67.1
Polish	9.5	30.5	30	35.4	6.3
Slovak	9.3	29.5	59	24	4.4
South Italian	11.7	21.4	56	53.9	14.6
Greek	4.1	4.9	25	26.4	7.7
Hungarian	8.8	27.8	64	11.4	8.6
Lithuanian	7.9	29.4	14	48.9	6.7
Scottish	15.3	36.5	11[c]	0.7	57.9

[a] Figure for years 1908 to 1910.
[b] Figure for persons more than age fourteen.
[c] Figure for English returnees.
Source: Joseph 1914, pp. 179–193.

Discrimination

Immigrant Jews faced many kinds of discrimination, ranging from street violence to subtle institutional exclusion. Jews were often the targets of anti-Jewish brutality meted out by neighboring Irish, Italian, and Polish youth gangs, occasionally abetted by Irish police. For example, in 1902, a rabbi's Lower East Side funeral procession was attacked by Irish bystanders who were hostile to the burgeoning Jewish population that was overtaking their neighborhood. The police responded by vigorously beating, arresting, and fining the funeral marchers. Allegedly their commander ordered his men to "club their brains out" (Howe 1976, p. 129).

Eastern European Jews also confronted extensive discrimination and prejudice in attacks by the media, in housing, and in employment. Such maltreatment can be attributed partly to nativist and antileftist sentiments—to which all immigrants were subject—as well as to specifically anti-Jewish campaigns (Lipset 1990; Miller 1977). An extreme incident in Georgia resulted in the lynching of Leo Frank in 1914. Jews also faced anti-Semitic diatribes, especially during economic downturns. While such bigotry is normally associated with marginal groups such as the Ku Klux Klan, some well-known anti-Semites were figures of national prominence, including the radio priest, Father Richard Coughlin, and automotive pioneer Henry Ford—who published Jew-baiting harangues in his newspaper *The Dearborn Independent* and authored *The International Jew* (Dinnerstein, Nichols, and Reimers 1990; Howe 1976).

As the children of immigrant Jews sought to move from the ethnic ghetto to mainstream American life, their efforts were hampered by pervasive social discrimination. Some policies established in the 1880s remained in effect through the late 1960s. Jews were generally prohibited from joining social and athletic clubs as well as from living in exclusive neighborhoods. After the 1920s, many private colleges and universities, especially those in the northeast, established quotas to limit the number of Jewish students present, regardless of their qualifications. For example, in the 1920s, Harvard president Abbott Lawrence Lowell decried the college's "Jewish Problem" and limited the fraction of Jews in attendance to 15 percent in order to "reduce anti-Semitism" and protect Harvard from the same fate among Gentiles as was suffered by "the summer hotel that is ruined by admitting Jews" (Takaki 1993, p. 306). A study made two decades later, in 1949, determined that "whatever way you adjust and reassemble the data, applications made by Jews to north eastern colleges are less often accepted than those of the Protestant or Catholic" (Ivy and Ross 1957, p. 136).

If admission to colleges and universities was difficult, getting accepted to graduate programs, especially in professional fields, was an even greater challenge. For example, the proportion of Jews enrolled in law, engineering, medicine, architecture, dentistry, commerce, fine art, and social work all dropped substantially between 1935 and 1946, even though the pool of Jewish applicants for such opportunities continued to grow. A 1945 article reported that while three out of four non-Jewish applicants were accepted into medical schools, only one out of every thirteen Jews could matriculate (Ivy and Ross 1957, p. 138). Due to the combination of quotas and the large number of Jewish applicants, in 1940, non-Jews had a one in seven chance of admission to Cornell Medical School, while Jews had only one-tenth of that likelihood for admission, one in seventy (Glazer and Moynihan 1963, p. 156). These statistics are born out in the following story, told by a relative of Steven Gold.

> I graduated from high school in New Haven in the late 1930s and applied to a private college in my home state, planning to major in biology. A representative of the admissions staff told me that if I would go into biochemistry, they would let me in. However, if I planned to apply to medical school, they would reject my application because Jews have almost no chance of getting into medical school, and so the college's record of medical school admissions would be adversely affected.

This same person enrolled in a state university and was finally able to attend medical school in the late 1940s after having made extensive personal contacts with medical school faculty members during his wartime service as a research biologist.

Even those Jews who, against such odds, successfully completed their doctoral training, often found employment in existing institutions unavailable. For example, Jews with Ivy League Ph.D.s were excluded from positions in college departments of English because, as eastern Europeans and Jews, it was felt that they could not partake in "the Anglo Saxon spirit" of the field (Howe 1976, p. 426).

While Jews were well represented in certain industries such as textiles, sales, entertainment, and the independent professions, they were systematically excluded from employment in elite firms and leading corporations. In the early 1950s, studies of the job markets in Los Angeles and Chicago found that between 17 and 20 percent of all job openings requested non-Jewish applicants, and the highest rates of exclusion were for white-collar and professional jobs, especially in law, insurance and accounting (Waldman 1956, pp. 211–214). In 1960, 8 percent of the country's college graduates were Jewish. However, they accounted for less than one-half of 1 percent of the executive personnel at the leading American industrial companies (Glazer and Moynihan 1963, p. 148). The experience of Harvard Law School professor Alan Dershowitz is telling. Upon graduation from Yale Law School in the early 1960s—first in his class and editor-in-chief of the prestigious *Yale Law Journal*—the jurist applied to "a few dozen" Wall Street law firms and was rejected by every one of them (Dershowitz 1991, p. 53).

As late as 1968, researchers for the U.S. Equal Employment Opportunity Commission determined that Jews were still systematically excluded from management positions in major corporations. While 75 percent of all Jews age 18 to 22 attended college, and 25 percent of recent Ivy League graduates and 15 percent of graduates of the Harvard Business School were Jews, three different studies revealed that only 4.5 percent to 8.9 percent of managers in corporations were Jewish. This pattern persisted even in New York City, where Jews constituted 50 percent of all college graduates at that time. Moreover, researchers found that this low representation could not be attributed to a lack of interest in such fields among Jewish students (EEOC 1968, pp. 667–671).

Despite the many forms of prejudice confronted by eastern European Jews, it should be noted that they never encountered the degree of hostility and discrimination experienced by non-Whites (Dinnerstein, Nichols, and Reimers 1990; Ivy and Ross 1957; Lipset 1990). This may be due to Jews' Caucasian appearance and their long presence in the United States (Sibley-Butler 1991).

The Rapid Social Mobility of Jews

Given that the context of the Jewish immigrant community in the early twentieth century was characterized by overcrowding, poverty, health problems, vice, and discrimination, how is it that the descendants of this lowly immigrant hoard "made it" to the top of the American social ladder so rapidly?

Explanations for ethnic status attainment (or lack thereof) fall into two types: those that emphasize culture, and those that stress structural and contextual factors. Cultural explanations assert that a group's value system and cultural outlook (for example, their family arrangements; gender roles; child-rearing habits; attitudes toward work, money, and education; norms for the distribution of resources; and time orientation) predispose the members of a group into patterns of behavior that will either foster or discourage their economic and social advancement. Structural and contextual explanations emphasize the importance of a group's resources and liabilities (for example, their possession of skill, education, capital, social networks, and legal status; demographic characteristics; and degree of organization) as well as the historical context of settlement (the availability of job opportunities, and the degree of host hostility encountered) as determining a group's ability to adjust to the host society. Scholars have marshalled both cultural and structural explanations to account for the mobility of eastern European Jews. We contend that the most accurate understanding of the Jewish experience is obtained by considering the combined effects of both.

Cultural Explanations

In both the popular mind and in much scholarly work, Jewish economic mobility is accounted for by Jewish cultural traits based in the Jewish religion, nurtured in the Jewish family, and reinforced through the Jew's historical role as social outsider. These include a high regard for education, a propensity for entrepreneurship, a Protestant work ethic that outstrips that of Protestants, parental dedication to and gratification from their childrens' achievement (called *nachis* in Yiddish), and a host of other values that fostered success in America (Chiswick 1991; Farber, Mindel, and Lazerwitz 1988; Glazer 1955, p. 31; Lipset 1990).

Thomas Sowell and Nathan Glazer each summed up this position well: "The internal values and traditions of the Jews were almost tailor-made for success in the American economy. . . . In short, the Jews had the social patterns and values of the middle class, even when they lived in slums" (Sowell 1981, pp. 93–94). "The Jewish workers violated most of the patterns of lower-class behavior, and were in many important ways indistinguishable from the non-Jewish as well as the Jewish middle class" (Glazer 1955, p. 33).

Structural Explanations

While there is some validity to this explanation for Jews' economic accomplishments in the United States, the stress on culture fails to acknowledge the more specific ways by which immigrant Jews were able to utilize a unique combination of skills and experiences, patterns of settlement, and social and economic opportunities, in conjunction with their cultural attributes, to achieve record economic mobility. To quote the authors of *The Transformation of the Jews*, "Those who have argued for the centrality of Jewish values as the explanation of rapid Jewish integration in America minimize the importance of social class and residential differences between Jewish and non-Jewish immigrants" (Goldscheider and Zuckerman 1984, p. 158).

A fuller understanding of why Jews achieved upward mobility more rapidly than the other southern and eastern European immigrants considers their motivation, demographic profile, human capital, as well as Jewish values, and can be summarized as follows. While most other immigrant flows of the turn of the century were made up of uneducated and unskilled males from rural areas who were temporary sojourners, Jewish migrants were more skilled, literate, and familiar with urban life. Because of conditions in eastern Europe, they came as permanent settlers, and in the company of their families.

Unlike other southern and eastern Europeans, Jews were refugees. They left not simply with the plan of working for a few years, but to permanently escape harsh violence and ethnic oppression. The "Great Immigration" of eastern European Jews from 1881 on was initiated by a series of anti-Semitic edicts, called the "May Laws," which concentrated Jews into "the pale of settlement," and further curtailed their already limited civil and economic rights. Eastern European Jews had long suffered violent, government sanctioned mob attacks, called pogroms, but after the 1880s, these became especially fierce, as suggested by the following report published early in this century.

> The congestion within the pale is the cause of terrible destination and misery. Fierce massacres occurred in Nizhniy-Novgorod in 1882 and in Kishinev in 1903. Many other pogroms have occurred, and the condition of the Jews has been re-

duced to one of abject poverty and despair. (Abrams 1911, p. 409)

Accordingly, millions left with no plans to return. "These laws and the extensions of them left most Jews no choice but to emigrate" (Farber, Mindel, and Lazerwitz 1988, p. 403).

Finally, eastern European Jews' mobility was shaped by the fact that they were received by an established group of coethnics—German Jews—who provided them with resettlement services and access to an economic niche (albeit in a paternalistic and sometimes exploitative manner).

Social and Demographic Characteristics

In terms of their levels of skill and literacy, permanence of settlement, and family composition, eastern European Jews more closely resembled the northern Europeans (of the "old immigration") that came in the early nineteenth century than the other southern and eastern Europeans (of the "new immigration") with whom they arrived (Joseph 1914) (see Table 1). These characteristics provided Jews with several advantages.

Skills and education gave Jews easier access to employment and self-employment. The presence of families offered permanence, stability, and resource yielding networks. Jews' stability is indicated by the fact that while 1 in 3 Poles and over half of all Slovaks, south Italians, and Hungarians returned to Europe, only 8 in 100 Jews did so. Patterns of chain migration—facilitated by Jews' extensive ties with relatives and former townsfolk—reduced the cost and effort involved in emigrating to and getting established in the United States. Their access to kin-based resources can be seen in that 62 percent of Jewish immigrants who arrived in the United States between 1908 and 1914 had their passage paid for by relatives. In contrast, among non-Jews, less than half that fraction (29 percent) received such aid (Kuznets 1975, p. 113).

In addition, they were accompanied by wives and children who provided them with a labor force for home-based manufacturing—an activity which involved a sizeable component of immigrant Jews. The large number of children under fourteen, school-age, that accompanied them (about 25 percent of the group) also gave the Jews a head start over other contemporaneous immigrants whose numbers in-

cluded few children (for example, only about 9 percent of Poles and Slovakians and 4 percent of Greeks were children under fourteen) (Joseph 1914, p. 180). These nationalities would have to wait for years before members would be exposed to the Americanizing and mobility fostering effects of an American education. Permanence of settlement and access to American schools also led Jews to learn English much faster than other immigrant groups. As Goldscheider and Zuckerman (1984, p. 168) pointed out "[w]ithin five years from immigration, two-thirds of Jewish industrial workers spoke English compared to 29 percent of all other immigrants."

Whether they hailed from small villages or large cities like Warsaw, Jews were familiar with the patterns of urban life—valuable resources for entrepreneurship—prior to migration. By contrast, other groups were rural in origin. Moreover, a large proportion of Jewish immigrants had worked in various kinds of factories and other industrial settings prior to migration. While 67 percent of all Jewish immigrants had worked in skilled trades prior to their emigration, between 75 and 80 percent of all Polish, Slovakian, and Italian immigrants had previously been unskilled laborers (Joseph 1914, p. 190). In fact, while Jews accounted for only 12 percent of all immigrants to enter the United States in the first decade of this century, they constituted over 29 percent of all skilled immigrants (Joseph 1914, p. 189).

German Jews

The economic mobility of immigrant eastern European Jews in the first decades of the twentieth century was also fostered by the help provided them by the established German Jewish population (Lipset 1990; Sibley-Butler 1991). German Jews, who entered the United States in the 1840s and 1850s, were socially, economically, and politically much more Westernized and secularized prior to their arrival in the States than eastern European Jews. German Jews quickly became concentrated in business and retail trades and made major progress in assimilating to middle-class American culture, politics, and language when the mass of eastern European Jews began to enter the country around 1880.

Out of sincere concern for their Jewish brethren and wary that the presence of these impecunious relatives—exotic in their dress, Orthodox in their religion, socialistic in their politics, and Yiddish-speaking—might arouse anti-Semitism, the German Jewish community sought to aid and Americanize the newcomers. Resettlement programs, which were extensive, yet sometimes heavy-handed and condescending, ranged from economic help and job placement to Jewish reform schools for wayward youth. As a result of friction, German and eastern European Jews feuded, and created parallel community institutions reflecting their different cultures and locations of settlement in Manhattan (Germans became known as tony "uptown Jews;" in clear contrast to the more humble, Russian "downtown Jews" of the Lower East Side). Yet faced with the common threats of anti-Semitism and the Holocaust, and inspired by Israel, the two groups eventually became united in numerous political, economic, and philanthropic endeavors (Farber, Mindel, and Lazerwitz 1988; Goren 1970).

German Jews also provided eastern European Jewish immigrants with an avenue to mobility, as by 1870 German Jews owned virtually the entire New York clothing industry, both retail and wholesale (Goldscheider and Zuckerman 1984, p. 160). About 50 percent of eastern European Jews had worked in this industry prior to migration (Joseph 1914). Since Jewish immigrant workers were already experienced in this type of work, employers prized them as they required little on-the-job training. Accordingly, Jews all but monopolized garment jobs, driving out natives and other immigrant groups until they abandoned them for more desirable positions (Kessner 1977). During the fifteen-year period from 1899 to 1914, the garment industry absorbed 400,000 workers, providing economic opportunities for a large number of these and other skilled immigrants (Kahan 1978).

Immigrant Jews also preferred to enter the garment industry because, as a heavily Jewish endeavor, it reduced the likelihood of contact with anti-Semites, allowed workers some ability to conform to their laws of religious observance, and offered a path for mobility based on coethnic cooperation and trust (Howe 1976; Kahan 1978; Kessner 1977).

Garments, Peddlers, and Entrepreneurs

Because much of the actual work of garment assembly was done in apartment-based sweatshops (a form of production introduced by immigrant Jews), the cost of setting up one's own enterprise was extremely low. Consequently many employees were able to save up the small investment needed to become self-employed. At the same time, the location of businesses in and near residential tenements permitted owners to put their family members, relatives, and recently arrived friends to work.

Moreover, because the garment industry was so heavily saturated by Jews, it was soon characterized by vertical integration and a multilevel system of ethnic-based cooperation that has now come to be referred to as an ethnic enclave (Portes and Manning 1986). Bonds rooted in language, religion, and common region of origin brought together workers and employers with suppliers, manufacturers, subcontractors, and retailers. While wages were kept low, flexibility and stability were maximized. Moreover, as long as the garment industry continued to expand—from the manufacturing of clothes to the broader area of retail sales—and new arrivals came in from overseas, avenues for mobility were ensured. As Kahan (1978, p. 241) underscored, "[w]ith older cohorts moving from the home industry sector from the sweatshop into the factories, into clerical positions, management and entrepreneurship, there existed a special dynamic pattern of economic adjustment for this group, the largest single occupational group in the Jewish immigration."

While the garment industry was the most important source of employment, Jewish immigrants—many of whom possessed useful skills—were extensively involved in other industries that were developing in the early part of the century, including construction, manufacturing, cigar and cigarette making, shoemaking, and food preparation (Joseph 1914, p. 188; Kahan 1978, p. 245).

Another low-investment business occupation favored by immigrant Jews was pushcart peddling. Waldinger (1992, pp. 15–16) highlighted that "[o]f the roughly 5000 pushcart peddlers in Manhattan at the turn of the century, approximately 60 percent were Jews. In turn, peddlers bought their merchandise from Jewish wholesalers and retailers, providing jobs for many other Jews." Patterns of economic cooperation and trust associated with Jewish families and communities also provided valuable entrepreneurial resources—especially parents' support for their childrens' entrepreneurship (Goldscheider and Kobrin 1980; Saracheck 1980). Finally, Jewish entrepreneurs were noted for their flexibility and willingness to accept low wages, develop new systems of production, and tolerate a lack of amenities (Kessner 1977).

In this original photo collage, the artist, Joanne Leonard, includes a letter written in 1942 that tells of a donation made, as a memorial to her grandmother, to the Joint Refugee Committee to further the work of anti-Fascist refugees trapped in concentration camps and refugees being resettled in Europe. She uses these family photos and letters to raise the issues of memory, memory loss, and history, particularly against the claim that the Holocaust, during which six million Jews died in Europe, did not happen or is exaggerated.

For some, this pattern paid off. In a relatively short time, many Jewish immigrants moved off of the bottom rung of the occupational ladder, the ethnically dominated occupations of home-based garment assembly and pushcarts, to the higher rung of retail trade, real estate, and other enterprises. "Indeed" stressed Kessner (1977, p. 65), "among the first generation immigrants it was not medicine, law or even their vaunted thirst for education that carried them forward. It was business."

Ethnic Organizations

While established German Jews provided their eastern European coethnics with organizations, activities, and jobs, the eastern Europeans themselves also brought with them a strong tradition of communal organization, experience with their governments prior to emancipation,[2] and extensive involvement with social movements and political activism following their emancipation in Eastern Europe (Katz 1971). Jews transplanted, modified, and invented a diverse array of associations, clubs, *landsmanschaften* (hometown lodges), synagogues, and mutual benefit societies which made up the communal basis of Jewish American life (Goldscheider and Zuckerman 1984).

The groups they created provided religious and moral guidance, education, political socialization, economic aid, health care, burial services, musical training, dancing lessons, and summer excursions—in sum, a social life and entertainment (Weisser 1985; Weinberg 1988). For example, the 1919 volume of the *American Jewish Year Book*, which sought to enumerate every Jewish organization in America, used close to 60 pages to list just New York City's entries. A particularly colorful instance of the Jewish organizational tradition is that of the Brownsville Boys Club, which was created by Jewish adolescents in Brooklyn who, without adult help, lobbied the city to provide them with playground space, funding, and their own building. A social historian who studied the club traces its young members' organizational skills to their having been raised by people themselves immersed in an activist milieu (Sorin 1990, p. 41).

Despite their Jewish constituencies, many immigrant organizations maintained a secular orientation, thus fostering the inclusion of the many nominally religious and nonreligious Jews together with their

Orthodox cousins (Schwartz 1988). The goals of Jewish organizations were extremely diverse—they were socialist, communist, and capitalist in ideology; American, European, and Zionist in national identification; cerebral and athletic in orientation; conservative and radical in cultural outlook. All offered a myriad of services, bases for mutual aid, and a place for the social incorporation of new arrivals.

Jews' accommodation with American life was facilitated through their growing involvement in American and western European innovations to their religious traditions including modern Orthodoxy, Reform, Conservative, and Reconstructionist Judaism. By joining such movements, eastern European Jews were able to maintain Jewish religious practice and communal involvement while simultaneously following lifestyles and ethical traditions that were compatible with those of the Christian society where they lived.

As time passed, Jews continued to build Jewish organizations such as philanthropic and political groups to advance Jewish interests on the national and international level, and a variety of Jewish clubs, camps, schools, colleges, universities, hospitals, retirement homes, and other institutions to provide for their own community needs, foster interaction, and socialize future generations.

Upon arrival, Jews were not as politically adept as the Irish, but many had experience with community activities and leftist movements prior to emigration. As they gained confidence, sophistication, and competence in the English language in the New World, they expressed their newfound political skills in various forms of activism, particularly in trade union movements and leftist parties.

Education and Mobility

While the "cultural values" explanation for Jewish mobility emphasizes the Jews' traditional love of education as a major reason for their mobility, the immigrant generation made only moderate progress in schooling. New York's public colleges became predominantly Jewish only after the immigrant generation had begun to establish themselves in industry and small business (Gorelick 1981; Sowell 1981). However, the social characteristics that distinguished Jews from other immigrants of the same era allowed them to make more money and live more stable lives. In turn, these advantages facilitated their childrens' educational achievements. Jewish children were able to attend school longer and often attended college (Goldscheider and Zuckerman 1984, p. 168; Joseph 1914). For example, relatively few Jewish daughters were removed from school to perform domestic duties, and were thus able to continue their education at higher rates than non-Jewish immigrant girls (Chiswick 1991; Weinberg 1988, p. 173).

The timing of the garment industry's rapid growth at the start of the century provided Jewish immigrants who had the appropriate skills with a viable economic foothold in America. Then, changes in American society that included the vast expansion of the public and service sectors of the economy, coupled with the growth of public higher education, offered their educated offspring ample opportunities to gain white-collar jobs. A further set of historical circumstances provided Jews, many of whom already valued a higher education, with unique opportunities to obtain it.

Prior to the 1920s, a college education was seldom a path to economic betterment, especially for non-WASPs. Rather, it served as a source of preparation for the Christian clergy (clearly not an option for Jews) and a kind of veneer for the cultural enrichment of the sons of gentlemen who had already achieved high social rank (Gorelick 1981). However, as the century progressed, an increasing number of jobs in government—as teachers, librarians, psychologists, principals, social workers, administrators, accountants, clerks, doctors, nurses, auditors, and engineers—were opened to those who could master the skills and score well on the civil service exams reformers had put in place to limit the patronage system controlled by political "machines" (Waldinger 1992). This movement had great effects in New York City, where a vast population, an excellent system of public schools and colleges, and a high standard of public service created an extensive demand for persons able to fill these positions.

While few early Jewish immigrants had made major strides in education, many in the second generation had taken advantage of opportunities for college. With jobs largely unavailable during the depression years of the 1930s, many young Jews stayed in school, gaining more education than they would have had access to had work not been restricted. In the decade of the depression, the proportion of Jews attending college (47 percent) was twice that of other denominational groups (23 percent) (Kahan 1986, p. 135). By 1934, about half the women college students in New York were Jewish (Weinberg 1988, p. 175). Following World War II, the G.I. Bill paid for veterans' continued education. Kahan (1986, p. 130) stressed that "a combination of the Depression period and the G.I. Bill raised the educational endowment and skill level of the American-born offspring of Jewish immigrants to the very top layer in American Society." Largely excluded from employment in major corporations, they entered the public sector in sizeable numbers. Summarizing the extensive involvement of Jews in public sector professions, Waldinger (1992, p. 21) asserted that, "Garments was the Jewish enclave of the past; Jewish mobility into the middle-class made teaching the Jewish niche of the mid-1960s."

While Jewish values—including an entrepreneurial outlook and a love of education—contributed to the rapid economic progress of eastern European Jewish immigrants, structural and historical factors also played a vital role in Jewish mobility. In comparison to others who immigrated about the same time, Jews were more literate, more skilled, more urban, and they came with their families to settle permanently. Upon arrival, they had the good fortune to receive resettlement aid and access to an economic niche by virtue of their connections with established German American Jews. Finally, their family patterns and economic stability allowed Jewish immigrants to take better advantage of educational opportunities.

Retaining Ethnic Solidarity in Mainstream America

Despite their rapid entry into mainstream America, Jews maintain behavioral patterns that distinguish them from non-Jews, confounding assimilationist sociological theories that suggest ethnicity will be lost as groups become largely middle class (Gordon 1964). For example, when compared to the established White population, Jews tend to more often live near fellow Jews, have relatively lower rates of intermarriage (to non-Jews), visit relatives more often, and retain more liberal to left politics and party affiliation (Lipset 1990, p. 19). [Of the forty-three Jews in Congress, thirty-five are Democrats (Hoffman

1993)]. Some of these patterns have recently begun to erode, yet the Jewish community maintains a high degree of viability. No longer can Jewish solidarity be accounted for by anti-Semitism, concentration in ethnic industries, residential segregation, and regional concentration. Jews now are leaders in the very institutions and corporations that once excluded them, live in neighborhoods that once maintained restrictive covenants against them, and are unflinchingly labeled as "Anglos" in political discourse. Once concentrated in the northeast, sizeable Jewish communities now exist in Texas, Florida, California, and Colorado (Kosmin 1991). Jews' accomplishments coupled with their continued viability as a community suggest that they have achieved social and economic mobility in America despite retaining their unique cultural, religious, communal, and economic patterns. Accordingly, their experience validates the theoretical formulations of cultural pluralism and situational ethnicity as a practical means of coping with life in the United States.

Currently the focus of much American Jewish life is secular and concerned with Jewish institutions, philanthropy, and the state of Israel. In recent years, many Jews, especially younger ones, appear to be displaying the patterns of assimilation that their group long resisted. For example, despite a continuing tendency towards liberalism, increasing numbers now vote Republican and have no connection with any sort of Jewish organization. Since 1985, the intermarriage rate has increased to what some see as an alarming level, approximately one in two (Kosmin 1991; Mayer 1991; Waxman 1990).

At the same time, however, Jewish viability in American society is being maintained through continuity, renewal, and immigration. Since the 1970s, several hundred thousand Jewish immigrants have settled in the United States from Russia, Israel, Iran, South Africa, and other nations (Gold 1992). While these immigrants' expression of their Jewishness is often quite different from that maintained by native Jews, they do bring youth and energy to American Jewish life. Moreover, as recent immigrants, they have revitalized neighborhoods, businesses, and institutions in areas such as West Los Angeles and Brooklyn that established American Jews have largely abandoned.

Many of today's Jewish immigrants are highly educated professionals, yet others retrace estab-lished patterns of immigrant adaptation. Today's Jewish immigrants, such as the Middle Eastern brothers who run Guess, Jordache, and countless smaller garment companies, have also found riches in the rag trades and other traditional Jewish enterprises—food, retail sales, construction, and real estate. Many also establish their own associations and storefront synagogues (Byron 1992; Gold 1992; Herman and Lafontaine 1983; Kass and Lipset 1982).

At the same time, parts of the native Jewish community have recently become involved in activities and lifestyles more Jewish than those of their youth. For example, a growing group of formerly secular *baal tshuvah* (returning to the faith) Jews have adopted Orthodoxy, reflecting a search for lost roots and spirituality in an affluent and secular society (Davidman 1991). A colorful example of these is Issac Bitton, a "former hippie and drummer" whose rock group Les Variations once performed with Gladys Knight and Curtis Mayfield. Currently Bitton—along with his eleven children—is an active member of the ultra-Orthodox Hassidic enclave of Crown Heights, New York (Noel 1993, p. 11).

Conclusion

At present, students of American Judaism debate the future of their group. The most pessimistic proponents (sometimes called "cataclysmists") suggest their disappearance, because with each additional generation the religious and institutional involvement of American Jews decreases; older, observant Jews die; and the intermarriage rate swells (Waxman 1990). A more optimistic stance asserts that predictions about the total assimilation of American Jews have been made since the 1870s, underscoring that while certain patterns of traditional Jewish behavior have diminished over time, American Jews do maintain many rituals and practices they find personally meaningful. The optimists also tend to see intermarriage in terms of the old proverb—not losing a daughter, but gaining a son. Rather than being a threat to Jewish survival, intermarriage also has the potential to incorporate non-Jews into the Jewish community (Goldscheider 1986).

Taking a more philosophical bent, tradition-oriented scholars and rabbis claim that regardless of the health of institutional Judaism, the spiritual dimensions of American Jewish life are suffering. They worry that because of American Jews' obsession with organized activities, the private and family basis of the religion—the Jewish home—has been undermined (Schulweis 1991). Similarly, other critics suggest that in constantly focusing upon the Holocaust and anti-Semitism as critical to Jewish self-definition, the community is raising a generation of Jews who are more concerned with their status as victims than with the ethical, spiritual, and religious foundations of Judaism (Glazer 1990; Seidler-Feller 1991). Suffice it to say that American Jews, like much of American society, are in a state of both flux and self-examination.

Notes

1. Yiddish is the German-based, Hebrew-scripted language specific to eastern European Jews.

2. Before the 1860s, most Jews of eastern Europe were not entitled to equal citizenship. Instead, they had fixed and delimited social rights as determined by contract with the Tsars. This arrangement, while extremely restrictive, did give Jewish communities some rights of self-government through administrative institutions or *Kahals* (Goldscheider and Zuckerman 1984, pp. 13, 40–41). Hence, eastern European Jewish immigrants who came to the United States had inherited a tradition of community self-control.

15

From Peasant to Worker: The Polish Immigrant in the United States

DOMINIC A. PACYGA

On 17 January 1915, Mary Dorula married Stanley Walkosz in the Church of the Sacred Heart in Chicago's Back of the Yards neighborhood. Father James Strzycki presided over the ceremony as the bridal party of twelve friends and family looked on. Both Dorula and Walkosz had immigrated to the United States from the Polish mountain village of Szaflary in the Austrian occupied province of Galicia. Walkosz, who was some nine years older than Mary Dorula, met his future wife while visiting his brother, Adalbert, in Chicago. He left his coal mining job in Pennsylvania to stay in the city and wed Mary. Walkosz found employment in the Chicago stockyards.

The Walkosz family grew quickly. Just after the couple's fourth wedding anniversary in 1919 Mary gave birth to her third child, Pauline. Eventually Stanley and Mary had seven children, five girls and two boys, all of whom were raised in the parish of the Sacred Heart. Most of their grandchildren would also be raised in the parish which was located just west of the Union Stock Yards. The history of the Walkosz family and of the parish of the Sacred Heart is largely representative of the history of Polish immigration to the United States before World War I (Walkosz n.d.).

Sacred Heart Parish was the third Polish Roman Catholic parish to be founded by the immigrants of Back of the Yards. In addition, three other Polish Catholic churches were located just to the northwest and northeast of the Union Stock Yards, in the neighborhoods of Bridgeport and McKinley Park. Poles flocked to the industrial district from the 1870s up to World War I. Father Francis J. Karabasz organized the parish in 1910. Previously the priest, who was born in the Polish settlement in Lemont, Illinois, served at various Polish parishes on the South Side. Karabasz's assignment saw him locate a parish in the poorest section of the Stockyard District (AC/Karabasz file). His future parishioners would, to a large part, come from the Polish mountains as Sacred Heart became the center for Polish Highlander community life in the city. Leaving their rural past behind, these mountaineers or *Gorale* became industrial workers in Chicago's vast complex of livestock pens, slaughterhouses, tanneries, and packing plants (Koenig 1980, 2:872; Sacred Heart Parish 1960, pp. 14–16).

With the exception of the Jews, at the turn of the century most of the migration from southern and eastern Europe was of peasant origin. Here are Polish immigrants just off the boat, the liner SS Touraine, on September 23, 1910. The women wear scarves tied around their heads as a show of modesty and decorum. (Theodore Wesley Koch Collection; Bentley Historical Library, The University of Michigan)

Polish Emigration

The account of Polish immigration to the United States is part of the history of the transformation of Poland from a rural east European society to an important center for the expansion of industrial capitalism. Poland, divided among three occupying powers and located on the edge of European industrial development, was transformed by the growth of the international capitalist economy in the years after the Congress of Vienna that took place in 1815. The result was the integration of Poland into that economy and the migration of much of the nation's rural peasant population to receiving areas in Europe, the Russian Empire, and the Americas (Pacyga 1991, chap. 1).

By the end of the nineteenth century Poles found themselves working in Polish factories, German and French coal mines, Siberian agricultural settlements, Brazilian farms, and all across the American industrial landscape. Large urban American Polish settlements appeared in Chicago, Detroit, Milwaukee, Brooklyn, Buffalo, Rochester, Utica, and across New Jersey, Connecticut, and Pennsylvania. Most toiled in industrial jobs, but a growing middle class also appeared in urban Polonia before World War I. Polish stores and businesses dotted the cityscape and a growing number of Polish American professionals served the needs of the Polish settlements. Poles also settled in rural settlements throughout the East and Midwest, especially in Connecticut and Wisconsin (Bukowczyk 1987, chap. 2).

Immigration from the Polish lands was not always easy. Poland had been partitioned at the end of the eighteenth century and again in 1815 after the Napoleonic Wars. The three occupying powers, Germany (Prussia before the 1871 creation of the German Empire), Austria-Hungary, and Russia often had conflicting attitudes toward emigration and even internal migration. Polish peasants were constricted by serfdom in Austrian Poland (called Galicia) until 1848 and in Russian Poland until 1864. Poles living in the Prussian occupied territories were freed from serfdom during the Napoleonic era in 1807 (Kieniewicz 1968, pp. 172–182). Even after *uwlaszczenie* (emancipation), the old precapitalist ways persisted in the countryside.

Economic Modernization

The process of modernization, of which emigration was but a part, was a slow and often painful one. Traditional Polish peasant society revolved around the four seasons and the land. The church, the community, and the family provided the pillars of rural society. The family provided the major work group, and land was held in its name. Polish peasants lived throughout much of the nineteenth century in isolated self-contained villages, but slowly outside influences began to transform the countryside. Still, even after the granting of land to the peasantry, the relationship between large landholders and small farmers remained one of inequality in all three of the Polish partitions. Of course, new economic and technological realities eventually did bring change to peasant society (Thomas and Znaniecki 1919, 1:98; Dobrowolski 1971, pp. 297–298). Immigration was definitely one of these developments.

The decision to emigrate was an outgrowth of new economic, political, and technological realities. While the resolution to leave the village, even if only temporarily, was an individual one, outside forces quickly came into play. What and where did opportunities for making a successful emigration exist? How did local authorities view emigration? What kind of transportation technology existed to allow migration abroad?

By the 1880s the Polish lands went through an economic transformation that entailed the loss of much of their traditional agricultural markets in the west and the arrival of industrial development in Polish cities such as Lodz and Warsaw. The emancipation of the peasantry proved vital for this economic metamorphosis. Peasants, now freed from their bonds to the land, faced a precarious future as independent farmers. Many of the emancipated peasants became nearly landless or lost all their property in the volatile capitalist rural economy. However, the new economic conditions also presented some Polish farmers with opportunities. Land was now available for purchase either from other peasants or from large landholders eager to rid themselves of some or all of their property. The major problem was how to get funds to purchase the land. Temporary emigration provided one possible answer (Kieniewicz 1968, pp. 172–182).

Prussian Poland provided the first major source of Polish emigrants. This process quickened after the creation of the German Empire in 1871. The 1880s witnessed the greatest overseas migration from Prussian Poland. The movement spread to Russian Poland in the 1890s and to Galicia at the turn of the century (Pacyga 1991, pp. 23–24). When Stanley Walkosz emigrated to the United States before World War I, he was already part of a well-established migration pattern. Indeed, seasonal migration had been part of Polish mountaineer culture for some time. The movement across the seas was a natural outgrowth of the traditional patterns of searching for work (Dobrowolski 1971, p. 294).

As did others, Poles moved in migration chains with pioneers sending for relatives and friends. The creation of Polish enclaves in the United States often took on a regional pattern, such as the creation of the Polish Highlander community in Back of the Yards which greeted Walkosz and Dorula on their arrival in Chicago. The chain migration pattern proved very successful in the creation of Polish colonies abroad. This migration system fit well into the expanding capitalist market and international transportation system (Gliwicowna 1936, p. 205).

While American historians talk about the railroads opening up the American West, in Europe the railroads spread throughout eastern Europe making once-isolated villages more accessible. This resulted in the expansion of the European emigration movement to the east European lands. Poland, divided politically and economically by her powerful neighbors, was in the midst of an agricultural adjustment to the capitalist market system. Hence it provided ready and willing participants in the ongoing transfer of population throughout the capitalist system.

The Journey

One of the major obstacles, however, to Polish emigration was the crossing of borders. Poles living outside of the German Empire first had to cross into Germany in order to reach the great ports of northwestern Europe. By the time Polish emigration reached its peak, the German ports of Hamburg and Bremen had the most embarkations for the United States. Therefore, the German Imperial government was instrumental in directing the flow of emigrants from eastern Europe to the United States. After 1892 the German authorities took a much more active part in regulating the movement across its borders. Germany instituted various checkpoints because of fear of the spread of cholera from eastern Europe to German territory. Five years later the German Reichstag passed further regulations that dealt with agents, ships, emigrant rights, and medical inspection. Furthermore, the increased complexity of American laws made it necessary to screen emigrants at the checkpoints (Taylor 1971, pp. 121–123).

The town of Myslowitz at the junction of the German, Russian, and Austrian empires provided one such border checkpoint. It was probably here that Mary Dorula left Galicia and entered German territory for the port of Bremen on her journey to Chicago. At Myslowitz emigrants had to pass a rudimentary medical examination and present their credentials. Once the emigrants entered the boarder station they could not leave except to board the train for the western ports (U.S. Senate 1911, pp. 29–31).

Some times the various imperial authorities opposed emigration, and at those instances Poles often had to sneak across the borders. Nevertheless, Polish emigration proceeded and emigrants made their way across the Atlantic to the United States. Emigrants flowed through the transportation systems created for the movement of goods and people across the European landscape.

The city of Bremen with its great port Bremenhaven was the point of embarkation for many Poles. Mary Dorula passed through this great port on her way to America. The young girl had her first taste of ice cream here as she waited for passage to New York and on to Chicago (Walkosz n.d.). The city seemed magnificent to emigrant eyes. One Polish traveler commented on the beauty of the city, claiming that every home looked like a palace. Bremen also proved to be expensive and sometimes dangerous for unsuspecting peasant travelers (Kula 1973, p. 245).

By the 1890s the trip to New York seldom exceeded twelve days. Few died on the crossing. Technological advances had made the trip fairly easy and quick. Discomfort for peasants who had never traveled at sea and overcrowding seemed to be the major problems faced by travelers. But the amount of return migration and travel again to America proved the point that the passage was no longer such a great obstacle.

After 1892 Ellis Island replaced the old immigrant checkpoint at Castle Garden on the tip of Manhattan in New York City. While the average immigrant stayed at Ellis Island for only two or three hours, others were detained for various reasons. For the great majority of Polish immigrants Ellis Island and New York City were temporary stops as they quickly moved on to their final destinations throughout industrial America (Abbott 1924, pp. 246–248).

Polish Immigration to Chicago

While the final destinations varied from small farms in Wisconsin to massive industrial neighborhoods in Pittsburgh, there were commonalties in the Polish experience in the United States. Generally speaking, the Polish diaspora was an industrial one. When Mary Dorula came to Chicago's South Side she found herself in one of the most infamous industrial neighborhoods in America. Ever since the publication of Upton Sinclair's *The Jungle* in 1905, the neighborhood had been identified as the quintessential industrial slum.

Communal Associations

The Stockyard District had emerged as a major industrial locale after the opening of the Union Stock Yards in 1865. By the turn of the century the small Polish colony was well established, with the Roman Catholic parish of St. Joseph at its institutional center. In 1906 Poles founded another parish, St. John of

God, just south of the stockyards. Four years later, Father Francis J. Karabasz took on the assignment to establish the parish of The Most Sacred Heart of Jesus. Like other Polish colonies, Back of the Yards saw the quick institutional development of its Polonia. The three parishes maintained schools and worked with the immigrant community to develop additional organizations. These three parishes were very Polish in character, as can be seen by the creation of Sacred Heart (Koenig 1980, 1:496, 515, 2:870; Wade 1987, p. 47). All of the lots except one that Karabasz bought for the location of a new multiuse building to serve as parish church, school, and hall were purchased from Poles. Bronislaw Kowalewski, a real estate agent and community leader from St. John of God Parish, helped Karabasz in this venture. The priest hired John Flizikowski to design the structure. The iron came from John Rybarczyk of South Chicago where Karabasz had earlier served as an assistant pastor. John Paczkowski provided the carpenters, while Albert Sikorski organized the masons. The Polish parish of the Sacred Heart of Jesus was a truly Polish venture from top to bottom.

Sacred Heart grew quickly. In 1925, 1,400 families lived in the parish and 1,519 children, including the children of Stanley and Mary Walkosz, attended the parochial school run by a Polish order of nuns, the Felician Sisters. Polish was the language of instruction in the parish school. The Felician Sisters reserved English for the "American" topics such as United States History. By this time the Polish Highlanders provided the backbone of the community (AC/Sacred Heart Correspondence File; Koenig 1980, 2:872; Sacred Heart Parish 1960, p. 47).

The creation and maintenance of a Roman Catholic parish was an extremely important symbolic, economic, and even political act for the young Polish American community. The parish was not simply a religious institution, it was a communal institution. Much of the life of the community revolved around it. Church steeples rose high all across the American industrial heartland challenging the smokestacks for control of the sky over cities like Chicago, Buffalo, Brooklyn, and Pittsburgh. In Chicago's Steel District, the Polish parish of St. Michael the Archangel, with its high Gothic bell tower, stood nearly surrounded by the smokestacks and foundries of Illinois Steel. The twin towers of St. Joseph Church and of St. John of God also stood out in sharp contrast to the smoke-belching chimneys of the packinghouses in

Back of the Yards. The Catholic parish was symbolic of the permanence and security of the Polish colony in industrial America. It also provided an important connection with the peasant past of the immigrants (Chalasinski 1935).

Religious practices in the Polish colonies of the United States replicated those in the Polish countryside. The Latin Tridentine Mass presented a familiar liturgy to the peasants turned industrial workers. In addition, the traditional holidays and rituals were reenacted in as close to a customary manner as they could be in the United States. Urban industrial realities often meant changes in traditions, but the essence of the celebrations was maintained. The church calendar closely resembled that used in the east European countryside. Christmas, Easter, Candlemas, and the Feast of the Assumption all were celebrated in customary Polish ways in Polish Chicago. The mass on All Soul's Day remained solemn, but because of the size of the Polish American parishes priests did not read out all the names of the dead on that day. Pre-Christian traditions of leaving food on grave sites and outside front doors also disappeared. The church further provided special devotions, vespers, and rosary hours as it did in Poland, reminding the immigrant community of its peasant and devotional past. The church bells marking the hour of prayer to the Virgin in the form of the Angelus prayer further reminded Polish immigrants of their homeland and religious and communal obligations (Obidinski and Zand 1987, pp. 62–72).

Practices changed because of new urban realities. In Poland the wedding celebration customarily lasted three days, but American work schedules prohibited such celebrations. Also families no longer had as much control over the selection of marriage partners as they did in Poland. Still, weddings, funerals, baptisms, and all the religious and public celebrations of the life cycle remained communal events of great importance.

Saloons provided public space for much of the communal celebrating. Almost all taverns in industrial neighborhoods had halls behind or above them that served as sites for the various celebrations of the life cycle and for meetings of church groups, labor unions, and ethnic fraternal organizations such as the Polish National Alliance, the Polish Roman Catholic Union, and the Polish Women's Alliance.

Wedding parties often attracted over 200 people who gathered in the saloon halls. They were costly affairs, in part paid for by the men of the community who purchased the right to dance with the bride by flinging silver dollars at plates. If the plate broke the dance was free; if not, the coin remained as a gift for the couple. The ritual eventually developed into simple gift-giving, but it remained popular throughout the first third of the twentieth century (Obidinski and Zand 1987, pp. 89–99).

Living Conditions

The churches, fraternal organizations, saloons, and communal celebrations with their rituals and merry-making stood in sharp contrast to the physical realities of the workplaces and the neighborhoods that Polish Americans found themselves in during the period of most intense immigration.

When Mary Dorula arrived in Back of the Yards the poverty and living conditions of the neighborhood stood in stark contrast to Chicago's magnificent downtown and the beautiful lakefront neighborhoods only a short streetcar ride away. Housing and sanitation presented a constant problem to residents. Mary McDowell, the head resident of the University of Chicago Settlement House, referred to the area as a frontier settlement with dirt streets, wooden sidewalks and buildings, and outdoor plumbing (USP/McDowell 1914a, p. 1). Considering the herds of animals being brought into the Union Stock Yards daily, the comparison was more real than symbolic. In 1905, Upton Sinclair, in his somewhat exaggerated account of local conditions, wrote of the area in much the same terms. Six years later another investigation repeated the charges of unhealthy living conditions among the cottages and two-flats of Back of the Yards (Breckinridge and Abbott 1911). Even a lawyer representing the meat packers in 1918 called for the ultimate destruction and rebuilding of the district as the only answer to the problems faced by its residents (*Chicago Tribune*, 6 March 1918).

Balloon-frame buildings provided the typical structure in Back of the Yards. This type of construction was a modification of traditional wooden buildings using lighter boards instead of timbers, and nails rather than mortise work. Closely fitted boards covered the outside of the house. These buildings were easy to build and cost much less than traditional wooden structures or those made of brick. In general, wooden two-flats and cottages filled the streets of Chicago's Stockyard District, but occasionally larger tenements and brick buildings stood mixed with the smaller buildings. Many of the buildings also contained storefronts, even along the residential side streets (Barrett 1987, p. 92; Duis 1976, p. 14).

A casual observer of the Back of the Yards in 1911 might not find the district to be congested. Most buildings occupied half or less of the city lots on which they stood, leaving room for front and rear yards. The neighborhood did, however, suffer from overcrowding. A survey of ten typical blocks in that year found an average of 208 residents per acre. Compared to other immigrant districts, this does not seem to be high, but given the actual size of buildings in Back of the Yards the living space proved to be inadequate. Lodgers were the cause of much of the overcrowding. According to the 1911 survey they made up 27 percent of the local population (Breckinridge and Abbott 1911, pp. 437–450). Many of these boarders were family members or friends from the villages of origin of the immigrants. The boarding system provided income for the family. It was also a form of women's work, allowing married women to contribute to the family income without leaving the home. In addition, the boarding system provided an important link in the migration chain from Poland and other east European countries to the Stockyard District. It was not simply an economic system forced by low wages, but also a communal response to the realities of migration and urbanization.

Vile odors permeated the district, a result of the meatpacking process. Smells from "Packingtown" filled the homes. "Bubbly Creek," as the west fork of the south branch of the Chicago River was known, supplied both an open sewer for the packinghouses and the northern boundary of the neighborhood. Large garbage dumps made up the neighborhood's western boundary helping to encircle the residential area with foul-smelling nuisances. The death rate for children living near the dumps was ferocious and the death rate for the neighborhood as a whole stood much higher than in Hyde Park, the fashionable lakefront neighborhood just to the east (Bushnell 1902, p. 38; USP/McDowell n.d., p. 2).

Other problems also plagued the district. Scum-covered ditches and open sewers were common. Houses without sewer or water connections existed in various sections of the neighborhood. At times the streets resembled canals. About 1910 street conditions improved, but many side streets remained unpaved, flooding during rainy seasons and creating dust storms during the summer months. The railroads that crisscrossed the neighborhood and served the stockyards also proved to be a problem. Hundreds lost their lives or limbs before the city mandated that the railroads elevate their lines passing through residential areas. Garbage collection was infrequent and alleys lay filled with refuse. The garbage became alive with worms and flies. Rats roamed freely through the alleys and yards (Bushnell 1902, p. 42; USP/McDowell 1914b, pp. 1–3).

While poor living conditions prevailed in Back of the Yards, they were not the worst in immigrant Chicago. Congestion was greater on the Polish Northwest Side and the Jewish West Side. In 1900 the death rate in the Stockyard District was lower than the city's as a whole, even though the Thirtieth Ward, a large part of which lay in Back of the Yards, led the city in deaths from consumption (pulmonary tuberculosis). The Stockyard District also accounted for a very high number of deaths from pneumonia. It ranked first among the city's wards in that category five times during the period 1894 to 1900. Conditions did not change much after the turn of the century (Chicago Department of Health 1894–1900, 1911; USP/Hedger 1908). Back of the Yards was not the worst area of the city, but its residents continued to suffer. Living conditions were as much a result of low wages, a lack of political power, and a lack of familiarity with American methods as they were of neglect or ignorance.

Working Conditions

Insufficient income proved to be the most serious barrier to changing living conditions in Back of the Yards and across industrial America. Various factors impacted on wages. Polish and other immigrants entered the Chicago packinghouses as unskilled workers. Before World War I the workday fluctuated greatly. Often men and women who left for their jobs in the morning had no way of knowing when they would return. The length of the workday depended on the availability of livestock at the Union Stock Yards. On days when producers placed fewer animals on the market workers worked a short day of four hours or less. When many animals flooded the pens the workday could drag on well into the night,

running twelve hours or more. Also seasonal shifts hit all the departments in the packinghouses. Unskilled workers in particular suffered from fluctuations in hours and days worked. In some departments unskilled workers were laid off for as much as one-third of the year. Most work, however, was steadier (Commons 1904, p. 13; Gross 1929, p. 14; U.S. Senate 1916, pp. 3214, 3521, 3464).

On the basis of the available data, we can see that the average packinghouse worker earned from $8.50 to $12.50 per week in 1910, the common wage in the neighborhood stood between $10 and $11 per week. From 1903 to 1915, the year Mary Dorula and Stanley Walkosz married, there had been a 10 percent increase in wages throughout the stockyards. The chief investigator of a report concerning packinghouse wages, however, claimed that this raise was inadequate. The period had seen not only a rise in wages, but also a rise in the cost of living. A comparison of prices in the Stockyard District in 1903 and 1910 showed an increase of 16.5 percent. In 1910, the average wage for a Polish husband in the Back of the Yards stood at $486.14 a year, at a time when it was estimated that the minimum amount necessary to support a family of five in the neighborhood was $800 per year. It was not surprising that many children went to work to help support their families. Also, working wives and lodgers played a decisive role in the survival of the family unit.

By 1916 the cost of living had spiraled as a result of World War I, leaving the Polish families of Back of the Yards far behind in their struggle for better living conditions. Mary Dorula and Stanley Walkosz, like their fellow Polish immigrants, found themselves near the bottom of the economic ladder in industrial America. Eventually their older children went to work at early ages in the stockyards or in other jobs that provided minimal incomes for boys and girls on Chicago's South Side. Mary, who only worked for a short time in the stockyards, took in laundry and boarders to help make ends meet. While Poles were legendary in Chicago as home owners, after World War I they enjoyed one of the highest rates of home ownership among ethnic groups in Chicago; the Walkosz family with seven children never purchased a house (Walkosz n.d.).

Family life cycles and community obligations also had a tremendous impact on incomes. Parochial schools charged a tuition of about 50 cents per month or about $5 per year. The parish church also had to be supported. In 1910, the average annual contribution to the local parish was $11.44. The Polish community took this obligation very seriously. A birth, marriage, sickness, or death could wipe out any savings these Polish immigrant families might have. Economic stability then became a major point of concern for Polish working-class communities, not only in Chicago, but across the country (Kennedy et al. 1914, pp. 9–14, 19–23; U.S. Senate 1916, pp. 3466–3467).

The initial response of the Polish immigrant colony had been an inward-looking one, a communal response. The parish, parochial school, Polish libraries, fraternal institutions, local businesses dedicated to ethnic usage, and even street gangs had all been part of this answer to the industrial American city. This response had been a necessary one. It gave Polish immigrants a sense of stability even in the constantly changing capitalist economic world. The communal response allowed Polish immigrants to lay down firm roots in the industrial United States and to become Polish Americans.

The Labor Movement

This response also provided a firm base for a wider reaction by these Polish and other working-class Americans to the living and working conditions that framed their lives. This extracommunal response included joining with other ethnic and racial groups in reacting to basic working-class issues. It lead not only to active participation in the labor movement, but also to political activity and to the creation of neighborhood organizations. All of these revolved around the desire to better their living and working conditions and to provide a better future for the community.

The first strike in the Union Stock Yards took place in 1869. Another walkout occurred in 1877 with the railroad strike and riots of that year. That was also the year that the first Polish family in the district, the Zulawski family, settled in Back of the Yards. The first modern organized job action, however, developed in 1886 when the Knights of Labor led, or rather misled, a work stoppage during the eight-hour strikes of that year. This 1886 strike would not be the last to sweep across the stockyards (McDowell 1970, pp. 336–337; Meyers 1929, pp. 336–337; Pelling 1968, pp. 63–66, 71).

Generally speaking, Poles and others of what was then called the "new immigration," were regarded as possible strikebreakers by union organizers. Polish immigrants filled the unskilled jobs in the packinghouses, especially after the 1886 strike. The craft unions often ignored those who were unskilled; and they, in turn, ignored the unions. During the riots surrounding the Pullman strike in 1894 the lack of organization became apparent as crowds ransacked railroad cars in the stockyards and railroad yards to the east of the neighborhood. Working-class anger was apparent everywhere across industrial America, but little effective leadership could be found, especially in the mass production industries such as meat packing and steel. Labor remained divided by trade and between skilled and unskilled workers, White and African American, immigrant and native born, old immigrant and new immigrant, male and female. The fragmentation of the working class seemed complete.

In the summer of 1900, the Amalgamated Meat Cutters and Butcher Workmen of North America, organized three years earlier in Cincinnati, established a local in the Chicago stockyards. Cattle Butchers Local 87 received its charter on 9 June, and the union came out from the industrial underground. The following year, the Amalgamated Meat Cutters moved its headquarters to Chicago from Omaha, and by 7 July locals were openly organizing in Chicago's packinghouses. The meat packers came to a quick agreement with the union which covered their skilled employees. Although the national union left the issue up to the locals, unskilled workers were not at first welcome in the Amalgamated Meat Cutters. Michael Donnelley, the president of the union, opposed this attitude and called for the organization of the unskilled immigrant workers. Donnelley understood that because of the technological and organizational changes in the production process, organizing unskilled workers was important for the unions. A few skilled workers could no longer shut a plant down as they had in 1886. For this reason the packinghouse union changed its attitude toward unskilled laborers. It would take further consciousness raising for the union to accept women and Blacks, but they did accept them when the reality of the employment structure in the Chicago stockyards made itself known. By 1900 women were playing an increasingly important role in the production of meat products. Blacks rapidly increased their presence in the stockyards after 1904. Mexican immigrants arrived in large numbers after 1920, further complicating the ethnic and racial mix in the packinghouses. Donnelley steered the union toward the acceptance of the unskilled, immigrants, women, and even African Americans into its ranks. The union attempted to create solidarity out of the fragmentation of the Stockyard District population. In fact, the six-week long 1904 strike was in large part in support of these unskilled workers who had been left out of the earlier agreement. Once again, however, organized labor came out a looser in its struggle for recognition from management (Brody 1964, pp. 14–15, 22–25, 35–41).

Polish Americans played a crucial role in this strike, making up a large percentage of the unskilled who stuck by Donnelley and the Amalgamated Meat Cutters. The Polish and other new immigrants proved crucial for support of the union in Chicago, just as they had two years earlier in the Anthracite coal mines of Pennsylvania (Greene 1968). After the defeat of the Amalgamated Meat Cutters and the other unions that supported it, a legacy of fear and despair covered the stockyards. Wages remained basically frozen until the outbreak of World War I in 1914 (Herbst 1930, p. 53).

At the time when Mary Dorula arrived in Back of the Yards in 1913, no unions represented the vast majority of packinghouse workers. The workers seemed permanently divided and unorganized. In that year the union had tried to reenter the stockyards, but for the most part workers shunned them. The memory of 1904 remained too bitter in the Stockyard District.

World War I soon provided a new set of opportunities for organized labor and the ethnic communities of Back of the Yards and industrial America. The fighting cut off European immigration, curtailment that was both a hardship and an opportunity for the Polish community. On the one hand, families were separated with no hope of reuniting until after the war. Polish Americans worried about their families, friends, and villages being caught up in the fighting along the Eastern Front. On the other hand, the stopping of the constant flow of new immigrants into industrial America gave the working class a stability it had not had before. Now new arrivals would not be pouring into the packinghouse, factory, and foundry jobs, serving to keep wages low. In turn, wartime production demands meant that there was a great demand for labor. The huge crowds that usually gath-

ered outside the packinghouse gates in search of work disappeared as the European conflict dragged on. When the United States entered the fighting in 1917, the work shortage became acute, giving organized labor an unprecedented opportunity to advance its position in the wartime industries. War production and the cutting off of immigration proved crucial to the revival of organized labor.

In July 1917, a major drive to organize the stockyards opened under the leadership of William Z. Foster and John Fitzpatrick. The campaign proved successful and resulted in the creation of the Stock Yard Labor Council (SYLC) which brought together the various unions interested in organizing the Chicago meat industry. The Stock Yard Labor Council provided a model for the organization of large mass production industries within the traditional American Federation of Labor framework. Polish workers flocked to the union under the leadership of John Kikulski and Alex Nielubowski. In 1918, a federally appointed wartime arbitrator, Judge Samuel Alschuler, granted an award which bettered working conditions in the stockyards and made the Stock Yard Labor Council a major force in the industry.

Once again, however, fragmentation hurt the movement, especially after the Chicago race riot of 1919. The problem of race proved to be a very difficult one for union leaders to overcome as African Americans entered Chicago and the industry in large numbers during the war. Also the leadership of the Stock Yard Labor Council and the Amalgamated Meat Cutters were soon at odds, causing a rupture in the labor movement. John Kikulski and Stanley Rokosz, both Polish union leaders, were murdered (Pacyga 1991, pp. 181–182, 245).

The strike that rocked the stockyards in December 1921 proved to be the end of the Amalgamated Meat Cutters as an important force in Chicago's meatpacking industry. Violence marked the industrial conflict, with police fighting primarily Polish and Lithuanian workers on the streets of Back of the Yards. Mary Walkosz hid a striker being chased by the police under the bed in her apartment. Her brother-in-law, Joseph, fought with mounted police during the strike. Mary's landlady hit a mounted policeman in the head with a brick as he was arresting her saloonkeeper son during the riot (Walkosz n.d.). The communal network that had been created by the Polish community supported the strikers in their battle to push back pay cuts and gain recognition from packinghouse management. The churches, fraternals, and small businesses of the neighborhood combined to help the workers. Despite this aid, the realities of the postwar period proved to be overwhelming for the union. The postwar depression, combined with the further fragmentation of the working class brought about by the 1919 race riot, created the basis for the defeat of the union. Furthermore, the postwar political reaction against organized labor and the American Left also provided a political atmosphere hostile to the Amalgamated Meat Cutters. In 1918 the federal government had intervened on labor's behalf; this would not be the case in 1921. By the end of January 1922 the union had lost the strike.

The packinghouse strike was only one of many similar labor conflicts throughout the United States during the World War I era. Polish workers struck in New York Mills, New York, in 1916, in the national steel strike in 1919, and in other similar stoppages across the country. By 1922 most of these labor organizations, with the important exception of the New York Mills unions, had met with defeat (Brody 1965; Pacyga 1991; Pula and Dziedzic 1990).

While the struggle for union representation during the World War I era was largely lost, the institutional base of the Polish community was not destroyed. As a result, the foundation for further labor struggles remained intact. Sacred Heart Parish continued to grow during the 1920s. At the end of that decade the Polish Highlanders Alliance of North America was founded, holding its first national convention at Sacred Heart Church and at Columbia Hall in Back of the Yards. Columbia Hall had been the headquarters of the big Polish Amalgamated Meat Cutters locals during the wartime organizational drive. The institutional web that the Polish community had created remained in place, allowing Polonia to act again. The communal organizations had proven to be important for the working-class institutions that tried to emerge during World War I. They would prove to be just as important during the Great Depression (Gromada 1982).

The Great Depression provided not only stiff challenges, but also opportunities. In Chicago's Stockyard District and in the South Chicago steel mill neighborhoods, labor unions appeared to once again organize workers. Also, neighborhood organizations appeared in both communities to reach out beyond narrow ethnic boundaries and reestablish the extra-

communal response seen during World War I. In Chicago's Stockyard District, the emergence of the Back of the Yards Neighborhood Council, and in South Chicago of the Russell Square Community Committee, proved crucial in combating the urban problems of blight and crime that faced these two largely Polish American communities. In both cases, Polish Catholic parishes—Sacred Heart in Back of the Yards and St. Michael the Archangel in South Chicago—played central roles. The institutional base laid down by the immigrant generation allowed the American-born generation to attempt to gain enough power to transform the community's economic and physical prospects. These two neighborhood organizations became very effective tools for bringing about change (Pacyga 1989; Slayton 1986).

Conclusion

The fact that Polish Americans maintained their sense of community was crucial in the development of the American working class. The communal response laid down an important foundation that provided Polish Americans with the ability to take part in an extracommunal response that included organized labor and grassroots neighborhood organizations. In turn, these organizations eventually provided the economic and neighborhood stability that led to generational upward mobility. By the third generation, Polish Americans were entering universities in large numbers and joining the professional middle class.

The institutional base laid down before 1914, however, had disintegrated as Americanization and suburbanization had its impact on these communities (Pacyga 1987). Stanley Walkosz died shortly after an industrial accident in 1948. He was injured when a row of lockers fell on him and pinned him as he worked as a janitor in a Chicago packinghouse. Mary Walkosz lived with her adult unmarried children in the tenement apartment on West Forty-Seventh Street where her husband had died until she moved with them in 1959 to a wooden three-flat owned by her oldest son Stanley. Mary died in 1966 and was buried alongside her husband in Resurrection Cemetery, a Polish Catholic burial place just southwest of Chicago. By that time the major meat packers had left Chicago. Her oldest daughter Angela died in 1961 of cancer, a disease that would kill three other

Weddings are among the ceremonial moments in which the community of those involved express and affirm their culture and their ethnicity. Here is the beautiful Polish wedding of the Dorula and Walkosz families on January 17, 1915. Both Mary and Stanley had emigrated from the town of Szaflary in the Podhale region of Poland. They were married at the Sacred Heart Church in Chicago and spent the rest of their lives in that Polish Roman Catholic parish. (Photo courtesy of Rita Walkosz Hurley)

daughters and her oldest granddaughter Hyacinth. The Union Stock Yards closed in 1971, the result of changes in transportation and meatpacking technology. Finally, in June 1990 Sacred Heart Parish closed its doors after celebrating its eightieth anniversary (Walkosz n.d.). When Sacred Heart closed, it was a parish shared by both Mexican and Polish Americans. In the 1970s and 1980s Back of the Yards itself became a primarily Mexican and African American neighborhood as those two groups came to dominate much of the old Stockyard District. The Archdiocese of Chicago in a move influenced by declining attendance and revenues, decided to close several of its former ethnic parishes in the neighborhood and combine the congregations. In 1990 over 60 percent of Chicago-area Poles lived in the suburbs. Three years later the last of the Walkosz family left Back of the Yards. Two of the grandchildren held doctoral degrees and one became a medical doctor. The family spread from Chicago to the West Coast and from Louisiana to Wisconsin. Assimilation, economic and educational mobility, and marriage within other ethnic groups all had their impact on the Walkosz family and on Polonia as the twentieth century came to an end.

16

The Civil Rights Movement: A Social and Political Watershed

ALDON MORRIS
CEDRIC HERRING

Give me your tired, your poor,
Your huddled masses yearning
  to breathe free,
The wretched refuse of your teeming shore.
Send these, the homeless, tempest-tost to me,
I lift my lamp beside the golden door.
—Inscription on the Statue of Liberty,
by Emma Lazarus (1883)

In August 1994, a burglar broke into Rosa Parks' apartment in Detroit, Michigan. When the burglar discovered that Mrs. Parks was home, he struck her in the face several times and escaped with $50. The news of the attack was beamed around the world. People in different corners of the globe were angered by the attack. An alert individual encountered the thief and immediately subdued him.

The thief had struck the woman who sparked the modern Civil Rights movement. On 1 December 1955 Mrs. Parks refused to relinquish her bus seat to a white man in Montgomery, Alabama. Legal statutes, commonly known as Jim Crow laws, mandated that Blacks and Whites sit in separate sections on the public buses. Thus, when Mrs. Parks refused to give her seat to a White man, she disobeyed one of the key laws undergirding the regime of racial segregation that was entrenched throughout the southern United States. In return for her defiance Parks was jailed like a common criminal. This was the same fate meted out to others who had previously violated Jim Crow statutes. But the reaction to Parks' case would be radically different. Montgomery's entire Black community would rise up in protest against racial segregation on the buses. Most scholars of the period consider this historic protest movement as the birth of the modern Civil Rights movement.

However, before grasping the significance of the Montgomery protest movement and the larger Civil Rights movement it spawned, one must first understand the social conditions that characterized the Black population throughout the first half of the twentieth century, for they contrast sharply with those of European immigrant groups. In an earlier study of the Civil Rights movement Morris (1984, pp. 1–4) dissected these conditions. It was found that the subjugation of Black communities was accomplished through a tripartite system of racial domination, consisting of economic, political, and personal oppression.

Economic oppression stemmed from the fact that Blacks were largely concentrated in the lowest-paying and dirtiest jobs the cities and rural areas had to offer. Thus, over 75 percent of Black men in the region toiled in unskilled jobs as cooks, porters, janitors, machine operators, and common laborers. Given that only about 25 percent of White males were so employed, these occupations were commonly referred to as "Negro jobs." Black women fared no better; approximately 50 percent of them were employed as domestics in contrast to less than 1 percent of White women. Another 20 percent of Black women worked in low-paying service occupations while only 10 percent of White women worked in such jobs. As to consequences of this economic inequality, Black people earned only about 50 percent of the median family income of Whites. From an economic standpoint it was clear that Blacks were separate and unequal.

Black inequality in the southern political arena was as apparent as that in the labor market. In fact, Blacks were systematically excluded from the political process. As a general rule, there were no Black officials in state, city, or local governments because of a variety of exclusionary measures Whites utilized to achieve Black disenfranchisement. White judges and juries decided the fate of Black people when they came into contact with the judicial system. Moreover, in addition to formal political exclusion, Blacks were victims of extralegal terror. Foremost among these were lynchings, where a Black person, typically male, was hung from a tree by rope until his eyes bulged out of their sockets and the victim slipped limply into death. At this point the corpse was either set on fire with gasoline or tied to the back of an automobile and dragged through the Black community as a warning to those who entertained ideas of rebellion. Thus, in the land of the free and brave, Blacks had no political rights that any White man was bound to respect. Blacks were without citizenship rights in the land proclaimed as a cradle of democracy.

The system of racial segregation was the third pillar of this tripod system of racial domination. As Morris (1984, p. 2) has shown:

Segregation was an arrangement that set Blacks off from the rest of humanity and labeled them as an inferior race. Blacks were forced to use different toilets, drinking fountains, waiting rooms, parks, schools, and the like. These separate facilities forever reminded Blacks of their lowly status by their wretched conditions, which contrasted sharply with the well-kept facilities reserved for whites only. The "colored" and "white only" signs that dotted the buildings and public places of a typical Southern city expressed the reality of a racial system committed to the subjugation of Blacks and the denial of their human dignity and self-respect.

In short, this encompassing tripartite system of racial domination was the social machinery placed around the necks of African Americans, sealing their fate at the bottom of American society. Thus, as recently as the 1950s, millions of American citizens were oppressed on the basis of skin color (Keith and Herring 1991). The longevity of this cruel system stemmed from the fact that it was backed by law, the iron fist of southern governments, and white supremacist organizations whose members attacked Blacks by night as they hid their identities underneath white sheets. These were the intractable forces that Rosa Parks and many others would eventually confront in broad daylight and, in so doing, transform them significantly.

Throughout history, human beings have rebelled against tyranny and oppression despite the risks and personal costs. Nevertheless, given such costs, people tend to first exhaust conventional channels before openly engaging in insurrectionary activities. Southern Blacks did likewise, despite the fact that most conventional channels were unavailable to them. Still many Blacks pleaded with the White majority for the right to vote, and during the two world wars millions of them voted with their feet by migrating to the North to escape the oppression and humiliation. Prior to the modern Civil Rights movement, the legal approach was the main strategy Blacks utilized in their attempt to overthrow racial domination. Indeed, it was the National Association for the Advancement of Colored People (NAACP) founded in 1910, that championed the legal approach. The NAACP won numerous legal battles, including its famous Supreme Court victory in the 1954 *Brown v. Board of Education* case, which declared that racially segregated schools were unconstitutional because they were inherently unequal.

However, the legal method alone was not capable of overthrowing the entrenched tripartite system of racial domination. Even when decisions were won, the association had limited means to enforce them in the face of a recalcitrant White population dedicated to maintaining White control and privilege. Additionally, the legal method was costly and slow because it was forced to proceed on a case-by-case basis. Moreover, the White majority often defeated the aspirations of the association by turning the legal method against the NAACP by bogging it down through countless legal maneuvers. As a result, by midcentury the NAACP had ripped away some of the legitimacy of the regime of racial domination, but the substance and consequences of the tripartite system of domination remained firmly in place. A more formidable force was required if the huddled masses of African Americans were to breathe free and unfettered by the tripartite system of domination.

When an oppressed group is unable to liberate itself through conventional means it has another option: the protest movement. Morris (1993a, p. 21) has defined a protest movement as "the deliberate multifaceted activities of an oppressed group that are directed specifically toward altering or destroying the system of domination that produces and manages the system of social inequality imposed on the oppressed." Moreover, in contrast to the peaceful modus operandi of the legal approach, the power of a protest movement derives from its ability to produce sufficient social disruption so that it provides the leverage enabling oppressed groups to initiate social change and empowerment. Therefore, given that a distinguishing character of a protest movement is its ability to produce social disruption, protest movements usually employ a combination of unconventional and conventional means. While members of the status quo usually view a protest movement as the irrational acts of hot heads, the protest movement is fundamentally a political phenomenon, because it is the vehicle through which oppressed groups engage in social conflict geared toward attaining the leverage and social resources necessary for empowerment.

It is extremely difficult for members of an oppressed group to create massive social disruption. One of the reasons for this difficulty is the fact that dominant groups usually repress such efforts through the use of violence, economic reprisals, and the incarceration of protestors. To overcome repression, social movements must be able to mobilize large numbers of people and resources to act collectively; they must develop social movement organizations that function as the vehicles through which social protest is formally planned and executed; and they must develop a repertoire of tactics and strategies that can be utilized during social conflict to force those in power to yield to movement demands.

Clearly, then, protest movements are complicated social phenomena usually crushed while in their embryonic stages by dominant groups. Nevertheless, throughout human history great social protest movements have sprung forward to confront tyrants and repressive dominant groups. Without such historic movements, the human community would be characterized by a great deal more oppression, exploitation, and human suffering than is currently the case. Thus, the protest movement was the only viable option for southern Black people in the 1950s if they were to challenge and transform the tripartite system of racial domination that had subjugated them for centuries. The African American community rose to the challenge. We now address that protest movement—the modern Civil Rights movement—which changed America and continues to alter its course. Our purpose here is not to describe or even analyze the majority of demonstrations and campaigns that constituted the modern Civil Rights movement. Such comprehensive accounts and analyses have been done elsewhere (Blumberg 1984; Branch 1988; Carson 1981; Garrow 1986; McAdam 1982; Morris 1984; and Sitkoff 1981). Rather, the goal here is to capture how that movement altered American racial relations fundamentally by selectively analyzing some of its pivotal campaigns of collective action.

Montgomery and the Black Protest Movement

When Mrs. Parks boarded that segregated bus in 1955 after a long day of work she was well aware of the injustice of paying the same fare as Whites but having to sit in the rear, and having to give up even the rear seat to a White person if all the seats in the White section were occupied. Mrs. Parks had disobeyed the law requiring segregation on the bus on other occasions. In fact, Mrs. Parks was accustomed to rebelling against racial segregation, given that she had served as the local secretary of the NAACP for years. She did not passively accept the rules underlying the regime of racial segregation. On 1 December 1955 she looked that White man squarely in the eyes and said "No" when he demanded her seat. The White bus driver immediately intervened on behalf of the White passenger, telling Mrs. Parks to make it light on herself by giving the White man her seat. Again she said "No," and remained in the seat with dignity and self-determination. The bus driver stopped the bus so Mrs. Parks could be arrested by the police and charged with violating a Montgomery segregation statute. The word swiftly moved through Montgomery's Black community that Mrs. Parks had been arrested for refusing to give her seat to a White man. This time, Mrs. Parks was not to stand alone, for within a matter of days the entire Black community would confront the White community and their policy of bus segregation.

The organized segments of Montgomery's Black community swung into action upon learning of Parks's arrest. One of the first to act was a group of Black women who were members of the Women's Political Council. They printed up leaflets declaring that it was time that the Black community boycott the city buses in light of this arrest. These Black women had begun the mobilization process, distributing leaflets throughout the Black community. The local president of the NAACP—E. D. Nixon—bailed Mrs. Parks out of jail and agreed with the members of the Women's Political Council that it was time to act. He contacted Montgomery's Black clergy about the arrest and the need for action. The Black clergy were the recognized leaders of the Black community, largely because they were directly connected to the Black masses through the churches, the most independent and resourceful institutions of the community. Through the churches, the Black masses could be organized to engage in collective action.

The proposal to boycott the buses was considered at a meeting of Black ministers from across the city. The ministers decided to support the proposed mass action. With this decision the mobilizational and organizational capacity of the developing movement grew substantially. Through the churches and ministerial leaders, large sums of money could be raised and the pulpit could be utilized as a major channel of communication capable of informing masses of people of the strategies and goals of the movement. The churches could also serve as the sites of rallies where the spiritual and emotional energies of the people could be mobilized for movement purposes.

Most movements need an individual who is designated as the leader and spokesperson of the movement. Individuals who are charismatic are often ideal for such leadership roles. Charisma is that quality which allows an individual to command respect, support, and allegiance from masses of people because of the leader's strong magnetic personality (Weber 1947). Charisma, therefore, is a dynamic relationship between leader and followers that enables ordinary people to accomplish extraordinary achievements.

The Reverend Doctor Martin Luther King, Jr., was chosen to be the leader of the Montgomery movement. This highly educated, twenty-seven-year-old minister was charismatic and possessed enormous oratorical skills. Through oratory, King knew how to stir people to action by dramatizing injustice and challenging them to make monumental sacrifices to achieve social change for themselves and for future generations. Because King was a minister he could play a major role in the mobilizational and organizational aspects of the movement, given that he was anchored in the church community. With King at the helm, the movement had solved the leadership dilemma.

The developing movement required a formal social movement organization to plan and execute mass action. The leaders of the movement decided that its social movement organization needed to function as an umbrella unit that consolidated the strength of the variety of Montgomery's existing Black political and social organizations. They formed such an organization called the Montgomery Improvement Association. It became an organization of organizations (Morris 1984). The movement now had a social movement organization capable of directing mass protest.

Finally, the movement had to decide on a tactic capable of disrupting the public bus system so severely that the city's White power structure would be forced to desegregate them. The leaders were aware that the majority of those who rode the buses were African Americans. Thus, if the movement were successful at preventing the entire Black population from riding the buses, it could destroy the economic foundation of the bus system and force the White power structure to concede to its demand of bus desegregation. The economic boycott was chosen as the tactic of the movement to produce the social disruption needed to ensure victory. Moreover, the boycott was also ideal because the decision was made that this movement would be characterized by nonviolent direct action.

The mass bus boycott began in early December 1955. Black people throughout the community refused to ride the buses. The movement organized a sophisticated car pool that transported Black people to and from work as well as other destinations. Some Black people walked to work. When one elderly woman was offered a ride as she walked, she refused, stating "my feets is tired but my soul is rested"

Overcoming the deep-seated segregation of the past necessitated the emergence of the Civil Rights movement. Martin Luther King, Jr., a Baptist minister and one of its major leaders, believed that non-violence as a strategy for protest was crucial to attaining the goals of the movement. (*U.S. News and World Report* Collection; Library of Congress)

(Raines 1977). Mass rallies were held in the churches throughout the Black community. At these rallies, money was collected to support the many activities of the movement. At these rallies, oratory and music were utilized to strengthen the resolve of the people as they experienced economic reprisals, arrests, and violence. Month after month Black people refused to ride the buses. As a result, the bus company experienced economic hardship so severe that they began to meet with movement leaders to solve the crisis. Nevertheless, racial segregation was so entrenched in the hearts and minds of the White population, they continued in their efforts to defeat the movement.

As the boycott wore on, much of the world watched on television, admiring the quiet dignity and determination of the protestors. The Black community, especially southern ones, watched with special interest, for they too desired to overthrow their systems of racial segregation. The mass protest endured for almost a full year because of the support it received from its leaders, the Improvement Association, its churches, and the marvelous mobilization and organization of the entire Black community. A solid protest movement had developed, generating power for an oppressed community.

Victory came on 13 November 1956 when the U.S. Supreme Court ruled that Alabama's state and local laws requiring segregation on the buses were unconstitutional. Rosa Parks and the Black community of Montgomery, Alabama, had set in motion a new force that would transform race relations in America. It was the force of a protest movement. Even as Black people walked for dignity and respect in Montgomery, additional mass bus boycott movements were springing up in other southern cities, utilizing the model and inspiration provided in Montgomery. The desire for freedom was now expressing itself through organized social protest.

Student Sit-ins

Throughout human history youths have played pivotal roles in protest movements. Young Black people proved to be no exception to this axiom during the Civil Rights movement. This was especially true for Black college students. College years are usually thought of as the period during which young people engage in valuable social experiments, learn, enjoy the freedom and fun derived from a carefree life unencumbered by employment and parenting responsibilities, and grow into mature adults eager to put into practice those American ideals of justice and fair play. Yet life for African American college students during the 1950s contrasted sharply with this idyllic image. They, too, were boxed in by the color line.

The first reality confronting Black college students in the South was the stark fact that they were restricted to matriculating in historically Black colleges. Because of the laws requiring racial segregation, Black students could not attend "White schools." Their entire educational experience from elementary to graduate school had to take place in colored schools. Black college students quickly discovered that most of the world beyond their Black campus was off limits to them because of their skin color. If they were allowed in the city's movie theater, they were required to go up in the balcony to the "colored section." They could not dine in White restaurants nor try on clothes in the department store, given that Whites would never buy garments that a "nigger" had worn, even momentarily. They could not utilize the best libraries in the city for they were White. And especially humiliating was the fact that they could not read nor eat at a lunch counter in a restaurant or five-and-ten store because those counters were reserved for Whites only. Thus, Black college students, like Black people generally, faced a harsh system of racial segregation despite the ideals they were being exposed to in the readings of the enlightened poets, Thomas Jefferson, Abraham Lincoln, and other romantic dreamers.

Therefore, Black college students took notice when the modern Civil Rights movement unfolded, because they knew their dreams and aspirations were intimately tied to the outcome of this new dynamic force. They were aware of the bus boycotts during the late 1950s, and some had participated in these conflicts. They were also aware that Martin Luther King, Jr., and his ministerial colleagues across the South had organized the Southern Christian Leadership Conference (SCLC) following the Montgomery bus boycott. The SCLC was a South-wide social movement organization whose goal was to organize protest movements across the South to overthrow the regime of racial segregation. Some Black students were involved in this unfolding protest movement and had established local activist groups on campuses.

However, by 1960 many Black college students were frustrated with the pace of the Civil Rights movement. These students wanted racial segregation overthrown immediately. In February 1960, Black college students stepped directly into the Civil Rights movement, quickening its pace and stiffening its resolve to topple the system of racial segregation. In 1960 the students developed what came to be known as the sit-in movement.

As previously stated, Black students abhorred the fact that they could not sit nor eat at lunch counters. On 1 February 1960, four Black students attending North Carolina A&T College walked into the local Woolworth store and took seats at its lunch counter. The employees, as well as the customers, were shocked at the students' obvious violation of an important law of racial segregation. The waiters trembled as they refused to serve the four Black students. One is reported to have said, "Sorry, but we don't serve niggers at the counter." One of the protesters replied, "That is fine, because we don't eat them, so a hot dog would be fine." The store closed that day rather than serve the Black students. The students returned the following day taking their seats at the counter. Day after day, they returned to the counter after being refused service. They were "sitting-in" at the lunch counter.

The White power structures of the South were shocked at this new development as it was broadcast throughout the region, and they vowed to crush it. It was too late. Black college students throughout the South took notice of the historic sit-in in Greensboro. Because these students were victims of the same injustice, and because many of them were already organized through campus activities, they were poised to act. Upon hearing the news from North Carolina, students at numerous Black colleges in the South organized themselves and walked into their local stores and "sat-in" at the lunch counters. The sit-in movement was under way. Indeed, between 1 February and 1 April 1960 Black students were involved in lunch counter sit-in demonstrations in approximately seventy Southern cities (Morris 1984, p. 195).

These sit-ins were dangerous business for the students. Many of them were beaten by White thugs. They were often jailed by the authorities for violating the law. As they sat-in they were routinely spat on, doused with mustard and ketchup, and burnt by cigarettes. Some were suspended by their colleges for participating in the protest. Others underwent excruciatingly painful conflict with their parents, who feared for their lives. Despite the costs, the students continued to sit-in at the lunch counters with books underneath their arms, collectively singing "Before I'd be a slave, I'd rather be buried in my grave and go home to my Lord and be free."

The sit-in demonstrations proved effective in many instances. Thus, by the summer of 1960, numerous cities had desegregated their lunch counters. Through the sit-ins the students had created sufficient disruption to force the White store owners and political elites to desegregate public facilities. The businesses lost money during the demonstrations, for White customers stayed away because of the threat of violence. Moreover, in many instances the Black community boycotted the White business community in support of the sit-ins. Then, too, during the sit-ins the resources of the local areas were strained because so much had to be directed to the police and other agents of social control to contain the demonstrations and keep them under surveillance. These southern cities also underwent a legitimacy crisis during the sit-ins, for the rest of the country and parts of the world watched on television as Black students were beaten and jailed simply because they wanted to sit at a lunch counter. As many watched, they wondered "Is this America?" All of these factors converged to generate social disruption of such a magnitude that social change became inevitable.

The sit-ins immediately energized the Civil Rights movement. One of the most important sources of energy was the participation of large numbers of students in the movement. The students became a distinctive force within the movement. Indeed, following the sit-ins they decided, often contrary to the wishes of SCLC and other older voices of the community, to formally organize their own South-wide social movement organization to join in the battle to overthrow the Jim Crow regime. The Student Nonviolent Coordinating Committee (SNCC) became the vehicle of change for many Black students. Indeed, members of SNCC went to numerous cities and rural areas of the South and organized Black communities, leading them into the struggle for desegregation and the gaining of the franchise for southern African Americans (Carson 1981; McAdam 1982).

The Overthrow of Jim Crow

By 1963 the Civil Rights movement was in full bloom. Its participants had developed a number of effective tactics, including the economic boycott, sit-ins, freedom rides, mass arrests, and mass marches to dramatize injustice. The movement had become entrenched throughout the South as it fought to eliminate racial inequality. Yet in 1963, racial segregation in the South was still a fact of life for African Americans. This reality led Martin Luther King, Jr., and many of his colleagues to conclude that in order for the regime of racial segregation to be overthrown, federal legislation that outlawed the entire system was required. They felt that this was especially the case, given that President John F. Kennedy and his

administration and the U.S. Congress were willing to tolerate massive racial inequality in the absence of Black protest. To achieve this national legislation, King and the SCLC targeted Birmingham, Alabama, one of the most racist and violently oppressive cities in the country, as the battleground on which to win this national legislation. Another goal of the Birmingham movement was to significantly desegregate Birmingham itself, for in so doing it would become clear to Black people throughout the South that change was possible. It was understood that in order for the Birmingham movement to be successful, it would have to produce prodigious levels of social disruption.

King and the SCLC launched the Birmingham movement in April 1963. Unlike many previous movements, the Birmingham strategists decided that multiple tactics utilized simultaneously were necessary to generate the magnitude of social disorder sufficient for victory. The tactical blueprint called for the use of an economic boycott by Birmingham's Black community that would cripple the White business community; mass arrests that would fill up Birmingham's jail which could house approximately 1,000 people; mass marches throughout the city that would tie up traffic and dramatize the enormous racial injustice; "kneel-ins" at White churches; and sit-ins and picketing. Additionally, the movement would have to endure the violence that was sure to be inflicted upon the demonstrators (Morris 1993a).

In early April 1963 the Black boycott of Birmingham's business community was implemented, thus officially launching the Birmingham campaign. Sit-ins in department stores and mass marches through the city followed. Kneel-ins at churches and the arrest of protestors, including King, on Good Friday further intensified the activities of the movement. On 2 May the movement sprung a surprise on the opposition. It activated thousands of elementary, high school, and college students who previously had been mobilized and trained by movement leaders to participate in demonstrations. This was a risky gamble, for it placed young children directly in the path of danger. The leaders answered the critics of the strategy by arguing that young people had the most to gain by a victory because otherwise their entire lives would be imprisoned by the walls of racial segregation and injustice.

On 2 May, successive waves of hundreds of grammar school and high school students marched

The Lincoln Memorial is one of America's major symbols of the struggle for racial equality. Here was the site of the major civil rights march on Washington on August 28, 1963, during which Martin Luther King, Jr. gave his famous "I Have a Dream" speech that expressed the goals of the movement. (*U.S. News and World Report* Collection; Library of Congress)

through the city and were arrested. But new waves of young children followed. At the end of the day 600 children had been arrested, bringing the total number of arrestees to nearly 1,000, thus filling Birmingham's jail. On 3 May, hundreds of additional youth attempted to march downtown where they would sit-in in department stores and demonstrate in front of City Hall. Bull Connor, a legendary racist who commanded the police and fire departments, attacked the children. They sicced vicious attack dogs on the children. The firemen of the city turned high-power fire hoses on the stunned children, tossing them around like pebbles in the eye of a hurricane. The students were beaten with billy clubs.

The world watched the horror on television. Despite the violence the children and other protestors just kept marching, picketing, and singing in disciplined waves. By 7 May the jails were full, city streets were swamped with demonstrators, the floors of department stores were clogged with young people

sitting-in, and the cash registers throughout the business community were silent, for protestors and the economic boycott ruled the day, not shoppers. Social order in Birmingham unravelled at the seams, unable to ensure that even the simplest of tasks could be accomplished.

Birmingham's White power structure knew the time had come to concede to movement demands. On 10 May Birmingham's business elites agreed to begin desegregating facilities and to upgrade Black employment. But the Birmingham movement generated another important development. Numerous Black communities, utilizing the Birmingham movement as a model, rose up in protest against the regime of racial segregation. Indeed, in the ten weeks following the Birmingham protest, there were at least 758 demonstrations, with 13,786 people arrested in seventy-five cities in the South alone (Morris 1993a, p. 633). The president was now forced to act, otherwise he would have the awesome task of attempting to govern a nation riddled with social disruption. Thus, President Kennedy concluded, "Now the time has come for this nation to fulfill its promise. The events in Birmingham and elsewhere have so increased the cries for equality that no city or state or legislative body can prudently choose to ignore them" (Morris 1993a, p. 633).

On 19 June 1963 President Kennedy sent a national civil rights bill to Congress that outlawed any form of racial segregation and discrimination. Before the bill passed Congress, it was amended so that it also prohibited discrimination on the basis of gender. Thus, many of the opportunities women now have can be traced to the sacrifices Black children made in the streets and jails of Birmingham in 1963. Leaders of the Civil Rights movement decided that the chances of the bill actually becoming law would be greatly enhanced if a massive march on Washington was conducted. On 28 August 1963 over 250,000 people from all over the nation converged on the Lincoln Memorial to demonstrate their support for civil rights and the passage of national civil rights legislation. It was at this historic demonstration that Martin Luther King, Jr., delivered his famous "I Have a Dream" speech. This impressive demonstration in the nation's capital deepened national support for the bill. As a result of widespread protest and this march, President Johnson signed the 1964 Civil Rights Act into law on 2 June 1964. With this stroke of the pen the Civil Rights movement had toppled the legal foundation of a segregated racist regime that had stood for nearly a century. The protest movement was proving to be a vehicle of empowerment for an oppressed group.

Selma and the Black Vote

The 1964 Civil Rights Act was a breakthrough, but it did not address the fact that even as recently as 1964, southern Black people did not have the right to vote. Thus, even with the Civil Rights Act, southern Blacks remained politically disenfranchised, unable to elect those who would govern them. The Johnson administration was in no mood in 1964 to entertain the idea of national legislation that would place the franchise in Black hands. The president felt that it was not feasible politically to consider such legislation, given the recent passage of the Civil Rights Act. The leaders of the Civil Rights movement disagreed,

arguing that in a democracy it is unjust to set the time table for another person's freedom. Once again, civil rights leaders came to the conclusion that a massive protest movement was needed to force the federal government to enact legislation that would seize the franchise for southern African Americans.

King and the SCLC decided to launch its protest in Selma, Alabama, to achieve the franchise on 2 January 1965. Selma was an excellent target because significant work to politicize and mobilize Selma's Black community had already occurred thanks to SNCC activists. The young people in SNCC believed that local Blacks had to organize themselves and lead their own social change efforts, rather than relying on a charismatic leader like King to come into town and save them as Moses had done in biblical times. In this spirit, SNCC workers had moved into Selma in February 1963 to assist the local Black community in organizing to achieve the franchise (Fairclough 1987, pp. 210–211). Thus, by 1965 a Selma movement already existed, comprised of local leaders, social movement organizations, tactical blueprints, and the clearly established goal of achieving the franchise for the local Black community.

Prior to 1965, SNCC and the local Black community had begun voter registration drives and marches to Selma's courthouse demanding the right to vote. In return for their efforts Black people who attempted to register were fired from their jobs and identified to the larger White community for further intimidation. Moreover, the police in Selma led by Dallas County Sheriff James Clark were resolute about how to deal with civil rights demonstrations. Such demonstrations were to be crushed by beating the participants with billy clubs, by shocking them with electric cattle prods, and by filling their respiratory systems with tear gas. By 1965 the scenario had been established whereby local Blacks continued to engage in protest for the franchise and the agents of social control continued to rebuff them with violence and economic reprisals. However, the preexisting movement had created the conditions for a confrontation of epic proportions.

On 18 January 1965, King and SCLC began their confrontation with Selma's White authorities by leading 400 Blacks to the courthouse to register to vote. They were not permitted to register. Such marches to the courthouse continued in the following days. The police began arresting demonstrators and closed down the registration office for a month. Nevertheless, the movement escalated the marches to the courthouse. By 3 February, Selma had definitely gained nationwide attention, for over 2,600 people—many of whom were children—were in jail. King himself was in jail, having been arrested on 1 February for participating in demonstrations.

As the jails filled in Selma, movement leaders organized major demonstrations in nearby Alabama cities, including Marian and Camden. They were all aimed at seizing the vote. A major development occurred in Marian during a night march of 400 people to the courthouse. Local police, joined by the Sheriff's Department, attacked the marchers. The actions of these agents of social control revealed the extent to which they were committed to maintaining the status quo. One participant described the scene:

The police, the deputies, and the troopers then chased Blacks off the streets. They turned all the lights out, shot the lights out, and they beat people at random." (quoted in Fairclough 1987, p. 239)

When a group of Blacks escaped into a cafe, they were hotly pursued by several state troopers. They began clubbing a Black woman. Her young son intervened only to be shot squarely in the stomach. Eight days later, the movement had to confront the fact that young Jimmy Lee Jackson died of his wounds.

Movement leaders decided to make sure that Jackson's death was not in vain. In his honor it was agreed that rather than make the courthouse the only major target, thousands of people would march from Selma to Montgomery—a fifty mile journey. In this manner, the movement would be honoring Rosa Parks and the birthplace of the movement and confronting George Wallace at the governor's mansion. During the march, the movement would dramatize to America and the world that Black people were disenfranchised. Governor Wallace, a legendary supporter of racial segregation, vowed to prevent the march and thus the confrontation in the capital.

On Sunday, 7 March, 600 marchers assembled at the Edmund Pettis Bridge headed toward Montgomery. As they emerged on the bridge, a scene reminiscent of those associated with apartheid in South Africa transpired. State troopers, many of them on horseback, rushed into the ranks of the marchers, beating them with billy clubs, trampling

them underneath galloping hooves, and spraying them with tear gas. The marchers retreated to their church with blood streaming down their faces and their eyes full of tear gas–induced tears.

The whole world watched this display of southern justice on the nightly news. The previous demonstrations in Selma had already predisposed President Johnson and some members of Congress toward the necessity of national legislation granting Blacks the franchise. The attack on the Edmund Pettis Bridge, known as "Bloody Sunday," further pushed federal lawmakers along this path.

King responded to the attack by promising that the march to Montgomery would proceed on Tuesday, despite legal injunctions to prevent it. He then sent out a call across the nation for religious leaders to converge on Selma and join the march to Montgomery. Thousands of people, many of them White, immediately traveled to Selma. On Tuesday, they joined the local marchers and headed toward the Edmund Pettis Bridge, en route to Selma. To their surprise, King only allowed them to kneel and pray and turn back to the church. King had made a secret deal with the Johnson administration not to march and risk further bloodshed. In exchange, the President was going to send national legislation to Congress pertaining to the franchise. However, the marchers were unaware of these subterranean understandings and began referring to the aborted march as "Turnaround Tuesday." That same evening, a group of White thugs attacked a group of White ministers as they walked in Selma. A White minister, James Reeb of Boston, was struck in the head. The wound was so severe that he died two days later. This development added to the pressure to pass legislation, especially given that a White person had been killed.

On 21 March the march to Selma resumed. Thousands of people from across the nation participated, including movie stars and other dignitaries. Several days later, the marchers arrived in Montgomery. As Wallace peeked through the window blinds of the governor's mansion, King triumphantly told the marchers that "We are on the move now and wild horses cannot turn us around." But when the dark came the great success of the day was shattered once again when a blast of shots shattered through a car window, killing a woman transporting marchers back to Selma. The victim was a white housewife, Viola Liuzzo, from Detroit, Michigan. She was murdered by members of the white supremacy group the Ku Klux Klan. Moreover, agents of the Federal Bureau of Investigations (FBI) were aware that the murder probably would take place, but did nothing to prevent it. This development created additional pressure on Congress to pass the federal legislation.

The social disruption created by the Selma movement was sufficient to achieve the franchise for southern African Americans. Indeed, a few days before the march to Montgomery, President Johnson "submitted to Congress a tough voting rights bill containing several provisions that movement leaders had earlier been told were politically too unpopular to be incorporated into legislative proposals" (McAdam 1982, p. 179). The Congress passed the bill and the President signed it into law on 6 August 1965.

In a decade, the Civil Rights movement had overthrown legal racial segregation and achieved the franchise for southern African Americans. The movement had also made a statement to all Americans that the protest movement could be utilized to promote democracy and to empower the disenfranchised.

Changes in the Status of African Americans

An important question is whether the Civil Rights movement had any lasting consequences. Below, we review some of the long-term changes that have occurred in the African American community as a result of the activities of the modern Civil Rights movement. Race relations and racial attitudes in the United States have changed dramatically over the past three decades (Allen and Farley 1986; Pettigrew 1985; Schuman, Steeh, and Bobo 1985; and Smith 1981). Although many of these changes are well documented, controversies persist about whether, in fact, race has declined in its significance as a determinant of social, economic, and political statuses and outlooks (Bobo 1983; Farley 1984; Herring 1989; Herring, House, and Mero 1991; and Thomas and Hughes 1986).

Through its several important victories, the Civil Rights movement affected the multiple forms of domination in some fundamental ways. For example, in response to the demand of the Civil Rights movement that discrimination be eliminated, Congress enacted Title VII of the Civil Rights Act of 1964. This act made it illegal for employers to discriminate against any individual because of race, color, religion, sex, or national origin. Also, as we have shown, because of the increasing power of the Civil Rights movement, the same government passed the 1965 Voting Rights Act which enfranchised the southern Black population. That same year, President Lyndon Johnson expanded equal opportunity efforts by issuing Executive Order 11246. It attempted to improve the economic and social statuses of protected-class individuals (that is, women and racial minorities) by requiring government contractors and subcontractors to take "affirmative action" to ensure that applicants and employees were not discriminated against. Under the order, government contractors and subcontractors were directed to identify, hire, and promote minorities and women in numbers roughly proportional to their availability in the labor market. The underlying premise was that if employers acted in a nondiscriminatory fashion, their work force would at some point reflect the composition of the populace that surrounded their establishments. These contractors were, therefore, expected to establish goals concerning the number of minorities and women they expected to hire and timetables for achieving them.

Between 1965 and 1973, many government contractors made good faith efforts to employ minorities and women (Weatherspoon 1985). There can be no doubt that these major victories were achieved by the Civil Rights movement, for, as we have demonstrated, it was that movement which waged general and specific protest campaigns for the purpose of overthrowing legal racial segregation, toppling lawful racial discrimination, and seizing the franchise for African Americans. Clearly the legal status of African Americans was advanced by the modern Civil Rights movement.

By the late 1970s, however, several societal forces were in place to question the very logic of the Civil Rights movement and to initiate measures aimed at reversing the gains of that groundbreaking struggle. These forces were able to move rapidly and without effective opposition because no counterforce with anything near the power and passion of the Civil Rights movement existed. Thus, by the 1980s, the Reagan administration promoted a major reversal in equal employment opportunity and affirmative action policies (Hudson and Broadnax 1982). The administration took the position that it "oppose[d]

court-ordered and court-sanctioned racial preferences for non-victims of discrimination" (Reynolds
1983, p. 260). It actively argued against the use of
preferential treatment and quotas in employment
practices. Some adversaries of affirmative action
went so far as to suggest that employers who tried to
implement equal opportunity programs would find
themselves in trouble if they established programs
that utilized set-asides or racial preferences (Hatch
1980).

Moreover, anti–affirmative action and anti–equal
opportunity policies were pursued even more vigorously during the Bush administration. The president
promptly vetoed a new version of the Civil Rights Act
on the grounds that it would require employers to establish quotas. Proponents of the legislation argued
that it would have broken no new ground and imposed no new burdens on employers, but simply restored the legal framework to what it had been prior
to a series of Supreme Court decisions in the late
1980s (Wilson, Lewis, and Herring 1991). The confrontations over affirmative action and other measures designed to achieve racial equality have triggered bitter disputes about the status of African
Americans (Herring and Collins 1995).

In part, disputes about the "progress" made by African Americans continue because there are data
that suggest that the socioeconomic position of African Americans relative to Whites has improved (Allen and Farley 1986; Farley 1984; Glazer 1975; Jaynes
and Williams 1989; and Wilson 1978), as well as data
that suggest that conditions for Blacks have deteriorated in recent decades (Allen and Farley 1986; Farley 1984; Hill 1981; Jaynes and Williams 1989; and
Murray 1984). Because of a history of progress and
retrenchment, it is debatable about whether African
Americans are any better off socially, politically, and
economically now than they were during the Civil
Rights movement. There are at least three different
views on this issue: (1) an optimistic view, (2) a pessimistic view, and (3) what has become known as the
polarization thesis. After providing overviews of the
basic claims of these perspectives, we will present
evidence that allows us to assess their accuracy.

The Optimistic View

The optimistic view argues that African Americans
have made significant progress toward racial equality and that racial discrimination has been signifi-

cantly reduced or even eliminated. Wattenberg and
Scammon (1973), for example, argue that a remarkable change took place in American society after the
1960s, as Blacks increased their educational attainment, moved into skilled blue-collar and white-collar
jobs, and began earning salaries close to those of
Whites. They even suggest that during the 1970s, a
majority of African Americans became middle class.
Similarly, Moynihan (1972) argued that young Black
husband-wife families living outside the South had incomes comparable to those of Whites.

While some proponents of this perspective acknowledge that gaps still remain between the earnings of Blacks and Whites, many of them attribute
such disparities to educational differences between
Blacks and Whites. Thus, they suggest that older
cohorts of African Americans who received poorer
educations earn lower returns for their educational
investments than their White counterparts. Newer
vintages of more equally educated young African
Americans, they claim, will experience similar earnings profiles to their White counterparts. So this
view holds out the promise that as African Americans improve their educational attainment relative to
Whites, their economic performance will also be
enhanced.

The Pessimistic View

In contrast to this optimistic perspective, there are
those who hold a far more pessimistic view. They
suggest that economic conditions for African Americans have not improved significantly since the Civil
Rights movement, and on many fronts things have
actually worsened. Hill (1981), for example, points
out that recessions continue to hit Blacks disproportionately, and that unemployment rates for African
Americans are typically twice those of Whites. Moreover, the number of African Americans living in poverty continues to rise at the same time that the number of Whites living in poverty decreases. And the
effects of racial discrimination on Blacks' incomes increase and accumulate over time (Thomas and Horton 1992).

According to this view, the seeming gains made
by African Americans on the occupational front have
been illusory for three central reasons: (1) economic
restructuring and technological breakthroughs
which have meant fewer opportunities for African
Americans, (2) a changing political climate which in

recent years has viewed equal opportunity policies as being very low priority, and (3) the cumulative effects of discrimination over the life course, especially among more highly educated Blacks. Thus, according to this perspective, most economic advancements made during the relatively prosperous times of the 1960s and 1970s were only temporary gains that have been cast away by changing political winds and changing demographic profiles.

The Polarization View

An intriguing argument which attempts to reconcile the discrepant claims about the changing status of Blacks vis-à-vis Whites is the "polarization thesis." Proponents of this perspective (for example, Wilson 1978, 1987) argue that some African Americans have taken advantage of the opportunities that resulted from the modern Civil Rights movement, while others actually have become worse off during the same period. They suggest that it is the competing, polarizing trends within the African American community that account for these seemingly contradictory findings about Black-White stratification patterns over the years.

William J. Wilson (1978), for example, argues that the Black community is becoming socially, politically, and economically polarized. He points out that a number of African Americans are completing college and moving into prestigious jobs that provide economic security, higher standards of living, and homes in the suburbs. These Blacks, he argues, have been able to take advantage of the opportunities that availed themselves as a result of the Civil Rights movement. On the other hand, many other Blacks are trapped in crime-infested, inner-city ghettos where schools are poor and where opportunities for employment and advancement are extremely limited.

Wilson's major argument is that as the Black community becomes more socioeconomically differentiated, race becomes a less important determinant of the life chances of individual African Americans than does socioeconomic status (class). The idea here is that as race declines in significance, Blacks will become more similar to Whites with the same socioeconomic position than they will be to other Blacks with vastly different socioeconomic statuses.

Below, we examine some of the changes in the status of African Americans. Not only do we examine trends as they relate to the social, political, and economic status of African Americans, but also, where possible, we examine the degree to which these trends have been similar or different for various segments of the African American community.

Has Race Declined in its Significance? A Look at Some Trends

Political Changes

The Civil Rights movement was successful in its effort to seize the franchise for millions of southern Blacks and to create favorable political conditions for the exercise of the franchise by Blacks outside of the South. The number of Black elected officials skyrocketed from 103 in 1964 (before the 1965 Voting Rights Act) to over 7335 by 1990. As Bobo and Gilliam (1990) point out, Black elected officials are to be found in the U.S. Congress, state legislatures, city councils, and school boards, and as mayors of major cities. Without a doubt, the Civil Rights movement ushered in a small-scale revolution in electoral politics, making it possible for significant numbers of African Americans to hold office for the first time since Reconstruction.

Some view the rise in the number of Black elected officials as an indication of the declining significance of race in American politics. But office-holding and real empowerment can be two different realities altogether (Morris 1992). The election of Black politicians has not automatically empowered the African American community, nor has it necessarily translated into social and economic equality. For example, while the number of Black elected officials topped 7300, by 1990 African Americans made up 0 percent of U.S. senators, 2 percent of the nation's governors, and 5 percent of the U.S. House of Representatives. Similarly, at the same time that Blacks have become more active in the electoral process, race has increased in its significance and surpassed class and all other characteristics as the primary organizing frame of American electoral politics (Huckfeldt and Kohfeld 1989). As the proportion of Whites who identify themselves as Democrats declined, the proportion of Black voters who identified themselves as Democrats increased from 57 percent in 1952 to more than 80 percent in 1988 (Herring 1989; and Tate 1992). Relatedly, the difference in the propor-

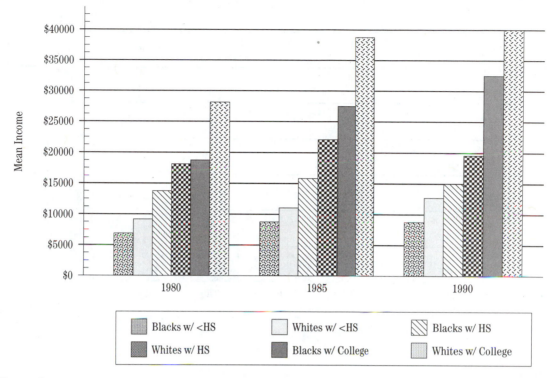

Figure 1
Mean Income of Males by Race and Education, 1980–1990

tion of African Americans and Whites voting for Democratic presidential candidates has increased from about 12 percentage points in 1948 to over 50 percentage points in the 1980s (Huckfeldt and Kohfeld 1989, p. 3). And one in five White voters report that they would refuse to vote for a qualified Black from their own political party for President of the United States (Williams 1990). Thus, many of "the racial divisions that have affected the vote ever since the 1960s are persisting features of party and electoral politics" (Gurin, Hatchett, and Jackson 1989).

Social and Economic Changes

In examining the patterns of change in the social and economic status of African Americans, there are some areas of dramatic and steady improvement. A notable example of this is in the area of educational attainment. African Americans steadily increased their levels of educational attainment, from an aver-

age of 5.8 years of education in 1940 to an average of 12.4 years in 1990. Accordingly, the educational attainment gap between Blacks and Whites had dwindled to less than half a year by 1990. But the impressive gains in education have not cashed into the kinds of rewards that one usually associates with educational credentials. For example, in 1948 the personal incomes of Black males were 54 percent of those of White males; by 1990, Black males still had personal incomes that were only 55 percent of those of White males. As Figure 1 shows, the bulk of these earnings differences persist even after we take education into consideration. Contrary to what the polarization thesis would lead us to expect, the gaps occurred for each educational attainment level. In 1990, for example, Black males with at least some college earned only 61 cents to every dollar that their White counterparts received. This translated into a disparity of more than $9,000. (Not shown are the parallel but larger differences that existed between Black women and White men.)

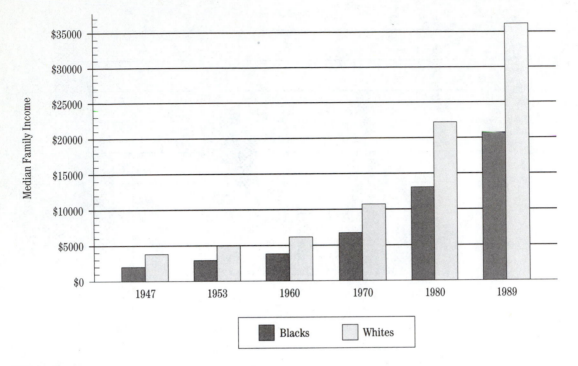

Figure 2

Changes in Median Family Income by Race, 1947–1989

Trends in family income reveal even more pessimistic news. As Figure 2 shows, in 1947 African American families earned about 51 percent as much as White families. By 1989, African American families still earned only 56 percent as much as White families. Even more telling is the fact that since the years immediately following the height of the modern Civil Rights movement, Black income as a percentage of White income fell from 61 percent in 1970. This trend clearly reflects a pattern of modest progress and subsequent retrenchment.

Changes in the occupational structure of the African American community suggest that Blacks have made progress in upgrading the kinds of jobs in which they are employed. As Figure 3 shows, however, this general improvement in occupation does not mean that the gap between Blacks and Whites has closed, as Whites have also substantially enhanced their occupational standing since 1940. Certainly, it is clear that much of this "upgrading" is due to the shift from goods production to knowledge production and the growing use of technology which displaced a large number of farm and manual laborers. In particular, the percentage of African Americans working in professional, managerial, technical, and administrative occupations increased from 4 percent in 1940 to 18 percent in 1990. There were, however, parallel changes among Whites in this upper echelon of jobs, as the percentage of Whites in professional, managerial, technical, and administrative occupations increased from 15 percent in 1940 to 31 percent in 1990. Therefore, the percentage gap between Blacks and Whites remained virtually constant over half a century. It is noteworthy that the largest relative gains for African Americans occurred in "transitional" occupations such as craft workers, operatives, and laborers.

The change from goods production to knowledge production and the increased use of technology also has had consequences for rates of Black unemployment and labor force participation. In the 1940s, "Black men were slightly more likely to be employed than white men, and Black women were significantly more likely to be employed than were white women"

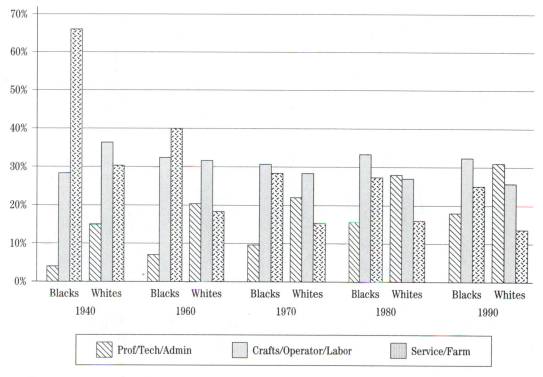

Figure 3

Percentage of Black and White Males in Selected Occupational Types, 1940–1990

(Jaynes 1990, p. 10). Since 1960, however, the Black unemployment rate typically has been at least twice that of Whites. A slight exception occurred in the early 1970s. Since that time, however, the ratio has approached 2.5 to 1. In 1992, the unemployment rate for Blacks males was 15.2 percent compared with 6.9 percent for White males.

Figure 4 shows that educational differences cannot account for the racial gap in unemployment rates. For all levels of education, unemployment rates of African Americans exceeded those of Whites with comparable levels of educational attainment. Each year, Blacks with college degrees had levels of unemployment that exceeded those of Whites with high school diplomas.

A final indicator of the social and economic status of African Americans is the poverty rate. On this measure, there was dramatic improvement since 1960, when more than 55 percent of African Americans compared with 18 percent of Whites lived below the poverty line. By 1970, these rates had fallen to 34 percent and 10 percent, respectively. Unfortunately, the pace of change did not continue after the 1970s, and during the 1980s the poverty rate actually increased for both Blacks and Whites.

Again, education cannot account for the racial differences in poverty. Figure 5 shows that African Americans at all levels of education are more likely than their White counterparts to be impoverished. For each of the periods examined, Blacks with college degrees are almost as likely to live in poverty as are whites without high school diplomas. These patterns provide little evidence for an optimistic interpretation, and they do not provide much in the way of support for the polarization thesis.

Conclusion

The modern Civil Rights movement affected American society in fundamental ways. It taught those Americans who were oppressed and excluded from

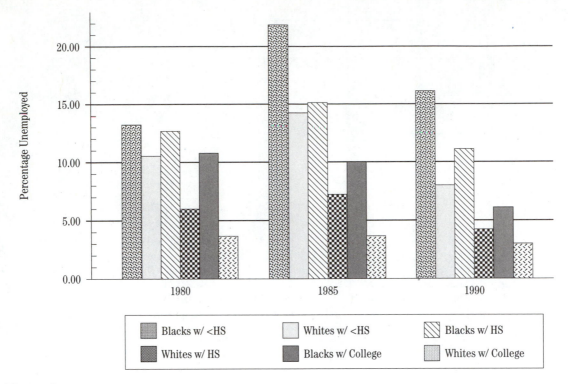

Figure 4

Percentage Unemployed by Race and Education, 1980–1990

the political process how to effectively pursue empowerment. It made the social protest option visible to the oppressed Black community and demonstrated how such protest could be successful. The movement also enriched the social protest option by providing it with new tactics and making such protest respectable. It also made a major contribution by making America a more democratic nation, for there can be no robust democracy in a society where people are oppressed and discriminated against on the basis of skin color.

What is often overlooked is that the Civil Rights movement has become a model of social protest for many other oppressed groups in America and throughout the world (Morris 1993b). In this sense, the movement opened up the political process and enabled White groups to achieve additional economic and social gains. Ironically, some of those gains made possible by the movement helped to maintain the relative differences between Blacks and Whites.

There is yet another achievement of the movement that remains hidden. It cleared the way for other contemporary non-White immigrant groups to come to America and gain an economic niche without having to fight the legally sanctioned racial discrimination that had stalled Black progress ever since the first African American was shackled in chains.

Nevertheless, it is clear that the modern Civil Rights movement did not solve America's racial problem. In the closing decade of the twentieth century, African Americans and many other non-White groups find themselves at the bottom of the social and economic order where they experience excruciating suffering. In many respects, the picture is one of gloom and doom, given the obstacles that continue to confront the African American community. Many conditions have deteriorated since the demobilization of the modern Civil Rights movement. The current economic picture of Black America is characterized by continuing subordination when compared

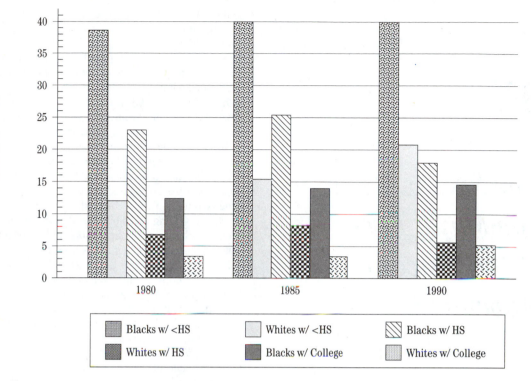

Figure 5

Percentage in Poverty by Race and Education, 1980–1990

with that of White America. The economic conditions of Blacks are heavily determined by the overall state of the American economy. Thus, when the economy is healthy, Blacks may experience absolute and even relative gains compared to Whites, as was the case in the three decades preceding the 1970s. But in the absence of political pressure or real economic growth, African Americans experience little progress and even decline.

The contemporary African American community is still significantly oppressed relative to the larger White community. In part, the current deteriorating economic and social circumstances found in the Black community can be laid at the doorsteps of that community. It seemed not to have sufficiently learned the lessons of the Civil Rights movement. One main lesson of that movement is that Black progress is intimately linked with social protest and that in its absence hard won gains can be reversed. Social protest will have to be utilized if the Black community is to seek true equality as it approaches the twenty-first century. Yet when the big picture is examined, White America bears the ultimate responsibility for racial discrimination. It profited from centuries of free slave labor and another three-quarters of a century of legal racial segregation. It benefited when the Civil Rights movement emerged because, to some degree, that movement enabled Whites to cast aside the tremendous burden of living the fiction that White skin automatically made some individuals superior. But most of all, the Civil Rights movement made America more democratic, and thus, benefited all of humanity. In that sense, the entire nation is indebted to Rosa Parks and the thousands of civil rights participants who put democracy and the principle of social equality above personal risks and narrow self-interests. The question now is whether America is going to complete the momentous task of eliminating racial discrimination and poverty generally. To accomplish this noble goal, all Americans will have to embrace its tired, it poor, its huddled masses yearning to breathe free.

17

Que Viva la Raza: *The Many Faces of the Chicano Movement, 1963–1971*

JAMES DIEGO VIGIL

In the spring of 1968, the Brown Berets and US (simply, us), militant bodies of the Chicano and Black communities, respectively, of Los Angeles, met in Elysian Park for the first efforts to join together and forge a coalition. The leaders of these two groups felt they faced similar exploitation and racism by the same oppressor and thus perceived a need for a coordinated thrust for revolutionary change. Ethnic and racial differences aside, the two organizations sought to broaden their separate movement activities into one that was based on class factors. Similar strivings toward building a broadly based class action against oppression were repeated countless times by other groups, but to no apparent avail.

By the spring of 1969, the Brown Berets were seen marching, in military boots and quasi-military safari jackets topped with berets of the appropriate color (one wag thought they resembled the Spanish Foreign Legion), to the spellbinding speech of Reis López Tijerina. Tijerina, the land rights leader from New Mexico, was at this moment rallying the crowd on the steps of the state capitol in Sacramento, California, with the famous refrain, *"Tierra. Tijerina. Tierra. Tijerina,"* repeatedly and militantly in the tone and style of a fundamentalist minister (which, indeed, Tijerina was). These words played on the deepest cultural emotions of the attentive and rapt audience composed mostly of Mexican American activists and students. They were whipped into a cultural nationalist frenzy at the precise moment when the Brown Berets marched—no, stomped—by; the sounds of boots from the pavement joined with the sounds of a possessed speaker. "Shades of Nuremberg rallies," crossed this observer's mind.

Had the Brown Berets, one of the most nationally recognized militant organizations of the Chicano movement, gone full circle? Not exactly. More to the point, the Berets represented what was an ongoing and constant dilemma within the broadly based Chicano Movement: the "contradiction" (as the popular word of the activist and radical 1960s would have it) of class interests diametrically in opposition to cultural nationalist strivings. Defined very broadly, class status implies that working and living conditions among different ethnic groups constitute a thread that connects them in their group interests. Cultural nationalism, in contrast, represents the unique, distinct, and shared historical and cultural experiences of a particular group—Mexicans in this instance. This either/or tug was the source of many an internal conflict among competing Chicano groups. It often sparked animosities between individuals in the same organization. Thus, much misspent energy led to nowhere, with activists spinning their wheels, as one "folk" social scientist once lamented.

What follows summarizes many of the key events, organizations, and leaders that characterized the Chicano movement (Vigil 1994). Woven into this discussion will be the elements that denote cultural nationalist versus class orientations and objectives. Any organization that reached out and attempted to develop ties with other ethnic groups or organizations in order to base actions on a common ground will be characterized as making a class effort. Cultural nationalist activities center around specific issues, such as bilingual education for the Spanish speaking. However, most of the organizations in the movement really showed some combination of the two ideologies, favoring one or the other in varying degrees, sometimes with a self-reorienting, "pragmatic" emphasis. The roots of this turmoil will be outlined and a reappraisal and reformulation will be suggested. A much better way to describe and assess organizational activities is through the prism of *ethnic* combined with *class*, or *ethclass* (Gordon 1964). The term ethclass captures the conjuncture of ethnic and class considerations in a social status system that places people in strata on the basis of race as well as socioeconomic criteria. An example of this would be Blacks that live in a segregated neighborhood, and only in that area are allowed social class status and mobility.

Organizations Prior to the Movement

Beginning with the *Mutualistas*, a nineteenth-century mutual self-help association that financed its members' major undertakings by pooling their resources, the motivations for most prominent Mexican American organizations centered on social and economic needs exclusive to the Mexican population. Quasi-military operations such as the "social banditry" in California and the cohesive "*Gorras Blancas*" in New Mexico (Acuña 1988; Vigil 1974) are important exceptions to this rule. Such organizations include, for example, LULAC (League of United Latin American Citizens) and the G.I. Forum, founded by World War I and World War II veterans,

respectively, in Texas (Márquez 1993). Even the CSO (the Community Service Organization), founded in the late 1940s in Los Angeles to formulate a political agenda, later became a type of modern day *mutualista*, providing social services to its members drawn largely from East Los Angeles. Later, MAPA (Mexican American Political Association), organized in 1960, followed in pushing for political change and representation (Gómez-Quiñones 1991). These organizations and others like them continue to this day and have expanded from their original goals to become more political, with some attempting to forge a national agenda.

Alongside these attempts to develop community organizations were recurrent efforts to organize Mexican American laborers. With problems rooted in job and living conditions, miners, farmworkers, and cattle ranch employees fashioned different types of protest and movement efforts. Sometimes these efforts involved more than just Mexicans; for instance, fellow miners and field laborers from White, Japanese, or Filipino background were often integrated into a more broadly based thrust for change and improvements. Early leaders of these groups foresaw that the rights of each ethnic group would be better served if coalitions with similarly affected ethnic groups were formed to make the cause stronger and capable of success. Throughout the Great Depression of the 1930s, there were many instances of this type of class-based organizing and camaraderie in various trades and occupations (Arroyo 1975). A recent biography on the times and long career of the deeply involved California activist, Bert Corona, clearly documents several decades of labor strife and other issues that pervaded Chicano communities (M. T. García 1994). Corona's role predates the official Chicano movement and shows, at least in this one life, how there has been a continual struggle for justice and improvements.

Budding a Movement: 1964–1968

No one writer has succeeded in identifying and assessing the early roots of what eventually became known as the Chicano movement (Rendon 1971). The Mexican American population was experiencing so many problems throughout the Southwest, as well as in smaller communities in the Midwest, that it is hard to pinpoint the precise impetus for social

Like all social movements, the Chicano movement had various thrusts. A major one was the farmwork-ers' struggle to organize as a trade union, the United Farm Workers. Cesar Chavez led *la causa* (the cause), in a non-violent fashion that, through the Mexican American Catholic tradition, also tied religion and politics. (Cesar Chavez Collection, Archives of Labor and Urban Affairs; Walter Reuther Library, Wayne State University)

change that characterized the 1960s. Economic and social inequity and injustice was widespread; many Mexican Americans suffered from low income, un-employment, poor housing, unequal education facili-ties and programs, segregation, and discrimination. Any or all of the above could be perceived as the mo-tivation for social action.

Indeed, it is precisely because there were multi-ple problem sources that the Chicano movement took on so many fronts and had so many thrusts. The Chicano movement in effect was simply the culmina-tion from several forces of all that came before, reaching a crescendo because the times were ripe for old scores to be evened. Ethclass considerations meshed with Mexican American cultural nationalism and class interests now included goals shared with

other ethnic populations, as if separate cultural na-tionalist movements joined to effect a class action.

It is important that we take *time*, *place*, and *people* (Vigil 1991) into account to adequately understand the Chicano movement. For example, place is central to the development of a people's ideology and politi-cal goals. Geography determines proximity to old cultural roots, for the U.S. government's acquisition of northern Mexico came in segments. Contact, con-flict, and change sequences unfolded in distinct ways in different regions. New Mexico was founded by Spanish-led settlers in 1598, an event that predates the English settlement of the United States. Similarly, south Texas has been colored by the Texas war of in-dependence, the War of 1846 to 1848, and a border history of conflict and friction, but also by close so-cial and commercial interaction. As compared to other parts of the Southwest, south Texas and New Mexico remained outside the total dominance of Anglo American society and culture for a longer period of time. California, Arizona, Colorado, and other places also have their special and distinct histories and cultural experiences.

Each of the leaders and organizations that gained attention and followers during the 1960s was preceded by incubatory developments that sometimes waxed and waned for decades. Mexicans in the United States are so varied in socioeconomic, cultural, and racial characteristics within the same location that when you include historical and regional considerations (such as the situation in the sixteenth century versus the nineteenth century, or the contrasts between Texas, New Mexico, and California), the heterogeneity and complexity increase tremendously (Vigil 1984). A particular time period or region often set the objectives and agendas that leaders and organizations had to follow.

Stealing Years: The 1960s

As one pundit said, "the Sixties is a decade that stole years from other decades" because so much happened, often simultaneously. Without question, it was an opportune time to start or accelerate all causes, for Black civil rights strivings, the just-initiated Great Society program of President Johnson, and the tumultuous Vietnam War made it so. Since the 1950s, national attention had been riveted on the challenges Blacks were making to established mores and practices. Serving as an example for other similarly if not equally subjugated ethnic and racial minorities, these engagements tended to energize people to join and participate. When the time was ripe, they did so, with convictions fed by long struggles of their own.

Additionally, the war on poverty renewed an age-old debate about the causes and consequences of poverty. With the government leading the way, billions of dollars were poured into poor communities, largely made up of ethnic minorities, to start job, education, health, training, social, and recreation programs (Jackson 1993). An unintended consequence of this attempt was to politicize poor communities to take matters into their own hands across several fronts. In essence, the state, in the name of the Great Society, fueled the already established small or large, weak or strong leaders and organizations carrying out their objectives (Evans, Rueschemeyer, and Skocpol 1985; Skocpol 1979).

The Vietnam War served as a multiple impetus. First, Chicanos began to realize that the enlistment and casualty rates were higher among them, as they generally were for all poor and working-class people. Second, the earlier hippie and countercultural ferment among mostly middle-class Whites, who eventually made up the bulk of the antiwar movement, were drawing public attention away from grassroots activities (Roszak 1969). As a result, they attracted more police to concentrate on the counterculture rather than on poor peoples. Many of these youth also actively participated in groups and organizations that reflected ethnic minority goals and concerns.

Tijerina and the Land Grant Movement in New Mexico

With the ingredients in place, the event that splashed onto the public's consciousness was the Tijerina-led *Alianza por Pueblos Libres* (Alliance for Free Communities), the land grant movement in New Mexico. Reis López Tijerina, an itinerant Protestant Spanish-speaking minister, had over several years organized a group of native *Hispanos* (descendants of the original settlers and land grant holders from the sixteenth century) to bring public attention to the land that had been stolen from them. Both individual and communal land grants from Spanish and Mexican rule had been guaranteed by the 1848 Treaty of Guadalupe Hidalgo (Griswold del Castillo 1990), the legal document that ended the war between Mexico and the United States. A parcel of land which (according to the original grantees) had belonged to a pueblo for grazing its sheep and had later fallen into the hands of the federal government and been incorporated into Kit Carson National Forest became one of numerous historical disputes and claims from the Alianza. Tijerina led a group of about 200 armed followers and reclaimed a portion of the national forest in the name of the pueblo, arresting two U.S. forest rangers in the process. This historically rooted but brazen act shook up the New Mexico establishment. Repercussions were also felt throughout the United States (Gardner 1970), even though the New Mexico situation was unique in that many Mexican Americans still held memories (and documents) of the land lost.

A subsequent courthouse raid and other altercations soon brought the issue of land rights to center stage (Nabokov 1969). Tijerina went on a barnstorming tour of the Southwest to remind Chicano leaders and groups that they were natives—not foreigners or "wetbacks." This incitement instilled in many individuals and groups a sense of righteousness in pursuit of their varied goals. It also generated and encouraged social change in other states where land might not have been an issue. Thus, the significance of this event must not be underestimated. Well advertised by the media, it combined an emotional fervor spurred by cultural nationalism, energized a people who were reminded that they had been raped and robbed, and solidified a strong class basis with regard to the loss of resources, linking them to Native Americans as "territorial minorities" (de la Garza, Kruszewski, and Arciniega 1973).

Chavez and the United Farm Workers Union

Within a few years, the Tijerina movement was effectively contained. It was a threat, but it also alienated many people because of the militant, quasi-military pose it struck. Not so the farmworkers struggle in California, however. Farmworkers of different shades and persuasions historically had fought for better working and living conditions. Left out of labor laws and agreements during the 1930s New Deal period, the average farmworker in 1964 still earned just under $9 a day. Gross farm income in the same year was $42.2 billion, making agribusiness a top moneymaker (Steiner 1970). Child labor was rampant, for all members of a family had to work if they were to survive. Migrant camps and the migratory nature of their life took heavy tolls on all family members, as inadequate housing, health care, nutrition, and education marked their lives. John Steinbeck's *Grapes of Wrath* forever seared into our hearts and minds the servitude nature of this type of existence. Long after Okies and Arkies, the subject of the book, found their way into other communities and occupations, Mexicans continued in the present, as they had in the past, to fill the places and positions of the farm laborer.

When Cesar Chavez gained national attention, he helped bring about changes and improvements for these workers. Chavez had grown up working the fields and experienced firsthand the many obstacles and demeaning situations that characterized this life. He had early on recognized that something had to be done, and when the opportunity presented itself he had dedicated himself to the migrant farmworker cause. Through a series of learning and training experiences with the Community Service Organization in the early years, and later under the tutelage of community organizing pioneers Saul Alinsky and Fred Ross, he gradually took over the helm of the farm laborers organizing committee which later merged with other fledgling farm laborer groups to form the United Farm Workers Union (UFW). Initially there was a large number of Filipinos in the union, but gradually most of the organization's leadership became Mexican (Dunne 1967); thus, it moved from a class- to an ethclass-based operation.

When the bracero program that brought millions of Mexican guest workers to work the fields over a twenty-two-year time span was ended in 1964, the conditions for organizing farmworkers were enhanced; employers could no longer simply use contracted labor as strikebreakers. Supported by most organized labor, including the AFL-CIO, the UFW initiated strikes and boycotts against farmers, singling out lettuce and grapes as the products to avoid. (Lettuce and grapes were high-profit, labor intensive crops in which California agriculture dominated the market.) The long-term *Huelga* (strike) was directed from headquarters in the central San Joaquin Valley town of Delano, California. Chavez and his aides meticulously planned and implemented the grape and lettuce boycott. Emphasizing nonviolence as a strategy, in the fashion of Martin Luther King and following the example of Mahatma Gandhi, Chavez on several occasions went on well-publicized fasts to demonstrate the nature and moral suasion of the farmworkers' cause. This approach attracted adherents and neutralized onlookers, as they became more impressed with tactics that showed deference and respect, rather than the outspoken and aggressive postures that were so common in the methods and actions of so many other contemporary organizations. Many progressive Whites, as well as many mainstream organizations, joined the cause; such individuals and groups manned the boycotters' picket lines at urban supermarkets.

Playing a strong role in this politicalization effort was *El Teatro Campesino*, a migrant worker–based theater group led by Luis Valdez (Broyles-Gonzales 1994), which spread the word through dramas and *actos* (acts or skits) that cultivated human sensibilities and reinforced morale in the face of grower-inspired violence. The boycotts were supported by many organizations and students, including leaders of the Catholic Church and such liberal politicians as the Kennedys (because of the Catholic connection and the nonviolence preached and practiced by Chavez). The Kennedys especially made it their cause and marshaled as much support as possible to ensure that the union was successful; Bobby Kennedy demonstrated this commitment symbolically when he joined Chavez in communion at a mass (Nelson 1966). The UFW strategy paid off. Public opinion and sentiment was solidly behind that segment of the working class that had been denied their rights thirty years before. As they negotiated and signed contract agreements with all the major growers, the UFW under Chavez' direction bettered the lives of farmworkers and their children.

After a few years, however, the growers countered by producing a company-like union to compete with the UFW. They also mounted a political campaign, contributing funds to local and state leaders who efficiently and effectively reversed the farmworker victories of the 1960s. Hard-won contracts lapsed and growers in the new political climate refused to negotiate new agreements. In 1993, Cesar Chavez died, but the struggle for the rights and privileges of farmworkers has not ended. It is now led by an experienced cofounder, Dolores Huerta, and by Arturo Rodríguez, Chavez' son-in-law. Organizers have renewed their effort in the fields and the UFW banner is again seen at the head of long, arduous protest marches. The union is still fueled and guided by the example and life of the late Cesar Chavez.

Highlighting Other Organizations

Tijerina and Chavez led organizations with a strong ethclass base in which ethnic and class characteristics blur and merge. Both of these thrusts gained the national limelight because they were solidly rooted in the historical experiences of the Chicano. They also served as exemplars throughout the Southwest, sparking other leaders and groups to stiffen their resolve and doggedly pursue their own agendas.

Education was among the issues that dominated discussions during the war on poverty, among Mexican Americans as well as other ethnic minority populations. In 1965 a group of educators in Los Angeles formed the Association of Mexican American Educators (AMAE), with the avowed purpose of eliminating barriers and policies that had thwarted educational performance and achievement among Mexican students. The AMAE also attempted to introduce new curriculum and learning programs based on language and cultural criteria relevant to the student population. The core of this group helped set the objectives and goals for educational change in the Southwest, although its activities were mostly conducted in southern California (Ericksen 1970). Although joined by only a small percentage of teachers in local school districts, the AMAE was strong enough to constitute a critical mass. Bilingual education, training and recruitment of minority faculty, curriculum changes to incorporate Mexican history and contributions, and an overall increase in resources in schools that traditionally were the poorest were among the major foci of the AMAE and its supporters. Other groups similarly hamstrung by inferior education also stood to benefit from their efforts. All poor and culturally different students who had been historically ignored would gain educationally and intellectually if these policy changes were effected.

In south Texas, meanwhile, a political spark was ignited to eventually mount a push for electoral representation nationwide. Crystal City, Texas, council members were mostly Anglo, four out of five in fact, but the town's population was 90 percent Mexican American. In the early years of the civil rights struggle, ethnic minorities were finally coming to realize the historical circumstances that kept them from power, especially the lack of political representation and resources that marked the Old South. Chicanos in Crystal City decided to organize and wrest this power from the small number of Anglos that had historically dominated the town; and, indeed, the entire border region with its long and deep legacy of racism, conflict, and oppression (Montejano 1987). In 1963 they succeeded in accomplishing this goal, and the victory became an example on which to base future actions, particularly consciousness-raising interactions and exercises. A core of activists from St. Mary's in San Antonio, Texas, used discussions and activities based on this experience to found a student group, MAYO (Mexican American Youth Organization). One of the leaders from the St. Mary's group was José Angel Gutiérrez, who methodically and expertly began the drive for a larger cultural nationalist political party, *La Raza Unida*. This local south Texas effort was soon to mushroom throughout key locations in the Southwest, as well as the nation.

At about the same time in Denver, Colorado, the Crusade for Justice led by Rodolfo "Corky" Gonzáles (an ex-prizefighter who had been a world-class contender) was well under way (Marin 1975). A strong cultural nationalist tendency characterized both the leader and the organization, for the time-place-people framework necessitated a unique approach to forcing change. Colorado had been a part of the New Mexico territory, and thus Mexican Americans there shared some of the same historical roots that made them feel assured in their demands for a place in the sun. Corky Gonzáles was very confident in his plans for bringing about change and he spoke with all the authority of an "American" (that is, one born in the United States) whose time had come. Economic development was the keystone to developments here, which included education and training, business entrepreneurship, political representation, and cultural awareness and involvement. This strategy was, in part, captured by Gonzáles' epic poem, "I Am Joaquín," which includes the lines:

I am Joaquín,
Lost in a world of confusion
Caught up in a whirl of a gringo society . . .
My fathers
    have lost the economic battle
and won
    the struggle of cultural survival.
And now!
    I must choose . . . (Gonzáles 1967)

Profoundly shaped by Gonzáles' cultural nationalism that emphasized community control of resources, many other leaders and groups, especially students, took up his call and path.

Student and Youth Groups

Finally, and not to be overlooked, were the student groups and organizations that surfaced. In south Texas, as noted above, there was MAYO. In Los Angeles, the first to frame an agenda of change in higher education was the Mexican American Student Association (MASA), which later changed its name to the United Mexican American Students (UMAS) at East Los Angeles College (Gómez-Quiñones 1978). Other areas of the Southwest generated their own student units with more or less the same objectives. All these groups supported what AMAE-like teachers' groups advocated, but they especially began the push for Chicano studies and higher education programs to recruit, retain, and graduate the professionals that were so desperately needed in the community.

Despite their general similarities, there were regional differences in the student groups that sprung up. Because the San Francisco Bay area was strongly influenced by early student activities at the University of California, the Chicano organizations in that region took a more radical line to ushering in change (Muñoz 1989). In fact, many joined with progressive Whites and other ethnic minority groups to form the Third World Liberation Front at San Francisco State, one of the first efforts at creating a "Rainbow Coalition." Berkeley students became particularly noted for one of the first intellectual organs, *El Grito* (a Quinto Sol Publication journal), that published literature, poems, and scholarly articles. Charting both activist and research paths for the future, this influential journal was headed by one of the few Chicano Ph.D.'s of that time, Octavio I. Romano-V., who successfully rallied students to the Chicano cause (Romano-V. 1968).

In addition to providing strong direction and helping recruit larger numbers on campus, Chicano students pushed for major changes at two levels: the community and the university. The latter was crucial, for there students were able to fill the tremendous voids in curriculum, students, and faculty. Chicano studies dates from this time period. It was relatively easy to start because part-time and nontenure track instructors were brought into the classrooms with little preparation. Those of us who experienced this situation told the running joke among ourselves that we entered the classroom with Carey McWilliams' *North From Mexico* in one hand and *El Grito* in the other. Administrators and entrenched faculty members worried little about this casual arrangement, seeing it as a temporary holding action until such a time as the movement would dissipate, which they were sure it would.

One area on which everyone seemed to agree was the need for student recruitment and learning programs, but even here the agreement was remarkably limited. Recruiting and retaining faculty members for the prized tenured, permanent faculty positions was another matter. Here the battle lines were sharply drawn, and even constant student pressure was unable to dislodge the ready rationales that academic spokespersons offered, such as that there were no "qualified candidates" available (Alvarez 1994). This struggle continues to this day, while Chicano and other such similar ethnic studies programs have managed to remain in place.

In the community, students also made their presence known. Joining community-based agencies and self-help programs, as well as protest groups like the United Farm Workers, students became skilled at balancing the demands of the classroom with those of the community from which they came. Many conferences and other organizational meetings that led to even more concerted plans and actions stemmed from the involvement of students. One group of students and community youth workers stands out in these developments. Young Chicanos for Community Action (YCCA) was a VISTA-sponsored unit, part of the government's Great Society interventions, that evolved into something quite different. They were greatly influenced by the more-militant example of the Black nationalists then rampant in American cities, particularly the Black Panthers. Under the leadership of David Sánchez, a model student from East Los Angeles, YCCA was supported in 1967 by different groups in starting a coffeehouse named the Pirana. This East Los Angeles location served as a meeting and rallying place for a sundry group of Mexican Americans from throughout the Los Angeles area who, in turn, attracted other participants to join in the political discussions and debates on the Chicano movement's future. The YCCA inspired the activists' metamorphosis into a formal militant group, the Brown Berets. The early Brown Berets (loosely defined as any Pirana attendee who donned a beret, purchased by someone who wanted everyone to look "militant") were a rather diverse group, and included a Puerto Rican and a Cuban as well as individuals who later became a film producer, a city councilman, and a university professor. The Brown Berets' militant stance attracted the attention of many disaffected street youths, especially gang members, who readily and easily redirected their fighting energies to established authorities like schools and law enforcement.

The Chicano Movement: Middle Years, 1968–1971

The Militant Phase: The Brown Berets

Each of the above thrusts was well under way by 1968, some gaining more public and media attention than others. In that year, an event occurred that gained more recognition for educational concerns in the city of Los Angeles and the state of California. A member of the Association of Mexican American Educators, Sal Castro, a high-school teacher in Los Angeles, had for years argued for major educational changes. School district officials had largely ignored and resisted his and others' calls for action with respect to the need of curriculum revisions, Latino faculty training and recruitment, and culturally sensitive learning programs for the primarily Mexican student populations in East Los Angeles schools. Meanwhile, at the Pirana coffeehouse, the transformation of certain members of the YCCA into the Brown Berets was moving in the direction of confrontation to force changes in the local public schools, joining to support Sal Castro's objectives. The Brown Berets, with the assistance of numerous student groups and leaders, helped forge an alliance for a previously unheard-of mass public display of civil disobedience (Muñoz 1989). The plans centered on students walking out of at least four overwhelmingly Mexican American high schools, an act that angered and threatened public officials and political leaders.

The event came to pass in March 1968 when 7,000 students marched through the streets of the barrios. These walkouts (or "blowouts," as the students referred to them) caught nearly everyone by surprise (Santana and Esparza 1974). Soon, students at other schools in the greater Los Angeles area, as well as elsewhere in the Southwest, staged additional blowouts. There was even a kindergarten walkout staged by the Crusade for Justice in Denver, Colorado. Such acts, considered defiant and out of character for an ethnic group that had always been considered humble, even docile, shook the establishment. Many of the still scarce Mexican Americans who had gained positions of influence and power were also shaken. Challenged and without a contingency plan, the dominant leaders' only response was to seek out and punish the culprits who had concocted this affair. Sal Castro, the Brown Berets, and others were singled out for retribution.

Within a couple of months of the walkouts, a grand jury returned an indictment that was as convoluted as it was contrived by the District Attorney's office. The decision centered on charging whoever "conspired" to plan a student walkout (itself a misdemeanor) with committing a felony (that is, conspiracy). Arrest warrants were issued for thirteen coconspirators, Sal Castro and the Berets being the main alleged plotters. The reaction from the Chicano community was immediate and unwavering. Thousands of people demonstrated around the "Glass House," the Los Angeles Police Department's main headquarters. For at least two years, the court case dragged on, and the cause of "the L.A. 13" reached even more eyes and ears. A sense of momentum had been built up and was not about to be stopped.

By 1969, another media-capturing incident continued the push for what at the time was considered radical educational change. For the third year in a row, the state department of education had hosted an annual conference focusing on the education of Mexican Americans. Titled *Nuevas Vistas* (New Horizons) and headed by a top state official, Eugene Gonzales, whose Spanish-surname helped to establish the importance of the gathering and to attract educators working within Mexican American populations, the conference hosted workshops, seminars, guest speakers, and the usual activities. That year, the keynote speaker was to be Governor Ronald Reagan, already an icon for the mounting conservative movement then under way in the state and nation. Some of the individuals named as conspirators in the previous year's school walkouts planned to attend the meetings and stage a massive walkout during Reagan's presentation. The disruption was poorly orchestrated, and most of the dinner attendees, mostly Mexican Americans, stayed in their chairs as the core protesters attempted to rally them by clapping in the Chicano movement style (a slow, gradually increasing cadence to a crescendo effect at the end) and walking to the podium where the governor stood. Instead, they wound up walking with only the companionship of an undercover police officer on each side. Fourteen protesters were arrested for the mild protest and were marshaled by undercover police officers to the paddy wagons awaiting their arrival.

Prior to the dinner walkout, some unknown persons had thrown flares into hallway closets on different floors of the hotel where the conference was being held. These premeditated acts of arson brought in the fire department and alerted officials that something might also be planned for the Reagan speech. The fires and the disruption of Reagan's speech made front-page news. Together, "the L.A. 13" and "the Biltmore 14" underscored how important the issue of education was to the Mexican Americans. As with the walkouts, a grand jury investigation to identify the conspirators in the events at the Biltmore Hotel followed. Nine were indicted, some of them already accused as part of "the L.A. 13," who now became "the Biltmore 9." More serious charges were leveled against them, for it was alleged they had planned the fires and disruptions without regard for the harm and damage they might cause.

Eventually the arrestees were vindicated, in no small part because it was learned that the police department had infiltrated the Berets. When the court cases were eventually settled (the Biltmore cases took over ten years because one of the alleged conspirators had jumped bail), the public discovered that entrapment was a strong possibility, which undermined the prosecution's case. More significant in the long run, however, was that a more militant stance had been taken and received enough support to encourage fence-straddlers to support educational change. Focusing on public education and problems similar to those experienced by other groups led to the Chicano efforts to broaden the exclusive, ethnocentric curriculum; increase the minuscule percent of minority students, faculty, and programs; and eliminate racial insensitivity, testing, and tracking. These problems also affected African Americans, for instance. So this ethclass series of challenges leaned strongly toward the class aspects that also affected other people.

The Brown Berets took a heavy toll with these two cause celebres. Police infiltration of the Berets had been a problem as early as the YCCA days at the Pirana, but turned more serious after these education-related and action-filled events generated more interest among youth, especially street gang members and other disenchanted people, in joining the Berets. Thereafter, even relatively innocuous events, like a protest march in front of a Safeway supermarket selling grapes in defiance of the UFW boycott, found the police monitoring the group. All of the key members were regularly harassed, tracked, and stopped for any pretext. Although the Brown Berets were to have other contributions and involvements, they eventually disbanded. Before they did so, their prime minister, David Sánchez, toured the Southwest and wrote a book on the experience, *Expeditions Through Aztlán* (1979).

The Denver Conference: 1969

While developments in Los Angeles took center stage, other regions in the Southwest were also capturing attention. Corky Gonzáles and the Crusade for Justice had made successful inroads in Colorado, building from the ground up to strengthen the community and its resources. Under Gonzáles' leadership, the Crusade now decided to host a conference for representatives and interests from throughout the Southwest (Muñoz 1989). Regional leaders had previously joined forces on an ad hoc basis, particularly at the 1968 Poor People's March in Washington, D.C., when Tijerina, Gonzáles, and others had united with Martin Luther King. However, the Denver Conference marked the first formal effort to bring all Mexican American political and activist agendas together.

Despite the clash of egos among the major movement leaders—Tijerina, Gonzáles (a relative newcomer), José Angel Gutiérrez (founder of the *La Raza Unida* party), and others—the conference resulted in some very positive outcomes (I. García 1989). Among them was a forum in which students and movement intellectuals could discuss organizational strategies to coordinate efforts on a broader front. Student organizations, for example, were in abundance and each bandied its own acronym and logo around to gain support for its acceptance as the common banner. Unity, however, was the goal of the day, and after long discussions, the conference attendees agreed to select a new student organizational title: *El Movimiento Estudiantil Chicano de Aztlán* (MEChA). It was also agreed that, upon returning to their home campuses, student leaders were to redefine their role in the context of MEChA's new objectives (Gómez-Quiñones 1978; 1991, p. 119).

Conference proceedings and deliberations also focused on the example of the Crusade in Colorado, emphasizing independence, self-reliance, and autonomy. Economic development and grassroots community-based work were considered the right path for the Chicano people, and Gonzáles helped frame such strategies. Following these paths with a decidedly cultural nationalist bent was a way to consolidate the movement by connecting different regional concerns and interests into a common foundation and building on a shared historical experience.

At this time an indigenous name for the Southwest was reintroduced. Historical and archaeological records, along with geopolitical logic, suggest that the original Indian inhabitants referred to the Southwest as *Aztlán*. As a coterminous geographical unit with deep cultural and linguistic legacies, modern northern Mexico and the southwestern United States are one (Wolf 1959). Tribes of early Americans traveled and migrated back and forth in the area. The last major migration was approximately 1,000 years ago when the Chichimecas (a group that included the Toltecs and Aztecs) moved south from what is now the United States. They called their homeland *Aztlán* (Vigil 1984).

The discovery of this name and history was an important event (*El Plan Espiritual de Aztlán* 1970). It meant that Mexicans were a people with roots, even though the McWasp (that is, middle-class, White, Anglo Saxon, Protestant) media instructed one otherwise. Mexicans were "illegal aliens" and "immigrants" according to reporters, or worse if you consider the household and colloquial use of "wetbacks" and "beaners." Giving legitimacy to their presence in the geographic region, Aztlán was also a way to create a nation state within a nation state, to rethink and reorganize the internal colonial system of domination that had been in place since the 1846 to 1848 conquest (Acuña 1988). This helped to galvanize support and generate participation for a combined regional thrust. It also showed the self-determination of the people, for now they used their own name for the region that includes the states of Texas, California, Arizona, New Mexico, Colorado, and parts of Nevada and Utah. As one observer remarked: In the animal world, it is kill or be killed; in the human world, it is define or be defined.

La Raza Unida: Where Culture Meets Class to Broaden the Base

Developing maturity and strength as each event unfolded, activists in south Texas began to realize that gaining political control of towns and districts populated by the bulk of the Mexican American citizens might have important national uses. Fresh from the victories at Crystal City and developments in other parts of the state, these young leaders under the leadership of Jose Angel Gutierrez began to think independently of both the Democratic and Republican parties (Shockley 1974). Their thoughts and plans were based on the fact that neither party consistently represented the interests of the Mexican people.

Beginning with efforts in Denver at the first Chicano Youth Liberation Conference in 1969, but consolidating these plans by the next gathering in 1970, the Texans advocated the establishment of an independent political unit, known as *La Raza Unida*. (*La Raza Unida* translates literally as the United Race, but its connotation refers to the historic idea of *una nueva raza*, "a new race" of peoples in the Americas who differ from—and are superior to—Europeans with their corruptions. The term *raza* also, not incidentally, means a ray of light.) Special nationalist experiences in south Texas, where restrictive racial codes and blatant segregation and discrimination practices held sway, engendered this separatist move. The idea of an independent party caught on, however, because other areas of the Southwest had similar social and political inequities. Gaining inspiration and momentum from the rallying speeches of Gutiérrez and Gonzáles, students and intellectuals of the *Movimiento* spread the word. Registration drives brought new voters into the fold and encouraged old voters to shift party allegiances. Since most Chicanos were Democrats, this third-party challenge primarily affected "the party of the working people" (I. García 1989).

For several years thereafter, even after the major energy of the Chicano movement had dissipated, *La Raza Unida* was still running campaigns, some of which played "spoiler" to mainstream candidates. Failing to convince most Chicanos that this was a viable way to challenge and revamp the two-party system and force its leaders to appreciate and heed the Chicano vote, *La Raza Unida* died for lack of "consensus." Mainly it missed connecting with the rest of the Southwest and the nation because the situation in south Texas was so special, with its largely rural, Mexican-majority communities in a former Jim Crow state. The very large proportion of the Mexican population living in urban areas simply failed to see the unifying thread of this political experiment. Moreover, third parties die hard in the United States,

as a long list of such bids lie strewn in America's historical past. Notwithstanding the fact that it meshed nationalist and class interests, it was just too ethnic Mexican to attract other similarly underrepresented working-class and ethnic and racial minority groups.

Nevertheless, it was a watershed event, as the first effort to contemplate and, for a short time at least, succeed in building a larger political consciousness among some elements in the Mexican American population. *La Raza Unida* was a natural outcome of the Chicano movement, a push for justice and equality that also entailed political issues and political organizing. Even the key leaders of separate southwestern regions for a time supported the goals of an umbrella third party. But undermining the premise of the party, *La Raza Unida* (note the relevance of time, place, and people) meant different things to different people. Tijerina in New Mexico was wrapped up in redressing the wrongs of the nineteenth century; Gonzáles in Colorado clamored for self-determination; and Gutiérrez in Texas wished to apply a local solution to a national problem. These differences in their aims and strategies were rooted in the distinctive histories of Mexican Americans in their respective areas. Each viewed the potential of a third party from a perspective shaped by local cultural interaction experiences. All the while each leader avoided publicly discussing class, cultural, and racial differences, the real contrasts and strains within *La Raza Unida*.

August 29

In East Los Angeles, on 29 August 1970, the defining event of the Chicano movement occurred. For several years Chicano observers had angrily voiced displeasure with the high casualty rates of Chicanos and other Latinos in the Vietnam conflict. Latino casualties were close to 20 percent in some branches of the armed services, far exceeding their proportion of the population. This disparity added insult to injury when considering other historically rooted injustices and liabilities. Rosalio Muñoz, a former student body president at UCLA and local activist, had received his induction notice in 1969. Refusing to abide by the legal obligation to perform military service, like tens of thousands of other American youth, Muñoz decided to resist the draft. With his protest, he sought to call attention to the fact that Chicanos were relegated to second-class treatment in higher education, but given front-line assignments in an unpopular war. Early in 1970, a loosely organized group of protesters led by Muñoz and the Brown Berets held a march in East Los Angeles and issued a short, mostly unpublicized, statement about this issue. Soon after, during and following the Chicano Youth Liberation Conference that spring, the impetus for a bigger, regional demonstration took off. Planning and discussions then transpired in various quarters to mount a national moratorium on the war.

When 29 August arrived, over 20,000 Chicanos, other Latinos, particularly Puerto Ricans, and progressive Whites massed for the march to cover several miles. The destination was to be Laguna Park in East Los Angeles, where speakers and group events were planned. After several hours of marching, the protesters reached their designated meeting place and began the rally. As it proceeded, little children played among the elderly, along with the more youthful, militant protesters. Unknown to this assembly, however, hundreds of law enforcement personnel from the Los Angeles County Sheriff's Department had been strategically placed in staging areas near the park. When an unrelated mishap more than a block away occurred, the sheriff's deputies were ordered to move in and disperse the crowd. Fighting broke out and spread rapidly, eventually encompassing the Whittier Boulevard shopping area of East Los Angeles (Morales 1972).

What began as a relatively peaceful, uneventful march and protest ended in a riot that caused millions of dollars of damage and the loss of three lives. Among those killed was Rubén Salazar, a *Los Angeles Times* columnist who had progressively become more concerned about Chicano matters. The strangeness of Salazar's death—he was shot in the head by a tear gas projectile fired from a sheriff's rifle, while he was sitting in a bar—added to the chaotic atmosphere. Millions of viewers watched the late news that day and were shown images that portrayed the marchers as rioters. Little was said about how the sheriff's actions instigated the riot, or how tranquil the gathering was before the outbreak.

Because of how this affair ended—in a negative, destructive turn of events outside their control—the organizers decided to plan another march. This they did on 16 September (Mexico's Independence Day). This, too, ended in a riot, as did another one arranged for January 1971. In short, organizational efforts and planning went for naught as the emotional energy and spirit of the movement rapidly unraveled under the purview of the coiled readiness of law enforcement.

August 29 was a day that would go down as the crucial moment when many Mexican American fence-straddlers and naysayers might have been convinced that a Chicano movement was needed, that it paid to be militant in demanding changes that were long overdue. Instead, the riot and its portrayal in the media turned off these uncommitted, undecided onlookers. A select group of community leaders, not particularly supportive of the movement anyway, spoke out against the marches as communist inspired and led. Police and the few Chicano political officials agreed with this assessment. In disarray and licking their wounds in retreat, the Brown Berets, Corky Gonzáles (who had been arrested during the riots because he had a gun and a large sum of money on him), and the other more vociferous leaders and groups lost both popularity and support (Acuña 1988).

Conclusion

The internecine struggle and competition between cultural nationalist and class-based leaders and groups was never really resolved. Each ideology had its strengths and weaknesses, but few adherents of each perspective were sufficiently flexible and resilient to allow for the search for common ground and compromise. Personalities also played a role. Vying for the leadership position atop all the separate movements and regions, each leader attempted to jockey and maneuver himself into power over all. Even when ideological views were close, personal attitudes and fragile egos interfered and kept a consensus from materializing. Cesar Chavez in California was the only leader who steered clear from such deliberations and continued to aim toward a wider national network. Perhaps the farmworker struggle was too localized and limited in its scope to garner wider and deeper interest and support from other movement forces and activities. Its ethclass-based approach combined the symbols of Mexicans, such as the *Virgen de Guadalupe* banner that led protests and marches and the plight of the migrant farmworker. This ethclass approach generated backing from many sources. Clear signals were sent to all that a life of exploitation and abject poverty was not to be tolerated. Anyone who had personally, or historically, experienced such situations and conditions was touched by the nonviolent appeal to our moral conscience. Other movement leaders failed to understand Chavez's strategy of reaching out and offering all observers a platform to step on.

Also, most Mexicans did not respond to the call for action and change because they were not sufficiently convinced that any one of the leaders or groups had the answer to what ailed Mexicans. Most of these onlookers were like everyone else in America; varying by race, class, culture, and, of course, time and place, there were as many different perspectives as there were people. As a Mexican proverb says, *Cada cabeza es un mundo* (Every mind is a world unto itself). Even those Mexicans who could be easily assembled along cultural nationalist lines felt compelled to follow their own noses to show personal autonomy and independence. In short, a movement cannot be built on a thousand heads.

Another confounding development leading to the demise of the movement was the federal government, often along with state and local governments. What the state giveth, the state taketh away. The Great Society of the 1960s was expressly aimed at low-income, poverty-stricken, minority populations. What started as a massive effort to combat and alleviate the worst effects of economic injustices and inequities, soon worked as a catalyst to awaken the political instincts, sentiments, and actions of ghetto and barrio dwellers and their supporters and sympathizers (Romo 1990). Community-based organizations intended to educate, train, and employ residents rapidly became political centers and pressure groups. These bodies joined the already strident civil rights units then under way; sometimes they were one and the same. This community buildup-turned-threat sparked a governmental reversal. The state intended to end poverty, not to subsidize the poor to end the state. Gradually the government apparatus began to cut back on distribution of resources. At the same time, its agencies sought to infiltrate and undermine any group whose expressed purpose was to overthrow the state, or minimally, to transform it into a truly democratic body. At this point the paramilitary arm of the state stepped in, in the guise of the FBI's COINTELPRO and other local and state undercover law enforcement agencies (Muñoz 1989).

Without doubt, the infiltration and tricks played on Chicano leaders and organizations by undercover agents had an effect on the movement (Acuña 1988). In addition to scaring off potential joiners and supporters, such tactics wore out key movement activists and allies. One of the more effective ways these agent provocateurs scuttled the methods and objectives of any organization was to "out militant" the militants of the unit. This often involved talking more radically about forcing change via violent means, including carrying and advocating firearms, bombs, and the like. This stratagem forced many individuals who were not that militant initially to feel they must measure up or be unworthy of calling themselves revolutionaries. Flushing out such persons helped thin the ranks of a number of groups of their most committed members, who felt they were in a "damned if you do and damned if you don't" situation. The infiltrators stretched organizational members so as to isolate and identify the main players, whom they regarded as the most threatening. Such activists, the undercover agents felt, needed to be tracked and taken out of the game. In time this worked, as one revelation after another of a "spy" and "egger-on" in their midst made many participants begin to second-guess everything that was discussed; with paranoia setting in, there was little to agree upon. This subterfuge of derailing a groups' objectives added to the normal "burn-out" process then taking its toll on activists, making them burn out faster.

José Montoya, one of the artist leaders of the Royal Chicano Air Force (RCAF) out of Sacramento, California, highlighted another development that weakened the Chicano movement: "The Chicano Movement went to the university to get a Ph.D.," as he put it. Indeed, many key participants and intellectuals that participated in movement activities, after studied involvement, operational depression, and provocateur-forced reflection and reconsideration, decided to pursue a life of research and writing, or to just drop out altogether, even though in their heyday they had attained some recognition as movement spokespeople. Their loss, especially within active student groups, took a toll on developments in higher education. Early successes in colleges and universities were not built upon because so few individuals were prepared or willing to make the necessary sacrifices to continue the struggle (Gómez-Quiñones 1991). What MEChA and other groups gained in 1971 was barely retained in subsequent years. The emergence of Chicano studies and the hiring of faculty to teach such courses was a small initial victory that did not develop into something bigger and stronger. Few student groups surfaced in the 1970s and early 1980s to reignite movement fires.

Not until the late 1980s and early 1990s were student groups resuscitated to force a bolder path of change. In large part this reclamation effort was spearheaded by a new, and large, batch of students of Mexican origin. Most of these leaders and participants came from the "1.5 generation"—that is, they are immigrants who arrived in the United States as children and were raised in a bilingual and bicultural setting and climate (Vigil n.d.). In many ways they appear to be oriented toward another thrust and renewal of a movement with long and deep roots. Among the first indications of this, for Los Angeles anyway, was the mobilization of students at UCLA in the spring of 1993 to push for a separate Chicano studies department. In part a student and community reaction to decades of neglect and undermining

of the interdisciplinary Chicano studies program and major, the protest was intensified by the timing of an announcement by UCLA Chancellor Young: He had rejected an enhanced status for Chicano studies shortly after the death of Cesar Chavez. A coalition of student, faculty, and community activists went on a hunger strike to convince the university's administrators that this change was long overdue. Public and media attention focused on the event, which ended with a partial victory for the activists. An important political and symbolic outcome of this peaceful confrontation was the naming of the department after a great leader: the Cesar Chavez Center. In effect, the fire and enthusiasm of the present generation of students, proud of their heritage and secure in their identity, was demonstrated in a public forum. Their message was delivered in a style reflective of the living example of a leader who practiced nonviolence and sacrifice combined with a decided determination to persevere over injustices and inequities.

18

American Indians and Political Protest: The "Red Power" Years

STEPHEN CORNELL

*"As I look around at the Indian situation,"
observed Robert Thomas in 1964, "it looks
like one big seething cauldron about ready to
explode." — quoted in Nabokov (1991)*

Robert Thomas was a Cherokee Indian, an anthropologist, and a political activist who knew what he was talking about, but his perception probably would have mystified most citizens of the United States. "A seething cauldron" was not the prevailing image of Native Americans in the early 1960s. For most Americans, bred on western movies or television shows and history texts that had little to say about Indians after the 1870s, Native Americans were a romantic part of the American past: Squanto teaching the Pilgrims to grow corn, Sacajawea guiding Lewis and Clark over the western mountains, Crazy Horse and Sitting Bull destroying Custer at the Little Big Horn. Tourists might still come across an Indian or two as they explored the back roads of the West, and take home silver earrings from Taos or a photograph of Navajo women tending sheep in the shadows of Monument Valley, but Indians were hardly prominent in popular consciousness, and the idea that Indian America was ready to explode would have been hard for most people to believe.

Then it blew up. From the early 1960s through most of the 1970s, a stream of unprecedented, often startling images of Native Americans poured out through the news media as the cauldron indeed exploded. Some of the events were sensational. In the Pacific Northwest in the mid-1960s, Indians defied both the law and the state police as they asserted their treaty-based fishing rights on the rivers of Washington State. Some were beaten and shot, and a running battle went on in the courts and on the stream banks for years. In 1969 a group calling themselves Indians of All Tribes seized Alcatraz Island, home of an abandoned federal prison, in San Francisco Bay, claiming it as Indian land. Their armed occupation, which lasted nineteen months, got media coverage around the world, and was followed by more than fifty forced Indian occupations of federal lands and buildings across the United States over the next five years. Then in 1973 a group of activists from the American Indian Movement, along with some Oglala Sioux traditionalists, seized the village of Wounded Knee on the Pine Ridge Sioux Reservation in South Dakota. Armed U.S. marshals, FBI agents, and Bureau of Indian Affairs police laid siege to the occupied village for a full ten weeks that riveted the attention of much of the nation.

These dramatic confrontations were only part of the story. In the 1960s and 1970s, Indians launched street demonstrations and protests across the country, went to court to pursue their claims, lobbied federal agencies and the U.S. Congress, turned out at the voting booth, and in a hundred different ways thrust their concerns before the government and the public. It was a political movement of major proportions: the mobilization of much of the Indian population in an extended and unprecedented struggle to regain control of their communities and their lives, and to reverse the powerlessness and poverty that had been theirs for a century (Cornell 1988; Day 1972; Nagel 1995).

Robert Thomas had been right.

The Native American Agenda

What lay behind this explosion of Indian political protest? It seemed sometimes as if everyone was making political noise in the 1960s. Alongside the Civil Rights movement, Black Power demonstrations, protests by farmworkers in California, an emerging Chicano movement, marches and demonstrations across the country against the war in Vietnam, the rising voice of feminism, the politically charged atmosphere of American campuses, and all the other turbulent events of the times, perhaps Indians were just jumping on the bandwagon, grabbing the opportunity to register some demands. In fact their grievances were long-standing, and their suddenly confrontational actions had a logic embedded in patterns of change in their own, tribal societies, in American society at large, and in the relationship between the two. What happened in the 1960s and 1970s was the logical outcome of changes in the Indian situation that made it possible for Indians to wage old political battles in new ways.

In the background of this new Indian political activism lay a long history of loss. First was the loss of homelands to the European invasion of North America, accompanied by an abrupt descent into stark and persistent poverty as the economic resources that had long sustained Indian societies were systematically stripped away. Second was the loss of power as the Indian nations, one by one, were brought under the tutelary administrative power of the United States, which took over the running of Indian communities and set out to transform them into a version of now-dominant American society. But the anticipated transformation never occurred. Instead, in the aftermath of these losses, by mid-twentieth century

many reservation communities had moved to the edge of disaster, with the worst unemployment rates in the country, health conditions resembling those of impoverished Third World nations, substandard housing, and a high incidence of social pathologies such as alcoholism and suicide (see the essay by Snipp in this volume; Brophy and Aberle 1966; U.S. Congress 1986).

Activist Indian politics reflected these concerns. To some degree, Indian political goals were the same as those of other ethnic and racial groups in the 1960s and 1970s: the battle for jobs and against discrimination, to improve health care and housing, to find a seat at the table of American prosperity. But significant parts of the Indian agenda were unique and reflected the distinctive history of Native Americans themselves. For many of them, and certainly for the bulk of the reservation population, the key issues were land, tribal sovereignty, and treaty rights. They were fighting for the preservation of a tribal land base (and, in some cases, the restoration of lands taken from tribes), to defend the principle of tribal sovereignty over that land base and over internal and external tribal affairs, and to gain recognition that treaties signed between the United States and the Indian nations were binding on the United States. Underlying these three concerns lay the common and distinctive themes that have driven much of Indian politics: not individual but *collective* survival and prosperity. While Indian politics has adopted a variety of causes, and various elements within it have often acted at cross-purposes, it has repeatedly held up the survival of Indian *peoples*—of nations within a nation—as its primary objective. In this it has departed from much of the politics of race and ethnicity in the United States, which has focused on the distribution of economic and political rewards within the American system and on breaking down barriers to individual achievement by members of minority groups, as through affirmative action programs. It has departed also from American political culture in general, a culture which vests rights in individuals, not groups, and which paints an idealized picture of an environment in which ethnic and racial identities function neither as the determinants of individual fortunes nor as the primary loci of individual allegiance, identity, and action.

These key objectives, clearly emerging in the 1960s and subsequent decades, were not new. For the most part, however, they had been ignored by the United States. During the early part of this century, Indians lacked the power to make their voices heard or to effectively place their concerns before the federal government or the public. About midcentury that began to change, and by the 1960s and 1970s, a newly mobilized Indian population had seized the initiative in Indian-White relations and was taking forceful action on its own behalf.

Precursors of the Activist Movement

The Indian activist movement provides a powerful illustration of the reciprocal relationship between incorporation and response in the relations between groups. The emergence of Indian activism demonstrates some important ways in which patterns of incorporation link subordinate groups to the dominant society and thereby shape the actions of those groups. It also illustrates the opposite: how the collective action of subordinate groups can transform patterns of incorporation.

The Indian political activism of the Civil Rights and "Red Power" years was built on a long history of resistance to the impositions of Euro-American society. From the earliest years of European contact, Indian nations had struggled in many ways to maintain their freedom—through expert diplomacy, elaborate trade relations, and armed warfare. Ultimately, exhausted by war, crippled by their collapsing tribal economies, or simply overwhelmed by the westward expansion of Euro-American society, Native Americans gradually came under the control of the United States. Eventually they found themselves confined to remnant lands called reservations, largely powerless to affect the major decisions, made in distant federal offices, that increasingly determined their lives.

The result was a hiatus in Indian political activity, a period lasting for most Indian nations from the beginning of their confinement to reservations well into the twentieth century. This period of political silence was a product of Indian circumstances. Their military and economic power was lost, and many Native Americans found themselves dependent on the federal government for support. The reservation system kept most Indian groups under close administrative supervision. Federal agents, who wielded practical power on the reservations, frequently bypassed Indian leadership. Even if there had been an opportunity to mobilize politically, few Indians had

the necessary expertise—gained through either education or experience—to operate effectively within the political system of the United States. In effect, the political arena was closed to them.

This hiatus was not total, however. Resistance survived in the daily patterns of living on numerous reservations, where many tribal members struggled to maintain both traditional religious practices and tribal social and political relations in the face of efforts by federal bureaucrats and religious missionaries to replace them with the individualism and Christianity of the dominant society. In addition, Native Americans asserted themselves in other ways. In 1911 a group of Indians—mostly professionals with roots as much in White as Indian society—founded the Society of American Indians, an organization devoted, according to its constitution, to "the honor of the race and the good of the country" (quoted in Hertzberg 1971, p. 80). The first national Indian political organization to survive for very long, its goals were fundamentally assimilationist. While wishing to support the welfare of Indian peoples, the Society of American Indians also sought an end to tribalism and a gradual integration of Indians, as individuals, into the American mainstream (Hertzberg 1971).

More indicative of what future decades would bring was the battle over land rights waged by the Pueblos of New Mexico in the 1920s. The federal government proposed legislation settling conflicting Indian and White land claims in New Mexico, favoring non-Indian interests and threatening Pueblo lands and water resources. Pueblo leaders claimed that the Bursum Bill, as the legislation was known, would "complete our destruction" (quoted in Sando 1976, p. 96). The All-Pueblo Council, an association of the Pueblo villages, along with a coalition of non-Indian organizations fought the legislation and won. Congress instead passed the Pueblo Lands Act, which placed the burden of proof of ownership on the non-Indian claimants. Unlike the work of the Society of American Indians, the focus of Pueblo action was squarely on land rights and the preservation of tribal communities. As such, it was both in the tradition of Indian resistance and a forerunner of the political future. Its substantial dependence on non-Indian support, however, also was indicative of the disadvantages Indians faced: few political resources and little access to the centers of power.

Events such as these were rare in these years. A large-scale social movement in favor of indigenous rights and tribal preservation would have to await the opening of the political arena, the acquisition of the resources necessary to sustain political battle, and the development of a broader group consciousness on the part of Native Americans that would allow this enormously diverse population—made up of numerous tribes with distinct histories and cultures—to join forces on behalf of a common political agenda.

Putting the Pieces in Place

A sequence of three major events—two of them related to Indian policy, one of them a worldwide upheaval—began to create sufficient conditions for effective political action. The policy events were the Indian Reorganization Act of 1934 and the so-called "termination" policy of the 1950s; the upheaval was World War II.

The Indian Reorganization Act (IRA), passed by Congress in 1934, was stimulated in part by a growing concern with the poverty of Indian reservations and in part by the New Deal's enthusiasm for communal structures. It signaled a shift in federal policy away from the breakup of reservation communities and lands and toward support for tribalism. It encouraged and endorsed the creation of constitutional tribal governments. In so doing, it formally reestablished federal recognition of Indian-selected, tribal leadership, and gave those leaders a federally recognized organizational structure in which to operate. In some ways, the change was modest: the Indian leaders who took over these new tribal governments had little power, for the governments themselves remained under the close supervision of the Bureau of Indian Affairs (BIA) in the Department of the Interior. But in endorsing tribal governments, the United States recognized tribes once again as political actors, and the tribal governments themselves eventually came to provide an organizational basis with which to further assert tribal control. In addition, the IRA represented a step away from the destructive individualism and assimilationism that had dominated federal policy for decades. While close federal supervision of Indian affairs continued long after the IRA had been passed, the legislation and the more supportive federal atmosphere attached to it formed a critical turning point. Thereafter, tribal objectives and tribal actions began to play a larger role in

Indian-White relations than they had in years (see Cornell 1988; Deloria and Lytle 1984; Taylor 1980).

The second major event was World War II. This had several important effects. First, large numbers of Native Americans joined the armed forces during the war. When the war ended and many of them returned to the reservations, they brought with them a new impatience with the political powerlessness of Indian communities as well as the experience necessary to take a leading role in creating change. In the late 1940s and 1950s these veterans moved into leadership positions in tribal government and in national Indian organizations, so much so that *The New York Times* reported, in 1952, that "a new, veteran-led sense of power is everywhere in Indian country" (Stevens 1952).

Second, job opportunities in wartime industries, combined with stagnant or deteriorating economic conditions on the reservations, prompted still more Indians to head for America's cities, beginning a massive migration that continued, partly with federal support through the government's relocation program, for the next few decades. The urban Indian communities that resulted eventually became the breeding ground of the radical wing of the activist movement.

Third, World War II and the urbanization it stimulated led to the growth of a supratribal political consciousness and identity within the Native American population. While such a consciousness existed prior to the 1940s and 1950s, it tended to be incidental and transitory. For example, as early as the 1600s, some eastern tribes attempted to band together against the Europeans so as to greatly increase their power. "Brothers," said a Narragansett Indian, addressing his own Indian enemies in 1642, "we must be as one as the English are, or we shall all be destroyed" (quoted in Armstrong 1971, p. 3). Other leaders likewise tried to build alliances among Native American tribes (see Dowd 1992; Spicer 1969, chaps. 1 and 2). While there were some successes, few such alliances lasted. They came to grief over the efforts of the colonial powers, and later the United States, to reinforce the divisions among Indian nations, and on the very real differences between those nations, which shared neither language nor culture nor, in some cases, clearly defined interests.

Until nearly the middle of the twentieth century, it was primarily tribal relations and the reservation environment that framed daily life for most Indians.

Both the administrative structure of Indian affairs—organized by reservation unit—and the indigenous organization of daily life—involving distinct tribal languages, cultural activities, and kinship relations—lent support to tribal identities and allegiance. A nascent supratribalism could be seen in pan-Indian activities such as powwows and in religious movements such as the Native American Church, but a comprehensive political consciousness rooted in daily experience as yet involved only a minority of Indians.

With World War II and urbanization, that began to change. In the cities Indians encountered bureaucracies that paid little attention to tribal differences, treating all Indians the same or as part of the larger community of the urban poor. Racial discrimination, likewise, tended to ignore tribal boundaries. In the armed forces or in the multitribal Indian communities that began to appear in many cities as the Indian migration continued, Native Americans from diverse backgrounds began to discover how much they shared, including their common differences from the non-Indian mainstream. Large numbers of them, newly experienced in the larger world of Indian-White relations, returned to the reservations, bringing with them that experience and the supratribal Indian consciousness it encouraged.

One practical result of these developments was the founding, in 1944, in Denver, of the National Congress of American Indians (NCAI). The seventy-five delegates from fifty or so tribes who attended the founding convention included a number of World War II veterans and others with extensive experience in the non-Indian world. Their organization was something new in Indian affairs. Open to any "Indian tribe, band, or community," the National Congress of American Indians (1944, p. 54) saw its mission as the preservation of tribal rights and tribal communities. It was explicitly supratribal: while it fought to preserve distinct Indian nations, it brought those nations together in a common political effort. It also became a source of political information relevant to member tribes, and an active lobby for Indian interests before Congress and the Bureau of Indian Affairs.

Perhaps the most important factor in the emergence of a supratribal political consciousness was the third event in this sequence that began to bring the pieces of the activist movement together: the "termination" policy. In the aftermath of World War II, eager to reduce federal spending and wary of anything that smacked of socialism (as some conservative politicians believed tribalism did), and wishing to solve "the Indian problem" once and for all, Congress reversed itself again. It abandoned the support for Indian communities that had characterized policy under the Indian Reorganization Act and, in a series of legislative acts, set out to dismantle the reservation system and integrate Indians, as individuals, into the mainstream of American life. It ended federal services to a small number of tribes, disestablished their reservations, and distributed tribal assets among individual tribal members—hence the name by which the policy came to be known: "termination." In addition, it allowed a number of states to assume jurisdiction over Indian reservations (Burt 1982; Wilkinson and Biggs 1977).

The economic results were disastrous. Tribes such as the Menominees, in Wisconsin, and the Klamaths, in Oregon, found themselves suddenly subject to extensive taxation, saddled with failing economies, rapidly losing their lands, and deprived of federal services that had been crucial to their well-being. States that initially favored the new policy turned against it as they discovered the full extent of the economic burdens they would have to assume when, in the absence of the federal government, the full weight of reservation poverty landed on their shoulders. The political results were even more far-reaching. Tribes across the country recognized in termination the greatest threat to their survival since the U.S. military campaigns of the nineteenth century. They also recognized that it was a threat all of them faced, and they responded accordingly. In 1954 the National Congress of American Indians convened an emergency conference in Washington, D.C., to voice virtually unanimous Indian opposition to the new policy. It was the first of a number of Indian efforts to change the course of federal action (Deloria and Lytle 1984, chap. 13). Indeed, Nancy Lurie (1971, p. 456) has argued that it was the policy of termination that finally made many Indians aware of "the diametrical opposition between Indian and white objectives," between the assimilationism of federal Indian policy and the preservation of Indian nations, and spontaneously produced "a common Indian opposition."

That opposition developed further at the American Indian Chicago Conference, convened in the summer of 1961 and attended by nearly 500 Indians from seventy tribes. D'Arcy McNickle, an Indian anthropologist who was there, later wrote (1973, p. 117) that the conferees had "a sense of being under attack, and it was this shared experience which drew them together." In their concluding document, drawn up for presentation to Congress and titled "A Declaration of Indian Purpose," they explicitly rejected termination and asserted the right of Indian peoples to choose their own futures and pursue them in their own ways (Lurie 1961; Witt 1968). The National Congress of American Indians played a significant role in the Chicago Conference, the largest multitribal gathering in decades and evidence of an emerging and increasingly political supratribalism. The 1961 Chicago Conference represented the coming together of a diverse but fairly well-established and experienced Indian leadership, products of the tribal governments that had emerged in the 1930s, of service in the armed forces and involvement in the larger society, and of the ongoing battle to stop or slow down termination.

A different set of political actors also was starting to emerge. Beginning in the mid-1950s, growing numbers of Indian students on campuses such as the University of New Mexico had begun to organize workshops, Indian clubs, and "youth councils" to discuss Indian education and other Indian issues. They sponsored their own youth conferences, including one in 1960 that brought together 350 Indians from fifty-seven tribes. Some of the leading figures in this emergent group showed up a year later at the Chicago Conference, but found themselves largely ignored by the more established Indian leadership. Partly in reaction, in August 1961, ten of them met in Albuquerque and formed the National Indian Youth Council (NIYC), with the conscious intention of generating a movement of political activism among younger Indians (Steiner 1968). While explicitly supratribal in its sense of common Indian interests, the council drew its membership largely from reservations, and placed the support of tribal identities and communities at the very center of its political agenda (Bonney 1977). But it was its political style—creative, confrontational, and impatient—that clearly broke from the past. "We were concerned with direct action," said Mel Thom, a Paiute Indian and one of the council's founders. "The younger Indians got together in the Youth Council because they didn't feel that the older leadership was aggressive enough. And we felt that Indian affairs were so bad that it was

time to raise some hell" (quoted in Steiner 1968, p. 40).

Raise some hell they did. The council was the first of a group of increasingly radical Indian organizations that dramatically altered the shape of Indian politics over the next two decades, rapidly seizing the initiative in Indian-White relations, and forcing the federal government increasingly to respond to their actions and concerns.

The Movement: The 1960s and Beyond

Open conflict first erupted in the Pacific Northwest, where a number of Indian tribes had been struggling for decades to assert their treaty-guaranteed rights to fish in northwestern waters. Over the years, individual Indians from a number of tribes had exercised those rights in defiance of state fishing regulations in Washington and Oregon. In 1954, however, the State of Washington arrested two members of the Puyallup Tribe for fishing repeatedly—and, according to the state, illegally—in the Puyallup River. An even split in the state supreme court led to the dismissal of the charges, which outraged non-Indian fishermen and at the same time encouraged other Indians in the area to exercise their treaty rights as well. With the fundamental rights issue unresolved, the state continued to arrest Indian fishermen. As the new cases worked their way through the courts, a major Indian protest began to build (see American Friends Service Committee 1970; Cohen 1986). Indians saw the fundamental issue as the gross violation of treaty rights. In addition, they complained that their nets, boats, and other equipment necessary to their livelihood were being confiscated by the state; others claimed to have been shot at while fishing and threatened by game wardens and non-Indian fishermen.

In 1963, an Indian organization called Survival of American Indians Association led a major demonstration at the state capitol building in Olympia demanding that the state observe Indian treaty rights. Indians from other tribes began to show up in support of the growing protest, as did members of the National Indian Youth Council, which staged its own protest in Olympia in 1964. With support from these organizations and some tribal governments, Indians began staging protest "fish-ins" on the Nisqually

River and other rivers as a direct challenge to the state governments. Some of these confrontations eventually turned violent. In 1965 Indians and Washington state officers clashed along the Nisqually River in "an emotion-charged battle of paddles, sticks, and stones" (American Friends Service Committee 1970, p. 110). Hank Adams, one of the protest leaders, was shot in 1971 while watching over a net for a friend on the Puyallup River (Cohen 1986, p. 81).

The fishing rights battle never seemed to end. Despite major court decisions in the 1970s in the Indians' favor, efforts by other fishing interests to overturn Indian fishing rights continued into the 1990s, and today the struggle goes on in the courts and the forum of public debate. Whatever its eventual resolution, the conflict marked a turning point in Indian protest, in both form and effect. Tactically, it represented a major shift toward more confrontational forms of action. Confrontation was by no means new in Indian affairs. In the late 1950s, for example, Tuscaroras in western New York and Lumbees in North Carolina, among others, had turned on occasion to civil disobedience in defense of treaty rights and their own communities. But the fish-ins and attendant demonstrations in the Pacific Northwest provided the largest, sustained example so far of a supratribal effort to challenge the prevailing power structure outside conventional channels of political action. It was controversial, not least among Indians themselves, including some of the tribes most directly involved (see American Friends Service Committee 1970, pp. 116–118), but it opened a new chapter in Indian politics. By initiating actions that directly challenged the states, Indians placed both federal and state governments on the defensive, forcing them to scramble in an effort to figure out how to deal with unexpected forms of Indian protest and with newly invigorated Indian concerns. For years Indians had been responding to federal actions. Now the power structure in Indian affairs had to respond to them. The tables had been turned.

These new, confrontational tactics were appearing elsewhere as well. In the early 1960s, for example, traditionalist Cherokees in Oklahoma openly defied state regulation of their hunting rights; Indians in California and Maine physically confronted loggers and road builders on their lands, forcing a halt to logging operations; Mohawk Indians in New York blocked a bridge between Canada and the United States, protesting violations of their treaty-guaranteed rights of free passage across the international boundary. In the hands of urban Indians, political activism took an even more aggressive turn (Nagel 1989). By the mid-1960s, Indians in a number of cities had begun street demonstrations over housing, health, and employment issues, and against federal neglect of urban Indian concerns. Then in October 1969, the San Francisco Indian Center, which had provided the San Francisco Indian population with a meeting place and an administrative center for programs, burned to the ground. In November a group of Indians calling themselves Indians of All Tribes, searching both for a place that could serve as a center for the community and for a way of making their concerns known, landed on Alcatraz Island in San Francisco Bay, site of a recently decommissioned federal prison, and demanded title to the island. The proclamation they read stated: "We, the native Americans, reclaim the land known as Alcatraz Island in the name of all American Indians. . . ." They pointed out that the island would make an ideal Indian reservation "by the white man's own standards," having a number of things in common with other reservations: "It is isolated from modern facilities, and without adequate means of transportation. It has no fresh running water. The sanitation facilities are inadequate. . . . There is no industry and so unemployment is very great. There are no health care facilities. The soil is rocky and non-productive and the land does not support game. . . . The population has always been held as prisoners and kept dependent upon others" (Indians of All Tribes 1972, pp. 40–41). The occupiers held the island for nineteen months before U.S. marshals finally removed them in June 1971 (see the detailed account in Fortunate Eagle 1992). While the original plans for the island—to develop an Indian educational and training center—never came to fruition, the incident served as a catalyst. It focused widespread media attention on Indian issues, and it provided Indians themselves with their own model of confrontational politics. Over the next five years there were more than fifty Indian seizures of land or buildings around the country (Cornell 1988, p. 190).

At about the same time, urban Indians in the Midwest founded what became the most radical of the new Indian organizations. In 1968, members of the large and growing Indian population of Minneapolis–St. Paul, concerned about discrimination against Indians in the city, formed a new organization that eventually came to be called the American Indian Movement (AIM). AIM began with local goals: to assist the city's Indians with problems such as housing, employment, and relations with city bureaucracies, particularly the police. One of its early activities was an Indian Patrol. AIM members tracked police movements in the Indian areas of the city at night, watching for harassment, monitoring arrests, mediating disputes between Indians and the police, and picking up inebriated members of the community and taking them home (Cohen 1973; Dewing 1985). Soon the organization widened its horizons. AIM members participated in the Alcatraz occupation, and organized demonstrations and occupations in other cities as well. By 1972 AIM claimed to have established sixty-seven chapters in the United States and two in Canada (on AIM generally, see Bonney 1977; Dewing 1985, p. 68).

As it grew, AIM began to turn its attention to reservation concerns. The urban Indians of AIM, wrote Gerald Vizenor—Anishinabe poet, journalist, and novelist—in 1973, "dreamed of returning to the ideal reservation as the new tribal warriors" (1976, p. 70). For most Indians, the reservations had always been the ultimate concern. Even for those with little, if any, reservation experience—which included many AIM activists—the historical and cultural dimensions of Indian identities were most profoundly associated with reservation communities. Furthermore, it was virtually impossible to make effective political use, especially in the courts, of the most distinctive aspect of the Indian situation—the treaty relationship—without a reservation base. Reservations were where tribes, as federally recognized actors, were located; they also were the homelands of the most traditional groups, groups who were most concerned with treaty issues.

In the early 1970s these two different constituencies—urban activists and reservation "traditionalists"—began to find common ground on the issue of treaty rights. Their convergence was tenuous, thanks to the substantial gap between reservation and urban worlds. Many reservation Indians were uncomfortable with the confrontational style of urban activists, their leftist politics, and their cultural posturing, and feared that urban groups would drain away resources now dedicated to reservation needs (Nagel, Ward, and Knapp 1988). On the other hand,

the activists offered not only a high level of energy, but a program of action on issues of direct relevance to reservation populations.

The two groups came together in the "Trail of Broken Treaties" caravan that traveled to Washington, D.C., in the fall of 1972. Planning for the caravan began on the Rosebud Sioux Reservation in August, and eventually involved a wide array of Indian organizations, including the American Indian Movement and the National Indian Youth Council (*Akwesasne Notes* 1974a, p. 2). In October, caravan participants left the West Coast, stopping at numerous reservations along the way, where they were joined by growing numbers of Indians. En route, they also put together a statement, known as the "Twenty Points," that set out a new framework for Indian-White relations. It proposed a reopening of the treaty relationship, allowing individual tribes to renegotiate their relationship with the United States, a thorough review of treaty violations, and the governance of all Indians according to treaty relations (*Akwesasne Notes* 1974a, pp. 63–88). The caravan's organizers had hoped that these points could be the basis of frank discussions with the federal government, but a crisis developed soon after the caravan's arrival in Washington in November, a week before national elections. Following a confusing altercation with security guards in the Bureau of Indian Affairs headquarters, caravan participants seized the building, where they remained for several tense days before a negotiated settlement persuaded them to leave. In the aftermath, the building itself was a shambles, numerous Bureau of Indian Affairs records were missing, and the "Twenty Points," unfortunately, had been largely forgotten (V. Deloria 1974; a detailed account is provided in *Akwesasne Notes* 1974a).

The most dramatic of the activist events was the 1973 takeover of the hamlet of Wounded Knee on the Pine Ridge Sioux Reservation in South Dakota. Wounded Knee is an infamous name in the history of Indian-White relations. In December 1890, during the final effort to force the various bands of the Sioux peoples onto reservations, a group of Sioux under a leader named Big Foot, being shepherded by the U.S. Army toward the village of Pine Ridge, camped one night on Wounded Knee Creek. In the morning, the army troops, who had closely surrounded the Indian teepees, ordered the Sioux to turn in their arms. The Indians refused, troops moved in to search the teepees, someone fired a shot, and without warning the heavily armed soldiers opened fire on the Sioux camp. Some 300 Sioux men, women, and children died (Utley 1963).

Eighty-two years later, on the night of 27 February 1973, Oglala Sioux traditionalists and American Indian Movement activists seized control of the hamlet of Wounded Knee, built on the massacre site, in protest against federal policies, against violations of the Fort Laramie Treaty of 1868, which recognized Sioux sovereignty over a vast expanse of the northern plains, and against what they viewed as a corrupt and federally controlled tribal government. Within twenty-four hours heavily armed federal agents had surrounded the village. A standoff that lasted for ten weeks, including nightlong firefights, on-and-off negotiations, and two deaths, captured media attention around the world (for day-by-day accounts from opposite sides of the conflict, see *Akwesasne Notes* 1974b and Lyman 1991).

The 1973 Wounded Knee incident was the highwater mark of militant Indian activism. There was more to come: additional occupations and protests took place over the next few years as more and more Indians mobilized in pursuit of their goals. It was not until after the Longest Walk, another march on Washington in 1978, that the more militant phase of the activist movement really came to a close. But it was Wounded Knee, on a huge reservation in rural South Dakota, that most successfully captured the attention of the world, and served notice that Indian-White relations would never be the same again.

Shaping a Movement: Identities, Opportunities, Resources

The Indian activist movement is an example of a social movement. Broadly, a social movement is the conscious, concerted effort of a group of persons, a collectivity, to change some aspect of a situation in which they find themselves. Social movements include everything from heading for the polling booth on behalf of a candidate to heading for the streets or the hills with guns, and range from boycotting a company's products to organizing a neighborhood trash cleanup.

Social movements sound simple: those with grievances get together and they act. But it is seldom easy to act effectively. Many persons and kinds of action are constrained in various ways. Getting together can be very difficult for those who are physically dispersed or whose freedom of movement or communication is restricted. Targets of action are sometimes inaccessible. Distant sources of power may be difficult to impress, and while workers can refuse to carry out their work, much depends on the nature of that work and the role it plays in the society. In a large industrial economy, a national strike by coal miners or autoworkers is likely to draw more attention than a national strike by orderlies in hospitals serving the ghetto poor. As Frances Piven and Richard Cloward (1979, p. 23) emphasized, "people cannot defy institutions to which they have no access, and to which they make no contribution." Collective action of one kind or another may also have unacceptable consequences. Few slave insurrections took place in the southern United States in the eighteenth and nineteenth centuries, in part because enslaved African Americans faced an overwhelming—and lethal—system of controls over their behavior. The insurrections that did occur often turned out to be suicidal (Genovese 1979). Some courses of action are constrained by the distribution of rights. Those who cannot vote, such as women in many parts of the United States prior to passage of the Nineteenth Amendment, or who have limited legal standing in the courts, such as slaves, lose the political options the polls and litigation represent. Thus economic and political relationships shape the opportunities groups have to pursue their grievances in effective ways.

Action also requires various kinds of resources: organizational networks to recruit participants; large numbers of voters for the polls; resources with which to bargain in negotiations; legal skills—or the money to buy them—if a case is going to court; and leadership to articulate grievances, coordinate events, and model behavior. Some kinds of action require few resources or very specific ones, others demand a lot. The most important point is that resources are differentially distributed; not everyone has the necessary assets to invest in a social movement. The distribution of resources is itself a constraint that makes some kinds of action more or less likely than others.

Finally, patterns of collective action depend on patterns of identity. When people act in concert they do so in the terms that make sense to them: as workers, or as women, or as farmers, or as African Americans, or as students. Consciously or unconsciously, they have to come to a conclusion about the appropriate basis on which to act.

These factors played a major role in the evolution of Indian political activism. Early in the twentieth century, American Indians had few political opportunities. The structure that incorporated them into the political economy of the United States—in particular, the closely controlled system of reservation administrative units—left few doors open to them. But even if Indians had had the opportunity to mobilize politically, few had the resources necessary to operate effectively within the larger political system. With few assets or opportunities, little other than local Indian political activity was possible.

The Indian Reorganization Act, World War II, and the termination policy, these were among the critical events that transformed the Native American situation. With the Indian Reorganization Act, the federal government opened the door to tribal assertions. The opening was modest, but by endorsing tribal governance it helped to put in place new political organizations that eventually became vehicles of political action, and it established tribes as legitimate political actors. A series of subsequent developments—World War II, the expansion of education, the urbanization of the Indian population, and federal programs that included urban relocation and the antipoverty programs of the 1960s—brought into Indian communities an array of resources, from dollars to organizational skills to political savvy, that could be applied to political organization and action. They produced as well a set of new constituencies oriented toward action who became increasingly supratribal in their politics and operated both within and outside the normal channels of tribal-federal relations: veterans, college students, and urban Indians. At the same time, the dramatic federal policy shift toward termination aroused major Indian opposition, prompting a concerted supratribal response that eventually took up a much broader array of issues.

Indians also were acutely aware of and affected by the political campaigns launched by diverse groups in the 1960s and 1970s that provided new models of collective action. Looking at the impact of the Civil Rights movement on Indian activism, for example, Sioux activist and scholar Vine Deloria (1974, p. 23) remarked that "[t]he obvious success of the marches and demonstrations in getting policies changed taught a very important lesson to many young Indians, who had seen their fathers and grandfathers

thwarted by the immense bureaucracy of the federal government." Faced with new opportunities and armed with new weapons, Indians took up some old issues and forced them into the consciousness of American society.

A fortuitous event further enlarged Indian political opportunities and the impact of Indian actions. In the 1970s the United States found itself faced with a major energy crisis as the cartel of world oil producers dramatically raised prices. Over the preceding decade, the country also had become increasingly aware of the finite nature of its other natural resources, from water to minerals to timber to fish. Certain Indian tribes, well supplied with some of these, suddenly found themselves sitting on vast riches. For the first time since the end of the great nineteenth-century land grab, Native Americans had structural power: they controlled something of value to the surrounding society (Ambler 1990; Cornell 1988; Schwartz 1976, p. 172). Their politics could no longer be dismissed as noise and nuisance. Now they could bargain from a position of strength.

The Federal Response

The Native American political activism of the 1960s and 1970s was unprecedented in twentieth-century Indian affairs, and it presented the federal government with a dilemma. On the one hand, to simply submit to Indian demands would be to endorse the extrainstitutional, confrontational tactics of the more radical parts of the Indian movement and to abdicate control over Indian affairs. On the other hand, to ignore Indian demands entirely would be to endorse a status quo that was increasingly indefensible. Reservations had gained wide public exposure as islands of poverty and repression in the midst of the wealthiest nation on earth; the termination policy, intended to simplify Indian-White relations, had had the opposite effect; and the activist movement not only had led to chaos in Indian affairs, but also was beginning to threaten significant political and economic interests. What to do?

The result, in the early to mid-1970s, was a policy potpourri that had three elements. The first was an organized effort to discredit and ultimately suppress the radical wing of the activist movement. The American Indian Movement, which had played a major role in much of the turmoil, was a primary target of this effort. The FBI succeeded in infiltrating the upper reaches of the organization, and along with Bureau of Indian Affairs and state and local police, maintained a campaign of surveillance, harassment, and indictment that lasted from the early 1970s into the 1980s and put many activists on the defensive in the courts (Johansen and Maestas 1979; La Course 1974; U.S. Senate 1976; Weyler 1982). This campaign led to few convictions, but it crippled much of the effort of AIM and other radical organizations, which had to pour time, energy, and money into their own defense.

The second element was symbolic investigation and reform. In 1975, with great fanfare, Congress established the American Indian Policy Review Commission. Indians were featured prominently on the commission and its staff, which carried out a series of studies on federal Indian policy, tribal government, resource management, the needs of urban Indians, and other topics. They sent to Congress a voluminous final report with concrete suggestions for program changes and new policy initiatives. The symbolic purpose of the commission was efficiently accomplished: it showed a response to Indian concerns at the policy level, and involved Native Americans themselves in that response. By the time the final report was issued, however, congressional and public attention had moved on; most of its recommendations were ignored.

The government also appointed a number of Indians to high positions in the Bureau of Indian Affairs and on advisory bodies of various kinds, and established a new position, filled by an Indian, as Assistant Secretary of the Interior for Indian Affairs. Again, the change was symbolically striking, suggesting growing Indian influence in policy-making and implementation. However, the organizational interests of these bodies and the power structure itself remained largely unaffected.

The third element of the federal response involved substantive change in the structure of Indian-White relations, and led ultimately to a shift in the pattern of incorporation linking Native Americans to the larger society. Despite years of federal controls, most Indian peoples had never abandoned the idea that they were sovereign nations, entitled by inherent right to govern their own affairs within the context of an encompassing—and variously defined—relationship with the United States. That sovereignty

had been recognized in treaties and upheld in the courts; in practice, however, it had been subordinated to the administrative supervision of the United States, enacted through the Bureau of Indian Affairs. Central to the Indian politics of the "Red Power" years had been demands for the practical recognition and implementation of that sovereignty—for tribal self-government, not as image but as reality. In the 1970s, the federal government responded to those demands. President Richard Nixon set the tone early in a 1970 message to Congress in which he repudiated the termination policy that had dominated the preceding two decades and proposed instead a policy of self-determination for Indian tribes (Nixon 1970). The policy became official in 1975 when Congress passed the Indian Self-Determination and Education Assistance Act, committing itself to the provision of "maximum Indian participation in the Government and education of the Indian people," and to "an orderly transition from Federal domination of programs for and services to Indians to effective and meaningful participation by the Indian people in the planning, conduct, and administration of those programs and services" (U.S. Congress 1976, pp. 2203, 2204). While the intent of the act seems clear, in practice it has been subject to diverse interpretations and to the resistance of those organizations, such as the Bureau of Indian Affairs, that stand to lose the most from enhanced tribal power. Twenty years later, however, it seems clear that the act began to provide the nearly powerless tribal governments of the Indian Reorganization Act period with some genuine muscle, and undermined some of the federal resistance to long-standing demands by Indians for enhanced control over their own affairs (Cornell and Kalt 1993; P. Deloria 1986).

This three-pronged response certainly served federal purposes. It strengthened those more moderate Indian political actors most closely linked to the current system of Indian-White relations—the tribal governments in particular—while suppressing the more radical ones most hostile to that system. At the same time, it also amounted to a significant change in the structure of incorporation itself, in the pattern of relationships linking Native Americans to the larger society. Commented one tribal planner in 1985: "In the 1960s, self-determination was a distant dream. We're a lot closer to it now" (quoted in Cornell 1988, p. 205). The partial realization of that dream is substantially a legacy of the Native American activist movement.

Conclusion

By the end of the 1970s the militant phase of the activist movement was winding down. Indian politics remained vibrant, but its style and focus were changing. In the 1980s, newly empowered tribal governments moved to the forefront of Indian political activity, taking control of reservation affairs, initiating their own programs of economic, political, and social development, and imposing their own agendas on Indian-White relations. Pursued mostly through established channels of political action and policy implementation, their activities were less visible and dramatic than the events of the two preceding dec-

ades. Nonetheless, they were significant: an increasingly assertive exercise of the sovereignty that the earlier social movement had struggled to achieve. That exercise has not always been successful. Many reservations today remain desperately poor, and tribal development programs have met with mixed success (Cornell and Kalt 1992; also Snipp, this volume). Many of the themes that motivated the activist movement—the stratified politics of Indian-White relations, the economic welfare of reservation and urban Indians, ultimate control of lands and resources—remain unresolved. A host of Indian constituencies and organizations, including many veterans of the militant activist years, continue to pursue them. But the social and political context of the struggle has changed. Thanks to the Indian activist movement, today it is not only Whites but Indians as well who are shaping the Indian future.

19

Borders and Bridges: Undocumented Immigrants from Mexico and Central America

LEO R. CHAVEZ

On the issue of immigration, public opinion in the United States rarely goes cold; rather it simmers most of the time and boils over occasionally, typically during periods of economic downturns (R. Simon 1985). In the early 1990s, the "immigration problem" was once again hot. In a time of entrenched economic recession and rising unemployment rates, as well as presidential election campaigns, it is not surprising that Americans were asked to "rethink immigration" (Brimelow 1992). In California, the governor blamed immigrants for the state's economic problems, fueling public sentiment "to narrow the gates to the Golden State" (Reinhold 1991, p. A1). At the national level, Pat Buchanan, a Republican presidential candidate during the 1992 primaries, regularly cited immigration as one of our biggest problems (*Newsweek* 1992, p. 33). In particular, Buchanan has warned that undocumented immigration puts the United States at risk "of not being a nation anymore" (Jehl 1992, p. A1). Such rhetoric resembles past anti-immigrant discourse, which placed immigration high on the list of threats to national security (Cornelius 1980).

Some pundits have also blamed Latin American immigrants for the riots that ravaged Los Angeles after the acquittal of the police officers accused of beating Rodney King. As two columnists in the *Los Angeles Times* argued:

> Weary conservatives and liberals have no shortage of explanations for the devastating Los Angeles riots. Yet a major factor has escaped serious discussion. It is immigration, currently running at unprecedented levels, that exacerbates the economic and social forces behind the riots. (Graham and Beck 1992)

Any discussion of undocumented Mexican and Central Americans must place these concerns with immigration and the characterizations of immigrants into historical context. Contemporary views of immigrants, even undocumented immigrants, need to be understood as the most recent manifestation of our nation's split personality when it comes to immigrants. On the one hand we make it possible for them to come to this "nation of immigrants," while on the other hand we fear and resent their perceived threats to our way of life, our jobs, our education system, and our health care system.

Many of these preoccupations, fears, and resentments suggest a limited understanding of why Mexicans and Central Americans migrate to this country and what they do when they get here. Most Americans have never met an undocumented Mexican or Central American. Such people are "known" through evening television newscasts, pronouncements by public officials, and perhaps in passing on public streets. The purpose of this chapter is to place undocumented Mexican and Central American migration within a historical and political economic context while at the same time letting them speak about their own lives. The reigning explanation of international migration places it within theories of world systems, international capital flows, metropoles, and peripheries. Peripheries are unbalanced by increased incorporation into the world system, leading to migration (Portes and Walton 1981; Sassen 1988). As an anthropologist, my job is to put a human face on these macrotrends (Marcus and Fischer 1986, p. 82).

Immigrants: Can't Live With Them, Can't Live Without Them

Because of America's history, "immigration" has become what anthropologists call a key symbol in American culture. Immigration is such a central and powerful concept that it is endowed with a multiplicity of referents and meanings; it raises highly charged emotions, which can often be contradictory to each other. As a consequence, America has had a love-hate relationship with immigrants. Even before we became a nation, anti-immigrant sentiments surfaced. Consider the following comment Benjamin Franklin made in 1751:

Why should the Palatine boors be suffered to swarm into our settlements, and, by herding together, establish their language and manners, to the exclusion of ours? Why should Pennsylvania, founded by the English, become a colony of aliens, who will shortly be so numerous as to Germanize us, instead of our Anglifying them? (Steinberg 1981)

At the same time that some colonists worried about immigration, other colonists believed that America needed more immigration and that immigration was being unfairly restricted by England. Indeed, the colonists felt so strongly about opening up America to immigration that this issue was one of the reasons they decided to separate from Great Britain, as stated in one of the key articles of the Declaration of Independence in 1776:

The present King of Great Britain . . . has endeavored to prevent the population of these States; for that purpose obstructing the Laws of Naturalization of Foreigners; refusing to pass others to encourage their migration hither.

After winning independence from England, America opened its doors to immigrants. The 1800s witnessed massive immigration from various European countries, as well as from China and Japan. Although the immigrants made possible the economic transformation of the industrial revolution, Americans became wary of their presence. On 15 May 1880, an editorial in the *New York Times* was one of many such articles sounding the alarm:

There is a limit to our powers of assimilation and when it is exceeded the country suffers from something very like indigestion. We are willing to receive immigrants just as fast as we can make them over into good American citizens. . . . [But] we are not in need of any more aliens at present. Foreigners who come here and herd together like sheep remain foreigners all their lives. We know how stubbornly conservative of his dirt and his ignorance is the average immigrant who settles in New York, particularly if he is of a clannish race like the Italians. Born in squalor, raised in filth and misery and kept at work almost from infancy, these wretched beings change their abode, but not their habits in coming to New York. . . . A bad Irish-American boy is about as unwholesome a product as was ever reared in any body politic. (R. Simon 1985, p. 186)

Such xenophobia, or the extreme, even irrational, fear of newcomers, may seem unwarranted from today's perspective. Of course, we now know that America did not become "lost" because of German, Irish, and Italian immigration. We have had an Irish American president, an Italian American sits on the U.S. Supreme Court. And, ironically, President Ronald Reagan was fond of joking that his grandfather was an illegal alien from Ireland.

Today, a century later, immigration is no longer predominantly from European countries. Asians and Latin Americans account for about 80 percent of contemporary legal immigration and most of illegal immigration. But the clarion calls sound familiar. In their book *The Immigration Time Bomb*, Richard Lamm, the ex-governor of Colorado, and Gary Imhoff, an ex-official of the Immigration and Naturalization Service, warn that

> At today's massive levels, immigration has major negative consequences—economic, social, and demographic—that overwhelm its advantages. . . . To solve the immigration crisis, we Americans have to face our limitations. We have to face the necessity of passing laws to restrict immigration and the necessity of enforcing those laws. If we fail to do so, we shall leave a legacy of strife, violence, and joblessness to our children. (Lamm and Imhoff 1985, p. 3)

Throughout American history, immigration policy has ensured a steady supply of immigrant workers and the American public has received them with mixed emotions. One person's threat is another person's labor force. Indeed, in certain regions of the country employers in industries such as agriculture and services (hotels, restaurants, landscaping, etc.) have come to rely on low-wage, immigrant workers. Even the threat of shutting off illegal immigration raises the specter of economic decline. Such was the case when the U.S. Congress passed employer sanctions as part of the monumental 1986 immigration law. By punishing employers with fines and even possible jail terms for hiring undocumented workers Congress hoped to stop undocumented migration by cutting it off at the source: jobs. However, due to loopholes in the law and inadequate enforcement, employer sanctions have not been effective in stopping undocumented immigration (Bean, Edmonston, and Passel 1990). Still, the possibility that immigrant workers might no longer be available made some employers uneasy. For example, the personnel manager at a luxury hotel in southern California stated that, in her opinion, if the employer sanctions law had worked there would have been

> . . . severe damage. A lot of properties would have a terrible situation. A majority of your staffing would suffer tremendously. Your food service in-

dustry would shut down, your hospitality [hotels] would suffer tremendously. Your pricing [would] go up. You'd see in this industry alone a lot of rooms closed. You'd see a lot of permanently closed properties. [But] no one has really suffered an impact other than creating additional paperwork flow. (L. Chavez 1992, p. 19)

Immigrants, even undocumented immigrants, have made important contributions to America's economic development, and they have been sought after, even actively recruited, for their labor, as shall be discussed below. At this point, however, it is well to consider alternative voices. There have been other histories on the land the United States now claims as its territory. Illegal immigration did not suddenly emerge during the twentieth century. Americans were once viewed with similar apprehension by the people whose lands they entered, often without permission.

An anecdote about an Indian delegation that visited Washington, D.C., is both humorous and telling. The leader of the Indian group told Alben Barkley, who was then vice president of the United States, "Young fellow, let me give you a little advice. Be careful of your immigration laws. We were careless with ours" (Lamm and Imhoff 1985).

Native peoples were not the only ones whose borders were crossed by illegal aliens. During the nineteenth century Americans routinely crossed into Spanish and then Mexican territory, creating one of the first large waves of undocumented immigrants into what is now the American Southwest. The warnings of José María Sánchez sound similar to modern day complaints about Mexican illegal aliens. In 1827, Sánchez, a Mexican boundary commissioner, was charged with examining the eastern boundary between Mexico and the United States. As a result of his examination, Sánchez became alarmed at the influx of Anglo-Americans to the Mexican side of the border:

> The Americans from the north have taken possession of practically all the eastern part of Texas, in most cases without permission of the authorities. They immigrate constantly, finding no one to prevent them. Repeated and urgent appeals have been made to the Supreme Government of the Federation [of Mexico] regarding the imminent danger in which this interesting territory is be-

coming the prize of the ambitious North Americans. . . ." (J. Chávez 1984)

Immigration as a "problem" often depends upon one's perspective. And undocumented immigrants themselves, as we shall observe below, have many reasons for coming, few of which have to do with collecting welfare, going to college, or obtaining free medical care.

The Public's Concern over Undocumented Immigration

Although a thorough discussion of the issues related to undocumented immigration would take us beyond the limits of this chapter, a few points must be addressed. One of the more persistent issues in the public discourse over undocumented immigration concerns the affect of immigration on citizen workers, especially Chicanos and African Americans. Research by academics and the government tends to conclude that immigration results in positive economic outcomes for the United States (J. Simon 1989). One study on Los Angeles, for example, found that immigration did result in Chicanos earning a little less than comparable workers in other parts of the country, but they also experienced steady job growth (Muller and Espenshade 1985). In other words, they may have made slightly less, but more people were working. The same study also found that African Americans did relatively better economically in areas that experienced immigration compared to areas with little immigration. However, because the evidence is inconclusive, public concerns about the effects of immigration remain high.

Undocumented immigrants are also characterized as placing a burden on publicly financed social services, especially the education system, public hospitals, and the welfare system. While there are costs to immigration, there are also benefits, including tax contributions. Undocumented immigrants, in particular, tend to pay more in taxes than they use in social services (Cornelius, Chavez, and Castro 1982; McCarthy and Valdez 1986; Muller and Espenshade 1985; J. Simon 1989). Complicating this issue, however, is that most of their tax contributions go to the federal government but the costs they incur are at the local and state levels. Hastily drawn studies commissioned by public officials often tend to exacerbate the problem. Such studies must rely on a number of assumptions about the undocumented population, assumptions which can easily exaggerate the costs of social services provided to undocumented immigrants (García y Griego and Chavez 1993; Rea and Parker 1992). Tax contributions can also be minimized by questionable assumptions and methodology, and by simply ignoring federal tax contributions. As a consequence, during periods of economic downturn undocumented immigrants are easily targeted by public officials who can use them in their battle for greater federal contributions to local treasuries.

Concerns about immigrants reached a new high in California in 1994. The lightning rod for the public debate was Proposition 187, a proposition on the November California electoral ballot. The proposition's ostensible goal was to deny all government-funded social services, particularly health care and education, to undocumented immigrants. Not surprisingly, the heated public debate over Proposition 187 came on the heels of a deep economic recession in California. Research has shown a correlation between negative public opinion toward immigrants, including scapegoating, and economic downturns, usually with an associated lag time (Cornelius 1980; R. Simon 1985).

The debate in California surrounding Proposition 187, and its passage by the voters, ignited a debate on immigration throughout the nation. Whether or not undocumented immigrants are denied health care and barred from public education will ultimately be decided in the court system. What is clear, however, is that those who voted for Proposition 187 were sending, at the very least, a symbolic statement about their concern over undocumented immigration, and even immigration in general. White ("Anglos" in local parlance) Californians, in particular, appear to be expressing sentiments of unease over immigration. Two out of three voting Whites in California (about 67 percent) voted for the proposition, a significantly larger proportion than the vote among African and Asian Americans (about half of each group voted for it) and Latinos (only about a 23 percent voted for it). The voting block provided by White voters ensured Proposition 187's passage. Importantly, even though Whites account for only a little over 50 percent of California's population, they account for about 80 percent of the voters, thus their views take on tremendous power.

Proposition 187, by targeting health care and education, aims its wrath entirely at reproduction: that is, women and children. Immigrant women and children typically use health services, and children are in school. The logic appears to be that if life is made difficult enough for the families of undocumented workers, then the families will return to their country of origin and the principal worker will also return and others will not come. However, there is absolutely no evidence that if you deny health care for women and children, and deny education for children, that it will do anything to reduce the economic magnet—jobs—that draws undocumented workers to the United States. It also ignores the displacement of peoples escaping the ravages of war, who may also be in the country illegally.

Why Mexicans and Central Americans Come to the United States

Mexicans and Central Americans have distinct patterns of immigration to the United States and often differing motives for leaving home (Chavez 1988). Most Mexicans migrate into territory that was once part of Mexico, and they have a long history of migration to the United States (Alvarez 1987). In contrast, Salvadorans, Nicaraguans, and Guatemalans began coming to the United States in large numbers in the late 1970s and early 1980s. Indeed, between 1980 and 1990, the number of Central Americans in the United States increased by 66.7 percent (U.S. Bureau of the Census 1991a, p. 4).

Although Mexicans and Central Americans have distinct histories of immigration, they all come from regions that are part of the United States' sphere of political and economic influence (L. Chavez 1990; Díaz-Briquets 1989; Pedraza-Bailey 1985). Mexicans migrate within a long-established labor market that extends beyond national borders. This international labor market just did not suddenly appear, as if by magic. U.S. employers and government policies of investment, trade, and use of Mexican labor helped to establish and nurture the patterns of Mexican immigration that we find today (Bustamante 1977).

The roots of Central American immigration can also be traced to U.S. involvement in the region. Salvadorans, Nicaraguans, and Guatemalans often migrated out of countries experiencing political turmoil and civil war (Marmora 1988). Our relationship to these countries did not begin when immigrants began arriving at our shores. Protecting our interests during the cold war meant supporting dictatorships in El Salvador, Guatemala, and Nicaragua (Palmer 1992). Such actions created a bridge between the United States and those nations, a bridge across which migrants traveled to the United States when political turmoil led to out-migration.

The Roots of Mexican Immigration

Because of the long history of Mexican immigration, a brief history of its origins and maintenance is in order (Massey et al. 1987). The presidency of Porfirio Díaz (1876 to 1911) laid the groundwork for both the Mexican Revolution and Mexican emigration to the United States (Cardoso 1980). One of Díaz's goals was to modernize Mexico. To accomplish this he invited foreign investment from Europe and the United States and separated the Mexican *campesino* (peasant) from the land, thus creating a mobile labor force for capitalist development. By the time his presidency came to end in the flames of the revolution, 5 million rural Mexicans had lost their rights to land. In villages and towns across the countryside, upwards of 98 percent of the farmers had no land to farm. During this time, the railroads were built, with U.S. financing, connecting the interior of Mexico with U.S. and European markets. The railroads also provided the rural labor force with a cheap means of transportation to Mexico's growing urban-industrial centers and to the northern border.

However, it was the growing demand for labor in the American Southwest that made migration attractive to Mexicans. In the early part of the twentieth century, New Mexico, Arizona, Colorado, and Oklahoma needed workers for their booming coal and copper mines. In California, the deserts of the Central and Imperial valleys were being transformed into rich and labor-hungry agricultural lands. A growing population and economy meant large-scale construction in expanding cities throughout the Southwest, but especially Los Angeles, San Diego, San Francisco, and Denver (Romo 1983).

In its search for a labor force to meet these new labor demands in the Southwest, the United States turned to international sources. Chinese immigrants were brought to work in the agricultural fields, mines, and railroads. Their immigration was virtually stopped with the Chinese Exclusion Act of 1882, a response to the "Yellow Peril" campaigns against the Chinese. The Japanese followed, but they too were characterized as a competitive threat due to their success in farming, fishing, and other economic endeavors. The gentlemen's agreement with Japan in 1907 closed the door on their immigration (Takaki 1989).

In the early 1900s, Mexicans became a preferred alternative labor force in the Southwest for a number of reasons (Cardoso 1980). Mexican culture was not so different from American culture, relative to Asian cultures. U.S. employers already had experience with Mexicans through their investments in Mexico. And Mexicans had a long history in the area; their presence was not new or exotic. In addition, Anglo-Americans commonly characterized Mexicans as indolent, passive, noncompetitive, inferior half-breeds who lacked ambition and who were satisfied with their lot in life (fatalists) (De León 1983). They were portrayed as people who would not become economic competitors with their employers. And finally, Mexicans were viewed as the quintessential temporary migrants who would return like "homing pigeons" to Mexico rather than staying permanently in the United States. With such characteristics, Americans viewed Mexicans as providing ample labor at little cost.

Indeed, so pervasive were these characterizations of Mexicans that, in 1911, the Dillingham Commission, which was established to study the immigration issue, argued that Mexican migration should be promoted as the best solution to the Southwest's labor problem (Portes and Bach 1985). It even went so far as to exempt Mexicans from the head tax for immigrants that was established under the immigration laws of 1903 and 1907. With the higher wages offered in the United States compared to Mexico, and with active recruitment campaigns by American employers, Mexicans with few opportunities in Mexico became attracted to jobs in *El Norte*.

By the 1920s, a pattern of migration had been established and the stage was set for the first large migration of Mexicans to the United States (Cardoso 1980). The postrevolutionary years in Mexico were chaotic and violent, especially in the countryside. Across the border, the United States needed labor. The American economy was growing, but perhaps more importantly for Mexicans, the United States

shut the door to low-skilled labor from Europe. The immigration laws of 1921 and 1924 severely restricted the immigration of southern and eastern Europeans. Mexicans, once again, became a suitable alternative for America's labor-hungry agricultural fields and factories. At this time Mexicans were even recruited to the Midwest, including Chicago, where the communities they established continue today.

Immigration from Mexico and the rest of the world came to a virtual stop during the Great Depression of the 1930s. In fact, many Mexicans returned to Mexico, some willingly, others unwillingly. Anti-immigrant, especially anti-Mexican, sentiments flourished during the early 1930s. President Herbert Hoover even blamed the depression on the presence of Mexican immigrants, providing another example of immigrant scapegoating during difficult economic times (Hoffman 1974). As a consequence, the Immigration and Naturalization Service routinely rounded up Mexicans and repatriated them back to Mexico, forcing them to take their U.S.–born children, who were American citizens, with them. Close to 500,000 Mexicans were repatriated during the depression (Hoffman 1974).

Although Mexicans were eschewed during the 1930s, the 1940s witnessed a renewed recruitment of Mexican labor. World War II ushered many American men and women out of the labor force and into military service. Many other women entered the workplace, but a labor need still existed. The United States turned to Mexico for unskilled and semi-skilled laborers who would work in the United States on a contract basis for a few months duration. Beginning in 1942, this program became popularly known as the "bracero program," *bracero* meaning "arms" in Spanish. Although the bracero program was to last only during the war years, its advantages as a ready source of cheap labor, especially in agriculture, proved irresistible, and it was continued until 1964.

During the course of the twenty-two-year bracero program, hundreds of thousands of Mexicans migrated to work in the United States, where they learned about the opportunities to be found in the American labor market and where they established contacts with American employers (Craig 1971). When the bracero program ended in 1964, the demand for the labor these workers provided did not vanish. Employers still needed the workers in their fields and on their ranches, only the workers could no longer migrate to work legally as braceros. Not surprisingly, the number of illegal workers rose dramatically after the termination of the bracero program (Reimers 1985). The system of employer-employee contacts, migration routes, and social networks continued to operate, only clandestinely. Same system, new rules.

The decades of the 1960s, 1970s, and 1980s witnessed a diversification in the use of Mexican labor, especially in the Southwest. Once working primarily in agriculture, undocumented Mexicans are now found in many urban and suburban jobs, performing work that pays low wages and offers few benefits. Such jobs are not generally attractive to U.S. citizens.

This brings us to the contemporary period. The motivations expressed by undocumented Mexicans, as well as Central Americans, reflect the historical realities outlined here. These motivations, however, need to be placed within an understanding of contemporary trends in undocumented immigration.

The Demographics of Contemporary Undocumented Immigration

Although many, if not most, undocumented immigrants come to the United States for relatively brief periods of time and then return home, some do settle and add to the existing population. Because undocumented immigrants are a clandestine population, making accurate estimates of their numbers is difficult. Some reasonable assessments are, however, available. For example, data from the 1980 census was used to estimate the number of undocumented immigrants in the country during the early 1980s at between 2.5 and 3.5 million (Passel and Woodrow 1984). This number was dramatically reduced by the approximately 3 million undocumented immigrants legalized under the 1986 immigration law, most of whom were Mexicans (about 75 percent), but Central Americans (about 5 percent were Salvadorans and 2 percent were Guatemalans) were also represented.

Undocumented immigrants come from many countries, including China, Ireland, Colombia, and so on. However, Mexicans and Central Americans make up a large proportion of the undocumented population. For example, Mexicans made up about 70 percent of the undocumented population in 1988, and Central Americans accounted for another 25 percent (Woodrow and Passel 1990). Undocumented Mexi-

In the contemporary period, due both to their illegality and their poverty, undocumented workers live and work under precarious conditions. Here a farmworker from Oaxaca, Mexico, works in northern San Diego County. He is next to plastic containers that are used to store pesticides and other chemicals, which farmworkers fill with water and use to drink and wash their dishes in. (Photo by Leo Chavez)

cans and Central Americans tend to be relatively young, most are between 18 and 34 years of age (Bean, Telles, and Lowell 1987; Borjas 1990b; Borjas and Tienda 1993, p. 717; Chavez, Flores, and López-Garza 1990; Warren and Passel 1987). Historically, most undocumented immigrants were males, but the proportion of females has increased so that most studies of community-based undocumented immigrants and newly legalized immigrants find that females account for over 40 percent of the population (Borjas and Tienda 1993; Chavez, Flores, and López-Garza 1990; Montes Mozo and García Vásquez 1988). Undocumented Mexicans and Central Americans have on average six to seven years of education (CASAS 1989; L. Chavez 1991).

More than 80 percent of undocumented immigrants live in five states: California, New York, Texas, Illinois, and Florida (Passel and Woodrow 1984). California attracts the largest proportion of undocumented immigrants of all nationalities. For example, Cornelius (1988) found that in 1987, California had approximately half (1.74 million) of the nation's undocumented immigrants. Not surprisingly, most undocumented immigrants from Mexico also choose California as their state of residence. California alone absorbs at least half of the total flow of undocumented Mexican immigrants (Cornelius 1988, p. 4). Of the approximately 3 million people legalized under the 1986 Immigration Law, most (55 percent) lived in California (CASAS 1989).

The number of undocumented immigrants who settle in the United States, rather than staying for a short time and then returning home, appears to be about the same at the end of the decade of the 1980s as it was at the beginning. Based on the 1980 census, Passel and Woodrow (1984) estimated that between 200,000 and 300,000 undocumented immigrants settled in the United States each year. Later, they examined data from the Current Population Survey and estimated that during the late 1980s about 200,000 undocumented immigrants settled annually in the United States (Woodrow and Passel 1990, p. 57). The similarity between the two estimates is important because it suggests that the monumental 1986 Immigration Law, which was designed to stem the flow of undocumented immigrants, had little effect on the number who settle in the United States each year. These estimates also suggest that the number who settle is much less than the millions of illegal aliens often said to come and, by implication, stay in the country.

Why Undocumented Mexicans and Central Americans Come to the United States: Their Own Stories

The reasons undocumented Mexicans and Central Americans give for migrating to the United States are most often very personal. They speak of economic need, personal tragedies, and unsafe environments. In 1986, I, along with some colleagues, interviewed undocumented Mexicans and Central Americans in San Diego, California, and Dallas, Texas (Chavez,

Flores, and López-Garza 1990). Table 1 summarizes their motives for migrating. Most Mexican men, many Mexican women, and even an important number of Central Americans migrated for economic reasons, such as to earn more money or to find a better job. Many women migrated for family reasons. Many Central Americans also cited reasons having to do with political turmoil in their countries. These summary categories of motivations for migration, however, do not capture the difficult decisions the immigrants had to make nor their emotional experiences upon leaving home and traveling to the United States.

Undocumented immigrants express many reasons for leaving home and migrating to the United States. I have grouped these reasons into a number of categories that have helped me to understand their experiences (L. Chavez 1992).

Migration as a Part of Family History   Enrique Valenzuela was raised on a *rancho*, which is a very small agricultural community usually consisting of only a few families, in the state of Puebla. His family's land depended upon rain to grow corn and beans. "If it rains, there is work for six months. If it doesn't rain, there is no work." During the periods of little or no work, his father migrated elsewhere to earn money, including the United States, where he worked under the bracero program for twelve years. Sometimes he'd get a contract for three months, six months, and then he would return to the *rancho*, to Mexico. And then the next year he would return [to the United States] again. He did that until the bracero program ended."

After the bracero program ended in 1964, Enrique's father continued to migrate to the United States as an undocumented worker, relying on contacts with employers that he made during the years he worked as a bracero. His father, however, always tried to discourage Enrique's interest in migrating to the United States, both in person and in his letters to Enrique. "My father had told me that it was very hard and that he suffered a lot. 'When you first come to this country you suffer,' he would tell me. He tried not to have me come here."

In 1970, Enrique, then 23, decided to take advantage of his father's presence in the United States and convinced his father to help him migrate. His experience reflects a number of themes found in the migration experiences of undocumented Mexicans. The

Table 1 Reasons for migration for undocumented Mexicans and Central Americans by sex in sample from San Diego, California and Dallas, Texas

	Undocumented			
	Mexican		Central American	
Motivations for migration	Men N=156 %	Women N=129 %	Men N=195 %	Women N=90 %
Familial motives				
Spouse brought	0.0	10.1	0.5	8.9
Parents brought	3.2	12.4	1.5	7.8
Babysit	0.0	3.9	1.0	3.3
Other family	9.0	11.6	4.6	10.0
Total familial motives	12.2	38.0	7.7	30.0
Economic motives				
Unemployed	12.8	9.3	5.1	2.2
Underemployed	3.8	7.0	4.6	7.8
Low wages	18.6	9.3	7.7	3.3
High wages in U.S.	12.2	10.1	8.2	6.7
Help parents	1.9	4.7	1.5	2.2
Other economic motives	9.0	2.3	4.6	0.0
Total economic motives	59.0	42.6	31.8	22.2
Political motives				
Escape personal conflicts, threats	0.0	0.0	4.1	1.1
Escape political conflicts, civil war	0.0	0.0	37.8	34.4
Escape threats from political/gov't org.	0.0	0.0	4.1	2.2
Total political motives	0.0	0.0	46.1	37.8
Other				
Curiosity/adventure	9.6	2.3	3.1	1.1
For a better life	5.1	4.7	4.1	1.1
Study	1.9	1.6	0.5	2.2
Other	12.2	10.9	6.7	5.6
Total other	28.8	19.4	14.4	10.0
Totals	100.0	100.0	100.0	100.0

Source: Chavez, et al. 1990.

need for labor in the United States and the resulting bracero program created the opportunity for Enrique's father's migrations, which laid a foundation for Enrique's own migration. His father's experiences provided Enrique with both a psychological and a social bridge between himself and the United States.

The story of Andrea Portrero provides another perspective and an even deeper historical relationship to the United States. Andrea was fifty-two years old in 1986. She is the matriarch of a large, three-generation, extended family living in the San Diego area. She has twelve children and ten grandchildren, seven of whom were born in the United States.

Andrea's links to the United States extend back two more generations, to her mother and her grandparents. Her mother was born in Los Angeles, California, in 1918. Andrea's grandparents were in Los Angeles at the time working for a railroad company making boxcars. In 1922, the family returned to Mexico to visit family in the town of Val Paraiso, Zacatecas. While there, her grandfather was killed. As a result of the tragedy, Andrea's mother and her children stayed in Zacatecas. Later, Andrea's mother's brother and sister moved to the United States and eventually became legal immigrants. As for Andrea's mother, she remarried in Zacatecas and did not return to the United States until recently, after her second husband died, to visit Andrea.

In 1969, Andrea and her family moved from Val Paraiso to Tecate, in the Mexican state of Baja California. In Tecate, Andrea washed and ironed clothes. Her two oldest daughters and her husband found work in an electronics assembly plant, one of the U.S.–owned *maquiladoras* that produce goods for sale in the United States. Two years after they arrived in Tecate, Andrea's family needed money, despite four members working. Andrea's maternal aunt had settled in Fresno, California, and offered to help Andrea find work. Andrea recounts the discussion she and her husband had about which one of them should seek work in the United States:

Well, he said, "Maybe I should go to the United States." I told him, "Look, you shouldn't quit your job. If you like, I'll go. I can work a few days and then come back." He said, "Well, if you want to go with your aunt and see what kind of luck you have, well go then." So that's why I went to my aunt's.

Andrea migrated to the United States to help support her family. Her grandparents' migration to the United States early in the twentieth century created the link that facilitated Andrea's own migration. Her family connection in the United States meant that she, not her husband, was best suited to migrate and find work. Andrea's story underscores the long historical relationship Mexicans have with the United States. Five generations of Andrea's family, spanning most of the twentieth century, have at some time lived and worked in the United States.

Target Earners   A general theme in the reasons undocumented immigrants give for leaving home is the need to earn money for a specific, or targeted, purpose. These "target earners" typically desire to earn a certain sum of money and then return home after a short period of time, anywhere from a few months to a couple of years. This is quite common among young, unmarried men and women who said they left to help their parents. Comments such as the following by two women were common: "My family needed money for planting and food" and "I came to work and send my parents money." Men occasionally said they came for similar reasons, such as the comment by the person who talked about his first trip to the United States, "My family was very poor and I came only for three or four months and then returned to help my father." These reasons suggest the importance of family in the decision to migrate. Sending a member to the United States, or even a nearby urban area in Mexico, is often part of an overall family strategy to find as many sources of income as possible, thus helping to ensure the family's economic survival. Family members working away from home are, in essence, agents of the family who send back, or bring back with them, needed resources.

Dissatisfaction with Local Economic Opportunities
Many undocumented immigrants who come to the United States to work say they did so because they were dissatisfied with their opportunities in the local Mexican economy. Comments of this type were the most common among both Mexican men and women. I should point out here that the overwhelming majority of Mexicans do not migrate to the United States, even during periods of economic crisis. Most Mexicans stay in Mexico and develop strategies for dealing with economic hard times rather than undertaking a migration that would mean leaving family, friends, and country for long periods of time.

Margarito was a man in his mid-thirties who was living in a squatters' settlement in northern San Diego County, known as Green Valley, in 1988. For Margarito, the underlying economic problem pushing him and others to migrate was that the wealth of Mexico does not trickle down to the peasant. As he said about his native Oaxaca, and Mexico in general, "Oaxaca has a lot of minerals, vegetation. There is silver, gold, trees, mountains. Mexico has a lot, plus oil, a lot of natural resources, fish. But the wealth stays with very few. It never goes to the farmers."

Single mothers often feel similar pressures. In January 1989, Isabel had been in the United States one month, living in makeshift housing of plywood, cardboard, and plastic built on a hillside in Carlsbad, California. She had left a daughter, age six, and a son, age three, in Oaxaca. She said:

I came here because there is no work over there [in Oaxaca]. Oaxaca has no factories, no large businesses to employ people. When you do find work it's very difficult. You work from nine in the morning to nine at night for little pay and it's hard to find another job. I was told that there were good wages here and that there was plenty of work for women. Right now I do housekeeping, but sometimes I do that and sometimes I don't. It's not stable [work].

The Immigrant's Dream   Another major theme in the reasons undocumented immigrants give for leaving Mexico has to do with what I call the "immigrant's dream," which is similar to the "American dream." Undocumented immigrants view the United States as the land of opportunity, a place where the streets are paved with gold, and a place where hard work and sacrifice can earn them upward mobility, perhaps not for themselves but at least for their children. They believe the United States offers them the opportunity to *progresar* (to progress) and *mejorar económicamente* (better themselves economically). They hope to *superarse* (to surpass their current situation or circumstances) both for themselves and their families. They believe jobs in the United States offer a chance to *subir* (rise). As one fellow put it, "I always wanted to be a mechanic and here I had better opportunities to rise (to that position)." Symbolic referents which stress upwardness and positiveness meta-

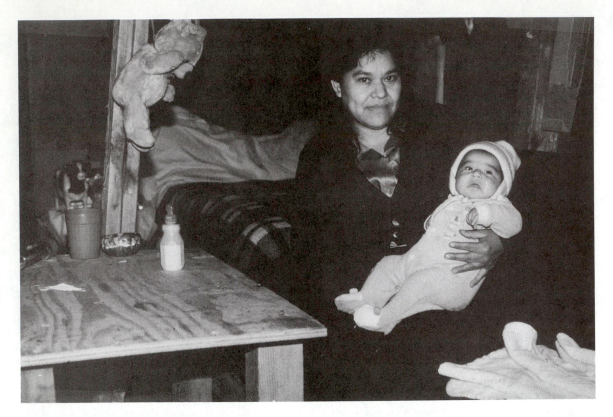

Though their unprotected lives often hang by the slim thread of luck, the undocumented often make enormous sacrifices for the sake of their children's future. Here this woman and her child are living in a makeshift wooden shack near Rancho Santa Fe, in California. (Photo by Leo Chavez)

phorically associate going north with social and economic improvement and mobility.

There is little in such affirmative declarations that might be construed as a fatalistic attitude or a resigned acceptance of one's lot in life. On the contrary, undocumented immigrants express a sense of optimism and hope for the future that has to be seen as part of their immigrant ethic. They come ready to accept sacrifices for a better tomorrow.

Family Reunification and Family Conflicts   Family motivations influence migration in a number of ways. Family members left behind sometimes migrate to join a family member already in the United States. In addition, many young people migrate to help their parents economically. This portrayal emphasizes the importance of the family, rather than just the individual, in the decision to migrate. However, family life is not always a model of unity and cohesion. Conflict

between family members can also be a powerful influence in the decision to migrate.

People who migrated because of family conflicts often felt a deep personal anguish. Beatriz Valenzuela's eyes moistened as she remembered her mother and the conflicts between her and her brothers. Beatriz was in her mid-forties and had the tired look of someone who regularly worked long days. She showed a sly sense of humor when she spoke. She was born and raised in Manzanillo, Colima, where her family owned a large bakery. Because of the family business, Beatriz was able to study through high school. But in 1964, when Beatriz was twenty years old, her father died, setting off a period of bitter conflict between her and her two older brothers, who believed they, alone, were entitled to their father's inheritance. Beatriz believed her brothers, especially her oldest brother, were taking advantage of her, since she was the youngest child. She believed that she, too, deserved some of the inheritance. This led to a period of intense conflict that eventually caused her to leave for the United States. As she said,

> I tolerated them [her brothers] for five years, but only for my mother, only for her. I would tell her that I wanted to go to another place or come here [the United States], and she would start to cry. When I couldn't stand it anymore, that's when I asked her for permission to come here.

Beatriz did not migrate to the United States for economic reasons. Family conflict drove her out of her home. Beatriz' memory of her trip from Manzanillo to Tijuana recalls the loneliness and fear felt by someone who had never left home before. "I was afraid and I would think of my mother. So the whole way I cried." At the time Beatriz told this story, it had been about seventeen years since she had last seen her mother.

Adventure and Curiosity   Undocumented immigrants also migrate to satisfy their curiosity about what it is actually like in the United States. As one woman said, "I came because I wanted to know San Diego." Or as a man said, "I wanted some adventure." Another woman said, "I came on vacation and I stayed." And for yet another woman, "I came to acquaint myself with the United States and then study English." Some found, however, that the adventure was more difficult than they expected, as this fellow found: "I came for adventure to the richest country, where we live like dogs."

Central Americans

Undocumented immigrants from El Salvador, Nicaragua, and Guatemala whom I have interviewed migrated to the United States for some of the same reasons as Mexicans, as Table 1 indicates. Few, however, had the historical connections or linkages to the United States that their Mexican counterparts did.

As important as economic and familial motives were, they were often overshadowed by the frequent references to the danger interviewees felt in their homelands because of political conflicts, a reason rarely found among Mexicans. Salvadorans, Guatemalans, and Nicaraguans frequently mentioned their fear of guerrilla activity and the dangers presented by the state's attempts to repress such activity. I would also suspect that the motives some interviewee's gave for leaving may have been economic on the surface but actually reflected the devastation to local economies caused by the turmoil these countries have experienced.

Although many examples could be presented, consider the case of María Favala, who expressed her distress over the daily dangers in El Salvador as the reason she and her husband left. In December 1988 María was living in makeshift housing in Green Valley. María was pregnant, twenty-two years old, and had left three children—ages five, three, and two—with her mother-in-law when she and her husband migrated from El Salvador. Seven years earlier, some men had burst into María's family's home. "They killed my father. They killed him at home in front of us." María also spoke of the other atrocities she has witnessed, "They come to take a man out of his home and the next day you'll find him dead on the road, with no head. They'll make a statue with a stick with the head on top in the middle of the road, and they'll leave the body thrown over to the side."

María and her husband had been in the United States for six months. They joined her husband's father, who had been in the United States for three years and who also fled El Salvador because of fear. Her father-in-law suspected his son's life was in danger after he was temporarily arrested in El Salvador, and so he sent word for María's husband to join him in San Diego. María agreed. "They were looking for my husband. They would leave us notes saying that if they didn't find him then they'd get his wife. So he was afraid. He said, 'If I leave, something will happen to you here.' So, that's why we came over here."

Conclusion

Violence, or the threat of violence, led many Central Americans to leave home. Migration under these circumstances, combined with their relatively recent migration, highlights a major difference between these undocumented immigrants and those from Mexico.

What is striking for Mexicans, in contrast, is not merely that they migrate primarily to work. What is important is how history, social relationships, and economic structures have converged to expand the conception many Mexicans have of where they can seek work. By this is meant that they are not restricted to jobs in their local economy. Their possible labor market, or where they can trade their labor for wages, includes the United States. The politically defined border between Mexico and the United States is not a hindrance to the operation of this international labor market, which relies on a long history of entrenched patterns of migration. Importantly, these patterns of migration did not appear suddenly nor without involvement by the U.S. government and American employers. If the history of Mexican, and more recently Central American, immigration teaches us anything it is that bridges once built are difficult to demolish.

20

Cuba's Refugees: Manifold Migrations

SILVIA PEDRAZA

El vino, de plátano; y si sale agrio, ¡es nuestro vino! —José Martí, "Nuestra América" (1891)[1]

Over thirty years of political migration brought close to a million Cuban immigrants to American soil, harboring distinct waves of immigrants as well as distinct refugee "vintages," alike only in their final rejection of Cuba. Each of the major waves of migration has been characterized by a very different social composition. To understand the changing characteristics of the exiles over time, we need to pay attention to the changing phases of the Cuban revolution. As Peter Rose (1981, p. 11) underlined, "refugees do not live in a vacuum. They are part of an intricate sociopolitical web that must be seen as the background against which any portrait of their travails must be painted and any dissection of their innermost thoughts and feelings must be pinned." This analysis focuses both on the nature of the decisions the refugees made to leave Cuba and the larger social structures that shaped the exodus. As C. Wright Mills (1961) said, the sociological imagination lies at the intersection of personal troubles and historical issues.

Including those who were born here, the total number of Cubans in the United States as of 1990 was approximately 1,042,433, 73 percent of which were immigrants.[2] The 1990 census data gives us, for the first time, the opportunity to look at the social and demographic characteristics of the distinct waves of the Cuban exodus.[3] The various waves of Cuban migration brought very different sets of social resources with them—such as their social class, race, education, family, institutional knowledge, and values. Over the course of more than 30 years of exodus, they also arrived in the United States at times when the social context that greeted them presented them with vastly different amounts of opportunity—such as economic growth or recession, government policy programs, a warm welcome, or a cold reception. Hence, Cuban immigrants from the various waves of migration have undergone rather different processes of incorporation into American society, as will be seen in their contrasting social outcomes. As a result, to portray Cubans in the United States by figures that describe them as a whole (for example, by their overall education or poverty level) masks the vastly different social realities which they represent. Those many social realities are the result of their manifold migrations.

Refugees as a Social Type

As E. F. Kunz explained, a refugee "is a distinct social type." The essential difference between refugees and voluntary migrants lies in their motivations:

> It is the reluctance to uproot oneself, and the absence of positive original motivations to settle elsewhere, which characterizes all refugee decisions and distinguishes the refugee from the voluntary migrants. (Kunz 1973, p. 130)

Hence, the key idea necessary to understand the refugee in flight is that of the "push" (cf. Lee 1966). While ordinary immigrants are more likely to be "pulled" by the attraction of the opportunity to fashion a better life, as Barry Stein (1981, p. 322) succinctly expressed it, "the refugee is not pulled out; he is pushed out. Given the choice, he would stay." Political exile is the last step of a process of profound political disaffection that, as Kunz (1973) stressed, is often accompanied by the refugees' fear for their safety given their interpretation of events and self-perceived danger.

To explain the enormous variance among refugees' experiences, Kunz (1973, p. 137) spoke of "vintages," or refugee groups that are distinct in "character, background, and avowed political faith." When dramatic changes in the society take place gradually, individuals react differently. Some oppose changes that others support, some call for compromises that to others smell of collaboration:

> As the political situation ripens for each, they will leave the country as distinct 'vintages' each usually convinced of the moral and political rightness of his actions and implicitly or openly blaming those who departed earlier or stayed on. (Kunz 1973, p. 137)

"Vintages" (defined by attitudes) may or may not be the same as waves (defined by timing). The Cuban political exodus holds both distinct waves of migration and "vintages."

The First Wave: Cuba's Elite

Nelson Amaro and Alejandro Portes (1972) portrayed the different phases of the Cuban political immigration as changing over time with the exiles' principal motivation for their decision to leave. With the unfolding of the Cuban revolution, over the years "those who wait" gave way to "those who escape," and they to "those who search." Bringing the analysis up to date, I have added "those who hope" and "those who despair." Overall, the Cuban migration is characterized by an inverse relation between date of departure and social class of the immigrants.

Typical of the first phase of the immigration were "those who wait." The Cuban exodus began with the triumph of the Cuban revolution in 1959 over the tyranny of Fulgencio Batista with the exit of the *Batistianos*. But at this time the majority of Cubans shared in the euphoria of the revolution's hard-won success. It was only when the revolution entered a more radical phase that the exodus of political immigrants really took force.

In this first wave, those who left were Cuba's elite. These upper and upper-middle classes were not tied to Batista's government but were bound to a political and economic structure that, Amaro and Portes underlined, was completely interpenetrated by the demands and initiative of American capital:

> These executive and owners of firms, big merchants, sugar mill owners, manufacturers, cattlemen, representatives of foreign companies and established professionals, were those most acquainted with the United States' political and economic guardianship of Cuba, under which they had created or maintained their position, and thus were the least given to believe that the American government would permit the consolidation of a socialist regime in the island. (Amaro and Portes 1972, p. 10)

Hence, amidst the economic and diplomatic war that ensued between Cuba and the United States (cf. Schreiber 1973), they decided to leave. The refugees of this first wave came to the United States driven by Cuba's overturning of the old order through revolutionary measures, such as the nationalization of American industry and agrarian reform laws, as well as by the United States' severance of diplomatic and economic ties with Cuba, all of which entailed serious personal losses. Maximiliano Pons[4] came to the United States in 1960 after Castro nationalized the American company he was working for. The son of Spanish immigrants, Catalan traders from Barcelona, like many of his social class Maximiliano was educated in the United States:

> I finally broke with my father when I went to college. I elected to go to Yale and not to Salamanca, where my father wanted me to go. I wanted to be an American very badly. I became an American aspirant. The United States had just emerged victorious from World War II. I was away at Yale then. Four years later I returned to Cuba, went to work for a leading American company, bought a house in the Marianao suburbs—the whole thing. Fidel turned me back into a Cuban. (Llanes 1982, pp. 53–54)

"Those who wait" characterizes those first refugees that came imagining that exile would be temporary, waiting for the inevitable American reaction and help to overthrow Cuba's new government. In this first stage the exile's political activity was intensely militant, supporting military counterrevolution against Cuba. Of these, the exiles' invasion of the Bay of Pigs in April 1961 was the largest and most tragic. This first phase of the Cuban exile ended with the fiasco of the Freedom Fighters' attempt to liberate Cuba from Castro's hold. In *Playa Girón* the 2506 Brigade fought against the Cuban rebel army, waiting for the air cover the United States had promised. They waited in vain (see H. Thomas 1977).

"Those who escape" constituted the second phase that was set on by the growing political turmoil when the Catholic church was silenced after denouncing the revolution (Alfonso 1984). The electoral system collapsed when jubilant crowds around Castro chanted, "*¿Elecciones para Qué?*" ("What do we need elections for?"); and Castro announced that he had always been a Marxist-Leninist and would be so until the day he died (cf. Thomas 1977). The exodus doubled. As Amaro and Portes (1972) noted, the inverse relationship between date of emigration and social class in Cuba began to show. Still largely a middle-class exodus, now it was more middle than upper: middle merchants and middle management, landlords, middle-level professionals, and a considerable number of skilled unionized workers who wanted to escape an intolerable new order.

The immigrants of the first two phases were not so much "pulled" by the attractiveness of the new society as "pushed" by the internal political process of the old. When the private universities and schools began to close in 1961, fear that the children would be educated by the state became prevalent. Miranda Martín, the daughter of a doctor who had initially sympathized with the revolution, remembered:

> We became aware gradually of the shift in Fidel's policies to the left, but *papi* didn't want to leave Havana. . . . My mother, on the other hand, was panicked. What will they do to the children in school? Will they force them to go to communist schools? Finally in August we left. (Llanes 1982, pp. 23–24)

Over 14,000 children came alone through Operation Peter Pan, sent by their frightened parents (Walsh 1971). "What began as a trickle," wrote Fagen, Brody, and O'Leary (1968, p. 62), "was, by the middle of 1962, a small flood." Data from the 1990 census show that of the 757,187 Cubans in the United States who immigrated after the revolution, 25 percent arrived during the first wave, 1960 to 1964 (see Table 1).[5] At this time the Cuban Refugee Program was initiated that assisted most of the refugees in Miami.

The higher class origin of these Cuban refugees has been well documented. This initial exodus overrepresented the professional, managerial, and middle classes—31 percent—as well as clerical and sales workers—33 percent. Likewise, the educational level of these refugees was remarkably high (Fagen, Brody, and O'Leary 1968, Table 7.1).

With breathtaking speed, in a couple of years the Cuban revolution had moved through distinct phases. Nelson Amaro (1977) captured the progressive stages as: first, democracy; then, humanism; followed by nationalism; thereafter, socialism; culminating in Marxism-Leninism. The United States' punitive policy—cutting the sugar quota, instituting a trade embargo, and backing the exiles' invasion of Cuba—no doubt aided the rapidity of this transition. Amidst this swift progression of stages, some refugees dissented at one point, some at others. To Castro, they were all the same: *gusanos*, or counterrevolutionary "worms."

Casal (1979) observed that although the "highly belligerent" counterrevolutionary movements of the first two phases never actively engaged all exiles, they did draw on the financial or moral support of most exiles who hoped for Castro's overthrow and for their own return to Cuba. Unable to reach their goal, the Cuban communities became disenchanted with such activities and withdrew their support. As Kunz (1973, p. 133) stated, when refugees realize "that the doors are closed behind" them, they begin to take the steps that change them from temporary refugees into exiles.

Kunz (1981, pp. 45–46) also distinguished various refugee types according to their ideological-national orientation while in exile. Kunz's labels for the various ways in which one can be an exile constitute an effort to delineate the types: the *restoration activists*, the *passive hurt*, the *integration realists*, the *eager assimilationists*, the *revolutionary activists*, and the *founders of utopias*. Revolutionary activists single-mindedly "subjugate matters of family and chances of long-term resettlement" to the purposes they set out to achieve, while eager assimilationists may engage in a "hyperactive search for assimilation and the achievement of material success" as a way to forget their guilt and their past. Still, it is important to realize, as Kunz proposed, that while these "solutions" to the problem of exile may be found side by side, most individuals "pass through these role-phases from their day of flight, progressing and regressing" throughout their exile careers (Kunz 1981, p. 46), passages that may well be facilitated by specific events.

The very life of Lourdes Casal attests to various exile role passages and to diametrically opposed ways of shouldering the felt sense of historic responsibility which exile can entail. In her attempt to express her strong identification with Cuba and to solve her personal problem of exile, Lourdes Casal lived through many exile lives. Casal was a formidable woman: a Black Cuban from the middle class, in the 1970s she was the founder of the magazine *Areíto* and the Antonio Maceo Brigade, for Cuban Americans open to or in sympathy with the revolution. Yet at the beginning of the revolution, in the late 1950s and early 1960s, she had been an active member of groups, such as the *Juventud Universitaria Católica* (Catholic University Students) and the *Directorio Revolucionario Estudiantil* (Students' Revolutionary Directorate), that fought against Batista's dictatorship but also actively opposed the communist direction in which the revolution was then headed. As a result of her travels to Africa and her return to Cuba, as well as the social movements then rippling through the United States, she underwent a pro-

found personal change and, in the early 1970s, became an active supporter of the Cuban revolution. When she learned that she had little time left to live, she chose to return and, in 1981, die in Cuba (see Institute of Cuban Studies 1982). As Peter Rose (1981, p. 11) highlighted, few social scientists have turned their attention to the sociology of exile, and even those few "rarely have delved into the social and psychological ramifications of those affected" by refugee migration and resettlement.

After the October Missile Crisis in 1962, the flights ceased, forcing the migration rate to slow down. The United States provided direct transportation only for over 1,000 Cuban exiles, who had been imprisoned in Cuba as a result of the Bay of Pigs fiasco, and their relatives. The Cuban government exchanged the prisoners for vital needs: medicine, medical and surgical equipment, food, and money.

During this period, Cubans arrived that had either previously stayed in other countries or had escaped Cuba illegally in boats and rafts to the shores of Key West. Francisco Mateo crossed the ninety miles between Cuba and Key West in a small rowboat:

> Some of the people who left from Mariel (in 1980) took twenty hours to cross the distance in a motorboat. We took twenty days in 1962, my family and me, in a boat with three oars and holes. . . . You tell me how eight people could leave on an eight-foot rowboat and expect to get anywhere. Across the Miami Causeway maybe, but not those stinking, treacherous ninety miles. God was with us. There is no other answer. (Llanes 1982, p. 19)

During this phase of the exodus, close to half of the arrivals were blue-collar workers, skilled and unskilled, and a large proportion were agricultural workers and fishermen (Casal 1979, Table 1). Cuba introduced food rationing and compulsory military service at this time, further spurring the exodus.

The Second Wave: Cuba's Petite Bourgeoisie

In the fall of 1965 a chaotic period ensued when hundreds of boats left from Miami for the Cuban port of Camarioca, where they picked up thousands of relatives to come to the United States. "Those who

Table 1 Number of Cubans in the United States, by Year of Immigration, 1990

Year of Immigration	Number of Cubans	Percent	Immigrated from 1960 to 1990 (%)
Born in the US	285,244	27.4	—
1987–1990	33,837	3.3	4.9
1985–1986	16,963	1.6	2.4
1982–1984	23,163	2.2	3.4
1980–1981	125,313	12.0	18.2
1975–1979	33,256	3.2	4.8
1970–1974	109,731	10.5	15.9
1965–1969	173,287	16.6	25.1
1960–1964	174,275	16.7	25.3
1950–1959	50,956	4.9	—
Before 1950	16,406	1.6	—
Total	1,042,433	100.0	100.0 (689,825)

Source: 1990 U.S. Census, Public Use Microdata Sample, 5 percent, weighted.

search" characterized this next major wave of the Cuban migration. In response to President Lyndon Johnson's "open door" policy that welcomed refugees from communism, the Cuban exodus was organized and concerted. For eight years, the U.S. and Cuban governments administered an orderly air bridge as the *Vuelos de la Libertad*, or Freedom Flights, daily brought Cubans from Varadero to Miami that the Cuban Refugee Program swiftly processed and resettled, dispersing them throughout the United States.

Kunz (1973) distinguished anticipatory refugee movements from acute ones. The joint policy of the U.S. and Cuban governments turned this initially acute exodus into a coordinated and orderly anticipatory refugee movement. Though for quite different reasons, the U.S. and Cuban governments have often "cooperated with the enemy," as Jorge Domínguez (1991) stressed. When the refugee airlift closed, thousands of flights had brought more than 250,000 persons. As Table 1 shows, 41 percent of Cubans who immigrated to the United States after the revolution came over during the years of the air bridge, 1965 to 1974.

Throughout this period, a Memorandum of Understanding regulated the immigrants' departure, giving the immediate family of exiles already living in the United States priority (J. Thomas 1967). Both countries compiled their "master lists"—one composed in the United States of those who claimed their relatives in Cuba, and one composed in Cuba of those who requested departure. Jointly, both governments decided who would emigrate and the migration proceeded through family networks. Cuba barred from exit young men of military service age, as well as professionals, technical, and skilled workers whose exit would cause a serious disturbance in production or delivering of social services, such as doctors (Clark 1975).

With this phase of the migration, the exodus of the upper and upper-middle classes largely came to an end. This wave of immigration was largely working class and *petite bourgeoisie*—employees, independent craftsmen, small merchants, skilled and semiskilled workers. Amaro and Portes judged (1972, p. 13) that over time the political exile increasingly became an economic exile as "those who search" searched for greater economic opportunities than were provided in a socialist society that instituted a new ethic of sacrificing individual consumption to achieve collective goals.

Without doubt, these were some of the leanest and most idealistic years of the Cuban revolution. To spread access to a basic education and health care, young, educated Cubans went to live in the countryside, working in literacy campaigns to educate the poor, illiterate peasants, and in public health campaigns to provide basic health care. At the same time, the impact of the hemispheric trade embargo imposed by the Organization of American States in 1964 resulted in a spare parts crisis and other profound economic dislocations (Schreiber 1973); the exodus drained technical and administrative skills; and Cuba failed in her attempts to cease being a sugar monoculture, industrialize, and diversify. In Amaro and Portes' (1972) view, increasingly the immigration ceased to be a political act and became an economic act. Yet their distinction missed the reality that while life in Cuba grew harsh for all, it turned particularly bitter for those who had announced their dissent by declaring their intention to leave. Those who applied to leave lost their jobs, were ostracized as enemies, and were forced to do hard labor in agriculture. Antonio Chacón applied to leave Cuba in 1962, but was unable to leave until 1966, by which time he was suffering from malnutrition, diabetes, and high blood pressure:

> We had applied for an exit permit. This meant that I would lose my job at the newspaper. We had planned for a few months of unemployment. It was unavoidable. . . . Then, slam. . . . The door closed and I was inside. Unemployed. We finally left in 1966. Can you imagine that? Four years knocking around doing "volunteer work" on weekends in order to get the food allowance. We lost our belongings. Everything we owned was sold or traded for food. We ended up living with my friend Jacobo, who took us in at great risk. I lost eighty pounds in those four years. (Llanes 1982, pp. 93–94)

The social transformations the Cuban revolution effected—political and economic—were so pervasive that they always "pushed" Cubans. America, in facilitating the migration, always "pulled" them. Moreover, the Cuban migration is unique in the extent to which both the U.S. and Cuban governments organized, concerted, and facilitated the exodus (Tabori 1972). Together, I argued, they set in motion a system of political migration that for many years proved beneficial to both. The loss of the educated, professional middle classes indeed proved erosive to the Cuban revolution, but it also served the positive function of externalizing dissent. At the same time, in the United States the arrival of so many refugees who "voted with their feet" also served to provide the legitimacy necessary for foreign policy actions during the tense years of the Cold War (Pedraza-Bailey 1985).

Now the Cuban community in the United States became increasingly heterogeneous, varying widely in their social class origin. The former social distinctions were perpetrated and reenacted in exile, often with little bearing to their life in America. Those who had belonged to the five most exclusive yacht and country clubs in Havana founded another in Miami, with nostalgia dubbed "The Big Five." Cubans of working-class origin remained outsiders to these attempts to recreate a golden past that seems to ever grow only more golden.

When the migration began in the early 1960s, 31 percent of the Cubans who arrived in the United States were professionals or managers. By 1970, only 12 percent were professionals or managers. More than half the arrivals, 57 percent, were blue-collar, service, or agricultural workers (Aguirre 1976, Table 2). While Cuban exiles are clearly heterogeneous, their celebrated "success story" obscures it. It particularly serves to obscure the many Cuban poor. Still, the Cuban poor have always been evident in many neighborhoods of *la Southwestcera*, as Miami's Southwest is affectionately called. But not only are they hidden from the view of Americans, Cubans also tend to hide them from themselves (cf. Domínguez 1975). Casal (1979, p. 116) emphasized the costs of the "success story": it prevents Cubans from getting "a clear picture" of their true situation; it desensitizes them and others to the hidden costs of "success;" and it isolates Cubans from other American minorities.

Cuban immigrants that arrived after the air bridge ended consisted of refugees that had first lived in Spain. Portes, Clark, and Bach (1977) found that these émigrés represented Cuba's "middling service sectors:" cooks, gardeners, domestics, street vendors, shoe shiners, barbers, hairdressers, taxi drivers, small retail merchants. They left Cuba during the period when Castro launched a new "revolutionary offensive" in Cuba, confiscating over 55,000 small businesses that were still privately owned (Mesa-Lago 1978), "pushing" out the little entrepreneur and his employees. By and large, the refugees of this "vintage" believed in the promises of the revolution until the Cuban government labeled them *parásitos*, or "parasites," and took over their small businesses.

With the economic transition to socialism effected, in the 1970s the Cuban government cast the shape of the political system: the new Cuban Communist Party held its first congress, a new constitution was declared, and Fidel Castro formally became President. In many ways, the old idealism and romanticism of the 1960s gave way to what Mesa-Lago (1978) called pragmatism. The failure of the mobilization of hundreds of thousands of Cubans all over the island to make the national goal of cutting 10 million tons of sugar in 1970 issued this new phase. Cuba reintroduced material incentives and wage differentials to promote greater economic growth; other mass organizations, such as the *Poder Popular*, or Organs of Peoples Power, took form. With this institutionalization, Cuba increasingly took on the features of eastern European communism (cf. Roca 1977).

For the vast majority of Cubans in the United States, throughout these years the issue continued to be life in America. Yet that very stability, and the cultural impact on the young who lived face to face with the social movements of "the sixties" in America, gave birth to an increased ideological pluralism, denser than that which had always existed though obscured by the uniform rejection of Cuba. As Casal (1979, p. 128) observed, "the Cuban community is not monolithic now (if it ever was)."

Among other splits, such as social class and waves of migration, the Cuban community is certainly cleft by age, by generations. Typically, immigrants experience a pronounced generation gap when parents raised in the old world confront their children raised in the new. But this gap reflects more than that; it is the difference between political generations that result from sharing a common location in a historical social process that subjected them to specific experiences during their youth—a stratification of experience that shapes a frame of reference for the future (Mannheim 1952). Among Cuban exiles, the gap between the political generations which came of age during certain critical periods of Cuban history (cf. Zeitlin 1966) and that which came of age in America, under the impact of the Civil Rights and anti-Vietnam War movements, is often a chasm.

It was fifty-five progressive young people who, in December 1977, first broke through nineteen years of hostility, abuse, and isolation. Grouped as the Antonio Maceo Brigade, their visit throughout the island left behind a profound mark. Cuba filmed it: *55 Hermanos* (55 Brothers and Sisters) captured their search for cultural identity; for some, for political identity. Widely shown in Cuba, it proved heartrending: evidence of the suffering that exile had brought both those who left and those who were left behind.

In 1978, a Dialogue took place between the Cuban government and representatives of the Cuban community in exile, as a result of which the Cuban government agreed to the release of political prisoners, to promote the reunification of families rent apart by the exodus, and to allow Cubans in the United States to visit their family and their homeland.

All at once, the counterrevolutionaries (*gusanos*) of yesterday respectfully became "members of the Cuban community abroad," the release of political prisoners began, and the return visits of Cuban exiles commenced. The Cuban community split into the opposing camps of those who supported and opposed the Dialogue; those who returned and those who refused to visit Cuba. Still, since that day, hundreds of thousands of Cubans have returned to Cuba every year—seeking the family they loved and the vestiges of the life they once led.

The Third Wave: Cuba's Marielitos

Since the flow of Cuban refugees had halted for many years, few expected the chaotic flotilla exodus in 1980. Initiated in April by those who asked for political asylum at the Peruvian Embassy, within days it grew massive. When this acute refugee exodus ceased the following fall, it had brought over 125,000 more Cubans to America, approximately 18 percent of all Cuban immigrants (see Table 1). This wave lacked order and process. From Miami, thousands of boats manned by relatives sped across the ninety miles of sea to Cuba's Mariel Harbor. At times they succeeded in bringing their families, other times they brought whomever angry officials put on their boats. Toward the end, this included Cuba's social undesirables: those who had been in prison (whether they had committed real crimes or had only succeeded in challenging the state), mental patients, and homosexuals.

In Cuba, these "antisocial elements," this *escoria*, or "scum," as the government called them, represented a large public slap in the face: no longer the immigrants of the transition from capitalism to communism, but the children of communism itself. In America they arrived in the throes of President Jimmy Carter's ambivalent government policy that both welcomed them "with open hearts and open arms" and sought to delimit the flow.

In the United States, after twenty years of celebrating the achievements of Cuban exiles, the press contributed to their damaging portrayal. It focused on the criminals, the homosexuals, the many Blacks: categories of people to whom Americans accord too little respect. Who were the *Marielitos*? Were they "scum?"

To dispel the more damaging and inaccurate portrayals, Robert Bach (Bach 1980; Bach et al. 1981/1982 pp. 33–35) studied their characteristics by sampling the *Marielitos* while still in the processing cen-

The migration of refugees whose exodus is motivated by both political and economic conditions in their country can result in waves of migration that are quite distinct in their social composition. Here is a family of Cuban refugees from the first wave of the Cuban exodus, being resettled by the Cuban Refugee Program, in Cleveland, 1962. (*Miami News* Collection; Historical Museum of Southern Florida)

ters and the refugee camps soon after their arrival. Among the most salient was their youth (most were young men, single or without their families) and the visibly higher proportion of Blacks than ever. Their former occupations showed that most were from the mainstream of the Cuban economy, hardly scum. Also salient was their overwhelmingly working-class origins—close to 71 percent were blue-collar workers. Mechanics, heavy equipment and factory machine operators; carpenters; masons; and bus, taxi, and truck drivers led the list of occupations (Bach et al. 1981/1982, p. 34). These characteristics, stressed Fernández (1982), suggested new generational strains may have developed from the more limited economic and political opportunities available to the

young when the older generation of Cubans who made the revolution held the key posts, as well as the burden of military service in Cuba and overseas shouldered by the young (cf. Díaz-Briquets 1983). "Those who hope" might well characterize this wave.

In the United Sates, the press focused inordinately on the criminal element. Indeed, there were many who had been in prison. According to the Immigration and Naturalization Service, of the 124,789 Mariel refugees, around 19 percent (or 23,970) admitted they had been in jail in Cuba. Of those who had been in prison, 5,486 were political prisoners, whereas 70 percent of those who had been in prison had been jailed for minor crimes or for acts, such as vagrancy or participation in the extensive black market, that were crimes in Cuba but not in the United States. The Cuban *Ley de la Peligrosidad* (Law of Dangerous Behavior) made some forms of dissent "antisocial" behavior, controlled by prison terms, such as participating in the black market (buying or selling clothes or food); desertion or dodging military service; refusing to work for the state, particularly in the cane fields; and trying to escape Cuba illegally (Bach et al. 1981/82 p. 46). Of those who had been in jail, the immigration service considered only 7 percent to be serious criminals—less than 2 percent of all the *Marielitos* (Montgomery 1981).

Given their youth, the *Marielitos* clearly constituted a different political generation, one whose coming of age was long after the early revolutionary struggle and sharp social cleavages that demanded enormous sacrifices but also affirmed the loyalty of many. Roughly half of the Mariel immigrants came of age during the late 1960s or the 1970s, when problems of freedom of expression became particularly acute for artists and intellectuals, such as the incident sparked by Heberto Padilla's poem expressing the marginality of those who were *"Fuera del Juego"* ("Out of the Game"). Moreover, deviance, particularly homosexuality, was scorned and dealt with by prison sentence. Comparisons with the years of Batista could no longer serve to promote the consent of a generation that scarcely could remember them.

The *Marielitos*, therefore, were a significantly different "vintage"—one whose lived experience (*experiencia vivencial*, as we say in Spanish) contrasted sharply with other "vintages." In particular, at the two poles of twenty years of emigration stand two "vintages" that at best can hardly comprehend one another and at worst may be, as Kunz noted (1973), hostile. Over time the dramatic changes the Cuban revolution effected progressed through distinct stages, and these stages interacted with the social characteristics of those affected to produce markedly different processes of political disaffection.

To put it simply, let me give two stark examples. A typical 1960 *émigré* was an older, white, male executive that would likely have become disaffected by the nationalization of American industry in the early years of the revolution. But a typical 1980 *émigré* was a young, Black, male bus driver that would scarcely have minded that nationalization. Instead, he might have spent many years believing in the professed goals of the revolution, until a bout of prison terms for his participation in the extensive black market of the 1970s promoted his disaffection. Mariano Medina was a Black Cuban and former army officer that fought in Angola. He spoke of the distance that separated him from the earlier exiles:

> I can now see that they feel no ill will toward me and may even want to help me, but they can't help me come to grips with the twenty years I've spent in Cuba. They don't understand how I feel. (Llanes 1982, p. 170)

Despite the willing help of many in the Cuban community, many others exhibited a defensive prejudice against the newcomers, who might tarnish their reputation. The first and latest waves of Cuban refugees in the United States live side by side but remain aloof from one another. For them, as Kunz (1973, p. 137) pointed out, the date of departure from Cuba "signifies the bona fide" of their "political credo." Thus, they tend to blame each other for having left too soon or stayed too late. And the Cuba they long for is not quite the same Cuba.

Oscar Handlin ([1951] 1973) wrote of the immigrants from Europe at the turn of the century, those who came to fashion America. He caught the sadness, despair, and nostalgia of every person that has been uprooted:

> Yesterday, by its distance, acquires a happy glow. The peasants look back . . . and their fancy rejoices in the better days that have passed, when they were on the land and the land was fertile, and they were young and strong, and virtues were fresh. . . . Alas, those days are gone, that they believed existed, and now there is only the bitter present. (Handlin [1951] 1973, p. 98)

All Cuban immigrants in America for many years missed Cuba. But one night in Key West, while speaking with four refugees from Mariel, the difference struck me. While fishing, they listened on the radio to a baseball game being played right then in their hometown in Cuba. The early refugees' nostalgia attached them to the Cuba they knew—*la Cuba de ayer*, before the revolution. The Mariel refugees' longing was for *la Cuba de hoy*, of the revolution.

In the last few years, Cuba's refugees from the fourth major wave of the exodus have fled Cuba on make-shift *balsas*. Here *balseros* signal to a rescue plane, in December 1993. Of their initial crew, five survived six days of cold, December seas, but one died along the way. (Photo by Hector Gabino. *The Miami Herald*)

The Fourth Wave: Cuba's Balseros

The Mariel exodus proved so traumatic, both for the United States and Cuba, that immediately thereafter the doors to further migration closed. However, in the mid-1980s both governments signed a new Migration Agreement that provided for the immigration to the United States of up to 20,000 Cubans and up to 3,000 political prisoners a year, as well as for the deportation of excludable *Marielitos* back to Cuba. However, in actual practice only around 2,000 visas were being given a year.

Cuba's economic crisis reached new depths when communism collapsed in eastern Europe, particularly in the Soviet Union, on whom Cuba had been enormously dependent for trade and economic subsidies. The impact of these losses has been devastating: a decline in the national product of one-half, and in investment of two-thirds from 1989 to 1993 (Mesa-

Lago 1994). As a result, Cuban industry has been paralyzed, public transport hardly operates, the sugar harvest was abysmally low, and electricity has become sporadic, with Havana suffering blackouts during which people rely on candles, if candles can be found. As Mesa-Lago (1994) explained, "because of the eroding value of the peso, health care, education, pensions and other free services—which used to be the pride of the revolution—are rapidly deteriorating." The economic crisis is so severe that in the fall of 1990 Castro himself declared it "a special period in a time of peace." Such a *período especial* was to have been temporary, but coupled with the United States' tightening of the embargo (the Torricelli Law) in 1992, the end of the crisis is not in sight.

Cuba has attempted to forestall the worst of it by opening the economy to trade, investment, and tourism, and by reintroducing the use of U.S. dollars, measures which, Castro himself emphasized, were not intended to reintroduce capitalism but to "save socialism." But these policies have not succeeded in turning the economy around. Abject need and hunger have now become the reality of Cubans' lives during this "special period."

At the same time, the dissident movement has grown and developed into a social movement. Despite different political thrusts (for example, democratic socialist, environmentalist, Christian democrat), all dissidents in Cuba have increasingly called for a new democratic opening, a liberalization of the political structures that Castro adamantly refuses (see Hidalgo 1994). Hence, both economic and political want now drive the new Cuban emigration, together with the ever-present desire for the reunification of families still rent apart.

The new Cuban exodus has taken several forms (cf. Rodríguez-Chavez 1993), with illegal emigration being the major one. Cubans have become so desperate that they leave on *balsas*—rafts, tires, or other makeshift vessels—risking death due to starvation, dehydration, drowning, or sharks. The *balseros*, as they are called, now risk the arduous crossing so regularly that since 1991 *Los Hermanos dal Rescate* (the Rescue Brothers) constantly patrol the sea in helicopters searching for them. According to the U.S. Coast Guard, 5,791 *balseros* managed to reach safety in the United States from 1985 to 1992. As economic conditions have worsened in Cuba, the numbers have risen dramatically. While in the year 1989 less than 500 *balseros* arrived, by 1991 the number had risen to over 2,000 and by 1993 to 3,656. In 1994, due to the crisis in August and September, over 37,000 Cubans were rescued at sea. Their gratitude to their rescuers knows no limits. Early one morning in August, Lizbet Martínez, a twelve-year-old girl with a long, blonde ponytail, climbed aboard a raft with her parents. She took aboard her most prized possession—her violin. When the U.S. Coast Guard rescued them, she played "The Star-Spangled Banner" on her violin for them—a plaintive melody of gratitude for those who had saved her life (Balmaseda 1994a).

When the *balseros* arrive in Florida, they are welcomed—briefly—as heroes. But so many have died tragically at sea. Claudia Pérez was fifteen months old when she died in her mother's arms—one hour before she was rescued. Raísa Santana died because she drank seawater, reserving the only drinking water left in their vessel for her son (*El Nuevo Her-*ald, 26 December 1993, p. 1). "Those who despair" constitute this last wave of migration.

August 1994 comprised yet another historic turn in Cuba. On 5 August, massive riots took place in the streets of the center of Havana, in which thousands of Cubans participated. Their behavior expressed the enormous material want that shapes their lives, their disdain for the privilege reserved for party members and foreigners, and the enormous wish for civil liberties that now permeates Cuban society (Rivas-Porta 1994).

Shortly thereafter, Castro gave orders to the Cuban Coast Guard not to discourage emigration from Cuba's shores. Immediately, thousands of *balseros* put out to sea in the hopes of reaching Miami. But an abrupt policy change made the Cubans unwelcome. Under orders from President Bill Clinton and Attorney General Janet Reno, the U.S. Coast Guard blocked their progress and directed them to Guantánamo Bay Naval Station where over 30,000 people lived in tents for the nine months during which they were allowed entry to the United States.

As a result of the crisis, a new migration agreement was signed in September 1994 that promised that the United States will now give at least 20,000 visas a year for Cubans to immigrate to the United States. However, in May 1995 another abrupt policy change allowed the refugees in Guantanamo to come to the United States at the same time that the United States signed another migration agreement with Cuba that stipulated that all *balseros* found at sea will be returned to Cuba. "Cuba bleeds," headlined Liz Balmaseda (1994b), "and the drops are called rafts." But the U.S. government now denies their claim to being refugees.

Cubans in the United States

Race

Though Cuba has always been a multiracial society, despite their differences, prior to Mariel both major waves of Cuban immigrants were predominantly White. Yet, while throughout the decade of the 1960s the occupational distribution of Cuban refugees became more representative of Cuban society, "paradoxically," said Benigno Aguirre (1976, p. 105), Cuban Blacks "participated less in it."

Table 2 Number of Cubans in the United States, by Race, and by Year of Immigration, 1990

Year of Immigration	Race				Total	
	White	Black	Other Race	Asian	N	Percent
1987–1990	84.0	2.6	13.0	0.4	33,838	100.0
1985–1986	85.8	3.3	10.9	—	16,963	100.0
1982–1984	76.8	5.6	17.0	0.6	23,163	100.0
1980–1981	77.3	6.0	16.3	0.4	125,313	100.0
1975–1979	74.8	5.6	18.5	1.1	33,256	100.0
1970–1974	84.1	1.7	13.9	0.3	109,731	100.0
1965–1969	82.4	1.5	15.6	0.5	173,288	100.0
1960–1964	90.7	1.9	7.3	0.1	174,275	100.0
Total	83.5	2.9	13.3	0.3	689,825	100.0

Source: 1990 U.S. Census, Public Use Microdata Sample, 5 percent, weighted.

The 1953 Cuban census put the proportion of Blacks at 27 percent. In Cuba, like much of the Caribbean, social class and race overlap in the extreme. But while the social class level of the Cuban migration dropped, for fifteen years the immigrants remained overwhelmingly White. Data from the 1990 census shows that 86.3 percent of the immigrants were White Cubans, 0.2 percent were Black Cubans, 0.3 percent were Asians (no doubt *Chinos Cubanos*), and 13.2 percent designated themselves as belonging to "other race."[6] In Cuba, as in the rest of the Caribbean, this usually corresponds to Mulattoes, or *Mulaticos*, as they are affectionately called. Charles Wagley (1968) described the social definition of the races in the Americas. In the South of the United States a dual racial classification was used—Black versus White—that was based on ancestry ("one drop of Black blood"). By contrast, throughout the Caribbean the social definition of race was based on phenotype buttressed by social status—"money bleaches," the Brazilians say. Moreover, three different racial categories were recognized—Black, White, and those who are mixed, variously referred to as *Mulatos* (Cuba), *Pardos* (Brazil), and *Trigueños* (Puerto Rico).

The differential migration of the Cuban races up to this time was quite explainable. Two different social processes, Aguirre (1976) concluded, were at work. At the outset, the revolution pulled the power out from under the upper classes that had deliberately excluded Blacks from their midst. The immigration proceeded through the chain of extended family and friends, further selecting Whites. In addition, the migration policy of the United States and Cuba contributed to Blacks being excluded as they gave priority to close relatives of Cubans already in the United States.

Moreover, Blacks in Cuba did benefit from the revolution. Cuba never had a "separate but equal" system of legal segregation; and Cuban culture was a "creolization" of White Spanish and Black African cultural traditions. Yet prerevolutionary Cuba excluded Blacks from the pinnacles of society: yacht and country clubs, the best vacation resorts and beaches, hotels, and private schools reserved for the elite.

One of the first acts of the revolution was to make these exclusive facilities public, available to all, regardless of color or wealth. In addition, the Cuban government promoted new opportunities for Blacks in employment and education. Richard Fagen et al. (1968, p. 20) noted that the race problem in Cuba was "a boon to Castro." The revolutionaries found it extremely useful for discrediting the old social order. With the "instant liberation" of Blacks "tens of thousands of disadvantaged Cubans were recruited into the ranks of revolutionary enthusiasts."

Indeed, as Table 2 shows, about 91 percent of the refugees who came over in the first wave, Cuba's elite, were White. But the proportion of Whites declined quite markedly during the second wave. From 14 to 19 percent of those who immigrated from 1965 to 1979 considered themselves as "other." The *Marielitos* had the lowest proportion White of any

Table 3   Number of Cubans in the United States, by Poverty Status, and by Year of Immigration, 1990

Year of Immigration	Poverty Status		Total	
	Above Poverty Line	Below Poverty Line	N	Percent
1987–90	61.3	38.7	33,838	100.0
1985–86	79.4	20.6	16,963	100.0
1982–84	83.8	16.2	23,163	100.0
1980–81	72.1	27.9	125,313	100.0
1975–79	71.3	28.7	33,256	100.0
1970–74	87.7	12.3	109,731	100.0
1965–69	88.0	12.0	173,288	100.0
1960–64	91.5	8.5	174,275	100.0
Total	83.5	16.5	689,825	100.0

Source: 1990 U.S. Census, Public Use Microdata Sample, 5 percent, weighted.

wave—77 percent—while 16 percent considered themselves "other" (most likely Mulattoes) and 6 percent considered themselves Black. By American standards, fully 22 percent were non-White.

Given the Cuban revolution's appeal to race, why such a large presence in recent years? As early as the 1970s, Geoffrey Fox (1971, p. 21) remarked that "almost all those emigrating today are among the poorer classes in Cuba, the very people in whose name the revolution was made," Blacks included. To study "the defections of the sans-culottes," Fox interviewed a few working-class émigrés in Chicago and concluded that both for White and Black workers the salience of race in the revolution created strain—Whites complained of favoritism, Blacks of tokenism. Moreover, although discrimination was eliminated, racial prejudice persisted in Cuba, attitudes which Cuban Blacks might have sensed as real, denying the changes effected. As Max Weber ([1922] 1946) pointed out, whatever their origins, ideas, once established, take on a life of their own and guide action.

Whatever role their race may have played in the decision to emigrate, Black Cubans find their steps uncertain in America. As Blacks, they are not fully accepted by Whites; while among Blacks, they are Cubans (cf. Dixon 1988).

Poverty

Looking at other indexes of structural assimilation (cf. Gordon 1964), as one might expect, these racial differences have consequences in America. The

1990 census puts the proportion of all Cubans in the United States whose incomes are below the officially defined poverty line as 16.5 percent—a figure that compares quite favorably to that of the total U.S. population. But when we consider Cubans by their race, we can see that while only 14 percent of White Cubans fall below the poverty line, 35 percent of Black Cubans and 23 percent of racially mixed Cubans fall below the poverty line—figures that compare most closely to the poverty rates among Black Americans and Puerto Ricans in the United States. Representations of Cubans as a homogeneous group by social and demographic profiles of the total Cuban American population mask these dramatic differences. Table 3 shows the proportion of Cubans who were poor in 1990. Among Cuba's elite who had immigrated during the first wave, and their children, it is rather negligible: about 8 percent.

By and large, this first wave of Cuban migration brought enormous social resources with them—resources of social class, race, education, training, values, and expertise consonant with those of an industrial capitalist society, and the intimate knowledge of American society many had as its closest neighbors to the south. These social resources were "translated," made valuable in America by the warm welcome they received when they came to the United States—the multifaceted Cuban Refugee Program of assistance (see Pedraza-Bailey 1985). Moreover, at the time of their arrival in the 1960s and early 1970s, substantial economic opportunities existed in the United States, particularly in Miami. Together, all of these opportunities allowed the development of a Cuban ethnic enclave in Miami (see Portes and Bach 1985; Rieff 1993; Portes and Stepick 1993) and a Cuban "success story" much loved by the American media.

By contrast, the poverty rates are much higher in the other waves of Cuban migration, from around 12 percent of those who immigrated during the air bridge (1965 to 1974) to 28 percent of the *Marielitos*. These immigrants brought with them far fewer social resources—of social class, race, education, and prior knowledge of American institutions and culture. They also arrived when the social context was not propitious—after the assistance provided by the Cuban Refugee Program had ended and when the United States was in a serious recession. Among the *balseros* who arrived most recently, 1987 to 1990, the proportion of poor rises to a dramatic 39 percent. It is unlikely that this is solely due to the recency of their arrival. Time, of course, will tell.

Gender

That immigration has a decided impact on the labor force participation of women is a recurrent finding of immigration research (Pedraza 1991). It is also a central issue in studies of Cuban immigrants (Pérez 1988; 1986). In contrast to the extremely low rates of labor force participation of women in Cuba prior to the revolution, when most women did not work outside the home, Cuban women who immigrated to the United States after the revolution have extremely high rates of labor force participation. Yolanda Prieto's (1987) study of Cuban women in New Jersey, a more working-class community than Miami, concluded that the major determinant of the massive entrance of these women into the labor force was their social class: these women were middle class either in their origin or, if working class, in their aspirations. Achieving the upward mobility of the Cuban family in the United States made women's work necessary and broke with the traditional Cuban notion that a woman's place is in the home, justifying the massive entrance of women into the labor force.

Indeed, Cuban women overwhelmingly saw work as an opportunity to help the family, rather than as an opportunity for self-actualization. Thus, Myra Max Ferree (1979) wrote that Cuban women were an example of employment without liberation. Cubans had apparently stretched the traditional view of women

existing for the family to include employment as part of that role, while implying no necessary change in values. However, Lisandro Pérez (1988) argued that generational differences should be taken into account. The first generation reared in the traditional culture might well view employment as instrumental, but the second generation, more American, might hold a different set of attitudes. Pérez (1986b) also showed that Cubans' relatively high family income is partly due to the high proportion of dual-income families, underscoring the central role of women's work in the Cuban "success story."

Occupations and Education

Among Cuban immigrants, educational and occupational attainment also vary by waves. Data from the 1990 census show that fully 25 percent of Cubans who immigrated during the first wave had graduated from college, while only 7 percent of the *Marielitos* had the same level of education.

Table 4 shows their occupations. Over one-third of the immigrants who came over in the first wave, both men and women, work as managers and professionals, and another very sizeable proportion work in white-collar jobs in sales, technical, and administrative support. Among the *Marielitos*, the most common occupations were operators, fabricators, and laborers. Next come precision, production, craft, and repair workers for the men (nearly one-quarter); and sales, technical, and administrative support for the women (over one-third).

The occupational insertion of the immigrants from the second wave also shows the uniqueness of the first. Among those who immigrated on the heels of the revolutionary transformation of Cuba, over one-third of the immigrants today—as well as then—hold managerial and professional occupations, while among those who immigrated during the period of the air bridge, the proportion of professionals and managers is half that.

Political Refugees or Economic Immigrants?

Differences abound, yet the questions the recent refugees posed were the same that for over thirty-five years have framed the debate over the meaning of the Cuban migration. Interpretations of the mean-

Table 4 Number of Cubans in the United States, by Occupational Attainment, Year of Immigration, and by Sex, 1990

Year of Immigration	Managerial, Professional	Technical, Sales, Adminis- trative Support	Services	Precision Production, Craft, Repair	Operators, Fabricators, Laborers	Farming, Forestry, Fisheries	N	Percent
Men								
1982–1990	10.8	16.2	19.2	20.2	29.5	4.1	23,688	100.0
1980–1981	8.9	18.7	15.6	23.0	30.2	3.6	57,206	100.0
1965–1979	19.3	25.1	11.7	21.2	21.1	1.6	109,119	100.0
1960–1964	37.8	27.8	7.3	13.7	12.3	1.1	58,100	100.0
Total	20.4	23.4	12.3	19.8	21.9	2.2	248,113	100.0
Women								
1982–1990	10.0	32.6	23.8	6.9	26.7	—	19,056	100.0
1980–1981	9.6	36.1	25.5	5.3	23.2	0.3	27,413	100.0
1965–1979	17.2	42.8	16.5	4.1	19.2	0.2	102,425	100.0
1960–1964	33.9	41.6	11.6	3.5	9.2	0.3	57,556	100.0
Total	20.2	40.7	17.0	4.3	17.6	0.2	206,450	100.0

Source: 1990 U.S. Census, Public Use Microdata Sample, 5 percent, weighted.

ing of the exodus once again polarized into two positions: at one pole, the immigrants were said to be a manifestation of the loss of legitimacy of the Cuban revolution, discrediting it; at the other pole, the immigrants were said to be propelled by the scarcity of consumer goods, merely embarrassing it (see Fernández 1982). Hence, at one pole the immigrants were seen as political refugees; at the other, as economic immigrants. Over 1 million persons for over one-third of a century: Are they political or economic immigrants?

As Figure 1 makes clear, two different axes determine the definition. First, the motivation of the immigrants serves to define them as political or economic immigrants—a sociological distinction. Since all societies are simultaneously and inextricably political and economic, in our perceptions, political and economic conditions are entangled. In a society in transition, political disaffection easily results when government policies to change the basic economic allocation dislocate people: they lose their economic, social, and ideological "place." Even in a stable society, lack of economic opportunities easily results in lack of trust for public leaders. In this sense, Cuba's refugees are, and have always been, both political

and economic. But when people grow politically disaffected, when they lose faith and trust in their government and its cause, they can no longer be disposed of as simply economic immigrants. Cuba's refugees are, and have always been, fundamentally political.

Second, the governments that regulate their exit and arrival define immigrants as political or economic immigrants—a legal distinction (see Pedraza-Bailey 1985). At one end, in the United States, Haitian refugees consistently encountered a hostile reception: the refusal to grant them amnesty, their interdiction at sea, and their deportation back to Haiti despite the political violence that reigned there. Haitians who were refused asylum can be seen as undocumented refugees. Until now, Cubans had always been recognized and welcomed as refugees. But during this last crisis, for the first time they were defined as aliens attempting to enter the United States by illegal means. With the end of the Cold War and under the resurgence of nativistic attitudes across the land, Cubans have now become undocumented refugees.

At the other end, in Cuba, over the course of time, all who left were labeled traitors and counterrevolutionaries, whether they were supporters of the *an-*

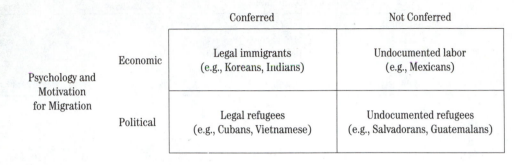

Figure 1
Typology of Migration

cien régime of Batista or, like Húber Matos, had fought against Batista in the hills of the Sierra Maestra, side by side with Fidel Castro for a nationalist and social-democratic revolution they felt Castro's espousal of communism had betrayed.

A society where the only choice possible is to "love it or leave it" provides too few choices. A truly democratic society is defined not only by its party structure, constitution, delegation of authority, and electoral representation, but principally by its capacity to tolerate and incorporate dissent. The Cuban exodus, now over one-third of a century old, has been driven not only by the trauma of revolutionary change in Cuba, and by the economic hardships caused both by the inefficiencies of the new economic system and by the isolation of the trade embargo, but also by Cuba's incapacity to tolerate dissent. The Cuban revolution's only solution to dissent has been to externalize it. Cuba has yet to provide political channels to express and incorporate the other dissenting voices.

Conclusion

Over the time span of the exodus, over one-third of a century, Cubans in the United States have been undergoing a profound attitudinal transition: from refugees to immigrants to ethnics. The first part of this transition in attitudes, from refugees to immigrants, has, under the impact of constant new immigration and the centrality of Cuba in the lives of the immigrants, been slow to take place (cf. Rieff 1993). But to the degree that Cubans have ceased to look back (like Lot's wife) to Cuba as the only source for the meaning of their existence and identity, to the degree that they have started to look forward and to carve their future in this country as Cuban Americans, to that degree it has been achieved (cf. Portes 1984).

The second part of this transition in attitudes, from immigrants to ethnics, is more inexorable because it corresponds to the demographic transition presently under way. Although quite young still, a second generation is now in our midst that was raised under American institutions and socialized in American schools, the great transmitters of tradition, culture, and values. Although these young Cubans' assimilation may have been delayed by their growing up in the Cuban enclave in Miami, like any people, they are the soil in which they rooted and grew. Moreover, they lack the felt sense of a Cuba they did not know.

In "The Agony of Exile," Rubén G. Rumbaut (1991a) underscored that the meaning of exile is different across generations. To the parents' generation, who made the decision to leave, exile represents a profound loss and a profound commitment. And it entails a worldview that will be defended. By contrast, to their childrens' generation, exile is an inherited circumstance. Typically, they are in solidarity with the family's predicament, but do not need to protect their parents' worldview. Their focus is on the future in the new society.

In between these two lies the "1.5 generation"— those who left at the dawn of their adolescence and, like Robert Park's (1928) "marginal man," are forever caught between two worlds, the land of their birth and the land that tended them. As one of them, it is my hope that the second generation that has now rooted in the United States will not be so American that they will lose touch with their history and culture, with their *Cubanía*. But as a sociologist I have to recognize that such may well be the price to be paid for shedding the pain of exile.

Notes

1. *The wine is from plaintain; and if it proves sour, it is our wine!* — *José Martí, "Our America" (1891).* Translated by Juan de Onís (1954).

2. Data for this article is from the 1990 U.S. Census, 5 percent sample, Public Use Microdata Sample, weighted. The author wishes to gratefully acknowledge the assistance of Lisa J. Neidert, University of Michigan, Population Studies Center, and Rubén G. Rumbaut, Michigan State University, Department of Sociology, in helping her obtain the 1990 census data for this article.

3. Close to 20 percent of Cuban immigrants arrived after the 1980 census.

4. Based on the collected life stories of hundreds of Cubans, José Llanes (1982) drew fifty-eight composite characters whose human dramas were representative of immigrants from the various waves. Names are fictitious.

5. The U.S. census precoded the variable on year of immigration by intervals: before 1950, 1950–1959, 1960–1964, 1965–1969, 1970–1974, 1975–1979, 1980–1981, 1982–1984, 1985–1986, 1987–1990, and born in the United States. Hence, while technically the Cuban exodus that belongs to the revolution should begin with the year 1959, when the revolution triumphed, it is not possible to disaggregate the data in that way. Fortunately, this precoding does not do too much violence to the data because, as Fagen, Brody, and O'Leary (1968) pointed out, it was really the nationalization of American industries in October 1960 that turned the emigration from a trickle into a flood. Likewise, the migration from Mariel is pretty well approximated by the years 1980–1981, although technically it took place in 1980.

6. In recent U.S. censuses, data on race, ethnicity, and ancestry are the result of self-identification by those answering the census questions.

21

Dominicans in the United States: First- and Second-Generation Settlement, 1960–1990

SHERRI GRASMUCK
PATRICIA PESSAR

Dominicans are one of the fastest growing of the new immigrant groups in America. The following story of one Dominican family illustrates a number of important features of the contemporary wave of Dominican immigration into the United States, namely that migration is a multifaceted process involving economic, political, and sociocultural factors. The Molina migration history begins with Rafael Molina. As a university student in Santo Domingo and committed to improving the lives of children around him, Rafael studied education. When Dominican president Juan Bosch was overthrown in a military coup in 1963, Rafael, as student leader at the Autonomous University in Santo Domingo, joined others in the street fighting that followed. During the evolving political crisis, which resulted in the defeat of Bosch supporters and U.S. occupation of the island, Rafael feared for his life and believed his political sympathies would interfere with his aspirations to become a school administrator. Despite his political hostility toward the United States, in the summer of 1965 Rafael nonetheless applied for, and easily obtained, a temporary resident visa for the United States.

In 1966 Rafael returned home briefly to marry Mercedes, his high school sweetheart, who was working as an elementary school teacher. The couple was reunited in New York the following year. By 1970 they had two young children, and Mercedes was anxious to begin work as a sewing machine operator so that they might save for the children's education. Back in the Dominican Republic, Rafael's father had died, leaving his mother a widow. It was decided that Rafael would sponsor Gertrudes' emigration so that she might care for his two children and free Mercedes to work outside the home. Rafael and Mercedes had agreed to send remittances to Gertrudes' sister, who was caring for Gertrudes' two youngest sons. By 1980, the middle son, Tomas, had graduated from the university in Santo Domingo, only to find that jobs in communications were few and starting salaries low. Rafael convinced a Colombian clothing manufacturer to hire his brother as an accountant, and Rafael, who was now a U.S. citizen, was able to sponsor Tomas's emigration. Given the negative educational experience of Tomas, Gertrudes concluded that it made no sense for her last son, Carlito, to complete school in the Dominican Republic and she sponsored his migration to the United States the following year so that he could enroll in a U.S. high school and eventually a university.

The range of motivations behind Molina family members' decisions to leave Santo Domingo for New York between 1965 and 1981 include fear of political persecution, desire for marital reunification, greater economic opportunity, provision of child care for relatives, and opportunities for higher education. The Molinas's story illustrates the range of influences behind the massive displacement of Dominicans from their island. First, the timing of the first migrant's departure, 1965, coincides with a period of political turmoil after a thirty-year dictatorship in the Dominican Republic and the appointment of a United States consulate eager to facilitate out-migration in the face of rising political tensions. Second, none of the migrants were unskilled agricultural laborers, but rather were, by Dominican standards, relatively educated urbanites. Third, it is possible to make sense of the movement of these five individuals only by reference to the meaning of family ties and the significance of the social networks available to the family. Like the Molina family, the bulk of the Dominican community living in the United States arrived during the past two and one-half decades. Prior to the early 1960s, emigration from the Dominican Republic was severely curtailed owing to the Trujillo dictatorship (1930 to 1961) which limited even the issuance of passports to Dominican citizens (Crasweller 1966; Frank Canelo 1982). After the assassination of Trujillo in 1961, Dominican migration to the United States rose sharply. The defeat of the popular revolt of 1965 and the United States military occupation of the Dominican Republic during the Johnson presidency contributed to a certain amount of politically motivated migration, such as the case of Rafael, and early on established New York as the preferred destination of Dominicans. For the most part, however, Dominican migration to the United States is motivated by a desire to advance economically, as was the case for most of the other members of Rafael's family, and is often accompanied by a keen sense of cultural longing for what has been left behind.

The Magnitude and Concentration of the Dominican Community

By 1990 Dominicans had become the fourth largest of the Latin American and Caribbean immigrant communities living in the United States, behind only Mexico, Cuba, and El Salvador, and fourteenth among all immigrant groups in the United States (U.S. Bureau of the Census, 1990a). The total number of Dominicans legally admitted to the United States between 1961 and 1989 was 451,324. In addition to the immigrants, approximately one-third of the Dominican community consists of second-generation Dominicans born in the United States. In the 1990 U.S. census, 520,151 persons identified themselves as having Dominican ancestry, of whom 347,858 were born in the Dominican Republic.[1] The number of Dominicans in the United States has grown substantially over the last decade, with approximately one-half of Dominican immigrants having arrived in the last decade. The Dominican population in the United States is youthful: the median age being 27 years. The 1990 U.S. census also shows an apparently female-dominant population—53 percent.[2]

Beyond legal immigration a substantial number of Dominicans have entered the United States surreptitiously. The most typical manner of entering illegally is to overstay a tourist visa or to enter with false documentation. While the extent of illegal entry is not actually known, the 1980 census revealed only approximately 11 percent of the Dominican ancestry population to be illegal residents.[3]

The impact of Dominicans in the United States is even greater than is suggested by its volume, given the concentration of Dominicans in the New York/New Jersey area. In 1990, 91 percent of Dominicans in the continental United States resided in only four states: New York (68.9 percent), New Jersey (11.1 percent), Florida (6.4 percent), and Massachusetts (5.1 percent) (U.S. Bureau of the Census 1990).

New York City is the undisputed capital of Dominicans in the United States. There Dominicans live in relatively concentrated neighborhoods, in Washington Heights, the Lower East Side, and in Queens, and there they have made their most distinctive social and cultural impact. A majority of the Dominican immigrants living in New York City in 1990 had arrived in the United States during the decade of the 1980s.[4] Beginning in the early 1980s Dominicans assumed greater visibility as a distinct Spanish-speaking community in New York with their own literary magazines, a cultural center *(Casa Dominicana)*, and the formation of the Dominican Studies Institute at City College (CUNY). Each year in August thousands of Dominicans gather to celebrate a Dominican Day Parade, where even politicians from the Dominican Republic are eager to be seen. The music and dance traditions of mergengue witnessed a growing audience in New York beyond the Dominican community, furthered most recently by the worldwide acclaim of the Dominican musical group,"4.40."

Social Characteristics of the Migrants

Social Origins

The story of the Molina family told above illustrates that although unemployment and poverty are widespread in the Dominican Republic, the bulk of Dominican out-migration has not drawn predominantly from the poorest and most marginalized sectors of Dominican society. Dominican migrants, especially those who left in the late 1960s and 1970s, tend to be overwhelmingly urban in origin and come from the relatively youthful, skilled, and educated sectors of Dominican society (Grasmuck and Pessar 1991). After the first several decades of Dominican out-migration, however, economic conditions severely deteriorated in the Dominican Republic with rampant inflation, declining real wages, and a continued weakening of the Dominican peso (Inter-American Development Bank 1992, p. 81). The steady decline in Dominican real wages throughout the 1980s made a grim contrast to the rising relative value of the dollar.[5] The "emigration option" became increasingly attractive to a broader sector of the Dominican population in the 1980s, cutting deeper into both the less skilled of the urban sectors and the more middle-class professionals. The fact that highly skilled Dominicans are emigrating in larger numbers is reflected in the fact that 11.3 percent of those who emigrated between 1985 and 1990 were college graduates, as compared to an average of 5.1 percent for the two prior decades. Thus, among Dominican immigrants we can detect a trend toward internal diversification typical of other immigrant streams, with growing numbers of both upper and lower strata immigrants joining the Dominican community. However, despite the fact that Dominicans are not from the poorest or most unskilled sectors of their home society, they have relatively low levels of education

compared to the overall U.S. population: for example, while 75.2 percent of the total U.S. population over the age of twenty-five has completed high school or more, only 41.5 percent of similarly aged Dominicans have.[6] These relatively low levels of education translate into low levels of occupational attainment in the United States.

Settlement: New York City, Jobs, and Family Structure

As the Molina family's case history also shows, the majority of Dominicans come to the United States with the intention of finding employment. Typical of the history of other new immigrant groups, Dominicans have moved into the lower occupations of the U.S. labor market and they rank among the lowest paid workers (Gurak and Falcon-Rodríguez 1987). The predominant occupations for Dominicans lie in the manufacturing, retail sales, and service industries (Grasmuck and Pessar 1991). Women tend to work more in the manufacturing sectors, especially in manual jobs in the garment industry in New York City. Men work in both manufacturing and service, especially in hotels, restaurants, and hospitals. While most immigrants' incomes lag behind the average income of the total U.S. population, Dominican income ranks at the bottom of even other immigrant groups. While the national median household income in the United States was $30,056 in 1990, the comparable Dominican figure was only $10,130 (U.S. Bureau of the Census 1990a). There are several factors important in understanding the economic situation of Dominicans: the economic climate of their place of settlement, New York City; their relatively low levels of education, as already mentioned; the high rate of single-parent households; and racial discrimination. That Dominicans are concentrated in New York City is very relevant to their economic situation. During much of the 1970s and 1980s the New York City economy underwent significant restructuring that resulted in a relative increase in the supply of low-wage jobs and a decrease in high-wage jobs. This increase in low-wage jobs has benefited the first generation of Dominican immigrants by making it possible to find employment. However, given the increased structural constraints of relatively few higher paying jobs in the region and overcrowded schools and apartments in the upper Manhattan community,[7] second-generation Dominicans may find their social mobility more circumscribed than was the case for earlier waves of immigrants. The fact that more than 40 percent of Dominican children live in households with incomes below the poverty line is already cause for alarm.[8] Indeed, recent census data reveal that being born in New York puts Dominicans at significantly greater risk than being an immigrant: the poverty rate for those born in the Dominican Republic is 30.4 percent, compared to 44.0 percent for those born in New York and 22.0 percent for those born in Florida (U.S. Bureau of the Census 1990a).

The declining New York City economy, however, is only part of the explanation for the relatively disadvantaged position of many Dominicans. The 1990 census, for example, reveals that Dominicans are lagging behind other immigrant groups also residing in New York City with respect to their participation in higher-level occupations. Table 1 compares Dominicans with the other four largest Caribbean immigrants arriving in New York City after 1965. Whereas approximately 50 percent of all Dominicans in the labor force work as either operatives or laborers or in personal service occupations, only 37 percent of the Jamaicans, 47 percent of the Colombians, 45 percent of the Haitians, and 40.2 percent of the Cubans are so concentrated. Similarly, only 10 percent of Dominicans occupy the three top occupational categories of managers, professionals, and semi-professionals or technical workers compared to 21.4 percent of the Jamaicans, 12.5 percent of the Colombians, 17 percent of the Haitians, and 16.3 percent of the Cubans.

Part of the explanation for the relatively low occupational attainment of Dominicans lies with Dominicans' relatively lower levels of education compared to other Caribbean immigrants in their place of settlement; more than 60 percent of Dominicans over twenty-five have not completed high school, compared to 33.7 percent of Jamaicans, 41.3 percent of Colombians, 35 percent of Haitians, and 52.7 percent of Cubans also residing in New York City (Table 1). This disadvantage in the labor market is further exacerbated by their comparatively low levels of English language acquisition. Almost half of the Dominican immigrants in New York City reported that they spoke no or very little English, which puts them at a disadvantage relative to the other recent Caribbean immigrant groups also residing in New York City (Table 1). Typically, language disadvantages decrease as immigrants lengthen their stay in the host society. This is also true of Dominicans and is expressed by the lower proportion of non-English speakers among the immigrants with the most time in the United States (Table 2).

In contrast to what has been said about a "declining stock of U.S. immigrants" (Borjas 1990a), the proportion of college graduates among adult Dominicans over twenty-five years old has increased among those arriving between 1985 and 1990—11.3 percent compared to an average of 5.1 percent for the two prior decades (Table 2). English acquisition is likely to be more rapid for these more-educated recent immigrants. Even more dramatically, we see that 27.1 percent of Dominicans born in the United States over the age of twenty-five are college graduates. To some extent this group represents a unique group who entered the United States before 1965. Nonetheless, their high educational levels compared with national averages illustrate the considerable diversification within the Dominican community. These more-educated Dominicans are often the frustrated voices raised in protest of media portrayals of Dominicans as the "losers" among new immigrant groups without attention paid to the diversity that exists within the community.

Compounding the educational and regional disadvantages faced by Dominicans is evidence of considerable family disorganization. The 1990 census revealed high rates of marital disruption in Dominican households: almost 40 percent of Dominican households with children under the age of eighteen in New York City were female householder families, which is considerably more than other Caribbean immigrant groups in New York (Table 1) and well over the national average of 16.5 percent or the foreign-born average of 10.8 percent (U.S. Bureau of the Census 1990a). Mother-only families have had substantially higher poverty rates than other groups in the United States for the past twenty years (Garfinkle and McLanahan 1986). High poverty rates of the Dominican population are concentrated among the female householders; whereas married Dominican householders have poverty rates somewhat comparable to national averages, 19.1 percent to 13.1 percent, respectively, more than half, or 52.4 percent, of Dominican female householders are living below the poverty line. These households also have the lowest levels of education and English acquisition of all Dominican households.[9] Moreover, it appears that being born in the United States to a female-headed

Table 1   Occupational Distribution, Education, English Fluency, and Labor Force Participation of Selected Caribbean Immigrant Groups in New York City, 1990

	Dominicans (%)	Jamaicans (%)	Colombians (%)	Haitians (%)	Cubans (%)
Occupation					
Managers	4.7	7.7	5.6	4.6	7.8
Professionals	4.0	10.2	5.0	8.6	7.1
Semiprofessionals/technical	1.3	3.5	1.9	3.8	1.4
Clerical	11.7	21.2	13.3	16.2	16.5
Operatives/labor	26.9	7.2	19.5	12.4	16.6
Personal service	22.6	29.5	27.8	32.8	23.6
Less than high school education	61.3	33.7	41.3	35.0	52.7
No or little English	47.7	0	33.9	19.4	44.5
Homeowners	8.6	32.0	19.4	33.6	30.7
Labor force participation	62.6	80.0	75.7	78.6	68.1
Female householders	39.2	32.7	20.1	31.2	15.7
Identifying as					
White	24.3	0.8	53.2	1.1	60.4
Black	25.2	98.4	3.1	97.8	13.4
Other	50.1	0.2	42.5	0.0	25.7
Asian	0.3	0.6	0.8	1.0	0.5
Total in NYC	206,719	104,916	62,650	66,178	24,933

Source: Public Use Microdata Samples of the 1990 Census, Five Counties in the New York City area; occupational breakdowns calculated by John Mollenkopf, CUNY Data Service, with the remaining statistics calculated by Rubén Rumbaut.

household drastically increases the chances of poverty for Dominicans, even more so than immigrant status (Table 3); fully 63.6 percent of Dominican female householders born in the United States are living in poverty, compared to 48.8 percent of the Dominican immigrants who are female householders. Interestingly enough, Dominican family disorganization is predominantly a "New York story" revealed in the fact that 41.0 percent of Dominicans households in New York City (PMSA) are female-headed compared to only 22.0 percent in New Jersey or 15.3 percent of Dominican households in Florida (U.S. Bureau of the Census 1990a). This reinforces our earlier point regarding the extent to which the depressed New York economy has especially negatively affected the second-generation of Dominicans residing there. The obstacles these single-parent households will face in the future are likely to be substantial. These are especially worrisome trends because dual-income families, not single-headedness, have been associated with the "success stories" of other immigrant groups, such as the Cubans (Pedraza 1992, p. 244).

Table 2   College Graduates and English-Speaking Ability by Year of Entry for Adult Dominican Ancestry Population Over the Age of 25

	College Graduates (%)	English Ability	
Year of Entry		"Not well" (%)	"Not at all" (%)
Born in U.S.	27.1	7.8	2.0
1985–1990	11.3	29.2	34.2
1980–1984	5.1	32.5	23.5
1975–1979	6.5	25.6	25.2
1970–1974	4.2	26.8	14.4
1965–1969	4.5	27.8	8.8
Before 1965	9.4	27.8	13.2

Source: Public Use Microdata Samples of the 1990 U.S. Census; calculations by Rubén Rumbaut.

Table 3 Poverty Rates by Household Structure, Citizenship Status, and Race of Dominican Ancestry Group

	Married Householders	Female Householders	All Dominican Householders
Citizenship status			
Born in U.S.	22.5	63.6	41.5
Naturalized citizen	13.9	40.6	24.7
Not citizen	17.5	48.8	31.1
Self-identified race			
White	11.9	49.1	26.6
Black	36.8	54.1	42.3
Mulatto/other	16.0	53.2	47.0

Source: Public Use Microdata Samples of the 1990 U.S. Census; calculations by Rubén Rumbaut.

Immigrants of Color

Considerations of race discrimination are also important to a full understanding of the obstacles confronting the Dominican population in the United States. Despite the fact that traditional sectors of the Dominican Republic have been reluctant to acknowledge their strong African ancestry, Dominicans are a mixed population with a relatively low percentage describing themselves as White compared to other Spanish-speaking immigrant groups in New York City (Table 1). The vast majority identified themselves as either Mulatto, specified in the census as Other, (50.1 percent)[10] or Black (25.2 percent), which is considerably more than for either Colombians or Cubans, the next two largest Spanish-speaking immigrant groups in New York City. Skin color as it turns out is also a very significant predictor of poverty among Dominicans. Table 4 compares the poverty levels of White, Black, and Mulatto Dominicans according to household status. Black and Mulatto Dominicans have strikingly higher poverty levels than White Dominicans—42.3 percent and 47.0 percent compared to 26.6 percent, respectively. More notable, however, is the fact that even married Black householders have more than double the poverty rates of their lighter-skinned counterparts—36.8 percent compared to 16.0 percent for Mulatto and 11.9 percent for White, married Dominican household-

ers. The economic advantages of being married, which we noted earlier, is apparently almost completely wiped out for Black Dominicans. To further complicate the picture we wanted to include considerations of immigrant status to see whether a kind of "triple jeopardy" was operating such that being a Black, female householder, and a nonnaturalized immigrant would maximize one's chances of living in poverty. Table 4 reveals that this is not the case but does dramatically illustrate the extent to which second-generation Dominicans compared to Dominican immigrants are disadvantaged. The highest poverty rates for all three color groups are among those female householders born in the United States, all above 60 percent, with Whites surpassing Blacks and Mulattos—66.1 percent, 60.4 percent, and 63.9 percent, respectively.

Self-Employment

Self-employment, especially in firms that cater to immigrant clients, is often an important part of the immigrant settlement process in places where immigrants abound (Bonacich and Modell 1980; Light 1972; Portes and Bach 1985; Waldinger 1986). The overwhelming majority of Dominicans are employed as wage laborers. Nonetheless, an increasing number are using their savings and skills to begin their own businesses (Guarnizo 1992; Portes and Guarnizo 1991). The overall rate of Dominican self-employment in New York City (6.3 percent) is comparable to other foreign-born groups in New York City.[11] Dominican businesses tend to be small and their market is often ethnically-oriented. Businesses associated with Dominicans include *bodegas*—small, walk-in stores that stock food specialties of the Hispanic Caribbean—neighborhood restaurants, subcontracting garment shops whose labor force is overwhelmingly Hispanic, and nonmedallion or "gypsy" cabs that service immigrant neighborhoods (cf. Waldinger 1986; Portes and Guarnizo 1991). As is usually the case for immigrant entrepreneurs, Dominican businesspeople concentrate in the most risky and often most physically dangerous niches of the local economy. As the headline of a recent newspaper article on Dominicans' growing role in supermarket ownership in New York attests, "Dominicans Thrive Where Big Chains Won't Go" (Myerson 1992).

Table 4   Poverty Rates by Race, Citizenship Status, and Household Structure among Dominican Ancestry Population

	White			Black			Mulatto		
	Born U.S.	Naturalized Citizen	Not Citizen	Born U.S.	Naturalized Citizen	Not Citizen	Born U.S.	Naturalized Citizen	Not Citizen
Married householder	17.4	11.0	9.3	41.4	33.2	29.6	19.5	6.6	17.6
Female householder	66.1	39.0	40.5	60.4	44.0	52.4	63.9	39.8	51.3

Source: Public Use Microdata Samples of the 1990 U.S. Census; calculations by Rubén Rumbaut.

Such immigrant-owned businesses are likely to multiply as the Dominican community matures. There are at least two reasons for this projected increase. First, as the community grows in number, it is likely to generate an increasing demand for ethnic-oriented goods and services; and second, there is a clear pattern whereby Dominicans who are employed in a Dominican-owned business go on to start their own business. According to one recent survey of Dominican business owners, more than one-third had begun their careers as employees in other businesses owned by Dominicans (Guarnizo 1992).

In summary, then, relatively low levels of education and English-language acquisition combined with the racial discrimination immigrants of color experience in the context of the increasingly declining labor market conditions of New York City, combined with evidence of considerable family disorganization, have disadvantaged the Dominican community even when compared to other new immigrant groups in New York City. Despite these collective difficulties, however, there is evidence of a growing class diversification within the community reflected in growing rates of self-employment and relatively high levels of education for the most recent immigrants.

Gender

Consistent with the national pattern for Latin American immigrant women of occupational succession into the least desirable jobs (Tienda, Jensen, and Bach 1984), the vast majority of Dominican women are employed as operatives in New York's declining manufacturing sector, as noted above. Gender discrimination contributes to the constraints and burdens Dominican women face while employed, or seeking employment, in the United States. First, women are rewarded even less than men for the education and skills they bring to the U.S. marketplace.

For example, according to the 1990 census, a Dominican male with a high school education earned an average annual income of $13,437 as compared to $8,177 for a Dominican woman with an equivalent level of education (U.S. Bureau of the Census 1990a). Moreover, women who arrive with professional job experience are less likely than their male counterparts to find similar employment in the United States.[12] Across the board, Dominican women earn significantly lower incomes than men. This is illustrated by the 1990 census figures which show that Dominican women earned about 60 percent as much as Dominican men, or an average annual income of $9,298 compared to $15,472. In fact, being female exerts a more depressive pull on earnings than does undocumented legal status.[13]

Despite the low wages of women, Dominican immigrants have been heard to say, "The Dominican Republic is a country for men, the United States is a country for women." Captured in this refrain is the perception, and often the objective experience, of Dominican women having increased their personal autonomy and authority as a result of emigration. In many cases this change is attributable to the fact that women more often work for wages in the United States and contribute substantial amounts to the family budget. As with many working women, Dominican women often perceive their wages as their private property and use them to negotiate greater gender parity with husbands. Women's gains include often successfully challenging men's sole control over the status of household head, transforming the mode of budgetary control from one based on male authority and privilege to a more egalitarian mode, influencing men to assist women in certain child care and household duties, and, more generally, greater social independence (cf. Grasmuck and Pessar 1991). Indeed, it is both women's access to greater opportunities for employment in the United States and the gains such employment represents for improved domestic relations that cause many to oppose their husbands' plans and strategies for a relatively rapid return to the Dominican Republic.[14] Women who seek to protect their personal gains by remaining in the United States often adopt a budgetary strategy aimed at consuming goods and services which will at once reduce the family's savings and make life in the United States more secure and stable. In contrast, men may actively pursue a plan for return by trying to save household income and limiting their own and the household's expenditures on commodities that will root the family securely in the United States. As men say, "Five dollars wasted today means five more years of postponing a return to the Dominican Republic."[15]

It appears, however, that these "empowering options" do not come without a cost. If household renegotiations over decision making and control are not successful and instead result in increasing rates of marital instability and single-headedness among women, as apparent from the statistics cited above, then gains in autonomy for women are likely to be accompanied by stark increases in poverty. The success of migration ventures for families is often dependent on the preservation of household bonds of solidarity, albeit renegotiated ones. In contrast, when household bonds are severed, the migration project often falters, because the individual resources of single members, especially women, are insufficient to sustain it. Indeed, the fact that public welfare assistance constitutes 11.1 percent of Dominican women's total income[16] may reflect the fact that a significant proportion of Dominican women with children may have accepted periods of unemployment and state dependency rather than toil with no child care in low-paying, insecure jobs that net little more than welfare payments. There is a painful irony associated with the fact that the newfound autonomy claimed by some immigrant women may come at the cost of the collective mobility project of their households, and in many cases lead to poverty.

Deep Ties to the Homeland

First- and second-generation Dominicans living in the United States represent almost 14 percent of the population of the Dominican Republic. This figure illustrates the dramatic impact this out-migration has had economically and culturally on the home society (Georges 1990). Although Dominicans have spent more than twenty-five years consolidating an important and visible community in the New York City area, it is difficult to underestimate the extent to which strong ties bond Dominicans living in the United States with those in the Dominican Republic. When someone in the Dominican Republic says, "If I don't get a job soon I too will have to go over there," no one doubts where that "over there" is. One recent study revealed that 77 percent of the households in the capital of Santo Domingo claimed to have at least one relative living in the United States, and fully 27 percent received financial support at least once a month from a family member living abroad (Lozano 1992, p. 19).[17]

First-generation Dominicans living in the United States often experience a great deal of social pressure to prove themselves, perhaps to compensate for the sacrifices their families have made in order to send them abroad. A popular refrain in the Dominican Republic spoken to someone about to depart for work in the United States, is, "If things go well for you over there, write."[18] The clear implication is that if things go poorly, the migrant should not add this failure to the historical record, and perhaps, more to the point, should think twice before returning. In this way, the successes of emigrants become mythologized while their misfortunes are minimized and hence experienced as doubly painful for those who actually suffer them. A recent antidote to this romantized version of the migration experience held by many Dominicans is the recently released film, *One-Way Trip (Un Pasaje de Ida)* by Dominican filmmaker, Agliberto Meléndez, which is about the clandestine journey of forty Dominicans on a boat trip to New York. Referred to by some as a kind of Dominican *El Norte*, this first full-length, feature film by a Dominican was of such interest to the public in the Dominican Republic that a black market immediately developed to respond to the intense public interest in the film (Minero 1993, p. 25). Further evidence of the cultural difficulties confronting Dominicans in the United States is a recent study of return migrants in the Dominican Republic that revealed that in contrast to the predominance of economic motivations behind the decision to leave the island, sociocultural preferences were the major factors behind the decision to return home (Guarnizo 1993, p. 13).

In the contemporary period, immigrants from Asia and Latin America predominate. Here a Dominican immigrant tends to his *bodega* (small grocery shop) where goods, such as newspapers and music, cater to the tastes of the Latin population. Signs both in English and Spanish announce that smoking is not allowed and credit will not be extended. (Photo by Steve Gold)

That many Dominicans find themselves concentrated on Manhattan Island and yet maintain strong ties and identification with their island of origin, Hispaniola, creates a sense of living permanently "between two islands." In the following two reflections, Julia Alvarez, a Dominican writer who came to the United States with her parents as a young girl in the early 1960s captures in her seemingly autobiographical novel, *How the García Girls Lost Their Accents*, how this sense of island identification often interacts with cultural alienation from the same home society. In the first passage, Yolanda and her sisters in adapting to American lifestyles during adolescence strive to distance themselves from their associations with life on the island:

We learned to forge Mami's signature and went just about everywhere, to dance weekends and football weekends and snow sculpture weekends.

We could kiss and not get pregnant. We could smoke and no great aunt would smell us and croak. We began to develop a taste for the American teenage good life, and soon, Island was old hat, man. Island was the hair-and-nails crowd, chaperons and icky boys with all their macho strutting and unbuttoned shirts and hairy chests with gold chains and teensy gold crucifixes. By the end of a couple of years away from home, we had *more* than adjusted. (Alvarez 1992, p. 108)

Yet when Yolanda returns to the Dominican Republic after a period of five years in the United States, she permits herself to confront her cultural alienation:

All around her are the foothills, a dark enormous green, the sky more a brightness than a color. A breeze blows through the palms below, rustling their branches, so they whisper like voices. Here and there a braid of smoke rises up from a hill-side—a *campesino* and his family living out their solitary life. This is what she has been missing all these years without really knowing that she has been missing it. Standing here in the quiet, she believes she has never felt at home in the States, never. (Alvarez 1992, p. 12)[19]

The transnational nature of the Dominican community is exemplified in the fact that Dominicans living in the United States and returnees have become significant investors in certain key industries in the Dominican Republic. These include housing, finance, tourism, and export processing. For example, by 1984 *Dominicanos ausentes* (a local term for Dominican emigrants) accounted for 60 percent of the total amount of that year's formal housing sales, or D.R. $240 million (U.S. $85 million) (Guarnizo 1992, p. 328). Some of the Dominican immigrant entrepreneurs we discussed earlier have taken advantage of investment and financial incentives within the export processing sector recently offered by the Dominican government. Availing themselves of the even cheaper labor offered in the Dominican Republic, a few have even established offshoots of their own manufacturing operations in New York. Also reflective of the depth of continued ties between the Dominican Republic and the Dominican community in the United States is the fact that until recently the political concerns of the Dominican community in New York were oriented toward the Dominican Republic (Georges 1984). In the late 1970s, for example, the National Association of Dominicans Abroad (ANDE), one of the principal Dominican organizations, sought as its main goal the right to vote in Dominican national and local elections. All the major political parties of the Dominican Republic have offices in New York. This "transnationalism" is reflected as well in Dominicans' comparatively low rate of naturalization as compared to other new immigrants. Only 16,000, or 7.8 percent, of the 204,000 Dominicans who were admitted for permanent residence in the United States between 1960 and 1980 had become U.S. citizens by 1980. This rate increased to 17.7 percent according to the 1990 U.S. census. By contrast, immigrants from many Asian countries had naturalization rates well above 50 percent during this time period (Warren 1988, p. 4). The 1990 census data reveal that among Dominicans resident in New York City, 21.7 percent were citizens.[20]

Many in the first generation still aspire to return home one day. However, as the community matures and a second-generation of Dominicans assert their distinctive interests, a notable change in political orientation has occurred (Georges 1993; Torres-Saillant 1989). The Dominican national constitution was amended in 1994 to permit dual nationality, which presumably reflects the desires of many Dominican immigrants to maintain their political rights on both sides of the border. Dominicans have begun to assert themselves in the political life of New York, as evidenced by several Dominicans choosing to run for elected office. Community-based efforts focusing on the discrimination Dominicans face as Spanish speakers and people of color has also produced a new Dominican leadership.

Cultural and Ethnic Identity

Issues of cultural and ethnic identity for Dominicans are likely to be prominent for the Dominican community in the future. Most Dominicans strongly identify as Dominican, rather than as Hispanic. As Dominicans emerge as a distinct ethnic group and vie for government recognition and resources, they find themselves in an ambivalent position with respect to New York's largest Hispanic bloc, the Puerto Ricans. While many Dominicans recognize that Dominicans and Puerto Ricans share common economic, educational, and social service needs, some Dominicans feel they are being held back in politics and community organizations by a Hispanic leadership that pursues a narrowly Puerto Rican agenda. According to Guillermo Linares, the first Dominican elected to New York's City Council, "We all sit at the table when it comes time to draw lines. When the time comes to determine who represents all of us, we no longer find ourselves at the table" (González 1992, p. B4).

To understand the significance of skin color and race identity and the dilemma they pose for Dominicans one must keep in mind the significantly different "race climates" of the Dominican Republic and the United States (Grasmuck 1992). Despite the obviously mixed phenotype of the majority of the population, many Dominicans in the Dominican Republic discount their African heritage and prefer to think of color in nationalistic terms, restricting the term "Black" for the residents of the other half of the island—the Haitians. Discussions of race, in fact, still remain a somewhat taboo topic in the Dominican Republic. One way of thinking of this is to consider the historically different significance of a drop of African blood on the two sides of the border. In the traditional Dominican context to be partly White (which includes most Dominicans) was to be non-Black. In contrast, race laws in the United States historically defined someone as "colored" if they had even a drop of African blood (David 1991). How then does one interpret this Dominican African ancestry? While race may be, in the words of Rubén Rumbaut, a "pigment of our imagination," it poses real challenges to those whose identities lie on both sides of the cultural divide. Dominican immigrants leave a society with a low level of public consciousness and debate surrounding color discrimination and enter a society highly politicized around race and race identity. The dilemma comes when Dominicans with African features or dark skin, regardless of their social sense of self, find themselves identified by many in the United States as Black and are discriminated against on that basis (rather than language, for example), and are often not prepared to interpret discrimination on these grounds. These issues will undoubtedly remain salient for second-generation Dominicans, who are likely to experience both color discrimination and racial pride in the United States in a way different from that experienced by their parents coming from the Dominican Republic.[21]

Conclusion

While the Dominican immigrant community is at once maturing and beginning to face the challenges posed by second-generation Dominicans, the community at the same time continues to receive a sizeable proportion of new members from the island and is likely to continue to receive them at current levels for the indefinite future. This is because, for one thing, deteriorating economic conditions in the Dominican Republic continue to provide a strong emigration incentive for those with the means to leave. In addition, the continued consolidation of the Dominican community in the United States attracts aspiring Dominicans who benefit from the social space and cultural protection afforded by the Dominican network in place in the United States. Rumbaut (1992) has pointed out that "American history is notable for its near mass-extinction of non-English lan-

guages." The linguistic pattern in the United States among the children of immigrants is that in all ethnic groups, without exception, children consistently prefer English to their mother tongue. Such a pattern is happening and can be expected to continue in the Dominican community.

While linguistic struggles between parents and children are part of an old immigration story, the Dominican case adds a new twist to an old tale. While the second generation of Dominicans is growing and maturing, the active interplay between homeland and host society continues. This makes the generational struggles among Dominicans different from those often depicted in literature where the first generation of immigrants are more cut off from regular contact with their homeland. Moreover, amidst this sustained flow of new Dominican immigrants, a third generation of Dominicans can be expected to appear very soon (that is, those whose grandparents emigrated in the mid-1960s, whose parents came of age in the mid-1980s, and who can be expected to reach adolescence in the mid to late 1990s). This third generation will step on stage, so to speak, at the same time that new Dominican immigrants continue to arrive in large numbers. If these patterns persist, and if the ties to Dominican society remain as strong as they are today, we can expect extremely complicated situations to arise in the future among these different generations living in the United States and their relations in the Dominican Republic. One can imagine, for example, the complexities for the third generation of Dominican Americans who speak little Spanish and yet are expected to have some kind of sustained contact with both a "home" society, which increasingly escapes them emotionally and linguistically, but to which their parents and grandparents maintain strong ties, and with new Dominican immigrants who speak little English. Dominican Americans will likely continue to struggle internally with these complex issues as they simultaneously consolidate their more recent efforts to gain social and political recognition in the United States.

Notes

1. The 1990 census reveals that 51.9 percent of all Dominican immigrants entered the United States between 1980 and 1990. Those entering during the last decade constitute 28.9 percent of the entire Dominican-ancestry population living in the United States (calculations provided by Rubén Rumbaut).

2. The 1980 U.S. census also showed an apparently female-dominant population. However, there is considerable evidence to suggest that the male population is subject to a significant undercount; Larson and Opitz (1988, p. 13) and Báez Evertsz and D'Oleo Ramírez (1985, p. 348), relying on data from the U.S. consul general in Santo Domingo, concluded that of the 220,000 persons who received immigrant visas between 1960 and 1980, 52 percent were male.

3. One study of undocumented Dominicans in New York concluded that although approximately 33 percent of Dominicans had been illegal at some point during their residence in the United States, only 17 percent were illegal at any given time because a high proportion of undocumented Dominicans eventually succeeded in regularizing their status (Pérez 1981). The 1980 census enumerated 170,689 persons who reported single Dominican ancestry or reported Dominican as one of multiple ancestries (U.S. Bureau of the Census 1983a, p. 14) with approximately 19,000 undocumented Dominicans, or approximately 11 percent (Warren 1988).

4. Of the 206,719 Dominican immigrants counted in the 1990 census, 58.7 percent had arrived in the 1980s, 28.9 percent in the 1970s, and 12.3 percent in the 1960s (U.S. Bureau of the Census 1990a; calculations by John Mollenkopf of CUNY Data Service).

5. The salary differential between the two countries, based on the annual average exchange rate and the minimum monthly salary, rose steadily from 287 pesos in 1974 to 1,794 pesos in 1987. By 1987 the minimum monthly salary for full-time work in the United States was six times higher than that of the Dominican Republic (Grasmuck and Pessar 1991, p. 46).

6. Similarly, college graduates constitute 20.3 percent of the U.S. population twenty-five years and older, but only 7.9 percent of the comparable Dominican age group.

7. The Washington Heights area of New York City, where many Dominicans reside, is one of the most densely populated areas of the city where some families double- and triple-up in cramped apartments. Moreover, a thriving drug trade in the area coupled with high rates of poverty create conditions ripe for conflict and violence (González 1993).

8. The 1990 census reports that 43.2 percent of children of Dominican ancestry live below the poverty line, compared to 17.9 percent for all children in the United States (U.S. Bureau of the Census 1990a, 5 percent Public Use Microdata Sample).

9. Only 3.7 percent of Dominican female householders have completed college compared to 9.7 percent of married Dominican householders. Additionally, fully 59.2 percent of Dominican female householders report that they speak very little or no English compared to 41.2 percent of married Dominican householders. This means that the difficulty of raising children on one salary is magnified by the fact that that salary is likely to be very low given the educational and English proficiency levels of this group.

10. The census categories did not actually include the choice of "mulatto" but the selection of "other" by 50 percent of the Dominican population can reasonably be interpreted to mean the wide percentage of the population which considers itself "mixed" or "indio" in the local vernacular.

11. The self-employment rate for foreign-born in New York City was 7.6 percent, compared with 6.3 percent for Dominicans, 9.2 percent for Cubans, 6.6 percent for Haitians, and 4.6 percent for Jamaicans (1990 Public Use Microdata Sample for the five counties making up New York City); calculations by John Mollenkopf of the CUNY Data Service.

12. In our 1981 survey of Dominicans in New York, while 23.1 percent of our female respondents had been professionals in the Dominican Republic, less than 3 percent were able to match this level of employment in New York. Among the male migrants, the process of channeling into the lower-skilled jobs, although similar, was less extreme.

13. Based on our 1981 survey, we found that the ordinal stratification of wages moved up from undocumented women, to documented women, to undocumented men, to documented men (Grasmuck and Pessar 1991). This does not imply, however, that undocumented Dominicans do not also suffer from vulnerability and discrimination. Yet contrary to conventional wisdom this is not a disadvantage that undocumented persons always "import" with them as a result of more disadvantaged backgrounds and lower skills than their documented counterparts. In fact, in the early 1980s we found that undocumented Dominicans in New York City were somewhat better educated and had held more prestigious jobs prior to emigration than did documented Dominicans. However, as the economic crisis of the 1980s in the Dominican Republic intensified, illegal immigrants increasingly came from the more popular and less-skilled sectors of the home society.

14. The gender tension over the decision to return home is dramatized in Julia Alvarez's novel, *How the García Girls Lost Their Accents*: "There was still some question in his mind whether or not he might move his family back. But Laura had gotten used to the life here. She did not want to go back to the old country where, de la Torre or not, she was only a wife and a mother (and a failed one at that, since she had never provided the required son). Better an independent nobody than a high-class houseslave. She did not come straight out and disagree with her husband's plans. Instead, she fussed with him about reading the papers in bed, soiling their sheets with those poorly printed, foreign tabloids. "The Times is not that bad!" she'd claim if her husband tried to humor her by saying they shared the same dirty habit" (Alvarez 1992, p. 144).

15. The greater aversion to return home expressed by many Dominican women compared to men in our study of Dominicans living in New York City in the early 1980s (Grasmuck and Pessar 1991) was also revealed in a recent study of return migrants in the Dominican Republic. Among such returnees, women showed higher reluctance to return than did men, although driven by social and familial pressures women ultimately did so (Guarnizo 1993, p. 16).

16. This compares with only 1.0 percent of total income of Dominican males coming from public welfare assistance (U.S. Bureau of the Census 1990a).

17. Remittances, the peoples' version of foreign aid, are impressive. By the 1980s, remittances from Dominicans abroad matched the combined contribution to the national GDP of the country's three traditional export commodities: sugar, coffee, and tobacco. Indeed, remittances lagged behind only tourism as the leading international revenue producer (Espinol Jacobo 1993).

18. "Si te lleva bien allá, escribe." This is the refrain of the popular merengue, *Elena* by the Dominican group "4.40." The song describes the misfortunes of Elena, a Dominican who leaves in a boat and encounters the drug-filled underworld of New York City.

19. Yolanda's positive experience of feeling at home in the Dominican Republic upon returning is contrasted with the internal monologues of another "Yolanda," who is a character in a novel by Pedro Vergés, another Dominican writer. Yolanda Martínez, a bored young woman recently returned from the United States, muses:

> In this damn country there was no Coney Island where one could have a good time on a *Saturday* (in English in the original), no 24-hour movies, or Hot Dog stands for when you go for a walk, or big stores to spend hours and hours just looking, there was nothing at all. A lot of dirt, a lot of filth, a lot of laziness, many soldiers, many thieves, many politicians, that is what there was. (Cited in Castro 1985, p. 503)

20. The Dominican naturalization rate of 21.7 percent in 1990 compares with 44.7 for Panamanians, 42 percent for Bajans and Cubans, 27 percent for Haitians, and 34 percent for Jamaicans. Calculated from the 1990 U.S. census by John Mollenkopf.

21. For example, a second-generation Dominican woman living in New York City recently told us of her pride in being Afro-Caribbean. She explained how she challenges her mother when the latter advises her daughters to revere their father's light-skinned, blue-eyed ancestors and to minimize their mother's much darker-skinned forebear.

22

Filipino Americans: Many and Varied

BENJAMIN V. CARIÑO

The Filipino population in the United States has grown rapidly over the years. In 1990, Filipinos in the United States numbered over 1.4 million, increasing by more than 80 percent from approximately 775,000 in 1980 (U.S Bureau of the Census 1991). Filipinos now constitute the second largest Asian American group, behind the Chinese who numbered 1.6 million in the same year. Analysis shows, however, that of these 1.6 million Chinese, as many as 200,000 are ethnic Chinese from Vietnam, and smaller numbers come from such countries as Cambodia, Laos, Malaysia, and Indonesia (Rumbaut 1991a). Thus, in terms of national origin, the Filipinos are the largest single Asian group in the United States today, and only Blacks, Mexicans, and Puerto Ricans (who share with Filipinos a common colonial history) are more numerous. Yet little is known about the Filipino population in the United States.

What are the major characteristics and human resource endowments of Filipino Americans? Are they achieving success in the host society, or are they encountering adaptation problems exacerbated by their invisibility? What is the role of contemporary Filipinos in America? This essay is precisely aimed at providing a portrait of the Filipino community in the United States. It will document the historical context in which such a community has grown, both in size and diversity, outlining in the process the changing patterns and trends in Filipino immigration to the United States. Factors that may have influenced the success of Filipino Americans in adapting to their new environment will also be examined. A major theme that is emphasized in this essay is that, like most ethnic groups in the United States, it is difficult to speak of a single Filipino community in America (see Alegado 1991). Filipino American communities are many and varied—in origin, in socioeconomic characteristics, and in their levels of incorporation into the economy and social structures of the host society.

This essay relies heavily on information available from previous works on the subject, as well as from census reports and statistics published by the Immigration and Naturalization Service (INS). In addition, this article draws on data from the Comparative Longitudinal Asian Immigration (CLAIM) Project, a joint undertaking of the University of the Philippines, Konkuk University, Seoul, and the Program on Population, East-West Center, and funded through a grant from the Sloan Foundation.

The Historical Context

For almost ninety years Filipinos have been part of the migratory stream of newcomers in search of a better life in the United States. Depending upon conditions in both the sending country and the receiving society, their numbers have "at times, surged heavily or, alternatively, ebbed to a trickle" (Posadas 1986–1987, p. xi). Indeed, Filipino immigration to the United States is an active phenomenon that has gone through three main historical phases (see East-West Population Institute 1990). To better understand these phases, it is important to take into account the country's long colonial experience.

The Philippine Colonial Experience

As argued elsewhere (Cariño 1987), the character and magnitude of Filipino immigration has been largely influenced by the country's colonial experience, first with Spain, and later with the United States. A major legacy of the colonial experience with Spain is the predominance of Catholics among Filipinos (about 84 percent), as well as their subordination to the dominant economic and political interests of the colonizers. As will be noted later in this essay, this may help explain Filipino Americans' acceptance of similarly dominant interests in the United States.

The colonial experience with the United States started when the Philippines was ceded to the United States by the Treaty of Paris at the end of the Spanish-American War in 1898. The Philippines thus became a U.S. territory, establishing the political, economic, and social linkages between the two countries (Pido 1986). This colonial experience had a pronounced impact on Philippine immigration to the United States (for more on this, see Cariño 1987). Among other things, it produced a pervasive Americanization of Philippine culture which is a major factor in the integration of Filipinos with the host society. This colonial heritage has been enhanced by a Philippine educational system that has been patterned after that of the United States and uses English as the medium of instruction (Abad and Eviota 1984). In part because of the similarity of the educational systems between the two countries, it has not been difficult for the U.S. economy to absorb surplus professional labor from the Philippines.

The colonial heritage likewise engendered strong business ties with the United States, an economic relationship that is regarded by many as one of dependency, and one that has persisted even with decolonization. The strong economic presence of the United States in the Philippines is evident in the fact that, in 1991, 17.2 percent of total foreign investments in the country can be attributed to the United States (NEDA 1992). By facilitating the movement of labor, this economic linkage has undoubtedly contributed to the volume of migration between the two countries.

The strong American military presence in the Philippines in the past is yet another legacy of the country's U.S. colonial experience. For a long time, the United States had, by agreement, legal jurisdiction over two major military base facilities in the Philippines that contained the largest U.S. naval and air force fleet in the Asian Pacific region. This substantial military presence has boosted migration flows from the Philippines in at least two ways. First, it facilitated international marriages between American military personnel and Filipino women, as evidenced by the numerous departures of Filipino women from the military base facilities in the country. Second, it similarly facilitaed the recruitment of Filipinos into the U.S. military (Cariño 1987). In 1970, for instance, there were more Filipinos serving in the U.S. Navy than in the entire Philippine Navy. In San Diego, which houses the third largest Filipino community in the United States, more than half of all Filipinos are employed by the U.S. Navy (Reimers 1985; Rumbaut 1991a). All these factors, coupled with the presence in the United States of an already substantial Filipino American population, allowed for a rapid response to the 1965 changes in U.S. immigration law.

Immigration Phases

The colonial experience of the Philippines provides the context for reviewing the major historical phases of Filipino immigration to the United States. In turn, these historical phases provide the broad perspective from which one can appreciate the growth and increasing diversity of the Filipino community in the United States.

The first phase started at the turn of the century with the recruitment of workers by sugar plantation owners in Hawaii and California (Lasker 1969). About 150,000 Filipino laborers for these plantations were recruited, mostly between 1907 and 1930, and about half went to Hawaii. The flow of these early

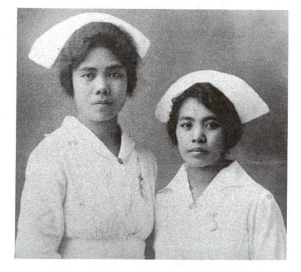

In the contemporary period, the immigration of Filipino nurses is a sizeable component of the professional immigration. Such a migration has long roots. Here two Filipino nurses are in the Firland Sanatorium, in 1931, shortly after arriving from the Philippines. They worked as nurses in Seattle, Washington, for over 20 years. (Filipino American National Historical Society)

Filipino immigrants was facilitated by a U.S. law that, until 1935, allowed Filipinos to enter the United States freely as U.S. nationals (Cariño et al. 1990).

The Philippines became an independent territory when it was granted commonwealth status in 1935. Consequently, Filipinos were no longer U.S. nationals and immigration to the United States was virtually halted. A national-origins quota system limited it to not more than 100 Filipinos annually in the next decade (Pido 1986). It was not until the Philippines gained its independence from the United States in 1946, following the end of World War II, that the number of Filipino immigrants rose substantially, beginning the second phase of immigration which lasted until 1965. The number of Filipino immigrants during this period, however, did not exceed 4,000 in any given year (Arnold, Minocha, and Fawcett 1987, Table 6.1). During the two decades from 1946 to 1965, over 34,000 Filipinos came to the United States (East-West Population Institute 1990).

The 1965 amendments to the U.S. immigration law eliminated restrictions based on country of origin, that virtually barred Asian immigration to the United States, and allowed entry primarily on the basis of kinship with persons already in the United States, or on the basis of occupational characteristics. This marked the beginning of the third phase of Filipino immigration to the United States. This landmark immigration reform dramatically increased the number of immigrants from Asia as a whole. The sudden upsurge in the number of immigrants has, however, been particularly pronounced in the case of the Philippines, rising from less than 16,000 during the preamendment 1961 to 1965 period to more than 325,000 in the 1986 to 1990 interval, more than a twenty-fold increase. From 1965 to 1990, the Immigration and Naturalization Service (INS) reported more than 1 million Filipino admissions to the United States (INS 1991). Indeed, since 1965, only Mexico sent more immigrants to the United States than the Philippines, underscoring the great import and impact of Filipino immigration.

The Current Situation

Sending Country Conditions

In the 1980s, the social, economic, and political conditions in the Philippines reinforced the contemporary patterns of international migration from the country. The Philippines has had a stagnant and, at times, even declining economy over the past decade or so, a fact that can only trigger further emigration in the future. The economic problems of the country were particularly severe during the period following the assassination in 1984 of a leader of the opposition, Senator Benigno Aquino, eventually leading to the downfall of the Marcos regime and the subsequent formation of the Corazon Aquino government.

The poor state of the economy during that period is vividly illustrated by the fact that in 1984 and 1985, the growth of the country's gross national product (GNP) was recorded at a negative 8.7 and 7.7 percent, respectively. These figures become even more alarming when viewed in the context of the high population growth rates in the country. For most of the 1980s, the country's population was growing at a rate of 2.6 percent. The perceived instability of the Aquino government—not unfounded given the several coup attempts against the regime—exacerbated the economic problems of the country as it discouraged investments, both foreign and domestic, and encouraged "capital flight."

Although there are signs of economic recovery, the country's economic problems continue to be serious, a trend that is aggravated by a severe power crisis, and a huge government budgetary deficit and external debt that totaled $29.6 billion in 1990. In 1991, the unemployment rate remained at a high annual average of 10.6 percent. From January to September 1992, reports from the National Economic Development Authority revealed that the growth of the country's GNP was a minimal 0.9 percent. The Philippine population growth rate, meanwhile, declined only slightly at 2.4 percent.

The Demographic Response

The severe economic crisis that has confronted the Philippines has obviously combined with historical factors and immigration policy reforms in the United States to trigger an even more massive flow of emigrants from the country in recent years. Indeed, migration (not just to the United States but to other parts of the world) has been observed by many as a common demographic response of Filipinos to social and economic problems (see Smith 1976). In fact, the country has now become a main source area for international migrants: those leaving home in search of a better life in another country through permanent residence or temporary employment. Data from the 1980 U.S. census show that about 65 percent of Filipino Americans were immigrants, while about 35 percent were born in the United States. Clearly, then, the rapid growth of the Filipino population in the United States can be attributed to immigration, which gradually rose to 71,279 in 1990, constituting 22.1 percent of total Asian immigration to the United States in that year (Table 1). Smaller flows have also moved such other countries such as Canada and Australia.

Just as significant is the growth in the number of migrant workers who seek temporary employment abroad. Following the formal adoption of the Overseas Employment Program by the Philippine government in the mid-1970s, the flow of temporary migrant workers from the Philippines to the Persian Gulf region and east Asia had become more prominent than the more permanent immigration streams to the United States and other Pacific Basin countries

Table 1   Immigration to the United States from the Philippines and Asia, Fiscal Years 1971–1990

		Philippines	
Years	Asia	Number	Percent of Asia
1971–1980	1,588,178	354,987	22.3
1981–1990	2,738,157	548,764	20.0
1985	255,157	53,137	20.8
1986	258,546	61,492	23.8
1987	248,293	58,315	23.5
1988	254,745	61,017	23.9
1989	296,420	66,119	22.3
1990	321,879	71,279	22.2
1971–1990	6,019,180	1,026,653	17.1

Source: U.S. INS, *1990 Statistical Yearbook*, Table 2. (Washington, D.C.: U.S. Government Printing Office).

(Cariño 1992). Since the institution of the Overseas Employment Program, the number of processed Filipino contract workers who sought temporary employment abroad increased almost twenty-fold over a sixteen-year period, from just over 36,000 in 1975 to almost 700,000 in 1991. The high level of labor flows in 1991 was obviously a reflection of the massive reconstruction and rehabilitation efforts in the aftermath of the Persian Gulf War earlier in the year.

The Socioeconomic Consequences

The massive migration flows from the Philippines have, on the one hand, been interpreted as a case of labor "oversupply." Quite a few scholars (see Arcinas 1987; Pernia 1976) have observed that the continuing out-migration flows from the country are basically a reflection of the inability of the economy to absorb the supply of labor, particularly certain high-level skills (for example, physicians and nurses). In other words, the problem is viewed as one in which the supply of skilled and professional labor has outpaced the economy's absorptive capacity. Available evidence reveals that the Philippines sent more professional workers to the United States in 1990 than any other country, and ranked second (next to Mexico) in the number of executives and managers (INS 1991). It should be noted that Mexicans are far more numerous than Filipinos following legalization of their status under the amnesty provisions of the Immigration Reform and Control Act (IRCA).

From a macro perspective, migration is viewed as a means of easing the worsening unemployment problem of the country. The country's high rates of unemployment, particularly during the mid-1980s (when labor migration flows were encouraged through the Overseas Employment Program), strengthens this argument. However, the problems of unemployment and poverty reflected the political crisis of the period as well as of the government's largely ineffective economic policies. In particular, the tight credit situation during the period, coupled with restrictions on the importation of raw materials, has resulted in the closure of many industrial firms and the subsequent massive layoff of workers.

At a more micro level, migration may be seen as an opportunity and mechanism for upward social mobility, both for the migrants and the families they leave behind. There is hardly any question that migration, through repatriated earnings, has had a favorable impact on household welfare in the Philippines. A survey of Filipino immigrants in Honolulu in 1982 revealed that almost three-quarters of the households surveyed sent remittances to relatives in the Philippines, averaging about $600 per household during a twelve-month period. Over a one-year period, these remittances totaled more than $373,000 from 623 households (Caces 1985, p. 148). In absolute terms, foreign exchange remittances from temporary migrant workers are even more substantial. Over a ten-year period from 1977 to 1987, repatriated earnings from overseas workers averaged $578.1 million, steadily increasing from $162.9 million in 1977 to $791.91 million in 1987 (Philippine Overseas Employment Administration 1988).

The contrary view is that the departure of individuals, especially of the economically active and productive segment of the population, is really just a reflection of the exploitative relationship between the more advanced receiving society and the poorer sending country. As noted in another study (Cariño 1987), this argument is often made in the context of the "world system" theory which, as a derivative of Marxist theory, views international migration as a movement of labor from peripheral to core nations caused by the advanced capitalist economies' search for cheap labor and their desire to perpetuate economic and political dependency relationships (Petras 1981; Portes 1981).

There is some evidence to substantiate this claim, but only with respect to the earlier waves of Filipino immigration. Evidence elsewhere shows, for instance, that the Filipino community on Oahu, Hawaii, generally stands below the American population on the island in terms of occupational status (Cariño 1981). However, 1980 census data reveal that the educational attainment and occupational status of Filipinos as a whole exceed the average of the native-born American population (Rumbaut 1991a; U.S. Bureau of the Census 1984).

The Diversity of the Filipino Community

As previously indicated, Filipino Americans today make up one of the fastest growing Asian American populations. Based on the 1980 census, nearly half of the Filipino population in the United States (46 percent) is in California; 17 percent is in Hawaii, 6 percent in Illinois, 5 percent in New York, and the rest are scattered in such states as New Jersey, Washington, and Texas. On the whole, Filipino Americans are known to have high levels of educational attainment relative both to the U.S. and Philippine populations (Fawcett and Gardner 1992). Along with the Asian Indians, Filipinos are also more proficient in the English language than other Asian groups in the United States, and are often regarded as the most "westernized" of Asian Americans. Compared with other immigrant groups, the Filipino immigrant population likewise contains a higher proportion of professionals, and for that reason tend to be less "clustered" and "visible."

There are, no doubt, subgroups within the Filipino community in the United States that occupy a disadvantaged position (in terms of various dimensions of socioeconomic status) vis-á-vis other ethnic groups in the host society. As noted earlier, this observation is particularly accurate in the case of Filipinos on Oahu, Hawaii. Analysis of an Office of Economic Opportunity (OEO) census in 1975 confirmed the common view that Filipinos on Oahu, when compared with other ethnic groups on the Hawaiian island, occupied the lower strata of the host society's social and economic life, reflecting their slower pace of adaptation (Cariño 1981). Factors contributing to this slower pace of adaptation include the unusual character of the Filipino age-sex structure and the consequent lack of a stable family life; their lesser experience in a competitive trading economy; and possible resistance to including Filipinos in programs aimed at rectifying inequalities of opportunity in the occupational life of the host country.

Aside from sociodemographic factors, and the Filipinos' relative lack of experience in a trading economy, the disadvantaged position of Filipinos in Hawaii has been attributed to their heavy reliance on interpersonal networks which offer only limited opportunities for newcomers who might obtain better positions under less restrictive labor market conditions. In the words of Caces (1986–1987), personal networks "unwittingly channel new immigrants to the societal segments with which they are most familiar. This is a rather narrow segment, as demonstrated in employment-seeking processes."

More significantly, racism has, in the view of many observers, served as a major obstacle to the socioeconomic advance of Filipinos in the United States. Indeed, many scholars have noted that the process of assimilation among Asian groups in the United States, as a whole, has been polarized along racial lines (Bonacich 1976; Daniels and Kitano 1990). A manifestation of this phenomenon among Filipinos is the reproduction of the Filipino national identity and ethnic community in the host society. In Alegado's (1991) view, two main factors have contributed to the persistence of the Filipino ethnic community: the continuing large influx of immigrants from the Philippines, which sustains the Filipino culture and identity, and the continued subjection of Filipinos to institutional discrimination and anti-immigrant prejudices.

Despite the commonality of race and nationality, however, the increasing size and diversity of the Filipino community in the United States precludes superficial generalizations about the group as a whole. "The old-timers"—the plantation-based, pre–World War II Filipino immigrants—were composed predominantly of single men who lacked the support of the traditional Filipino family and kinship system. They contrast sharply with the migrant flows of professionals and Filipino families of the post-1965 period. The few who arrived in the intervening years between the war and the 1965 immigration reform offer only a tenuous link between the two larger immigration streams.

Socioeconomic disparities can also be observed between the foreign-born and the increasing number of U.S.–born Filipinos. Studies have shown that local-born Filipinos generally have higher income and occupational status than immigrant Filipinos, reflecting the advantage of the former in being reared in the American culture and being more familiar with the economic opportunity structure of the host society (Cariño 1981; Cabezas, Shinagawa, and Kawaguchi 1986–1987). In this connection, it is noteworthy that some scholars have observed that foreign-born Filipinos in Hawaii, in their efforts to participate in various aspects of the social and economic life of the host society, often have to contend with resistance from an unlikely source—U.S.–born Filipino Americans. Such resistance is often manifested in cleavage, avoiding association, and, at times, social conflict (Okamura and Coller n.d.). However, other scholars have also observed that Filipino immigrant students are now performing better than U.S.–born Filipinos in terms of educational attainment (Portes and Rumbaut 1990).

Finally, socioeconomic disparities also exist among immigrants entering through various admission categories. To document these disparities, data taken from two surveys that form part of a larger comparative, longitudinal study of Filipino and Korean immigrants are presented in Table 2. These two surveys are the Philippines Immigrant Predeparture Assessment (IPDA) Survey, which entailed face-to-face interviews in Manila with 2,077 new immigrants in 1986; and the Status Adjusters Survey (SAS) in 1986. By definition, "status adjusters" are those who apply for an immigrant visa (or a "green card") while they are in the United States as nonimmigrants. In 1986, 25 percent of the cohort of Filipino immigrants age eighteen to sixty-nine were status adjusters (Cariño et al. 1990). A total of 1,565 Filipinos participated in this mail survey.

As shown by the evidence presented in Table 2, socioeconomic disparities exist at various levels. At one level, variations can be observed across immigrant visa categories, namely the *immediate relatives* of U.S. citizens (spouses and minor children) who, by law, are exempt from the numerical limitations; *family preference* immigrants, which includes other relatives of U.S. citizens and the spouses and unmarried adult children of resident aliens; and *occupational preference* immigrants, which consist of workers needed by the U.S. economy. Both family preference and occupational preference immigrants are subject to numerical limitations.

In particular, sociodemographic disparities can be discerned between immigrants who enter the United States on the basis of occupational skills and those who are admitted under the family reunification provisions of the immigration law. On the one hand, the occupational preference immigrants are more likely to be women in their prime working ages, more professional in occupational background, and characterized by higher levels of educational attainment and occupational skills. On the other hand, family-based immigrants (especially immediate relatives) tend to be relatively older, much less professional in occupational background, and generally stand below the occupational preference immigrants in terms of educational attainment (Cariño et al. 1990). Family preference immigrants (as opposed to immediate relatives) are somewhat closer to occupational preference immigrants in terms of the socioeconomic indicators examined. The results of a recent study confirm these observations (Liu et al. 1992). The authors of this study attribute the diversity that exists among Filipino Americans to a phenomenon of "dual chain migration." The first chain derives from family connections to Filipinos immigrants prior to 1965 and is referred to as "relative-selective." The second chain stems from the highly qualified immigrants who entered the United States through the occupational categories within the preference system and is referred to as "occupational-selective." The authors are quick to note that, given the bias of the 1965 immigration policy reform for family reunification, the persistence and lengthening of both chains will depend primarily on family-based immigration in the years to come.

The composition of the immigration stream in terms of the major admission categories has also changed dramatically in recent years. Up to the mid-1970s, about half of the admissions under the new law were on the basis of occupational skills. In the 1980s, however, a majority have been admitted on the basis of the family reunification provisions of the law. This change in the pattern of admissions has clearly contributed to the increasing diversity of the Filipino community in the United States.

At another level, "nonimmigrant admissions" also contribute to the growing diversity of the Filipino American community. By admitting students and workers as temporary nonimmigrants, the United States is also unwittingly accepting immigrants because many such students and workers eventually find a way to adjust their status. As shown by the evidence presented in Table 2, the status adjusters, when compared with the predeparture immigrants, have generally superior educational and occupational qualifications. Also, a higher proportion of status adjusters are women, a pattern that can be attributed to the many fiancees of U.S. citizens who achieve immigrant status through marriage.

Thus, beyond the commonality of race and ethnicity, diversity seems to have characterized the experience of Filipino Americans and, to a great extent, the Filipinos' sense of themselves in the host society (Posadas 1986–1987). Indeed, the reality of being a Filipino in the United States has meant, in different times and places, plantation work, privileged student status, the postwar brides, and the more professional flows—doctors, nurses, and teachers—with their relatives and extended families.

Conclusion

This provided a profile of the major socioeconomic and human resource characteristics of Filipino Americans. It also documented the historical context in which the Filipino community in the United States has grown, both in size and diversity. The factors that may have influenced Filipino immigration were likewise examined. In the process, the possible consequences of such processes for both the sending country and the immigrants were highlighted.

A major theme that has surfaced in this analysis is the rapidly increasing magnitude and diversity of the Filipino community in the United States. On one hand, the rapid growth of the Filipino American community in recent years can be attributed to the worsening social and economic situation in the Philippines which, combined with historical factors and immigration policy reforms in the United States, have served as a powerful "push" for the massive and continuing immigration of Filipinos. On the other hand, the sources of diversity are many and varied: historical factors, place of birth, recency of migration, modes of entry, and the U.S. system of immigrant preferences and numerical limitations.

Because of the increasing size and diversity of the Filipino community in the United States, the opportunities for meaningful participation in the socioeconomic life of the host society are not the same for various groups among Filipino-Americans. As ob-

Table 2   Socioeconomic Characteristics of Predeparture Immigrants and Status Adjusters by Immigrant Visa Category, IPDA and SAS, 1986

	Total Number	Mean Age	Percent Female	Mean Years of Education[d]	Percent with Professional or Technical Jobs
Predeparture immigrants					
Immediate relatives	1,105	40.5	66.6	9.8	21.6
Family preference	510	33.4	44.9[c]	12.1	25.5
Occupational preference	114	36.0	58.4	13.5	58.4
Others	48	42.0	33.3	11.3	—[e]
Status adjusters[a]					
Immediate relatives	1,429	35.2	78.7	12.4	35.5
Family preference	29	33.5	75.9	12.4	77.8
Occupational preference	89	36.1	74.8	14.6	83.1
Others	11	38.4	18.2	14.6	—[e]
Predeparture immigrants and status adjusters[b]					
Immediate relatives	2,534	39.2	69.6	10.5	25.1
Family preference	539	33.4	52.6	12.2	38.5
Occupational preference	203	36.0	64.1	13.7	64.7
Others	59	41.1	29.5	12.1	—[e]

[a] Excludes from the analysis twelve cases from whom data were missing. (See text for explanation of "status adjuster.")

[b] Average and percent figures represent weighted average of predeparture immigrants and status adjusters.

[c] Includes 136 fiancees (K-1 visas).

[d] For predeparture immigrants, exludes 271 cases for whom data were missing; for status adjusters, excludes 16 cases for whom data were missing.

[e] No data available.

served in many studies (see Alegado 1991; Cabezas, Shinagawa, and Kawaguchi 1986–1987; Forman 1991), however, what seems common among many Filipino Americans, as with other immigrant groups in the United States, is that forms of racial discrimination, directly or indirectly, have to some extent hampered the economic success of Filipinos in the host society.

For Filipinos, however, a subordinate status is a familiar experience. Filipino immigrants come from a country that is beset by severe social and economic problems. They have also suffered under centuries of colonial domination and from the development of an oligarchical leadership composed mostly of the social and economic elite of the country. This may help explain the apparent vulnerability of Filipino Americans to structural discrimination and subordination. The observation of one study aptly describes

the situation of Filipino immigrants in the United States:

> Even when made aware of their subordination, the immigrants may subjectively compare degrees of subordination: what they left behind versus what they find here. Any structural discrimination they encounter thus is familiar. In the Philippines, they saw subordination by dominant economic and political interests; in the United States, they may see themselves subordinated by similarly dominant interests. (Cabezas, Shinagawa, and Kawaguchi 1986–1987, p. 17)

In this connection, one important question that emerges concerns the possible effect of subordination on the socioeconomic advance of the Filipino community in the United States. It seems evident, in this regard, that Filipino Americans share this pre-

Table 3   Occupational and Family Preference Immigrants from the Philippines: 1966–1990

		Visa Category			
	Total	Occupational Number	Preference Percent	Family Number	Preference Percent
1966–1970	61,688	30,530	49.5	31,090	50.5
1971–1975	96,357	49,606	51.5	46,610	48.5
1976–1980	98,480	19,035	19.3	78,605	79.8
1981–1985	96,972	18,470	19.0	78,431	80.9
1986–1990	98,097	19,676	20.1	78,266	79.8

Sources: 1966–1977: INS (*Annual Report*, Table 7A); 1978–1979: INS (*Statistical Yearbook*, Table 7A); 1980–1981: INS (*Statistical Yearbook*, Table 5); 1982–1985: INS (*Statistical Yearbook*, Table 21); 1986–1990: INS (*Statistical Yearbook*, Table 5).

dicament with many other ethnic groups. As noted in another study (Cabezas, Shinagawa, and Kawaguchi 1986–1987), the socioeconomic advance of Filipino Americans partly lies in the recognition of the dominant societal structures that produce subordination, and in the willingness to eliminate or reshape such structures. But beyond such recognition, the success of Filipino Americans in coping with various problems in the host society may also depend to some extent on their ability to submerge internal divisions and small group loyalties in favor of ethnic unity and an overall Filipino identity.

These trends and patterns of Filipino immigration and the characteristics of the Filipino community in the United States must be taken into account in policy decisions by both the sending and receiving societies. For the Philippines, a major concern must be the continued departure of the highly skilled segment of the population, reflecting as it does the flaws in the country's manpower and development policies. For the United States, the concern must not only be focused on the increasing volume of Filipino immigration, but also on its increasing diversity and how such diversity can mesh with the host society's social and economic system (see Cariño et al. 1990).

23

The Entrepreneurial Adaptation of Korean Immigrants

PYONG GAP MIN

Before 1965, Asian nationals had been virtually barred from entry into the United States. Laws passed in the early 1920s restricted immigrant visas almost entirely to persons from European countries. But the liberalization of U.S. immigration law in 1965, which repealed that discriminatory national-origins quota system, made possible the mass migration of people from Asian countries to the United States. Largely as a result of this liberalized immigration law, many new immigrant groups are now visible in Los Angeles, New York, and other major U.S. cities. Among them, Koreans form one of the most rapidly growing minority groups. The Korean population in the United States grew from less than 100,000 in 1970 to nearly 1 million in 1993, achieving an extraordinary tenfold increase in a span of two decades.

It is, however, due to their unique mode of economic adjustment, rather than their fast growth, that Korean immigrants have captured the interest and attention of the American public and social scientists. The post-1965 Korean immigrants have earned their living mainly through self-employment in small businesses. In Los Angeles, New York, and other large cities, Korean merchants are visible in almost every inner-city neighborhood. The visibility of Korean merchants to the American public has also been heightened by their conflicts with Black residents. These conflicts between Korean merchants and Black customers have been covered by the local media in many cities. The black boycott of two Korean stores in Brooklyn, New York, which started in January 1990 and came to an end in May 1991, received national media headlines. However, the 1990 Brooklyn boycott was not the climax of Korean-Black tensions. The American public witnessed the outburst of Korean-Black conflict in the Los Angeles riots in April 1992. Approximately 2,300 Korean-owned stores in South Central Los Angeles and Koreatown were burned and/or looted during the riots.

Why are Korean immigrants concentrated in small business? Why are many Korean stores located in Black neighborhoods? What factors contribute to Korean-Black conflicts? What are the effects of Koreans' business-related intergroup conflicts on the Korean community? Answering these significant questions is central to understanding Korean immigrants' adjustment in America. This chapter on the new Korean immigrants intends to answer these and other questions relating to Korean immigrants' commercial activities in America. It consists of five interrelated sections. The first section covers the immigration and settlement of Koreans in the post-1965 era. The second section discusses the factors that have contributed to Korean immigrants' concentration in small business. The third section is concerned with the effects of Koreans' commercial activities on their economic mobility. The fourth section examines Korean immigrants' business-related intergroup conflicts, with special emphasis on Korean-Black conflicts. The final section focuses on the positive effects of business-related intergroup conflicts on Koreans' ethnic solidarity.

Immigration

The Influx of Korean Immigrants

Approximately 7,200 Koreans who came to Hawaii to work on sugar plantations between 1903 and 1905 made up a small, first wave of Korean immigrants to the United States. However, the major waves of Korean immigrants came only after the passage of the new immigration law in 1965 and its full implementation in 1968. The 1970 census first classified Koreans as a separate group, and found the number of Koreans in the United States to be less than 70,000. The majority of Americans knew about Korea and Koreans before 1970 mainly through their memories of the Korean War and movies like MASH which reflected the misery of the war. With the exception of Honolulu, the locus of the old Korean enclave, Korean Americans were invisible in American cities before 1970.

As shown in Table 1, the number of annual Korean immigrants steadily increased after 1970. It reached the 30,000 mark in 1976 and maintained that or a higher level throughout the 1980s. Next to Mexico and the Philippines, South Korea was the third largest source country of immigrants to the United States in the 1970s and 1980s. The large influx of Korean immigrants during this period led to a phenomenal increase in the Korean American population. The 1990 census estimated the Korean population to be a little less than 800,000. When those legal and illegal Korean residents not counted in the census and those Korean immigrants who came after 1990 are considered, the total Korean population as of 1993 may be close to 1 million.

Table 1 Koreans Immigrated and Naturalized, 1965–1989

Years	Koreans Immigrated	Koreans Naturalized
1965–1969	17,869	6,982
1970–1974	93,445	14,716
1975–1979	148,645	48,884
1980–1984	162,178	67,646
1985–1989	175,803	73,407
Total	597,940	211,635

Source: Immigration and Naturalization Service. *1965–1989 Statistical Yearbook* (before 1977 it was called *Annual Report*). Washington, D.C.: U.S. Government Printing Office.

Motivations and Causes of Mass Exodus

Like most other Asian immigrants, Korean immigrants chose the path of U.S.–bound migration mainly because they were attracted to the prospect of a higher standard of living and a better quality of life in this country. The aspiration for a higher standard of living was particularly important for those Korean immigrants who came in the 1970s, when the standard of living in South Korea was very low. In 1975, the United States maintained a standard of living approximately ten times as high as in South Korea. South Korea improved her economic conditions significantly in the 1980s as per capita income increased from $843 in 1977 to $2,199 in 1987 (Korean National Bureau of Statistics 1984, p. 451; 1987, p. 467). Therefore, the economic motivation may have been less important for those Korean immigrants who came here in the late 1980s. However, the United States was and still is a better place than South Korea in terms of quality of life. For example, Seoul, from which more than half the new Korean immigrants came, is worse than most American metropolitan cities in terms of pollution, traffic congestion, crime, and housing conditions.

Another important motivational factor for the mass migration of Koreans to the United States has much to do with the difficulty of sending children to colleges and universities in South Korea and the relative advantages of a college education in the United States. Strongly influenced by the Confucian cultural tradition, Koreans place great emphasis on their children's education. According to a nationwide survey,

90 percent of the parents in South Korea expected four years of college or higher education for their sons and 70 percent expected the same level of education for their daughters (Korean National Bureau of Statistics 1987). However, a college education is very expensive in South Korea. Moreover, only a small fraction of candidates can get admitted to colleges and universities due to strict quotas of freshmen imposed by the Education Ministry. Many parents choose U.S. immigration in order to give their children a second chance for a college education. Many others do so to relieve the financial and psychological burdens of sending their children to higher educational institutions in South Korea. These push and pull factors are important for understanding the motives and patterns of Korean immigration. However, the close military, political, and economic connections between the United States and South Korea have also made significant contributions to the large trans-Pacific migration of Koreans to the United States. In fact, these structural factors, rather than Korean immigrants' motivations, seem to be central in answering the question of why South Korea has been the second largest Asian source country of U.S. immigrants next to the Philippines. In particular, the U.S. military involvement in South Korea since the Korean War has directly influenced the mass migration of Koreans to this country by providing a context for intermarriages between Korean women and U.S. servicemen. *Annually* more than 3,000 Korean women married to U.S. servicemen immigrated to the United States in the 1980s. The U.S. military and political involvement in South Korea has also indirectly influenced the mass migration of Koreans to the United States by heightening the aspirations of Koreans, particularly middle-class Koreans, for a higher standard of living. Koreans have been greatly exposed to American lifestyles through the presence of U.S. servicemen and the U.S. media, as well as the diffusion of American popular culture. Many Koreans seem to have chosen U.S.–bound migration in the early 1970s because of their exposure to the affluence of American life.

Characteristics of the New Korean Immigrants

Whereas the pre–World War II Asian and White immigrants came from rural and low-status backgrounds, the new Asian immigrants are characterized by urban, white-collar, middle-class backgrounds. Korean immigrants, like many other Asian immigrants, are drawn from the urban, middle-class segments of the Korean population. The vast majority of them lived in Seoul and other major metropolitan cities prior to immigration (Park, Fawcett, and Gardner 1990). Using high school and college graduation rates as indicators, Korean immigrants received higher levels of education than the U.S. general population, although South Korea is far behind the United States in the overall educational level. For example, the 1990 census indicates that 33.5 percent of Koreans who immigrated to America in the 1970s and 1980s completed four years of college education in comparison to 20.3 percent of the U.S. general population (U.S. Bureau of the Census 1993a). Most significant, the new Korean immigrants generally held white-collar and professional occupations in South Korea, with only a small fraction engaged in farming and blue-collar occupations (Hurh and Kim 1988; Min 1989; Park, Fawcett, and Gardner 1990). The U.S. government revised the 1965 Immigration Act in 1976 to discourage the immigration of aliens based on occupational skills. This revision made it difficult for highly educated Korean professionals to immigrate to the United States. Moreover, due to improved economic and social conditions in South Korea, and the negative publicity of Korean immigrants' adjustment problems in the United States, middle-class Koreans, particularly professionals, have little incentive to immigrate. A survey recently conducted in Seoul shows that low-class Koreans, rather than middle-class Koreans, reveal a strong preference for immigration to the United States (Yoon 1993). Thus, Korean immigrants who came to the United States recently are likely to be lower in socioeconomic status than those who came here in the 1970s. This speculation is supported by survey studies (Hurh and Kim 1988; Min 1989). In addition, the overall socioeconomic background of Korean immigrants is likely to be even lower in the future.

Another important characteristic of post-1965 Korean immigrants is their heavy Christian background. Christianity is not the major religion in South Korea, with only a little more than 20 percent of Koreans affiliated with Christian churches (Korean National Bureau of Statistics 1987). However, the majority of Korean immigrants are found to have attended Christian churches in South Korea. For example, in a 1986 survey of Korean immigrants conducted in Seoul (Park, Fawcett, and Gardner 1990, p. 60), 54 percent of the respondents reported that they were affiliated with Protestant (41.6 percent) or Catholic (12.3 percent) churches in Korea. Many Korean immigrants who were not Christians in Korea have become Christians in the United States. Thus, some 75 percent of Korean immigrant families are affiliated with Korean immigrant churches (Hurh and Kim 1990; Min 1989). Max Weber (1952) and his followers (Kennedy 1962; Stone 1974) linked the Protestant religious ethos to the spirit of capitalism. It may not be accidental then that Korean immigrants, heavily Christian, are also heavily entrepreneurial.

The Concentration of Korean Immigrants in Small Businesses

The High Self-Employment Rate

Korean merchants are found in almost any neighborhood in Los Angeles, New York, and other major Korean communities. Based on their casual observations, the American public seems to believe that Korean immigrants are business-oriented people. Survey studies conducted in major Korean communities tend to support the validity of their casual observations. A study conducted in 1986 showed that 45 percent of the Korean workforce in Los Angeles was self-employed (Min 1989). A study based on interviews with Korean married women in New York City revealed an even higher self-employment rate (Min 1992). Forty-nine percent of Korean married women in New York City and 62 percent of their husbands were found to be self-employed, thus the overall self-employment rate of married Koreans reached 56 percent. Since another 30 percent of the New York City respondents were employed in Korean-owned businesses, only 14 percent were employed in non-Korean firms. This means that Korean immigrants are highly segregated in their economic activities.

Types of Major Korean Businesses

When minority/immigrant groups are overrepresented in small business, they are concentrated in a few or several business lines. Korean immigrants are heavily concentrated in the grocery/green grocery retail, retail of manufactured goods imported from South Korea and other Asian countries, dry cleaning service, and a few other business lines.

The retail grocery is the most common type of Korean-owned business in the United States. According to the Korean American Grocers Association of southern California, there were some 2,800 Korean-owned grocery/liquor stores in Southern California as of 1990, which accounted for 25 percent of the total independent grocery/liquor stores in that area. Atlanta, Philadelphia, Washington, and Seattle Korean communities are also heavily concentrated in the retail grocery area. In the New York Korean community, there are also a large number of fruit and vegetable stores in addition to many grocery stores.

Another type of business common to major Korean communities is the retail of manufactured goods imported from South Korea and other Asian countries. These items include wigs, handbags, clothing, jewelry, hats, and shoes. This type of business has been made possible mainly because of the radical increase in Korean exports to the United States since the early 1970s. As will be discussed shortly, many Korean immigrants took advantage of their native language and connections with South Korea and established import and wholesale businesses dealing in Korean-imported fashion items in the 1970s. They distribute Korean-imported merchandise mainly to Korean retailers.

Another popular Korean business line is the dry cleaning service. The dry cleaning business is attractive to Korean immigrants partly because it is very suitable for a family business involving husband-wife coordination and partly because it involves shorter hours of work and a lower level of physical strength than other types of retail businesses. Whereas Korean retail stores selling groceries, liquor, and Asian-imported manufactured goods are heavily concentrated in Black and Hispanic neighborhoods, Korean dry cleaning shops are heavily concentrated in White, middle-class areas.

There are 700 Korean-owned garment manufacturing businesses in Los Angeles and 350 in New York. More than 95 percent of Korean garment manufacturers subcontract work from White manufacturers. A large number of Chinese and Latino immigrants in Los Angeles and New York also engage in garment subcontracting (Bonacich 1993; Waldinger 1985). Immigrants dominate garment subcontracting mainly because they have easy access to new immigrants as a cheap labor source. In addition, there are 1,500 Korean-owned manicure shops and 700 fish retail stores in New York, but these businesses are not popular Korean businesses in other Korean communities. There are also many Korean-owned gas stations and fast food franchise stores in Los Angeles.

Major Reasons for Koreans' Entry into Small Businesses

Historically, immigrants have shown a higher self-employment rate than native-born Americans (Light and Sanchez 1987; Min 1987). The tendency of immigrants to make occupational adjustments in small business can be explained by their language barrier and other disadvantages for employment in the general labor market. Like other immigrant groups, Korean immigrants turn to small business due to their labor market disadvantages. It is important to note that most Korean immigrants, although highly educated, have a serious language barrier—a more serious language barrier than Filipino and Indian immigrants who used English in schools prior to immigration. In a survey of Korean merchants in Atlanta, disadvantages in the general labor market and the opportunity for economic mobility through self-employment were found to be the two most commonly cited reasons for Koreans' entry into small business (Min 1984).

Factors Facilitating Koreans' Commercial Activities

Although Blacks and other minority groups face severe disadvantages for employment in the general labor market, they are severely underrepresented in small business. This suggests that Koreans' disadvantages for employment in the general labor market cannot completely explain their high self-employment rate. Whereas other disadvantaged minority groups have disadvantages for small business as well as in employment in the general labor market, Korean immigrants have some advantages for small business. Therefore, in order to explain Koreans' exceptionally high self-employment rate in a more satisfactory manner, we need to consider what advantages Korean immigrants have for the establishment and operation of small business.

Korean immigrants work exceptionally long hours to survive as small entrepreneurs, and their readiness to work long hours for a moderate amount of income is an asset to commercial activities. Does Korean immigrant merchants' hard work have a cultural basis? In explaining Korean immigrants' commercial activities, Kim (1981) points out that Confucianism, which has had great effects on Korean culture, is similar to Protestantism in its emphasis on work ethic, future orientation, and frugality. Confucianism also places great emphasis on family ties, and Korean immigrants' strong family ties help them with the establishment and operation of small business (Min 1988b). Thus, the Confucian cultural tradition and the Protestant background of Korean immigrants partly contribute to their commercial activities, although situational and structural factors seem to be more important than these cultural factors for Korean immigrants' commercial activities.

Most Korean immigrants work harder here in the United States than they did prior to immigration. Moreover, most Korean immigrant women work outside the home for the family business or for the accumulation of start-up capital, whereas few married women in South Korea participate in the labor force. These findings suggest that Korean immigrants' commercial activities, based on long hours of work and unpaid family labor, are more of a reaction to their immigrant situation than a reflection of their values brought from Korea (Min 1988a, 1988b). This has an important implication for the understanding of Korean and other immigrants' commercial activities. Many Americans tend to think that Korean immigrants' cultural values regarding work ethic and family ties are suitable for small business. Yet it seems to be more accurate to say that Korean immigrants turn to small business due to their disadvantages in the general labor market, and that they have to work long hours and mobilize family resources in order to survive as small business owners. Not only Koreans, but also many other immigrants, specialize in labor-intensive small businesses. Their long hours of work and dependence on family labor seem to be largely survival and coping mechanisms in an alien environment.

As previously noted, Korean immigrants are characterized by their middle-class background. Class resources associated with their middle-class background are also important facilitating factors for their commercial activities in the United States. Unlike earlier White and Asian immigrants who usually came here penniless, post-1965 Korean immigrants

Some immigrant or ethnic groups are so concentrated in small businesses that they have come to be known as a "middleman minority." In the United States at the turn of the century, Jewish, Italian, Greek, Chinese, and Japanese immigrants were all middleman minorities. Today, foremost examples are Koreans, Arabs, Iranians, and Cubans in Puerto Rico. Here, a Korean immigrant is selling American baseball caps in Los Angeles, 1993. (Photo by Steve Gold)

have brought a significant amount of money from Korea which provides them with start-up capital. In a predeparture survey conducted in Seoul, 1,986 Korean male immigrants reported that they planned to bring an average of $18,400 to the United States (Park, Fawcett, and Gardner 1990). Korean immigrants' class resources are not limited to material resources. Korean immigrants of middle-class background have brought with them nonmaterial resources conducive to their commercial activities, including the knowledge, motives, attitudes, and values required for successful entrepreneurship (Light and Bonacich 1988; Min 1988b; Yoon 1991).

Another factor facilitating Korean immigrants' commercial activities in the United States is the trade relations between the United States and South Korea. In the late 1960s and 1970s, the late President Chung Hee Park, taking advantage of cheap labor sources in South Korea, developed an export-oriented economy (Kim 1981, pp. 81, 86; Light and Bonacich 1988, chaps. 2, 3, and 4). As a result, South

Korea has made a great increase in her exports to developed countries since the late 1960s. By virtue of the traditional close relationship between the two countries, the United States has been the largest recipient of Korean merchandise. Korean exports to the United States have radically increased since 1970 when a massive immigration of Koreans to the United States started. Many Korean immigrants, taking advantage of their ethnic background, have been able to establish import businesses dealing in Korean-imported items such as wigs, handbags, shoes, jewelry, and clothing. Korean importers distribute Korean-imported merchandise to Korean wholesalers who distribute it to Korean retailers and peddlers, thus establishing the vertical integration of Korean businesses. However, the increase in labor costs in South Korea and a series of devaluations of the U.S. dollar since the 1988 Seoul Olympic Games have made it unprofitable for Korean importers to import merchandise from South Korea. Thus, many Korean importers in the United States have recently turned to Taiwan, Hong Kong, China, and other Asian countries.

Small Business and Economic Mobility

Researchers interested in ethnic mobility and urban occupational structure have long assumed that small business can be one main avenue for the economic mobility of minority groups, especially immigrant groups (Glazer and Moynihan 1963; Light 1972; Waldinger, Ward, and Aldrich 1985). Research by Portes and his colleagues based on interviews with Cuban immigrants in Miami supported this assumption concerning the relationship between ethnic mobility and small business (Portes 1987; Portes and Bach 1985; Wilson and Portes 1980). They showed that Cuban enclave workers in Miami—self-employed business owners and employees working for coethnic businesses—did as well economically as those in the primary labor market. It would be an interesting issue to examine whether small business has helped Korean immigrants to achieve economic mobility.

First of all, self-employed people, whether immigrant or native born, tend to underreport their actual incomes (Boissevain 1984, p. 29). Therefore, census data undercount Korean immigrants' incomes to a greater extent than members of other groups who have lower rates of self-employment than Koreans. In order to examine the economic effects of self-employment more accurately, randomly selected Korean immigrants in Los Angeles and Orange counties were personally interviewed in 1986. The median household income for all U.S. households in 1986 was $24,897 (U.S. Bureau of the Census 1988b, p. 442), and over 70 percent of Korean immigrant households in Los Angeles reached that income level that year. As expected, self-employed Koreans earned substantially higher household incomes than employed Koreans. Approximately 19 percent of self-employed workers, in comparison to 36 percent of employed workers, reported their annual household income to be below $25,000. In contrast, 43 percent of entrepreneurial respondents chose $50,000 or more as their household income in comparison to only 20 percent of employed respondents (Min 1989). Whereas self-employed and employed respon-

Table 2 Comparison of Business and Nonbusiness Families in Other Indicators of the Standard of Living

	Business Families (%)	Non-Business Families (%)	Level of Significance
Home ownership	49.6	28.4	$p < 0.000$
Living in a good community[a]	30.7	21.7	$p < 0.05$
Expensive house[b]	15.1	2.6	$p < 0.005$
3 bedrooms or more	51.5	31.3	$p < 0.000$
3 cars or more	29.8	16.8	$p < 0.000$
Private school	12.1	9.3	$p < 0.1$

[a] White middle- or upper-class areas.
[b] Housing price $400,000 or more.
Source: Min (1989), based on personal interviews with 518 Korean immigrants in Los Angeles and Orange counties.

dents significantly differed in family income, they had similar numbers of workers per family.

Other indicators such as home ownership, the number of bedrooms, and the number of cars suggest that Korean business families maintain a higher standard of living than nonbusiness families (see Table 2). Operation of a small business involves long hours of work and other social costs (see the last chapter of Light and Bonacich 1988). Nevertheless, self-employment in small business helps many Korean immigrants to achieve the economic mobility that is difficult to achieve through employment in the general labor market.

A comparison of Korean workers employed in Korean firms with those employed in non-Korean firms revealed that Koreans working for coethnic businesses have disadvantages in terms of wages and fringe benefits (Min 1989). Moreover, Korean employees working in the general economy, compared to those working for coethnic firms, were found to have big advantages in human capital investments as reflected in their educational level, fluency of spoken English, and length of residence in the United States. However, most Koreans working for Korean firms expect to start their own businesses and thus consider gaining business training and information to be the main benefit of their employment in Korean-owned businesses. As noted above, Korean immigrants who have their own businesses maintain a higher standard of living than those employed, and new Korean immigrants acquire business experience and training mainly through their employment in Korean-owned businesses. Accordingly, it can be argued that Korean businesses have positive effects on the economic well-being of the Korean community as a whole.

The Middleman Minority Role and Business-Related Intergroup Conflicts

Those minority groups that specialize in small business but lack political power often play a middleman minority role. Middleman minorities are usually found in societies rigidly stratified between two racial groups and are usually brought to the host society by the ruling group to distribute its products to minority members (Blalock 1967; Bonacich 1973; Hamilton 1978; Rinder 1959; Zenner 1991). There is no evidence that Korean immigrants have been brought here by the U.S. government or White corporations to distribute corporate products to minority customers. However, Korean immigrant merchants serve disproportionately large numbers of Black and Hispanic customers. Since there are some structural factors that encourage Korean immigrants to enter small business in minority neighborhoods, Korean immigrant merchants can be considered to play a middleman role in the United States.

Korean Merchants' Concentration in Minority Neighborhoods

Korean merchants serve all three groups of customers: White, Black/Hispanic, and Korean. Yet they depend on Black and Hispanic customers to a greater extent than is expected by chance. For example, although Blacks constituted 25 percent of the population in the four Atlanta counties where most Korean businesses were located, Black customers made up 44 percent of the total customers for Korean-owned businesses in Atlanta in 1982 (Min 1988a, p. 69). Only 9 percent of New York City Koreans live in Brooklyn, yet more than 25 percent of the New York City Korean businesses are located in this borough which is heavily populated by Blacks. Starting in 1985, Koreans in Los Angeles developed swap meets,

"a new form of retailing, in which individual sellers lease booths in a large building and sell their wares independently" (Chang 1990). Approximately 10,000 Koreans in southern California engage in indoor and outdoor swap meet businesses, and 80 percent of Korean-owned swap meets are located in Black/Mexican areas (my interviews with the president of the Korean Swap Meet Store Owners' Association in Los Angeles in 1992; see also Chang 1990). In addition, a large number of Korean-owned grocery/liquor stores, gas stations, and dry cleaning shops in Los Angeles are located in Black and Hispanic neighborhoods. Knowledgeable Korean experts indicate that some 70 percent of the businesses located in many Black neighborhoods in New York and Los Angeles are owned by Korean immigrants.

Why Are Korean Businesses Concentrated in Minority Neighborhoods?

There are two major types of Korean businesses in minority neighborhoods, and some structural factors encourage Korean immigrants to operate both types of businesses in Black neighborhoods. One type of Korean business heavily concentrated in Black neighborhoods is the grocery/liquor business. Predominantly White neighborhoods have major grocery chain stores, and therefore do not need Korean-owned small grocery stores. There are, however, few grocery chains in low-income Black neighborhoods. Due to the low spending capacity of the residents, high crime rates, and vandalism, big grocery chains are unwilling to invest in these areas (Light and Bonacich 1988; Min 1988a, p. 73). Moreover, many people in low-income Black neighborhoods do not have their own cars, and thus they depend on small neighborhood grocery stores for major grocery shopping. Accordingly, independent Korean grocers in low-income Black neighborhoods do not encounter as much competition as those in predominantly White areas. This suggests that racial segregation in urban residential patterns has created small business niches in low-income minority areas that have been filled by Korean and other immigrant merchants (Min 1988a).

The other major type of Korean business in minority neighborhoods is the fashion business that deals in Korean- and Asian-imported items such as wigs, handbags, hats, clothing, shoes, and so forth. Korean immigrants specialize in the fashion business mainly because they have easy access to Korean suppliers and importers, and Blacks and Hispanics are the major customers for these fashion stores. In addition to their easy access to coethnic suppliers, Korean retailers in minority neighborhoods have another advantage for operating the fashion business, and this advantage has much to do with the structural change in the U.S. economy. In the contemporary postindustrial economy, fashion cycles have become very short (Waldinger 1985). Independent Korean store owners can respond to fashion changes more effectively than department stores. Whereas big department stores, going through several channels for merchandise order, have difficulty getting new fashion items quickly, independent Korean store owners can get hot items the next day from Korean suppliers. This suggests that the restructuring of the U.S. economy in the late 1970s and early 1980s has created the opportunity for independent business owners to successfully compete against large department store owners in low-income minority areas (Chang 1990). In fact, the first Korean swap meet in south central Los Angeles was established in 1985 using a closed Sears building.

Black Hostility Toward Korean Merchants

Middleman minorities specializing in businesses in minority neighborhoods have received hostility and rejection from minority customers in different forms. For example, in the Indian riot in Duban in 1949, African residents attacked Indian stores, homes, and persons because they considered the Indian store owners and renters to be mainly responsible for their economic hardship (Hunt and Walker 1974). In the United States, many Jewish stores in Black neighborhoods became targets of systematic burning and looting in the riot of 1943 and the riots of the 1960s (Bender 1969; Capeci 1985; Cohen 1970). Korean merchants in Black neighborhoods, like middleman merchants in other societies, have been subject to all forms of hostility, rejection, and violent reactions by Blacks. Korean ghetto merchants in Los Angeles, New York, Philadelphia, Baltimore, and other cities have received blacks' hostility and rejection in several different forms: verbal and physical assaults, press attacks, murder, arson, boycotts, and looting.

Black hostility against Korean merchants has been most severe in New York and Los Angeles, the

two largest Korean communities in the United States. Five major Black boycott movements against Korean stores have occurred in New York since 1981. The boycott of two Korean produce stores in Brooklyn, New York, in January 1990 drew national media headlines (Min 1990, 1991). In the 1980s, Koreans in Los Angeles had maintained better relations with Blacks than those in New York. However, Korean-Black tensions were heightened in the spring of 1991 when a fourteen-year-old Black girl was shot to death in a Korean-owned grocery store while struggling over an unpaid bottle of orange juice with the female owner. Blacks were more angry about the court decision than the incident when Mrs. Du, who shot the black girl, was given a five-year probation in November 1991. Unfortunately, five months later, the jury verdict in the case of four White police officers accused of beating a Black motorist, Rodney King, of innocent sparked riots in Los Angeles. Approximately 2,300 Korean stores in south central Los Angeles and Koreatown became targets of destruction and looting in the Los Angeles riots in April 1992, which resulted in property damage of more than $350 million (see Table 3). Korean immigrants absorbed 45 percent of the total property damage incurred by the Los Angeles riots.

Many Korean stores were the targets of arson and looting in the Los Angeles riots mainly because a disproportionately large number of Korean stores were located in south central Los Angeles' Black and Hispanic neighborhoods. As previously indicated, the majority of grocery, liquor stores, and gas stations in many Black neighborhoods in south central Los Angeles were owned by Koreans. Yet there is evidence that Black rioters selectively targeted Korean-owned stores. A postriot investigation by the FBI indicated that Black gangs, who were responsible for planning the 1992 riots, consciously targeted Korean-owned stores for arson and looting (*Korea Times Los Angeles* 1992a). The Anti-Violence Coalition, a Los Angeles–based civil rights organization, monitored programs by FPKK, a major Black radio station in Los Angeles (90.7 FM), and reported that the radio station instigated Blacks to push Korean merchants out of Black neighborhoods and encouraged Blacks to target Korean-owned stores after the riot started (*Korea Times Los Angeles* 1992b).

What factors have contributed to Black hostility toward Korean merchants? Korean-Black conflicts commonly start with an altercation between Korean

Table 3 Damage (Looted or Burned) Statistics of Korean-Owned Businesses in Los Angeles During the Riots as of May 6, 1992 (compiled by *Radio Korea*)

Categories of Businesses	Number	Estimated Costs ($)
Car repair shops	61	9,707,800
Beauty salons	39	5,607,300
Dry cleaners	82	21,269,000
Electronic stores	60	16,085,000
Furniture stores	21	4,215,000
Gas stations	39	4,792,000
Jewelry shops	49	6,316,000
Liquor stores	187	41,812,000
Grocery markets	273	66,873,850
Swap meets	336	54,941,300
Restaurants	93	10,797,900
Clothing stores	222	34,458,734
Video shops	29	4,192,500
Others	376	65,894,810
Total	1,867	346,962,394

Source: Korea Times Los Angeles, English edition, 11 May 1992.

merchants and Black customers in Korean-owned stores. As analyzed elsewhere (Min 1993a), Korean merchants' language barriers, their cultural differences from Black customers, and mutual prejudice all contribute to the interracial tensions felt in business transactions between Korean merchants and Black customers. However, these social-psychological variables cannot explain Blacks' long-term boycotts of Korean stores and their targeting of Korean stores for arson and looting. Boycotts of Korean stores are usually organized by Black nationalists who consider outsiders' commercial activities in Black neighborhoods as economic exploitation. Thus, the Black nationalist ideology emphasizing the economic autonomy of the Black community and the perception of Korean merchants as economically exploiting Blacks are also important contributing factors to the black hostility toward Korean merchants. In addition, the exaggerated, unbalanced, and even biased media coverage of Korean-Black conflict has deteriorated the already strained relations between the two communities. Finally, inner-city Black economic problems and Korean immigrants' disadvantages for economic adjustments have provided structural conditions for Blacks' hostility toward Korean

merchants. Although the larger system is responsible for Black economic problems, Korean immigrants trapped in small business in Black neighborhoods bear the brunt of Blacks' economic frustrations because they are easy targets.

Business-related Intergroup Conflicts and Ethnic Solidarity

Members of a group tend to achieve internal solidarity when they encounter threats from the outside world. Korean merchants have encountered threats from different outside interest groups: Black customers, White suppliers, White landlords, government agencies, and labor unions. These business-related intergroup conflicts have, in turn, contributed to the solidarity of the Korean community.

Effects of Black Hostility on Ethnic Solidarity

Of all forms of business-related intergroup conflicts, Black hostility in the forms of armed robbery, physical violence, murder, arson, boycott, and looting has undoubtedly been the most serious type of threat Korean merchants have encountered. Therefore, Korean-Black conflicts have had the most pronounced effects on Korean internal solidarity. A series of conflicts with Blacks made Korean merchants more aware of the potential collective threat in this country from Blacks and have thus contributed to the fostering of solidarity among Korean merchants. By establishing local business associations, Korean merchants in Black neighborhoods tried to solve problems with Black customers and residents collectively. But Black boycotts of Korean stores in New York and other cities and the destruction of Korean stores in the Los Angeles riots threatened Korean immigrants' economic survival itself. Thus, they have led not only Korean ghetto merchants, but all Koreans, to be concerned about their common fate and marginal status in America.

While a number of both minor and major incidents of business-related conflict with Blacks have enhanced Koreans' solidarity, two recent events, the 1990 Black boycott of Korean stores in New York and the destruction of so many Korean stores in the 1992 Los Angeles race riots, have most significantly affected the Korean community in terms of their ethnic solidarity and political consciousness. This section will examine how the destruction of many Korean stores during the 1992 Los Angeles riots solidified the Korean community.

Koreans in Los Angeles were angry about the scale of destruction that the Korean community suffered during the riots. However, they were more outraged by what they believed to be the White conspiracy behind the destruction. In their view, the U.S. media intentionally focused on Korean-Black conflicts during the riots in order to vent Blacks' frustrations on Korean merchants. Angry about the biased media coverage, a number of Koreans sent letters of protest to the U.S. and Korean media. For example, in her letter to the editor of The *San Francisco Chronicle*, a Korean American woman wrote:

> We Korean Americans are angered by the wrong message, delivered to the public by the media coverage. The horror we have witnessed during the past few days is not because of animosity between blacks and Koreans. It is not a black vs. Korean issue. It is an American issue. This is a time when every citizen, every resident to America needs to stop and think. (Chey 1992)

On 2 May, Koreans in Los Angeles held a solidarity and peace rally at Ardmore Park in Koreatown, just the day after the riots came to an end. Some 30,000 Koreans from all over the Los Angeles area participated in the rally, the largest Korean meeting ever held in the United States (Sunoo 1992). Koreans also showed solidarity in their efforts to help the victims of the riots. In order to help the Korean victims of the riots in Los Angeles, Korean Americans all over the United States participated in fund-raising campaigns, which raised approximately $4.8 million (*Korea Times Los Angeles* 1992c).

Many Koreans in Los Angeles and other cities felt that Koreans became innocent victims of Black-White racial conflict in the 1992 riots mainly because of their lack of political power. Thus, the Los Angeles riots have heightened Koreans' political consciousness. Keenly aware of their need for improving political power, more and more Korean immigrants have applied for naturalization since the riots, and more of them have registered to vote. For example, a study based on county voter registration records estimated that 2,400 Koreans in Orange County, California, reg-

istered to vote in 1992, accounting for 30 percent of the Koreans who had registered since 1983 (*Korea Times Los Angeles* 1992d). In the 1992 elections, Jay Kim, representing the forty-first district in California, was elected to the U.S. House Representatives, becoming the first Korean member of Congress in American history. That same year, three other Koreans were elected as members of the state legislatures in Washington, Oregon, and Hawaii. They were elected largely because they received the support of conservative, White voters. It is, however, true that Korean support, not only as voters but also as helpers in the election campaigns, also contributed to their success.

The Los Angeles riots also heightened second-generation Koreans' sense of ethnic identity and ethnic solidarity. It was the first major event in Korean American history that provided 1.5- and second-generation Koreans with an opportunity to think about their common fate as Korean Americans. Many young Koreans suspected that the police did not care about protecting Koreatown, mainly because Korean Americans are a powerless minority group. Immediately after the riots, many young Koreans showed their quick response by writing articles in major English dailies in California and the English section (weekly) of *Korea Times Los Angeles*, attacking the media bias and the police inactivity in protecting Koreatown. Thomas Chung, a political science major at UCLA, responded to an interview by a *Korea Times* reporter:

> Korean Americans felt abandoned by the police and the whole justice system. It was really hard for me to watch the news because this is where I grew up and it was on fire everywhere. The culmination of the whole thing is just anger. Before, I felt alienated because I wanted to support the African American community, but the Korean American community was targeted purposely, and I felt like we were being used as scapegoats to channel their anger. On Saturday, when I went to the Koreatown rally, I was so glad I was there. After that, I realized we all have to pitch in. (Ha 1992)

A large number of 1.5- and second-generation Koreans participated in the solidarity rally held at Ardmore Park, Koreatown, on 2 May 1992, which made the rally a truly multigenerational rally. They realized that the riots exposed the political weakness of first-generation Koreans and that it is their obligation to protect the interests of their parents' generation. The Korean-American Coalition in Los Angeles, the Coalition of Korean-American Voters in New York, and several other 1.5- and second-generation Korean American associations in other cities held meetings in late 1992 in order to establish a national coalition organization similar to the Japanese Citizens' League.

Business-related Conflicts with Other Interest Groups and Ethnic Solidarity

Korean retail store owners generally depend on White wholesalers and manufacturers to supply their merchandise. They have been subject to some discrimination from White suppliers in terms of the quality of merchandise, price, the speed of delivery, and even parking allocations. However, Korean retailers have not passively accepted unfair treatment by White suppliers. Korean merchants in each business specialty area have established a trade association and used collective strategies to protect their interests against White suppliers. For example, Korean grocers in each of the major Korean communities established the Korean-American Grocers Association, and several local Korean grocers' associations established a national association in 1988. Both local and national Korean grocers associations organized grocery shows in which many grocery manufacturers and wholesalers paid large sums of money in order to advertise their grocery items. Korean produce retailers in New York, through their association, have organized six boycotts of White suppliers since the late 1970s, and every boycott was successful in making the targeted wholesaler accept their demands (Min 1993b).

A very small proportion of Korean merchants have their own buildings for business operations (only about 10 percent), and the others depend on White landlords. Koreans' dependence on White landlords puts them in a position vulnerable to exploitation. Typically, when Korean-owned businesses become successful, their landlords raise rents two or three times within a short period of time. Many Korean store owners have to keep their stores open long hours partly to cover escalating rents. However, the Korean merchants' conflict over economic interests with White landlords has also helped to solidify

them. For example, in the 1980s Korean business leaders actively lobbied the New York City mayor and council for the proposal of a new law that would regulate commercial rents (*Korea Times New York* 1985, 1992). They took a leading role in organizing three major multiethnic demonstrations in New York City, asking for legislation to regulate commercial rents. In the 1989 mayoral election, Korean business leaders supported David Dinkins over the incumbent mayor Ed Koch, because Dinkins promised to take some measures for rent regulation (*Sae Gae Times* 1989).

Conclusion

Post-1965 Korean immigrants are generally drawn from the middle class of Korean society. Although well educated, they have a language barrier and face other disadvantages for employment in the general labor market. However, they find niches for small business in minority neighborhoods where big corporations are reluctant to invest. Thus, Korean immigrants are highly entrepreneurial, and a significant proportion of Korean-owned businesses are located in Black and Hispanic neighborhoods.

For many Korean immigrants, small business is the main avenue through which they achieve economic mobility. However, their commercial activities have effects on their noneconomic lives. Like middleman minorities in other societies, Korean merchants specializing in business in minority neighborhoods have received hostility and rejection from minority customers in several different forms. This has been the case particularly in Black neighborhoods. Of all, boycotts and arson/looting are the two most severe forms of hostility. Many Korean stores in New York and Los Angeles were targets of long-term Black boycotts during the 1980s and 1990s. Yet the hostility of Blacks toward Korean merchants was most relentlessly expressed during the Los Angeles riots in April 1992. Approximately 23,000 Korean-owned stores became targets of destruction by Black and Hispanic rioters.

Koreans' business-related intergroup conflicts are a topic of great sociological interest. Yet even more sociologically interesting are the positive social functions of such intergroup conflicts on Koreans' ethnic solidarity. First of all, a series of conflicts with Blacks has made Korean store owners aware of the potential collective threats they face in this country and has thus contributed to the fostering of solidarity among Korean merchants. In many Korean communities, various Korean trade and local business associations have been established mainly to deal with the conflicts relating to Koreans' involvement in Black ghetto businesses. The long-term boycotts of Korean businesses in New York and Los Angeles and the 1992 Los Angeles riots led not only Korean business proprietors, but also other Koreans, to be concerned about their common fate and marginal status in the United States. The destruction of many Korean stores in Los Angeles, in particular, heightened second-generation Koreans' ethnic identity and political consciousness. Moreover, Korean merchants' conflicts with White suppliers, White landlords, and government agencies have also enhanced their ethnic solidarity. Korean trade associations have successfully organized several boycotts of White suppliers, and they have learned important political skills via their efforts to handle business-related intergroup conflicts.

The majority of second-generation Koreans, who do not have serious language difficulties, are likely to find white-collar and professional occupations in the general labor market. In addition, American-born Koreans are not attracted to labor-intensive small businesses. Thus, there will be a radical intergenerational transition in occupations from self-employment to employment in the general labor market. However, as long as the current scale of Korean immigration (a little less than 20,000 per year) remains unchanged, a large proportion of Korean Americans will continue to work in a segregated Korean ethnic economy. And as long as Korean Americans are economically segregated, their business-related intergroup conflicts and reactive solidarity will continue to be dominant features of their adjustment to American life. Moreover, the victimization of many Korean merchants during the Los Angeles riots, as a major historical event in the Korean community, will continue to have effects on the ethnic identity of second- and third-generation Koreans.

24

A Legacy of War: Refugees from Vietnam, Laos, and Cambodia

RUBÉN G. RUMBAUT

The image is dramatic and seared into the American national consciousness: It is 7:53 A.M. on 30 April 1975, as the helicopter lifts off the roof of the American embassy in Saigon, carrying the last remaining Marines away from the surrounding panic to the safety of the U.S. Seventh Fleet in the South China Sea. As the helicopter disappeared from view, two decades of American involvement in Indochina came abruptly to an end. But the end of one form of involvement was followed by the beginning of another.

Two decades later the images are less dramatic and clear-cut, but they are no less extraordinary: Tra Minh La, who had escaped from Vietnam with nine other family members on a sixty-foot wooden boat in 1979, became the manager of the $22 million Kim Son Restaurant that opened in 1993 in Houston's old Chinatown. Not far away another refugee from Vietnam, K-Bres, a fifty-eight-year-old former village chief of the highland Ma tribe, who was imprisoned for several years in a Vietnamese reeducation camp, spent most days in Houston rummaging through trash bins, collecting aluminum cans to survive—the cans earned him $40 a month to supplement the food stamps and federally subsidized housing that he and his wife, Ka-Loih, received (their refugee cash assistance had expired a few months before); he had to compete with five other residents for cans, and despite the stench stored the aluminum inside his apartment to guard against theft. At the Massachusetts Institute of Technology, Tue Nguyen, a twenty-six-year-old Vietnamese boat refugee, set an MIT record in 1988 by earning his seventh advanced degree, a doctorate in nuclear engineering, just nine years after arriving in the United States, and landed a job at IBM designing technology for the manufacture of semiconductors. Another boat refugee from Vietnam, Vu Thanh Thuy, a former war correspondent who drifted for days in the Gulf of Siam with her husband and two small daughters and survived harrowing attacks by Thai pirates, became in 1987 the first recipient of the National Organization for Women's "21st Century Woman Award" at a ceremony in New York City. In Santa Barbara, a preliterate Hmong woman from the Laotian highlands, recently converted to Christianity, asked her pastor if she could enter heaven without knowing how to read. In Chattanooga, Tennessee, a twelve-year-old Cambodian girl, Linn Yann, placed second in a 1983 regional spelling bee (ironically, she missed on "enchilada"); her story was made into a Disney TV movie, *The Girl*

Who Spelled Freedom. More Americans are likely to have read an occasional newspaper account of a Vietnamese high school valedictorian than they are to recall the 1989 massacre of Cambodian and Hmong children at Cleveland Elementary School in Stockton, California, by one Patrick Purdy, a racist drifter who fired over 100 rounds from an AK-47 assault rifle into a crowd of children before shooting himself. Only a few films, such as *The Killing Fields* in the 1980s and *Heaven and Earth* in the 1990s, have brought images of the war to American audiences from the point of view of Cambodian and Vietnamese protagonists. And none has told a tale such as that of a middle-aged Hmong refugee, who arrived in San Diego in 1980 after spending five years in refugee camps in Thailand, and who had this to say about his situation a few years later (as translated and quoted in Rumbaut 1985, p. 471–472):

Some nights the sleep hardly comes to me at all. . . . In our old country, whatever we had was made or brought in by our own hand, we never had any doubts that we would not have enough for our mouth. But from now on to the future, that time is over. . . . We are not born to earth to have somebody give us feed; we are so ashamed to depend on somebody like this. . . . Any jobs they have require a literate person to get. We have the arms and legs but we can't see what they see, because everything is connected to letters and numbers. . . . In this country everything is money first. You go to the hospital is money, you get medicine is money, you die is also money and even the plot to bury you also requires money. These days I only live day by day and share the $594 for the six of us for the whole month. Some months I have to borrow money from friends or relatives to buy food for the family. I'm very worried that maybe one day the welfare says you are no longer eligible for the program and at the same time the manager says that I need more money for the rent, then we will really starve. I've been trying very hard to learn English and at the same time looking for a job. No matter what kind of job, even the job to clean people's toilets; but still people don't even trust you or offer you such work. I'm looking at me that I'm not even worth as much as a dog's stool. Talking about this, I want to die right here so I won't see my future. . . . How am I going to make my life better? To get a job, you have to have a car; to have a car you have to have money; and to have money you have to have a job. . . . Language, jobs, money, living and so on are always big problems to me and I don't think they can be solved in my generation. So I really don't know what to tell you. My life is only to live day by day until the last day I live, and maybe that is the time when my problems will be solved.

The Newest Asian Americans

In the years following the end of the Vietnam War in 1975, over 1 million refugees and immigrants from Vietnam, Cambodia, and Laos arrived in the United States. Together with their American-born children, by 1990 they already represented more than one out of every seven Asian Americans, adding significantly not only to the size but to the diversity of the Asian-origin population in the United States. They are the newest Asian Americans, and the story of their migration and incorporation in America differs fundamentally from that of other Asian-origin ethnic groups.

To be sure, except for persons of Japanese descent, the overwhelming majority of Asian Americans today are foreign-born, reflecting the central role of contemporary immigration in the formation of these ethnic groups. But unlike the others, most of the Indochinese[1] have come as *refugees* rather than as *immigrants*. Unlike post-1965 immigrants form the Philippines, South Korea, China, India, and elsewhere in Asia whose large-scale immigration was influenced by the abolition of racist quotas in U.S. immigration law, the Indochinese have entered outside of regular immigration channels as part of the largest refugee resettlement program in U.S. history, peaking in 1980 and continuing ever since. As refugees from three countries devastated by war and internecine conflicts, they have experienced contexts of exit far more traumatic than practically any other newcomers in recent times, and they have had no realistic prospects of return to their homelands. Moreover, their reception as refugees reflects a different legal-political entry status, a status that facilitates access to a variety of public assistance programs to which other immigrants are not equally entitled. The American *welfare* state has shaped their incorporation far more than any other immigrant group in U.S. history, even as their exodus and resettlement were themselves complex, unintended consequences of the intervention and ultimate failure of U.S. foreign policies and of the American *warfare* state. Indeed, the Indochinese case underscores the need to attend carefully to historical contexts, and particularly to the role of the state and of war itself, in explaining specific types of migrations and ethnic group formations.

The Vietnamese, Laotians, and Cambodians do not share a history of several generations in America as do the Chinese and Japanese—a history marked early on by harshly discriminatory treatment and official exclusion—nor of a half-century of direct U.S. colonization as do the Filipinos. At first they could not be resettled into coethnic communities previously established by earlier immigration, since they essentially were nonexistent prior to 1975; and in the resettlement process they were more likely to be dispersed throughout the country than other large immigrant groups. Unlike recent Asian immigrant flows, most notably those from India, which have been characterized by large proportions of highly educated professionals and managers, the Indochinese flows, with the notable exception of the "first wave" of 1975 evacuees from South Vietnam, have been characterized by far larger proportions of rural-origin and less-educated peoples than any other Asian immigrant group in decades. There are also significant contrasts with other large refugee groups: for example, unlike refugees from Cuba and the former Soviet Union, who are among the oldest populations in the United States, the Indochinese comprise the youngest populations in the United States, with median ages of less than twenty for all groups except the Vietnamese, partly a reflection of high levels of fertility. All of these particular sociodemographic characteristics and contexts of exit and reception have shaped their modes of adaptation to the American economy and society.

While as refugees of the Vietnam War they generally share a common history and experiences that distinguish them from other Asian American groups, the various Indochinese ethnic groups—Vietnamese, Khmer (Cambodian), lowland Lao, Hmong, Mien and other Laotian and Vietnamese highlanders, and ethnic Chinese from all three countries—also differ from each other in equally fundamental ways. As will be elaborated in this chapter, they have different social backgrounds, languages, and culture, often adversarial histories, and they reflect different patterns of settlement and adaptation in America. They range from members of the elite of the former U.S.–backed governments to Vietnamese and Chinese "boat people," survivors of the "killing fields" of Cambodia in the late 1970s, and preliterate swidden farmers from the highlands of northern Laos. And within each of these ethnic groups there are major differences, especially by social class of origin, between different "waves" or cohorts of arrival, and by gender and generation. Tens of thousands of Amerasians—children of Vietnamese mothers and American fathers who served in Vietnam during the war—have also been resettled in the United States under a special law enacted in 1987 (the "Amerasian Homecoming Act"); much discriminated against and stigmatized as *bui doi* (children of the "dust of life"), they too form yet another distinct and poignant legacy of the war.

This chapter cannot consider each of these points in detail, but will aim rather to provide an overview of the most salient patterns. We begin, however, with a brief discussion of the war that led to the formation of Vietnamese, Cambodian, and Laotian communities in America. One of the ironies of the war that took America to Vietnam, and of the war's expansion into Cambodia and Laos, is that a sizable part of Vietnam, and also of Cambodia and Laos, has now come to America.

A Legacy of War: Indochinese Refugees in Historical Perspective

The Indochinese refugees are a product of the longest war in modern history—the thirty-year Vietnam War (1945 to 1975) and its metastasis into Laos and Cambodia in the 1960s and early 1970s (see Karnow 1991). An immensely complex conflict that still creates bitter controversy and whose full significance will continue to be assessed and debated for years to come, the war was a tragedy of staggering proportions for Americans, Vietnamese, Cambodians, and Laotians alike. With the exception of the American Civil War a century earlier, the Vietnam War became the most divisive event in U.S. history. By war's end about 2.2 million American soldiers had served in Vietnam—their average age was nineteen, five to seven years younger than in other American wars—and almost 58,000 died there or were missing in action, their names memorialized in a wall of polished black granite dedicated in 1982 in the nation's capital. The war also cost the United States over $120 billion from 1965 to 1973 alone, triggering a postwar inflation and an economic chain reaction that shook the world economy. The war defined an entire generation of young people in the 1960s, polarized the American electorate into "hawks" and "doves," and led to President Lyndon Johnson's early retirement from politics in 1968 and to the Watergate scandals of the Nixon administration in the early 1970s. "The first war that the U.S. ever lost" produced a "Vietnam syndrome" whose political ramifications still affect the formulation of American foreign policy—for example, President George Bush promised "no more Vietnams" before launching Operation Desert Storm in 1991—and the trajectory of national elections—for example, the controversies over the Vietnam-era draft status of former Vice President Dan Quayle in 1988 and of President Bill Clinton in 1992.

The war also produced a massive refugee population for whom the United States assumed a historic responsibility. Not coincidentally, Vietnam represents at once the worst defeat of U.S. foreign policy in the cold war era and the leading example (with Cuba) of the functions of U.S. refugee policy; Vietnamese (and Cubans) admitted as political refugees into the United States have served as potent symbols of the legitimacy of American power and global policy. The circumstances of the U.S. withdrawal from Vietnam, the dramatic fall of Saigon and its aftermath—and indeed, the extent to which such refugee flows were a dialectical consequence of U.S. foreign policy in an era of East-West superpower rivalry— also provided added moral and political justification for significantly expanded domestic refugee programs, which totaled some $5 billion in cash, medical assistance, and social services to primarily Indochinese refugees from 1975 to 1986 alone (Rumbaut 1989b; cf. Zolberg, Suhrke, and Aguayo 1989; Zucker and Zucker 1987).

If the war divided America, it devastated Vietnam, Laos, and Cambodia. During the period of U.S. involvement after the defeat of the French at Dien Bien Phu in 1954, it is estimated that over 4 million Vietnamese soldiers and civilians on both sides were killed or wounded—a casualty rate of nearly 10 percent of the total population. The total firepower used by the United States on Vietnam exceeded the amount used by the United States in all its previous wars combined, including both world wars. In South Vietnam alone, about a third of the population was internally displaced during the war, and over half of the total forest area and some 10 percent of the agricultural land was partially destroyed by aerial bombardment, tractor clearing, and chemical defoliation [research on the long-term health effects on the local population of the dumping of more than 11 million gallons of the toxic defoliant Agent Orange is still fragmentary, but see Korn (1991) and Schecter et al. (1995)]. In Laos the war exacted its greatest toll on the Hmong, an ethnic minority from the rural highlands who had fought on the U.S. side against the Pathet Lao; before the fall of Vientiane about a third of the Hmong population had been uprooted by combat, and their casualty rates were proportionately ten times higher than those of American soldiers in Vietnam. In Cambodia, whose fate was sealed after the expansion of the war in 1970, as many as a quarter of its people may have died during the horror of the late 1970s. The war shattered the region's economy and traditional society. A tragedy of epic proportions, the "war that nobody won" left these three countries among the poorest in the world. In an economic ranking of 211 countries in the mid-1980s, Vietnam was ranked 202nd (with a per capita national annual income of roughly $130), Laos 208th ($100 per capita), and Cambodia 211th (the world's poorest at $50 per capita) (Rumbaut 1991a).

Since the end of the war in 1975, over 2 million refugees are known to have fled Vietnam, Laos, and Cambodia. The refugee exodus was shaped by complex political and economic factors. As is true of refugee movements elsewhere, the first waves of Indochinese refugees were disproportionately comprised of elites who left because of ideological and political opposition to the new regimes, while later flows included masses of people of more modest backgrounds fleeing continuing regional conflicts and deteriorating economic conditions. Vietnamese professionals and former notables were greatly overrepresented among those who were evacuated to American bases in Guam and the Philippines under emergency conditions during the fall of Saigon; Laotian and Cambodian elites, by contrast, were much more likely to have gone to France (the former colonial power in Indochina, where French-speaking Indochinese communities, particularly in Paris, had developed as a result of more than half a century of previous migration). Among the first to flee on foot across the Mekong River into Thailand were the

Hmong, but they were the least likely to be resettled by Western countries at the time; most were to languish in Thai camps for years thereafter. In Vietnam and Laos, meanwhile, several hundred thousand persons with ties to the former regimes were interned in "reeducation camps" (many years later, beginning in 1989, over 50,000 of those former Vietnamese political detainees would be resettled in the United States under special legislation). In Cambodia the cities were deurbanized as the population was forced into labor camps in the countryside; the capital of Phnom Penh became a ghost town practically overnight. But the exodus of the 1975 refugees was only the start of an extraordinary emigration.

A massive increase of refugees beginning in late 1978 was triggered by the Vietnamese invasion of Cambodia, which quickly ended three years of Khmer Rouge rule; the subsequent border war between Vietnam and China in early 1979, which accelerated the expulsion of the ethnic Chinese *petit bourgeoisie* from Vietnam; a new guerilla war in the Cambodian countryside, already wracked by famine and the destruction of the country's infrastructure; and the collapse of both the Chao Fa guerilla resistance against the Pathet Lao and the new system of collective agriculture in Laos, compounded by mismanagement and natural catastrophes. Hundreds of thousands of Cambodian survivors of the Pol Pot labor camps fled to the Thai border along with increased flows of Hmong and other refugees from Laos; about 250,000 ethnic Chinese from North Vietnam moved across the border into China; and tens of thousands of Chinese and Vietnamese "boat people" attempted to cross the South China Sea packed in rickety crafts suitable only for river travel, many of whom drowned or were assaulted by Thai pirates preying on refugee boats in the Gulf of Thailand. By spring 1979 nearly 60,000 boat people were arriving monthly in the countries of the region. These events led to an international resettlement crisis later that year when those "first asylum" countries (principally Thailand, Malaysia, and Indonesia) refused to accept more refugees into their already swollen camps, often pushing boat refugees back out to sea, where many perished—Malaysia alone pushed some 40,000 out—or forcing land refugees at gunpoint back across border minefields (U.S. Committee for Refugees 1985, 1986b, 1987). In response, under agreements reached at the Geneva Conference in July 1979, Western countries began to absorb significant numbers of the refugee camp population in Southeast Asia.

By 1992, over 1 million had been resettled in the United States, 750,000 in other Western countries (principally Canada, Australia, and France), and many others still languished in refugee camps from the Thai–Cambodian border to Hong Kong. Harsh "humane deterrence" policies and occasional attempts at forced repatriation sought to brake the flow of refugees to first-asylum countries, with limited success. After 1979 the number of boat refugee arrivals declined, but never dropped below 20,000 annually (until abruptly coming to a halt in 1992), exacting a horrific cost in human lives: it has been estimated that at least 100,000 boat people, and perhaps over twice that number, drowned in the South China Sea (U.S. Committee for Refugees 1987). Beginning in the 1980s, an Orderly Departure Program (ODP) allowed the controlled immigration of thousands of Vietnamese directly from Vietnam to the United States, including two groups with a unique tie to the war: Amerasians and former reeducation camp internees. By 1992, over 300,000 Vietnamese had immigrated to the United States through the ODP, including 161,400 in the regular family reunification program, 81,500 Amerasians and their accompanying relatives, and 61,000 former political prisoners and their families (U.S. Committee for Refugees 1993; U.S. GAO 1994). But both the Amerasian program and refugee processing for former political prisoners were to end by 1995. In Laos and Cambodia, meanwhile, refugee flows had virtually ended by the early 1990s, with the focus shifting to the voluntary repatriation of refugees still in camps in Thailand and elsewhere.

Indeed, an entire era was coming to a close, while a new phase of the Indochinese diaspora was opening. The end of the cold war in 1989, the collapse of the former Soviet Union in 1991, United Nations–supervised elections in Cambodia in 1993 that sought to end its long-running civil war, and the end of the U.S. trade embargo against Vietnam in February 1994 were but the most remarkable events of a compressed period of extraordinarily rapid and fundamental changes in international relations that has transformed the nature of Indochinese refugee resettlement in the United States. In this post–cold war context, the United Nations High Commissioner for Refugees proclaimed the 1990s as the "decade of repatriation." Already most of the ODP family reunification cases in the 1990s have been leaving Vietnam as regular immigrants, not as refugees, a pattern likely to become more pronounced over time; and the flows from Laos and Cambodia to the United States have been, respectively, sharply reduced and virtually terminated. For some first-generation Indochinese adults exiled in America, the new developments in their homelands may permit the possibility of return or of establishing business and other linkages between their native and adoptive countries; but for a sizable and rapidly growing second generation of young Vietnamese, Laotian, and Cambodian Americans now rooted in communities throughout the United States and speaking accentless English, a new era was dawning in which the legacy of war will likely increasingly recede in practical importance. Theirs is an American future.

The Indochinese in America: A National Perspective

The research literature that has accumulated on Indochinese Americans is surprisingly large, especially when compared to that of larger groups with much longer histories in the United States. As state-sponsored immigrants, their refugee status has provided them not only with greater access to various forms of government assistance, but in some respects it may have also made them an "overdocumented" population in comparison with other immigrants. Indeed, the 1975 refugees in particular may be the most closely studied arrival cohort in U.S. history. In this section we summarize data from various national-level sources to describe their patterns of immigration, settlement, and socioeconomic progress.

Immigration History and Population Size

Among Asian Americans, the Indochinese constitute the most recently formed ethnic groups. According to the Immigration and Naturalization Service (INS), the first recorded Vietnamese immigration to the United States occurred in 1952, when eight immigrants were admitted; the first Cambodian immigrant arrived in 1953, and the first Laotian in 1959 (U.S. INS 1991). As late as 1969, fewer than 200 Cambodians and Laotians combined had immigrated to the United States, and the total from Vietnam amounted to little more than 3,000—mostly university students from elite families as well as diplomats and war brides who had come in the late 1960s in the wake of the rapid expansion of U.S. involvement in Vietnam. As Table 1 shows, in the early 1970s the number of these pioneer immigrants quintupled to nearly 15,000 from Vietnam, while increasing only slightly to nearly 300 each from Cambodia and Laos. Thus, when Saigon fell in April 1975, the Vietnamese in America numbered around 20,000, while the number of Cambodians and Laotians was still negligible.

About 130,000 refugees, nearly all from South Vietnam, were resettled in the United States during 1975. A small number arrived from 1976 to 1978, bottoming out in 1977, but a massive new inflow began in late 1978 in the context of the international refugee crisis described earlier. As Table 1 shows, about 450,000 Indochinese refugees arrived en masse from 1979 to 1982 alone, peaking in 1980 (the record year in U.S. refugee resettlement history) when 167,000 were admitted. Compared to the 1975 "first wave," this new wave of refugees was not only much more numerous but also much more heterogeneous: they included the "boat people" from Vietnam, the survivors of the Pol Pot period in Cambodia, and the lowland Lao and the highland Hmong, many of them coming from rural backgrounds, with little education, knowledge of English, or transferable occupational skills, and having endured prolonged stays in refugee camps overseas. What is more, the timing and context of their entry into the United States further complicated their reception: the peak year of their arrival (1980) coincided both with the crisis of the tens of thousands of Mariel Cubans and Haitians arriving in Florida in small boats and with the highest domestic inflation rates in memory, followed during 1981 and 1982 by the most severe economic recession since the Great Depression. The confluence of these events in turn contributed to an accompanying political climate of intensifying nativism, racism, xenophobia, and "compassion fatigue."

Since 1982, Indochinese arrivals have oscillated between 40,000 and 80,000 annually. Vietnamese *refugee* admissions, which totaled over 650,000 from 1975 to 1992, have been supplemented by a substantial though little noticed flow of over 170,000 nonrefugee Vietnamese *immigrants* who arrived in the United States during the same period—the latter in-

Table 1   Arrivals in the United States from Cambodia, Laos, and Vietnam, 1952–1992: Refugees, 1975–1992; Nonrefugee Vietnamese Immigrants, 1975–1992; and pre-1975 Immigrants

Fiscal Year	Post-1975 Refugee Arrivals			Nonrefugee Immigrants	Total Arrivals
	Cambodia	Laos	Vietnam	Vietnam[a]	
1992	193	7,272	26,841	45,580	79,886
1991	199	9,232	28,450	33,764	71,645
1990	2,323	8,719	27,714	28,271	67,027
1989	1,916	12,432	22,664	15,880	52,892
1988	2,805	14,556	17,654	4,391	39,406
1987	1,539	15,564	23,012	3,635	43,750
1986	9,789	12,869	22,796	6,068	51,522
1985	19,097	5,416	25,457	5,134	55,104
1984	19,851	7,291	24,818	5,244	57,204
1983	13,114	2,835	23,459	3,290	42,698
1982	20,234	9,437	43,656	3,083	76,410
1981	27,100	19,300	86,100	2,180	134,680
1980	16,000	55,500	95,200	1,986	168,686
1979	6,000	30,200	44,500	2,065	82,765
1978	1,300	8,000	11,100	2,892	23,292
1977	300	400	1,900	3,194	5,794
1976 (+TQ)	1,100	10,200	3,200	4,201	18,701
1975	4,600	800	125,000	3,038	133,438
Subtotal (1975–1992)	147,460	230,023	653,521	173,896	1,204,900

Period	Pre-1975 Immigrant Arrivals				Total Arrivals
	Cambodia	Laos	Vietnam		
1970–1974[b]	286	292	14,661		15,239
1960–1969	98	69	3,167		3,334
1952–1959	6	1	219		226
Subtotal (1952–1974)	390	362	18,047		18,799
Total (1952–1992)	147,850	230,385	845,464	(included)	1,223,699

[a] Totals include 55,985 Amerasians from Vietnam admitted as immigrants in FY 1989–1992.
[b] Totals include 98 Cambodians and 96 Laotians who entered as immigrants in FY 1975.
Source: Compiled from records maintained by the Statistics Division, U.S. Immigration and Naturalization Service; U.S. State Department; and U.S. Office of Refugee Resettlement. Refugee arrivals from 1975 to 1981 are rounded to the nearest hundred; the totals for 1976 include a transition quarter (TQ) as a result of fiscal year changes.

cluding persons coming to the United States from other countries and from Vietnam through the Orderly Departure Program, among them the young Amerasians and their accompanying relatives mentioned earlier. By 1992, as detailed in Table 1, total Indochinese arrivals numbered 1,223,699: 147,850 Cambodians (12 percent), 230,385 Laotians (19 percent), and 845,464 Vietnamese (69 percent). Of that total, 86 percent entered as refugees (the remainder as immigrants), and four out of five arrived in the United States only since 1980. U.S. government agencies collect these data only by nationality, not ethnicity, so it is not possible to determine the proportion of ethnic Chinese or other minority groups among them; however, the U.S. Office of Refugee Resettlement has estimated that of the 213,519 Laotian arrivals from 1975 to 1990, 92,700 were highlanders, primarily Hmong.

The immigration statistics do *not* include their children born in the United States, which a study of their patterns of fertility estimated at nearly 200,000 by 1985 (Rumbaut and Weeks 1986). Allowing for natural increase, and adjusting for mortality (which is low since this is a very young population) and emigration (which is negligible), Bouvier and Agresta (1987) projected a 1990 Indochinese population of over 1,300,000, including 859,600 Vietnamese, 259,700 Laotians, and 185,300 Cambodians. But the 1990 U.S. census enumerated a Vietnamese population of only 614,547 (well below even the number of actual *arrivals* from Vietnam since 1975—see Table 1); a Laotian population of 239,096 (including 90,082 Hmong); and a Cambodian population of 147,411. Taken together, the 1990 census count of these Indochinese groups totaled just over 1 million; well *below* all available projections. What accounts for such a huge disparity, especially among the Vietnamese? The gap is too large to be explained by a census undercount, which probably did not exceed 5 percent among the Vietnamese in particular. One clue is that the 1990 census also counted a Chinese population of 1,645,472—well *above* what had been expected [Bouvier and Agresta (1987) had projected a 1990 Chinese population of 1,259,038]. It appears that sizable numbers of ethnic Chinese from Vietnam, Laos, and Cambodia indicated their ethnicity as Chinese in response to the appropriate census question. Earlier research had estimated that ethnic Chinese comprised up to 25 percent of total Vietnamese arrivals, and up to 15 percent of total Cambodian arrivals. Applying these proportions to the 1990 census figures would yield an additional 200,000 from Vietnam and 20,000 from Cambodia, bringing the total Indochinese population more closely in line with both immigration and natural increase data.

Patterns of Settlement

The 130,000 (mostly Vietnamese) refugees who arrived in the United States in 1975 were sent initially to four government reception centers—at Camp Pendleton, California; Fort Indiantown Gap, Pennsylvania; Fort Chaffee, Arkansas; and Eglin Air Force Base, Florida—where they were interviewed by voluntary agencies and matched with sponsors throughout the country, including individuals,

church groups, and other organizations. U.S. refugee placement policy aimed to disperse the refugee population to all the fifty states in order to minimize any negative impacts on receiving communities ("to avoid another Miami," as one planner put it, referring to the huge concentration of Cuban refugees there), and indeed the 1975 Indochinese refugees were more significantly dispersed than other immigrant or refugee populations. One study (Baker and North 1984) found that the refugees were initially placed in 813 separate zip code areas in every state, including Alaska, with about two-thirds settling in zip code areas that had fewer than 500 refugees and only 8.5 percent settling in places with more than 3,000 refugees. Less than half were sent to the state of their choice. Despite this general pattern of dispersal—shaped by government policy, the availability of sponsorships, and the relative absence of family ties and previously established ethnic communities in the United States—areas of Indochinese concentration nonetheless began to emerge, particularly in California, and to grow rapidly as a result of secondary migration from other states. Significantly, by 1980, 45 percent of the 1975 arrivals lived in a state other than the one where they had been originally sent; the proportion in zip code areas with fewer than 500 refugees had dropped to 40 percent, while those residing in places with more than 3,000 more than doubled to 20 percent; and the refugee population living in California had doubled from about 20 percent to 40 percent, where they were concentrated in southern California metropolitan areas (Los Angeles, Orange and San Diego counties) and to the north in the "Silicon Valley" city of San Jose.

As the much larger waves of Indochinese refugees began to arrive in the late 1970s and especially during the 1980s, their patterns of settlement continued to be shaped by the factors noted above, especially by the social networks that were becoming increasingly consolidated over time. Government policies and programs (such as the Khmer Guided Placement Project, dubbed the "Khmer Refrigerator Project" by Cambodians because of its frostbelt locations) sought the dispersal of refugees without family ties away from high impact areas, while most others were reunited with family members already residing in areas of high concentration. Remarkably, by the early 1980s about a third of arriving refugees already had close relatives in the United States who could serve as sponsors, and another third had more distant relatives, leaving only the remaining third without kinship ties subject to the dispersal policy (Hein 1993).

In addition, different localities of concentration emerged for the different ethnic groups, with the largest Cambodian community in the nation developing in the Long Beach area of Los Angeles County, the Lao in San Diego, and the Hmong in Fresno and surrounds in California's agricultural San Joaquin Valley (in the Fresno telephone directory, the Vangs—one of but two dozen Hmong clan names—are as numerous as the Joneses). By 1990 the largest Vietnamese concentration in the United States was found in Orange County, with its hub in the communities of Santa Ana and Westminster ("Little Saigon"); among recent Orange County home buyers, the Nguyens outnumbered the Smiths 2 to 1. Los Angeles, San Jose (where the Nguyens outnumbered the Joneses in the phone book fourteen columns to eight), San Diego, and Houston followed in rank order. But Indochinese Americans continue to reside in every state of the nation, and their patterns of settlement differ in some significant ways from those of other Asian Americans.

While California is home for 12 percent of the total U.S. population, 39 percent of the Lao—and of all Asian Americans, for that matter—live there; the degree of concentration in California is even greater for the Vietnamese (46 percent), the Cambodians (46 percent), and the Hmong (52 percent). By contrast, while over 20 percent of Asian Americans reside in New York/New Jersey and Hawaii (compared to 10 percent of the U.S. population), less than 4 percent of all Indochinese have settled in those states. After California, the Vietnamese are most concentrated in Texas (11 percent), with sizable communities in Houston and Dallas (which began to be formed by the 1975 cohort, attracted by employment opportunities), and along the Gulf coast (especially of shrimp fishers). Remarkably, the Vietnamese are already the largest Asian-origin group in Texas as well as in the contiguous states of Louisiana, Mississippi, Arkansas, Kansas, and Oklahoma. After California, another 17 percent of all Cambodians are concentrated bicoastally in Massachusetts and Washington, and despite their comparatively small numbers they are the largest Asian-origin group in Rhode Island. The Lao are the most dispersed among the Indochinese groups, and they do not predominate in any state; but the Hmong are the most concentrated, with another 37 percent located in the contiguous states of Minnesota and Wisconsin, where they are by far the largest Asian-origin groups—an extraordinary development considering that prior to 1975 there had been virtually no immigration from Laos to America.

Social and Economic Characteristics

Table 2 presents a summary of demographic and socioeconomic characteristics from the 1990 census, comparing the main Indochinese ethnic groups to each other and to the total U.S. and Asian-origin populations. These data underscore the significant differences between the various Indochinese ethnic groups and between the Indochinese and other Asian Americans. All of the Indochinese groups are much younger than other Asians or the total U.S. population, reflecting their much higher levels of fertility. American and Asian American women average just under 2 children born per woman aged 35 to 44 (an approximate measure of completed fertility), compared to 2.5 for the Vietnamese, 3.5 for the Lao and Cambodians, and 6.1 for the Hmong (the highest in the country). Thus, despite the recency of their arrival, over a third of the Hmong (35 percent) are already U.S.–born; amazingly, their median age was under thirteen years (compared to thirty-three for the American population), and that of U.S.–born Hmong Americans was just five years.

These indicators vividly demonstrate the dynamics of new ethnic group formation through immigration and rapid natural increase, and underscore the socioeconomic importance among the Indochinese of families with a high proportion of dependent children. While less than half of American households have children under age eighteen, over two-thirds of the Vietnamese, over four-fifths of the Lao and Cambodian, and 90 percent of the Hmong households consist of families with minor children. The structure of these families is a key social context shaping the adaptation of these recently resettled groups—including their efforts at collective pooling or "patchworking" of economic resources amid constant tension over changing gender roles and intergenerational conflicts (Kibria 1993). Indochinese groups, and Asian Americans generally—who are preponderantly foreign-born, as Table 2 shows—exhibit a smaller proportion of single-parent female-headed households than the U.S. norm; the main exception are Cambodian refugees, whose higher rate (25 per-

Table 2 Social and Economic Characteristics of Vietnamese, Cambodian, and Laotian Groups in the United States, Compared to the Total U.S. Population and the Total Asian-Origin Population, 1990

	U.S.	Asian	Vietnamese	Cambodian	Laotian	Hmong
Total persons	248,709,873	6,876,394	593,213	149,047	147,375	94,439
Nativity and immigration						
Born in the U.S. (%)	92.1	34.4	20.1	20.9	20.6	34.8
Immigrated pre-1980 (%)	4.5	27.9	30.5	9.4	16.3	15.7
Immigrated 1980–1990 (%)	3.5	37.8	49.3	69.6	63.1	49.5
Age						
Median age	33.0	29.2	25.6	19.7	20.5	12.7
Median age (U.S.-born only)	32.5	14.7	6.7	4.7	5.4	5.2
Family contexts[a]						
Fertility (per woman 35–44)	2.0	1.9	2.5	3.4	3.5	6.1
Female householder (%)	16.0	11.5	15.9	25.4	11.3	13.6
With own children <18 yrs. (%)	48.2	59.2	69.0	83.8	82.8	90.1
Children <18 with 2 parents (%)	73.0	84.6	76.6	71.0	82.6	86.2
English (persons over 5)[b]						
Speak English only (%)	86.2	24.6	6.2	4.0	3.2	2.6
Does not speak "very well" (%)	6.1	39.8	60.8	70.0	67.8	76.1
Linguistically isolated (%)	3.5	25.1	42.1	54.7	51.5	59.8
Education (persons over 25)						
Less than 5th grade (%)	2.7	7.1	11.4	40.7	33.9	54.9
High school graduate (%)	75.2	77.6	61.2	34.9	40.0	31.1
College graduate (%)	20.3	37.7	17.4	5.7	5.4	4.9
Postgraduate degree (%)	7.2	14.4	4.4	1.6	1.5	1.5
Employment (persons over 16)[c]						
In labor force (%)	65.3	67.4	64.5	46.5	58.0	29.3
Unemployed (%)	6.3	5.2	8.4	10.3	9.3	17.9
Of those employed						
Upper white-collar (%)	26.4	31.2	17.6	9.8	5.0	12.8
Lower white-collar (%)	31.7	33.3	29.5	23.3	15.2	18.9
Upper blue-collar (%)	11.4	7.8	15.7	17.2	19.8	13.9
Lower blue-collar (%)	14.9	11.9	20.9	30.0	43.9	32.1
Income[d]						
Median family income ($)	35,225	41,583	30,550	18,126	23,101	14,327
Per capita income ($)	14,420	13,806	9,033	5,121	5,597	2,692
Below poverty (%)	13.1	14.0	25.7	42.6	34.7	63.6
Receives public assistance (%)	7.5	9.8	24.5	51.1	35.4	67.1
Own home (%)	64.2	48.3	40.1	19.7	24.0	11.1

[a] Fertility: children ever born per woman aged 35 to 44.

[b] Linguistically isolated: a household in which no person age fourteen or older speaks English only or very well.

[c] UW-C: professionals, executives, managers; LW-C: clerical, sales; UB-C: repair, craft; LB-C: operators, fabricators, laborers.

[d] Percent of persons below the federal poverty line; percent of households receiving public assistance income.

Source: U.S. Bureau of the Census, *1990 Census of Population: Asians and Pacific islanders in the United States*, CP–3–5 (1993). These data are based on a sample (not the 100 percent tabulations) and are subject to sample variability; group sizes may differ from Table 2.

cent) reflects the disproportionate presence of many widows whose husbands were killed during the Pol Pot period of the late 1970s (one study in San Diego found that more than 20 percent of Cambodian women were widowed; Rumbaut 1989a). Among the Vietnamese, the proportion of children under eighteen living at home with both parents is slightly above the U.S. norm of 73 percent, but well below that of Asian Americans generally (85 percent)—partly a reflection of the sizable number of unaccompanied refugee children from Vietnam and of youths (disproportionately males) who escaped with other relatives or adult guardians.

As the most recently arrived Asian Americans, the 1990 census made clear—not surprisingly—that the substantial majority of the Vietnamese, Cambodians, and Laotians did not yet speak English "very well;" in fact, about half of all Indochinese households were classified by the census as "linguistically isolated" (see Table 2 for a definition). While 38 percent of all Asian American adults were college graduates and 14 percent also had postgraduate degrees—about double the respective levels of attainment of the U.S. population (20 percent and 7 percent)—all of the Indochinese groups were much less educated on average, particularly the non-Vietnamese groups, as detailed in Table 2. Only about a third of the refugees from Laos and Cambodia were high school graduates, while higher proportions had less than a fifth-grade education, underscoring the rural origins and severe social class disadvantages of many refugees in these ethnic groups.

Relative to the U.S. population, Asian Americans as a whole also showed higher rates of labor force participation, lower unemployment, and a greater percentage of professionals and managers among those employed; but the profile for each of the Indochinese groups was precisely the opposite in each of these indicators. Employed Indochinese were twice as likely to have jobs as operators and laborers (significant levels of downward occupational mobility have been noted among the earlier arrivals), and their levels of self-employment were significantly below those for other Asian Americans and the U.S. population, as were family and per capita incomes and rates of homeownership. Moreover, as shown in Table 2, poverty rates for the Indochinese groups were two to five times higher than for the U.S. population, and the disparity in welfare dependency rates was even greater. Approximately one-fourth of the

Vietnamese fell below the federal poverty line and received public assistance income, as did one-third of the Lao, about half of the Cambodians, and two-thirds of the Hmong (the latter are probably by far the highest rates in the country). By comparison, poverty rates for the U.S. and Asian American general populations were about the same (13 to 14 percent), with fewer than a tenth of households relying on public assistance. The diversity of these socioeconomic profiles underlines the widely different social class origins, age and family structures, and modes of incorporation of Asian-origin immigrants and refugees, and the senselessness of "model minority" stereotypes. These census data, however, tell us little about the equally significant differences *within* ethnic groups, especially between different "waves" or cohorts of arrival, and of the dynamics of their socioeconomic progress over time. For that we turn to other nationwide and local-level data sources.

Occupational and Economic Progress Over Time

One useful longitudinal data source on the economic progress of the 1975 to 1979 refugees comes from their federal income tax returns—a data source unique among all immigrant groups in the United States. Indochinese refugees who arrived between 1975 and late 1979 were issued Social Security numbers in blocks through a special program in effect at the time, and as a result it is possible to obtain annual aggregate data from the Internal Revenue Service on incomes received and taxes paid by these cohorts. For example, we can compare the median adjusted gross income received annually during the 1980s by the first wave of refugees who came in 1975 (who, as noted above, numbered 130,000, almost all Vietnamese) to the median income of the 1976 to 1979 arrivals (a much more heterogeneous group, including about 60,000 Vietnamese, 49,000 Lao and Hmong, and 9,000 Cambodians—most of whom arrived in 1979 and were thus about four years behind the 1975 cohort). The data show clearly the economic progress of these groups over time; worth noting is the fact that since 1985 the median income of the 1975 refugees has surpassed that of all U.S. tax filing units. As would be expected, the incomes of the 1976 to 1979 cohort still lagged noticeably behind the incomes of the 1975 cohort, but the gap between them has been closing rapidly: their income ratio (of the

1975 to the 1976 to 1979 cohort) was 1.62 in 1982 but had dropped to 1.24 by 1988. Overall, their income from wages more than doubled during the 1980s; still, as late as 1988 over a quarter (28.6 percent) of individual W-2 forms were under $5,000, while less than a fifth (19.6 percent) of W-2 forms were over $25,000. In 1988, more than 10,700 tax returns reported income from self-employment, totaling over $103 million. All together, these Indochinese Americans reported more than $2.2 billion in annual income in 1988 and paid $218 million in federal income taxes (U.S. ORR 1991).

From the beginning of the Indochinese refugee resettlement program in 1975, the federal government has funded annual surveys of representative national samples of this population. Reported each year by the Office of Refugee Resettlement (ORR) in its *Annual Report to the Congress*, these are the only national survey data available that provide a comprehensive picture of their occupational and economic adaptation over time; the results, however, are not broken down by ethnicity but are reported for the Indochinese refugee population as a whole. In general the data show increasing rates of labor force participation and, once in the labor force, decreasing unemployment rates over time in the United States. Arrival cohorts reflect very low rates of labor force participation and high rates of unemployment, especially during their first year in the United States, when most refugees are enrolled in English language (ESL) classes and job training programs while receiving cash and medical assistance. However, more recent cohorts show only small increases in labor force participation rates over time, staying in the 30 to 38 percent range (about half the U.S. rate of 66 percent), unlike earlier cohorts who moved from first-year rates in the 20 to 30 percent range to 40 to 50 percent by their second and third year in the United States. In particular, the 1975 to 1978 cohorts have shown the highest rates of labor force participation, exceeding the rate for the U.S. population within their first four years in the country; and they also have had lower unemployment rates than the U.S. average (except during the recession of 1981 to 1982, when their unemployment rates almost quadrupled). In part, these patterns reflect the relative handicaps of different arrival cohorts, especially their level of education at arrival. For example, the annual surveys have documented a decline in the educational levels of adult refugees over time: 1975 arrivals averaged 9.4

years of education, those arriving from 1976 to 1979 averaged 7.4 years, 1980 to 1984 arrivals averaged 6.8 years, and 1985 to 1989 arrivals about 5 years. The 1990 to 1992 cohorts (which included many former reeducation camp internees from Vietnam) reversed this trend, however, averaging eight years of education (Haines 1989; U.S. ORR 1985, 1992).

The promotion of "economic self-sufficiency" is a principal goal of the U.S. refugee program, as enunciated by the Refugee Act of 1980; that is, to ensure early employment and minimal reliance on public assistance [such as Aid to Families with Dependent Children (AFDC) and Medicaid]. Eligibility for the latter and levels of benefits vary widely among the states, however. For example, during the mid-1980s in the two largest states of Indochinese settlement, monthly AFDC benefits for a family of four ranged from $591 to $734 in California, but only $141 to $221 in Texas; indigent two-parent families with dependent children were eligible for AFDC in California but not in Texas; and indigent adults without dependent children were eligible for local general assistance in California but had no such "safety net" in Texas. As a result, the initial decision to resettle refugees in one state or another involves not only a choice of possible destinations but of possible destinies as well. Particularly with extraordinarily needy populations, such as many Indochinese refugees, the availability or unavailability of different "safety nets"—that is, their placement in segmented state welfare systems—significantly affects not only their subsequent occupational and economic adaptation, but all other aspects of their lives as well.

A recent national study of refugee public assistance utilization during the 1980s found that 18 percent of all Indochinese refugee households were economically self-sufficient after their first year in the United States, some of them never having used public assistance. Of those who did, 41 percent had exited from public assistance programs within their second year in the United States, as had 57 percent by their fifth year. However, there were significant differences between refugee households in California (home to over 40 percent of the total Indochinese population) and those in the rest of the country (the study grouped the other forty-nine states together, although there are major differences among them in their programs of public assistance). In California, only 7 percent of the Indochinese households were financially independent after their first year and only

An Air America helicopter crewman helps board Vietnamese evacuees amid the fall of Saigon in April 1975, which today remains a sad and bitter memory for thousands who fled their homeland. With the end of the Vietnam War, two decades of American involvement in Indochina came abruptly to an end. Yet one of the ironies of the war that took America to Vietnam, and of the war's expansion into Cambodia and Laos, is that a sizeable part of Vietnam, and also of Cambodia and Laos—over one million refugees—has since come to America. (United Press International/Bettmann Archive)

18 percent had exited from public assistance by their second year, compared to 26 and 57 percent, respectively, outside Califonia; by the end of five years, only 26 percent of California refugee households had left public assistance, compared to 75 percent outside California (Bach and Argiros 1991). The following situation is illustrative: in California in 1986, an eligible refugee family of six could qualify for monthly AFDC cash benefits of $941, plus medical coverage for the family through Medicaid, and could be eligible for a maximum monthly allotment of $382 in food stamps. To equal that level of cash assistance alone, the family's principal wage-earner would need to have had a full-time job at $5.50 per hour, though such a job would be unlikely to provide any health care benefits for the family. For that matter, in that year a family of six with *both* spouses working forty hours a week, fifty-two weeks a year, at the minimum wage of $3.35 per hour, would still fall below the federal poverty line.

If residence in California is significantly associated with a higher level of reliance on public assistance, what are the determinants of such reliance among refugees *within* California? A study of welfare dependency among Indochinese groups in southern California found that the strongest determinant was the number of dependent children in the family; the proportion of family income coming from public assistance grew by about 10 percent for each dependent child. Welfare dependency significantly *decreased* over time in the United States, however, and *increased* with age and poor health status (Rumbaut 1989b). Indeed, concern over medical care coverage is often a decisive consideration for large refugee families who continue to remain on public assistance (including Medicaid). Though it keeps them below the poverty line, they would rather stay on welfare than risk low-wage jobs that provide no health care insurance at all. Cambodian refugees we interviewed in San Diego in the 1980s, for example, referred to their MediCal stickers (as Medicaid is called in California) as being "more valuable than gold" (Rumbaut et al. 1988).

The Indochinese in America: A Local Perspective

Since the late 1970s a large number of community surveys, ethnographies, and epidemiological and clinical studies by sociologists, anthropologists, psychologists, psychiatrists, and other researchers have reported on many aspects of the Indochinese refugee experience in settings throughout the United States. While they lack the generalizability of national survey data, and use different samples and methodologies that often preclude comparisons across studies, they nonetheless have greatly added to the richness and depth of our available knowledge for specific ethnic groups. For instance, qualitative studies include ethnographic field work on Indochinese communities and families from California (Gold 1992) to Philadelphia (Kibria 1993); case histories of refugee adults (Freeman 1989) and youth (Rumbaut and Ima 1988a); and edited collections of papers on each of the Indochinese ethnic groups (Haines 1985). The results of major quantitative community surveys have been collected by Haines (1989). One of these was a cross-sectional study of Vietnamese, Chinese-Vietnamese, and Lao 1978 to 1982 arrivals residing in Boston, Chicago, Houston, Seattle, and Orange County, California (Caplan, Choy, and Whitmore 1991; Caplan, Whitmore, and Choy 1989); another, the Indochinese Health and Adaptation Research Project (IHARP), was a longitudinal study of Vietnamese, Chinese-Vietnamese, Cambodian, Lao, and Hmong adults and children in San Diego based on representative samples of 1975 to 1983 arrivals (Rumbaut 1989a). These two are among the few to focus specifically on the ethnic Chinese from Vietnam. Some illustrative findings from the IHARP study are presented below, focusing on aspects of the refugee experience that are missing from census and other official data. The respondents, men and women ranging in age from eighteen to seventy-one, were interviewed at length in their native languages in 1983 and again a year later. Results are presented by ethnic group and for three key cohorts of arrival (1975, 1976 to 1979, and 1980 to 1983, the latter being the most numerous). First we will look briefly at their migration experiences and their adaptation and mental health after arrival in the United States, and then touch on pregnancy outcomes among Indochinese mothers and the educational progress of their children.

Migration, Adaptation, and Mental Health

Table 3 summarizes basic information on the social background and migration process of the refugees. About 90 percent of the Hmong and 55 percent of the Cambodians came from rural areas, while the Chinese and Vietnamese were overwhelmingly from urban sectors in South Vietnam. These differences are reflected in their levels of premigration education: the Vietnamese were the most educated (9.8 years), followed by the Chinese (6.6 years), the Cambodians (4.9 years), and the Hmong (1.7 years). Two-thirds of Hmong adults had never attended school and were preliterate, their language lacking an alphabet until the 1950s when missionaries in Laos developed a written notation for what had been until then only an oral tradition. There were also very significant social class differences by cohort of arrival. The 1975 refugees were much more likely to come from highly educated professional and managerial classes, while less educated farmers, fishers, and manual laborers predominated among the more recently arrived. Vietnamese and Hmong men included high proportions of former military officers and soldiers, while the ethnic Chinese—a largely segregated "middleman" minority of merchants from Saigon's Cholon ("Large Market") area, which had been referred to as the largest Chinese city outside of China after Singapore—were least likely to have had any prior involvement with either the military (ARVN) or the South Vietnamese or American governments during the war. Indeed, very few Chinese-Vietnamese cited "past associations" in their motives to flee.

A distinction often made between refugees and other classes of immigrants revolves around their different motives for migration and the traumatic nature of their flight experiences. Refugees are said to

Table 3 Social Background Characteristics and Contexts of Exit of Indochinese Refugees in San Diego County, by Ethnic Group and Year of Arrival in the United States (IHARP Adult Sample, $N = 500$)

	Ethnic Group				Year of Arrival		
	Vietnamese	Chinese	Cambodian	Hmong	1975	1976–1979	1980–1983
Educational background							
Years of education	9.9	6.6	4.9	1.8	11.9	6.5	5.2
High school graduate (%)	47.1	19.3	13.3	2.8	74.4	23.8	15.6
Knew some English (%)	39.5	13.2	5.8	1.8	66.7	12.8	13.1
Never attended school (%)	1.9	12.3	23.3	67.9	2.6	19.2	29.4
Rural background (%)	5.1	4.4	55.0	89.9	5.1	36.0	39.1
Occupational background							
Professional/managerial (%)	25.6	11.2	5.3	3.0	38.7	15.4	7.3
Military (%)	25.6	6.1	15.9	31.3	35.5	20.8	17.9
Clerical (%)	7.5	1.0	1.8	2.0	9.7	4.0	2.3
Sales (%)	18.8	38.8	14.2	2.0	6.5	20.1	18.6
Blue-collar (%)	10.5	27.6	8.0	2.0	3.2	7.4	15.2
Farmers, fishers (%)	10.5	14.3	54.0	59.6	3.2	32.2	37.6
Motives for exit[a]							
Number of "political" motives:	3.3	2.9	5.1	4.2	2.7	3.9	4.0
TARGETS	2.6	2.3	4.5	3.9	2.1	3.3	3.5
REBELS	0.7	0.6	0.5	0.3	0.6	0.6	0.5
Number of "economic" motives:	0.5	0.7	2.1	0.4	0.2	0.6	1.1
VICTIMS	0.1	0.2	1.7	0.2	0.1	0.2	0.7
SEEKERS	0.4	0.5	0.4	0.2	0.1	0.4	0.4
Migration events							
Fled without family (%)	13.4	11.4	29.2	19.3	25.6	19.8	15.9
Gave bribes to exit (%)	32.7	71.7	19.3	21.3	13.2	35.7	39.0
Feared would be killed (%)	73.2	73.7	80.7	92.7	18.4	86.6	83.0
Assaulted in escape (%)	30.6	36.8	25.2	25.7	0.0	24.4	36.7
Violence events in exit	2.1	1.9	3.1	2.5	0.9	2.2	2.7
Years in refugee camps	0.6	0.9	2.1	2.9	0.2	1.2	1.9
Mental health status, 1984							
Sleep problems (%)	21.7	23.9	55.8	61.5	23.1	36.6	42.7
Appetite problems (%)	14.6	11.4	42.5	22.0	5.1	18.0	27.0
Positive well-being (%)	50.6	31.0	15.0	35.5	56.4	35.1	30.8
Demoralization moderate (%)	28.8	43.4	46.7	34.6	30.8	35.7	39.8
Demoralization severe (%)	20.5	25.7	38.3	29.9	12.8	29.2	29.4

[a] Data refer to the number of migration motives reported by the respondent, classified as follows:

TARGETS: *Forced* relocation, to new economic zone, into reeducation camp; imprisoned prior to exit; *fear* of arrest or harm from new regime; past political involvement with old regime, armed forces; association with U.S. government, military, or CIA; drafted to fight in Cambodia; loss or confiscation of personal property or wealth; general harrassment.

REBELS: Protest communism, lack of freedom; refusal to join cooperative; other political-ideological reasons.

VICTIMS: Starvation, famine, lack of health care, harsh or poor economic conditions, inability to make a living.

SEEKERS: Seeking better future, education, prospects for children; family reunification; other miscellaneous reasons.

be motivated to flee by fear of persecution ("political" motives), while immigrants are defined by their aspirations for better material opportunities and self-advancement ("economic" motives). IHARP respondents were asked to state all of their motives for leaving the homeland; over fifty different reasons were given, ranging from fear of repression or imprisonment in reeducation camps, to past associations with the former regime and ideological opposition to communism, to desires for family reunification, better education for their children, and an improved standard of living. Some of these reasons may be defined as political in nature, others as economic or social. Often both kinds of reasons were cited by the same respondent, making the usual distinction between refugees and nonrefugees simplistic and misleading.

Their exit motives were classified into four main types, as shown in Table 3. Two involved more clearly political motivations: (1) specific perceptions and experiences of fear or force, past political associations, and related motives (labeled TARGETS), and (2) explicit forms of protest and ideological reasons (REBELS). The other two types involved more clearly socioeconomic considerations: (3) harsh material conditions of famine and other dismal economic conditions (VICTIMS), and (4) miscellaneous "pull" motives, such as seeking a better education for the children or family reunification (SEEKERS). Cambodians reported by far the most TARGET and VICTIM reasons for flight (reflecting their life-threatening experiences during the holocaust of the late 1970s) and hence both more political *and* economic exit motives. The Hmong also reported many TARGET motives, the Vietnamese the most REBEL motives, and the Chinese the most SEEKER or economic "pull" motives, as well as the fewest past associations with the former regime. Despite this diversity of motives in the refugees' decision to leave, by far more TARGET motives were reported overall (3.3 per person) than any of the other three motive types: REBELS (0.6), VICTIMS (0.5), and SEEKERS (0.4). By this classification, far more political motives (3.8) than economic motives (0.9) were reported, underscoring the qualitative difference in modes of exit between refugees and conventional immigrants, but also the fact that economic and social as well as political factors were interwoven in the decision to flee.

Other extraordinarily stressful exit experiences of the refugees are detailed in Table 3. Most feared they would be killed during their escape, except for the 1975 refugees. The Cambodians suffered the greatest number of family loss and violence events, followed by the Hmong and the Vietnamese. The Chinese—and more recent arrivals generally—were most likely to have left together as a family, to have given bribes to exit, and to have been assaulted in the escape (often by Thai pirates). Once they reached a country of first asylum, the Hmong stayed in refugee camps far longer than any other group before being resettled in the United States, followed by the Cambodians, the Chinese, and the Vietnamese. Taken together, such differences in the migration events experienced by these refugee groups help explain why the Cambodians and the Hmong had a significantly higher number of chronic health problems (physical symptoms lasting six months or longer) whose onset occurred between their exit from their homeland and prior to their arrival in the United States.

Although not shown in Table 3, the refugees' social background characteristics were reflected in their socioeconomic position in San Diego as they struggled to rebuild their lives. The same ethnic group rank order was mirrored in their levels of English literacy, employment and labor force participation in the local economy, income and welfare dependency, although all groups were progressing gradually if at different rates over time. English ability increased over time; it is primarily a function of level of prior education, and secondarily of (younger) age and longer time of residence in the United States. The biggest differences in labor force participation and unemployment rates were seen between the 1975 first-wave refugees and later arrivals, reflecting the national survey data reviewed above. The most recently arrived refugee families reported very low annual incomes, and by 1984 about two-thirds of the 1976 to 1979 arrivals and over four-fifths of the 1980 to 1983 arrivals in the sample still had incomes that fell below the federal poverty line, compared to a poverty rate of about 15 percent for the general U.S. and local populations. It was one of the respondents in the IHARP study, a middle-aged Hmong refugee who had arrived in 1980 after spending five years in refugee camps in Thailand, who was quoted at some length at the beginning of this chapter, vividly and eloquently expressing the complexity of the economic and related psychosocial problems faced by some of these recently arrived groups.

The measure of mental health status reported in Table 3 was based on a screening scale used by the National Center for Health Statistics in a major national survey of the general American adult population (Link and Dohrenwend 1980). It found that 74 percent of Americans scored in the "positive well-being" range, 16 percent in the "moderate demoralization" range, and 9.6 percent in the "severe demoralization" range (indicative of "clinically significant distress"). But in 1984 the respective prevalence rates for the Indochinese refugees were 34 percent "positive well-being," 38 percent "moderate demoralization," and 28 percent "severe demoralization" (see Table 3). That latter figure was three times the level of severe distress found for the general American population; a year before in 1983 the corresponding refugee rate had been four times higher than the U.S. norm. The demoralization rates were highest for the Cambodians—who had experienced the most traumatic contexts of exit—followed by the Hmong, Chinese, and Vietnamese.

The process of psychological adaptation is temporally as well as socially patterned. The first several months after arrival in the United States tend to be a relatively hopeful and even euphoric period, but during the second year, a period of "exile shock," depressive symptoms reach their highest levels, followed by a phase of psychological recovery after the third year. The general pattern was described succinctly by an elderly Cambodian widow (quoted in Rumbaut 1985, pp. 469–470):

> I was feeling great the first few months. But then, after that, I started to face all kinds of worries and sadness. I started to see the real thing of the United States, and I missed home more and more. I missed everything about our country: people, family, relatives and friends, way of life, everything. Then, my spirit started to go down; I lost sleep; my physical health weakened; and there started the stressful and depressing times. But now [almost three years after arrival] I feel kind of better, a lot better! Knowing my sons are in school as their father would have wanted, and doing well, makes me feel more secure.

What, among all of the stressors reviewed above, affected refugee mental health the most? The San Diego study found that, in 1983, the principal predictors of demoralization were *prearrival* factors: the number of TARGET motives reported, an index of family loss and separation, and a rural background. A year later, however, the effect of these prearrival stressors had receded and current difficulties, primarily being unemployed, emerged as stronger predictors of depressive symptoms. Past losses and events seemed to heal with time and recede in importance as present demands and challenges grew in psychological significance. Over time the "refugees" generally became more like "immigrants," as contexts of incorporation supplanted contexts of exit in their compelling psychosocial effects. In addition, by 1984 a significant predictor of lower distress and greater satisfaction was an attitudinal measure of biculturalism; measures of monocultural styles (traditionalism or assimilationism) showed no effect on psychological outcomes. That is, refugees who adopted an "additive" acculturative strategy, adapting to American ways while retaining their ethnic attachments and identity, appeared to reduce psychological distress over time. This finding points to the importance of creativity and flexibility in the acculturative process (Rumbaut 1991b, 1991c; Vega and Rumbaut 1991).

Infant Mortality, Children's Educational Progress, and Some Paradoxes of Acculturation

Another dimension of the Indochinese adaptation process may be gleaned from a study of the infant mortality rates of all ethnic groups in San Diego County, based on a linked data set of all live births and infant deaths recorded in the metropolitan area from 1978 to 1985 (Rumbaut and Weeks 1989; Weeks and Rumbaut 1991). The Indochinese ethnic groups appeared to be at high risk for poor infant health outcomes. After all, they had come from a background of high fertility and high infant mortality, often with preexisting health problems, and had levels of unemployment, poverty, welfare dependency, and depressive symptomatology that greatly exceeded U.S. norms. Lack of English proficiency limited their access to health care, and indeed Indochinese pregnant mothers exhibited the latest onset of prenatal care of all ethnic groups in San Diego.

Remarkably, however, the Indochinese overall were found to have much lower infant mortality rates (6.6 infant deaths per 1,000 live births) than the San Diego County average (8.5), and two refugee groups actually exhibited the lowest infant death rates: the Vietnamese (5.5) and the Cambodians (5.8). Only the Hmong (9.1) had a higher infant mortality rate than non-Hispanic whites (8.0), though still much lower than Blacks (16.3), and vastly lower than their own infant death rate prior to their arrival in the United States (104). It is also worth noting that a common characteristic of the ethnic groups with below-average infant death rates—including other Asians (Japanese, Chinese, and Filipinos) and Hispanics (mostly of Mexican origin)—is that they were composed of populations with a significant proportion of immigrants. Several behavioral factors were associated with these positive outcomes, particularly the nearly universal absence of tobacco, alcohol, and drug abuse among pregnant Indochinese women—and among immigrant women generally—in contrast to U.S.–born groups. One implication of these findings is that "subtractive" acculturation—that is, a process of "Americanization" that involves the learning by immigrant women of bad habits in the U.S. milieu—may have negative consequences for infant health.

Other evidence bearing on the future prospects of the coming generation of Indochinese Americans comes from a study of their educational attainment in San Diego schools (Rumbaut and Ima 1988a). Data on academic grade point averages for the entire cohort of high school seniors, juniors, and sophomores in the district ($N = 38,820$), including nearly 2,400 Indochinese students, were compared for students who spoke English only and for students who spoke a primary language other than English at home (virtually all of them immigrants or children of immigrants). The latter are in turn classified by the schools as either Fluent English Proficient (FEP) or Limited English Proficient (LEP). Among all groups in the school district, the Indochinese had by far the highest proportion of LEP students (over two-thirds overall, and over 90 percent of the Cambodians), reflecting the fact that they were the most recently arrived immigrants. By contrast, less than a third of the Hispanics (mostly of Mexican origin) were classified as LEP, and less than 15 percent of the East Asians and Filipinos.

Despite the language handicap, however, the academic GPA of Indochinese students as a whole (2.47) significantly exceeded the district average (2.11) and that of White Anglos (2.24). The highest GPAs were found for immigrant Chinese, Korean, Japanese, and Vietnamese students. In fact, with the main exception of Hispanics, all of the non-English immigrant minorities were outperforming majority Anglo students; falling below the district norm were Mexican Americans and African Americans. Even the Hmong, whose parents were largely preliterate peasants from the Laotian highlands, and the more recently arrived Cambodians, who were mostly rural-origin survivors of the "killing fields" of the late 1970s, were outperforming all native American majority and minority groups. Significantly, for all ethnic groups without exception, English monolinguals (usually U.S.–born) exhibited lower GPAs than their bilingual FEP coethnics (usually foreign-born); this was particularly clear among the East Asians, Filipinos, and Indochinese. Indeed, educational achievement appeared to decline from the first (immigrant) generation to the second and third generations. These findings run counter to the conventional assumption that the more Americanized immigrants become, the greater will be their success in the competitive worlds of school and work. Instead, the implication is that Americanization processes, all other things being equal and to the extent that they involve "subtractive" rather than "additive" forms of acculturation, may be counterproductive for educational attainment.

But all other things are never equal except in mathematical models, and acculturative processes always unfold within concrete structural and historical contexts. Exactly why and how the immigrant ethic that appears to yield the positive outcomes reviewed above, often despite significant disadvantages, erodes over time in the United States remain, at present, unanswered questions. It is also unclear whether "additive" adaptations, such as fluent bilingualism, can be sustained beyond one generation in the United States. In the end, the complex processes of assimilation to different sectors of American society will vary for different types of second-generation Indochinese Americans located in different types of familial, school, and community contexts—from the inner cities to the suburbs to diverse ethnic enclaves—and exposed to different types of role models and forms of racial discrimination. Bilateral relations between the United States and the countries of origin are also likely to affect not only future immigration flows but also entrepreneurial opportunities and the very nature of institutional life within established refugee communities. Many outcomes are possible. The future of Vietnamese, Laotian, and Cambodian Americans will likely be as diverse as their past, and will be reached by multiple paths.

Conclusion

Vietnamese, Laotian, and Cambodian Americans now form a sizable and diverse component of the Asian-origin population in the United States. They are the newest Asian Americans, most having arrived only after 1980; and they are also among the fastest growing populations in the country as a result of both the largest refugee resettlement program in American history—a legacy of the nation's bitterest and most divisive war in this century—and of fertility rates that are among the highest of any ethnic group in the United States. They differ from other Asian Americans and indeed from most other immigrant groups in significant ways, especially in the contexts of exit and reception that have shaped their refugee experience. They differ greatly from each other as well, and, if the studies reviewed here are any indication, the Vietnamese, Cambodian, Lao, Hmong, and

ethnic Chinese generations now coming of age in America will differ still again from that of their parents. In their diversity they are writing yet another chapter in the history of the American population and society, and in the process they are becoming, quintessentially, Americans.

Note

1. "Southeast Asian" is sometimes preferred over "Indochinese" mainly to avoid any connection to the usage of the latter term during the period of French colonial rule in Vietnam, Laos, and Cambodia. Southeast Asisan, however, is a broad and imprecise term both geographically and historically, covering as it does a vast region and countries as diverse as Thailand, Burma, Malaysia, Indonesia, Brunei, Papua New Guinea, and the Phillippines, nore of whom share the fateful history of U.S. involvement during the "Indochina" war nor of special U.S. sponsorship of refugees who fled after the collapse in 1975 of U.S.-backed governments in Saigon, Vientiane, and Phnom Penh. To avoid the cumbersome repetition of each of the nationalities and ethnic groups being here considered while retaining those more precise geographic and historical meanings, Indochinese will be used in this chapter to refer collectively to refugees from the three countries of Vietnam, Loas, and Cambodia. In any case, it should be noted that persons from those countries do *not* identify ethnically either as Indochinese or Southeast Asian.

25

New York as an Immigrant City

FREDERICK M. BINDER

DAVID M. REIMERS

No American city as been as ethnically and racially diverse as New York. From its founding as a Dutch colony in the 1620s to the present, it has been home to an enormous variety of newcomers. Until World War II the city's inhabitants were largely European and African in origin. However, recent decades have witnessed the arrival of immigrants from all parts of the world, and today America's largest metropolis claims residents speaking over 100 languages. In the 1990s New York is truly a global city demographically; with the possible exception of Los Angeles, it is matched by no other large American city in its diversity. As a consequence of this amazing mix of people, New York's history has been marked by continual racial and ethnic conflict, often frustrating its attempt to seek social harmony.

Past Migration

The "Old" European Immigration

Dutch colonizers quickly eliminated Native Americans as a significant force in the city's life, but the Dutch West India Company was not so successful in attracting Dutch settlers to build its empire in the New World. As a result, the company welcomed others to New Amsterdam. Among the first settlers were French-speaking Walloon families, Scandinavians, Italians, Flemish, English, a few Germans, Quakers, and the city's first Jews. These last came from Recife, Brazil, when the Dutch were forced to surrender that colony to the Portuguese, who brought with them the feared tribunal of the Inquisition. New Amsterdam's Governor Peter Stuyvesant himself was prejudiced against Jews, but the West India Company, which had important Jewish backers, insisted that he permit them to remain in the town, though they were not granted full rights as citizens (Goodfriend 1992). The acceptance of a few Jewish refugees was typical of New York's attitude toward settlers during the colonial era. Because of the need for labor, first the Dutch and then the British allowed persons from all over Europe to settle in colonial New York. By 1643, less than twenty years after the town's founding, Father Isaac Jagues, a French Jesuit, claimed to have heard seventeen languages spoken on the streets of New Amsterdam. He was not enthusiastic about what he observed: "The arrogance of Babel has done much harm to all men; the confusion of tongues has deprived them of great benefit" (Archdeacon 1976, p. 32). As in the case of Jews, permission to settle sometimes came grudgingly. Stuyvesant also disliked English Quakers, whom he considered to be troublemakers lacking the true faith of the Dutch Reformed Church. Catholics had an especially bad time in colonial New York, and although a Roman Catholic served as the colony's governor briefly in the 1680s, they did not worship openly and enjoy full civil rights until after 1750.

Labor shortages also account for the appearance of another group: African slaves. Lacking qualms about slavery, Hollanders imported the first slaves as early as 1626, two years after the city's founding. Their numbers remained small until 1655 when the ship *Witte Paert* docked at New Amsterdam with a cargo of 300 captives. By the time the English seized New Amsterdam in 1664, Africans made up about 20 to 25 percent of its population. Most labored for the West India Company, but some were bonded to individuals who owned them. With the introduction of slavery came slave codes and customs that limited the life of slaves. Dutch restrictions were loose, however; some slaves received their freedom for a special service. However, this freedom was not automatically passed on to their children (McManus 1966).

Whatever plans the Dutch had for an empire in North America abruptly ended when the English seized the colony in 1664, and again in 1673 after the Dutch briefly retook it. Like the Dutch, the British needed laborers and did little to discourage diversity. A small but steady stream of Europeans and New Englanders found their way to the city that had been renamed New York in 1664. Among the most prominent newcomers were French Huguenots, Calvinists seeking religious tolerance in the New World (Butler 1983; Goodfriend 1992). As non-Dutch immigrants arrived, Dutch influence began to wane. By the end of the seventeenth century, Hollanders accounted for only about half of the city's citizens; by the time of the American Revolution, scarcely one-sixth.

At the outset tensions existed between the Dutch and their new conquerors, but on the whole the English pursued a tolerant policy toward most Protestants, including the Dutch Reformed. The Anglican Church replaced the Dutch Reformed Church as the established one, but the latter's communicants worshipped openly as did other Protestants. English replaced Dutch as the official language; English laws replaced Dutch statutes and the Dutch style of archi-

tecture declined. The English Naturalization Law of 1740 granted citizenship to all Protestants, including Quakers, and even to Jews. As early as 1695 Jews had established a public synagogue in a rented room, naming it "Shearith Israel." In 1730 the congregation erected its own building, the first such to be built in mainland America (Goodfriend 1992).

Life was not so kind to Black New Yorkers. Continuing to face labor shortages, the English imported African and West Indian slaves. By the middle of the eighteenth century, Blacks made up about one in five of the city's residents. Moreover, slave codes became tighter under the English. Periodic slave revolts and rumors of more to come brought harsh reprisals from the English masters. Following the 1712 slave uprising, the city indicted thirty-nine Blacks for murder or as accessories to murder and convicted twenty-three of them. The court sentenced to death all those convicted. While most were hanged, three were burned to death, one by slow fire (Scott 1961).

The American Revolution ushered in an era of tolerance. The Anglican Church lost its established position, and Catholics in the city began to worship openly upon the dedication of their first church building in 1785. The new state also removed civil restrictions on both Catholics and Jews, who could now vote and hold office. Until the 1830s this growing tolerance of nationality and religious groups combined with low rates of immigration to lessen ethnic tensions. The French Huguenots, who adapted to English ways quickly, had struggled in the eighteenth century to maintain their French Church, but most of their congregants ultimately became Anglicans (later Episcopalians), and frequently intermarried with English and old Dutch families. The Dutch and English themselves often intermarried, and some descendants of the original Dutch settlers became Anglicans. While the Dutch language continued to be used in the Reformed churches after 1776 and served as the principal tongue in a number of upstate villages, on the streets of New York City it was heard with declining frequency. Indeed, New York seemed to be entering a period of homogeneous Americanization in the early nineteenth century, with its population being increasingly native-born (Rosenwaike 1972).

The American Revolution also brought better times for African American New Yorkers, though it by no means ended racism. Inspired by the Enlightenment ideas of equality, several prominent New

Yorkers emancipated their slaves and in 1785 banded together to found the New York Manumission Society. The society labored for the next fifteen years to persuade the legislature to end slavery. Their efforts bore fruit in 1799, when the lawmakers passed a bill for gradual emancipation. Not content with this cautious approach, antislavery advocates finally succeeded in winning legislative approval for the total eradication of slavery in the state, effective in 1827 (White 1991).

The end of slavery for Black New Yorkers did not bring equality. Black males did not win parity at the ballot box until enactment of the 15th Amendment in 1869. Moreover, African Americans faced segregation in most public conveniences, churches, and schools. New York did not pass racial civil rights laws covering public accommodations and schools until the late nineteenth century, and even then the enforcement of such measures was lax. Throughout the nineteenth century Blacks found employment in the lowest paying occupations. While a small professional Black elite also emerged, most Black women who worked found employment as domestics and Black men as unskilled laborers. No racial ghetto existed yet, as Blacks frequently lived on the same blocks, and even in the same buildings, as Whites. But Black quarters were usually run down and located in the most unhealthy sections of the city (Scheiner 1965, pp. 15–44). Furthermore, the proximity of Black and White residences bespoke of common poverty rather than racial harmony.

The years after 1830 were momentous in New York City's ethnic history. New York's position as the nation's premier port and industrial center attracted millions of immigrants from Europe. While Scotch-Irish, Welsh, Scots, and English from Great Britain continued to settle there between 1830 and 1890, Germans and Irish Catholic immigrants arrived in the greatest numbers and made the most dramatic impact on the city's social life.

Economic conditions in Europe coupled with the lure of America drew Germans who mostly entered through New York's Castle Garden at the tip of Manhattan. By the time of the Civil War they made up 15 percent of Manhattan's population and almost 10 percent of neighboring Brooklyn. While many Germans were peasants or farmers who found jobs as unskilled laborers, many others possessed skills useful in an urban and commercial economy. Thus they became prominent as cabinetmakers, shoemakers, ma-

chinists, tailors, and practitioners of many other crafts (Nadel 1990). By the late nineteenth century a German business elite had emerged, some of whom mingled with the city's older Anglo elite. Included among the successful Germans were Jews who arrived with little or no funds. By the 1880s a substantial Jewish middle class and elite had their own organized life; they thought of themselves as being both Jewish and German. The most prominent among the German Jews were department store owners, some of whom began life in America as rural peddlers, storekeepers, and bankers (Sachar 1992).

Germans lived in most of the city's wards, but the main area of settlement was *Kleindeutschland*, below 14th Street on Manhattan's East Side. Their shops featured goods for German immigrants and German was the language of the streets. They also published their own newspapers and insisted that the Lutheran and Catholic churches they attended be staffed by German priests and pastors (Dolan 1983; Nadel 1990).

New Yorkers watched uneasily as the German population grew. Irish church leaders did not want Germans to have their own priests, and strait-laced Protestants did not approve of the German beer halls, opened to gaiety on Sundays. In the late nineteenth century the rising German Jewish entrepreneurs encountered a renewed anti-Semitism, as they were barred from membership in prominent social clubs, some of which had previously welcomed them (Sachar 1992, pp. 98–102).

Yet the German presence was not nearly as tension-producing as was the Irish. To begin with, the Irish were the largest foreign-born group in the city; after 1840 their rapid growth and presence could scarcely be ignored. The Irish were also among the city's poorest residents. The famine Irish arrived penniless, at times on the verge of starvation. While conditions improved in Ireland after 1850, they remained so bad that hundreds of thousands of poor Irish still headed for America, with New York being a main attraction.

Arriving with few urban skills and penniless, what else could Irish men do but unskilled labor, working on the docks, digging up streets, and finding whatever casual jobs were available? When they reached for better jobs they sometimes encountered signs reading "No Irish Need Apply." Women, who often emigrated as single persons, found employment as domestics. Unskilled employment paid poorly and

Irish immigrants lived in tenements, often crowded into filthy unsanitary buildings and rooms. No wonder that disease and crime flourished in such Irish neighborhoods as the infamous Five Points slum, near the present-day City Hall. The impoverished Irish also comprised a disproportionate number of the residents of the almshouse. Not surprisingly, Irish women domestics at times felt somewhat privileged because they lived with families in better accommodations (Ernst 1965).

New Yorkers considered these poor immigrants a threat, the carriers of poverty and disease. They also disapproved of the saloons that served as the centers of Irish social life. Protestant temperance advocates believed that Irish immigrants were prone to excessive drink, which they viewed as one of society's major social ills. Some Protestant reformers established missions in Irish districts, to convert the immigrants to Protestantism and to teach them their brand of cleanliness and temperance (Pernicone 1973).

It was not only poverty that brought disapproval of the Irish from native New Yorkers. Perhaps as important was their Catholicism. Protestants watched uneasily as Catholic churches spread rapidly in the city. As noted, New Yorkers had disestablished the Anglican Church and granted Catholics full civil rights, but many still believed Catholicism to be nothing more than a superstitious faith, run by evil Popes and Jesuits from Rome. The anti-Catholic Know-Nothing movement, which flourished between 1840 and 1860, found many adherents in New York City who believed some of the fantasies of anti-Catholic bigots; in the 1850s the Know-Nothings briefly captured City Hall, though no nativist legislation was forthcoming (Spann 1981).

Catholics and Protestants also quarreled over the nature and treatment of religion in the publicly supported schools. When Archbishop John Hughes asked for state support for Catholic schools to counter what he considered the Protestant and anti-Catholic orientation of the private state-supported Public School Society, Protestants insisted the schools were nonsectarian and that no public funds should go to denominational schools. The ultimate consequence of the controversy was the establishment of a true public school system in New York City, which was the sole recipient of state school funds (Pratt 1961).

While the Irish had little money, they did have numbers. Their arrival during a time when New York

State was enacting universal White manhood suffrage enabled them to use the ballot and politics to gain power and employment in the public sector or in industries, such as construction, dependent upon governmental beneficence. At first Tammany Hall wanted nothing to do with those "Papists," but the hall soon changed and opened its doors to the Irish. The Irish loyally served and ultimately took over the hall and the Democratic Party. By the turn of the twentieth century they were controlling city wards and electing mayors. Tammany Hall politicians were not bothered by rules and regulations; immigrants were quickly and illegally naturalized so that they could vote; voters were either paid to vote correctly or intimidated; and ballot boxes were stuffed or lost (Erie 1988).

Many old-stock New Yorkers were shocked by Tammany Hall's practices and its Irish supporters. They also disapproved of saloons, which were often the centers of political activity and were scandalized by political corruption. Graft at city hall was common, and politicians and police took bribes to avoid enforcing Sunday closing laws and shutting down the city's many houses of prostitution. Reformers periodically put together coalitions to throw the (Irish) rascals out, but Tammany's power did not end until after 1933 (Erie 1988).

Irish workers often found themselves in conflict with Black New Yorkers. As unskilled immigrants, they competed with Blacks for low-paying jobs, such as along the city's docks. Antagonism between Black and Irish workers led to physical clashes, the most notorious occurring during the draft riot of 1863. White New Yorkers, and especially the Irish, resented the Civil War draft, which allowed the rich to pay a substitute to avoid military service. Some also opposed a war for the abolition of slavery and they feared that free Blacks might take their jobs if they were drafted. Opposition to conscription erupted into violence in July 1863. For days the city was out of control, and Black citizens were chased, attacked, and even lynched by angry Whites. The Black orphanage on Manhattan's 43rd Street was burned by a mob. Not until federal troops arrived did calm return (Bernstein 1990).

The city's first Asian community also dates from this period. Chinese immigrants were small in number during the nineteenth century, but the outlines of Manhattan's Chinatown were formed by the 1870s. In 1870, 300 people lived there; after twenty years the number reached 2,000. By 1890 New York's Chinese set the social pattern that was to continue for the next sixty years. Due to immigration laws that prohibited wives from joining their husbands who had migrated before restrictions were passed, Chinatown was largely a bachelor society that grew slowly before World War II. Discrimination limited Chinese employment to laundry work, restaurants, or small tourist shops. Most New Yorkers viewed Chinatown as an exotic tourist attraction and had no thought of integrating the Chinese population with New York's institutional life (Kwong 1987, pp. 13–14).

After 1890 the great migrations from Germany and, to a lesser extent, Ireland, slowed. During World War I immigration from Europe fell drastically, and the immigration restriction acts of the 1920s put a ceiling on large-scale European arrivals. Some Germans continued to emigrate during the 1920s, as did Irish following the achievement of Irish independence and the end of the civil war in Northern Ireland in 1922. But the Great Depression further cut into the numbers; only 528,000 immigrants entered the United States during the 1930s, of whom about 100,000 settled in New York City.

New York's German and Irish Americans gained greater status in the city, especially as they moved up the economic and social ladder. Assimilation was virtually forced on Germans during World War I as things German came under severe attack. New York's German institutions changed their names, and the circulation of German newspapers dropped. The Germania Life Insurance Company, for example, became the Guardian Life Insurance Company. Yet the weakening of German American culture was partly a matter of time and generation. As immigration slowed and the second and third generations became educated in the public schools and worked in the larger city economy, German American institutions lost support. After 1900 *Kleindeutschland*'s residents dispersed as the most prosperous moved to the outer boroughs and to the suburbs (Nadel 1990).

The "New" European Immigration

While immigration from Ireland and Germany dropped, that of southern and eastern Europe soared between 1890 and 1924. Italians, Jews, Poles, Slavs, Greeks, Norwegians, and Swedes became the new voices of the city (Rosenwaike 1972). Those years also witnessed the beginnings of the city's

Arab community, which centered first in Manhattan and then Brooklyn (Orfalea 1988, pp. 76–80).

But the two major groups changing the city's demography after 1890 were eastern European Jews and Italians; by World War I they made up the largest ethnic groups in the city. The city sported Jewish neighborhoods and several sections know as "Little Italy." In New York City's rapidly changing neighborhoods few city wards were composed entirely of a single nationality, but immigrants created their own cultural life by building a dazzling array of institutions suited to their needs. Synagogues, Italian Roman Catholic parishes, fraternal societies, theatres, coffee shops, musical clubs, and political associations delineated the city's landscape. City blocks came alive with Italian *fests*, such as the celebration of the Madonna of 115th Street in Italian East Harlem (Glazer and Moynihan 1963; Orsi 1985; Rischin 1962).

Immigrants tended to congregate into distinct occupations. Italian and Jewish street peddlers were ubiquitous, and both groups worked in the city's exploding garment industry, though Jews were predominant. Italians, often illiterate, labored in unskilled jobs, helping to construct the city's subways and other building projects. Jewish women were more apt than Italian women to work outside the home, especially in the garment industry. Yet both Italian and Jewish married women made garments at home for the clothing industry, and both took in boarders as a source of income. For a fortunate few, white-collar jobs were a possibility, but more likely the better skilled and literate immigrants opened shops catering to immigrant needs and tastes (M. Cohen 1993; Glenn 1991; Kessner 1977).

During these years of rapid city growth, the new immigrant groups kept largely to themselves. Italians did experience conflict with the Irish-dominated Catholic Church that looked down on the Italian's form of Catholicism. Jews, quicker than Italians to naturalize as citizens and enter politics, experienced Irish and Tammany resistance to their political aspirations. Older generations of New Yorkers denigrated the newcomers because of their different ways. Yet the primary task of the new immigrants was to make a living. And for those with limited English, job choices were few.

By the 1920s, at the same time that Jews moved through the city's public schools and aspired to enter colleges, anti-Semitism increased. Both Columbia

Yesterday, as well as today, the Statue of Liberty in New York's harbor remains a major symbol of this nation of immigrants. Here two women and a boy, at the turn of the century, stand in awe at the sight of the statue in the distance. (Theodore Wesley Koch Collection; Bentley Historical Library, The University of Michigan)

and New York universities resorted to Jewish quotas after 1920, and the city's medical schools tightened entrance requirements for Jews. In white-collar employment, one scholar reported that Jews were excluded from most of the city's general office work (Sachar 1992).

The Internal Migration of Blacks

No immigrant group, however, experienced anything approaching the hostility facing Black Americans. Blacks had begun moving to New York City and other northern cities after 1880, but with the drastic cut in European immigration during World War I, this migration stream rapidly increased. Initially African Americans found homes throughout the city, but after 1900 racial segregation in housing became the rule. Harlem had been a German, Irish, and Jewish middle-class neighborhood before a real estate boom collapsed just after 1900. Realtors then turned to newly arriving Blacks to fill their empty apartments, and Harlem quickly became an all-Black neighborhood, the center of African American life in the city (Osofsky 1966). It was there that the Harlem Renaissance, a rich outpouring by Black novelists, poets, and musicians, flourished in the 1920s.

As Harlem became a Black ghetto it also became a slum, with crowded housing in substandard apartments, poor health standards, overcrowded schools, and high unemployment. Black New Yorkers, regardless of their education, largely labored in low-paying jobs, as they had done for generations. For women, this meant domestic work; for men unskilled jobs such as porters and janitors. The Black elite usually served only the Black community by opening small businesses such as funeral parlors or beauty shops or by working as teachers and clergymen. Blacks were largely left out of the city's politics and public sector. The first Black policeman was not appointed until after 1910, and he became a tourist attraction as Whites came to see the "Nigger cop" (Greenberg 1991).

No incident revealed the political weakness of Black New Yorkers more than the riot of 1900, during which Irish police beat Blacks at will. Calls for an investigation yielded little. An internal police investigation cleared the police of charges of brutality, in spite of eyewitness testimony to the contrary (Osofsky 1966, pp. 46–52).

Migration Lull

Ethnic Conflict

Ethnic conflicts and tensions were heightened during the Great Depression. In sharp contrast to the boom of the 1920s, which saw the great expansion of Brooklyn, Bronx, and Queens neighborhoods, the city's residents experienced high unemployment and an uncertain future. As many as a third of New Yorkers were unemployed at times, and in Black neighborhoods the situation was desperate as half of the families lacked a bread winner.

Desperation in Harlem exploded into a riot in 1935 largely directed at White-owned stores and the police. Mayor Fiorello La Guardia's task force investigating the riot revealed the extent of the city's racism that penned African Americans into rundown housing, limited their employment, and segregated them in schools and other public institutions of poor and second-class quality. Yet little change occurred after the inquiry (Greenberg 1991).

However, Black leaders took the initiative and organized to improve their lot. Picketing forced modest changes in the hiring practices of several stores on Harlem's 125th Street, where African Americans had been confined to janitorial work. The Rev. Adam Clayton Powell and the National Association for the Advancement of Colored People (NAACP) prodded the city's transit system to hire Blacks for positions such as bus drivers (Greenberg 1991).

Jewish, German, Italian, and Irish New Yorkers also found themselves pitted against one another during the depression years. The German-American Bund, centered in New York City, staged rallies, complete with Nazi uniforms, in Madison Square Garden. Such demonstrations angered a Jewish community cognizant of Hitler's anti-Semitic policies. Some Jewish organizations called for boycotts of German goods, a call that prompted attacks from a few German New Yorkers. Actually, the German supporters of Hitler and the American Bund were mostly recent immigrants and few in number (Bayor 1988).

Father Charles E. Coughlin, the anti-Semitic radio priest, had more New York followers than did the Bund, but they were usually among the poor Irish rather than the city's Germans. The most outspoken cleric supporting Coughlin was Father Edward Lodge Curran. Coughlin's followers organized the Christian Front to promote anticommunism. Supported largely by Irish New Yorkers, the front attacked not only communism but Jews as well. Coughlin was eventually plagued by a shortage of funds and silenced by the church and the federal government during World War II (Bayor 1988).

Hostility between Irish and Jewish New Yorkers on occasion erupted into street fighting. In the ethnically mixed Washington Heights district of upper Manhattan, the Christian Front had a significant number of supporters. Even after the front came under government attack during the war, tensions existed in the neighborhood, and on more than one occasion Irish adolescent gangs attacked Jewish youths. Similar incidents occurred in the South Bronx during the 1940s (Bayor 1988).

On the political battlefield Jews and Italians pushed aside Tammany Hall and supported the reform administration of Fiorello La Guardia, who was mayor from 1934 to 1945. Under his leadership, expansion of the civil service led to a decrease in Irish on the city payroll. In addition, La Guardia appointed a number of Jews to administrative positions. That the feisty mayor made ethnic appeals was part of the city's politics. La Guardia, who spoke seven languages, was the ideal political candidate for New

Yorkers. Once, while campaigning, he was accused by his opponent of being anti-Semitic. The Episcopalian La Guardia, whose mother was Jewish, challenged his accuser to discuss the issues, the debate to be "Entirely in the Yiddish Language" (Mann 1959, pp. 156–157).

Foreign affairs aggravated ethnic tensions. The city's Roman Catholics generally favored a Franco victory in the Spanish Civil War, but New York's Jews worried about the rise of fascism and opposed Hitler's and Mussolini's aid to Franco. While some Italian Americans were critical of Mussolini, most initially approved of him, which brought them into conflict with Jews and Blacks who were upset by Italy's invasion of Ethiopia (Bayor 1988).

America's entrance into World War II in 1941 brought an end to debates about Hitler, Mussolini, and Spain. Yet the real shift in New York's nationality and religious conflicts did not occur until after World War II. New York State and City pioneered in legislation and action combating racial and religious discrimination. In 1945, responding to pressure by the American Jewish Congress and other ethnic organizations, the state enacted the nation's first law banning religious and racial discrimination in nonsectarian educational institutions. It then banned housing and employment discrimination and established a commission, the State Commission Against Discrimination, to enforce the law. The city enacted similar laws and established the City Commission on Human Rights.

Ethnic Tolerance

Compliance with the antidiscrimination laws was not immediate, nor was funding adequate for the agencies responsible for enforcement. Yet the racial and religious barriers for equal opportunity in employment and access to education and housing began to fall after World War II. By the 1960s Jewish quotas were eliminated at most universities and medical schools that had established them during the 1920s. Firms that had refused to hire Jews or Catholics shifted policy, and even in the large investment banking houses and Wall Street law firms changes were noted by the 1980s (Sachar 1992). In 1983, for example, an old WASP firm, Morgan Guaranty Trust Company, named Boris S. Berkovitch as vice president. Another Jew, Edward Miller, became head of Manufacturers Trust Company, the nation's fourth largest

bank. Italian Americans Anthony P. Terracciano and Peter C. Palmieri became vice chairman of Chase Manhattan Bank and vice chairman of the Irving Trust Company, respectively, during the same decade (Dinnerstein and Reimers 1988, p. 157). Even some of the old-line exclusive clubs began to open their doors. And in the political life of the city, both Jews and Italians increasingly won both municipal and state elections. Jewish candidates were usually liberal Democrats, while Italians often voted Republican or were less willing to support the liberal agenda.

The greater tolerance among New York's religious and European nationality groups was part of a national trend and marked a dramatic change in the history of religious and ethnic tensions. Higher educational standards, a growing economy, and declining racism and bigotry stemming from the recent war against fascism all played roles in explaining the shift. Also important was the fact that the third and fourth generations of the descendants of the "new immigration" did not seem so different to Americans of northern and western European ancestry. The later generations had learned English, no longer spoke the language of their grandparents, served loyally in the nation's military, and in many other ways seemed to participate in an emerging European-American culture.

The inevitable also happened: growing intermarriage among Americans of varied European descent. Indeed, even among Italians and Jews, who had seen little intermarriage among first-generation immigrants, intermarriage was becoming common after 1970. As a result, the traditional neighborhoods of these people began to disappear and the institutions which supported them weakened.

This is not to argue that all of New York City's citizens of European background were merging into a general White culture; only that many were. By midcentury it was difficult to find distinctive French Huguenot, English, German, or Dutch culture in New York City. Ethnic events like the annual Von Steuben Day Parade were largely symbolic of the past. Scandinavians were not far behind in their fading ethnic identity. As the editor of the century-old *Nordisk Tidende*, down to only 5,000 weekly issues in 1989 from 20,000 in the 1940s, put it, "The situation of the paper mirrors that of the Norwegian community. The population is steadily diminishing. I guess it will eventually assimilate and vanish, because many who still live here are older, their children have moved out of Brooklyn and there's no more Norwegian immigration" (*New York Times*, 17 March 1991).

Clearly New York still had several Jewish, Italian, and Irish communities. Among Jews, 70,000 or so Hasidim, located mostly in Brooklyn, lived apart from other New Yorkers, including other Jews. The city also had a thriving Orthodox community, centered in Brooklyn's Boro Park section. In the 1980s two newer Jewish communities emerged: the Soviets and the Israelis, each in the process of defining their identities. Other Jews were active in Conservative or Reform temples, while still others were scarcely observant and often were as at home in the gentile world as in the Jewish (S. Cohen 1988).

New York's Irish also inhabited several communities. By 1980, reflecting trends in intermarriage, half of those identifying themselves to the census said they were only part Irish. Most New York Irish had little interest in affairs in Ireland and they no longer lived in predominately Irish neighborhoods or read Irish American newspapers. They were also apt to marry non-Irish. If they attended the St. Patrick's Day Parade, it was largely a symbolic participation. Yet some Irish remained concerned about events in Northern Ireland, and a few, mostly first or second generation, organized groups like NorAid to assist the Republican cause there. The first and second generation was more apt to attend Irish American cultural events, like the *feis*, begun in the 1930s, and patronize the Irish Arts Center or Irish Repertoire Theater. A new Irish immigration, largely undocumented at first, in the 1980s served to bolster Irish-American New York culture (Reimers, forthcoming).

Italian ethnicity was most apt to be found in working-class neighborhoods like Bensonhurst in Brooklyn or the Belmont section of the Bronx. Bensonhurst also received a considerable number of Italian immigrants after World War II, and their presence reinforced the Italian culture of their community.

New immigration of Europeans was important to other groups as well. A substantial Greek immigration in the 1960s and 1970s created a first-generation Greek neighborhood in Astoria, Queens. Polish refugees from the initial collapse of the Solidarity Movement in 1979 headed for Greenpoint in Brooklyn and brought new Polish life to that older, heavily Polish community. As one resident of Greenpoint put it, "If it hadn't been for the emigres, we would have lost much of our language and traditions" (*New York Times,* 22

June 1984). Soviet Jews created another ethnic community in Brooklyn's Brighton Beach section.

New York's White community changed in still another way after 1945. Rising prosperity, improved transportation, and rapid suburban development opened the way for a significant White exodus. Whites made up nearly 95 percent of the city's population in 1940. Over 400,000 left in the 1940s, another 1.2 million in the 1950s, and 500,000 in the 1960s. Another 1 million moved in the 1970s and 1980s. The 1990 census found that only about half of New Yorkers were White.

Contemporary Migration

The city's population declined after 1950, but the drop would have been substantially greater without the in-migration of southern Blacks, Puerto Ricans, and new immigrants from Latin America and China.

Black Migration: Internal and International

The great migration of Blacks from south to north that occurred around World War I continued in the 1920s, but slowed during the Great Depression. After 1940, African Americans once again moved to New York City and they were about one-quarter of the city's population by 1980. During the 1970s the immigration slowed, and few southern Blacks settled in New York during the 1980s. The Black population grew, however, largely due to natural population increase and the arrival of Blacks from the Caribbean, especially Jamaica, Guyana, and Haiti. That migration dated from the 1790s, although the numbers were not especially large. By 1930 foreign-born Blacks accounted for about one-sixth of the city's Black population. Few new immigrants arrived during the 1930s, and changes in the immigration laws in 1952 placed severe limits on English-speaking West Indian immigration. Under the liberal immigration laws of 1965 these restrictions were removed, and Black West Indians once again entered New York City in large numbers. The 1980 census revealed that of the 300,000 non-Hispanic Caribbeans in New York, 80 percent had arrived after 1965. By 1990 one-quarter of Black New Yorkers were foreign-born, largely English-speaking from Jamaica, Trinidad, Tobago, Barbados, and Guyana (Kasinitz 1992).

Whether native- or foreign-born, the city's Black residents quickly learned about New York's historic patterns of racism. The Civil Rights movement of the 1960s did not leave New York untouched. While the conflicts in the South monopolized newspaper headlines and television news broadcasts, African Americans in New York organized demonstrations and protests to deal with their problems. Both the city and the state responded with legislation and programs to combat racial discrimination, and the lives of Black New Yorkers changed considerably after the 1960s.

Prior to that time they had been underrepresented on the city's public payroll and largely omitted from its politics. By the 1970s Blacks were overrepresented in public employment, although they tended to hold lower-paying positions and to be concentrated in health services and welfare positions rather than in city services, like the Irish-dominated fire department. In electoral politics, African Americans represented a growing segment of the city's voters. As such, they began to win municipal, state, and congressional elections. Finally, in 1990, David Dinkins, after winning a very close election, became the first Black mayor in the history of New York City (Mollenkopf 1992; Sleeper 1990).

The City University of New York also responded to Black pressure and adopted an open admissions plan effective in 1970 that substantially increased Black enrollment. In addition, picketing at some construction sites won better jobs for African Americans. Desegregation of the public schools was more difficult to achieve because of early resistance by the board of education and the White exodus. By 1990 only about 20 percent of the city's public school children were White. Despairing of desegregation, some Black activists urged community control, and they supported a decentralization plan implemented in the late 1960s. Debate over the direction of the schools erupted into a Black-Jewish conflict in 1968, pitting the largely Jewish teachers union against the local, Black-controlled school board (Sleeper 1990). By the early 1990s more Black (and Hispanic) teachers and administrators were being employed by the schools, but many problems such as old buildings in need of repair, low scores on achievement tests, high drop-out rates, and violence in the schools remained.

Educational problems represented but one difficulty facing Black New Yorkers. The unemployment rate for Blacks was consistently higher than for Whites, and Blacks were more apt to live in poverty.

And after declining during the immediate postwar years, during the 1980s the poverty rate increased. Blacks also encountered a segregated housing market, and violence was not unknown when African American families attempted to buy or rent in predominately White neighborhoods. Nor was racial violence limited to housing incidents (Rieder 1985). Conflicts between the police and Black New Yorkers periodically made headline news, and several serious incidents of White mobs killing Blacks occurred during the 1980s. In 1989, a Black youth was killed by a predominantly Italian American mob in Brooklyn's Bensonhurst, and two years later African Americans and Hasidic Jews clashed in ugly demonstrations in the Crown Heights section of Brooklyn. Black New Yorkers were also pitted against Korean grocers during the late 1980s and early 1990s in several Black neighborhoods, featuring Black boycotts of Korean stores and charges of racial insensitivity by both sides (Sleeper 1990).

Unemployment, poverty, riots, killings, and conflicts over schools and city programs all indicated that improvement in the nation's race relations after World War II had by no means brought equality. New York, as did the rest of the nation, moved forward, but fell short of finding racial justice and peace. The incidents and conflicts of the 1980s prompted some Black and White observers to see a deterioration of race relations following the progress of the civil rights era of the 1960s and 1970s.

Puerto Rican Migration

West Indians were only one of the major immigrant groups arriving in New York after World War II. Puerto Ricans had been present in the city before 1940, but not in large numbers. Though they were American citizens, their experience in the city resembled that of many immigrant groups. Their status as citizens enabled them to migrate from Puerto Rico to the mainland freely, and tens of thousands did so during the 1940s and 1950s. After 1960 the migration slowed, and some Puerto Ricans even returned to the island. Puerto Ricans were the city's largest Hispanic group, though their proportion of that population has now declined to about 50 percent. They were also the poorest of the city's Latinos, with over 40 percent living below the poverty line. They, too, faced discrimination and segregation but were unable to use their citizenship status effectively in poli-

In New York City at the turn of the century, immigrants lived in their own ethnic neighborhoods where they managed to make a living surrounded by the comforts of their own language, customs, music, and food. Here, census taking begins in New York City as the enumerator asks questions of the immigrant woman who sold fish in the market. (Theodore Wesley Koch Collection; Bentley Historical Library, The University of Michigan)

tics. Because of their poverty and lower educational achievements Puerto Ricans were less apt to vote and organize (Rodríguez 1989).

Latin American and Asian Migration

Dominicans made up the city's second largest Hispanic group, and by 1990 were the largest immigrant group in the city (Grasmuck and Pessar 1991). The city also claimed Cuban, Salvadoran, Guatemalan, Honduran, and other Latin immigrants, mainly from the Caribbean. From elsewhere in the Caribbean French- and Creole-speaking Haitians arrived in substantial numbers (Laguerre 1984).

Caribbean people dominated the immigration flow after 1970, but Asians also contributed a substantial number of newcomers after immigration poli-

cies changed in 1965. Chinese were the largest group, bursting the bounds of Manhattan's old Chinatown and spreading into Queens and Brooklyn. Taiwanese generally settled in Queens, while immigrants from Hong Kong and China lived in Manhattan (Chen 1992; Kwong 1987). Asian Indians, a "brain drain" migration of highly educated physicians, engineers, and entrepreneurs, added to the Asian influx, as did Koreans who operated many small businesses in addition to the ubiquitous greengrocer shops that dotted New York's streets (Kim 1981).

New Europeans

After World War II, European refugees settled in New York, but overall European immigration dropped in the early 1970s. Then events in Europe drove thousands to seek a new home in New York City. Refugee flows from Poland and the Soviet Union were most noticeable; and even after the collapse of communism, immigrants from Poland and the former Soviet Union continued to migrate. They were joined by many young Irish who were leaving Ireland because of the poor economic conditions there (Reimers, forthcoming).

Altogether, the new immigration was striking, the most diverse in the city's history. Coming from over 100 countries, the latest arrivals swelled the foreign-born population and added to the city's historical diversity. For example, in 1990, the census reported that among the older European groups there were 838,000 persons of Italian descent, 535,846 Irish, 296,809 Poles, and 330,230 Russians. Of the newer groups, the census reported 332,713 persons of Dominican ancestry, 84,454 Colombians, 391,744 non-Hispanic West Indians, 61,722 Mexicans, 94,590 Asian Indians, and 238,191 Chinese (Department of City Planning 1993, p. 8; U.S. Bureau of the Census, 1990c). After little immigration during the years of the Great Depression and World War II, New York was becoming an immigrant city again. The census recorded over 900,000 immigrants settling in New York during the 1980s, and by 1990 immigrants made up 28 percent of the city's population, not counting undocumented aliens (Department of City Planning 1992, p. 6).

In 1992, 120,000 immigrants settled in the city, with the Dominican Republic, the former Soviet Union, China, Jamaica, Guyana, the Philippines, Poland, Ireland, Trinidad and Tobago, and Colombia being the top ten sending nations (*New York Times*, 13 June 1993).

Neighborhoods were being changed by the latest wave of newcomers. As pressure grew for housing in Manhattan's Chinatown, Chinese bought buildings in Little Italy and on the Lower East Side, once the center of the city's Jews. In 1974 *The Forward*, the famous Jewish radical newspaper, sold its building to a Chinese group. By 1987 it had declined from a high of nearly 250,000 readers to only 20,000 (*New York Times*, 25 May 1987).

Ethnic succession perhaps peaked when Jewish customers reported that the best bagels in New York could be purchased at a Queens store operated by Thai immigrants. One customer remarked of the ba-

gels sold in this store run by seven Thais, "They are really very good; crispy, chewy, lots of stuff on top. Some days you can't get a table in there . . . are there Thailand Jews?" (*New York Times*, 21 Feb. 1991).

Conclusion

The new immigrants are clearly visible in the city's economy, schools, streets, neighborhoods, and culture. Half of the city's eighty foreign-language newspapers, frequently sold at newsstands run by Asian Indians, were established between 1970 and 1990 (*Daily News*, 10 Dec. 1990). A city overwhelmingly Christian and Jewish was witnessing the spread of new religious groups. New York's first Hindu temple was founded in 1977, and by 1991 had seven full-time priests (*New York Times*, 11 June 1991). Buddhists and Sikhs appeared along with Hindus, and the city's Islamic population grew rapidly. The largest Mosque, which can accommodate 1,000 worshippers, opened in 1991, and by then Muslims could send their children to ten Islamic schools (*New York Times*, 3 Oct. 1992, 4 May 1993).

The new immigrants' impact upon politics was just emerging in 1991 when Dominicans elected one of their own to a city council seat. The post–World War II immigrants arrived at a time vastly different than prior waves. Immigrants of the past did not encounter such a wide range of public services, nor did they have civil rights laws to protect them. But they also encountered a diversity of jobs in the rapidly growing manufacturing sector. As the city restructured its economy after 1945, many immigrants found jobs in a rapidly growing service sector and in small business niches in the economy (Waldinger 1986 and Winnick 1990). Aggravated by the national economic recession, New York experienced another round of economic difficulties in the early 1990s. How these may impact future events remains unclear. It is clear, however, that just when conflicts among White ethnic groups abated and when New Yorkers of European descent were becoming less significant in the city's life, Blacks, Hispanics and Asians were becoming more so. As immigrants and ethnics, they will no doubt continue to shape New York City's future as they constructed her in the past.

26

Los Angeles: Explosive Diversity

MEHDI BOZORGMEHR

GEORGES SABAGH

IVAN LIGHT

On a balmy spring day in 1992, Los Angeles exploded into thirty-six hours of arson, rioting, and looting. The fury erupted after a jury handed down a not guilty verdict in the trial of four officers charged with beating a black motorist. That verdict flew in the face of well-publicized evidence—a clandestine videotape of the officers brutally beating the driver. All of the officers charged were Anglo. As television viewers around the world watched in horror, mobs of looters plundered stores, with arsonists torching the premises. The 1992 Los Angeles riot was the worst in the U.S. history since the New York City draft riot of 1863. Moreover, the 1992 riot was unique to Los Angeles,[1] whereas L.A.'s Watts riot in 1965 preceded comparable uprisings in several other American cities. The "L.A. 2000" epilogue's warning of the *Blade Runner* scenario: "the fusion of individual cultures into a demotic polyglotism ominous with unresolved hostilities"[2] had become a reality. According to the *New York Times*, "In two centuries, the City of Angels has gone from a Spanish village, to a Mexican garrison, to a Midwesterners' retirement and agricultural haven, to the epitome of the post-war suburban ideal, to a polyglot capital of the Pacific Rim, to what some people see as a 'Blade Runner' nightmare."[3]

The Los Angeles riot spectacularly publicized the long-standing conflict between Blacks and Koreans. In all, 2,073 stores were affected, with a median damage of $75,000, and total damages of $359 million. Korean-owned businesses suffered 73 percent of the total damages sustained by business (Ong and Hee 1992). More than 14,000 people were arrested during the riots, but only 5,633 were arraigned in Los Angeles Municipal Court. Of those arraigned, 51 percent were Latino, 36 percent were Black, and 11 percent were White.[4] The ethnic and racial diversity of the arrestees shows that rioters by no means included only Blacks, although they were overrepresented. On the other hand, some evidence suggests that the riot's focus on Korean firms was indeed a product of the Black/Korean conflict (Light, Har-Chvi, and Kan 1994). First, Los Angeles' Hispanic communities had not had a prior public quarrel with

The research for this chapter was partly funded by the Ethnic Los Angeles grant from the Russell Sage and Mellon foundations to Roger Waldinger and Mehdi Bozorgmehr. We thank Roger Waldinger for his ideas and Claudia Der-Martirosian for her help with data analysis.

the Korean merchants in their neighborhoods as had the African Americans (Cheng and Espiritu 1989; Light and Bonacich 1988, chap. 12). Second, fire damage to stores was much heavier in the African American neighborhoods of Los Angeles than elsewhere in the city, suggesting that the motive of revenge was more significant in these areas. Third, 68 percent of the African American public interpreted the riot as "a protest," whereas only 39 percent of Hispanics did so (Bobo et al. 1992, Table C14). Finally, after the riot, Korean merchants negotiated protection payments with black street gangs[5] whom they openly suspected of having set the fires that burned their stores.[6] The Korean/Black conflict and the diverse ethnic background of the arrestees underscores the impact of immigration and ethnic change on Los Angeles. However, rapid ethnic change in Los Angeles has produced less publicized conflicts among Asians, Anglos, and Hispanics about political representation, multiculturalism in schools, use of English on business signs, and other issues (Horton 1989). Los Angeles, between the two riots (1965 to 1992), was ethnically transformed by immigration more than any other American city.

Immigration to Los Angeles

A striking geographical feature of today's immigration is its Pacific Coast destinations (Portes and Rumbaut 1990). Whereas New York City, Chicago, Detroit, and other Atlantic Coast or midwestern metropolitan regions received the bulk of immigrants before 1965, Los Angeles has become the major port of entry for new immigrants. Of the 3.3 million foreign-born persons who arrived in the United States between 1975 and 1980, 17 percent settled in Los Angeles and 12 percent settled in New York.[7] This pattern has persisted to the present. Of the 2.3 million persons admitted legally to the United States between 1987 and 1989, 18 percent chose Los Angeles and 13 percent selected New York. The addition of undocumented workers only widens the intermetropolitan gap. In marked historical contrast, of the 3.5 million who arrived in the United States before 1950, only 5 percent settled in Los Angeles and 14 percent settled in New York City.

Before 1960, Los Angeles was a magnet for interstate migrants from middle western and southern states. Some of these migrants were the descendants of Europeans whose first settlement areas had been along the eastern seaboard (Davis 1990, p. 114). "Here was the city that, only thirty years before, had been known as 'Iowa's seaport' because of all the transplanted midwesterners who had settled there, now serving as the magnet, the destination of choice, for people from all over the Third World" (Rieff 1991, pp. 131–132). Unlike New York, the international presence was "virtually nonexistent" in the 1960s in Los Angeles (Soja, Morales, and Wolff 1983). The massive influx of immigrants has made Los Angeles a multicultural and multiethnic metropolis. In less than thirty years, Los Angeles has surpassed New York in absolute number of immigrants (2,895,066 versus 2,082,931) and in relative proportion of its foreign-born population (34 percent compared to 28 percent). In 1990, about one out of every four foreign-born persons (5 million out of 19.8 million) in the United States lived either in greater Los Angeles or New York City. Indeed, other than California and New York, *no state* exceeded Los Angeles or New York City in the number of immigrants. But, in contrast to a decade ago, Los Angeles is now the nation's premier immigrant metropolis. Moreover, unlike New York City, which has always been multicultural and multiracial, Los Angeles has recently become so after two decades of unprecedented ethnic change. The new immigration has already introduced a large population of American children whose parents were born abroad. Higher birth rates among most foreign-born ensure that immigrants and their ethnic descendants will comprise an even bigger portion of Los Angeles' population by 2000. In that year, they will no longer be minorities in a statistical sense (Waldinger and Bozorgmehr, forthcoming).

Immigration has also reinvigorated population growth in Los Angeles. Los Angeles' pre-1960 history of population growth was impressive, but in the 1970s, before the immigrants arrived, there was a marked slow down in this growth. Southern California's huge and immensely profitable weapons industry declined as the Vietnam War wound down. Although peace benefited the people of the United States, then as now peace hurts the Pentagon-dependent economy of Los Angeles.[8] Happily for the region's prosperity, Ronald Reagan's presidency reinvigorated the international arms race, threatening world destruction, but bringing new defense contracts to Southern California, and immigrants reversed the region's decline of population. As a conse-

Table 1   Population of Major Ethnic Groups, and Population by Place of Birth in Los Angeles County, 1960–1990

Ethnic Group and Place of Birth	Percent Distribution			
	1960	1970	1980	1990
Major ethnic groups				
Whites (non-Hispanic)	80.5	67.1	52.9	40.8
Hispanic	9.6[a]	18.3	27.4[b]	36.4[c]
Black	7.6	10.8	12.6	11.2
American Indian/Eskimo	0.1	0.4	0.6	0.5
Asian/Pacific Islander/other	2.2	3.4	6.5	11.0
Total	100.0	100.0	100.0	100.0
Place of birth				
Born in California	34.5	39.5	47.7	40.8
Born in other states	51.8	43.3	37.1	27.0
Foreign-born	9.5	11.3	22.3	32.2
State of birth not given	4.2	5.9	—	—
Total	100.0	100.0	100.0	100.0
Total population	6,038,771	7,032,075	7,477,503	8,863,164

[a] Spanish-surname population based on sample.
[b] Excludes Hispanic Blacks.
[c] Excludes Hispanic Blacks, Hispanic American Indians, and Hispanic Asians and Pacific Islanders.
Sources: U.S. Bureau of the Census. *U.S. Census of Population: 1960.* Vol. I, *Characteristics of Population. California.* Washington, D.C.: U.S. Government Printing Office, 1963; U.S. Bureau of the Census. *U.S. Census of Population: 1960. Persons of Spanish Surname.* Washington, D.C.: U.S. Government Printing Office, 1963; U.S. Bureau of the Census. *1970 Census of Population.* Vol I. *Characteristics of the Population. California.* Washington, D.C.: U.S. Government Printing Office, 1973; U.S. Bureau of the Census. *1980 Census of Population.* Vol. I, *Detailed Population Characteristics. California.* Washington, D.C.: U.S. Government Printing Office, 1983; U.S. Bureau of the Census. *1990 Census of Population and Housing.* Summary Tape File 1A and Public Use Microdata Sample (PUMS).

quence, with a population of 14.5 million in 1990, Los Angeles is now not only the second largest metropolis of the United States after New York, but is also the most rapidly growing (Sabagh 1991).

Ethnic Change

In general, immigration is most difficult to absorb when immigrants are numerous, and of different ethno-racial stock than the settled population. Both conditions are met in Los Angeles. In 1960, 80 percent of Angelenos were non-Hispanic Whites, 50 percent had been born in other states, and only 10 percent were foreign-born (Table 1). The new immigration of the 1970s and 1980s radically altered this picture by adding to the population of existing ethnic groups, such as Mexicans, and creating new ethnic groups such as Central Americans, Vietnamese, and Iranians, who had not lived in Los Angeles before.

By 1990, non-Hispanic Whites became a demographic minority, accounting for only four out of every ten Angelenos. By contrast, the demographic share of Hispanics increased spectacularly from 10 percent in 1960 to 36 percent in 1990,[9] with the group nearly matching Anglos as the largest in Los Angeles. Between 1960 and 1990, Asians were rapidly catching up with Blacks as the next largest minority group in Los Angeles.[10] In 1960 there were about two Asians for every eight Blacks, but by 1990 there were as many Asians as Blacks.[11] The rapid growth of the Asian population, including Koreans, as compared to a nearly stagnating Black population, contributed to the explosive violence of Blacks against Korean immigrant entrepreneurs in Los Angeles.

Ethnic diversity increased between 1960 and 1990, but most of this occurred before the 1980s (Table 1). Despite sensational exaggerations in the press, Los Angeles is not unique in this respect. For instance, in 1990, New York was even more diverse than Los Angeles, but their components of diversity were different. Compared to Los Angeles, New York had many more immigrants from the Dominican Republic, Haiti, Jamaica, and Guyana. Caribbean immigrants also contributed to the diversity of the Black population in New York.

The most remarkable change is the increase in the foreign-born Angelenos from 9.5 percent in 1960 to 32.2 percent in 1990, more than a threefold increase (Table 1). In other words, in 1990, three out of every ten Angelenos were immigrants. Even more striking, however, is the shift in immigrants' origins from Europe to the Third World. In the thirty-year period, the number of immigrants from Europe and Canada declined from 375,000 to about 250,000 (Table 2). Immigration from the former Soviet Union is an exception. The number of Russian immigrants declined by half between 1960 and 1980, but it more than doubled in the 1980s because of Jewish and Armenian refugees. At the same time, there was a spectacular growth in the number of immigrants from Latin America, Asia, and Africa from nearly 200,000 to about 2.5 million. In 1960, immigrants from the Third World were only half as numerous as immigrants from Europe and Canada. By 1990, their numbers were ten times greater. This ratio soared to twenty-seven times more for immigrants who arrived in the 1980s.

From 1970 to 1990, Latin America, and more particularly Mexico, contributed the most to the immigrant population of Los Angeles. In 1990, there were 1.7 million persons born in Latin America, of whom nearly 1.2 million were born in Mexico (Table 2). Asians ranked next in importance, numbering more than 750,000 in 1990. While there was a sharp rise in the number of immigrants from Africa, this number was still very small even in 1990.

Diversity increased among immigrants from specific Third World regions, particularly between 1970 and 1990. In 1980, the overwhelming majority of immigrants from Latin America had been born in Mexico. In spite of vast Mexican immigration in the 1970s and 1980s, their share of foreign-born Latinos decreased steadily from 1970 to 1990. As a result of political terror in Guatemala, and civil war in El Salvador, the foreign population from these two countries increased eight-fold between 1980 and 1990 (Table 2).

While Asians have lived in southern California since 1850, the Asian population has burgeoned un-

der the impact of recent immigration. To the Japanese and Chinese groups who comprised the bulk of Asian residents in the 1960s, the new immigration added a much more diverse Asian-origin population. Long southern California's largest Asian-origin group, the Japanese lost their historical preeminence in the 1980s. In that decade, the Japanese were the slowest growing Asian group; the Chinese were the most rapidly growing. Because Japan was so prosperous in this period, Japanese had no desire to leave their homeland, and Japanese immigration was minimal. As a result, the Japanese population fell from first to fourth rank in size, the Chinese rose from third to first rank, and the Koreans rose from fourth to third. Los Angeles has no "Little Manila" Filipino enclave. Nonetheless, despite their residential invisibility, Filipinos were the second most numerous immigrant group in both 1980 and 1990.

Immigration has made an important contribution to the population growth of Filipinos, Chinese, and Mexicans. Immigration accounts for about half of the growth of the Mexican population, and about three-quarters of both the Chinese and Filipino population increase. For other groups, such as Iranians and Vietnamese, immigration accounts for most of their population growth by 1990.

Cultural Impact of Immigration and Ethnic Change

The increasing diversity of residential streets and business districts, stores and services, sights and sounds, cultural symbols, business signs, entertainment, and languages are all cultural impacts of immigration. In the 1950s, Whites from Iowa held annual picnics at MacArthur Park in downtown Los Angeles. They served corn and baked ham and organized square dances. Today, MacArthur Park hosts a multicultural and multilingual population of homeless vagrants, gang members, peddlers, Central American shoppers, and Latina mothers and infants. Tacos and tortillas have replaced corn and baked ham. In the 1950s, there were two small Asian immigrant enclaves (Little Tokyo and Chinatown), and somewhat larger Mexican enclaves in East L.A. and the San Fernando Valley. By 1993, these original enclaves expanded, and ethnic villages have been created by the immigrants of the 1970s and 1980s. The new ones include Little Saigon, Koreatown, Chinese enclaves in the San Gabriel Valley including Monterey Park, and

Table 2   Number of Immigrants by Continent, Region, or Selected Countries of Birth in Los Angeles County, 1960–1990

Continent, Region, and Selected Country of Birth	Number of Immigrants (in 1,000s)			
	1960	1970	1980	1990
Europe	291	262	217	199
Former USSR	44	32	34	51
Canada	84	68	56	47
China/Hong Kong/Taiwan	9	17	43	138
Japan	17	19	33	45
Korea	1	NA	51	116
Philippines	9	NA	73	168
Other Asia	17	NA	140	315
Total Asia	53	96	340	782
Mexico	118	207	698	1,174
Guatemala and El Salvador	2	NA	45	321
Other Latin America	17	95	171	214
Total Latin America	137	302	914	1,707
Africa	2	NA	18	30
All other	3	12	8	37
Total	614	772	1,587	2,853
Country of birth not reported	3	21	78	133
Total immigrants	617	793	1,665	2,986

Note: NA means not available.
Sources: See Table 1.

Salvadoran Pico-Vermont enclaves. In spite of the existence of such ethnically homogeneous localities, however, there has been an increase in the ethnic diversity of most areas in Los Angeles. This diversity was brought about by the flight of Whites who were replaced by Asians, Blacks, and Hispanics. Nevertheless, between 1970 and 1990, segregation decreased noticeably among African Americans and Asians, but increased slightly among Hispanics. The rise in segregation among Hispanics can be partly attributed to the rapid increase in their numbers. Although Blacks did not experience in-migration in the 1980s, they are highly segregated in Los Angeles (Massey and Denton 1993).

The Los Angeles Black community is more diverse than suggested by the sensational media images of crime, riots, and family collapse. Black Los Angeles actually contains a newly enlarged and prosperous middle class, beneficiaries of government employment and policies, as well as an underclass in deepening poverty. The bifurcation only increases the class segregation of the Black population that Franklin Frazier (1957) lamented long ago. While Blacks still suffer high racial segregation, the decrease in segregation over time is quite striking. At the same time, employment opportunities for Blacks have become increasingly tied to skill level, education, and the deconcentration of manufacturing and service jobs in the metropolitan area. Many Blacks in the manufacturing sector have lost their jobs because the factories that once hired them have closed or relocated (Johnson et al. 1992).

Another impact has been an explosion of ethnic and immigrant businesses, the most visible of which are restaurants. Although its air is polluted, its freeways often congested, and its streets unsafe, Los Angeles's food is excellent, varied, and inexpensive. Prior to the 1970s, Chinese, Japanese, and Mexican restaurants were already common, but subsequently there has been a rapid growth in the number of Korean, Thai, Vietnamese, Filipino, Indian, Central American, Iranian, and Armenian restaurants. The quality, low price, and diversity of ethnic cuisine in Los Angeles creates a dietary cosmopolitanism that reflects a cultural melting pot.

Although Spanish is the mother tongue of most Los Angeles students, over eighty languages are now spoken in Los Angeles schools. Between 1977 and 1987, the Hispanic student population in Los Angeles County schools increased from 30 to 48 percent and the non-Hispanic White segment dropped sharply from 47 to 28 percent. Asian students increased from 5 to 10 percent in the same period (Sabagh 1991, Table 6). Starting in the 1960s, the trends for Hispanic and non-Hispanic students are much more pronounced in the city's schools. Expanded ethnic populations in schools and colleges creates a pressing need for developing bilingual and multicultural education. At the university level, it encourages criticism of Eurocentric education and the emergence and consolidation of Chicano and Asian studies programs. Demand for a full-fledged Chicano studies department in addition to the existing center at UCLA was brought to the attention of the nation in televised demonstrations and hunger strikes in 1993.

The proliferation of foreign languages is not limited to schools and college campuses. A myriad of languages are often heard on Los Angeles streets. Foreign language traffic schools have mushroomed to meet the needs of immigrant drivers. Driving a car is a virtual necessity in Los Angeles, where, since the decline of the once-famous street railway system, public transportation has gone into shocking eclipse (Light 1988). Because people must drive their cars everywhere, they are at risk of traffic citations, which can be removed from one's driving record by attending traffic schools offered in English or in foreign languages. A recent California "Traffic Violator School Licensing List" included the following foreign language traffic schools in Los Angeles County: one in Armenian, seventeen in Chinese (Mandarin), one in Hebrew, seven in Korean, four in Japanese, four in Persian, one in Russian, fifty-eight in Spanish, and nine in Vietnamese.

Predictably enough, there has also been a proliferation of immigrant-owned and oriented media in Los Angeles. In the 1950s, Los Angeles had no Spanish-language television channels, only one Spanish-language daily, and a few Spanish-language radio stations. Today Los Angeles has three Spanish-language UHF television channels, seven Spanish-language radio programs, and two daily Spanish-language newspapers. UHF television channel 18 devotes its entire programming to broadcasts in Arabic, Armenian, Cambodian, Hindi, Korean, Mandarin, Persian, Tagalog, and Vietnamese. News programs from China, Taiwan, Korea, and Vietnam can be seen daily on Los Angeles television stations. There are also radio programs in many of these languages. According to the *Los Angeles Times* (15 April 1991, p. F12), "the radio dial is becoming a mirror of the mix of Southern California."

Economic Change

Recent immigration to Los Angeles has occurred under economic circumstances that resemble and differ from what existed during the Great Migration of 1880 to 1914. Then as now, Los Angeles was a nonunion town whose cheap labor was a major industrial attraction. As did earlier immigrants, newcomers of the 1980s entered a rapidly growing economy—indeed, one that, thanks to extravagant military spending, expanded at a much faster rate than the rest of the U.S. economy in the Reagan/Bush era. But they also entered a civilian economy undergoing major structural change, in which the region's high-wage manufacturing industries relentlessly contracted, in part, because of the military spending that brought prosperity to others. Although Los Angeles is the largest manufacturing center in the country, job expansion in today's Los Angeles occurs in services and in low-wage manufacturing industries rather than in the high-wage smokestack industries of the industrial era. Although the prospects for successful immigrant adaptation depend on the regional economy, the complex transformation in southern California's economic structure makes it difficult to predict which of the immigrant groups that comprise Los Angeles' population are best situated to move ahead. The general prognosis is, regrettably, unfavorable, because without the resumption of the cold war, defense spending will not rescue Los Angeles again. Moreover, high-wage jobs in the civilian sector are declining in number as a result of economic restructuring; only low-wage jobs are growing. In effect, the good jobs are moving away from Los Angeles, leaving behind the worst jobs.

Manufacturing employment in Los Angeles increased from 881,000 in 1969 to 925,000 in 1987, a peak year for military spending in the Reagan era. Then it decreased to 895,000 in 1991.[12] During the same period, manufacturing employment declined sharply by about 50 percent in New York and Chicago. By the mid-1980s, Los Angeles had become the prime industrial metropolitan area of the United States. Also there has been a sharp decline in automobile, steel, and tire production. These heavy industries are unionized and they pay high wages and good benefits. The recent decline in manufacturing employment, however, is indicative of a restructuring of the civilian economy that undermines the manufacturing base. According to Sassen (1990, p. 288), "the need for centralized management and control feeds growth in cities such as New York and Los Angeles." As compared to Chicago and Los Angeles, however, New York is "qualitatively and quantitatively different" in that it has a higher incidence of finance, investment, real estate, and business, with more firms in sectors that are international in character (Sassen 1990, p. 302).

The last few years have witnessed a radical restructuring of manufacturing employment in Los Angeles. Producing commercial as well as military airplanes, the aircraft industry has always been an important sector of the Los Angeles economy. This high-technology industry is very dependent on military expenditures, and the end of the cold war, while very beneficial for the people of the earth, damaged southern California's lucrative weapons procurement industry. After the Vietnam War ended, and military expenditures declined in consequence, employment in the aircraft industry in southern California fell from about 203,000 in 1959 to about 90,000 in 1977, a disaster locally blamed on President Jimmy Carter. With the Reagan and Bush administrations' reanimated cold war vision in Washington, defense industry employment rose to about 130,000 in 1986 (Scott 1990). With the end of the cold war, reductions in military expenditures adversely affected employment in the Los Angeles aircraft industry. It has declined much more rapidly than local industrialists and real estate boosters would like, if less rapidly than military security would warrant (Peltz 1993; see note 8). Because of military cutbacks, aerospace employment must continue to decline from its peak of 375,000 in 1988.[13] Happily, some of this decline has been compensated for by the rise in civilian high-tech industries, including civilian aircraft. In southern California, employment in civilian high-tech industries increased more than fourfold between 1959 and 1986 (Scott 1990, p. 1578).

The region's major manufacturing expansion has been small firms specializing in garment, apparel, food processing, and furniture manufacturing. All these industries rely on the cheap and unskilled labor of immigrants. In Los Angeles County, between 1981 and 1988, manufacturing employment declined by 4 percent. In the same period, employment in apparel and textile products, where immigrants concentrate, increased by 17 percent.[14] According to Scott, "the heart of the country's clothing industry, which traditionally includes sweatshop jobs, has moved from New York to Los Angeles."[15] Another cause of the growth of low-skill, low-income jobs is immigration itself. According to Portes and Rumbaut (1990, p. 238), "the presence of foreign workers, many undocumented, has been central to the survival of the garment and footwear industries in New York and Los Angeles."

There has also been a growth in the underground cash economy. Transactions in this economy cannot be measured directly because no one reports them. However, the city's ubiquitous street vendors are a visible sign of its thriving informal economy. In 1992, the City of Los Angeles contained 3,000 street vendors, most of whom were Hispanic immigrants.[16] The vendors fought for five years to change the law before the city council finally voted in 1994 to legalize street vending on a trial basis.[17]

Types of Immigrants and Economic Change

Each type of immigrant has a different impact on the Los Angeles economy. Although most new immigrants come from the Third World, they differ among themselves in their education and in their reasons for leaving their homelands. Economic immigrants and refugees are two different types of newcomers. Each has characteristically different motivations for migration. Economic immigrants relocate in response to economic opportunities in the country of destination; for example, employment at wages superior to what they were earning in their homeland even if those wages are very low by American standards. The immigrants are further differentiated in terms of temporary labor migrants and more permanent immigrant professionals and entrepreneurs (Portes and Rumbaut 1990). As might be expected, temporary labor migrants have less education and work experience than do the other two types of economic immigrants. Refugees leave their country of origin because of famines, revolutions, and wars (Gold 1992, Pedraza-Bailey 1985).

Most immigrants in Los Angeles are economic immigrants, but some are refugees, exiles, or asylees. The number of refugees in Los Angeles County increased from 59,344 in 1980 to 164,477 in 1990.[18] After the Vietnam War, most refugees in southern California were Indochinese. However, like the rest of the immigrant population, this refugee segment of the Los Angeles population has become increasingly heterogeneous in the last two decades. In 1990, 31 percent of refugees were from world regions other than Southeast Asia as compared to only 15 percent in 1980. If data were available on the motivation for leaving the country of origin, there would be many more refugees among these immigrants. For example, on the basis of motivation, 40 percent of a random sample of Iranian immigrants in Los Angeles were classified as refugees, as compared to a mere 24 percent who had the legal status of "refugee or asylee" (Bozorgmehr and Sabagh 1991, p. 125).

Immigrants also differ markedly in human capital, commonly measured by level of education. Data on the number of years of schooling show great variation among immigrants (Table 3). With about 3 percent completing four or more years of college, Mexican and Central American immigrants had the lowest level of schooling; immigrants from Africa and Asia (most notably Indians, Filipinos, Chinese, and Iranians) had the highest.[19] Among major Asian groups, the Vietnamese had the lowest level of schooling. Surprisingly, Europeans were not the most highly educated immigrants (Table 3). As expected, immigrant groups with higher educational credentials are much more likely to be in professional and managerial occupations. The proportion of immigrants in these two top occupations varied from roughly a low 7 percent among Mexicans and Central Americans to 40 percent and higher among immigrants from Europe, Canada, Japan, China/Hong Kong/Taiwan, the Philippines, India, Iran, Israel, and Africa. Not surprisingly, immigrant groups with a high concentration in the top two occupations had relatively few manual workers, especially unskilled and semiskilled. By contrast, 59 percent of Mexicans were manual workers, of whom 41 percent were unskilled.

Central Americans were also highly concentrated in manual jobs.

Table 3 shows wide variations in self-employment among immigrants in Los Angeles. With a rate of about 32 percent, Koreans had the highest self-employment rate, followed by Iranians and Israelis (about 27 percent) and Lebanese (24 percent).[20] Rates of self-employment are in the 15 to 25 percent range for Europeans, Soviets, Canadians, and most Asian groups. The level of self-employment among immigrants reflects, in part, their ability to rely on coethnic networks, social capital, the money they have, and premigration entrepreneurial experience.

Combined with a fairly high rate of self-employment among some Asian and Middle Eastern immigrants, the presence of even some manual workers among these immigrants has favored the growth of ethnic economies in Los Angeles. *Ethnic economies* consist of the ethnic self-employed, ethnic employers, and their coethnic employees. Los Angeles possesses the largest ethnic economies of any metropolitan area in the United States. According to Mike Davis (1990, p. 104):

> The *internationalization of class formation* has recently made a quantum leap in Los Angeles beyond any other North American city. The most WASPish of big cities in 1960, Los Angeles now contains more polyethnic diversity than New York, with a huge manual working class of Latinos and a growing rentier stratum of Asian investors. On an elite level, Japanese capital, arriving like a *tsunami* in the early 1980s, has already become a discreet but major player in the city politics. Meanwhile the great waves of Chinese, Korean and Armenian middle-class immigrants, augmented by Israelis, Iranians, and others, have made Los Angeles the most dynamic center of ethnic family capitalism on the planet.

Ethnic economies are an important adjunct to the economic mobility of all groups (Light et al. 1993). But some groups benefit more from ethnic economies than others, and only in selected cases has self-employment emerged as the dominant mode of adaptation among recent immigrant groups.

When immigrants become self-employed, they contribute to the growth of ethnic economies in Los Angeles, and immigrant entrepreneurs mainly em-

Table 3   Socioeconomic Characteristics of Persons 25 to 64 Years Old, Hispanic and Non-Hispanic Foreign-born by Selected Regions or Countries of Birth, Native-born, and Total Population in Los Angeles County, 1990[a]

Regions and Countries of Birth	Percent with Four Years of College or More	Percent Self-Employed	Managerial Professional Occupations	Manual Workers[b] Skilled	Manual Workers[b] Unskilled
All persons	23.9	11.2	32.5	11.6	15.7
Native-born	28.1	11.1	39.5	10.0	9.9
Foreign-born	17.5	11.4	20.8	14.1	25.2
Europe and Canada					
Europe	29.5	18.3	47.3	11.8	7.1
Former USSR	27.7	20.5	33.4	18.5	13.2
Canada	29.5	16.9	45.6	7.4	7.3
Asia					
Japan	37.9	14.7	41.0	5.6	4.8
Philippines	55.8	5.1	39.9	7.1	7.8
China/Hong Kong/Taiwan	42.8	16.4	42.4	5.7	10.3
Korea	37.6	31.8	30.2	10.6	10.6
Vietnam	13.1	11.8	24.4	15.4	19.2
Thailand	37.0	11.7	29.6	10.3	11.6
India	60.4	14.8	54.1	3.9	6.4
Iran	42.5	26.5	41.7	10.2	7.6
Israel	29.6	27.2	42.8	15.8	6.5
Lebanon	26.7	23.5	34.8	17.6	7.1
Other Asia	25.8	18.7	32.5	12.1	13.0
Africa	45.8	15.1	44.8	6.0	9.1
Latin America					
Mexico	2.8	6.8	7.0	18.4	40.7
Guatemala	3.4	8.2	6.3	17.3	29.5
El Salvador	3.0	7.2	6.5	15.3	32.4
West Indies	17.6	11.9	27.5	12.3	15.4
South America	21.3	14.1	28.5	12.9	15.8
Other Latin America	10.0	8.1	16.1	13.9	22.4

[a] Includes only countries with 2,500 or more employed males ages 25 to 64.

[b] Unskilled includes semiskilled.

Source: U.S. Bureau of the Census. *1990 Census of Population and Housing*, Public Use Microdata Sample (PUMS).

ploy labor migrants. Sometimes these are coethnics, but many low-wage, small-firm immigrant industries, especially the garment industry, have developed because immigrant entrepreneurs hire immigrant laborers from other countries. In assessing the expansion of garment sweatshops in Los Angeles, Light (1988, p. 62) has argued that "although most workers were Mexican women, Chinese from Hong Kong and Taiwan, Koreans and Latin Americans were their principal employers." In such cases, the immigrant entrepreneurs provide the jobs that attract immi-

Racial and ethnic inequality long sustained and endured can suddenly explode, calling for attention. Here, even the palm trees in Los Angeles look like they were ignited during the April–May 1992 riots. (AP/Wide World Photos)

grant laborers without creating new jobs for native-born workers.

Uneducated migrants work for the lowest possible wages, partly because these wages are higher than what they would have received in their homeland. They are attracted by the opportunities for employment in Los Angeles even though these opportunities are in the dirtiest, smelliest, and worst-paid sectors of the Los Angeles economy. In turn, their presence in Los Angeles contributes to the growth of this sector because employers know that abundant cheap labor is available. Many of these labor migrants are undocumented and subject to employer exploitation. Employers threaten to report undocu-

mented workers to the Immigration and Naturalization Service (INS) ("la Migra") if workers object to substandard wages and working conditions. In this way, unscrupulous employers reproduce conditions of semiservitude in a formally free country. Most undocumented workers hail from Mexico or Central America, and because of the geographical proximity and the varied possibility for illegal entry, they go back and forth between Los Angeles and their homeland, a movement often prompted by immigration raids. As Massey et al. (1987) have shown, however, many of these immigrants end up settling in Los Angeles, thus expanding the permanent pool of unskilled workers.

The massive immigration to Los Angeles in the 1970s and 1980s coincided with increased employment in this metropolis. Thus, the estimated number of wage and salary workers in nonagricultural establishments in Los Angeles increased by 45 percent, from 2,896,800 in 1970 to 4,222,900 in 1988.[21] This growth in employment is caused by net international migration, by net internal migration, and by those entering the labor force between 1970 and 1988. Espenshade and Goodis (1985, p. 15) estimated that of the growth in total employment in Los Angeles County of 645,200 persons, about two-thirds were the result of the employment of immigrants who arrived between 1970 and 1980. Mexican immigrants contributed nearly half of this addition to employment. As labor migrants, however, they contributed two-thirds of employment in manufacturing and in eating and drinking establishments (Espenshade and Goodis 1985, p. 18). In 1980, about two-thirds of workers in low-wage manufacturing were immigrants, and about two-thirds of these immigrant workers were Mexicans. With their lower educational levels, Mexican workers had to seek employment in the low-wage sector of the Los Angeles economy, and contributed to the growth of this benighted sector.

Increased poverty is another consequence of labor migration and the expansion of low-wage employment in Los Angeles County. Although some immigrant newcomers are moving ahead, especially by the poor living standards of their homelands, increased immigration has increased poverty and widened income differentials. The percentage of foreign-born Hispanics who were below the poverty line (that is, annual income less than $11,102 in 1986 dollars) increased from 20.3 percent in 1969 to 26.1 percent in 1979 and to 44.6 percent in 1986–1987. Since their share of the total employed population increased from about 5 percent in 1969 to 22 percent in 1986–1987, the level of poverty among all workers doubled from 8 percent to 18 percent in this period (Sabagh 1991). Growing poverty may be due to such structural factors as the decline of core manufacturing industries and the growth of the low-wage sector. But while structural changes are important, they clearly affect Los Angeles's multiethnic population unequally, with some groups bearing the brunt of change through wage depression and unemployment.

Conclusion

For many years, Bostonians, New Yorkers, Chicagoans, or San Franciscans snickered at the mention of Los Angeles as a metropolis. To them, Los Angeles was not really a city but an endless string of suburbs with no urban core, or "suburbs in search of a city." The beaches, the sun, the palm trees, the doughnut-shaped drive-ins, the cults and fads, and Hollywood all added to the perceived unreality of Los Angeles. Los Angeles was considered unique only because it was oddly different from eastern and midwestern cities. To some extent this image of Los Angeles reflected a by-gone reality. Los Angeles County was, and still is, a vast metropolis covering about 4,000 square miles, nearly four times the size of New York. An interminable succession of one- and two-story residences, vast tracts of vacant land, and small shopping centers spread out over this space. The absence of high-rise buildings further contributed to the unmetropolitan appearance of Los Angeles. Los Angeles was the first American city to adopt the automobile as a means of urban transport. The automobile facilitated urban sprawl, which in turn made the automobile and the freeways indispensable.

The ubiquitous White internal migrants from the Midwest and the East Coast also made Los Angeles distinctive. In the 1940s, these migrants accounted for nearly three-quarters of Los Angeles's population growth, and their share was even higher in the 1920s and the 1930s. Speaking of the southern California region, McWilliams (1946, p. 227) asserted that "the volume and velocity of migration . . . account for most of the unique features of the region's cultural landscape."

In spite of the major transformations of the last three decades, Los Angeles is still unique among major American cities. It has an informal lifestyle, elevated health and fitness consciousness, mild climate, and beaches. But this idyllic picture is marred by urban sprawl, freeway jams, and the ever-present threat of fires, floods, and earthquakes. To this day, the automobile reigns supreme despite costly and prolonged attempts at building a mass transit system. Starting in the 1960s, however, Los Angeles and other giant American cities became much more alike in physical landscape. Los Angeles acquired an urban core with its appropriate skyscrapers, whereas New York, Chicago, and Boston were ringed by sprawling suburbs that replicated Los Angeles's earlier image. The flood of immigrants from the Third World replaced the earlier American internal migrants in Los Angeles, bridging the gap in cultural landscape with other immigrant cities.

In effect, immigration has transformed Los Angeles from a regional to an international metropolis in one generation. Los Angeles is now a highly cosmopolitan city as a result of massive labor migration from Mexico and other Third World nations, as well as immigration of entrepreneurs and professionals from east Asia and the Middle East. Partly reflecting this polarity in types of immigrants, the Los Angeles economy has become polarized between a high-wage, high-tech, financial sector, still heavily dependent on the Pentagon, and a low-wage, small-firm, "sweatshop" sector. Immigrant professionals work for the aerospace and military companies; immigrant entrepreneurs own small firms in which they employ labor migrants. The Los Angeles 2000 Committee task force warns that Los Angeles is "rapidly becoming a bimodal society" with a widening disparity between "high skill, high paying jobs and low skill, low paying jobs." As a result "there is a widening gap between the rich and the poor in the quality of health care, education, and housing."[22]

Neither economic restructuring nor immigration is unique to Los Angeles. Restructuring characterizes major world cities everywhere. The largest metropolitan regions of Europe, such as Berlin, Frankfurt, London, Madrid, and Paris, have also received a substantial labor migration from the Third World. As a result, they too are becoming increasingly diverse. In this sense, Los Angeles parallels other world cities in economic change and population growth through immigration. Although qualitatively similar to what is happening elsewhere, the case of Los Angeles is also quantitatively worse. For one thing, the end of the cold war has not had the devastating economic impact elsewhere that it has had in Los Angeles. For another, most world cities have not recently experienced nearly so much immigration, nor so much from the Third World, in such a short time span. In effect, the good jobs sector of the Los Angeles economy has contracted more and the immigrant population of Los Angeles has expanded more than have the economies and populations of the other world cities undergoing restructuring. Obviously these antagonistic trends cannot continue infinitely without profoundly changing a city's charac-

ter. In a sense, the riot of 1992 brought the consequences of contradictory economic and population trends to the attention of the whole world.

Los Angeles will not close down, but it will never be the same, and the process of change is emotionally and politically troublesome. The results of a *Los Angeles Times* poll in early 1993 showed that about two-thirds of Angelenos thought "there were too many immigrants in Los Angeles" (11 July 1993, p. A1). With a low voter turnout, the Anglo vote was critical in electing the current Los Angeles mayor, Richard Riordan, over Michael Woo, his Chinese American challenger.[23] Apparently, the Anglos responded favorably to Riordan's campaign motto: "tough enough to turn L.A. around." Even California's governor and state senators have placed curbing immigration, both documented and undocumented, at the top of their political agenda. It might be easier, however, to turn Los Angeles around than the Los Angeles–bound immigration.

A devastating earthquake rattled Los Angeles on 17 January 1994. Measuring 6.7 on the Richter scale, the quake killed 61 and injured 9,000 persons, seriously damaged 30,000 houses, and caused at least $13 billion in structural damage.[24] In terms of damage, this earthquake was the worst natural disaster in the U.S. history. President Clinton offered the city an $8.6 billion aid package. While the Congress has not opposed this aid package, it has required relief agencies to prevent illegal immigrants from receiving assistance beyond the initial ninety days.[25] Under congressional pressure, the INS conducts random checks of long-term aid recipients, and the Federal Emergency Management Agency (FEMA) requires proof of lawful residence before extending housing assistance beyond three months.[26] Opponents of the federal restrictions see an irony in that "the message of the restrictions is ugly, painting a nation that has been generous with disaster aid around the world as petty and miserly at a time when people within its own borders are in need."[27] This drastic measure is not so surprising, however, when seen against a background of calls by some local, state, and federal officials to cut off health and education benefits to illegal immigrants and for greater border security to stem the tide of illegal immigration.

Ironically, the earthquake played a short-lived positive economic and social role for Los Angeles. The earthquake gave the economy, especially the large construction industry, a boost. Regardless of class, ethnic, and racial backgrounds, all Angelenos experienced a recurring trauma during the earthquake and its jarring aftershocks.[28] This shared traumatic experience temporarily overcame the interethnic and interracial bitterness caused by the Los Angeles riots.

The rising anti-immigration sentiments among many Angelenos and Californians, particularly Whites, has found its most dramatic expression in Proposition 187, alias "Save Our State," that was submitted to the California electorate in November 1994. This proposition "makes illegal aliens ineligible for public social services . . . , and public school education at elementary, secondary, and post-secondary levels." Furthermore, "it requires various state and local agencies to report persons who are suspected illegal aliens to the . . . United States Immigration and Naturalization Service."[29] This proposition was passed by a wide margin, not just in California, but also in "multicultural Los Angeles County, [which] voted in favor of Proposition 187 by a 12-point margin."[30] While Whites are a demographic minority in Los Angeles, as we have shown in this chapter, they make up the vast majority of the electorate. According to the *Los Angeles Times* exit polls, 67 percent of White Californians voted for this proposition as compared to only 23 percent of Latinos. Surprisingly, nearly half of African Americans and Asians also approved Proposition 187. These voting patterns suggest a continued polarization of ethnic relations in California, especially in Los Angeles where most undocumented workers are found. But Los Angeles and California are not unique in this respect, though they have the dubious distinction of being the first to ride the waves of anti-immigration emotions. In a poll taken in the United States, outside of California, just over half of those polled favored a law in their state "that would eliminate education, health, and welfare benefits for illegal immigrants and their children."[31] Unsurprisingly, support for such a law was higher in states with a large number of illegal immigrants. Legislation against illegal immigrants, including Proposition 187, will be decided in the California courts, where it is currently bogged down, and ultimately in the Supreme Court. One unanticipated consequence of the passage of Proposition 187 has been a surge among Latinos, who do not want to be outvoted or excluded by law, in acquiring U.S. citizenship.

Notes

1. A similar incident took place in Miami in 1990 when a Latino police officer shot and killed a Black motorcyclist and his passenger, but the ensuing riots were put down by a massive show of police force (Portes and Stepick 1993).

2. Los Angeles 2000 Committee, *LA 2000: A City for the Future* (Los Angeles: Los Angeles 2000 Committee, 1988), p. 86.

3. Timothy Egan, "California Dreaming—Once More, With Feeling: L.A. Tries to Do the Right Thing," *New York Times*, 20 February 1994, sect. 4, p. 1.

4. Liberman, Paul, "51% of Riot Arrests Were Latino, Study Says," *Los Angeles Times*, 18 June 1992, p. B2.

5. Reflecting the ethnic and racial diversity of the city, even gangs in Los Angeles are not all Black. According to Davis (1990, p. 316), "aside from the 230 Black and Latino gangs which the LAPD have identified in the Los Angeles area, there are also 81 Asian gangs, and their numbers are also rapidly growing." Most Asian gangs restrict their activities to shaking down coethnic merchants for protection payments. Los Angeles also has a "Mexican Mafia" known as "M" (eme in Spanish) that makes unsuccessful efforts to control the lifestyle violence of young pachuchos in the barrio (Jess Katz and Robert Lopez, "Mexican Mafia's Impact on Gangs Questioned," *Los Angeles Times*, 28 September 1993, p. B1).

6. Patrick Lee and Penelope McMillan, "Skepticism Greets Meeting Between Blacks, Koreans," *Los Angeles Times*, 2 May 1992, p. D1.

7. U.S. Bureau of the Census, *1980 Census of Population. Detailed Population Characteristics. United States Summary* (Washington, D.C.: Government Printing Office, 1984); and the companion volumes for the states of California and New York.

8. James F. Peltz, "Southern California's Aerospace Industry is Down But Not Out," *Los Angeles Times*, 26 September 1993, p. D1.

9. In 1960, the population data for Hispanics were based on Spanish surnames and are not strictly comparable to data from the 1970, 1980, and 1990 censuses based on answers to a "Spanish origin" question. It is likely that the true proportion of Hispanics in Los Angeles was higher than that reported by the census because of the undercount.

10. According to Muller and Espenshade (1985, p. 15), the "fourth wave" of immigrants to California is characterized by the predominance of Asians and Hispanics.

11. In view of greater census underenumeration among Blacks than among other groups, the percentage of Blacks may be somewhat higher than the figure shown in Table 1.

12. *Los Angeles Times*, 27 July 1991, p. A1.

13. *Los Angeles Times*, 30 September 1991, p. D9.

14. U.S. Bureau of the Census, *County Business Patterns, 1988 California* (Washington, D.C.: U.S. Government Printing Office, 1990); and U.S. Bureau of the Census, *County Business Patterns, 1981 California* (Washington, D.C.: U.S. Government Printing Office, 1983). The number of apparel establishments increased by 16.7 percent while that of other manufacturing establishments remained almost the same.

15. *UCLA Daily Bruin*, 1 August 1991, p. 6.

16. *Los Angeles Times*, 26 January 1992, p. E1.

17. James Rainey, "Vendors Cheer as Legalization Wins Final OK," *Los Angeles Times*, 5 January 1994, p. A1.

18. California Employment Development Department, *Annual Planning Information. Los Angeles-Long Beach*, June 1991, p. 9.

19. The presence of a significant number of college students among some of these groups may account for this high number and for the discrepancy between educational and occupational achievement.

20. Some immigrant subgroups such as Jews from Iran have a much higher rate of self-employment than the highest rate given in Table 3 (Bozorgmehr 1992).

21. California Department of Finance, *California Statistical Abstract 1971* (Sacramento: Office of State Printing, 1971); and California Department of Finance, *California Statistical Abstract 1990* (Sacramento: Office of State Printing, 1990).

22. Los Angeles 2000 Committee, *LA 2000: A City for the Future* (Los Angeles: Los Angeles 2000 Committee, 1988), 10.

23. Richard Simon, "Anglo Vote Carried Riordan to Victory," *Los Angeles Times*, 10 July 1993, p. A25.

24. Jeffrey Rabin and Chip Johnson, "5.3 Aftershock Rattles Southland," *Los Angeles Times*, 21 March 1994, p. A1; and B. Drummond Ayers Jr., "Los Angeles Is taking Rapid Road to Recovery," *New York Times*, 17 March 1994, p. A7.

25. Seth Mydans, "Los Angles Debates Quake Aid for Illegal Immigrants," *New York Times*, 6 February 1992, p. A12; Alan C. Miller and Glenn F. Bunting, "House Unit Acts to Bar Quake Aid for Illegal Immigrants," *Los Angeles Times*, 2 February 1994, p. A1; and Rich Connell, "Quake Brings Immigration Issue to National Forefront," *Los Angeles Times*, 3 February 1994, p. A1.

26. Alan Miller, "INS to Randomly Check Quake Aid Applicants," *Los Angeles Times*, 19 March 1994, p. A1.

27. Seth Mydans, "Los Angeles Debates Quake Aid for Illegal Aliens," *New York Times*, 6 February 1994, p. A12.

28. In the month after the earthquake, more than 74,000 people received mental health counseling assistance through FEMA's Project Rebound-Earthquakes and the Red Cross program (see Gordon Dillow, "Post-Disaster Mental Health Efforts Debated," *Los Angeles Times*, 21 March 1994, p. A1).

29. California Voter's Handbook.

30. Patrick J. McDonnell, "State's Diversity Doesn't Reach Voting Booth." *Los Angeles Times*, 10 November 1994, pp. A1, A34.

31. December Times Mirror Center for People & The Press, "Public Expects GOP Miracles: 66% of Democrats Want Clinton Nomination Challenged." News release, 8 December 1994.

27

Miami Spice: The Ethnic Cauldron Simmers

GUILLERMO J. GRENIER

LISANDRO PÉREZ

Hurricanes and heat keep Miami in the news. Even before Andrew ripped through southern Florida on 24 August 1992, Miami was a city used to the national attention brought about by championship football, social turmoil, and epic natural disasters. Consider this: at various points between 1980 and 1990, anyone living in Miami was at least vaguely aware that the Contra war in Nicaragua was being plotted in a restaurant in the area of the city known as Sweetwater, that the fads and fashions sparked by the TV series *Miami Vice* were based on more than a modicum of truth, that on any given day in the early 1980s one might witness a street shootout of the "cocaine cowboys" in a nearby shopping center, that the desperate plight of the Haitian boat people unearthed the deep racism still underlying social relations in the region, and that the chaotic flotilla of 125,000 Cuban refugees (*Marielitos*) in 1980 would create further problems for the already stressed social fibers of the city. Meanwhile, the Miami Hurricanes were winning national football championships at a monopolistic rate, and Miami earned its place in history by hosting Super Bowl XX as well as four major riots in a ten-year period. Miami is hot, with no sign of cooling off.

These episodes illustrate the unparalleled dynamics of what is arguably the most international American city. What makes Miami distinct is not its large number of foreigners. Other cities such as New York and Los Angeles have many more immigrants. The uniqueness comes in how immigrants in Miami see themselves and their potential for success. Do not give the Cubans, Nicaraguans, or Haitians the traditional gospel of starting at the bottom rung and queuing up for the benefits of society that perhaps their children will enjoy if they play by the rules. The rules are different in Miami. Here, immigrants want their piece of the pie now and, what's more, many of them get it, beyond the rhetoric of the American dream. With over 50 percent of the population of Hispanic descent and another 23 percent composed of various groups of Blacks from either the Caribbean

We thank our colleagues in the Department of Sociology and Anthropology at Florida International University. Their series of articles on Miami for the American Sociological Association's *Footnotes*, welcoming ASA members to Miami, provided us with a great deal of material for this chapter.

islands or from the old South, multiculturalism is Miami. Its manifestations are evident as sociologists try to answer the questions that have always guided the discipline: Who rules? Who benefits? How do newcomers fit in?

In the Beginning: A Brief History of Miami

Greater Miami has experienced a quarter-century of profound changes that have transformed the city from a midsized tourist haven into a socially and economically complex metropolis. The history of Miami since the early 1960s has been driven by one particular phenomenon: immigration. Dade County's history can be divided into two basic chapters: before the immigrants and after the immigrants.

From its inception at the turn of the century until World War II, Miami was a frontier city; an American city but not a typical one. Newer, less traditional than other urban centers, it was neither fully southern nor northern. With a large transient population and a large proportion of inhabitants who were first-generation migrants, it lacked a consolidated socio-political structure and a coherent elite. As late as the 1980s, newcomers of the appropriate class background could rapidly become members of the business and civic elite. There is the case of a married couple who, in the 1980s, moved to Miami from Texas, with political connections to a Texas congressman, business interests in South America, northeastern elite schooling, and development interests in Miami. They quickly became proteges of Miami's preeminent business and civic leader. Within a only few years, one of them was elected president of the Greater Miami Chamber of Commerce. This absence of an established civic leadership allowed some immigrants to grab the reins of leadership in significant arenas of county life, as was the case for some members of the Cuban middle and upper classes arriving soon after the 1959 Cuban Revolution.

The very creation of Miami hinged on the actions of northern retirees. In the 1880s, Henry Flagler, millionaire partner of John D. Rockefeller in Standard Oil, moved to Florida for his health and found a new career as a railroad and real estate magnate. A decade later he was persuaded to extend his rail line from Palm Beach to Biscayne Bay by Julia Tuttle, a

Miami is now home to immigrants from all of Latin America and the Caribbean, so much so that some have called it a major Latin American capital. Here, young men from Haiti, who have just arrived in Miami on this boat, are waiting to be unloaded. (Photo by Patrick Farrell. *The Miami Herald*)

widow from Cleveland who had relocated to a 600-acre tract on the northern bank of the Miami River where it flows into the bay. It is the core of downtown Miami today.

The railroad provided the necessary transportation link to fuel the local economy. In those early days, Bahamian immigrants provided the local unskilled labor, outnumbering Black Americans who came to Miami primarily from northern Florida and Georgia. When Miami was incorporated in 1896 with a petition bearing 368 signatures, only about 200 residents of the locality were White. The new city's promoters solicited the remainder of the signees among the local Black residents, primarily immigrants from the Bahamas.

The enfranchisement of Blacks did not last long. The city officials elected as a result of the process of incorporation were White. Since the Bahamians considered themselves the social equals of Whites, it was important for the new city leaders to impress upon local Black citizens the appropriate behavior standards of "southern Negroes."

Decades before Disney World, the state's laissez-faire policies gave northern entrepreneurs license to create fantasy real estate and resort ventures. Venetian and Spanish-style housing developments sprang up, while the indigenous mangrove coast was plowed away to make room for luxury hotels and wide beaches. Without the trees to hold the sand, the ocean washed it away. The hotel proprietors arranged for sand to be pumped back onto the beaches.

Miami experienced dizzying growth between 1910 and 1925. Its population grew from 5,000 to 146,000 and the development of Miami Beach began full throttle upon completion of a causeway across the bay. By the early 1920s, the planned residential community of Coral Gables, surrounded by lush, tropical vegetation, was constructed and the doors of the University of Miami were opened.

The boom came to an abrupt end in 1926, when a major hurricane devastated the city, followed by years of financial crisis and the Great Depression. By the mid-1930s, the local economy had begun to recover and a significant influx of Jewish working-class and middle-class migrants from cities in the Northeast led to the proliferation of distinctive small hotels and apartments in what became the Art Deco district of south Miami Beach.

World War II began the next development boom that picked up steam in the 1950s. Many servicemen and women stationed in south Florida during the war decided to stay. Refugees from northern winters opted for Florida on an ever-larger scale, and the popularization of air travel brought Miami vacations within reach of growing numbers of Americans across the country. Most of the large hotels of Miami Beach (including the Fontainebleau, opened with great fanfare in 1954) date from this era. Between 1940 and 1950 the population of metropolitan Miami (Dade County) nearly doubled, rising from 268,000 to 505,000; by the 1960s it had reached nearly 1 million.

As Miami acquired the dimensions of a large metropolis, its economy began to diversify, and it took on an increasingly important role as a port city in the commerce between the United States, the Caribbean, and Latin America. Miami's image as a tropical paradise for tourists had faded by the 1960s. Eventually, tourists to Florida preferred Walt Disney World, the fantasy creation near Orlando in Central Florida. Meanwhile, those seeking sun and sea vacations increasingly found them further south in the Bahamas, Jamaica, Puerto Rico, and Yucatán.

Even before the massive arrival of Cubans in the early 1960s, the influx of northern "Anglos" (as non-Hispanic Whites are called in Miami) had slowed to a trickle, becoming negative after 1970. The next wave of Miami's development would be spurred by migration from Latin America, especially from Cuba.

Before the 1960s, Miami's population consisted largely of Black and White southern migrants and their descendants, transplanted northerners (including many Jews), and Bahamian and other Caribbean Blacks and their descendants (Mohl 1983, 1986). In the past twenty-five years, however, a very large number of Latin Americans and a substantial Haitian population added to the mix with dramatic impact.

Cubans fleeing Castro's Cuba, Miami's principal immigrant group, began arriving in significant numbers in the early 1960s, following the 1959 Cuban Revolution and the failure in 1961 of the Bay of Pigs invasion. The U.S. government encouraged and aided the flow (Masud-Piloto 1988). While significant numbers of Cubans settled in New York and New Jersey, Miami was the preferred destination of the vast majority, making Cubans Miami's most visible minority. Their numbers were considerably increased by the Mariel boatlift of 1980, which brought 125,000 Cubans to the United States, the majority of whom settled in Miami. In 1990, there were 953,400 persons of Hispanic origin or descent in Dade County, representing slightly more than 49 percent of the entire metropolitan population. Cubans accounted for about 59 percent of Miami's Hispanics (U.S. Bureau of the Census 1990b).

Southern Florida now has the highest proportion of foreign-born residents of any U.S. city. The two largest incorporated areas within Greater Miami occupy, nationally, the first and second positions in the percentage of their residents who were born in a foreign country. Hialeah, a working-class suburb of Miami that has received large numbers of Cubans since the 1960s, is by far the most populated area by immigrants in the entire nation: 70.4 percent of its population was reported in 1990 to be foreign-born. The city of Miami, second in the nation, had a proportion of 59.7 percent foreign-born.

In contrast to the growth of the foreign-born and Latin population, the White non-Hispanic population has been decreasing, in relative terms, since 1960, and in absolute terms since 1970. From a peak in the 1950s of 85 percent, it is expected to decline by the year 2000 to around 30 percent of the total, barely more than the figure for Blacks. The decline is evident throughout the metropolitan area, including the suburbs and Miami's elite residential areas.

Miami has always been a service-centered economy, but after the arrival of the Cubans, the focus of those services shifted from tourism to Latin America. Beginning in the 1960s, Miami came to displace New Orleans as the country's principal trade entrepôt with that region. By 1980, 100 multinational corporations had their Latin American headquarters in Miami and Miami stood second only to New York as an international banking center. By the mid-1980s, Miami International Airport was the ninth-busiest airport in the world in passengers and the sixth largest in air cargo tonnage. About 160,000 workers, one-fifth of its labor force, were directly or indirectly employed in airport and aviation activities. By the late 1980s, Miami's industrial profile was similar to other newer American cities in which the economy was led by services, wholesale trade, finance, insurance, and real estate.

The existence of the Cuban enclave played a pivotal role in Miami's economic transformation. Cubans frequently headed the import and export companies, the banks that financed the transactions, and the transportation and service companies that moved goods and services. Over 25,000 Hispanic businesses exist in Dade County, the vast majority of which are small enterprises that have been described as the true engines of Miami's economic growth (Satterfield 1987). But many of the most powerful economic corporations are also Latin-owned and operated. While Miami holds only 5 percent of the U.S. Hispanic population, it has close to half of the forty largest Hispanic-owned industrial and commercial firms in the country. By the mid to late 1980s, 40 percent of Miami's banks were owned by Latins (Botifol 1985), as were numerous Latin insurance companies, shipping firms, and innumerable import and export establishments. Some of the most important developers are Latins, one of which became the first Latin to head of the Greater Miami Chamber of Commerce.

The city's emergence as the "capital of the Caribbean" has reinforced these trends in immigration. Miami now is the most desired migration point of many Latins, especially the elite and middle classes of the Caribbean and Central America. In the late 1980s, as the United States abandoned the Contra war (which had largely been based in Miami), a broader, working-class flow emerged, first from Nicaragua, then from other Central American nations as well as Colombia, the Dominican Republic, and Puerto Rico. Meanwhile, the frustration of democracy in Haiti increased migration pressures there, despite U.S. policies designed to repress the Haitian flow.

This transformation of Miami's demographic profile got under way precisely when Black Americans were beginning to achieve the civil rights that had been denied them since the city's founding at the end of the last century. The new White immigrants, the Cubans, soon received the benefits of being a minority. While Blacks have experienced some progress, the White Cuban immigrants leaped over Blacks, quickly garnering the political and economic gains that still elude Black Americans. Meanwhile, the response by many Anglos has been to abandon the city. "White flight" decreased the number of non-Latin Whites in Dade County by 24 percent between 1980 and 1990.

Much of this decline is attributable to the decrease in the elderly population, once so characteristic of the greater Miami area. In 1980 the elderly comprised 52 percent of the population of Miami Beach, the highest proportion for any city of comparable size in the United States. Although Miami Beach had been founded by wealthy White Anglo-Saxons, with anti-Semitic restrictions in housing, hotels, and private clubs that persisted for several decades, the retirement wave made it the most Jewish of American cities as well. By 1950, the population of Miami Beach was half Jewish; by 1970, the proportion had reached 80 percent.

All this began to change in the 1970s, in the context of a stagnant tourist economy and the diversion of the retirement flow to other parts of Florida. By 1980, the Jewish component of the population had declined to 66 percent, falling even more steeply thereafter to reach 40 percent in 1990. The decline of the elderly component of the population was just as rapid, dropping to 30 percent in 1990.

Much of the old Miami still persists. The largest firms are still owned and managed by Anglos. Most television, radio, and print media are in English, and even most of those that use Spanish are owned by Anglos. Coral Gables, the elite city within the broader metropolitan area, retains and Anglo majority, while Miami Beach and North Miami continue to

As in most major cities filled with immigrants, ethnic enterprises flourish in Miami. Here this *carnicería* (butcher shop) expresses well the family underpinnings of many ethnic enterprises in its name, *Los Doce Hermanos* (The Twelve Brothers and Sisters), as well as the immigrants' deep attachment to their homeland, as the letters that spell Cuba issue from the Cuban flag. (Photo by Brian Smith, *The Miami Herald*)

be heavily Jewish. Miami Beach holds a small group of octogenarian Stalinists, unrepentant Communist Party members from the 1930s. Black Americans, Miami's first minority, continue to struggle. How does it all work together? Let's look at some of the communities that make up Miami.

The Creation of Cuban Miami

The rise of Cuban Miami as the largest concentration of Cubans in the United States—and the third largest Hispanic community in this country, after New York and Los Angeles—effectively starts in 1959 with the exodus from the island that resulted from the social transformations effected by the Cuban Revolution. Prior to that time, Miami, largely because of its youth and weak economic structure, was never the principal destination of Cuban immigrants to the United States. Sizable Cuban communities thrived in New York, Key West, New Orleans, and Ybor City, on the outskirts of Tampa, in the nineteenth century. New York was still the premier destination for migrants from the island in the period between World War II and the rise of Fidel Castro's government.

This pattern of migration started changing with the creation of rail and highway links between Miami

and Key West and their extensions to Havana by way of regular ferry service. Air service between Miami, Key West, and Havana dates back to the 1920s and represents a pioneering effort in the history of passenger aviation. Those transportation links served to make Miami the principal staging area for the increasingly close relationship developing between Cuba and the Florida peninsula. In the year 1948, for example, Cuba led all countries in the world in the volume of passengers exchanged with the United States (U.S. Immigration and Naturalization Service 1948).

While Miami did not have the employment opportunities Cuban immigrants required during the first half of the twentieth century, it did receive those seeking refuge from the shifting fortunes of the island's turbulent political history. Two deposed Cuban Presidents, Gerardo Machado (in 1933) and Carlos Prío Socarrás (in 1952), made their homes in Miami. A prominent Cuban politician of the 1940s built Miami's baseball stadium. Even Fidel Castro spent time in Miami in the 1950s.

As leader of the 1959 Revolution, Castro initiated a process of change that, in its rapidity and pervasiveness, alienated large sectors of the Cuban population (Fagen, Brody, and O'Leary 1968, pp. 100–101) and contributed to the creation of Cuban Miami (see the essay by Pedraza in this volume).

The concentration of Cubans in Miami, starting in the mid-1970s, was first preceded by their intentional dispersion throughout the United States. The Cuban Refugee Program was established in February 1961 as a federal effort to provide assistance in handling the large influx of people coming from Cuba. One of its purposes was to ease the demographic and economic pressures that the influx was exerting on south Florida. A resettlement program was established through which families arriving from Cuba were given assistance if they immediately relocated away from Miami. According to a Cuban Refugee Program Fact Sheet (U.S. Department of Health, Education, and Welfare 1978), 300,232 persons were resettled away from Miami between February 1961 and August 1978, approximately 64 percent of all Cubans then arriving in the United States. The bulk of the resettled Cubans went to New York, New Jersey, California, and Illinois. In 1965, 42 percent of the Cuban population in the United States lived in Dade County; in 1970, 40 percent. Resettlement reached its peak in the late 1960s. Thus, contrary to the experience of most immigrant groups for whom concentration in a city or region formed part of the process of adjustment to the United States, Cubans underwent a process of intentional dispersion early in their history of immigration.

In the mid-1970s, however, a process of concentration in Miami was well under way as many resettled Cubans decided to make their home in southern Florida. By 1980, the U.S. census found about 52 percent of all U.S. Cubans living in Greater Miami (Pérez 1985, p. 3). The concentration of Cubans in south Florida increased during the 1980s as the majority of the *Marielitos* settled in Miami, where they could find employment in a familiar cultural environment. By the 1990 census, 57 percent of all U.S. Cubans lived in Dade County. The process of concentration in Miami is likely to continue in the 1990s due to the aging of the large middle-aged cohort that left Cuba during the early 1960s and were resettled away from Miami.

In the mid-1990s, the "open arms" U.S. policy toward Cuban migrants shifted dramatically. The massive exodus of rafters in the summer of 1994, initiated by the Cuban government's decision not to impede their departure, prompted the United States to intercept vessels on the high seas and send their occupants to detention camps outside the United States. The attorney general vowed that the detained Cubans would never enter this country. In May 1995, in the interest of ending the indefinite detention of more than 20,000 Cubans at the Guantánamo Naval Base, the United States announced that the detainees would be processed for admission to the United States. However, recognizing the prevailing anti-immigration sentiment, the move was balanced with an unprecedented announcement: future rafters would be intercepted and returned directly to Cuba, under the terms of an agreement reached with the Cuban government. The only way Cubans will be allowed to enter the United States from Cuba will be through the normal visa application process conducted by the U.S. Interest Section in Havana. At least 20,000 people will presumably be admitted every year through that process.

Cuban migration to the United States will continue, but the exile era has officially ended. Cubans are no longer recognized as refugees by the U.S. government. The newly established policy for their admission into this country does not differ in any essential way from the policy toward other nationalities: one enters only through the regular visa process; those attempting to enter without authorization will be returned. In effect, there has been, for the first time, a normalization of one aspect of U.S.–Cuba relations.

Colonizing Dade County

Today persons of Cuban origin represent Miami's largest ethnic group. Cubans account for 56 percent of Greater Miami's foreign-born population and a majority of all Hispanics in the area. About 29 percent of Dade County's 1990 population was of Cuban birth or descent.

The demographic importance of the Cuban presence in Miami is evident in a myriad of ways. The "Cubanness" of the area is manifested not only in demonstrable terms, such as economic activities and cultural events, but also in its more intangible "ambience." Rieff, a New Yorker who has written on Miami, noted that Cubans have largely taken control of the "atmosphere" of the city (1987b, p. 71).

More than thirty years after the start of the exodus, today Cubans have a sense of "rootedness" in Miami. Miami is the capital and mecca of Cubans in the United States. As Rieff (1987a, p. 224) expressed it, "Cubans are probably the only people who really do feel comfortable in Dade County these days. . . . Miami is their town now." Nearly 70 percent of the Hispanic population of Greater Miami, Cubans can be found throughout most of the metropolitan area. Generally the heaviest concentrations of Hispanics within the county are found along a belt running west from downtown (which is located east of Biscayne Bay) all the way to the western edges of the metropolitan area.

Since the 1960s the settlement of Cubans along this belt has proceeded from east to west, emanating largely from the area known as "Little Havana," which is located within the city of Miami, some twelve to fifteen city blocks directly west of downtown and stretching west along Calle Ocho for about fifteen city blocks.

Middle-income Cubans, and especially professionals, are now hard to find in Little Havana. As with so many other immigrant groups, their upward mobility has taken them to more suburban areas. The current residents of Little Havana are likely to be blue-collar and service workers, the elderly, the poor, and recent immigrants, including other Latin American immigrants (Longbrake and Nichols 1976, p. 42).

As one moves north or south of the "Hispanic belt" the proportion of Hispanics declines, so that the northern and southern edges of the county contain some of the lowest concentrations of Latins. An exception is the city of Hialeah, in the northwestern portion of Greater Miami, as more than two-thirds of its population is now Hispanic. Hialeah contains many of the region's manufacturing plants. During the late 1960s and the 1970s, its predominantly White blue-collar population was rapidly replaced with Cuban blue-collar families (Longbrake and Nichols 1976, pp. 39–40).

Hispanic-Anglo segregation is not as high as one might expect, unlike Hispanic-Black segregation. The latter two populations exhibit considerable spatial distance. Blacks in Miami tend to live in fairly confined areas, segregated from both Hispanics and Anglos. Those areas are located primarily—although not exclusively—in the northern half of the city of Miami, in unincorporated zones in northern Dade County, and in the city of Opa-Locka, also in the north. Pockets of Black settlements in the southern half of the county, in Coconut Grove, and in the Richmond Heights and Perrine area in southern Dade, are all zones and neighborhoods with few Hispanics.

The Cuban Enclave: OriginsR and Consequences

Miami's Cuban community is regarded as the foremost example in the United States of a true ethnic enclave. Portes and Bach (1985, p. 203) define an ethnic enclave as "a distinctive economic formation, characterized by the spatial concentration of immigrants who organize a variety of enterprises to serve their own ethnic market and the general population." The basis of the enclave is a highly differentiated entrepreneurial activity. Miami is the metropolitan area in the United States with the highest per capita number of Hispanic-owned businesses (O'Hare 1987, p. 33). The community's entrepreneurial base was established largely by those Cuban immigrants, especially those that arrived in the first wave in the early 1960s, who possessed the complex of skills and attitudes that eventually made possible their entry into a wide range of self-employment (Portes 1987). This is the process Portes and Bach (1985, p. 203) call "the

successful transplantation of an entrepreneurial class from origin to destination during the first waves of the migration."

A strong and diversified entrepreneurial activity is responsible for the enclave's most important overall feature: institutional completeness. Cubans in Miami can, if they wish, literally live out their lives within the ethnic community. The wide range of sales and services, including professional services, available within the community makes that possible.

Three factors have been identified as promoting Cubans' economic and political activity in Miami: (1) structural factors arising from the human capital Cubans brought with them and their geographical concentration in Miami (Portes and Bach 1985); (2) the role of the U.S. government in providing aid to the arriving Cuban refugees (Pedraza-Bailey 1985); and (3) the creation of a collective Cuban-American political identity arising from the interplay of the U.S. state and Cuban exile counterrevolutionary organizations.

The favorable reception by the U.S. government entailed a "direct line" for Cuban exile leaders to the centers of political power in Washington. Unlike other immigrant and ethnic minorities who struggled painfully for years or even generations to gain access to the corridors of power, Cuban leaders accessed it almost from the start. This window of opportunity greatly boosted the Cubans in the 1960s and the Miami economy in general. Waves of Cuban immigrants stimulated demand and received substantial subsidies from the federal government. With a high rate of labor force participation, especially among women, the Cubans also contributed significantly to productivity growth. The 1973 to 1974 recession even had an indirect benefit for Cubans. Construction, one of the growth sectors during the boom, almost completely collapsed, spurring many American residents to flee Dade County. In the subsequent recovery of the late 1970s a de facto segmentation of the industry emerged, with Latins becoming the leaders in home construction and Anglos maintaining dominance in large-scale commercial construction. This division reflected an even more important development within the community, the emergence of the Cuban enclave.

Without a doubt, not all Miami Cubans are rich and powerful businessmen. Even the fact of business ownership is somewhat misleading. Of the nearly 25,000 Latin-owned and operated businesses in 1982, only 12 percent had paid employees and together they generated a total of only 18,199 paid jobs (Díaz-Briquets 1984), a number only slightly higher than the number of Latins in Dade County who belong to unions. Moreover, most of the Latin employment growth during the 1970s was directly attributable to population growth. Latins (as well as Blacks) were underrepresented in Miami's fastest growing industries, especially financial services, which were dominated by Anglos (Cruz 1990, p. 17). The "White flight" abandonment of Dade County that began in the 1970s was class selective. Anglo laborers and production workers left in great numbers, while executives and managers remained or moved into Dade County. Although Latins (and Blacks) expanded their representation in the professional and executive occupations during the 1970s and 1980s, Anglos still outnumbered them.

To understand Cubans in Miami, it is important to trace the consequences of the structural organization of the enclave. Foremost is the well-studied influence of economic adjustment, but the enclave also affects areas such as acculturation, interethnic relations, and political participation.

Economic Adjustment

Most of the research by Portes and his associates on the Miami enclave deals precisely with its relationship to the relatively successful economic position of Cuban immigrants. The enclave insulates the immigrant somewhat against the usual processes of the secondary labor market. In contrast to Mexican immigrants, who must join the open labor market in peripheral sectors of the economy throughout the country—what many economists have called the secondary labor market—many recent Cuban immigrants enter the U.S. labor market largely through the large number of businesses in Miami that are owned or operated by members of their own group that arrived earlier. While wages may not be higher in the enclave than they would be in the larger primary labor market, ethnic bonds provide for informal networks of support that facilitate the learning of new skills and the overall process of economic adjustment, thereby blurring the usual differences that exist between the primary and secondary labor markets. The enclave's positive implications for economic adjustment have maintained the Cubans' socioeconomic position relatively high in comparison

with many other immigrant groups (Portes 1981, pp. 290–295; 1982, pp. 106–109).

Acculturation

The completeness of the enclave also has the effect of slowing down the process of acculturation, for it tends to insulate immigrants from the dominant society and culture, allowing for the retention of their culture of origin. Using language as one indicator of the degree of acculturation, studies have found that most Cubans use only Spanish at home and in many of their daily activities (Díaz 1980, pp. 48–50). One of the extraordinary results from the 1990 census on Greater Miami is that the proportion of all persons who live in a household in which Spanish is spoken is greater than the proportion of the population residing in homes in which only English is spoken.

The institutional completeness of the enclave has made Spanish a public language in Miami, one that is not confined to the intimacy of the family or the peer group. It is the *lingua franca* for conducting a wide range of business and personal matters beyond one's family and friends. The extensive use of Spanish is one reason why language use is such a critical—and often explosive—issue in Miami, one that is the frequent battleground for interethnic conflicts. It is not a coincidence that the "English-Only" movement was born in Miami (see the essay by Portes and Schauffler in this volume).

The retention of the language and cultural patterns of the country of origin are also attributable to the fairly recent arrival of most Cubans to this country. The process of acculturation among Cubans in Miami is also retarded by the perception many Cubans have of themselves as reluctant migrants, compelled to leave their country, but who expect to return to Cuba someday. Consequently they have little desire or motivation to assimilate into this society. Moreover, there have been periodic waves of fresh arrivals from Cuba, of which the largest and most recent was the Mariel boatlift. The new arrivals renew and reinforce Cuban culture within the immigrant community.

Interethnic Relations

The insulation of the immigrant within the enclave may have positive implications for the initial process of economic adjustment, but it poorly serves interethnic communications and understanding in Miami.

In a metropolitan area in which the various ethnic groups—especially Blacks—are spatially segregated, the existence of an institutionally complete community among Cubans makes it even less likely that Hispanics and Blacks can create the basis for a better understanding and a common agenda. Of course, this social distance is due to many causes, but the ability of the Cubans to live largely within their own community is undoubtedly one of them. Compared to American Blacks, Cubans in Miami have a far greater likelihood of being able to work with coethnics, shop in stores owned and operated by coethnics, and obtain professional services from coethnics.

The social distance created by the enclave has contributed to the changing perception of Cubans by the Black community in Dade County. While recent evidence indicates that during the 1960s and early 1970s Black Miami viewed Cubans as possible allies in the battles for minority empowerment in Dade County, by the 1980s this view had been replaced by one of clear antagonism.

Political Culture

During the 1980s Cubans in Miami were able to establish pivotal local power, exercised through an increasing number of elected officials and organizations such as the Cuban American National Foundation, the Latin Builders Association, the Hispanic Builders Association, and the Latin Chamber of Commerce. By the late 1980s, the City of Miami, Hialeah, Sweetwater, West Miami, and Hialeah Gardens all had Cuban-born mayors. The city manager and the county manager were Cubans. Cubans controlled the City Commission and constituted more than a third of the county delegation to the state legislature. After Claude Pepper died, a Cuban woman, Ileana Ros-Lehtinen, won his U.S. House of Representatives seat. Most recently, Lincoln Díaz-Balart, a Cuban-born politician elected to the state legislature in the mid-1980s, joined Ros-Lehtinen as the second Cuban American from Miami in Congress. Nowhere else in America, nor even in American history, have first-generation immigrants so quickly and so thoroughly appropriated political power.

Perhaps for this reason, no aspect of Cuban Miami attracts more national and international attention than its politics. Even after more than thirty years, Miami remains a community of exiles, largely preoccupied with the political status of the homeland. At the same time, however, Miami's Cubans demonstrated a strong participation in the U.S. political system at the local and state levels.

The Second Generation

Although the enclave favors the retention of the culture of the homeland, delaying the process of acculturation, it is unlikely that Cubans in the United States will be an exception to the usual intergenerational shift toward greater acculturation and assimilation. English is the principal language among Cubans who have lived all or most of their lives in the United States. Sharp intergenerational differences exist among Cubans in Miami, with alienation between parents and children usually found in Cuban families with interactional problems (Kurtines and Miranda 1980, pp. 181–182; Szapocznik, Scopetta, and Tillman 1978, pp. 42–44). An important focus of these intergenerational tensions are the conflicting value orientations with respect to dependence and independence (Bernal 1982, p. 197). Cuban culture foments the continued dependence of children on their parents, even in the teenage years and beyond. Cuban children, however, are more likely to have internalized the norms of independence commonly found in U.S. society. One adaptation that reduces intergenerational tension is "biculturality," by which each generation adjusts to the other generation's cultural preferences: "parents learn how to remain loyal to their ethnic background while becoming skilled in interacting with their youngsters' Americanized values and behaviors, and vice versa" (Szapocznik and Hernandez 1988, p. 168).

Intergenerational shifts will also be set in the very structure of the enclave. Coming up rapidly behind that large cohort of first-generation entrepreneurs that have created the enclave are those Cuban Americans born in the United States or who arrived as children from the island. Largely educated in this country, the influentials in that group are less likely than the older generation to be entrepreneurs and more likely to be professionals. A rapid change may well take place in the years ahead in the very economic basis of the Cuban community, from a community dominated by "first-generation" entrepreneurs to one of "second-generation" professionals. Such a shift will alter the Cuban community's relationship with the rest of the city. The advent of that dramatic intergenerational shift is already evident in the participation of Cubans within the traditional "Anglo" institutions that hold true economic power in Miami: banks, law firms, insurance companies, real estate, advertising, professional services, and public bureaucracies.

Removed somewhat from their Cuban origins, the new generation of Miami Cubans will more likely think like immigrants rather than exiles, will have a new agenda, and will easily find common ground and solidarity with other ethnic groups in the community, especially the growing number of other Latins.

Black Miami

During the 1980s, Miami's periodic outbursts of racial violence seemed curiously out of step with the evolution of urban America. On four separate occasions during the decade, major riots engulfed predominantly Black neighborhoods of the city, resulting in numerous deaths and hundreds of millions of dollars in property damage. The 1992 riot in Los Angeles, however, served notice. Miami was not a unique holdover from the 1960s, but rather an early warning for a fragile social order in which traditional racial divisions are exacerbated by the new patterns of international migration.

The grievances of Miami's Black population are as old as the city itself. While Black workers played an integral role in the construction and staffing of Miami's early tourist economy, the immigrants were subjected from the outset to a strict southern segregationists ethic. Blacks were forced to reside in "Colored Town" (later Overtown), just north of downtown Miami. When Overtown proved too small to handle the population, a new Black neighborhood, Liberty City, was constructed to the northwest, complete with a wall to separate it from its White neighbors. Portions of this wall can still be seen along N.W. 12th Avenue, between 62nd and 67th Streets.

More remarkable, perhaps, is how long this practice of apartheid endured. In 1960, Miami still scored about ninety-nine on the residential segregation index (where 100 signifies perfect segregation). At

that point, Black workers in Miami Beach hotels and restaurants were still obligated to carry identity cards and return to their homes across the bay each night.

In the 1960s, the Civil Rights movement did away with the formal institutions of segregation. But just at the point where new opportunities for Miami's Black minority began to appear, the city was transformed by the sudden arrival of Cuban refugees. The rapid economic advance of this group, fueled in part by government programs aimed at national minorities and in some instances displacing Black participation in the labor market, provided a new focal point for discontent.

The aspirations of Black Miami confronted two debilitating conditions: weak community leadership and an unresponsive political system. The first was a by-product of the typical 1960s era urban renewal programs, especially freeway construction. Like many southern cities, Miami's segregationist practices had allowed Overtown to emerge as a vibrant center of small businesses, andprofessions catering to the local Black population and culture. Urban renewal virtually destroyed Overtown, displacing much of the Black middle class to newly desegregated suburbs or away from Miami altogether.

Miami's metropolitan political system also offered little possibility for redressing Black concerns. As constituted by a "good government" reform in 1957, it combined a strong county government, administered by a nonpartisan commission elected at large and an appointed manager, with twenty-six municipal governments (the largest being the city of Miami) retaining some local autonomy but subordinated to standards set by the county. More than half of Dade County's population and territory fell outside of these municipalities. With some 60 percent of metropolitan Miami's Black population residing in unincorporated Dade County, and much of the remainder located in the city of Miami with its large Hispanic majority, the chances of effective Black political representation were minimal.

The riots of the 1980s thus crystallized a widespread anger in Black Miami over both its failure to keep pace economically with other social groups and its lack of political voice. As in the 1960s, the response of city elites was to create a series of economic and social programs designed to shore up Black neighborhoods. The Greater Miami Chamber of Commerce created a $5.4 million program, all of it from private money, to create jobs and businesses in Liberty City. In 1981, community developers, the county, and the city tried to attract firms to a business park in the heart of Liberty City. In 1984 the city of Miami tried to develop a shopping center in Overtown, seeking to fill it with Black-owned businesses. A Black-controlled community business development organization opened a shopping center in 1985 on the border between Liberty City and "Little Haiti" (Reveron 1989).

The task of rejuvenating the Black community was daunting. In the early 1980s there were only 3,885 Black-owned businesses in Dade County that generated $110 million annually in sales and receipts. Only 11 percent of the Black-owned businesses had employees, who totaled 2,323 in 1982 (Hartley and Martinez 1990). Because of this lack of infrastructure, all efforts at business development encountered difficulties. The Liberty City park had no firms in 1989, and the Overtown shopping center had only two stores in its nine spaces. The shopping center on the border of Liberty City and Little Haiti, however, was a success, providing more than 130 jobs (Reveron 1989). In a gesture which might well symbolize what it takes to develop successful businesses in Miami's Black communities, its developer, Otis Pitts, was awarded a MacArthur Foundation "genius" award in 1990 (Viglucci 1986).

When Martin Luther King visited Miami in 1966, he noted Miami's racial hostility and warned against pitting Cuban refugees against Blacks in competition for jobs (Porter and Dunn 1984, p. 13). While competition for jobs is difficult to document, in the area of governmental contracts with minorities, Cubans clearly prevailed over Blacks. Between 1968 and 1980, the Small Business Administration (SBA) cumulatively dispersed 46.6 percent of its Dade County loans to Hispanics and 6 percent to Blacks. Following the 1980 riots, nearly 90 percent of the SBA loans went to Hispanics or Whites.

Not surprisingly, at the end of the 1980s, there were still few Black businesses in Dade County. Only 1.4 percent of Dade's Black population owned businesses in 1988, and only one Black Miami business was among 300 firms on six separate lists of top firms in *Black Enterprise* magazine in June 1988 (Castro and Yaney 1989; Miami-Dade Chamber of Commerce and Greater Miami Chamber of Commerce 1990). In 1991, Black businesses in Dade County numbered about 5,400. The total of all persons employed in those businesses was 6,208, or 4 percent of the Black population.

Haitian Miami

While Miami's racial and ethnic profile is often portrayed as a tripartite division between Anglos, Hispanics, and Blacks, it is important to note the ethnic diversity within Black Miami as well. The 1990 census classified just over 19 percent of metropolitan Miami's 2 million residents as Black (non-Hispanic). Of this group, however, about 60 percent are U.S.–born African Americans, and the rest largely Caribbean immigrants.

Haitians represent the largest nationality group among Black Caribbean immigrants. They number approximately 70,000 in the Greater Miami area. The first group of Haitians to arrive in a boat on south Florida shores landed at Pompano Beach on 12 December 1965. Having shot up Haiti's Presidential Palace, the immigrants qualified for and were given political asylum. When, in December 1972, the second boatload of Haitians arrived in south Florida, again at Pompano Beach, they were not granted the status of political refugees.

The only ones who came forward to offer help to the new wave of Haitians were members of the National Council of Churches and a few established resident Blacks. The Blacks who tried to help were led by the Reverend James Jenkins of Liberty City's Friendship Missionary Baptist Church and a small group of Black ministers called the Black Baptist Alliance. They provided food, shelter, and clothing. Since the Haitians were not recognized as refugees, they were ineligible for welfare or work permits.

By the spring of 1974 there were about 400 Haitians settling into an area a few miles north of downtown Miami that was becoming known as "Little Haiti." With their numbers growing, it became increasingly difficult for the Baptist Alliance to carry the burden alone. Consequently, in 1974 the National Council of Churches sent a fact-finding team to south Florida to look into the human service needs of Haitians. The end result was the establishment, by the National Council of Churches, of the Christian Community Service Agency to act as a conduit of church funds to assist the Haitians. This organization remained the primary source of support for the Haitians, who continued to arrive in such numbers that by 1978 it was becoming increasingly clear that Miami had a problem—a very serious problem. Hundreds and eventually thousands of undocumented Haitians were slipping illegally into the country and into Miami's Little Haiti.

The Haitian refugees tend to be semiskilled, have some education, and have lived in urban areas in Haiti. Yet because Haiti is the least developed nation in the western hemisphere, to be semiskilled and have some education and urban experience still leaves one at a disadvantage in the United States. The refugees had completed an average of only four to six years of formal education prior to arrival and less than 5 percent had graduated from high school. Their native language is Creole. Few spoke English. Many were sick. Even more of them were disoriented, vulnerable people who had risked their lives for a chance in America.

There are also significant numbers of Black Bahamians, Jamaicans, and others from additional Caribbean islands. While the numbers of these immigrants are significantly less than those of Latins, they have contributed to the relative and absolute growth of Miami's Black population. By the early 1980s, Greater Miami became one of only sixteen metropolitan areas in the United States with more than 300,000 Blacks. Between 1970 and 1980, Dade's Black population grew by 47 percent, a growth rate exceeded only by Atlanta.

To a degree the Haitians mirror the Cubans, in that an early wave of middle- and upper-class refugees from the Duvalier regime in the 1960s and 1970s was followed by a poorer wave of immigrants in the 1980s and 1990s. There is also Little Haiti, a section of Black Miami in which Haitian shops, restaurants, and other small businesses, along with a large residential area, are concentrated. Nonetheless, while Haitians have sought to emulate the success of the Cuban enclave, the reality of their situation, including U.S. immigration policies to keep them out rather than welcome them and their social experience in being treated as Blacks, has been very different.

The relations between Haitian and native-born Blacks in Miami have often been strained by perceptions of cultural differences and economic competition, familiar to students of American immigration history. Recent developments may be working to narrow the gap, however.

Black Miami Unites

In June 1990, Nelson Mandela visited Miami during his triumphal U.S. tour. Unlike the civic honors bestowed upon him at his other stops, his official Miami reception was decidedly cool, the product of Cuban and Jewish outrage at his refusal to disavow his cordial relations with both Fidel Castro and Yasser Arafat. The mayors of Miami Beach and Miami, and the Metro Dade Commission, refused to honor or meet with Mandela, much to the anger of local Black leaders. Then, a week later, a large Haitian demonstration in front of a Little Haiti store, where a customer had argued and fought with a Cuban proprietor, was violently dispersed by Miami police.

In the wake of these events, a group of Black professionals and community leaders headed by lawyer H. T. Smith organized a convention boycott, calling on national organizations with meetings scheduled in Miami to take their business elsewhere. The demands of the Boycott Coalition were an apology to Nelson Mandela from Miami's elected officials, an investigation into police conduct in the Haitian demonstration, a series of economic measures to promote Black economic interests, especially in the tourist industry, and the reform of Miami's political system to provide greater Black representation.

The boycott had a quick impact. Before the end of 1990, thirteen organizations, including the American Civil Liberties Union and the National Organization for Women, had canceled their Miami meetings. It is estimated that the boycott cost Miami over $60 million in convention-related business before it was lifted in 1993.

Official response to the boycott demands was slow to materialize but began to accelerate in 1992. The Greater Miami Convention and Visitors Bureau established a scholarship program with local corporate sponsorship at Miami Dade Community College and Florida International University, under which Black students are provided with training culminating with a degree from FIU's School of Hospitality management and subsequent management level employment in a local tourist industry firm. As of late 1992, thirty-six students were enrolled in the FIU program.

Meanwhile, the mayor of Miami Beach and the Metro Dade Commission issued retroactive statements honoring Mandela. The Cuban American mayor of the city admitted there might have been errors made in handling the situation but did not issue an apology.

Most importantly, Miami's political system underwent a major reform. A federal court ruling in late 1992 ordered Dade County to immediately replace its at-large commission with a single member district system in order to guarantee effective minority representation. Following the establishment of boundaries for thirteen districts, the local election process was initiated on 16 March 1993. The result was a Metropolitan Commission composed of four Blacks, six Hispanics, and three Anglos: a representative commission after all these years.

Conclusion

In the 1990s, Miami will reach a crossroads as its various ethnic communities are restructured. Both Cuba and Cuban Miami will see fundamental changes. It is likely that we will see the reestablishment of normal transportation, communications, and commercial links between Miami and Cuba. It is also likely that midway through the decade change will come to Cuba—with or without Fidel Castro. If changes in Cuba are in the direction of democratization, they will produce a warming in U.S.–Cuba relations. The basic conditions, therefore, that created and shaped Miami's exile community will finally be altered, after more than three decades.

Just as Castro will pass on in the 1990s, so will his contemporaries here, the exile generation. In 1990, that large cohort of exiles will be between the ages of fifty and sixty-nine, in the twilight of their lives, and their economic and political (and ideological) influence will wane, rapidly making way for a new generation.

A new generation of Miami Black Americans, newly arrived and native-born, will also be faced with new challenges. For the first time in Miami's history, its Black community sees some rays of hope as the city's political system offers a voice for Miami's Black community. Since the same system will also increase Hispanic representation on what until now has been an Anglo dominated commission, turning this newly representative system into an effective vehicle for reconciling the interests of Miami's divergent communities is perhaps the city's most critical challenge at the close of this century.

In Miami's development we now sense its emerging nature as a city on the edge, as Portes and Stepick (1993) argue, a city in the throes of an economic, social, political, and cultural transformation that is forcing us to reevaluate our ideas of how urban centers develop and how immigration impacts on this development. Ultimately the differences within the communities and class and ethnic differences will influence the development of the urban social environment of Miami. If "White flight" continues as the wave of largely poor immigrants from Latin America and the Black Caribbean continues to increase, Miami's success may already have seen its best days. Cities abandoned to minority populations, no matter how successful a portion of that minority is, do not fare well.

Yet if the emerging political and social structures in the city become representative of its multicultural core, then Miami promises to be one of America's great cities. Its diversity serving as a bridge—not so much between the United States and the rest of the world, but between the power centers of diverse communities emerging within urban America. If the conflict between newcomers and immigrants finds no resolution in Miami, then we will have to learn from its failures. One way or another, Miami points toward the future.

28

Washington, D.C.: The Changing Social Landscape of the International Capital City

ROBERT D. MANNING

My people came [involuntarily] and planted the great [elm] trees, these Hispanics come [voluntarily], and enjoy the[ir] shade . . ."
—African American school teacher explaining the roots of tensions between Blacks and Latinos in Washington, D.C.

The inauguration of Bill Clinton as the forty-second President of the United States provided attendees and television viewers alike with a panoramic snapshot of the world's most famous capital city. The festivities, held on the Mall, included such majestic images as the Capitol rotunda, Jefferson Memorial, Supreme Court, Library of Congress, Smithsonian museums, Washington Monument, and the glorious fireworks display over the Lincoln Memorial. This awe-inspiring cluster of political institutions, cultural treasures, and historic monuments is complemented by many other imposing off-mall sites such as the White House, John F. Kennedy Center, Pentagon, Arlington National Cemetery, Vietnam Memorial, and the infamous Watergate complex.

For most Americans, these picturesque scenes of familiar political and cultural icons reinforce the longstanding view of Washington, D.C., as the "Federal City." That is, a politically and financially dependent appendage of the national government that lacks its own social history or economic dynamism. This common misconception characterizes the District as a small, urban anachronism that is inhabited primarily by transitory politicians, armies of pencil-pushing (White, middle-class) bureaucrats, and special-interest lobbyists. From this perspective, the District is not a "real" city but simply an urban repository for the administrative machinery and cultural patrimony of U.S. society: a virtual theme park of American cultural and political history. Hence, its significance is portrayed as largely symbolic—the floral centerpiece of Western democracy. This is a particularly poignant irony. In 1994, District residents (two-thirds African American) are still denied such basic political rights as self-governance and full representation in the U.S. Congress; although opponents of D.C. statehood emphasize its small demographic base, its population was greater than ten predominantly White states as recently as 1970.

The focus of this essay is the changing social and economic "landscape" of the emergent metropolis of Washington, D.C.; the once dominant urban core—the District of Columbia—and its surrounding suburban areas are both examined as complementary

The research assistance of Enrique Soto, Susan Barlow McMillan, Patrick Lemmon, and Anita Butera is gratefully acknowledged.

parts of the larger metropolitan whole. The objective is to illustrate how structural changes in the U.S. economy are profoundly influencing the ongoing "social construction" of urban America. In the case of Washington, D.C., which is presented as a prototype of the "postindustrial" city, it features the metamorphosis of a previously biracial provincial "Southern town" into a multicultural, international metropolis (Manning 1995, 1996). This process, which includes the recent arrival of "new" immigrant groups, has produced more complex intergroup relations and heightened political tensions as old and new racial and ethnic groups negotiate the increasingly contested social "terrain" of the postindustrial metropolis. Accordingly, then, it will be argued that Washington, D.C., has transcended its historical "exceptionalism" and now constitutes a harbinger of the future of multicultural relations in American society (Manning and Pedersen 1992; Manning 1995, 1996).

During the spring of 1991, the dual strains of cultural and economic change culminated in two days of violent protest and civil disobedience in the Mt. Pleasant neighborhood of the District—the first social insurrection since the devastating riots of 1968. This seemingly unexpected "civil disturbance," which pitted primarily young Central American immigrants as well as some local White and Black residents against a largely African American police force, presaged the Rodney King inspired uprisings by nearly one year (Manning 1995; Manning and Pedersen 1992; U.S. Commission on Civil Rights 1993). In a sense, the Mt. Pleasant riots signified the maturation of Washington, D.C., as a major American metropolis; it is the seventh most populated metropolitan area in the United States. Before recounting the circumstances that precipitated this incident, it is necessary to review the historical, social, and economic conditions that previously defined the delicate web of race relations in the District.

Race, Class, and Nationality in the "Federal Town," 1790 to 1945

The location of the capital of the thirteen American colonies was fraught with controversy as both regional rivalries and personal (financial) interests stymied the selection process. In 1790, this political impasse was broken by Alexander Hamilton and Thomas Jefferson, enabling George Washington to

select the site of the "federal town" that still bears his name (Bowling 1991); the District of Columbia was formally incorporated in 1871 with the addition of Georgetown and Washington County to the original eastern Potomac River site. This compromise is not insignificant. It ensured that Washington would assume a distinctly southern character and thus openly embrace the chattel property rights of slaveholders (Abbott 1990; Green 1967; Landis 1948). Hence, unlike other northern cities, African Americans played a major role in the social and economic history of the city from its inception.

White Town/Black Labor

Upon its founding, Washington was a peculiar city, both in terms of its social composition *and* its economic base; in 1800, the U.S. census reported that over one-fourth (29 percent) of the Washington population of 14,093 was comprised of African slaves and a few Black freedmen. Unlike other East Coast cities (New York, Boston, and Philadelphia), it was neither a center of manufacture or of commerce. This profoundly influenced race relations in the District by reinforcing the local economy's dependence on cheap African slave and free Black labor. That is, Washington could not compete with other nearby cities for European workers since the rising productivity of industrial manufacturing and high profits of commercial trade translated into greater wages for manual laborers.

The traditional northern European immigration of the early and mid-nineteenth century (from England, France, Scotland, Germany, or Ireland) was *not* followed—in Washington, D.C.—by the arrival of large numbers of southern and eastern Europeans in the late nineteenth and early twentieth centuries. Instead, the "new" immigrants from such countries as Italy, Greece, and Russia generally bypassed the District for higher wages in the rapidly urbanizing metropolises of the Northeast and Midwest. As a result, rather than participate in the larger regional labor system of immigrant "sojourners," local employers recruited less costly Black freedmen from Virginia, Maryland, and the Carolinas. According to Constance M. Green, in her classic study of race relations in Washington, D.C., this pattern contrasts sharply with other nearby cities such as Philadelphia, where the arrival of new immigrant groups (for example, the Irish) dramatically curtailed the employment opportunities of "colored workingmen." Green contends that "Negroes in Washington looked upon [European laborers] as enemies, but *foreign immigrants in the capital were too few in the last decades of the century* to be a determining factor in the local labor market" (1967, p. 134, emphasis added).

Throughout the antebellum period, the bulk of the manual and craft occupations in Washington were performed by African slaves and Black freedmen. These included carpenters, bricklayers, stonemasons, wheelwrights, coopers, blacksmiths, plasterers, cabinetmakers, printers, shoemakers, cab drivers, draymen, and domestic servants. The enormous contributions of this primarily male workforce, whose origins have been traced largely to West Africa (Angola, Senegambia, Gold Coast), are mirrored in its rapid population growth and relative prosperity; by 1860, African Americans numbered 14,307 (19 percent of the District's population) and more than three-fourths had purchased their freedom. These trends are instructive. They illustrate the unique economic and social opportunities available to African Americans in the District. For example, the proportion of freedmen in the District's African American population increased more rapidly than in any other slave state—from 39 percent in 1820 to 78 percent in 1860—compared to only 6 percent for the entire South at the onset of the Civil War (Berlin 1974). Even so, slavery was not abolished in the capital until 16 April 1862 and District slave owners were financially compensated for the loss of their "property" rights.

Favorable socioeconomic opportunities continued throughout the nineteenth century for African Americans in the District (Green 1962, 1967; Landis 1948). As the political power and economic prosperity of the United States blossomed, fueled by the escalation of European immigration to the industrial Northeast and Midwest, the consolidation of the American administrative state accelerated the expansion of Washington, D.C. Beginning with the Civil War, federal military and administrative bureaus were centralized in the District at an unprecedented scale. The rapid growth of the federal workforce— from 2,199 in 1861 to 13,124 in 1881—precipitated the concomitant increase in the District's population from 61,357 in 1860 to 147,491 in 1880; federal civilian workers rose from 9 to 20 percent of the total District labor force during this period (Abbott 1990).

The expansion and concentration of the federal bureaucracy in Washington, D.C., was ensured by the enactment of the landmark Pendleton Civil Service Act of 1884 and the establishment of the Interstate Commerce Commission in 1887; federal civilian employment climbed to 20,834 in 1891 and to 29,044 in 1901. Furthermore, the local economy experienced a post–Civil War "boom," propelled by public contracts (especially military), a large contingent of resident soldiers, and the construction of a modern infrastructure including roads, bridges, water lines, and sewers. As Table 1 shows, this ambitious urban expansion, which helped to insulate the District from the national and international economic downswings of the 1890s, resulted in the tremendous growth of the District's population through the turn of the century.

For African Americans, whose social (working "caste") status was fundamentally conditioned by southern race relations, this period of bountiful economic opportunities highlights a historical irony: the relatively low wages of Black-dominated manual, craft, and domestic service occupations effectively discouraged the participation of European immigrants in the local labor market. In 1900, for instance, the U.S. census lists only 638 employed Italian men and very few performed unskilled jobs that characterized their work experience in other northern cities (Gillette and Kraut 1986). The African American community, by comparison, grew rapidly—from 48,377 in 1880 to 86,702 in 1900. In fact, between 1890 and 1910, the District featured the largest population of African Americans in the nation—even larger than New York City—and produced its own vibrant cultural "renaissance" led by such native luminaries as jazz great Duke Ellington. This unique pattern, which coincides with the enormous escalation of European arrivals in the late nineteenth century, is illuminated by comparing the African American and immigrant populations, respectively, of other northern cities in 1900 (cf. Lieberson 1963; Lynch 1973): Washington, D.C. (31 percent versus 7 percent), Baltimore (16 percent versus 13 percent), Philadelphia (5 percent versus 23 percent), New York (2 percent versus 37 percent), Chicago (2 percent versus 35 percent), Pittsburgh (5 percent versus 25 percent), and Boston (2 percent versus 35 percent).

During World War I, the "closed" labor system of the southern Black Belt temporarily acceded to the labor demands of northern industry (Manning 1989, chap. 5). The resulting Great Black Migration is responsible for the dramatic growth of African American communities in cities throughout the Northeast and Midwest, as rural Black laborers were finally permitted to enter the more remunerative industrial workforce (Lynch 1973; Marks 1989; Spero and Harris 1931). Ironically, the District was not a major beneficiary of this new labor migration pattern. The District's Black community grew at only a modest rate through the mid-1930s and accounted for about one-fourth of the city's total population; southern migrants from outside of Maryland and Virginia comprised only one-fifth of the African American population (Green 1967, chap. IX). As a result, the size of the District's Black community fell from first in 1910 to third in 1920 and then to fifth in 1930 behind New York, Chicago, Philadelphia, and Baltimore.

In synthesis, the historic prominence of the African American community in Washington, D.C., contrasts sharply with the experience of other northern cities at the turn of the century. The Black population of the District increased during the peak of European immigration to the industrial North and paralleled the steady decline of its respective White, foreign-born population. This pattern underscores the exceptionalism of Washington, D.C.; in the urban North, African American communities generally date from World War I, while European immigrants were the primary source of unskilled labor. Nevertheless, the District was a predominantly White city; African Americans did not exceed one-third of the city's population until after World War II.

Prosperity and Poverty: A Tale of Two Cities

The early twentieth century witnessed the continued expansion of centralized administrative, information collection and dissemination, and war coordinating bureaucracies—especially during World War I and its immediate aftermath. This is mirrored in the virtual explosion of the federal civilian labor force—from 38,911 in 1910 to 94,110 in 1920—which accounts for the unprecedented increase (from 25 percent to 40 percent) of this component of the total District workforce. Not surprisingly, the population of Washington, D.C., continued to increase rapidly (see Table 1). This growth was primarily due to the influx of native Whites from the northeastern states, Maryland, and Virginia (Abbott 1990; Green 1967);

African Americans increased moderately, accounting for 27.1 percent, while foreign-born Whites steadily declined to 6.1 percent of the District's population in 1930.

Favorable employment opportunities and intensifying demographic pressures profoundly affected race relations in the District during the early twentieth century. Overall, this period witnessed the systematic erosion of the social and economic fortunes of the African American community through: (1) the attack on the prosperous Black middle-class, which included skilled craftsmen, white-collar professionals, and small businessmen; (2) institutionalization of racial occupational segmentation in the private and public sectors; and (3) success of public and private policies that promoted residential segregation and Black concentration in the inner city. Together, these trends fundamentally circumscribed the social, political, and economic opportunities of the African American community of Washington, D.C. (Frazier 1969; Green 1967; Landis 1948). The following examples are illustrative of the "new nadir" of race relations in the District after World War I.

The relatively high socioeconomic status of the District's Black population at the turn of the century was mirrored in its impressive levels of skilled employment, high rates of entrepreneurship, and residential dispersion throughout the capital. In the 1890 census, for instance, over one-fourth of all African Americans residing in Washington, D.C., were reported as self-employed; nine out of ten barbers in the city were Black. Only two decades later, self-employed African Americans were being forced into the wage labor market while Black tradesmen were being relegated to unskilled and semiskilled "helper" positions. By 1940, just 1.4 percent of employed Black men were listed by the U.S. census as business proprietors or managers, while the proportion of Black barbers in the District had plummeted to about 30 percent (Frazier 1969).

The deteriorating occupational status of African Americans, which pervaded both private and public sectors, became official U.S. government policy following the election of Woodrow Wilson as President in 1912. This institutionalization of "racial privilege" had a threefold impact: (1) an overall hiring preference for Whites; (2) occupational definitions that relegated Blacks to the lowest manual and white-collar positions; and (3) wage discrimination that offered higher compensation for "White" occupations.

Table 1 African American and Foreign-born Population of Washington, D.C., 1870–1990[a]

Year	D.C. total (number)	African American (number)	Decennial change (percent)	African American (percent)	City rank[b]	Foreign-born (percent)
1860	73,492	14,307[c]	—	19.1	7	—
1870	109,092	35,455	222.8	32.5	3	12.6
1880	147,491	48,377	36.4	32.8	3	9.7
1890	230,402	75,572	56.2	32.8	1	8.0
1900	278,785	86,702	14.7	31.1	1	7.0
1910	331,389	94,446	8.9	28.5	1	7.4
1920	438,112	109,966	16.4	25.1	3	6.7
1930	487,336	132,068	20.0	27.1	5	6.1
1940	664,064	187,266	41.8	28.2	4	5.1
1950	803,144	284,313	51.8	35.4	5	4.9
1960	763,891	411,737	44.8	53.9	5	5.1
1970	756,276	537,712	30.6	71.1	5	4.4
1980	638,333	448,906	−16.5	70.3	6	6.4
1990	606,900	399,604	−10.9	65.1	8	9.7[d]

[a] Until 1980, U.S. census reports did not specify the race and ethnicity of the foreign-born population of the District of Columbia. Hence, in this table, the category "African Americans" includes Latinos of African heritage. Also, it is generally agreed that African Americans (particularly males) and members of new immigrant groups tend to be underreported in official U.S. government surveys.

[b] The ranking of the District of Columbia in relation to the U.S. cities with the largest African American populations (see Lynch 1973).

[c] The African American population included 3,176 slaves and 11,131 freedmen in 1860 (see Berlin 1974).

[d] This is a low estimate due to the substantial undercount of Latinos and other recently arrived immigrants. In the Washington, D.C., suburbs, for example, the foreign-born comprise 12.8 percent of the enumerated population.

Source: U.S. Bureau of the Census, *Current Population Reports*, 1973, 1984, 1992.

The confluence of these patterns is exemplified by the employment practices of the Federal government—the largest employer in the District.

In 1940, over one-third of the District's labor force was employed by the federal government, yet it hired only 16 percent of all African American workers compared to 40 percent of all White workers; Blacks comprised 28 percent of the District's population. In terms of occupational opportunities, the prospects for Blacks were even bleaker. In 1938, 90 percent of all African Americans employed by the federal government were confined to the least desirable custodial jobs. Even in the white-collar professions, college-educated Blacks were restricted to the lowest status positions. In fact, the U.S. State Department hired Blacks in Washington only as chauffeurs, messengers, and janitors; between 1924 and 1940, not a single "Negro" was employed by the department above the custodial level—a significant policy shift from the late nineteenth century when many Black professionals held prominent positions. Lastly,

wage discrimination virtually ensured the poverty status of African Americans. A "Negro job" was ten to twenty times as likely as a "White job" to offer an annual income of under $1,000 in 1940 (Frazier 1969; Green 1967; Landis 1948).

The dependence of the Washington, D.C., economy on the U.S. government reached its zenith during the New Deal; by 1940, almost one out of every two civilian workers (44 percent) in the District was a federal employee. This "pencil sharpener revolution" engendered a complementary demand in building trades, manual labor, and personal services. Again, like the preceding periods of rapid urban expansion, foreign workers—especially from Europe—did not satisfy this escalating demand for manual laborers. In fact, even if a shift in "race preference" had occurred, the dramatic decline in immigration during the Great Depression would have severely hindered such a recruitment strategy. Overall, total immigration to the United States plummeted during the Great Depression and then rose only moderately in the 1940s. Similarly, in the District the incoming flow of unskilled immigrant workers dwindled to a trickle; between 1930 and 1940, the District's foreign-born populace registered a net increase of only 4,100 (INS 1992c). Not surprisingly, District employers encouraged a resurgence in Black migration, especially from the Lower South states of the Carolinas, Alabama, and Georgia (Abbott 1990; Green 1967; Williams 1988). This resulted in the more than doubling of the African American population of Washington, D.C., between 1930 and 1950. In the process, this population movement constituted the catalyst for the profound demographic changes that began to crystallize in this period.

The D.C. "Black Belt": The Social Construction of Urban Apartheid

The racial segregation policies promulgated by the U.S. federal government during the Wilson presidency begot the more pernicious forms of institutional racism that characterize the more recent African American experience. During the 1920s, the social, political, and economic rights of African Americans were systematically eroded and juridically redefined. The most striking feature of the White backlash was the concerted campaign to concentrate the widely dispersed African American population—both within and outside the District—into the racially contained neighborhoods of the inner city. This explicit private *and* public policy of segregating the nation's capital, which, not incidentally, was highly profitable for local business interests (Gale 1987; Jackson 1985; Landis 1948), underlies the twin social and economic pillars of contemporary "Black D.C."

The economic and political exigencies of World War II generated new opportunities for the African American population. Wartime military mobilizations, coupled with the enormous growth of the U.S. government workforce, resulted in a sudden shortage of labor in the District; between 1940 and 1950, the federal civilian labor force rose from 139,770 to 223,312—its largest absolute decennial increase. Of course, the escalating demand for both blue-collar and white-collar workers was not satisfied by the sluggish flow of immigrant workers; the foreign-born population increased by only 5,500 in the 1940s. Instead, the wartime labor shortage inspired the temporary dismantling of gender and racial occupational barriers. The U.S. government's attack on "Jim Crow" policies began with the Ramspeck Act of 1940 and other reforms of the Civil Service regulations. In 1941, mounting pressure by national Black leaders to increase African American employment in the more skilled occupations of the war industries culminated in President Roosevelt's Executive Order 8802, which forbade racial discrimination in federal employment and established the Fair Employment Practices Committee (FEPC). These policies, which signaled the end of the South's exclusive control over its racially subjugated workforce, provided the political framework and economic impetus for the second large-scale migration of African Americans from the rural South.

In the District, the influx of southern migrants contributed to the tremendous growth and increased sociocultural diversity of the Black population during the 1940s. The irony of this newly earned economic and geographic freedom is that African Americans were not granted the right to choose their residential neighborhoods after arriving in the District. Instead, both old and new Black residents found that the historically negotiated pattern of geographically dispersed African American communities was being unilaterally replaced by an aggressive policy of inner-city containment; their housing options were increasingly limited to the "Black crescent" of downtown Washington. This general trend followed earlier episodes of the involuntary dismantling of existing Black neighborhoods in the outer urban districts or in the Maryland and Virginia suburbs and their relocation to the demographically compressed inner-city "Black belts" (Landis 1948; Weaver 1948). The displacement of the long-standing Black community of Georgetown in northwest Washington during the 1920s and 1930s attests to the profitability of the process to White "renovators" and realtors, as well as the success of revising the social history of this now racially homogeneous enclave of professional and political elites (Lesko, Babb, and Gibbs 1991).

Undeniably, the demographic growth of the District's metropolitan area illuminates distinctly different residential patterns by racial group. What is particularly noteworthy is that the process of racial enclosure in the inner city was inextricably tied to the growth of the metropolitan suburbs. On the one hand, African Americans disproportionately clustered within the compressed inner urban zones; comprising less than one-third of the city population in

1940, Blacks accounted for almost two-thirds of the net increase of 60,000 in the inner city during the 1930s. On the other hand, Whites were moving en masse to the rapidly expanding, racially homogeneous neighborhoods of the outer urban zones and the expanding Maryland and Virginia suburbs. Between 1930 and 1940, four-fifths of the net District metropolitan increase of 290,000 occurred outside of the inner-city zones; Whites constituted 85 percent of the net growth in the outer urban zones and 95 percent in the suburbs. Hence, middle- and working-class Blacks were forced to compete with each other for the limited, aging housing of the central city, while White residents were encouraged to enter the racially exclusive and relatively cheaper housing market of the newly constructed suburbs.

By 1947, this racial containment strategy had successfully concentrated about two-thirds of all urban District Blacks within a 2.25-mile radius of the downtown business center (Landis 1948). These residential "Black belts" were clustered within a crescent-shaped rampart that was bounded by the Navy Yard in the southeast to McMillan Reservoir in the north and around to Rock Creek and the Potomac on the west. Not surprisingly, the all-important White residential and commercial enclave lies within the African American residential bulwark and includes the business district, a fashionable hotel and diplomatic corridor, George Washington University, D.A.R. Constitution Hall, and of course the White House—all dependent upon the availability of low-wage labor (Landis 1948). In order to ensure the stability of this tenuous social geography, the federal and District governments promoted passive segregation policies such as restrictive covenants for preventing property transfers to specified groups (primarily Blacks and Jews);[1] denying construction permits and/or financing for African American housing developments outside the Black "zones"; condemning Black-owned residential property for the "public interest" (such as for parks, highways, government buildings, and White housing); permitting suburban communities to voluntarily reject participation in public housing programs; and manipulating federal mortgage loan (FHA, VA) programs to curb the flow of African Americans toward the suburbs.

The latter two policies are crucial to understanding the contemporary patterns of urban "apartheid" in general and Washington, D.C., in particular.

First, in the Washington, D.C., metropolitan area, the predominantly White suburbs have consistently refused to support subsidized housing projects within their jurisdictions. This has increased pressure on the District to solicit federal resources for expanding their public housing programs. Not surprisingly, the construction of low-income residential projects in the District has reinforced existing patterns of racial segregation. As the president of the Federation of Citizen Associations asserted during his U.S. Senate testimony on the National Capital Housing Authority's (NCHA) proposed mandate in 1944, ". . . public housing must be continued on a segregated basis, [or the White] people of Washington are going to, by a very great majority, oppose it" (cited in Landis 1948, p. 41). Hence, from the start, public housing has played a central role in the strategy of racial containment; in 1940, 83 percent of the 3,259 permanent low-rent units operated by the NCHA were occupied by African Americans.

The second factor concerns the planned growth of the metropolitan suburbs. From its inception in 1934, the Federal Home Administration's (FHA) mortgage insurance programs were intended to stimulate the construction industry as well as the demand for consumer durables (ovens, refrigerators, washing machines, automobiles) by offering easy financing terms for new homeowners in the White suburbs. This New Deal program, which implicitly affirmed existing discriminatory lending practices (neighborhood "redlining") by private financial institutions, severely handicapped the purchase or even the remodeling of older homes in the inner cities. As a result, FHA and other subsidized mortgage loan programs (VA for military veterans) accelerated the suburban expansion of metropolitan Washington; between 1934 and 1960, Maryland (Prince George's, Montgomery) and Virginia (Fairfax) suburban counties received more than seven times as much mortgage insurance as the District. Significantly, the long-term impact of these programs was clearly understood by federal housing officials. In 1939, only five years after its creation, the FHA perceptively assessed the impact of its programs on the racial composition of the city:

> . . . the "filtering-up" process, and the tendency of Negroes to congregate in the District, . . . [means that] eventually the District will be populated by

Negroes and the surrounding areas in Maryland and Virginia by White families. (cited in Jackson 1985, p. 213)

From Biracial City to Multicultural Metropolis, 1945 to 1970

Pax Americana: Cold War Foundations of the International Capital City

The post–World War II era witnessed the consummation of America's political and economic preeminence in the cold war order: *the Pax Americana*. The United States emerged as the dominant power of the international capitalist economy and, accordingly, Washington, D.C., became the undisputed capital of the "Free World." As a result, the major international financial and political institutions of the "new world order" were founded in the District during the late 1940s and early 1950s. These include such prominent international organizations as the World Bank, International Monetary Fund, Import-Export Bank, International Development Bank, and various offices of the United Nations. With the centralization of this bureaucratic apparatus, the United States embarked on a campaign to rebuild the war-ravaged economies of Europe and Asia as well as guiding the development of the Third World—especially the newly independent countries of Africa, Asia, the Middle East, Central America, and the Caribbean.

The new leadership responsibilities of the United States required the rapid expansion of such national bureaucracies as the Agency for International Development, Department of Commerce, Department of State, U.S. Information Agency, and later the Peace Corps. Of course, this prominent role in international relations—especially in the containment of Soviet and Chinese expansionism—provided the justification for the sustained growth of the Department of Defense (Pentagon), national military services (Air Force, Army, Marines, Navy), domestic and international security forces (National Guard, FBI, CIA), National Air and Space Administration (NASA), and the burgeoning private economy of consultants ("beltway bandits"), contractors, and defense-related corporations. The latter are particularly important due to their concentration in the nearby suburbs; these

new white-collar "armies" are responsible for the overnight emergence of suburban "edge" cities such as Crystal City and Reston in Virginia and Rockville and Gaithersburg along the I-270 corridor in Maryland (Garreau 1991).

The expansion of the international bureaucratic and diplomatic corps was a key factor in the postwar dynamism of the local District economy. It not only enshrined Washington, D.C., as the political capital of the new world order, but it also increased the cultural and national diversity of its white-collar workforce as foreign delegations and their personal staffs competed for prestigious assignments in the new international capital. This phenomenon contributed to the District's unique "culture of politics" and precipitated the revitalization of its aging downtown mansions and adjacent brownstones as Massachusetts Avenue was transformed into Embassy Row. This postwar shift in international capitals, from London to Washington, was also accompanied by a rapid growth in the number of foreign students enrolled in local universities (American, Georgetown, George Washington, Howard, Johns Hopkins, and Maryland); most were sponsored by their governments to study foreign relations, economic development, political science, international economics, or U.S. politics. Although only a relatively small number of foreign students remained in the District during the 1950s, they played a crucial role in forging the social infrastructure of their respective ethnic/national communities in subsequent decades.

Metropolitan Sprawl: Urban Decline and Suburban Domination

Ironically, the "take-off" of the Washington, D.C., economy in the post–World War II period is distinguished by the rapid diversification of its private sector. The most prominent sources of new employment include federally sponsored research and development (National Institutes of Health), design and engineering for defense projects (Rockwell), state and federal consulting contracts, provision of business supplies and office maintenance services, the proliferation of trade and professional lobbying associations, the growth of financial and investment services, an enormous expansion of office and residential construction, and an escalating demand for personal services (hotels, restaurants, entertainment, retail shops, health care) by government employees as well as visiting tourists and businessmen. Not incidentally, this economic diversification trend coincides with the meager growth of U.S. government employment during the 1950s (2.7 percent); the proportion of the total District workforce engaged in federal civilian employment shrank from 36 percent in 1950 to 30 percent in 1960. Nevertheless, the local labor market became increasingly bifurcated into highly skilled and educated white-collar workers and unskilled service and manual laborers and construction workers.

The most striking feature of this economic boom period is the rapid growth of the Maryland and Virginia suburbs. Between 1950 and 1970, the Washington, D.C., metropolitan area mushroomed from 183 to 523 square miles, and its suburban population swelled from 661,000 to 2,154,000 (U.S. Bureau of the Census 1982). This trend underlies the profound political and economic shift in the center of power of the postindustrial metropolis as new immigrants and racial and ethnic minorities confronted the invisible barriers of entry into the booming suburban growth "poles." Indeed, a distinguishing characteristic of the postindustrial political, informational, and service metropolis is the declining population of the inner city, with its increasingly dense concentration of impoverished racial and ethnic minorities and the extraordinary dynamism of relatively affluent and predominantly White suburbia. Not surprisingly, as the exemplar of the postindustrial metropolis, the District has been in the forefront of these sociodemographic trends (Manning 1995, 1996; Manning and Butera 1994). In the late 1940s, the population of the District of Columbia peaked at over 870,000 and, by 1947, the city accounted for almost three-fourths (72 percent) of the total metropolitan Washington, D.C., population. Although African Americans increased substantially in the 1940s (52 percent), they still comprised only 30 percent of the city's total population in 1947—a slight decline from the 33 percent of 1880. At the same time, the Washington, D.C., suburbs featured 313,000 Whites and only 26,000 Blacks. Three years later, in 1950, the District accounted for only half (55 percent) of the overall District metropolitan population, as the Maryland and Virginia suburbs nearly doubled to 661,000 during this brief period. Over the next two decades, the District's share of the metropolitan population fell sharply, to less than two-fifths in 1960 (38 percent) and then to about one-fourth (26 percent) in 1970.

Clearly this trend reflects the profound shift in the demography of the postindustrial metropolis due to the decline of the District's population (–6 percent) compared to the explosive growth of the suburban population (226 percent) during this twenty-year period. Admittedly the increasingly successful attacks on Jim Crow segregationist policies contributed to White flight from the District; in 1954, two major judicial decisions *(Brown v. Topeka Board of Education* and *Bolling v. Sharpe)* provided the legal basis for implementing racial integration in the District (Williams 1988). Even so, the exodus of 300,000 Whites from the city during the 1950s and 1960s accounts for only a modest proportion of the net White suburban increase of nearly 1.5 million (U.S. Bureau of the Census 1982).

The continuation of the racial enclosure policies in the central city is mirrored in the social geography of Washington, D.C. In 1947, only 2 percent of all Blacks in the metropolitan Washington, D.C., area resided outside of the District. Although African Americans experienced some mobility in the following decades, with one-fifth of all metropolitan District Blacks living outside of the District versus almost four-fifths of the White population in 1960, this gap widened in the following decade; in 1970, the proportion of African Americans residing in the suburbs rose to 25 percent compared to 90 percent for Whites. In contrast, the District experienced a dramatic shift in its racial composition. Between 1950 and 1960, African Americans increased from 284,000 to 412,000 while the White population fell from 519,000 to 352,000; more Blacks moved into the District (126,000) than out to the suburbs (87,000) during the 1960s. This resulted in African Americans—for the first time—comprising an absolute majority of District residents by the end of the 1950s. This trend continued in the following decade with Blacks rising to 538,000 and Whites falling to 209,000 in 1970. As a result, African Americans constituted an overwhelming 71 percent of the District's population in 1970, compared to a mere 8 percent of the suburban Washington, D.C., population.

The Black Metropolis and the New Immigrants

The new political and economic forces that profoundly influenced the emergence of the "Black Metropolis" are both local and national in their origins.

20th Anniversary of the Historic 1963 March on W

ugust 27, 1983   *We Still Have A Dream*
JOBS PEACE FREEDOM

The aspirations raised by the Civil Rights Movement remain only partly fulfilled. On the 20th anniversary of the 1963 March on Washington, many congregated once again on the steps of the Lincoln Memorial to march for jobs, peace, and freedom—"We Still Have a Dream." (Photo by William Clark. U.S. Department of the Interior, National Park Service)

First, for African Americans, this period featured the collapse of the Old South "racial state" which released millions of rural Black workers and their families from a declining agricultural economy and enabled them to pursue the economic opportunities of northern, midwestern, and western cities. However, the resistance of predominantly White trade unions to the influx of cheap labor, and the overall stagnation of the industrial (goods-producing) sector since the 1960s, ensured that most African American migrants would not participate in the postwar prosperity. Instead, the overwhelming majority were relegated to unskilled, low-wage occupations—especially at the bottom or "secondary" labor market of the low-wage service sector.

Second, the low occupational status of African American workers reinforced the existing patterns of urban confinement in the downtown Black belts; the influx of new Black arrivals did not push older residents into the more desirable neighborhoods of the

outer urban zones or the suburbs as posited by ecological models of ethnic succession. Third, the discriminatory policies of the federal FHA and VA mortgage insurance programs enabled millions of White families throughout the United States to abandon their urban brownstones and purchase single-family detached homes in the suburbs. This is illustrated by the enormous disparity in approved FHA mortgages for the District (8,038) versus its suburbs (44,432) between 1934 and 1960 (Jackson 1985, p. 211).

The rapidly changing social composition of the District in the post–World War II period contrasts sharply with the historic stability of the biracial city; remember, the proportion of Blacks actually declined between 1880 and the late 1940s. In fact, even during this period of urban population decline, the District remained an unpopular destination for immigrants. Overall, the number of foreign-born residents fell from 39,000 in 1950 (5 percent) to 36,000 in 1970 (4 percent); this absolute decline reflects the small number of immigrants arriving in the District as well as their relative ease in moving to the metropolitan suburbs.

Ironically, the powerful forces of social change that produced the Civil Rights movement also contributed to the dramatic increase in the magnitude and national diversity of the new immigration. That is, the demands for racial equality (which produced the Civil Rights Act of 1964 and the Voting Rights Act of 1965) were extended to include a "color blind" immigration admission process through the enactment of the Immigration and Nationality Act amendments of 1965. This resulted in a shift away from traditional European groups and toward those from Asia, Latin America, the Caribbean, and most recently Central America and Africa (Bryce-Laporte 1980; Portes and Rumbaut 1990). Nevertheless, the flow of "new" immigrants to the District was such a modest trickle in the 1960s that Hispanics and Asians were not even separately specified in the 1970 census reports on Washington, D.C.

Even before the influx of the post-1965 immigrants, the frustrated social and economic aspirations of African Americans in the central city Black belts achieved political expression in the form of large-scale urban riots. This tumultuous period of the late 1960s not only exacerbated White flight to the suburbs, but it also led to the abandonment of traditionally ethnic-owned (especially Jewish) businesses in the inner city—the future foothold of the "new" immigrant entrepreneurs. The "discovery" of the degrading conditions of Black urban poverty, moreover, led to the expansion of federal social and welfare services and new education and job training programs as the Civil Rights movement consolidated the political gains of poor African Americans (Piven and Cloward 1993).

This "war on poverty" campaign was conducted through the legislative initiatives of President Lyndon Johnson's Great Society programs and addressed such issues as housing segregation and occupational discrimination—especially in the public sector, including the federal bureaucracies. However, by improving the basic living conditions of the urban poor (Whites and Blacks), the expansion of the social "safety net" enabled many urban African Americans to abstain from the lowest-paying, least desirable jobs of the burgeoning service sector. It is this phenomenon that underscores the fortuitous timing of the new immigration. On the one hand, it stabi-

lized the urban labor market following the abrupt departure of African Americans from dead-end, unskilled jobs in the late 1960s, and, on the other, it provided the low-wage "fuel" for the suburban "growth machine" of the 1970s and 1980s.

The Postindustrial Metropolis

The Contemporary Social Landscape: The Changing Complexion of Multiculturalism

The tremendous growth of the District suburbs during the 1960s (from 1.2 million in 1960 to 2.2 million in 1970) continued uninterrupted over the next two decades. The population of the Virginia and Maryland suburbs increased from 2.4 million in 1980 to 3.3 million in 1990, while the District of Columbia registered a sharp drop from 756,000 in 1970 to 638,000 in 1980, and then slowly declined to 607,000 in 1990. Over this twenty-year period, the suburbs grew an astounding 54 percent, whereas the District experienced an absolute population decline of 20 percent; in 1990, the 85 percent suburban proportion of the District's metro area population was exceeded only by Atlanta (U.S. Bureau of the Census 1992d).[2] It is this demographic pattern, combined with the containment of the Black population within the District, that underlies the emergent "shortage" of manual workers in the District metropolitan labor market. That is, the traditional source of blue-collar craft and unskilled labor—African Americans—was available as evidenced by the rising unemployment levels of Black workers in the District; southern Blacks continued to migrate to the District in the 1970s as the District's suburbs remained largely inamicable to their residential aspirations. Hence, the historic legacy of housing segregation, together with the lack of an efficient mass transit system for delivering workers from the District to work sites in the suburbs, effectively limited the employability of African Americans in the outlying zones of the metropolitan area.[3]

A potential labor crisis was averted with the resumption of large-scale immigration. During the post–World War II period, legal immigration to the United States escalated from 1.0 million in the 1940s to 2.5 million in the 1950s, and then climbed sharply to 3.3 million in the 1960s, over 4.4 million in the 1970s, and then jumped to more than 6.3 million in the 1980s. In contrast to historical settlement patterns, however, the new immigrant groups prefer to live in the District's suburbs rather than in the central city. In 1970, for example, immigrants in the suburbs (4.5 percent) accounted for about the same proportion as in the city, (4.4 percent) but in absolute terms, constituted almost three times the foreign-born population of the District; African Americans, in comparison, accounted for 8 percent of the District's suburban population. Overall, this pattern reflects the moderate decline of the immigrant population in the District (11 percent) during the 1960s versus its rapid growth in the suburbs (55 percent). During the 1970s, the arrival of the new immigrant groups reinforced this pattern as the number of foreign-born residents in the city rose by 21 percent, whereas those in the suburbs increased by nearly 120 percent (U.S. Bureau of the Census 1982). Not incidentally, the ongoing diversification of the local and regional economy intensified the demand for distinctly different labor forces: highly educated white-collar professionals, blue-collar manual and skilled craft workers, and low-skilled clerical and service workers.

During the 1970s, the foreign-born population of the District experienced remarkable growth, and in the process reversed a seventy-year trend; the proportion of immigrants residing in the District fell from 7.4 percent in 1910 to 4.4 percent in 1970 before beginning its rapid ascent to 6.4 percent in 1980 and 9.7 percent in 1990. Furthermore, the sudden expansion of the District's foreign-born population was accompanied by a radical departure from its traditional national origins—away from the previous predominance of European immigrants. In 1994, the new foreign arrivals were primarily from Central America (El Salvador, Guatemala) and South America (Bolivia, Peru, Colombia), the Caribbean (Puerto Rico, Cuba, Haiti, Jamaica, Trinidad), Southeast Asia (China, Korea, Indochina), and Africa (Ethiopia, Ghana, Nigeria). As a result, the metropolitan Washington, D.C., area has undergone an extraordinary transformation—especially from a historical vantage point—from being a hostile to a hospitable destination for incoming immigrant groups. This is revealed in the reported settlement patterns of legal foreign residents. In 1991, according to the U.S. Immigration and Naturalization Service (INS), the Washington, D.C., metropolitan area ranked tenth among the top fifty urban metropolises in terms of the number of new immigrants that intended to reside in the respective urban area.

During the 1980s, the District continued to experience substantial change in its race and ethnic diversity. Although the total minority component of the District dipped slightly, to 73 percent, it features substantial internal flux: African Americans (65 percent) declined moderately, whereas Asians (2 percent) and especially Hispanics (5 percent) increased substantially. This reflects the continued Black flight to the suburbs (net decline of nearly 50,000), an influx of poor Central American refugees from wartorn El Salvador and Guatemala, and the arrival of Asian immigrants (primarily Koreans and Chinese); many of the latter are pursuing small business opportunities throughout the city.

For the first time since 1870, the heterogeneous foreign-born population of the District now exceeds 10 percent (when undocumented immigrants are included). This phenomenon is poignantly illustrated by the proliferation of fishermen along the Potomac River. The elegant bass boats and sophisticated fishing gear of White sportsmen constitute a stark contrast to the crude rods and handmade "trot" lines of numerous Central American, Vietnamese, and Cambodian fishermen who can be seen standing under the bridges, on the rocks, and along the river banks. For these immigrants, fishing is not simply a leisure activity. Rather, the allure of sportfish such as striped or large-mouth bass is secondary to the successful catch of perch, sunfish, or catfish for the evening meal (Symmes 1992).

Undeniably, the most remarkable sociocultural change is occurring in the suburbs. The combined racial and ethnic component rose from 24 to 31 percent during the 1980s; these figures belie the fact that the suburbs include substantially more minorities than the District due to its much larger population base. Between 1980 and 1990, the Census Bureau (1992d) reports that the three major racial and ethnic minority groups increased by 459,000, or an astonishing 83 percent, in the D.C. suburbs. This represents more than three times the growth rate of native Whites (24 percent); the total is considerably higher when undercounted immigrants (particularly Hispanics) are considered. Although the number of African Americans rose by 231,000 to 19 percent, this population gain was outpaced by the combined increase of Hispanics [116,000 (6 percent)] and Asians

[112,000 (6 percent)]; the heterogeneous Hispanic-origin community (Cubans, Guatemalans, Salvadorans, Bolivians, Peruvians) climbed more rapidly in the northern Virginia suburbs, while the diverse Asian community (Chinese, Indian, Korean, Filipinos, Vietnamese) increased more rapidly in the Maryland suburbs.[4]

Significantly, both immigrant groups achieved numerical parity during the 1980s and, in some localities, even exceed the native Black population. In northern Virginia, for example, Hispanics (14 percent) outnumber African Americans (11 percent) and Asians (7 percent) in Arlington, while Asians (9 percent) outnumber African Americans (8 percent) and Hispanics (6 percent) in Fairfax County (Cohn and Hsu 1991). Their rapid population growth rates and patterns of ethnic succession, moreover, have produced interesting crosscultural syncretisms. For instance, in the suburban Maryland community of Adelphi (Prince George's County), a well-known Chinese restaurant has adjusted to the loss of its previously White, middle-class market by attracting a largely African American, working-class clientele on weekdays and hosting a Hispanic nightclub on weekends; popular Latino bands attract Hispanic immigrants to the "King Kong" from the District, Virginia, and the Maryland suburbs with their pulsating *musica viva!* A personal favorite is a Vietnamese-owned restaurant in Greenbelt, Maryland. The "East Meets West" offers traditional Vietnamese dishes on one side of the restaurant and Mexican cuisine on the other. Here, one can enjoy lemon grass soup with an *enchilada* under the watchful eye of Mexican *mariachis*; all entrees are served by Vietnamese waitresses.

In conclusion, these examples highlight the enormous contribution of the new immigrants to the cultural diversity of the Washington, D.C., metropolitan area in general and the suburbs in particular. Overall, the combination of African Americans, Hispanics, and Asians accounts for slightly more than one-half of the net growth of the District area's population during the 1980s. When national origin is considered, the net increase of 216,000 immigrants accounts for nearly one-half of the total growth in racial and ethnic minorities. Over seven times as many immigrants reside in the suburbs (426,000) in comparison to the central city (59,000) in 1990. This fundamental change in the predominant settlement pattern of immigrants and native minorities mirrors the profound transformation of American social and economic life. That is, the ongoing shift in the socioeconomic dynamism of the postindustrial metropolis—from the central city to the outlying "edge cities" of the suburbs (Garreau 1991).

The Contemporary Economic Landscape: Urban Decline and Suburban Opportunity

The changing economic structure of the Washington, D.C., metropolis has profound implications to contemporary patterns of inequality in general and racial and ethnic inequality in particular (Manning 1995, 1996; Manning and Butera 1994). Indeed, based on the high education and skill demands of the U.S. postindustrial occupational structure (Silvestri and Lukasiewicz 1991) and the disproportionate concentration of the most lucrative jobs outside of the central city, it is clear that metropolitan suburbs offer the most desirable employment opportunities. For racial and ethnic minorities residing in the central city, the key to overcoming the labor market rigidities of the postindustrial metropolis is the penetration of suburbia.

As previously discussed, successful legal challenges to racially discriminatory education, employment, and housing policies have only recently fostered the growth of a vibrant African American middle class in the Washington, D.C., suburbs. Between 1970 and 1980, the proportion of African Americans residing in the District's suburbs rose from 25 to 47 percent and thus approached parity with District Blacks (401,000 to 445,000). This new Black migration continued in the 1980s and, by 1990, the African American population of the Maryland and Virginia suburbs accounted for almost two-thirds (62 percent) of the total metropolitan Black population (U.S. Bureau opf the Census 1992d). This represents a dramatic reversal of historic residential patterns: African Americans in the suburbs (632,000) substantially outnumber those in the District (395,000). See Table 2. Even so, the arrival of African Americans has not been warmly received by the existing suburban communities and has precipitated new patterns of White flight to the outer suburbs. This illuminates a defining feature of the African American experience in the post–Civil Rights period: resegregation of suburbia into racially homogeneous communities (Alba and Logan 1993; Massey and Denton 1988, 1993; Dent 1992; Gale 1987).

Table 2 Multicultural Change in Metropolitan Washington, D.C.: Race, Ethnicity, and Foreign-born (1970, 1980, and 1990)[a]

Area	1970		1980		1990	
	(No.)	(%)	(No.)	(%)	(No.)	(%)
District of Columbia						
White	209,272	27.7	164,244	25.7	166,131	27.4
Black	537,712	71.1	445,154	69.8	395,213	65.1
Hispanic	—[b]	—[b]	17,679	2.8	32,710	5.4
Asian	—[b]	—[b]	6,636	1.0	10,724	1.8
Other[c]	9,526[b]	1.2	4,620	0.7	2,112	0.4
Foreign-born	33,482	4.4	40,549	6.4	58,887	9.7
D.C. suburbs (Maryland and Virginia)						
White	1,957,709	90.9	1,847,572	76.3	2,293,002	69.1
Black	176,640	8.2	400,926	16.6	631,809	19.1
Hispanic	—[b]	—[b]	75,701	3.1	192,076	5.8
Asian	—[b]	—[b]	75,512	3.1	187,137	5.6
Other[c]	19,506[b]	0.9	22,878	0.9	12,650	0.4
Foreign-born	96,226	4.5	209,395	8.7	425,562	12.8
Washington D.C. metropolitan area (totals)						
White	2,166,981	74.5	2,011,816	65.7	2,459,133	62.7
Black	714,352	24.5	846,080	27.6	1,027,022	26.2
Hispanic	—[b]	—[b]	93,380	3.1	224,786	5.7
Asian	—[b]	—[b]	82,148	2.7	197,871	5.1
Other[c]	29,032[b]	1.0	27,498	0.9	14,762	0.4
Foreign-born	129,708	4.5	249,944	8.2	484,449	12.4

[a] Single African American and Latino men have been historically undercounted. This issue is especially important in the enumeration of the Hispanic-origin population due to the reluctance of undocumented immigrants to participate in the U.S. Census. Also, this data does not "double count" Hispanics by race, that is, Afro- and White Latinos are classified solely as Hispanics.

[b] In 1970, the population of Hispanic and Asian minorities was so small in the metropolitan Washington, D.C., area that the U.S. Census reported these groups in the combined "Other" category.

[c] This category includes those respondents that report themselves as distinct from the racial categories specified by the U.S. Census. These are people that identify their race or ethnic/cultural background as "mixed," such as products of interracial marriages.

Source: U.S. Bureau of the Census, *1970, 1980, and 1990 General Population Characteristics*. Washington, D.C.: U.S. Government Printing Office (1992).

The increasingly divergent urban-suburban opportunity structures shed light on only part of the picture of social inequality in the postindustrial metropolis. The key question is the income distribution of each racial and ethnic group within these spatially distinct areas. Overall, African American families are more than three times as likely as White families to live in poverty and over three times less likely to earn over $75,000 in 1989; Hispanics are slightly better off, followed by Asians. In terms of median household income, Whites ($46,000) earn over $21,000 more than Blacks ($25,000) in the District with Hispanics ($26,000) and Asians ($30,000) relatively close behind. Although the suburbs evidence substantially lower rates of poverty and much higher incomes, the most striking pattern is the consistently wide disparity in household income across all race and ethnic groups; the exception is the doubling of Asian income (over $70,000), which exceeds that of Whites in the suburbs. On average, median household income in the suburbs increases by about $14,000 for Hispanics, $21,000 for Native Americans and Blacks, over $20,000 for Whites, and a remarkable $40,000 for Asians (U.S. Bureau of the Census 1993b).

Consequently, for the majority of African Americans and especially the most disadvantaged in the District, the prosperity of the 1980s proved to be largely beyond their grasp. Urban poverty not only increased during the decade but it bulged into the "inner" suburban communities—especially in Prince George's County, Maryland, and northern Virginia (Alexandria and Arlington). In 1989, the median income of all households in the District ($31,000) represented only 61 percent of the adjacent suburbs ($50,000); the entire metropolitan area was $45,000 and includes Montgomery County, Maryland—one of the five most affluent counties in the nation—at over $70,000. Not unexpectedly, those District residents who were unable to escape the declining fortunes of the central city were more than three times as likely (16 percent) as their suburban neighbors (5 percent) to live in poverty (Grier 1992). It is this widening postindustrial "social divide" which underlies the escalating tensions that characterize the political landscape of the American metropolis.

The Mount Pleasant Riots: The Social Construction of the Postindustrial City

On 5 May 1991, the Washington, D.C., neighborhoods of Mount Pleasant, Adams Morgan, and Columbia Heights erupted into two days of riots and "civil disturbances." Attacks on District police, looting of commercial businesses, igniting of small fires, and systematic vandalism occurred after an African American policewoman shot a visibly intoxicated Salvadoran immigrant while arresting him for public drinking.[5] The perception that the incident typified an enduring pattern of police abuse of racial and ethnic minorities prompted spontaneous acts of violent protest by Latinos (primarily Salvadorans), which were later joined by African Americans and even some Whites from the three contiguous neighborhoods.

Curiously, more media attention initially focused on the "culture of public drinking" rather than on the deteriorating economic and social conditions that were exacerbating the plight of community residents. These latter factors are particularly important to the new Central American immigrants. Not only were Black/Latino tensions intensifying over scarce jobs and affordable housing, but desperation within the Latino community was increasing due to the tenuous legal status of recent arrivals, their limited access to local social services, and the unyielding demands for financial support from relatives in their native communities. For those Salvadorans and Guatemalans who left their war-torn societies for the "streets paved with gold," the collapse of the regional construction industry and subsequent decline in the local service sector precipitated an unexpected social battle for economic survival.

By the spring of 1991, the polarizing strains of the economic recession had severely aggravated racial and class divisions in the Mount Pleasant area (U.S. Civil Rights Commission 1993; Lane 1991). The rapid influx of Latino immigrants and the accompanying decline in the native African American population intensified the latent tensions between these minority groups; between 1980 and 1990, the African American population fell from 55 to 35 percent, while Whites rose from 30 to 35 percent and Hispanics rose from 14 to 27 percent (U.S. Bureau of the Census 1992). Furthermore, Latinos residing in the District were more likely to be recent immigrant men with few social contacts, little formal education, and poor knowledge of English, which severely hampered their entrance into more desirable suburban employment networks. Also, children of earlier immigrants were becoming increasingly frustrated by the gap between their rising aspirations and declining socioeconomic realities.

This situation exacerbated frictions over the dwindling number of available jobs among the District's culturally diverse workforce. Additionally, it fostered greater conflict outside of the city as African Americans and Latinos began competing with Whites for increasingly scarce jobs in the suburbs. Together with the ongoing expansion of Latino business and employment networks, as well as intensifying economic pressures to reduce production costs through subcontracting, these trends have potentially important implications. They suggest that an emerging feature of multicultural relations in the postindustrial metropolis is the rise of intergroup tensions in suburbia—especially in White, working-class communities experiencing rapid sociocultural change. This is demonstrated by the growing incidence of racially and ethnically motivated "hate crimes." For instance, a twenty-two-year-old White, working-class "Montgomery County redneck" was convicted in 1993 of attacking two young Black women in the blue-collar, suburban community of Wheaton, Maryland. The judge condemned the crime as a "savage attack" and "lynching by burning." The young man doused one woman with lighter fluid and threatened to ignite her. He was sentenced under a 1988 State of Maryland hate crime statute (Jennings 1993).

The District government, like all other major cities, is caught between the proverbial rock and hard place in responding to the social demands of the urban poor and the economic realities of the postindustrial metropolis. That is, the District government has become increasingly incapable of mediating disputes between poor people over limited access to jobs, housing, and public services while at the same time addressing the opposing interests of White, middle-class "gentrifiers" whose primary objective is to preserve the historic charm and rising property values of Mount Pleasant. Sadly, those minority groups that are most in need of government assistance tend to lack the necessary political influence to ensure that the municipal government responds to their needs—especially at a time of diminished public resources.[6]

It should not be surprising to note that the District government's official policy response to the plight of its politically impotent Latino residents has been largely symbolic. In 1993, the mayor filled the vacant directorship of the Office of Latino Affairs (OLA), modestly increased the OLA budget after successive years of reductions, offered an alcohol and substance abuse program for Latinos, and redirected existing recreation resources for Latino adolescents. Moreover, community groups were excluded from the long awaited police sensitivity training program; one ranking District police official stated, "we do not want [the training program] to devolve into a sociology class." These actions led the frustrated chairman of the District's Latino Civil Rights Task Force, a community advocacy group, to declare in 1993: "Maybe what we need [to get a serious policy response from the mayor] is another riot." Indeed, newly elected Mayor Marion Berry slashed the OLA budget in 1995 and it now operates with only a skeleton staff.

Conclusion

The historic transformation of Washington, D.C., from a biracial city to a multicultural metropolis, offers important insights into future patterns of social inequality and racial and ethnic relations. The emerging "multicultural majority" of U.S. central cities, as well as the widening "social divide" between urban core and suburban "edge" cities, appear to be defining characteristics of postindustrial American society. Although the urban destiny of the vast majority of new immigrants has shifted dramatically from the central city to the booming suburbs, these patterns have ominous implications for the urban poor, especially during the current period of U.S. industrial restructuring.

As the political and economic infrastructure of American cities continue to deteriorate, the economically disenfranchised may have little choice but to remain in the urban core of the postindustrial metropolis and struggle over the dwindling resources of the city government. This does not imply, however, that the future of multicultural relations in the postindustrial metropolis should be construed as inevitably bleak. Rather, for those groups that remain in the District, the possibility exists that new neighborhoods may emerge that cross-cut the traditional social axes of economic status and cultural and national background. This is illustrated by the ongoing social struggles and political coalitions being forged in the multicultural, multiclass community of Mount Pleasant, which may offer a potential harbinger of the social process of constructing the postindustrial city of the future.

Notes

1. The U.S. Supreme Court ruled in 1948 (*Shelley v. Kraemer*) that racial covenants, as legal devices for prohibiting Black occupancy, were "unenforceable as law and contrary to public policy." Even so, the Federal Housing Administration (FHA) did not decide until the following year that it would not insure mortgages on real estate with such restrictions. Furthermore, FHA policies failed to challenge the covenants already in effect and even offered a one-year grace period for builders or speculators to file new ones (Jackson 1985, chap. 11).

2. In 1990, according to the U.S. census, the top ten metropolises with the largest proportion of suburbanized residents were Atlanta (86 percent), Washington, D.C. (85 percent), St. Louis (84 percent), Pittsburgh (82 percent), Miami (82 percent), Boston (80 percent), Detroit (77 percent), Minneapolis–St. Paul (74 percent), Seattle (74 percent), and Cleveland (72 percent).

3. The Metropolitan Washington, D.C. subway, or "Metro" system was designed to transport workers and tourists from the suburbs to the central city. The first phase of the system, the "Red Line," was opened in 1977 and served the needs of those traveling from Capital Hill to Dupont Cirlce. In the 1980s, Metro service was expanded to the Maryland and northern Virginia suburbs; most subway stations in African American neighborhoods were not completed until the late 1980s and early 1990s. Hence, the contemporary crisis of the District's mass transit system is twofold. It does not provide cheap and efficient transportation for African Americans to travel from the District to job sites in the suburbs and it fails to deliver adequate service to the fastest growing segment of the metropolitan workforce: suburb-to-suburb commuters. As a Metro official acknowledged, "The [population] growth and jobs are occurring in areas that we have the most difficulty serving."

4. Unlike most U.S. cities, the Latino community of Washington, D.C., is not dominated by a single national group; the largest is from El Salvador and it accounts for much less than half of the Hispanic-origin population. In fact, at least twenty nationalities were represented at the 1995 D.C. Hispanic Festival. This cultural diversity contrasts with the predominance of individual Latino groups in other major U.S. cities in 1990: Los Angeles (67 percent Mexican), Miami (62 percent Cuban), Philadelphia (76 percent Puerto Rican), Chicago (65 percent Mexican), and New York (50 percent Puerto Rican).

5. Daniel Gómez, the Salvadoran dishwasher whose wounding sparked the Mount Pleasant riots, was acquitted in November 1992 of attempted assault on a District police officer. Conflicting evidence failed to conclusively corroborate the prosecution's contention that the knife wielding suspect was shot *before* being handcuffed. As a result, the trial ended with a hung jury and the prosecution decided not to retry to case.

6. Most Latino residents of the District are either undocumented, in an ambiguous legal status ("temporary residence"), or only recently "regularized" their immigrant status and therefore are ineligible to vote in District elections.

29

The First Americans: American Indians

C. MATTHEW SNIPP

By the end of the nineteenth century, many observers predicted that American Indians were destined for extinction. Within a few generations, disease, warfare, famine, and outright genocide had reduced their numbers from millions to less than 250,000 in 1890. Once a self-governing, self-sufficient people, American Indians were forced to give up their homes and their land, and to subordinate themselves to an alien culture. The forced resettlement to reservation lands or the Indian Territory (now Oklahoma) frequently meant a life of destitution, hunger, and complete dependency on the federal government for material needs.

Today, American Indians are more numerous than they have been for several centuries. While still one of the most destitute groups in American society, tribes have more autonomy and are now more self-sufficient than at any time since the last century. In cities, modern pan-Indian organizations have been successful in making the presence of American Indians known to the larger community, and have mobilized to meet the needs of their people (Cornell 1988; Nagel 1986; Weibel-Orlando 1991). In many rural areas, American Indians and especially tribal governments have become increasingly more important and increasingly more visible by virtue of their growing political and economic power. The balance of this chapter is devoted to explaining their unique place in American society.

The Incorporation of American Indians

The current political and economic status of American Indians is the result of the process by which they were incorporated into Euro-American society (Hall 1989). This amounts to a long history of efforts aimed at subordinating an otherwise self-governing and self-sufficient people that eventually culminated in widespread economic dependency. The role of the U.S. government in this process can be seen in the five major historical periods of federal Indian relations: removal, assimilation, the Indian New Deal, termination and relocation, and self-determination.

As with other ethnic groups in America, what form their political incorporation should take is a difficult issue for many Native American tribes. Here, Seminole Indians of the Dania, Florida, reservation cast ballots in the August 21, 1957 election to decide whether or not to adopt a tribal form of government. Such a government would have a charter forming a business corporation to represent the tribe. (Library of Congress)

Removal

In the early nineteenth century, the population of the United States expanded rapidly at the same time that the federal government increased its political and military capabilities. The character of Indian-American relations changed after the War of 1812. The federal government increasingly pressured tribes settled east of the Appalachian Mountains to move west to the territory acquired in the Louisiana purchase.

Numerous treaties were negotiated by which the tribes relinquished most of their land and eventually were forced to move west.

Initially the federal government used bargaining and negotiation to accomplish removal, but many tribes resisted (Prucha 1984). However, the election of Andrew Jackson by a frontier constituency signaled the beginning of more forceful measures to accomplish removal. In 1830 Congress passed the Indian Removal Act, which mandated the eventual removal of the eastern tribes to points west of the Mississippi River, in an area which was to become the Indian Territory and is now the state of Oklahoma. Dozens of tribes were forcibly removed from the eastern half of the United States to the Indian Territory and newly created reservations in the west, a long process ridden with conflict and bloodshed.

As the nation expanded beyond the Mississippi River, tribes of the plains, southwest, and west coast were forcibly settled and quarantined on isolated reservations. This was accompanied by the so-called Indian Wars—a bloody chapter in the history of Indian-White relations (Prucha 1984; Utley 1984). This period in American history is especially remarkable because the U.S. government was responsible for what is unquestionably one of the largest forced migrations in history.

The actual process of removal spanned more than a half-century and affected nearly every tribe east of the Mississippi River. Removal often meant extreme hardships for American Indians, and in some cases this hardship reached legendary proportions. For example, the Cherokee removal has become known as the "Trail of Tears." In 1838, nearly 17,000 Cherokees were ordered to leave their homes and assemble in miltary stockades (Thornton 1987, p. 117). The march to the Indian Territory began in October and continued through the winter months. As many as 8,000 Cherokees died from cold weather and diseases such as influenza (Thornton 1987, p. 118).

According to William Hagan (1979), removal also caused the Creeks to suffer dearly as their society underwent a profound disintegration. The contractors who forcibly removed them from their homes refused to do anything for "the large number who had nothing but a cotton garment to protect them from the sleet storms and no shoes between them and the frozen ground of the last stages of their hegira. About half of the Creek nation did not survive the migration and the difficult early years in the West" (Hagan 1979, p. 77–81). In the West, a band of Nez Perce men, women, and children, under the leadership of Chief Joseph, resisted resettlment in 1877. Heavily outnumbered, they were pursued by cavalry troops from the Wallowa valley in eastern Oregon and finally captured in Montana near the Canadian border. Although the Nez Perce were eventually captured and moved to the Indian Territory, and later to Idaho, their resistance to resettlement has been described by one historian as "one of the great military movements in history" (Prucha 1984, p. 541).

Assimilation

Near the end of the nineteenth century, the goal of isolating American Indians on reservations and the Indian Territory was finally achieved. The Indian population also was near extinction. Their numbers had declined steadily throughout the nineteenth century leading most observers to predict their disappearance (Hoxie 1984). Reformers urged the federal government to adopt measures that would humanely ease American Indians into extinction. The federal government responded by creating boarding schools and the allotment acts—both were intended to "civilize" and assimilate American Indians into American society by Christianizing them, educating them, introducing them to private property, and making them into farmers. American Indian boarding schools sought to accomplish this task by indoctrinating Indian children with the belief that tribal culture was an inferior relic of the past and that Euro-American culture was vastly superior and preferable. Indian children were forbidden to wear their native attire, to eat their native foods, to speak their native language, or to practice their traditional religion. Instead, they were issued Euro-American clothes, and expected to speak English and become Christians. Indian children who did not relinquish their culture were punished by school authorities. The curriculum of these schools taught vocational arts along with "civilization" courses.

The impact of allotment policies is still evident today. The 1887 General Allotment Act (the Dawes Severalty Act) and subsequent legislation mandated that tribal lands were to be allotted to individual American Indians in fee simple title, and the surplus lands left over from allotment were to be sold on the open market. Indians who received allotted tribal lands also received citizenship, farm implements, and encouragement from Indian agents to adopt farming as a livelihood (Hoxie 1984, Prucha 1984).

For a variety of reasons, Indian lands were not completely liquidated by allotment, many Indians did not receive allotments, and relatively few changed their lifestyles to become farmers. Nonetheless, the allotment era was a disaster because a significant number of allotees eventually lost their land. Through tax foreclosures, real estate fraud, and their own need for cash, many American Indians lost what for most of them was their last remaining asset (Hoxie 1984).

Allotment took a heavy toll on Indian lands. It caused about 90 million acres of Indian land to be lost, approximately two-thirds of the land that had belonged to tribes in 1887 (O'Brien 1989). This created a another problem that continues to vex many reservations: "checkerboarding." Reservations that were subjected to allotment are typically a crazy quilt composed of tribal lands, privately owned "fee" land, and trust land belonging to individual Indian families. Checkerboarding presents reservation officials with enormous administrative problems when trying to develop land use management plans, zoning ordinances, or economic development projects that require the construction of physical infrastructure such as roads or bridges.

The Indian New Deal

The Indian New Deal was short-lived but profoundly important. Implemented in the early 1930s along with the other New Deal programs of the Roosevelt administration, the Indian New Deal was important for at least three reasons. First, signaling the end of the disastrous allotment era as well as a new respect for American Indian tribal culture, the Indian New Deal repudiated allotment as a policy. Instead of continuing its futile efforts to detribalize American Indians, the federal government acknowledged that tribal culture was worthy of respect. Much of this change was due to John Collier, a long-time Indian rights advocate appointed by Franklin Roosevelt to serve as Commissioner of Indian Affairs (Prucha 1984).

Like other New Deal policies, the Indian New Deal also offered some relief from the Great Depres-sion and brought essential infrastructure development to many reservations, such as projects to control soil erosion and to build hydroelectric dams, roads, and other public facilities. These projects created jobs in New Deal programs such as the Civilian Conservation Corps and the Works Progress Administration.

An especially important and enduring legacy of the Indian New Deal was the passage of the Indian Reorganization Act (IRA) of 1934. Until then, Indian self-government had been forbidden by law. This act allowed tribal governments, for the first time in decades, to reconstitute themselves for the purpose of overseeing their own affairs on the reservation. Critics charge that this law imposed an alien form of government, representative democracy, on traditional tribal authority. On some reservations, this has been an on-going source of conflict (O'Brien 1989). Some reservations rejected the IRA for this reason, but now have tribal governments authorized under different legislation.

Termination and Relocation

After World War II, the federal government moved to terminate its long-standing relationship with Indian tribes by settling the tribes' outstanding legal claims, by terminating the special status of reservations, and by helping reservation Indians relocate to urban areas (Fixico 1986). The Indian Claims Commission was a special tribunal created in 1946 to hasten the settlement of legal claims that tribes had brought against the federal government. In fact, the Indian Claims Commission became bogged down with prolonged cases, and in 1978 the commission was dissolved by Congress. At that time, there were 133 claims still unresolved out of an original 617 that were first heard by the commission three decades earlier (Fixico 1986, p. 186). The unresolved claims that were still pending were transferred to the Federal Court of Claims.

Congress also moved to terminate the federal government's relationship with Indian tribes. House Concurrent Resolution (HCR) 108, passed in 1953, called for steps that eventually would abolish all reservations and abolish all special programs serving American Indians. It also established a priority list of reservations slated for immediate termination. However, this bill and subsequent attempts to abolish reservations were vigorously opposed by Indian advocacy groups such as the National Congress of American Indians. Only two reservations were actually terminated, the Klamath in Oregon and the Menominee in Wisconsin. The Menominee reservation regained its trust status in 1975 and the Klamath reservation was restored in 1986.

The Bureau of Indian Affairs (BIA) also encouraged reservation Indians to relocate and seek work in urban job markets. This was prompted partly by the desperate economic prospects on most reservations, and partly because of the federal government's desire to "get out of the Indian business." The BIA's relocation programs aided reservation Indians in moving to designated cities, such as Los Angeles and Chicago, where they also assisted them in finding housing and employment. Between 1952 and 1972, the BIA relocated more than 100,000 American Indians (Sorkin 1978). However, many Indians returned to their reservations (Fixico 1986). For some American Indians, the return to the reservation was only temporary; for example, during periods when seasonal employment such as construction work was hard to find.

Self-Determination

Many of the policies enacted during the termination and relocation era were steadfastly opposed by American Indian leaders and their supporters. As these programs became stalled, critics attacked them for being harmful, ineffective, or both. By the mid-1960s, these policies had very little serious support. Perhaps inspired by the gains of the Civil Rights movement, American Indian leaders and their supporters made "self-determination" the first priority on their political agendas. For these activists, self-determination meant that Indian people would have the autonomy to control their own affairs, free from the paternalism of the federal government.

The idea of self-determination was well received by members of Congress sympathetic to American Indians. It also was consistent with the "New Federalism" of the Nixon administration. Thus, the policies of termination and relocation were repudiated in a process that culminated in 1975 with the passage of the American Indian Self-Determination and Education Assistance Act, a profound shift in federal Indian policy. For the first time since this nation's founding, American Indians were authorized to oversee the affairs of their own communities, free of federal intervention. In practice, the Self-Determination Act established measures that would allow tribal governments to assume a larger role in reservation administration of programs for welfare assistance, housing, job training, education, natural resource conservation, and the maintenance of reservation roads and bridges (Snipp and Summers 1991). Some reservations also have their own police forces and game wardens, and can issue licenses and levy taxes. The Onondaga tribe in upstate New York have taken their sovereignty one step further by issuing passports that are internationally recognized. Yet there is a great deal of variability in terms of how much autonomy tribes have over reservation affairs. Some tribes, especially those on large and well-organized reservations have nearly complete control over their reservations, while smaller reservations with limited resources often depend heavily on BIA services.

Indian Country

As their name suggests, reservations (along with tribal trust lands) represent the last remaining landholdings subject to the control of American Indians; these lands are known as "Indian Country." About one-half of all American Indians live in these places, and most are located west of the Mississippi River. Since their creation, reservations have been places marked by severe economic distress. In 1928, the Meriam Report first documented the economic hardships of reservation life, and since then, studies continue to show that little has changed (Snipp 1989). American Indian reservations were and continue to be among the poorest places in America.

Welfare dependency has been a fact of life in Indian country since tribes were first interned on reservations and forced to depend on military rations for survival. Since then, this dependency has become considerably more complex and manifest in federal programs, such as the "war on poverty" efforts or the projects sponsored by the Comprehensive Employment and Training Acts (CETA) of the 1960s and 1970s. The abrupt termination of many of these programs in the conservative era of the 1980s reminded tribal leaders about the uncertainty of federal largesse and the need for achieving financial independence.

Instead of simply abandoning these destitute places, urban and reservation Indians alike view measures to revitalize and improve reservation life as essential. For reservation Indians, reservation projects that provide jobs and improve the quality of life make it possible to stay in the community, near not only family but also close to significant spiritual and cultural events, making it possible to follow traditional cultural practices. Reservations are also important because only on reservations are Indian communities allowed some measure of self-government. Furthermore, many urban Indians maintain close ties to family and friends on reservations, as reservations are places for cultural renewal. In some instances, they are places to return to when jobs and opportunities become available.

Indian Country has an unparalleled importance to American Indians because it represents the last remaining land base for Indian communities. For many American Indians, the land has spiritual as well as monetary value. As a result, efforts to develop tribal lands involve complex considerations such as how to profitably develop tribal lands without compromising cultural values. Nonetheless, reservation and trust lands represent a considerable resource that American Indians are beginning to develop for the benefit of their communities.

Natural Resources

There are 278 federally recognized American Indian reservations ranging in size from less than 100 acres to the largest reservation of 16 million acres—the Navajo—covering parts of Arizona, Utah, and New Mexico; an area about the size of West Virginia. These reservations account for most of the 56.2 million acres of Indian lands supervised by the federal government. The four major types of natural resources include agricultural land, timber, water, and mineral resources.

Since the late nineteenth century, the federal government has encouraged American Indians to adopt agriculture. Yet for most of the twentieth century, agriculture has been a declining industry in the American economy. Furthermore, the allotment policies actually caused declines in Indian agriculture (Carlson 1981). Not surprisingly, there is a long history of failed attempts to establish tribal farms and livestock herds. Some tribes suffered disastrous land losses during allotment. Other tribes, such as those in the Southwest had practiced agriculture for centuries, but they refused to adopt non-Indian technolo-

gies and their collective farm systems were also disrupted by allotment.

Since the late 1800s, non-Indians have been responsible for agricultural production on tribal lands. For example, when members of the Ute tribe in Utah refused to become farmers, Indian agents leased their lands to nearby Mormon farmers. This is typical of how agricultural land has been managed on most reservations. One study even found that non-Indians cultivated the most productive farmland while Indians were more likely to control less productive grazing land (Levitan and Johnston 1975).

The productivity of tribal land is a serious problem. Not surprisingly, reservations were established in places that have very little productive land. For example, the BIA classifies less than 1 percent of all reservation lands as highly productive and less than 5 percent of the giant Navajo reservation as highly productive farmland. For the average reservation, about one acre of productive agricultural land per resident (Summers n.d.).

Despite these problems, there are notable exceptions. For example, the Passamquoddy tribe in Maine used funds from land claims settlements to acquire and develop a high-quality blueberry farm that supplies gourmet markets, premium hotels, and Ben and Jerry's Ice Cream. The Ak Chin reservation south of Phoenix has a large and profitable agribusiness with over 10,000 acres of cotton and alfalfa in production (White 1990).

Although agricultural production is not widespread on many reservations, timber production is more common. Timber is often cut and processed outside the reservation, but a growing number of reservations have built mills to produce finished lumber. One of the oldest of these mills was established on the Menominee reservation in Wisconsin.

Historically, the BIA has had primary responsibility for overseeing tribal forests and for the harvesting and sale of tribal timber. In 1989, 237 federal reservations possessed nearly 16 million acres of forestland with potentially harvestable timber. Perhaps more significant is that in 1989, 149 reservations held about 6 million acres in commercial forests, not an inconsequential resource as its total harvested value is estimated at $158 million.

In the last twenty years, the BIA has been subject to numerous complaints as well as congressional investigations regarding its management of tribal forests. These complaints alleged fraud and mismanagement, particularly "sweetheart" deals between BIA employees and lumber companies, improper accounting, and incompetent resource management (Richardson and Farrell 1983). As a result, tribes are now considerably more involved in the management of their timber; and some tribes have instituted specialized forestry programs.

Water resources are also very important to reservations. The *Winters* doctrine, stemming from a 1908 Supreme Court decision over tribal water rights, guarantees that tribes have prior claims on water destined for their reservation. In the arid western United States, this gives reservations a powerful claim on a scarce and vital resource. This is perhaps most important for reservation development, especially in water-intensive projects such as agriculture. However, developing water for lease or sale off the reservation is tangled in the complex legal web of water rights, and has not been extensively pursued. Nonetheless, it is possible for tribes with extensive water rights, such as the Navajo, to lease their water to the arid cities of the Southwest, just as large growers have found it more profitable to lease their water than to use it for farming. Water also can be used to produce revenue in other ways; for instance, the Salish-Kootenai in Montana sell hydroelectric power from a dam on their reservation. In addition to water, the minerals available on reservation lands run the gamut from gravel to zinc and copper to energy resources such as uranium, coal, petroleum, and natural gas. Needless to say, the immense potential wealth associated with energy resources has attracted the most attention. By some estimates, 40 percent of all uranium and 30 percent of the strippable coal west of the Mississippi is located on tribal lands (Jorgensen et al. 1978).

Despite the enormous real and potential value of these resources, those tribes possessing coal and petroleum are not significantly wealthier than other tribes (Snipp 1988). Some tribes, such as the northern Cheyenne in Montana, view mining as a violation of their sacred relationship with the land. This traditional belief is frequently a source of conflict among those tribal members who adhere to traditional beliefs and tribal members who wish to develop the resources. A more important reason is that for many years, the BIA failed to exercise proper oversight in the process of making leases. Bureau of Indian Affairs oversights caused millions of tons of coal to be

sold at prices far below market value (Richardson and Farrell 1985; Snipp 1988).

Congressional inquiries and complaints by tribal leaders resulted in revamping BIA leasing procedures. The tribes also became more proactive in the negotiation of lease agreements, as with the creation of the Council of Energy Resource Tribes (CERT), which was formed in 1975 for the purpose of increasing tribal involvement in lease negotiations and for providing technical assistance to aide tribes in negotiations (Ambler 1990). Since then, many old leases have been renegotiated and tribal involvement has considerably improved the prices received for energy resources. However, more recently, a sluggish world market for coal and petroleum has dampened the earlier enthusiasm for exploiting these resources. The tribes have not had much direct involvement in mining the resources on their reservations. The capital required for mining operations vastly exceeds tribal resources. However, some tribes such as the Crow have investigated limited partnerships with energy companies, as well as imposing severance taxes on extracted resources.

Human Resources

Across the nation, education is a resource that is sorely lacking among American Indians. Indians lag far behind Whites by any measure of educational attainment. American Indian youth drop out at alarmingly high rates. In 1990, about 18 percent of American Indians ages sixteen to nineteen were not in school and had not completed high school, compared with 14 percent of Blacks and 10 percent of Whites. Surprisingly, among adults twenty-five years and older, American Indians are *more* likely to have graduated from high school than African Americans. This may be due to American Indians eventually completing their schooling with a GED instead of a regular high school diploma.

American Indians also seldom complete four or more years of college. In 1990, only about 9 percent of American Indians had completed four or more years of college, compared to 11 percent of African Americans and 22 percent of Whites. High drop-out rates and disrupted school experiences are undoubtedly major contributors to this problem. Moreover, American Indians are concentrated in two-year colleges aimed at vocational education. In 1984, for example, about 55 percent of American Indians in college were enrolled in two-year programs (Center for Education Statistics 1987, Table 2).

Although these statistics are for the nation as a whole, they are certainly indicative of reservation conditions. Education, training, and work experience are the best indicators of the human capital reserves belonging to reservations. The low levels of educational attainment bespeak the limited human resources for reservation development. The shortage of job opportunities in reservation communities further exacerbates the shortage of human capital as the best educated, most able-bodied workers have to leave the reservation for employment elsewhere.

Many tribes have decided to address this problem by establishing tribal colleges. The first tribal college was established in 1968 by the Navajo tribe. During the next twenty years, another twenty-three tribal colleges were established on reservations across the western United States. Most of them are small, two-year community college programs. A few such as Sinte Galeska College in South Dakota have a limited number of four-year programs. Almost all of these institutions depend heavily on federal funding and struggle to maintain the facilities and personnel adequate to meet accreditation standards. Because the students attending these schools have few economic resources, revenue from tuition or property taxes hardly exists.

Although relatively new, tribal colleges have the potential to play a key role for human resource development on reservations. The students attending these colleges are typically older and they frequently have very poor academic preparation; they are returning to upgrade their basic skills, obtain vocational training, or acquire a GED. The students who typically attend tribal colleges are not being diverted from educational opportunities elsewhere. They are students who would not be attending college under most circumstances, except that a tribal college gives them an opportunity to do so (Carnegie Foundation 1989). In 1989, an estimated 4,400 full-time students were enrolled at the twenty-four tribal colleges (Carnegie Foundation 1989). The advent of tribal colleges is a development of potentially profound importance. These institutions are training persons who in the past would have been labeled "hard-core unemployed." Their location also means they are able to coordinate program curriculum with

projects designed to meet special needs on specific reservations.

Tribal Development

The presence of economic poverty amid the abundance of resources found on some reservations is a long-standing paradox. This can be partly explained as a consequence of misguided policies such as allotment and bureaucratic mismanagement by the BIA. However, a complete answer is much more complex. More importantly, many tribes are actively trying to improve economic conditions in their communities in a variety of ways; some are based on conventional development strategies, others are unconventional in their approach.

Conventional Development Strategies

Since the 1950s, and even earlier in some areas, federal officials and tribal leaders adopted more or less "textbook" models for economic development. Like other rural communities, tribes have tried to attract industry by emphasizing a low-wage workforce, non-existent taxes, or by building infrastructure such as roads or industrial parks. Unlike other rural communities, they also have tried to start up their own businesses in construction, light manufacturing, agriculture, and a hodge-podge of other activities. There also have been a variety of efforts to encourage entrepreneurship. Many of these efforts reached a peak under the Economic Development Administration (EDA) and the Small Business Administration (SBA). Federal cutbacks in the early 1980s significantly reduced these programs, although some continue to offer assistance. Currently, most conventional development strategies focus on tribal operations or individual entrepreneurs.

The activities of tribes in promoting economic development are divided between attracting industries from outside the reservation and their own business start-ups. Attracting industries owned by non-Indians is a strategy that has been a mixed success. Often the goal of attracting outside industries is job creation, making labor-intensive and often low-wage industries most appealing. Light manufacturing, such as electronics assembly plants, has been located on a number of reservations. However, this strategy is problematic for at least two reasons. One is that the jobs created by such firms are typically low-skill, low-wage jobs with few benefits. Management jobs are seldom filled by Indians. While these firms have the virtue of providing employment, they do not often yield a significantly improved standard of living. Making the welfare poor into the working poor can be considered an improvement, but it is a small one.

A second problem with such industries is that it pits reservation workers against workers in developing nations. Like other rural communities, a number of reservations have watched local industries join the exodus overseas to obtain cheap labor. However, some tribes have been able to resist this trend. According to Chief Philip Martin of the Mississippi Choctaw, his tribe has been successful with light manufacturing because they can offer superior workmanship to compensate for lower overseas labor costs.

Tribally owned businesses face other dilemmas. One crucial problem is deciding whether a tribal business will operate to maximize employment, or whether it will seek to become an efficient, highly competitive enterprise. In theory, there is no necessary conflict between these goals. But in practice, such a conflict often exists when decisions about lay-offs or dismissing incompetent employees must be made.

The decision to choose between jobs or profits is often made more complicated by another problem—tribal politics. Like other communities, there are often disagreements among American Indians about the best course of action for tribal government. In connection with economic development, disputes may arise over the types of development, who is involved, and the disposition of jobs and revenues. Decisions that are politically astute may be disastrous for tribal enterprises. Some tribes have attempted to deal with this conflict by establishing business committees separate from the tribal government. This is intended to distance business decisions from tribal politics, but too often it merely shifts political disputes from the arena of tribal government to the business committee. Cornell and Kalt (1990) argue that political development is a necessary antecedent to economic development. For tribal governments to successfully undertake complex economic development projects, they must be able to exercise a great deal of administrative expertise as well as have the political stability to carry out long-range plans. Cor-

nell and Kalt (1990) echo others (for example, Vinje 1985) when they note that economic development projects must be consistent with tribal culture and lifestyles, and especially with the political culture of the tribe. Tribes accustomed to diffuse, highly decentralized decision-making processes will not accommodate economic development projects organized with a highly centralized management plan.

Unconventional Development Strategies

Unconventional development strategies are so named because they are development strategies based on the special legal and political status of American Indians—hence they are not options for economic development by non-Indians. This approach to economic development has become more common in the late 1970s and 1980s, possibly because there has been less federal support for conventional development projects. This approach has also been called the "legal road to economic development" (Olson 1988), and central to this strategy is the doctrine of tribal sovereignty.

Tribal sovereignty is a concept implicit in much of the preceding discussion because it is a central organizing principle in federal-Indian relations. The legal theory behind tribal sovereignty dates back to the founding of the United States and early decisions of the Supreme Court (Barsh and Henderson 1980). Briefly, tribal sovereignty means that by treaty and other agreements tribes have reserved certain legal rights of self-government. This provides tribal governments with a measure of self-rule subject to the authority of the federal government and exempt from most state and all local authority. With only a few exceptions, tribal governments have most of the same powers as state governments.

The so-called "legal road to economic development" exploits the powers of tribal sovereignty, treaty rights, and other legal agreements for the sake of developing a market niche for tribal enterprises. The ability to use tribal sovereignty for creating a market niche is crucial for having a successful enterprise. And indeed, there are a number of successful operations stemming from treaty rights, land claim settlements, and the use of tribal sovereignty to create a market niche.

Two of the best known developments stemming from treaty rights settlements are located in Maine and in the Puget Sound region of Washington State.

In 1975, the Passamaquoddy and Penobscot tribes of Maine won a major court victory and a ruling that these tribes might be eligible to claim up to two-thirds of the state. After protracted negotiations, a federal task force concluded negotiations with these tribes with a settlement of $82 million. With this settlement, the tribes purchased lands, established investment portfolios, and initiated economic development projects. The Passamaquoddy invested a full third of their settlement in economic development projects such as a construction firm, cement factory, and blueberry farm. These projects were meant to produce income for tribal services and provide jobs for tribal members. Some of these projects have been more successful than others, but they have been sufficiently capitalized and well managed that they are counted as successes by the tribe (White 1990).

The state of Washington in the 1950s and 1960s was the site of protracted struggles over Indian fishing rights. These struggles culminated in the court case of *U.S. v. Washington*, also known as the Boldt Decision. In 1974, federal judge George Boldt rendered a verdict that treaties signed with the Puget Sound tribes entitled them to 50 percent of the salmon harvested in this region each year in perpetuity. This was a major victory for these tribes, who shortly afterwards initiated economic development projects based on fishing. The Lummi and the Quinault in particular have vertically integrated aquaculture programs with fish hatcheries and fish processing plants. These tribes play key roles in Puget Sound conservation efforts and have a major stake in improving fisheries production. Furthermore, these activities are also a major source of tribal employment and revenue.

Because tribal sovereignty gives tribal governments the right to legislate for themselves, another direction in the legal road to economic development has been for tribes to make available goods and services restricted by local and state laws. In the mid-1970s, retail tobacco shops with products exempt from local and state taxes began appearing on Indian land across the United States. These are typically small shops with only a few employees and are operated either by individuals or tribal governments. Little systematic data exists about these shops, but anecdotally they are usually profitable operations, although the revenue they produce is not spectacular. They are nonetheless controversial because state and local governments resent the loss of tax revenues and local merchants complain about unfair competition.

However, compared to gambling, the controversy and revenues generated by tobacco shops are negligible. In 1978, the Florida Seminole won in court the right to operate a high-stakes bingo hall on their reservation. This enterprise was enormously successful and produced revenues in the millions of dollars. The success of the Seminole was quickly noticed by other tribes, and within a few years reservations across the country were engaged in high-stakes bingo. By the mid-1980s, many tribes were testing the legal waters by offering other types of gambling explicitly outlawed outside their reservations. After a series of legal tangles and complaints by state governments and by tribal officials, Congress enacted the Indian Gaming Act of 1988.

The Indian Gaming Act set the ground rules for tribes desiring to have gambling on their reservation and created the Indian Gaming Commission to oversee reservation gambling. The law establishes different classes of gaming and permits reservation gambling if the games offered are not fundamentally different from the gambling permitted outside the reservation in activities such as state lotteries. This law represents a mixed victory for the tribes because while they do not have the unconditional right to offer gambling, it does institutionalize gambling and protects it from state and local interference.

The future of reservation gambling is unclear. It has been a huge success for some tribes, especially those close to large urban areas. On reservations in remote places, it has been less successful and has attracted mainly a reservation clientele instead of wealthier non-Indian players. Reservation gambling also has become more competitive between reservations and from outside the reservation as more states liberalize their restrictions on gambling. These developments may eventually shrink tribal revenues from gaming, but in the short run they have been an enormous cash resource for many tribes.

A handful of tribes have also passed legislation skirting state and local zoning and environmental protection laws. However, these measures have proved as controversial on the reservation as they have been outside the reservation. A proposal to place a nuclear waste facility on the Pine Ridge reservation produced a heated conflict, as did a proposal for toxic waste that was summarily rejected by the Mississippi Choctaw. The Mescalero Apache are currently considering a nuclear waste storage facility, but this is also becoming controversial on and off the reservation.

Urban American Indians

Tribal resources and tribal development efforts, despite their undeniable importance for reservations, are largely outside the interests of urban American Indians. This is because few, if any, urban Indians live on tribal lands, receive services from a tribal government, or participate in tribal politics. In 1990, about 51 percent of all American Indians were city dwellers. As a result, urban American Indians are a very important part of the American Indian population.

The Urbanization of American Indians

In small numbers, American Indians have lived among Euro-Americans in towns since the founding of colonial settlements (Szasz 1988). However, beginning in the late nineteenth century, American Indians slowly began to become more urbanized, this process gradually accelerated throughout the twentieth century. Several factors hastened their urbanization.

Intermarriage with non-Indians—historically at relatively high rates—certainly influenced their residential choices. Some of these mixed-race families stayed near tribal communities, but in other instances Indian spouses, especially wives may have followed their partners to cities. The ubiquitous racism and antimiscegenation laws in place in many areas of the country at the turn of the century probably limited such movement. However, the children of such marriages, being less identifiable as Indian, probably found it easier to "pass" into Euro-American society. By the early 1920s, there were sufficient numbers of urban Indians, including many of mixed ancestry, that pan-Indian interest organizations began to appear, two such organizations were the Teepee Order and the Indian Association of America (Hertzberg 1971).

The federal government's actions also have hastened the urbanization of American Indians. The court decision of *Standing Bear v. Crook* in 1879 established that federal authorities did not have the ab-

solute right to forcibly confine Indians to reservations. American Indians were thus free to leave their reservations, and reservations no longer could serve to isolate the Indian population from the American mainstream.

Efforts to force the assimilation of American Indians into Euro-American society also had a predictable impact on American Indian urbanization. The allotment acts and subsequent land losses meant that to survive some Indians had to take up wage labor. The need to find wage labor probably encouraged the drift toward cities. Moreover, the Indian boarding schools also encouraged "outings" in which Indian students were placed in middle-class Euro-American homes. This exposure also may have led to some migration toward cities.

However, a major impetus for urban migration was that of the first and second World Wars. As the United States entered World War I, American Indians became became part of the war effort, from producing food to buying war bonds. More significantly, approximately 8,000 to 10,000 American Indians served in the military (Prucha 1984, p. 771). Again, there is little documentation, but participation in the war gave these men exposure to mainstream urban America as well as the skills to cope with cultural expectations of white society—prerequisites for urban migration.

Americans Indians still remained concentrated in rural areas after World War I. In 1930, when over half of the United States population lived in urban locales, barely 10 percent of the Indian population lived in urban areas (Snipp 1989). However, World War II and events afterward were major forces affecting the urbanization of the American Indian population.

World War II involved an unprecedented number of American Indians in the military. Aprroximately 25,000 American Indians served in World War II, in combat as well as in technical support positions. For tribes with a strong warrior tradition, the rate of enlistment was high, and the volunteers enjoyed considerable status in their communities (Hagan 1979, p. 158). Nearly twice as many American Indians were involved in the war effort at home, working in defense plants and related activities (Hagan 1979, Prucha 1984). After the war, many did not return to their rural origins.

The experience of World War II exposed American Indians to urban American culture and to the industrial society outside their reservations; it provided many with new job skills and work experience, and it was an impetus for many to leave their communities while they fought or worked in wartime industries (Prucha 1984). Equally important, the GI Bill provided an unprecedented number of American Indians with the opportunity to attend college. Out of necessity, many of these graduates settled in cities to find employment. The GI Bill also trained a generation of Indian leaders, many of whom returned to their reservations or became active in urban Indian social issues (cf. Ambler 1990; Snipp 1989; Weibel-Orlando 1991).

Postwar federal policy also played a role in moving American Indians to cities. As part of its effort to abolish the reservation system, the federal government's relocation programs resettled thousands of American Indians in cities. In doing so, it established a vital link for patterns of chain migration between cities such as Los Angeles and western reservations. An entire generation of urban Indian migrants to Los Angeles can be directly or indirectly linked to the use of this city as a relocation center (Price 1968; Weibel-Orlando 1991).

Considerable disagreement exists about the costs and benefits of urbanization for American Indians. When it was first proposed, one of the principle justifications for urban relocation was that economic opportunities would be greater in urban areas (Fixico 1986). However, the economic benefits of urban relocation appear to be questionable. Some evidence suggests that urban relocation improved the economic circumstances of participants (Clinton et al. 1975, Sorkin 1978) while other studies are less sanguine (Gundlach and Roberts 1978; O'Brien 1990; Snipp and Sandefur 1988).

Coping with Urban Life

The large influx of American Indians to urban areas prompted numerous studies in the 1960s and early 1970s (Thornton et al. 1982) of how American Indians coped with the alien culture and lifestyle they found in cities. Cultural assimilation is a central theme in this literature; the stress of urban adaptation is another.

The experience of urban American Indians contradicted the assimilationist expectations. Unlike other urban immigrants, American Indians did not readily adopt Euro-American culture, rather tribal culture and ethnic identity persisted. One possible reason is that racial discrimination posed barriers that denied entry into the dominant culture (Ablon 1964, 1965, 1971; Chadwick and Stauss 1975; Chadwick and White 1973; Guillemin 1975; Roy 1962; Stauss and Chadwick 1979). American Indian resistance to assimilating Euro-American ways has caused social scientists to rethink their expectations of the anticipated dissolution of Indian culture toward the apparent persistence and vitality of cultural traditions in urban environments (Vogt 1957). One explanantion for this persistence of tribal culture in cities is that they are able to find in ordinary urban landscapes the opportunity for cultural expression (Guillemin 1975). For example, the danger and physical demands of high steel construction work provide Mohawk men with opportunities to exhibit the bravery and risk-taking traditionally expected of them (Blumenfeld 1965).

In cities, social interaction and cultural expression take place in bars, pow-wows, Indian Centers, and churches (Weibel-Orlando 1991). Powwows are tribal events transported directly from reservation traditions. On reservations, powwows are held in ceremonial locations, while in cities they are most often sponsored by informal groups or pan-Indian organizations and held in gyms or public auditoriums. Similarly, urban Indian churches resemble reservation missions (Weibel-Orlando 1991), and bars and Indian Centers provide urban analogs to the tribal headquarters and nearby bars of reservation life (Weibel-Orlando 1991).

Also important in the persistence of Indian ethnicity in cities has been the emergence of pan-Indianism—a supratribal ideology that unites the interests of American Indians by virtue of their common heritage, independent of the varying social and political agendas of particular tribes such as the Sioux or the Navajo (Thomas 1965). In cities, pan-Indianism is particularly important as a unifying force because urban Indian populations come from a multitude of tribes. Since 1950, the spread of pan-Indianism has been extremely important as an organizational basis for urban Indian social life (Cornell 1988). With a few exceptions, community events and social gatherings for urban Indians are typically pan-Indian affairs—all tribes are welcome. Furthermore, the ideology of pan-Indianism itself has been an important resource for mobilizing political action locally (Nagel 1986; Weibel-Orlando 1991) as well as for building support for Indian issues across the nation (Cornell 1988).

What form their economic development should take is just as thorny an issue for Native Americans on reservations. Here is the facade of a gambling casino under a sign featuring the Native American face usually found on a nickel. (Photo by C. Matthew Snipp)

Pan-Indianism, therefore, has been a very positive result of urbanization.

However, a very negative result is that the transition to urban life may be partly responsible for problem drinking among urban Indians. While alcohol abuse and related health problems are a leading killer of American Indians on reservations and cities alike (Snipp 1989), the move to urban areas is for many American Indians an intensely alienating and anomic experience (Graves 1971). It removes them from the tight bonds of reservation life—traditions, family, friends—and places them in alien, disorienting city environments.

This difficult transition promotes heavy drinking and related problems such as incidents with police (Ablon 1965; Dozier 1966; Ferguson 1968; Graves 1971). Drinking among urban Indians may also differ from that of reservation Indians because urban Indian bars are an important venue for affirming ethnic identity within the community (Ablon 1965; Weibel-Orlando 1991). Furthermore, in cities where substantial status differences exist within the Indian community, shared drinking in a recognized Indian bar may serve as a leveling mechanism for building ethnic solidarity. In fact, shared drinking experiences among urban Indians may be "a way for one successful in the larger society to demonstrate to fellow Indi-

ans that he is still 'Indian'" (Thornton et al. 1982, p. 43). Local Indian leaders report that drinking in the local Indian bar is important for staying in touch with their constituents (Weibel-Orlando 1991). Of course, excessive alcohol consumption leads to encounters with police and to arrest rates that are considerably higher than for Blacks or Whites. American Indians are also disproportionately represented in prison populations; American Indians receive harsher treatment in the judicial system and do not effectively utilize resources such as legal assistance (Hagan 1976).

To many American Indians, many elements of city life fundamentally contradict the ethics of tribal culture and lifestyles. Many, if not most, are able to reconcile these conflicts. But there are many American Indians who struggle unsuccessfully with these dilemmas; they extract a toll manifest in substance abuse, mental disorder, family conflict, and involvement with the criminal justice system—personal "costs" of urban life.

Conclusion

Though small in number, American Indians have an enduring place in American society. Growing numbers of American Indians occupy reservation and other trust lands, and equally important has been the revitalization of tribal governments. Tribal governments now have a larger role in reservation affairs than ever in the past. Another significant development has been the urbanization of American Indians. Since 1950, the proportion of American Indians in cities has grown rapidly. These American Indians have in common with reservation Indians many of the same problems and disadvantages, but they also face other challenges unique to city life.

The challenges facing tribal governments are daunting. American Indians are among the poorest groups in the nation. Reservation Indians have substantial needs for improved housing, adequate health care, educational opportunities, and employment, as well as developing and maintaining reservation infrastructure. In the face of declining federal assistance, tribal governments are assuming an ever-larger burden. On a handful of reservations, tribal governments have assumed completely the tasks once performed by the BIA.

As tribes have taken greater responsibility for their communities, they also have struggled with the problems of raising revenues and providing economic opportunities for their people. Reservation land bases provide many reservations with resources for development. However, these resources are not always abundant, much less unlimited, and they have not always been well managed. It will be yet another challenge for tribes to explore ways of efficiently managing their existing resources. Legal challenges also face tribes seeking to exploit unconventional resources such as gambling revenues. Their success depends on many complicated legal and political contingencies.

Urban American Indians have few of the resources found on reservations, and they face other difficult problems. Preserving their culture and identity is an especially pressing concern. However, urban Indians have successfully adapted to city environments in ways that preserve valued customs and activities—powwows, for example, are an important event in all cities where there is a large Indian community. In addition, pan-Indianism has helped urban Indians set aside tribal differences and forge alliances for the betterment of urban Indian communities.

These alliances are essential, because unlike reservation Indians, urban American Indians do not have their own form of self-government. Tribal governments do not have jurisdiction over urban Indians. For this reason, urban Indians must depend on other strategies for ensuring that the needs of their community are met, especially for those new to city life. Coping with the transition to urban life poses a multitude of difficult challenges for many American Indians. Some succumb to these problems, especially the hardships of unemployment, economic deprivation, and related maladies such as substance abuse, crime, and violence. But most successfully overcome these difficulties, often with help from other members of the urban Indian community.

Perhaps the greatest strength of American Indians has been their ability to find creative ways for dealing with adversity, whether in cities or on reservations. In the past, this quality enabled them to survive centuries of oppression and persecution. Today this is reflected in the practice of cultural traditions that Indian people are proud to embrace. The resilience of American Indians is an abiding quality that will no doubt ensure that they will remain part of the ethnic mosaic of American society throughout the twenty-first century and beyond.

30

The Health of the African American Population

DAVID R. WILLIAMS

Health is an important and desirable personal and social resource. It determines the quantity and quality of life, and has an important effect on an individual's ability to capitalize on the opportunities available in society. The health of the American population is distributed unevenly across race. In 1990, the life expectancy at birth in the United States was 75 years. However, it varied from 69 years for African Americans to 76 years for White Americans (National Center for Health Statistics 1993). Moreover, life expectancy for African Americans has worsened in recent years. For every year between 1985 and 1989 the life expectancy for both African American men and women declined from the 1984 level (National Center for Health Statistics 1993). These racial disparities in health are not new. In 1900, life expectancy at birth for White Americans was 48 years compared to 33 years for non-Whites, who at that time were mainly Blacks. Thus, over the course of this century the health status of both Blacks and Whites has improved, but a gap in health between the two groups has persisted.

Table 1 indicates how pervasive racial disparities in health are. It presents the death rates for Blacks and Whites, men and women, and the Black-White ratios for the fifteen leading causes of death in the United States, where a ratio greater than 1.0 indicates that the death rate for Blacks is higher than for Whites. With the exception of pulmonary disease and suicide, deaths of all types are higher for Blacks than for Whites. The differences are small for some conditions, such as atherosclerosis, but substantial for others, such as homicide and AIDS.

Types of Explanations for Racial Disparities in Health

Traditionally explanations for differences in health between the races have focused on biological differences between the two groups. In the nineteenth century, medical research attempted to document that Blacks were biologically inferior to Whites and, therefore, were more susceptible to a host of illnesses (Krieger 1987). Most medical research is no longer so blatantly racist, but it still views most racial

Preparation of this paper was supported in part by the National Institute on Aging, grant no. AG-07904.

Table 1 Death Rates for the Fifteen Leading Causes of Death Per 100,000 Population by Race and Sex in the United States, 1988

	Men			Women		
Cause of Death	Black	White	Black/White Ratio	Black	White	Black/White Ratio
1. Heart disease	286.2	220.5	1.30	181.1	114.2	1.59
2. Cancer	227.0	157.6	1.44	131.2	110.1	1.19
3. Cerebrovascular disease (stroke)	57.8	30.0	1.43	46.6	25.5	1.83
4. Accidents	69.0	49.9	1.38	22.2	18.8	1.18
5. Pulmonary disease	26.0	27.8	.94	10.0	14.5	.69
6. Flu and pneumonia	28.0	18.0	1.56	13.4	10.7	1.25
7. Diabetes	19.8	9.6	2.06	22.1	8.4	2.63
8. Suicide	11.8	19.8	.60	2.4	5.1	.47
9. Liver disease and cirrhosis	20.7	12.1	1.71	9.3	5.0	1.86
10. Kidney disease	13.3	5.2	2.56	10.3	3.3	3.12
11. Atherosclerosis	4.6	4.0	1.15	3.3	3.0	1.10
12. Homicide and legal intervention	58.2	7.7	7.56	12.7	2.8	4.54
13. Septicemia	12.2	4.5	2.71	9.0	3.5	2.57
14. Perinatal conditions	18.9	6.2	3.05	15.3	4.8	3.19
15. AIDS	29.3	10.3	2.84	6.3	.7	9.00

Note: These death rates are age adjusted for differences in the age structure of the Black and White populations.
Source: National Center for Health Statistics, 1991. *Vital Statistics of the United States, 1988*, vol. II—Mortality, part A. Table 1-6, p. 10; Table 1-8, p. 12; and Table 1-37, p. 365.

variations in disease as due to underlying differences in biology. However, attributing racial disparities in health to innate physical differences between the races maintains the historic implication of superiority and inferiority. More recently, a second type of explanation focusing on lifestyle or health behavior differences between African Americans and White Americans has gained considerable currency. Although they appear very different, the biological explanation and the lifestyle explanation share the common view that racial differences in health are due to factors that reside inside of the individual, and can therefore be successfully addressed by medical interventions that target individuals. In contrast, the view presented here sees racial variations in health as primarily due to differences in the social location of groups. Thus, effective solutions to reduce these social disparities depend on broad-based interventions in the social conditions that shape the daily reality of people's lives. Before reviewing the evidence for the social embeddedness of health, the limitations of the dominant types of explanations will be considered.

The Limits of Biological Explanations

At the heart of the biological explanation for racial disparities in health is the conceptualization of what race is. The biological approach views racial taxonomies (Black, White, Asian, American Indian, and so on) as meaningful classifications of real genetic differences between human population groups. This approach assumes that race is a valid biological category, and that the same genes that determine what race an individual belongs to also determine the number and types of health problems that individual will have (Krieger and Basset 1986).

This view is deeply entrenched in our society and reflected in the approach of much medical research, yet it is seriously flawed. First, the concept of race and its attendant racist beliefs about the superiority of some groups and the inferiority of others developed in the late eighteenth century, long before modern scientific theories of genetics existed. The classification of the human population into separate "races" developed in the historical context of slavery and imperial colonialism, and served not only to classify human variation, but also to provide a rationale for the exploitation of groups that were regarded as inferior (Montagu 1965). Second, racial classification schemes are arbitrary and reflect changing social and political conditions across societies and in the same society over time. For example, a review of the racial classification schemes utilized by the U.S. Bureau of the Census between 1850 and 1980 indicated that no single set of racial categories had been used in more than two censuses during this period, with most used only once (Martin, DeMaio, and Campanelli 1990). Mexicans, for example, were classified as a separate race in the 1930 census, reclassified as White in the 1940 census, and designated as the "Spanish surnamed" population in the southwestern states in the 1960 census.

Likewise, racial classification is arbitrary and inconsistent across and within societies (F. Davis 1991). The racial categories used in the United States are different from that of some other countries. In the United States, Blacks were defined according to ancestry by a legal standard that used the "one drop rule," which defined an individual as Black if any known ancestor was African. In contrast, racial definitions in the Caribbean and Latin America, where slavery and mixed marriages and unions also existed, reflected considerations of ancestry, social status, and phenotype. The resulting racial categories were more fluid and recognized gradations between Black and White. Many fair-skinned persons of mixed ancestry (such as Lena Horne, W. E. B. Dubois, A. Phillip Randolph, and Booker T. Washington) who were classified as "Black" in the United States would have received a different racial label elsewhere (F. Davis 1991).

Our current views of race have been shown to be without scientific basis. There is more genetic variation within races than between races, and our current racial categories do not reflect biological distinctiveness. Regardless of geographic origin or race, all human beings are identical for 75 percent of known genetic factors (Lewontin 1982). Moreover, some 95 percent of human genetic variation exists within racial groups, with relatively small and isolated populations, such as Eskimos and Australian aborigines, contributing most of the between-group variation (Lewontin 1974). Stated another way, if an epidemic were to wipe out the entire population of the world except for Black persons in Africa, the world's population would then be entirely Black, but would still contain some 95 percent of all known genetic variation. Thus, our keen awareness of what racial group we belong to says much more about our society than about our biological makeup (Krieger and Basset 1986).

Proponents of biological explanations like to point to diseases that are clearly linked to the physical characteristics used to define race or that appear to be more frequent in some racial groups than others. An example of the former is the kind of skin cancer that more frequently affects light-skinned people; an example of the latter is sickle cell anemia, which is more common in Blacks than in Whites. However, sickle cell anemia is more prevalent not only in African Americans, but also in all peoples that lived in regions (the Mediterranean, Africa, and Asia) where malaria was endemic (Polednak 1989). Sickle cell anemia appears to be a protective, genetic adaptation to malaria produced by the interaction of biology with environmental conditions. Sickle cell anemia also illustrates that genetics is not static, but evolves over time as human groups adapt to environmental conditions. Thus, even when biological differences are found between two racial groups that live under different environmental conditions, these differences may not be due to innate physical differences, but to acquired ones that reflect the consequences of different living conditions over time.

Moreover, diseases that have a clear genetic component account for only a minuscule part of racial differences in health. Sickle cell anemia is the only potentially fatal disease that is linked to being Black, and Cooper and David (1986) indicate that only three-tenths of 1 percent of the total number of excess deaths among Blacks is clearly related to sickle cell anemia. Thus, differences in biology are not the primary cause of racial variations in health and disease.

The Lifestyle Explanation

Health practices—better nutrition and eating habits; diminished tobacco, alcohol, and drug abuse; more exercise; and the use of conflict resolution strategies—play an important role in maintaining and improving health. It is estimated that almost half of all deaths in the United States are linked to unhealthy behavior or lifestyle (U.S. Department of Health, Education and Welfare 1979). In comparison, 20 percent is due to environmental factors such as exposure to toxic substances, 20 percent to genetics, and 10 percent to inadequate medical care. Research also indicates that the health of the population can improve more through increases in healthy behavior than if an overnight cure were found for heart disease or cancer (Olshansky 1985). If changes were made in risky health behaviors, the death rates for Black Americans would be substantially reduced (U.S. Department of Health and Human Services 1985).

Proponents of the lifestyle explanation tend to regard health practices as psychological in nature. Typically health behaviors are assessed at the individual level, and only individual causes and consequences are attended to. John Knowles, a former president of the Rockefeller Foundation, illustrates this approach when he stated (1977, p. 58) that we are "born healthy and made sick as a result of personal misbehavior." He indicated that individuals are ultimately responsible for their own health and illness and are therefore under a "moral obligation" to preserve their health, because one person's poorly chosen health practices can become another's "shackle in taxes and insurance premiums." This approach also has a tendency to "blame the victims" for the particular health problems they face.

The Primacy of Social Structure

In contrast to both the biological or lifestyle explanation is a third approach that views health as a product of the socioeconomic, political, and cultural situations of social groups (Williams, Lavizzo-Mourey, and Warren 1994). Racial differences reflect distinctive histories and specific conditions of life that affect the risk of disease. An adequate understanding of racial differences in health must consider these larger social structures and processes and identify the ways in which they affect the health of individuals and groups. Health behaviors and the risk of disease must be understood in the context of the living and working conditions in which they emerge. This approach is illustrated by considering the relationship between socioeconomic status and disease, the effects of racism on health, and the consequences that the massive migration of African Americans from the rural South to the urban North had on their health.

Socioeconomic Status and Health

Socioeconomic status, whether measured by poverty, earnings, wealth, education, or occupational status, is one of the strongest known determinants of variations in health. During this century, despite general improvements in health, social and economic change, advances in health care technology, and improvements in the delivery of medical services, the socioeconomic status differential has not much narrowed over time. For example, since the implementation of the National Health Service in Great Britain in the 1940s, the health of all groups in society has improved, but the higher socioeconomic status groups have experienced greater improvement than their poorer peers such that the gap in health status has actually widened (Marmot and McDowall 1986; Wilkinson 1986). These socioeconomic differences in health are fairly universal and have been documented in western European countries, Australia, New Zealand, Japan, Canada, and throughout the Third World (Haan and Kaplan 1986; Marmot, Kogevinas, and Elston 1987; Williams, Wilson, and Chung 1992). Trends in the social distribution of certain diseases over time provide a further illustration of the impact of socioeconomic status on adverse changes in health. In the history of many diseases, even when an illness was initially more common among the higher classes, over time it became more prevalent among the less affluent (D. Williams 1990). For example, in the 1950s the rates for heart disease were higher among the rich. But as the social conditions that led to greater risk of heart disease were identified and knowledge of them became widespread, persons of higher social status changed their behavior (quit smoking, reduced dietary fat, exercised regularly) more rapidly than their less prosperous peers. Thus, the relationship between social status and heart disease changed from a positive one to an inverse one. Similarly, most of the initial AIDS patients in the United States were White, middle-class men, but most new cases of AIDS are concentrated among impoverished Blacks and Hispanics. Persons of higher socioeconomic status are more aware of health risks and more likely to initiate behavioral change because they command more resources to facilitate such change.

African Americans have lower levels of all socioeconomic status indicators than do White Americans. These differences are large and persist in spite

Table 2 Average Annual Percent of Persons Reporting to be in Fair or Poor Health by Household Income, Respondent Education, and Place of Residence for Blacks and Whites: United States, 1985–1987

Education	Household Income Less than $20,000		Household Income More than $20,000	
	Whites	Blacks	Whites	Blacks
Less than 12 years	33.1	38.8	16.1	20.5
12 years	15.2	17.9	6.8	9.6
More than 12 years	9.2	13.2	3.7	5.9
Total	16.6	19.0	5.1	7.6
Place of residence				
Urban	15.8	18.1	5.0	7.4
Central city	16.3	18.7	5.5	8.5
Not central city	15.5	16.3	4.8	5.7
Rural/small town	18.3	22.7	5.5	9.1
Number (in thousands)	61,029	15,089	113,919	9,079

Source: Peter W. Ries, *Health of Black and White Americans, 1985–87*, Vital Health Statistics 10(171). (Hyattville, Md.: National Center for Health Statistics, 1990), Table A, p. 5; Table 16, p. 55.

of the progress that Blacks have made in recent years (Jaynes and Williams 1989). For example, one-third of all Blacks are poor, compared to only one-tenth of all Whites. Thus, the higher rates of ill health and mortality among African Americans must be understood as part of a universal phenomenon in which poverty is associated with higher rates of disease and death (D. Williams 1990). At the same time, it must be remembered that race is an imperfect indicator of poverty. Although Blacks are disproportionately poor, two-thirds of all Blacks are not poor, and two-thirds of all poor persons in the United States are White.

The top panel of Table 2 presents the relationship between two indicators of socioeconomic status (household income and years of formal education) and self-assessed health for Blacks and Whites in the National Health Interview Survey between 1985 and 1987. The measure of health included here, the respondents' subjective report of their health, is one of the most robust indicators of health and is strongly related to objective measures of health. For example, persons who assess their health as "good" or "excellent" live longer than those who report it as "fair" or "poor." We can see that there are large disparities in self-assessed health by household income and level of education. For persons living in households with a total income of less than $20,000, 16.6 percent of Whites reported their health to be fair or poor, com-

pared to 19 percent of Blacks, while in households with incomes greater than $20,000, the comparable numbers for Blacks and Whites are 5.1 and 7.6 percent, respectively. Thus, the differences by income and education are much larger than the differences by race. At the same time, it is instructive that even when education and income level are held constant, Blacks still have higher levels of ill-health than Whites. This suggests that although most of the racial differences in health are accounted for by socioeconomic status, race also has an effect on health that is independent of its relationship with socioeconomic status. Racial discrimination is one critical factor that might operate independently of socioeconomic status, and/or interactively with socioeconomic status, to maintain an association between race and health.

Racism

As noted earlier, racist ideologies provided the rationale for differential treatment of social groups regarded as inferior. This treatment has also often been institutionalized as the consequence of organizational policies and procedures. Discrimination affects a broad range of social outcomes for individuals belonging to different racial groups, with those affected often unaware of it. Race, for example, transforms measures of socioeconomic status such that socioeconomic status indicators are not truly equivalent across race. On average, there are racial differences in the quality of education. Black students are disproportionately allocated or "tracked" into low-ability and noncollege preparatory groups that are characterized by a less demanding curriculum and lower teacher expectations (Jaynes and Williams 1989). Thus, the skills and knowledge that a student has upon completion of high school differs for high school graduates of different races. National data also consistently reveal that Whites have higher income returns on education than Blacks (Jaynes and Williams 1989). For both Blacks and Whites, the more years of formal education one has, the higher the income. However, when Blacks and Whites are compared at the same educational level, Whites earn more than Blacks. For example, the average annual income of a Black male college graduate is 74 percent that of his White counterpart (Jaynes and Williams 1989). Similarly, even after controlling for job experience and education, Blacks who are employed are more likely than their White peers to be exposed to occupational hazards and carcinogens (Robinson 1984).

In addition, the purchasing power of a particular level of income varies by race. Studies have documented that Blacks have higher costs of food, rent, automobiles, and auto insurance than Whites (King and Williams, in press). The prices of groceries at neighborhood stores, the cost per square foot of rental housing, and the premiums for auto insurance are all higher for residents of inner cities (where Blacks are disproportionately concentrated) than for those who live in more suburban locations. Moreover, a carefully executed study of auto dealerships found that compared to the purchase price of a new car offered to White men, White women had to pay a 40 percent higher mark up, Black men a 200 percent higher mark up, and Black women a 300 percent higher mark-up (Ayres 1991).

Institutionalized racism also affects the quality of services received across a broad range of institutions, including health care. Louis Sullivan, former secretary of the Department of Health and Human Services, stated that "There is clear, demonstrable, undeniable evidence of discrimination and racism in our health care system" (Sullivan 1991, p. 2674). A recent review of the available studies by the Council on Ethical and Judicial Affairs (1990) of the Ameri-

can Medical Association documents extensive racial disparities in the receipt of medical care. Compared to African Americans, White Americans are more likely to receive the following treatments: coronary angiography, bypass surgery, angioplasty, chemodialysis, intensive care for pneumonia, and kidney transplants. These differences remain, even after adjustments are made for the severity of the illness, the income of the person, and their insurance status. The data on kidney transplants are especially compelling: Whites are more likely to be placed on waiting lists for kidney transplants (Council on Ethical and Judicial Affairs 1990) and to wait only half as long as Blacks (Sullivan 1991).

Recently attention has also been given to environmental racism—that is, the disproportionate exposure of some racial groups to environmental toxic exposures. As it turns out, race is the strongest predictor of the location of hazardous waste facilities in the United States, even after adjustment for social class (Commission for Racial Justice 1987). Other evidence indicates that inner-city residents are five times more likely to be exposed to air and water pollution than their suburban peers, and that predominantly Black, poor, rural areas are also disproportionately exposed to health-threatening toxic materials from nearby industrial plants (Bullard and Wright 1987). Lead poisoning is also a major problem that disproportionately affects the health of minority children in the United States (Reed 1992). High levels of lead exist in the older buildings of impoverished neighborhoods where many African Americans are forced to live.

A handful of studies have examined how the experience of racism or racial discrimination affects health. In a study of Black and White women, Krieger (1990) found that Black women who passively experienced racial discrimination were four times as likely to have high blood pressure as those who talked with others about it or who took other action in response to the unfair treatment. Yet Black women were six times more likely than Whites to respond passively to unfair treatment, suggesting that they, probably accurately, perceived themselves as having little control over these encounters. Recent analyses of data from the National Study of Black Americans also indicate that having experienced racial discrimination recently, as well as having experienced it in the work setting, predicted lower levels of health (Williams and Chung, in press). Prior research suggests that at least two responses to the experience of racism may be especially predictive of adverse health consequences. An internalized denial of racial bias appears to have negative health consequences. Krieger (1990) found that Black women who reported that they had experienced no incident of racial or gender discrimination were two to three times as likely to have high blood pressure as those who said they had experienced unfair treatment. A second potentially health-damaging response is the oppressed minority's acceptance of the dominant society's racist ideology. Taylor and Jackson (1990) found in a study of Black women that internalized racism—belief in the innate inferiority of Blacks—was related to higher alcohol consumption. Similarly, analyses of data from the National Study of Black Americans document that internalized racism (measured as the respondents' endorsement of negative stereotypes about Blacks) was related to lower levels of well-being and physical health (Williams and Chung, in press).

Migration

The mass movement of African Americans from rural and urban centers in the South to the large industrial cities of the North was one of the most massive migrations in American history, and was quantitatively larger than the migration of many other ethnic groups to the United States (Lemann 1992). This dramatic migration had a profound impact on the living conditions of the Black population. It also held important consequences for their health.

In 1940, 50 percent of all Black Americans lived in the rural areas of the South, and an additional 25 percent resided in more urban southern communities (Jaynes and Williams 1989). Today's Black population is overwhelmingly urban and disproportionately represented in the inner cities of the largest metropolitan areas. In 1990, 47 percent of African Americans lived outside of the South, 57 percent (17 million) were residents of inner cities, and only 15 percent resided in rural areas (F. Wilson 1992). In the North and West, 77 percent of the Black population live in inner cities.

For each of the three decades between 1940 and 1970 there was a net out-migration of 1.5 million Blacks from the South to the North (Jaynes and Williams 1989). As Blacks migrated to the cities, they

were concentrated, by segregation and discrimination, into the least desirable neighborhoods (Massey and Denton 1993). A pervasive "web of discrimation" involving the actions and inactions of local and federal governments, financial institutions, and real estate companies were used to entrap Blacks in inner-city ghettos (Jaynes and Williams 1989). Municipal ordinances, restrictive covenants, and federal housing policies mandated segregation. Blacks were not allowed to buy or rent in certain White residential areas. Real estate agents steered Blacks out of White areas by quoting higher prices or directing them only to Black residential areas. Local banks routinely engaged in a practice called "redlining," that is, denying a loan on the basis of race where the applicant was attempting to buy a house in an integrated neighborhood. This usually involved either charging higher rates to make a particular loan unattractive or cutting off all conventional mortages in an area. School board policies often designated distinct attendance zones for White and Black children, and White neighborhood organizations were vigilant and sometimes violent in enforcing segregation.

Despite the reality of residential segregation and discrimination, Blacks who moved to the North were relatively successful, at least initially. These Black migrants were younger and had higher levels of education than those who remained behind in the South (Jaynes and Williams 1989). At the same time, they had lower levels of education than northern-born Blacks. Newly arrived Black immigrants were charged more rent and were paid lower wages than Whites in the North (Lemann 1992). Yet they earned considerably more even at low-paying jobs in the manufacturing economy of the North than they had earned as tenant farmers and sharecroppers in the South. Thus, these Black migrants probably saw the glass as half full rather than half empty. Compared to native-born Blacks in the North, they worked longer hours, had lower unemployment rates, more stable families, lower poverty rates, lower utilization of welfare, and higher income returns on their educational attainments (Farley and Allen 1989; Jaynes and Williams 1989). In fact, although southern birth limited the achievement of White men who moved to the North, it appeared to have had the opposite effect for Blacks (Farley and Allen 1989).

Primarily due to this migration of African Americans from low-wage agricultural employment to higher-wage manufacturing jobs, the economic status of the Black population as a whole improved considerably between 1940 and the 1960s. However, the economic gains from migration were relatively short-lived, as by the late 1960s they had come to an end (Jaynes and Williams 1989). At least two reasons account for this. First, it appears that the children of Black immigrants did not achieve the same economic success as their parents. African Americans who migrated from the South to the North may eventually have experienced a considerable amount of frustration as their jobs provided them with little opportunity for advancement, and racial segregation and discrimination in the housing market frequently kept them trapped in undesirable and deteriorating housing (Lemann 1992). This frustration may have been even more keenly felt by their children, whose reference group was not their Black counterparts who remained behind in the South, but their White peers in northern cities. Discouraged by their restricted opportunities and truncated options, the children of Black migrants may have been less motivated than their parents to work hard at jobs that did not hold out any real advancement.

A second and more important reason for the declining success of immigrants was the migration of jobs from the large metropolitan areas in the Northeast and Midwest. A major exodus has taken place of low-skill, high-paying, blue-collar factory jobs from the inner cities to the suburbs (Wilson 1987). This has led to a concentration of poverty and high unemployment in large northern industrial cities, trapping substantial portions of the African American population in the same geographic locations that attracted their parents or grandparents with the promise of economic opportunity (Massey and Eggers 1990).

Effects of Migration on Health Status

The mass movement of African Americans earlier this century had profound implications for their health. First, the Black migration disproportionately distributed the African American population to areas where living conditions are hostile to life and health. The lower panel of Table 2 shows the relationship between health, as self-assessed, and place of residence by level of household income for Blacks and Whites. A central city is the largest city in a large urban area. As we can see, irrespective of race or place of residence, higher-income persons are less likely to be in poor health than their low-income counterparts.

Also, at both income levels and in all residential settings, Blacks consistently have higher rates of poor health than Whites. However, for both Blacks and Whites, those who live in central cities have higher levels of ill health than those who live in the suburbs. The higher level of ill health among those who live in rural areas may be a selection effect. Persons of all races who did not migrate are more likely to be older and in poorer health than the typical migrant who is young, adventurous, and in good overall health.

The overall picture of the health of Blacks in central cities masks the ever-worsening conditions in some inner cities. A recent study documented that Black males between the ages of twenty-five and forty-four in Harlem, one of the poorest areas of New York City, are six times more likely to die than White males in the United States (McCord and Freeman 1990). Cardiovascular diseases, cirrhosis of the liver, homicide, and cancer are the main causes of excess deaths. Moreover, the life expectancy of Blacks in Harlem is lower than that of persons in Bangladesh, which is currently categorized by the World Bank as one of the poorest countries in the world.

Life in poor inner-city environments can include poor nutrition, poor education, crime, traffic hazards, substandard and overcrowded housing, low-paying jobs, unemployment and underemployment, and a lack of health insurance and access to basic health services. The Black urban poor encounter these conditions more frequently than their White counterparts. The Black poor have been increasingly concentrated in depressed inner-city neighborhoods with bad living conditions, while the White urban poor are more evenly dispersed throughout the city, with many residing in relatively safe and comfortable neighborhoods away from the inner city (W. Wilson 1987). Thus the experience of stress, even in a given city, may be qualitatively different for Blacks and Whites.

Stress in poor urban environments can lead to illness. Harburg and colleagues (1973) characterized neighborhoods in Detroit as either high or low in stress. High-stress neighborhoods were those where the median level of income and years of formal education completed were low, and levels of residential instability (percent of residents with less than five years tenure), marital breakup, and crime were high. Low-stress areas had the opposite conditions. Persons living in high-stress areas had higher levels of high blood pressure than those in low-stress areas, an association that was stronger among Blacks than Whites. Moreover, although national data reveal that Blacks are twice as likely as Whites to have high blood pressure, the blood pressure levels for Black and White males in low-stress neighborhoods did not differ.

Recent research has also indicated that community violence is health damaging. Persons who live in inner-city housing projects are twice as likely to experience violence as other persons (Gabarino et al. 1992). This violence affects the victims, those who witness it, and those who are forced to live with the consequences. The repetitive nature of witnessing traumatic events may have a cumulative effect, especially on children. Each homicide also means fewer social and material resources to support children and families, and domestic violence often precedes homicide. Physical and sexual abuse are thus prevalent stresses in the lives of the urban poor, and persons who experience sexual and physical assault are more likely to develop substance abuse and mental health problems (Bell et al. 1988). Living in these conditions, in urban war zones where physical safety is not assured, can also lead persons to live in a state of heightened vigilance and takes its toll on the human organism. A recent review noted that three studies of racial differences in blood pressure have documented that although average daytime blood pressure levels were similar for Blacks and Whites, blood pressure declined less among Blacks than Whites overnight, such that Blacks had higher blood pressure levels even while they were asleep (Williams, Lavizzo-Mourey, and Warren 1994).

Change in Health Behaviors

The internal migration of the African Americans also affected their health by changing their lifestyles in ways that lead to higher risks of disease and death. Health behaviors are important causes of the heavy burden of disease among Black Americans. The federal government estimates that there are 60,000 excess deaths—that is, deaths that occur for Blacks that would not occur if they had the same mortality experience as Whites—in the African American population every year (U.S. Department of Health and Human Services 1985). The government also indicated that smoking cigarettes and drinking excessively played a large role as underlying causes of the excess deaths in the Black population.

Along with the great migration and urbanization of Black Americans came a dramatic rise in their use of alcohol and tobacco. The current rates of smoking and alcohol abuse among Blacks represents an important historic shift in the social distribution of these behaviors (Williams 1991). In the 1930s death rates from lung cancer were only 50 percent as high among Blacks as among Whites, and up through the 1950s Blacks were much less likely to smoke cigarettes than Whites (Cooper and Simmons 1985). Up through 1955, mortality rates from cirrhosis of the liver (an indicator of alcohol abuse) were higher for Whites than for Blacks. Since then, the increases in cirrhosis mortality rates have been higher for Blacks than for Whites. Moreover, ecological data reveal that the highest increases for Blacks occurred in urban areas with large migrant populations (Herd 1985). Some studies have also found that rates of alcoholism are higher among Blacks who migrated from the South (Boone 1985). The migration of Blacks may have affected their drinking and smoking behavior in a number of ways. First, the great migration of Blacks shifted a considerable portion of the Black population from the relatively "dry" rural South, where social life revolved around churches and family associations, to the "wet" areas of the urban North, where taverns, nightclubs, and their associated alcohol use were an integral part of social life (Herd 1985). Second, the theologically conservative Protestant churches to which most Blacks belonged tended to frown on alcohol consumption, but their traditional authority and influence was reduced in the urban North (Frazier 1966).

Third, life in urban settings produced feelings of alienation, powerlessness, helplessness, and meaninglessness, thus creating the need for individuals to mask these feelings or obtain temporary relief from them by consuming tobacco or alcohol. Social and economic deprivation create adverse working and living conditions from which people will attempt to escape. The amount of stress one experiences in work, neighborhoods, and family, as well as the resources to cope with it, vary with social status (Williams and House 1991). African Americans, like other persons of low socioeconomic status face more stress and also have fewer options for dealing with it. People under stress are more likely to drink and smoke because these are socially approved ways that provide some temporary relief from the personal suffering that is induced by poor living conditions. Seeman and Anderson (1983) found that an individual's sense of powerlessness is directly related to how often and how much that person drinks and whether drinking problems develop. In addition, stressful life experiences also went hand in hand with increases in both powerlessness and drinking problems, setting a vicious cycle into place. Not surprisingly, national data reveal that the sale of alcoholic beverages increases during economic recessions and during periods of increased unemployment (Singer 1986). By implication, efforts to reduce the consumption of alcohol should focus not just on the individual but should also seek to improve the conditions under which people live and work that give rise to the need for alcohol in the first place.

Finally, enterprising economic interests saw the newly arrived Black population as a potential market and heavily targeted this vulnerable population for advertising campaigns. One of the consequences of the rise in socioeconomic status of Black Americans in Northern cities was the development of a Black consumer market. Civil rights scholars have noted that White retail merchants frequently played a crucial role in prodding White-controlled social and political institutions to accommodate the demands of Black activists (Jaynes and Williams 1989). Since the 1950s African Americans have been special targets of the tobacco and alcohol industries (R. Davis 1987; Levin 1988; Singer 1986).

Hacker, Collins, and Jacobson (1987) provide an excellent description of the strategies used by the alcohol industry to target the African American community. Alcohol products have been closely tied to the music, sports, and cultural events that are important to the values and tastes of African Americans. Image advertisements promote education, fatherhood, Black history and Black culture. The alcohol industry has also employed some of the best-known Black celebrities (such as Alex Haley, Lou Rawls, Wilt Chamberlain, and Patti Labelle) to promote their products. In addition, they provide substantial support for Black History Month and the United Negro College Fund, as well as a large number of social, religious, educational, athletic, and business programs for Black Americans. Some alcohol producers, such as Coors, have even linked continued economic support of the African American community to increased sales of the company's products.

The alcohol industry has also developed products exclusively for the African American market, such as malt liquors (beer with a higher alcohol content). Moreover, the saturation level of alcohol advertising in the Black community is higher than in the predominantly White market. This can be seen both in the advertising in the major Black magazines as well as in the outdoor media. Seventy percent of billboards in the United States contain advertisements targeted to Blacks, with cigarettes and alcohol, in that order, being the two most heavily advertised products. This heavy bombardment of images that revolve around alcohol is combined with the greater availability of alcohol in African American urban communities, availability that also leads to greater alcohol consumption (Singer 1986). Retail establishments for the sale of alcohol are more common in minority neighborhoods than in more affluent communities (Rabow and Watt 1982), a concentration that reflects the cooperation of government with large-scale economic interests, because in every state retail outlets for alcoholic beverages are licensed.

The effects of the migration of Blacks to northern cities on their health may not be all negative. The influx of large numbers of Blacks to northern industrial cities, combined with the successes of the Civil Rights movement, also provided a substantial political base and platform for representatives of the African American population to command national attention and influence in the halls of power. Since the differential distribution of power in our society affects the distribution of desirable goods, increased political power can lead to improvements in health. LaVeist (1992) has recently documented an inverse association between Black political power and postneonatal mortality rates—that is, the more political power among Blacks the better the survival of their newborn children. He suggests that the political empowerment of Black Americans may lead to more community-level political participation, increases in Black employment, and increases in the overall quality of life which may in turn lead to improved levels of health and well-being.

Adaptive Resources: The Family and the Church

An analysis of the health of African Americans would be incomplete without consideration of their strengths and health-enhancing resources. An exclusive focus on vulnerability and risk gives a distorted view of the struggles of a disadvantaged group. The evidence on the mental health status of African Americans provides somewhat of a paradox. Compared to Whites, Blacks tend to have higher rates of the more mild indicators of mental health problems (psychological symptoms and distress), which tend to disappear when taking socioeconomic status into account (Vega and Rumbaut 1991), but they have similar or lower rates of the more severe indicators of mental illness (psychiatric disorder).

Table 3 presents both current and lifetime rates for Blacks and Whites of the most commonly occurring psychiatric disorders in the United States. These data come from the largest study of psychiatric disorders ever conducted in the United States (Robins and Regier 1991). Conducted by researchers at five sites between 1980 and 1983, this study interviewed a sample of almost 20,000 Americans including both treated and untreated persons. The overall pattern is consistent. Rates of depressive disorders and alcohol and drug abuse are very similar for Blacks and Whites. Rates of schizophrenia are slightly higher for Blacks, but the difference is not significant when differences in socioeconomic status are controlled for. Recent data from the first study to use a national probability survey to assess psychiatric disorders in the United States are even more striking (Kessler et al. 1994). In this study of over 8,000 adults, Blacks had rates of mental illness that were similar or lower than those of Whites. Lower rates for Blacks than Whites were particularly pronounced for the affective disorders (depression) and the substance use disorders (alcohol and drug abuse).

Thus, although African Americans confront a broad range of social conditions that are risk factors for mental illness, they do not have higher rates of suicide (as seen in Table 1) or higher rates of mental illness than Whites. These findings emphasize the need for renewed attention to identify the cultural strengths and health-enhancing resources of disadvantaged groups. All social groups have cultural resources that facilitate their efforts to gain control over their environment. In this respect, two social institutions—the family and the church—stand out as crucial in the African American community.

Strong family ties and an extended family system are important resources that may reduce some of the negative effects of stress on the health of Black

Table 3   Rates (per 100) of Psychiatric Disorder for Blacks and Whites

	Current		Lifetime	
	Black	White	Black	White
1. Affective disorders	3.5	3.7	6.3	8.0
2. Alcohol abuse	6.6	6.7	13.8	13.6
3. Drug use history	—	—	29.9	30.7
4. Drug abuse	2.7	2.7	5.4	6.4
5. Schizophrenia	1.6	0.9	2.1	1.4

Source: Lee N. Robins and Darrel A. Regier (eds.), *Psychiatric Disorders in America: The Epidemiologic Catchment Area Study* (New York: Free Press, 1991), Table 3–6, p. 41; Table 4–3, p. 59; Table 5–1, p. 85; Table 6–3, p. 121; Table 6–12, p. 133.

Americans. A large body of data indicates that supportive social relationships are among the most powerful determinants of health (House, Landis, and Umberson 1988). This evidence indicates, for example, that social isolation is as bad for one's health as smoking. In addition to providing emotional support, social ties can provide information and practical assistance that enable individuals to pool and share their resources for survival. An emphasis on the strength of the Black family contrasts with the more negative image of the weak or nonexistent Black family that is routinely presented in the U.S. media. This image focuses on the high rate of female-headed households. Currently, two out of three births to Black women are to unwed mothers, but being an unwed mother is not necessarily synonymous with being the head of the household. One small ethnographic study conducted at the time of the 1970 census found that only 12 percent of officially classified "female-headed" households did not have a stable male cohabitant (Hainer et al. 1988). In this study, adult male cohabitants who were present in the majority of Black female-headed households and performed traditional domestic responsibilities were omitted in the enumeration of those households because their presence was deliberately hidden from the census enumerators. Residents may well have had what they perceived to be good reasons (such as protecting welfare benefits) for hiding the presence of these men.

Given the marginalized economic status of Black males and the low-wage jobs available to poor and working-class women, there is a strong relationship between poverty conditions and births to unwed mothers. Nonetheless, in the face of these serious social and economic challenges, the Black family is resilient and resourceful. For many Black persons, the concept of family is not restricted to individuals related by blood or to persons resident in the household. Families are composed of those who interact regularly in helping each other share resources, responsibilities, and crises (Stack 1974). These families operate across households and across generations to help each other survive. These patterns of supportive exchanges are not limited to the poor. Middle-class Blacks are also involved in the provision of material support to poorer members of their extended family.

It must be remembered that family structures and processes do not arise out of thin air or simply out of skewed family values. Instead, they are shaped by the larger social environment. Marital status, for example, is linked to larger economic processes. Unemployment, declines in income, and high job turnover are all associated with increased rates of marital dissolution; the number of female-headed households decline when male earnings rise and rise when male unemployment increases (Bishop 1977). The linkage between family structure and economic conditions is readily evident if the Black population of the United States is compared to the population of African ancestry in the English-speaking Caribbean. Both populations share a common heritage of slavery, but those outside of the United States currently face even more severe economic conditions. Not surprisingly, the percent of births to unwed mothers of Black populations in these Caribbean countries is even higher than that of the Black population in the United States. For example, 85 percent of all births in Jamaica and 73 percent of those in Barbados are to unwed mothers (Encyclopaedia Britannica 1993).

A recognition of the strengths of Black families should not be used to romanticize them as if they were a panacea for a broad range of adverse living conditions. While these networks of mutual aid and support do facilitate survival, they are also likely to provide both stress and support. Moreover, it is likely that cutbacks in government social services over the last decade have increased the burdens and demands on the support services provided by the Black family. At the present time we do not know how well the Black family is coping with these new challenges in the face of economic decline and worsening socioeconomic conditions.

The Black American church has been the most important social institution in the Black community. It emerged during the days of slavery as the only organization that was not controlled by the White power structure. These churches soon became centers of spiritual life as well as social and political life. They provided a place for Blacks to carry out professional and administrative roles that they were denied by the larger society, and were an important base of support for the Civil Rights movement (see the essay by Morris and Herring in this volume). Recent studies of African American churches document that, as in the past, they are still involved in providing a broad range of social and human services to the African American community (Caldwell, Green, and Billingsley 1992; Chang et al. 1994; Lincoln and Mamiya 1990). They also serve as important conduits of material resources to the community.

The clergy of all races also play an important role as providers of psychological and mental health support to their parishioners. A recent study of the African American clergy in New Haven, Connecticut, documented that Black ministers respond to a variety of needs that go beyond their role as spiritual advisor (Chang et al. 1994). This study found that most Black ministers spent much time counseling parishioners on issues ranging from sexual abuse to unemployment. The three most common problems that the clergy dealt with were marital and family problems, drug- and alcohol-related problems, and financial problems. This study also documented that the scope of these Black ministers' counseling ministry extended far beyond their own parishioners, with 85 percent reporting that they counsel persons from other religions and other religious denominations.

The New Haven study showed that the Black clergy also play an important role as a gatekeeper to the mental health system. Contrary to beliefs that religious leaders disparage professional models of counseling in favor of spiritual care, 47 percent of the African American clergy reported that they had referred parishioners to a community agency for mental health problems. Almost 44 percent of the clergy in this study reported that a referral had been made to them by an outside health professional or agency. Thus, at least some African American churches and mental health and social service agencies create networks of information and exchange in which institutional services are made accessible in church-based contexts of ongoing personal and community support.

The Black American church is also an important source of social integration and support. Congregation-based friendship networks function as a type of extended family and provide supportive social relationships to individuals as they go through the life cycle (Taylor and Chatters 1988). In addition, the African American population is arguably the most religious group in the industrialized world (Gallup Report 1985), and this high level of religious involvement on the part of Black Americans may also reduce the adverse consequences of stressful living conditions and promote their psychological well-being. Griffith and colleagues (Griffith, English, and Mayfield 1980; Griffith, Young, and Smith 1984) indicate that participation in Black church services can provide therapeutic benefits equivalent to those obtained in formal psychotherapy. The expression of emotion and the active congregational participation that is characteristic of some African American churches can promote a collective catharsis that enhances the reduction of tension and the release of emotional stress (Gilkes 1990).

Conclusion

African Americans continue to suffer high levels of disease, disability, and death. As we have seen, health must be understood within its larger social context. Health is inextricably tied to the social, political, cultural, and economic conditions under which people live. Larger social structures create stressful living conditions and working environments, and shape the nature of the response of social groups. The social distribution of health and disease reflect the convergence of a broad range of quality of life issues. As such, it is related to the struggle to make ends meet, to make sense out of life, and to cope with alienation and powerlessness.

The evidence presented here indicates that the health problems of African Americans are not primarily caused by underlying deficits in genetics or medical care. It follows that medical solutions alone will be limited in their capacity to improve health (D. Williams 1990). Thus, effective efforts to improve the health of groups, such as Black Americans, must consider changing the fundamental social conditions that created them in the first place.

31

The Wages of Race: Color and Employment Opportunity in Chicago's Inner City

MARTA TIENDA
HAYA STIER

As the second largest U.S. city at the turn of the century, Chicago was a booming industrial center on its way to becoming a major distribution center as well. Rapid industrial growth meant job growth—employment opportunities—which served as a magnet for southern and eastern European immigrants and northbound Blacks seeking better destinies than their southern origins had yielded. The dimensions of the "Great Migration" which relocated thousands of unskilled laborers from southern plantations to northern ghettos have been amply documented. Suffice it to recall that between 1916 and 1920 approximately 60,000 southern-born Blacks migrated to Chicago (Grossman 1991; Johnson and Campbell 1981). Chicago's ethnic landscape during the roaring twenties and the depression years provided the substance for Robert Park's (1914) classical notions about the cycle of race relations, assimilation, and along with his colleagues Burgess (1923) and Wirth (1938), the very conception of the city. Before the Depression, Chicago was Carl Sandburg's "City of Big Shoulders"—the "stormy, brawling, husky" place where the world's butchers, toolmakers, wheat stackers, railroad and freight handlers lived and labored.

As the end of the century approaches, Chicago is a different city—a place where opportunity has been dimmed as industrial restructuring eliminated thousands of semiskilled and unskilled jobs that paid family wages. Like other northern cities (that is, Detroit,

This research was supported by grants from the Rockefeller Foundation and from ASPE of the Department of Health and Human Services to the Institute for Research on Poverty of the University of Wisconsin. Our work was also supported by a fellowship at the Center for Advanced Study in the Behavioral Sciences, and we appreciate the generous support of the John D. and Catharine T. MacArthur Foundation for this opportunity. We gratefully acknowledge comments from Ronald Mincy, Christopher Tilly, and Harry Holzer. We appreciate technical assistance from Adelle Hinojosa, Safia Khan, and Frank Norfleet in preparing the final manuscript.

The data on which this article is based were collected under the auspices of the Urban Poverty and Family Structure Project of Chicago conducted under the direction of William J. Wilson, who provided generous access to the Urban Poverty and Family Life Survey of Chicago and the Social Opportunity Survey.

How well a group did in the United States had a lot to do with the manner in which it became incorporated in the labor force, a process that in itself was historically contingent and depended upon a multitude of factors, economic and cultural. Here a Mexican American is working in an industrial manufacturing plant, a job not easily attained by many minorities. (Photo by Steve Gold)

Cleveland, Philadelphia, New York, and St. Louis), Chicago's industrial decline is mirrored in its demographic decline; the city has lost nearly 1 million people since midcentury. During the 1980s Chicago fell from second to third largest city, surpassed by Los Angeles. Yet Chicago remains a destination for thousands of immigrants who continue to diversify the ethnic terrain. Contemporary immigrants destined for Chicago and other major cities originate from Mexico, Central America, Korea, and other parts of Asia rather than Europe. The 1990 census indicated that Blacks comprised nearly two-fifths of the city's population; Hispanics (overwhelmingly Mexican and Puerto Rican) another fifth; and Whites, many of them European ethnics, most of the remaining two-fifths.

Against this backdrop, it is unsurprising that Chicago was a core city for the Civil Rights movement, for Jessie Jackson's Operation PUSH, and for William J. Wilson's (1987) "Truly Disadvantaged." That poverty and joblessness has become more concentrated and devastating for Chicago's Black population suggests that race matters as much at the close of the century as it did at the beginning. This study is an investigation of just that question—how does color influence job opportunities in the city of "Big Shoulders?" Accordingly, in this chapter we examine differences in labor force activity among parents residing in economically and ethnically diverse neighborhoods. Our general aim is to identify and evaluate the circumstances that sustain inequities in employment opportunities within a single labor market.

Comparisons among Blacks, Whites, Mexicans, and Puerto Ricans who resided in Chicago's inner-city neighborhoods, and between the inner-city sample and a national sample of parents with similar background characteristics, serves several objectives that have been neglected in prior research.

First, we broaden the population at risk of exclusion from the labor market to include Hispanics. This is an important extension in light of recent evidence that Hispanic poverty has been less responsive to economic recovery than White and Black poverty, and theoretical arguments about possibly distinct mechanisms undergirding persisting poverty among Blacks and Hispanics (Moore 1989; Tienda 1989; Tienda and Jensen 1988). Second, we provide a national benchmark against which to compare the allegedly deviant labor force behavior of inner-city residents. Finally, we evaluate the relative merits of structural and individual explanations for the pervasiveness of joblessness in inner-city neighborhoods while appraising the salience of color in shaping employment opportunity in Chicago.

A brief overview of relevant theoretical considerations serves to frame the empirical analyses of race and ethnic differences in labor force activity. Following a description of the data sources, we first compare rates of labor force activity among race and gender groups for samples of Chicago and U.S. parents. Subsequently we evaluate how color and opportunity influence employment experiences in Chicago by assessing empirically the relative importance of individual and structural factors on labor force activity. We supplement our statistical analyses of labor force activity and wages with a textual analysis about social opportunity in Chicago, emphasizing race and ethnic differences in perceptions of opportunity and competition along color lines as voiced by inner-city residents themselves.

Theoretical Considerations

Various perspectives have been marshaled to explain widened race and ethnic economic inequities. One set of explanations focuses on the opportunities to work, including inequities produced by employer discrimination, while others emphasize individual differences in skills or willingness to work. Although these explanations of race and ethnic differences in inner-city joblessness have limitations and they are not mutually exclusive, it is instructive to discuss them separately for heuristic purposes. Evaluating their relative merits can provide useful insights about why race and ethnic differences in labor force activity have widened.

Opportunities and Ethnic Inequality

A dominant structural explanation for the rise of inner-city joblessness focuses on the decline of well-paying manufacturing jobs in old industrial centers (see Kasarda 1985; Wilson 1987). Wilson proposed that "social isolation" from mainstream work norms and behavior sustains chronic labor market inactivity in ghetto poverty neighborhoods. Yet this explanation does not resolve the puzzle of large race and ethnic differences in employment behavior among inner-city populations residing in a single labor market.

"Queuing theory" provides another structural perspective of ethnic differentials in employment (Hodge 1973). Simply stated, when job opportunities requiring low to moderate skills (education) decline, ethnic employment competition intensifies for two reasons. First, employers can be more "choosy" about which workers to hire, allowing their preferences and prejudices to play themselves out differently from a situation of labor scarcity. Second, the changing ethnic composition of neighborhoods resulting from differential migration of minority workers can reconfigure the extent and nature of job competition along color lines (Hodge 1973; Tienda, Donato, and Cordero-Guzmán 1992).

Several studies have documented that immigrants participate in the labor force at a higher rate than natives (Borjas and Tienda 1987), but researchers do not agree whether immigrants displace native minority workers. Aggregate econometric evidence indicates that the labor market impacts of recent immigrants are benign, but studies of specific firms and communities reveal intense competition between immigrants and native minorities (Tienda and Liang 1993). At issue for the contemporary debates about pervasive inner-city joblessness is whether native minorities both perceive competition from immigrants and experience lower employment rates than comparably or less-skilled immigrants.

Finally, discrimination against people of color could also give rise to race and ethnic differences in joblessness even in the absence of skill differences because it implies unequal treatment of entire groups. For example, a recent study showed that the employment returns to education came to depend increasingly on race and ethnicity between 1960 and 1985, particularly among women with less than a high school degree (Tienda, Donato, and Cordero-Guzmán 1992). However, this study did not consider whether higher minority joblessness was associated with group differences in willingness to work.

Individual Differences and Ethnic Inequality

The human capital explanation for the disadvantaged labor market standing of minority workers is essentially one that focuses on workers' differences in skills. Assuming that education has uniform exchange value, then the highest labor force participation rates should correspond to groups with the most education. The skill deficit explanation of rising inner-city joblessness also has been tied to the "spatial mismatch hypothesis" (Holzer 1991; Kasarda 1985). Simply put, the rapid decline of unskilled and semiskilled jobs from old industrial centers has produced a major imbalance between the supply of workers available and the demand for a highly educated workforce. Stated as a variant of the spatial mismatch hypothesis, the skill deficit interpretation maintains that increased inequities between White and non-White workers reflect the failure of minorities to keep pace with the rising educational requisites of new jobs.

An alternative individualistic explanation for pervasive inner-city joblessness suggests that weak labor force attachment is coupled with a preference for welfare (Murray 1984). Although several critics have challenged this interpretation, few studies have investigated the willingness of inner-city dwellers to work (Tienda and Stier 1991; Van Haitsma 1989). One variant of the weak attachment explanation is that jobless inner-city residents require higher wage rates than employers are willing to pay, given their skills and prior employment experience. Reservation

wage refers to the minimum hourly pay required by workers in order to accept a job offer. Applied to race and ethnic differences in labor force activity, this implies that jobless minority workers demand higher compensation as a condition of accepting a job than similarly skilled nonminority workers. This argument has been summoned to explain the anomaly of high minority unemployment coupled with high rates of labor force activity among undocumented workers. However, few studies have examined empirically whether people of color actually demand higher wages than their skill characteristics warrant. We evaluate this claim in some detail below.

Data

The primary data source for our empirical analyses is the Urban Poverty and Family Life Survey (UPFLS) of Chicago, which was conducted by the National Opinion Research Center (NORC) during 1986 and 1987 (William J. Wilson, principal investigator). The UPFLS is based on a sample of parents aged eighteen to forty-four in 1986. Because the study was conceived as an investigation of family life in poor neighborhoods, cost considerations dictated a sampling strategy that deliberately overrepresented poor neighborhoods. Using a 1980 family poverty rate of 20 percent or higher as the cutoff point to designate poverty tracts, 39 percent (n = 332) of the 843 census tracts within the Chicago city limits so qualified.[1] Among the remaining tracts, 40 percent recorded family poverty rates ranging from 5 to 19 percent, and an additional 21 percent of Chicago tracts had 1980 family poverty rates below 5 percent. Completed interviews were obtained from 2,490 respondents, including 1,186 Blacks, 368 Whites, 484 Mexicans, and 453 Puerto Ricans.[2] The overall response rate of 79 percent varied from 75 to 83 percent among ethnic groups.

The UPFLS is based on a single interview, but contains several retrospective sequences on life experiences, including work behavior, marriage, fertility, education, and welfare participation. Extensive information about respondents' family backgrounds, current employment status, reservation wages (as defined above), income sources, and attitudes toward work is also available. The UPFLS was supplemented by the Social Opportunity Survey (SOS), an in-depth, open-ended interview conducted with a small subset (N = 161) of respondents who participated in the main survey. Although not a random sample of the general survey, respondents interviewed for the SOS represent the race, ethnic, and gender variation of the general survey. Responses to the SOS provide richly textured information that aids in interpreting the statistical results from the main survey.

Finally, we use the 1987 National Survey of Families and Households (NSFH) to produce baseline tabulations of labor force activity against which to compare the employment and unemployment rates of the Chicago sample. A national benchmark is important to establish the bounds of generalizability of the Chicago Survey, which excluded individuals residing in very low poverty areas.[3] To maximize comparability with the UPFLS, we selected all parents aged eighteen to forty-four from the national survey.

Race and Ethnic Differences in Labor Force Activity

Table 1 compares labor force participation rates for Chicago and the United States. The "core city" refers to residence within city boundaries, as in accordance with U.S. census guidelines, while the noncore city refers to suburban fringes and nonmetropolitan areas. Within Chicago, separate tabulations for "mixed poverty" and "ghetto poverty" neighborhoods, respectively, refer to census tracts with family poverty rates of 5 to 20 and 30 percent or more.

These tabulations show that the labor force activity of parents depends on ethnicity and gender, and within Chicago's inner city, neighborhood context as well. Three broad generalizations are warranted from these results. First, in the aggregate, labor force participation rates are uniformly lower among Chicago parents than parents nationally, with gaps approaching 10 percent for men, and exceeding 15 percent for women. Likewise, unemployment rates also were appreciably higher in Chicago than in the nation as a whole, 9.4 versus 3.2 percent for men, and 5.1 versus 4.5 percent for women.

Second, while race and ethnic differences in labor force activity are clearly evident in both samples, Chicago's Blacks and Puerto Ricans experience extreme labor market hardships.[4] Chicago's Black inner-city

men participated in the labor force at a rate 11 percent below Black men nationally, compared to a 5 percent point difference for White men and a 2.3 percent gap for Mexican-origin fathers. The narrow labor force activity gap for Mexican-origin men reflects the high immigrant composition of the Chicago sample relative to the national sample, combined with the higher labor force participation rates of immigrant compared to native-born men (Bean and Tienda 1987).[5] Among mothers, the lowest rates of labor force activity correspond to Puerto Ricans, whose rates hovered around 35 percent in both samples.

Racial disparities in unemployment were also apparent in both samples. Black parents experienced the highest unemployment rates. Chicago's Black fathers were three times more likely than White fathers to be unemployed in 1987, but this compared favorably with the national Black-White unemployment ratio of 3.6. For women, the ratio of Black-White unemployment was approximately 2 in Chicago, compared to 3.5 nationally.

Third, residence-specific differentials in labor force activity rates were substantial between mixed and ghetto poverty neighborhoods in Chicago. In the mixed poverty neighborhoods, 91 percent of fathers were in the labor force compared to 79 percent in ghetto poverty neighborhoods. Black fathers are largely responsible for the low average male participation rate in ghetto poverty areas. Mothers residing in mixed poverty neighborhoods participated in the labor force at a rate of 56 percent compared to 43 percent among those residing in ghetto poverty neighborhoods. Furthermore, in mixed poverty neighborhoods, Black and White women exhibited the highest rate of labor force activity, but in ghetto poverty areas, Mexican origin women did. Unemployment rates of Black fathers were 50 percent higher in ghetto versus mixed poverty areas, but unemployment rates of Hispanic fathers did not vary appreciably from neighborhood poverty status.

Making Sense of Ethnic Inequality: Voices from the Inner City

Robert Hodge's (1973) conception of the labor market as an ethnic queue is helpful in understanding race and ethnic differences in employment statuses and provides some insight into the emergence of

Table 1   Labor Force Participation and Unemployment Rates of Parents Ages 18 to 44 by Ethnicity and Gender: Chicago's Inner City and United States, 1987

| Sex and Ethnicity | Chicago | | | | | | | | United States | | | | | | | |
| | Total Inner City | | | Mixed Poverty Neighborhoods | | Ghetto Poverty Neighborhoods | | | Total | | | Core City | | | Non-Core City | |
	[N]	LFP	UNEM	LFP	UNEM	LFP	UNEM	[N]	LFP	UNEM	LFP	UNEM	LFP	UNEM
Men	[811]	86.1	9.4	91.0	7.0	78.9	13.0	[1796]	96.2	3.2	96.1	4.1	96.4	2.3
Black	[308]	82.6	12.9	88.4	10.0	76.5	15.9	[334]	94.1	8.4	94.3	11.8	93.7	2.6
White	[127]	91.3	4.3	93.0	5.5	84.9	0.0[a]	[1321]	96.6	2.3	96.3	2.2	96.8	2.3
Mexican	[228]	94.8	1.7	96.5	1.7	89.6	1.8	[118]	97.1	4.5	98.3	5.1	—[b]	—[b]
Puerto Rican	[148]	84.4	7.6	85.6	7.3	82.1	8.2	[23]	90.1	13.3	91.5	14.5[b]	—[b]	—[b]
Women	[1523]	50.0	5.1	56.0	4.5	42.9	5.8	[3387]	65.8	4.5	66.5	4.7	65.0	4.1
Black	[719]	50.5	6.4	59.8	6.2	42.8	6.5	[793]	73.7	10.3	72.5	10.2	76.4	10.4
White	[237]	55.9	3.3	58.4	4.1	45.7	0.0[a]	[2296]	65.2	2.9	67.2	2.3	63.4	3.4
Mexican	[261]	52.8	1.9	52.4	1.5	54.2	3.7	[227]	60.0	9.9	58.0	9.5	77.1[c]	13.1[c]
Puerto Rican	[306]	35.4	1.3	37.4	1.6	31.9	0.6	[71]	34.8	4.6	32.9	5.1	—[b]	—[b]

[a] Based on unweighted cell sizes of 30 men and 58 women.
[b] Based on unweighted cell sizes of 20 or less observations.
[c] Based on weighted cell size of 30 or less observations.
Source: UPFLS (1987) and NSFH (1987).

wage inequities. The basic principle of queuing is that in the process of matching workers to jobs, the most desirable workers are hired first and the least desirable workers are hired last. Thus, when labor demand shrinks, increases in joblessness will be greatest for workers at the bottom than for workers ranked higher in the queue (Hodge 1973; Lieberson 1980; Reskin and Roos 1990). In a totally meritocratic society, skills would be the primary basis for determining individuals' rank in a hiring queue; in practice, gender, race, and national origin are also decisive. Distortions in rankings based purely on skills result because employers have imperfect information about prospective workers productivity, organizational procedures often protect and reward others with seniority, and employers discriminate against entire groups of workers (Hodge 1973).

We believe that the worsened labor market position of Black and Puerto Rican inner-city workers results partly from displacement through competition with new sources of labor. Responses to the open-ended interviews from the Social Opportunity Survey lend support to this interpretation, and reinforce our contention that the significance of color increases when the supply of jobs at a given skill level contracts and the supply of immigrant workers increases. These responses provide richly textured substance about how opportunity depends on race and national origin.

Jack's comments in response to a question about who gets ahead are especially illuminating about how color shapes opportunity in Chicago.[6] Jack is married, Black, and forty-four years old. He has five kids to support. He lives in an extremely poor, all-Black neighborhood and has never been on welfare, though he lost his house when the steel mill shut down. To the question on who gets ahead, he responded:

It's still the same old thing: Whites get ahead much quicker than Blacks. Still the same, that hasn't changed, 'cause on this job right now, I can see that. But there's nothing you can do about if you want keep your job, you gotta just lay dead and try to make it. . . . I can say this much: a man gotta do what he gotta do. If he gotta work, he gotta work: it's as simple as that. (emphatically) *You got to work*. If you don't do that, you gonna rob, and steal and I can't do that, 'cause what would I do in jail with 5 kids and a wife, you know? So I *have to work*.

When probed about who is least likely to get ahead, he replied:

It would go between the Puerto Ricans and the Blacks. They, they catch hell too. 'Cause a lot of them work out there with me. I don't know why, I don't know: I wish I knew why. I don't know why it's like that. They're just saying "that's the way it is." It's like a set pattern.

A Mexican respondent was equally blunt in linking job opportunities to ethnicity. Says he: "The good jobs go to the gringos!"

Many respondents—too numerous to report—spontaneously answered questions about opportunity and access to good and bad jobs with references to race and ethnicity. Some acknowledged that color could be used to reserve slots for minorities, but there was general agreement that the best jobs go to Whites, and especially educated Whites, while the worst jobs go to Blacks and Latinos. "Someone has to do the dirty work," replied a White woman in response to several questions about employment opportunities. Having networks and connections was another recurrent theme in responses to questions about access to good and bad jobs, but no one clarified how one gets connections in the first place, except that the rich and White folks have them.

Belinda, a thirty-nine-year-old, married, Black woman with six kids who is on public aid, sees this quite clearly:

Well, like, I've been here all my life, you know, and it's a White-ruled system, you know. If you're White, you've got a better chance. And I've personally experienced that as far as moving. . . . You know, I've went on jobs and seen a White person

get hired and I was told there was no job. The job wasn't even, didn't even require skills, you know, things like that.

Several respondents, especially Blacks, lamented the better times they experienced in the past. Letonya, a forty-one-year-old mother of two, legally separated, and on strike from a certified nurse's aid job, articulated this position:

Now they don't have the jobs that they had, even back in the sixties . . . before, I don't think it's as many people are from other countries was over here like it is now, you know. And it seems to me they're sort of moving out the Americans (laughs). I mean taking up the jobs where the peoples, Americans used to have. 'Cause I had worked for quite a few in my life, I worked in a hospital until four years ago and they laid us off. And it was a lot of foreigners over here you know. And when they come over here, they work for little or nothing, you know what I'm saying. . . . *Americans understand that they couldn't live off with what little money they were paying.* (authors' emphasis)

Some would say Letonya had a higher reservation wage than immigrants, meaning that she demanded higher wages than other workers with similar characteristics. However, a more perceptive interpretation, particularly in light of the empirical evidence provided below showing that Blacks have lower reservation wages than similarly skilled Hispanics or Whites, is that Letonya understands what it takes to live in Chicago, and appropriately questions a decline in her economic status.

Letonya was not the only respondent who blamed her weakened labor market position on illegal immigrants. Several respondents implicated immigration as a major reason for the narrowing of economic opportunities among Chicago's native residents. In contrast to several econometric studies based on national data which show little competition between native minorities and recent immigrants, Chicago's inner-city parents perceived acute competition and job displacement from recent immigrants, and especially undocumented immigrants.

According to Sabrina, a forty-two-year-old widow in a visiting union, who has seven kids and a history of chronic unemployment, opportunity is lessened by the presence of immigrants:

NO. They're bringing too many foreigners over here from other countries for any of us to get ahead or do what it's our right to do, and that's put us down.[7]

Jane, a married mother of four who holds a GED and has been on welfare since 1972, thinks only the bad jobs go to illegals. Her response supports the view that undocumented migrants take the jobs U.S. natives refuse:

Well, the bad jobs are usually for the poor Spanish people that come here illegally and they'll work for 2 dollars an hour . . . because they're afraid, you know, that they're . . . and they get mistreated and . . . stuff. Wetbacks they call 'em, which I don't see why they do. Call 'em Spanish or whatever, but . . .[8]

Yet other respondents felt that unauthorized immigrants were taking both good and bad jobs. For example, Ann, a thirty-year-old mother of four who has never been on welfare and is currently married, doesn't believe there are enough good jobs for everybody:

'cause the illegal aliens . . . they're taking all the jobs, 'cause its cheap labor. They can work for $3.00 an hour, whereas they're looking for $8 or $9 an hour. . . . I'm not prejudiced against any illegal alien, but they shouldn't come here, and be taking our jobs and what belongs to us.

Susan, a disabled welfare mother of three, also agrees:

I don't know, I might sound . . . like I'm prejudice or something, but I think they should go back to where they belong and let the White people work and make a living. . . . The Mexicans should go to Mexico and the Puerto Ricans should go back to Puerto Rico and the Blacks to Africa.[9]

These responses from Chicago's inner-city poor illustrate that competition for jobs is not simply a matter of Blacks versus Whites, or undocumented migrants versus native minorities. We were particularly struck by the responses of Puerto Ricans and Mexicans indicating job competition between them, intense discrimination against Puerto Ricans, and more generally, perceived rankings of Mexicans as workers preferred over Puerto Ricans. Dolores, a forty-two-year-old Puerto Rican welfare mother of four, has never been married and has been on public aid since 1969, following a highly unstable employment history between the ages of fifteen and twenty-four. She has lived in Chicago thirty-three years, and in her current neighborhood sixteen years. It is not lack of effort that keeps her jobless:

I go down to . . . I would love to work there, you make good money there. Every time I go they say they are not hiring. Every time I go I sign an application. They tell me I already have a lot of them there, but I say give me another. . . . I go down to the next factory where they make plastic flowers and apply there too. And to the next one and like that. . . . My daughters tell me they see help wanted signs, but when I go over they say no. My daughters tell me they want Mexicans and that I should just say I'm Mexican. Because they'll work for less and all that. But I say, "But they'll ask me for my papers!" But if you want the job, maybe you have to do that. Because I have some Mexican friends they go and get the job just like that. Because it's a $5 an hour job but they would only pay the Mexicans like $3.50.

When probed if she would work for $3.50, she replied:

I wouldn't mind. I used to work for less, for eighty cents an hour. I wouldn't mind, but then when I go they say they aren't hiring. But the other ones get the jobs and then I don't want to go back, you know.

While arguments about differences in the labor market standing of Mexicans and Puerto Ricans based on national data lend themselves quite readily to structural interpretations that emphasize the decline of blue-collar operative jobs as being responsible for the declining economic status of Puerto Ricans, this is only part of the story. In particular, it is difficult to argue that geography is a major factor in the differential labor force activity of Mexicans and Puerto Ricans living in poor Chicago neighborhoods. Also, a recent study provided strong evidence that the position of Puerto Ricans in New York was undermined by recent immigrants (Tienda and Donato 1993). Therefore, we wish to emphasize the importance of intense discrimination as a factor responsible for the declining labor market status of Puerto Ricans.

Explaining Race and Ethnic Inequality in Labor Force Participation

Our theoretical discussion identified differences in skill and willingness to work as two plausible reasons for race and ethnic variation in labor force activity. If educational deficits were the primary reason for high joblessness among inner-city minorities, we would expect lower education levels in Chicago than the United States, and among ethnic groups, the lowest educational levels for Blacks. In fact, as Table 2 shows, the educational characteristics of Chicago's Blacks and Whites are generally similar to those of their national counterparts. Hispanic parents were far more educationally disadvantaged than Black inner-city parents, averaging seven to ten years for Mexicans and Puerto Ricans, respectively. Yet the highest labor force activity rate (and lowest unemployment rate) corresponds to Mexican, not White or Black fathers.

Despite the similar mean levels of completed schooling among the Chicago and the national samples, the proportion of high school graduates revealed greater discrepancies. For example, between 55 and 58 percent of Chicago's Black parents were high school graduates, compared to about 80 percent of the Black parents nationally. Among Mexicans, 35 to 38 percent of Chicago parents reported having graduated from high school, compared to 43 to 63 percent of mothers and fathers nationally. The greatest discrepancy in high school graduation rates corresponds to Whites, as over 90 percent of the national sample compared to two-thirds of the inner-city sample claimed high school diplomas.

Race and ethnic differences in receipt of transfer income (that is, public aid) show relatively low rates of welfare utilization among White and Mexican parents, particularly when contrasted with the welfare participation rates of Black and Puerto Rican parents. Less than 10 percent of White mothers nationally were on public aid at the time of the survey, compared to 30 percent of White mothers in Chicago. By contrast, over half of Chicago's Black and Puerto Rican mothers were on aid in 1987, compared to between one-quarter and one-third of their national counterparts. Despite their severe educational disadvantages, Chicago's Mexican parents made relatively low use of transfer income. Ineligibility for means-tested benefits (because of immigrant status) may partly explain the low rates of welfare utilization among Mexicans, but their high rates of labor force activity, even at low wages, also may obviate their need for transfer income.

Skill, Color, and Labor Force Activity

A more succinct assessment of the relationship between ethnicity, education, and work activity is possible by examining simultaneously several determinants of labor force participation. For this we estimated a model that allowed us to evaluate the unique influence of individual, family, and neighborhood characteristics on the likelihood that respondents obtained a job. These results are illuminating and we summarize their lessons below. First, there are striking similarities in the determinants of labor force activity for both mothers and fathers, despite some noteworthy differences. For example, marriage increases the probability of labor force participation for both men and women, and the presence of young children has no influence on labor force activity of either mothers or fathers. Also, disability lowers the odds of labor force participation for both mothers and fathers. However, the presence of other adults increases the probability that mothers will work, but not fathers. This result has a parallel based on national populations (Tienda and Glass 1985).

Second, two findings are particularly noteworthy for our hypothesis about the relative importance of color and skill in determining labor force outcomes.

Table 2 Selected Demographic and Socio-economic Characteristics of Respondents by Ethnicity and Gender: Chicago's Inner City and United States, 1987 (Means and Percents)

	Blacks		Whites		Mexicans		Puerto Ricans	
	Men	Women	Men	Women	Men	Women	Men	Women
Chicago								
x̄ Education	12.1	11.9	13.1	12.1	7.0	7.2	9.7	9.9
(s.d.)	(2.1)	(1.9)	(2.9)	(2.6)	(3.6)	(3.8)	(3.0)	(2.8)
High School Graduates (%)	54.8	58.0	67.1	63.6	38.3	35.6	31.0	38.7
Employment								
Never worked (%)	9.9	20.7	0.0	8.6	0.9	17.0	5.4	28.6
Currently working (%)	69.7	44.1	87.0	52.6	93.1	50.9	76.8	34.1
Transfer income								
Ever received aid (%)	46.1	81.3	13.6	45.3	9.6	29.6	35.5	75.3
Currently on aid (%)	22.3	52.0	6.2	29.7	2.6	13.6	11.6	53.7
[N]	[338]	[689]	[126]	[238]	[229]	[260]	[145]	[309]
U.S. cities								
x̄ Education	12.7	12.5	13.8	13.3	10.9	9.6	—[a]	9.9
(s.d.)	(2.5)	(1.9)	(2.6)	(2.4)	(3.6)	(3.5)	—[a]	(3.2)
High School Graduates (%)	79.7	79.7	91.7	90.2	63.6	43.1	—[a]	39.3
Employment								
Never worked (%)	0.0	12.5	0.0	4.8	0.0	19.2	—[a]	40.5
Currently working (%)	83.3	62.1	94.0	65.3	90.6	48.9	—[a]	27.7
Transfer income								
Ever received aid[b] (%)	9.2	36.1	7.3	15.3	8.6	31.6	—[a]	46.2
Currently on aid (%)	3.6	27.5	1.7	7.4	2.9	16.4	—[a]	34.0
[N]	[213]	[529]	[613]	[1095]	[101]	[201]	[20]	[62]

[a] Too few cases.
[b] Received since 1982.
Source: UPFLS (1987) and NSFH (1987).

One is that high school graduation status significantly increased the labor market activity of inner-city mothers, but had no influence on the market activity of inner-city fathers. This indicates that education is a necessary, albeit insufficient condition for labor market success. Another key finding is that race and ethnic differences in labor force activity persist, even after standardizing our sample for differences in education, family status, and neighborhood characteristics. Specifically, Black fathers were significantly less likely, and Mexican fathers more likely, to participate in the labor force than their sta-

tistically equivalent White counterparts. Among mothers, significant ethnic differences were obtained only for Mexican-origin women, whose labor force activity exceeded that of White mothers with similar educational, family, and personal characteristics.

Finally, consistent with prior theorizing about the pervasiveness of joblessness in ghetto poverty areas, our results clearly show that labor force activity is significantly lower in neighborhoods with poverty rates in excess of 39 percent. It is unclear whether this result is obtained because residents of ghetto

Table 3 Average Wages and Selected Working Conditions of Parents Ages 18 to 44 in Chicago's Inner City by Ethnicity and Sex, 1987

Sex and Employment Status	Black	White	Mexican	Puerto Rican
Men, employed				
Hourly wage rate	$8.9	$14.6	$7.9	$8.0
Union (%)	47.0	25.0	52.1	38.4
High risk of job loss (%)	48.1	28.3	33.6	31.7
[N]	[225]	[111]	[213]	[118]
Men, jobless				
Reservation hourly wage	$5.8	$9.3	$7.2	$6.2
Want regular job (%)	89.9	76.3	100.0	82.2
Days/week willing to work	5.3	5.0	5.4	5.2
[N]	[83]	[16]	[15]	[30]
Women, employed				
Hourly wage rate	$7.3	$12.0	$5.2	$6.1
Union (%)	35.8	14.2	24.3	17.0
High risk of job loss (%)	71.6	73.0	58.6	31.1
[N]	[314]	[125]	[133]	[103]
Women, jobless				
Reservation hourly wage	$5.5	$6.6	$5.3	$5.3
Want regular job (%)	90.9	58.3	74.7	75.4
Days/week willing to work	5.1	4.5	4.8	4.9
[N]	[405]	[112]	[128]	[203]

Source: UPFLS (1987).

poverty areas refuse to work, or because they experience greater barriers obtaining jobs. This possibility is addressed next.

Reservation Wages and Working Conditions

The assumption that jobless adults do not want to work has been the basis of great controversy in policy arenas, despite evidence that many, if not most, would accept a job if one were offered (Tienda and Stier 1991). All respondents who were jobless (that is, both unemployed and not looking for work) when interviewed were asked whether they wanted a job and how many days they would be willing to work per week. In addition they were asked a series of questions about the salary and wages they required to accept a job.[10] These items are used to represent reservation wages, which refer to the minimally acceptable wages needed to accept a job.

Table 3, which summarizes this information, reveals that 90 to 100 percent of jobless Black and Mexican fathers indicated a willingness to work for over five days a week, compared to 82 percent of Puerto Rican and 76 percent of White fathers. In other words, among the jobless, it was White men rather than minority men who were least likely to report a "willingness" to work. A similar pattern was obtained for women, except that the ethnic and racial differences were even more pronounced. That is, over 90 percent of jobless Black women reported they wanted a job, compared to 75 percent of Hispanic women and less than 60 percent of White women. Moreover, Black women indicated they were willing to work more days per week, on average, compared to White women. Ethnic differentials in job security reveal that about half of employed Black men perceived themselves to be at high risk of losing their job (due to plant closings or employment reductions), compared to one-third of Hispanic men and just over one-fourth of White men.

Comparisons between actual wage offers and reservation wages reveal pronounced race and ethnic differences. Among the employed, average wage rates of men exceed those of women by 20 (Blacks and Whites) to 50 percent (Mexicans). For both sexes, there is a clear wage hierarchy among the employed, with the highest wages going to Whites, the lowest to Hispanics, and Blacks between these extremes. This pattern of race differences in wage rates is surprising in light of extensive empirical evidence that Hispanics outearn Blacks (Bean and Tienda 1987; Tienda and Jensen 1988). However, the results reported earlier (Table 2) indicate that employed inner-city Black parents may be a highly selective population group because a smaller share of them have jobs (see percent currently working). We consider the implications of such selectivity below.

The reservation wages of jobless workers are systematically lower than the actual wages of employed workers, although the wage gaps vary along gender and ethnic lines. For example, Whites have the highest reservation wages among both fathers and mothers. In absolute terms, it is Black, not Mexican fathers who request the lowest pay in order to accept a job. Despite their higher educational credentials, it is Black fathers who experience the highest rate of joblessness. However, the ratios of reservation wages to actual wages are virtually identical for White and Black men—approximately 0.65. Reservation wages of Mexican men are most in line with the actual wages of their employed counterparts, reaching a ratio of 0.9. Among women there is relatively little ethnic differentiation in wages desired to accept a job. Surprisingly, the reservation and actual wages are most similar for Hispanic women and least similar for White women.

On balance, the tabulations reported in Table 3 do not support claims that minority workers are less willing to work than nonminority workers, nor is there any evidence that minority workers expect higher wages than Whites as a condition of accepting employment. However, whether the reservation wages of jobless parents are realistic in light of the wage opportunities available to other parents from Chicago's inner city cannot be answered from the results reported in Table 3. This is because employed and jobless parents may differ in ways that are systematically correlated with wage offers. Therefore, it

is incorrect to compare directly the reservation wages of jobless individuals with the actual wages of those employed. To address this problem, we estimated the wage rates that men and women with comparable skills and personal characteristics would receive should they enter the labor market by deriving expected (that is, predicted) wages for both the jobless and the employed.[11] These results are summarized in Table 4, and can be readily interpreted as follows.

Expected (predicted) wages, which indicate the approximate offers jobless respondents can expect in the market given their skills and experience, are generally lower than actual or reservation wages, but the deviations vary along gender and ethnic lines. Of interest among the employed is the ratio of minority to White expected wages, which hover around 65 percent for men. However, among employed women, the ratio of expected wages reveals no gap for "statistically" similar White and Black women, a modest wage gap (8 percent) for Puerto Ricans and a substantial gap of 27 percent for Mexican-origin women. Thus, among employed parents, but especially fathers, our results show that *color matters*. Discrimination appears to maintain the color line between wages of minority and nonminority workers.

Of particular relevance for evaluating the hypothesis that inner-city residents do not want to work are the ratios of expected to reported reservation wages. The ratio of expected to reported reservation wages provides some support for claims that minority reservation rates are high relative to available market opportunities (Holzer 1993). However, two noteworthy qualifications are in order. First, the magnitude of the hourly wage disparities is small by most standards, about 10 percent. Second and far more striking are the minority-White disparities in expected wages, which range from 0.52 to 0.65 for men and 0.78 to 0.84 for women. A discrimination story is more consistent with these results than a reservation wage story not only because of the large racial disparities in expected wages, but also because reservation wages are "hypothetical" until a job offer is actually tendered. For Black and Puerto Rican men, the lack of job offers, not low wages, is the more salient aspect of limited opportunity.

Also pertinent for evaluating whether inner-city residents want to work are the ratios comparing the expected wages of employed and jobless parents. If the expected wages of jobless workers exceed those

Table 4 Comparison of Actual, Expected, and Reservation Wages of Parents, Ages 18 to 44 in Chicago's Inner City by Gender and Ethnicity, 1987

	Black	White	Mexican	Puerto Rican
Men, employed				
(1) Actual wages	$8.32	$12.80	$7.35	$7.72
Ratio to Whites	0.65	—	0.57	0.60
(2) Expected wages	$6.91	$10.60	$6.53	$6.81
Ratio to Whites	0.65	—	0.62	0.64
Men, jobless				
(3) Reservation wages	$5.55	$8.66	$6.89	$5.98
(4) Expected wages	$4.98	$9.60	$6.21	$5.55
Ratio (4) to (3)	0.90	1.10	0.90	0.93
Ratio (4) to (2)	0.72	0.90	0.95	0.81
Ratio to Whites	0.52	—	0.65	0.58
Women, employed				
(1) Actual wages	$6.77	$10.63	$4.74	$6.06
Ratio to Whites	0.64	—	0.45	0.57
(2) Expected wages	$5.97	$6.15	$4.46	$5.63
Ratio to Whites	0.97	—	0.73	0.92
Women, jobless				
(3) Reservation wages	$5.30	$5.36	$4.65	$4.11
(4) Expected wages	$4.55	$5.40	$4.29	$4.22
Ratio (4) to (3)	0.86	1.01	0.92	1.03
Ratio (4) to (2)	0.76	0.88	0.96	0.75
Ratio to Whites	0.84	—	0.79	0.78

Source: UPFLS (1987).

of their employed ethnic counterparts, then this would lend credence to arguments that unrealistic reservation wages are partly responsible for the low employment rates of inner-city workers. Once again, there is no evidence that this is the case for any of the groups compared. In fact, Black inner-city parents exhibit the lowest expected wage ratio between jobless and employed parents—about 0.75 for both sexes, even though Black parents have the highest rates of joblessness. For White and Mexican parents, the ratio of expected wages for jobless relative to employed parents ranges from 0.9 to 0.95, and Puerto Ricans fall between these extremes with ratios ranging from 0.75 to 0.81.

Overall, these results indicate that the reservation wages of jobless inner-city parents are not greatly out of line with those of comparably skilled workers. Stated more strongly, there is mixed support for claims that jobless parents are unable to find work because the minimum wages they require are unrealistically high relative to their skill endowments. However, there is a clear penalty for color—that is, expected wages of minority parents are but a fraction of their statistical White counterparts. The lower educational attainment of Hispanic women explains their lower wage rates relative to White women, but schooling deficits are not an obvious explanation for the wage gaps between Black and White inner-city mothers. And despite the role of unions in protecting wages of semiskilled and unskilled workers in the past, union membership is not an obvious explanation for the higher wage rates of White relative to minority inner-city parents, because White workers report lower rates of union membership than minority parents, yet their wages are appreciably higher.

The economic precariousness of the firms employing minority parents may also be partly responsible for the observed differences in wage rates coupled with strong statistical discrimination against Black and Puerto Rican men and women. That is, even if Wilson's (1987) claims are correct that urban job opportunities contracted during the 1980s, thereby resulting in higher joblessness for inner-city workers, this does not explain why employment options declined disproportionately for people of color, and particularly those residing in extremely high poverty areas.

Conclusion

Our emphasis on discrimination is not intended to discount the importance of explanations of persisting race and ethnic differences in wages which emphasize the surplus of laborers willing to work at low wages. Although we have taken into account individual differences in education and experience, there probably remain unmeasured skill differences (English proficiency, for example) among the groups. Ultimately, shifts in both supply and demand curves set wages, hence the various individual and structural interpretations reviewed at the outset are pertinent for understanding persisting wage disparities along color lines (Holzer 1993). Our approach considered both supply and demand factors that have been summoned to explain increased labor market inequities along race and ethnic lines.

By way of a tentative conclusion we propose that discrimination against people of color residing in the inner city, but especially Blacks, was heightened during the 1980s as the share of low-skill jobs declined. Even though Black jobless parents reported the lowest reservation wage rates and received the lowest wages among those with a job, employers are increasingly reluctant to hire them, particularly when alternative sources of immigrant labor are readily available. Stated differently, employers seem to prefer Mexican and White workers over Blacks, and to a somewhat lesser extent Puerto Ricans. Not only does color matter, but it has an exchange value that fluctuates according to general economic conditions and the availability of alternative sources of unskilled labor. But we want to be clear that, while the market may very well set prices, it is employers, not the market, who discriminate.

Notes

1. Officially there were 856 census tracts within the city boundaries, of which 843 retained the same boundaries used in 1970. All of the sampled tracts were taken from the subset with constant boundaries.

2. That the White sample fell well below the target reflects the low share of White parents residing in high-poverty neighborhoods.

3. The NSFH includes 13,017 respondents interviewed between March 1987 and May 1988. It includes a main sample of 9,643 respondents representing the noninstitutional U.S. population aged nineteen or older. In addition, several population groups were oversampled by a factor of two, including Blacks, Puerto Ricans, and Chicanos, single parents, persons with stepchildren, cohabiting persons, and persons who were recently married.

4. Unfortunately, it is not possible to tabulate separately Puerto Rican men in the NSFH. Although the participation rates for Puerto Rican women are lower than those of other minority women (Bean and Tienda 1987; Tienda 1989; Tienda, Donato, and Cordero-Guzman 1992), the small number of Puerto Ricans in the national sample warrants caution in interpreting the actual estimates.

5. The Chicago sample is approximately 85 percent foreign-born versus approximately 25 to 30 percent for the national population, depending on the enumeration of undocumented immigrants. The overrepresentation of immigrant Mexicans in the Chicago sample reflects the higher residential concentration of recent arrivals in poorer ethnic neighborhoods compared to native-born persons of Mexican descent.

6. All names used are fictive.

7. This was in response to a question about opportunity in Chicago.

8. This was in response to a question about who gets the bad jobs in Chicago.

9. Apparently the GED she received didn't provide a balanced account of U.S. labor history.

10. The exact wording of the question is: 6. A. What would the wage or salary have to be for you to be willing to take a job? *If R responds "Minimum Wage," re-ask A. Probe if necessary:* Is that per hour, day, week, or what?

11. These results and the technical procedures used are available from the authors. We use the term expected wages to indicate the average wage workers should expect given their education and labor force experience.

32

Language Acquisition and Loss Among Children of Immigrants

ALEJANDRO PORTES
RICHARD SCHAUFFLER

Where linguistic unity has broken down, our energies and resources flow into tensions, hostilities, prejudices, and resentments. These develop and persist. Within a few years, if the breakdown persists, there will be no retreat. It becomes irrevocable, irreversible. Society as we know it can fade into noisy babel and then chaos. — U.S. English policy statement

The United States is a rather unforgiving country when it comes to language use. Unlike other nations around the world where the use of more than one language is tolerated, North American culture demands a very rapid shift to English from those who are not native speakers. Other aspects of an immigrant group's culture may survive, but its native language is generally lost in America between the first and third generations. In this chapter, we examine findings from a study of children of immigrants that provides a first look at how the process of language adaptation is taking place among the "new" second generation—that which emerged in the wake of accelerated U.S.–bound migration during the last two decades.

While much research has been done comparing the experiences of recent immigrants to those who arrived at the turn of the century, not much is known about the adaptation process of their offspring. The sociological literature that exists on the second generation is based on the experience of children of the prior wave, mostly European pre–World War I immigrants. There is little reason to think that the adaptation experience of today's children of immigrants will perfectly resemble those of their earlier counterparts because of differences in the national composition of these immigration flows and major changes in the American economy and society.

Contemporary immigration was triggered by the 1965 Immigration Act, as well as by subsequent changes in American asylum and refugee policies. The overall direction of the new policies was toward greater universalism, eliminating previous discriminatory racial barriers and opening the doors of the country on the basis of uniform criteria. Since 1965, occupational skills, family reunification, and fear of political persecution have been the guiding criteria of U.S. immigration policy. As is well known, the result has been a rapid increase in immigration and an equally rapid shift in its origins from Europe to the Third World. Most of today's children of immigrants thus have parents who came from Latin America and Asia and about half are phenotypically non-White (Portes and Zhou 1993).[1] Cultural characteristics and skin color can affect significantly these youths' still unexplored process of adaptation to American society.

A key aspect of the process is, of course, the extent and rapidity with which children of immigrants shed their parent's languages in favor of English.

The size of the new immigration has given rise to sizable language enclaves in several U.S. cities and parallel nativistic concerns about these enclaves' resilience. In particular, the growing use of Spanish by Latin American immigrants has triggered increasingly gloomy assessments of the future prospects of linguistic unity, as illustrated by our opening quote. Miami is one of the cities most directly affected by recent immigration and the growing use of Spanish. The fact that our second-generation sample comes from that city allows us to examine directly the current scholarly and public concerns about the future of language assimilation in America.

Language Assimilation in Historical Perspective

The current debate over official language policy is the latest expression of a cyclical trend in the history of the country since its colonization. Generations of assimilated immigrants who "dropped the hyphen" and considered themselves just plain Americans have looked suspiciously upon later arrivals as the source of potential cultural disintegration. This was true even in the period prior to the Revolutionary War, when Benjamin Franklin complained in a letter written in 1751 that German immigrants in Pennsylvania "will shortly be so numerous as to Germanize us instead of our Anglifying them, and will never adopt our Language or Customs, any more than they can acquire our complexion" (Franklin 1959, p. 234).[2] The notion of "one nation, one language" has been often idealized as a state of linguistic perfection to which the nation should return. The biblical version of this myth is the threat of the fall from a monolingual Eden to a multilingual Babel in which the profusion of languages is associated with the decline of public morals. This idea was amplified by philosophers of the seventeenth and eighteenth centuries, including President John Adams, who contended that "language influences not only the form of government, but the temper, sentiments, and manners of the people" (Adams [1780] 1856, 7: 249–251).

During the colonial period, the notion that the country and its citizens were defined by a common language was justified on two grounds. First, along with incipient American nationalism came the idea that American English both reflected and constituted the democratic and rational nature of the country. American English was explicitly contrasted with

European languages, thought to embody the aristocratic and irrational traditions of the Old World. Second, the acquisition and use of English was seen as the litmus test of citizenship.[3] Lacking a common culture or common history, the use of English came to be seen as an essential part of "real" Americanism.

The two rationales were related insofar as the ability to think logically, seen as necessary for a democracy, was only possible on the basis of fluency in English. This perceived necessity for "Good English" has taken many forms throughout American history. In 1902, New Mexico's statehood was delayed until, in the words of one senator, "the migration of English-speaking people who have been citizens of other States does its modifying work with the Mexican element" (Baron 1990, p. 8). Nebraska banned teaching any foreign language to students below the ninth grade in 1919. Formal "Good English" campaigns were organized from 1918 to the early 1920s, in which language loyalty oaths like the following were designed for children to achieve the goal of "Good English" (Knowles-Robbins 1918, pp. 163–176):

I love the United States of America. I love my country's flag.

I love my country's language. I promise:

1. That I will not dishonor my country's speech by leaving off the last syllables of words.
2. That I will, as a good American, say "yes" and "no" in place of an Indian grunt "um-hum" and "nup-um" or a foreign "ya" or "yeh" and "nope."
3. That I will do my best to improve American speech by avoiding loud, rough tones, by enunciating distinctly, and by speaking pleasantly, clearly, and sincerely,
4. That I will learn to articulate correctly as many words as possible during the year.

In this and other ways, in schools and public life, monolingualism was linked to the idea of democracy, national unity, and allegiance to the country. Although many parents of upper- and middle-class backgrounds encouraged their children to learn Latin, French, or German, bilingualism on the part of immigrants was frowned upon. In part, this may have

While in general the United States has been a language graveyard, new immigrants have sought to preserve their languages, because it was those in which they functioned best and were skilled, and also because doing so upheld their cultures and their identities. Here a Spanish sign on a store window in Miami in 1963 displayed the hours the business was open and announced that English (rather than Spanish) was spoken. (Photo by Fraser Hale. Historical Museum of Southern Florida)

been due to the association of non-English languages with the existence of large ethnic communities which lay beyond the pale of the English-speaking population, out sight but never out of mind.

The Shifting Implications of Bilingualism

During the early twentieth century, opposition to bilingualism derived strength from the then dominant scientific wisdom. Academic studies in the fields of education and psychology argued that bilingualism created failure and mental confusion, and damaged the psychological well-being of immigrant children. Two schools of thought existed at the time: one

which argued that lower intelligence caused the failure of children to acquire English, and another which argued the opposite. The first school (low intelligence/low English) based its conclusions on beliefs about genetic differences between races, arguing that heredity limited the ability of immigrants to learn. The second school (low English/low intelligence) based its conclusions on beliefs about environmental factors, in particular the use of a foreign language at home. Intellectual failure was imputed by this school to the "linguistic confusion" of these children exposed to two languages.

It was not until 1962 that these views were convincingly disproved by a methodologically sound study of the effects of bilingualism. French- and English-speaking children in Canada were studied by Peal and Lambert (1962) who demonstrated that, if social class was taken into account, true bilingualism was associated with higher scores on a variety of intelligence tests (Cummins 1981; Lambert and Tucker 1972). True bilinguals, defined as those who could communicate competently in two languages, were shown to enjoy a greater degree of cognitive flexibility and an enhanced ability to deal with abstract concepts than their monolingual peers. Instead of creating "confusion," having two symbols for each object enhanced understanding.

Subsequent studies have consistently supported the positive association between bilingualism and scholastic achievement. An analysis of a national sample of high school students in the United States, for example, found a positive correlation between academic achievement and bilingualism among Hispanic youth (Fernandez and Nielsen 1986). More recently, a study of San Diego high school students also showed significant differences in performance between true bilinguals and monolinguals, as well as between true bilinguals (defined by the local school system as fluent English proficient) and semibilinguals (defined as limited English proficient). Again, true bilingualism was shown to be a positive intellectual influence associated with higher scholastic achievement (Rumbaut and Ima 1988b).

Despite the demonstrated advantages of bilingualism, the United States is unique in the rapid rate at which other languages are abandoned in favor of English. In no other country are foreign languages extinguished with such speed (Lieberson, Dalto, and Johnston 1975). The typical pattern is for the first generation to learn enough English to survive economically; the second generation continues to speak the parental tongue at home, but English at school, at work, and in public life; by the third generation, the home language shifts to English which effectively becomes the mother tongue for subsequent generations.

This pattern has held true of all immigrant groups with the exception of some isolated minorities like the German-speaking Old Order Amish and Russian-speaking Old Believers. Spanish-speakers are an exception to this general pattern only in the sense that language assimilation is slowed down by the continued stream of monolingual Spanish immigrants into the same geographical areas. The delay

brought about by this concentration extends only to the first or foreign-born generation (Fishman 1988; Veltman 1983). So far, there is no evidence of seriously delayed linguistic assimilation among children of Spanish-speaking immigrants, although not enough research has been conducted on the topic. The empirical results presented in the following sections speak directly to this question.

Recent theoretical developments in the sociology of immigration have emphasized the role of social and cultural contexts in the process of adaptation. These developments lead to specific hypotheses concerning language assimilation among second-generation youth (Portes and Rumbaut 1990, chap. 6): Children growing up in sociocultural contexts where the native English-speaking majority is dominant or where immigrants from other linguistic backgrounds are most numerous will experience a faster process of home language loss and a rapid conversion to English monolingualism; those raised in contexts where there is a large conational immigrant concentration will have a greater probability of language retention. This is especially true if the immigrant concentration is a diversified ethnic "enclave" where a host of economic, social, and cultural activities are conducted in the parental language. In such instances, there is a clear economic incentive to retain proficiency in that language, along with greater ease in learning it and practicing it within the community. The predicted outcome is widespread bilingualism. Children growing up in large but impoverished immigrant communities occupy an intermediate position. There is less incentive for home language retention, especially if the use of this language is stigmatized as a symbol of lower status by the surrounding community. In these instances, the most likely outcome is limited bilingualism with less common usage and lower preference for the parental language than among offspring of more economically successful groups. Figure 1 summarizes these prediction.

Political Repercussions

Today, and despite ample research evidence on the potential benefits of bilingualism, considerable animosity remains among the native English-speaking population about immigrants' retention of their languages. In the 1980s, the so-called "English Only" or "English Official" movement gathered momentum,

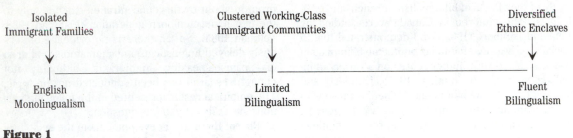

Figure 1
Predicted Second-Generation Language Outcomes by Context of Reception

presenting its concerns in the form of a policy question: Should English be declared the official language of government? Lobbying organizations such as U.S. English have placed a national version of an official language law, the so-called English Language Amendment (ELA) to the U.S. Constitution, before Congress every year since 1981. By now, almost every state has debated a state-level version of this law with eighteen having passed such legislation,[4] Advocates of English-only decisively oppose bilingualism in general, and bilingual ballots and bilingual education in particular, as a threat to the preservation of national unity. In their view, schooling and civic transactions conducted in a language other than English help preserve language enclaves that are both apart from and hostile to mainstream culture. Bilingualism is seen as contributing to a decline in the gross national product and increases in civil strife, as well as a threat to national security.[5]

Miami, the site of our study, is home to more foreign-born residents on a proportional basis than any other American city. Cuban immigrants have built a large and diversified ethnic community, originally known as Little Havana, which also serves as a cultural resource for other Latin American immigrants. Haitians are seeking to do the same in Little Haiti, a neighborhood which lies directly adjacent to Liberty City, Miami's principal African American section. Many native-born Whites have reacted to the immigrant influx and the emergence of the Cuban enclave by leaving the city or by militantly supporting the English-only movement. The result has been a debate over language more acrimonious than in other American cities (Portes and Stepick 1993).

In 1973, county commissioners voted to declare Dade County officially bilingual. Seven years later, however, a grassroots-led referendum repealed that ordinance and replaced it with a new one stipulating

that public funds could not be used to teach languages other than English or "promote a culture other than the culture of the United States" (Boswell and Curtis 1984, p. 121). In 1984, that ordinance was amended to legalize the use of languages other than English in tourism, emergency preparedness, and hospital information services for patients. In early 1993, the newly elected Dade County Commission, where Cuban Americans now comprise a plurality, rescinded the antibilingual ordinance mandating that public notices and brochures be printed in Spanish and, in certain cases, in French Creole, as well as in English. The decision triggered an immediate spate of lawsuits by opponents who argued that the county could not countermand the English-only amendment to the state constitution, passed two years earlier (Stewart 1993).

There is little doubt that foreign languages, particularly Spanish, are alive and well among first-generation immigrants in Miami. This pattern follows closely that established by other large immigrant groups in the past. Italian, Polish, and Jewish communities created by turn-of-the-century immigrants retained their home language for several decades (Glazer 1954). The key question is, however, the extent to which the extraordinary concentration of foreign-born groups in Miami is altering this historical trend by delaying the transition to English among second-generation youth and fostering their preference for foreign languages. We address this question next.

Language and the Second Generation in South Florida

The data set on which the following results are based comes from a survey of 2,660 eighth- and ninth-grade students in Miami (Dade County) and adjacent

Ft. Lauderdale (Broward County) schools. Included in this sample are children of the most diverse national origins, although, reflecting the composition of the immigrant population to the area, the largest contingents come from Cuba, Nicaragua, other Latin American countries, Haiti, and the West Indies. The survey defined "second generation" as youth born in the United States with at least one foreign-born parent (55.7 percent of the sample) and children born abroad but who had lived in the United States for five years or more (44 percent). The sample is evenly divided between boys and girls and the average age of respondents is 14.8 years. The sample included both inner-city and suburban schools and targeted schools where children of particular immigrant groups were known to concentrate. Ft. Lauderdale schools were selected to provide a counterpoint to Miami in terms of lower concentrations of second-generation youth. In total, twenty-two schools in both counties were included in the final sample.

In order to understand the language patterns of foreign-origin youth, we examined certain characteristics of the children themselves, their parents, and their families that have been shown in previous studies to influence the process of linguistic assimilation. These include age, sex, national origin, length of residence in the United States, education of the father and the mother, and the family's overall socioeconomic status. We explore how these characteristics affect the children's proficiency in English, their knowledge of the parental language, and their overall linguistic preferences. Our measure of English language proficiency is constructed from the students' reported ability to speak, understand, read, and write English.[6] Students chose their ability rating from four categories (Not at All, Not Well, Well, and Very Well), which were converted into an overall proficiency score. Other related questions asked about retention of the parental language, knowledge of any other foreign language, and language preferences. Table 1 presents our findings.

The first question of interest is the extent to which second-generation youth learn English. On this point, the evidence is unequivocal. For the sample as a whole, 73 percent report that they are able to speak, understand, read, and write English "very well" and an additional 26 percent "well." This leaves the sum total of those knowing little or no English at just 1 percent. Table 1 presents cross-tabulations of English proficiency with eight characteristics. In order to facilitate understanding these results, we discuss only those that are statistically significant.[7] Under this criterion, age, national origin, length of U.S. residence, parental education, and class self-identification are all significantly related to English proficiency. It is important to note that such differences exist only between the "well" and "very well" categories, signaling relative minor variations in English knowledge. Predictably, length of U.S. residence has the strongest association with English proficiency. Slightly over half of foreign-born children arriving recently report knowing English very well; the figure climbs to more than 80 percent among the native-born.

National origin also has a strong correlation with English ability. In this panel, the large Cuban-origin group is divided into those attending Latin-oriented private schools in Miami and those attending public schools. Differences between both groups on English knowledge are minimal, since over 70 percent of each category report knowing English very well. Highest proficiency is associated with children of European and Asian origin, grouped in the "Other" category, and with those of West Indian parentage. The latter result is directly attributable to the fact that most West Indian parents come from English-speaking countries such as Jamaica, Trinidad, Grenada, and the Bahamas. Second-generation Nicaraguans have the lowest English proficiency. This result is a direct consequence of the relative recency of Nicaraguan-origin migration. Very few of our Nicaraguan respondents are U.S.–born and most have been in the country less than nine years.

Weaker but still significant associations with father's education, mother's education, and family's reported socioeconomic status follow a predictable pattern. In every case, the higher the parental position, the better the command of English. The association with age is counterintuitive, however, since older children indicate less proficiency. This pattern is due to the tendency of more recent immigrant youth to enter school at grades lower than the respective native-born age cohort. In this sample, older students generally come from non-English-speaking countries and are among the most recent arrivals. Nicaraguan-origin youth form a sizable proportion of this group.

The key story in the first panel of Table 1 is the overwhelming dominance of English knowledge among children of immigrants and its strong positive association with length of residence in the United

Table 1 Language Patterns among Second-Generation Youth in South Florida, 1992 (*continued next page*)

| | English Proficiency | | | | Foreign Language Proficiency | | | | Language Preference: "I generally prefer to speak..." | | | |
	Not at All/Not Well %	Well %	Very Well %	$p<$[a]	Not at All/Not Well %	Well %	Very Well %	$p<$[a]	Other English %	Language %	$p<$[a]	Totals N[b]
Sex				n.s.[c]				n.s.			n.s.	
Boy	1.3	29.0	69.7		34.8	37.8	27.4		82.1	17.9		1286
Girl	1.2	23.3	75.5		33.2	33.8	33.0		81.0	19.0		1374
Age				0.001				n.s.			0.001	
< 13	0.6	19.7	79.7		33.3	38.4	28.3		87.7	12.3		513
14	1.2	23.7	75.0		32.2	36.9	30.8		80.8	19.2		1206
15	0.7	31.5	67.9		35.7	32.9	31.3		80.4	19.6		750
> 16	5.8	36.6	57.6		39.3	32.5	28.3		74.3	25.7		191
National origin				0.001				0.001			0.001	
Cuban, private school	1.2	27.3	71.5		11.0	51.2	37.8		93.6	6.4		172
Cuban, public school	0.6	24.0	75.4		27.8	39.7	32.5		81.8	18.2		968
Nicaraguan	3.1	42.0	54.9		21.9	36.4	41.7		74.3	25.7		319
Other Latin American	1.0	27.1	71.9		27.1	39.1	33.9		76.0	24.0		676
Haitian	4.4	23.5	72.1		70.6	19.1	10.3		87.5	12.5		136
West Indian	0.5	15.2	84.3		75.9	14.7	9.4		89.0	11.0		191
Other	0.5	18.0	81.5		61.4	22.8	15.9		89.4	10.6		189
Length of U.S. residence				0.001				0.001			0.001	
9 years or less	2.7	41.3	56.0		22.2	34.8	43.0		71.6	28.4		589
10 years or more	0.9	29.0	70.1		35.0	35.1	29.9		79.3	20.7		686
U.S.-born	0.9	18.1	81.0		38.4	36.5	25.1		86.9	13.1		1385

States. A very different story emerges when we consider retention of their parental languages. The same set of questions about English language knowledge was used to measure knowledge of the parental language or another foreign language. The index of foreign language proficiently was built in an identical manner to the measure of English knowledge that we reported above. The second panel of Table 1 presents cross-tabulations of this variable with the same set of individual and parental traits.

The bottom row of this panel indicates that one-third of the sample are already English monolinguals. The absolute number of such cases (902) far exceeds the number of children of West Indian and

other English-speaking nationalities, indicating a rapid loss of parental language among other immigrant groups.[8] Yet a comparable proportion of respondents report knowing their parental language "very well" and, hence, it is important to examine possible determinants of this difference. Among the set of potential predictors in Table 1, national origin has by far the strongest effect on home language retention. There is a clear difference between Latin American nationalities, on the one hand, and Haitian, West Indian, and Asian/European nationalities, on the other. Reported English monolingualism among West Indian-origin students and some students grouped in the "Other" category is again a straight-

Table 1   (cont.)

Characteristic	English Proficiency				Foreign Language Proficiency				Language Preference: "I generally prefer to speak. . ."			Totals N^b
	Not at All/Not Well %	Well %	Very Well %	$p<^a$	Not at All/Not Well %	Well %	Very Well %	$p<^a$	Other Language %	English %	$p<^a$	
County				n.s.				0.001			n.s.	
Dade	1.3	26.8	71.9		31.7	37.0	31.3		81.4	18.6		2467
Broward	1.6	16.1	82.4		62.7	20.2	17.1		82.9	17.1		193
Father's education				0.001				n.s.			n.s.	
Not high school graduate	1.8	30.6	67.6		30.4	37.0	32.6		78.5	21.5		605
High school graduate[d]	1.4	26.8	71.8		36.5	35.0	28.5		81.3	18.7		1321
College graduate	0.7	21.0	78.3		32.3	36.1	31.6		84.5	15.5		734
Mother's education				0.001				n.s.			n.s.	
Not high school graduate	1.5	30.6	67.9		33.4	35.0	32.1		76.8	23.2		647
High school graduate[d]	1.6	25.7	72.8		35.4	35.6	36.5		83.1	16.9		1415
College graduate	.3	22.1	77.6		31.2	29.4	31.4		82.9	17.1		598
Family socioeconomic status[e]				0.001				n.s.			0.001	
Poor	5.9	41.2	52.9		23.5	35.3	41.2		29.4	70.6		17
Working class	2.1	34.2	63.8		32.1	36.6	31.3		23.5	76.5		527
Lower middle class	1.4	26.7	71.9		33.4	36.2	30.3		18.9	81.1		900
Upper middle class	0.7	21.9	77.4		35.3	35.0	29.7		15.8	84.2		1216
Totals	1.3	26.1	72.7		33.9	35.8	30.3		18.5	81.5		2660

[a] Pearson's Chi Square probability.
[b] Absolute frequencies.
[c] Differences not statistically significant.
[d] Includes those with some college.
[e] Respondent's self-report.

forward consequence of their parents being English speakers. The same is not the case, however, among Haitian-origin youth whose home language is French or Creole. Nearly three-fourths of this group reports little or no knowledge of parental languages and only 10 percent declare themselves proficient in either.

The opposite is true among Latin American groups where home language loss affects only about one-fourth of respondents and drops to 11 percent among Cuban-origin students in private schools. Retention of the parental language, overwhelmingly Spanish, is in part a consequence of the recency of some migrant flows, such as the Nicaraguans. More importantly, however, it reflects the presence of a large and diversified ethnic community where Spanish is the language of intercourse for all kinds of transactions. Private school Cuban youth are the children of middle-class exiles who represent the core of this enclave economy. It is not surprising that they have the lowest propensity to give up Spanish. Combined with the pattern of responses in the first panel of the table, these results indicate that Cuban and other Latin American–origin youth in south Florida are mostly bilingual.

The originally Cuban and now pan-Latin enclave is located in Miami (Dade County). Hence, it is possible to predict that retention of Spanish will be greater among second-generation youth in Dade County than in adjacent Broward County (Ft. Lauderdale) where no similar phenomenon exists. This expectation is borne out by our results. Community of residence has the second strongest association with home language retention, with Dade County residents being almost twice as likely to be bilinguals (knowing the parental language "well" or "very well") as those living in Broward.[9] The very strong influence of community of residence is counteracted, however, by the passage of time. As shown in the second panel of Table 1, the longer the child has resided in the United States, the stronger the tendency toward English monolingualism. Among more recent arrivals, over 40 percent report full command of a foreign language, a figure that falls to just one-fourth among the native-born.

Parental education and family status have essentially no association with home language retention. This surprising result may be attributable to the contradictory effects of these variables. On the one hand, better-educated and higher status parents facilitate greater exposure of their offspring to American culture, and hence, accelerate their language assimilation. On the other hand, these parents have the greatest resources to promote retention of their own languages, such as sending children to bilingual private schools. Impressionistic evidence gathered by our project indicate that immigrant parents are consistently in favor of English language acquistion, but not at the cost of giving up their mother tongue.

Overall, these findings provide strong support for the hypotheses summarized in Figure 1: Children of relatively isolated immigrants—such as those living in Broward County, or Asians and Europeans grouped in the "Other" category—experience a faster language transition toward English only; offspring of relatively prosperous and highly concentrated immigrants, such as Cubans, are far more likely to retain the home language. The theoretical model does not do so well, however, in the case of

second-generation youth coming from clustered but poorer immigrant groups. Haitian-origin students, who best fit this characterization, are as prone to give up their parent's language as those from much less clustered groups. Nicaraguans, who also form a numerous and relatively poor community, are much less likely to have shifted to English monolingualism in the second generation, but this is the result of their relative recency in the country. The more complex analysis of the following section will clarify the relative explanatory weight of these various factors.

A final variable of interest is the youths' attitude toward speaking English *versus* speaking the parental language: Just because children of immigrants report high proficiency in English does not guarantee that they will use it, given the choice. The evidence on this point is presented in the last panel of Table 1. Preference for English is overwhelming: 80 percent of the entire sample endorses this alternative. Length of U.S. residence is strongly and positively correlated with English preference, but even among the most recent arrivals more than 70 percent opt for English over their home languages.

National origin is significantly associated with language preference, but the trend here differs from those found previously. Children of Haitian and West Indian parents, as well as those grouped in the "Other" category, lean strongly toward English in a fashion congruent with their weak retention of other languages. Cubans, however, also have a very strong preference for English, in particular those attending private schools. Despite their greater reported knowledge of their parent's language, over 90 percent of Cuban-origin youth prefer English. This result means that even among youth educated in Latin-oriented schools at the core of an ethnic enclave, linguistic assimilation is proceeding with remarkable speed. Somewhat lower attachment to English is found among Nicaraguans and other Latin Americans, a probable consequence of their recent arrival, but even among these groups three-fourths endorse their new country's language over native Spanish.

The significant and negative association of age with English preference corresponds to the pattern observed for actual knowledge of English, and is partially attributable to the older age of many recently arrived youth. Family socioeconomic status also has a significant and positive association with English preference. A 14 percent difference separates those coming from poor families from those reporting

Table 2   Least-Square Regression of Knowledge of a Foreign Language (Logged) on Selected Predictors ($N = 2,659$)

Independent Variable[a]	r[b]	b[c]	β[d]	τ[e]
Sex (female)	.033	7.133	0.096	5.79
Age	−.008	−0.161	−0.003	0.22
Length of U.S. residence	−.150	−9.075	−0.197	10.73[f]
County (Broward)	−.246	−14.579	−0.102	5.87[f]
Father's education	−.033	−0.575	−0.021	1.08
Mother's education	−.048	−0.291	−0.011	0.53
Father's occupation	−.013	−0.021	−0.013	0.70
Mother's occupation	−.041	0.001	0.001	0.08
Family socioeconomic status	−.045	−0.422	−0.010	0.57
National origin:				
Cuban, private school	.115	46.858	0.309	13.92[f]
Cuban, public school	.158	32.107	0.414	12.62[f]
Nicaraguan	.119	27.229	0.237	9.03[f]
Other Latin American	.115	29.327	0.343	11.34[f]
Haitian	−.144	−0.805	−0.005	0.23
West Indian	−.409	−31.436	−0.218	9.92[f]
Constant	—	198.152	—	—
R			0.563	
R^2			0.313	

[a] Variables coded according to their labels. Highest code in parentheses.
[b] Product-moment zero-order correlations.
[c] Regression coefficients taken to the fifth significant digit and multiplied by 100.
[d] Beta weights.
[e] T-ratios.
[f] $p < 0.001$

themselves as upper middle class. But again, these are differences in the context of overwhelming language assimilation. An eloquent indicator of this fact is the absence of differences between students in Dade and Broward schools. This result says that, regardless of whether second-generation children live in an English-only environment or in one where foreign languages are commonly spoken, their preference for the language of the land is the same.

What Makes a Bilingual?

The principal difference observed in this children of immigrants sample pertains to home language retention rather than English acquisition. In other words, the central question is not whether immigrant youth learn and prefer English, but the extent to which they retain some command of the parental foreign language. When they do, the second generation will be bilingual; when they don't, it will join the monolin-

gual majority. As seen in the preceding section, there is some confusion about the relative effects of national origin, length of U.S. residence, and place of residence as significant predictors of foreign language knowledge. To clarify them, we conducted a more complex statistical analysis in order to establish unambiguously the relative effects of different individual and family characteristics on language retention. The results are presented in Table 2. We outline in this section a nontechnical summary of these findings.

They reveal clearly the nature of the forces arrayed for and against bilingualism in the second generation.[10] The most significant factor on the probilingual side is national origin. Latin American nationalities display, without exception, a much greater probability to retain the parental language. Cuban-origin students in public schools are, for example, almost a third (32 percent) more likely to know Spanish than other nationalities are to know their respective foreign language, controlling for other variables. The corresponding figure for Cubans in private school is 47 percent, and that for Nicaraguans is a smaller but still significant 27 percent. The strong negative effect of West Indian origin on bilingualism (–31 percent) is again interpretable as a direct result of monolingual English-speaking parents.

The most significant factors on the promonolingual side of the equation are time in the United States and residence outside of Miami. These factors correspond to the temporal and spatial dimensions of language assimilation. Each five-year period of U.S. residence reduces foreign language proficiency by approximately 9 percent net of other variables. Living in Ft. Lauderdale rather than in the Miami enclave reduces proficiency in this sample by almost 15 percent. The single unexpected finding is the effect of gender: girls have a slight, but significant, advantage over boys in parental language retention.

The results of this analysis tend to fit the predictions about second-generation bilingualism summarized in Figure 1: Children of immigrants living outside areas of coethnic concentration give up their parental languages faster and the process is further accelerated by the passage of time; those raised in internally diversified ethnic communities have a much greater chance of retaining some command of their parental language. The Miami economic and cultural enclave represents the key factor accounting for the endurance of Spanish in this sample. The finding that French or Haitian Creole do not do nearly as well suggests that, contrary to our original expectations, youth from concentrated but impoverished immigrant communities experience the shift toward English monolingualism as fast as isolated immigrant groups.

Conclusion

English is alive and well in Miami. This is the American city most heavily affected on a relative basis by recent immigration and, hence, the one where the demise of English dominance, so feared by nativist organizations, should be most evident. Our results indicate that such fears are unfounded. Second-generation youth not only report widespread competence in English, but also demonstrate on unambiguous preference for it in everyday communication.

Children raised in the core of the Spanish-speaking Miami community (those attending bilingual private schools) are actually the most enthusiastic in their preference for the language of the land. These results indicate that, contrary to nativist fears, what is at risk is the preservation of some competence in the languages spoken by immigrant parents. Given that foreign language knowledge is a valuable intellectual asset, the rapid transition toward monolingualism represents a loss to the individual and to the community as a whole.

Based on an analysis of U.S. historical data, Portes and Rumbaut (1990, chap. 6) noted that even highly educated immigrant parents did not stand much of a chance of transmitting their language to their children if they did not have strong social supports. The overwhelming dominance of English monolingualism would effectively neutralize any such efforts. This prediction is confirmed in these results by the insignificant effects of father's and mother's education on parental language retention and by the heavy influence of place of residence. Immigrant parents' dreams of communicating with their children and grandchildren in their native language will not come true, for the most part. By the same token, nativist fears that the nation will fragment into a multilingual babel will prove unfounded. Only in places where immigrant groups concentrate and manage to sustain a diversified economic and cultural presence will their languages survive past the first generation. In the absence of policies promoting bilingualism, even these enclaves are likely to be engulfed in the course of two or three generations by the irresistible force of language assimilation.

Notes

1. The figure is arrived at by adding the percentage of Latin American immigrants estimated to be non-White on the basis of the 1990 census figures on the racial composition of the Hispanic population to Asian and Black immigrants.

2. Franklin imagined the Germans to be of darker hue than the English settlers. The general attitude of European colonists toward the indigenous languages of the New World was even more contemptuous (see Dillard 1985; Greenblatt 1976; and Laird 1970). Subsequent educational policies and political regulations aimed to eradicate Native American tribal languages, including forced removal of children from their families and homes at an early age (Berrally and McCarty 1990); (see also Leap 1981). Similarly, European colonists sought to prevent Africans arriving in the United States as slaves from maintaining their languages. For a more extensive discussion of the suppression of African languages and the influence of African languages on English, see Bailey, Maynor, and Cukor-Avila (1991), Herskovits (1990), Marckwardt (1980), Mintz (1970), and Veltman (1983).

3. For a full exposition of these developments see Baron (1990).

4. As of mid-1993, Arizona, Arkansas, California, Colorado, Florida, Georgia, Hawaii, Illinois, Indiana, Kentucky, Mississippi, Nebraska, North Carolina, North Dakota, South Carolina, South Dakota, Tennessee, and Virginia had passed such laws; Kansas, Minnesota, Wisconsin, Michigan, Ohio, Pennsylvania, New York, New Jersey, and Massachusetts were considering passage. Delaware, Iowa, Maryland, Missouri, New Hampshire, Texas, Utah, and West Virginia all defeated versions of this law (Baron 1990, p. 201).

5. This argument is presented in a 1989 report from the Council for Inter-American Security, known as *Santa Fe II* (Crawford 1986). Counterarguments regarding GNP and civil strife can be found in two papers by Fishman and Solano (1989, 1990).

6. Previous studies (Fishman 1969; Fishman and Cooper 1969; Fishman and Terry 1969) indicate that self-report measures are both reliable and valid.

7. Because of the large sample, a strict criterion was adopted that defined as significant only differences with a probability level of 0.001 or less. Large sample sizes make even trivial differences statistically significant. The $p < 0.001$ level indicates a one-in-a-thousand chance that the observed differences would be a result of chance in a random sample.

8. The cutoff points for the language categories in this case are slightly below those in the first panel of Table 1, where the strong rightward skew of the English language proficiency index forced us to classify in the "well" and "very well" categories only those reporting very high proficiency. The more balanced foreign language proficiently index allows other cutoff points. If the same cutoffs were used, the proportions of those reporting themselves as English monolinguals (left-most column of Table 2) would increase by 7.4 percent, and those declaring full command of a foreign language (right-most column) would decline by 15 percent. The same patterns of association discussed in the text would be obtained, however.

9. The small Broward County sample is a consequence of the greater dispersion of second-generation students in its schools and the consequent difficulty of interviewing them. A supplementary survey, conducted in the fall of 1992, doubled the size of the Broward sample. These data were not available at the time of this analysis.

10. The dependent variable, the foreign language proficiency index (FLPI), was transformed logarithmically so that regression coefficients in the second column of the table can be interpreted as percent change in this index. National origin is represented by a series of dummy variables with "Other" as the excluded category. The corresponding coefficients are interpretable as the net effect of each nationality in comparison with the (predominantly European and Asian) "Other" category.

33

Optional Ethnicities: For Whites Only?

MARY C. WATERS

This paper reviews the current meaning of ethnicity for the descendants of nineteenth- and early twentieth-century European immigrants to the United States and contrasts that experience with the identities of people with non-European origins—the descendants of earlier forced immigrants and conquered peoples and the growing number of voluntary immigrants from non-European countries. The paper proceeds as follows. First the proposition that ethnic identity is optional for most Americans of European background is put forth. Empirical evidence that this is the case is reviewed. The social and historical forces that allow ethnicity to be an option are described.

The experience of non-Whites in the United States is then contrasted. Non-Whites have much more limited options with regard to their ethnicity because of particular historical and social circumstances in the United States. Using the example of current relations on college campuses between Blacks and Whites, I trace the influence that different degrees of options have on everyday encounters between people and the everyday social psychological consequences of failing to recognize this key difference between race and ethnicity.

Ethnic Identity for Whites in the 1990s

What does it mean to talk about ethnicity as an option for an individual? To argue that an individual has some degree of choice in their ethnic identity flies in the face of the common sense notion of ethnicity many of us believe in—that one's ethnic identity is a fixed characteristic, reflective of blood ties and given at birth. However, social scientists who study ethnicity have long concluded that while ethnicity is based in a *belief* in a common ancestry, ethnicity is primarily a *social* phenomenon, not a biological one (Alba 1985, 1990; Barth 1969; Weber [1921] 1968, p. 389). The belief that members of an ethnic group have that they share a common ancestry may not be a fact. There is a great deal of change in ethnic identities across generations through intermarriage, changing allegiances, and changing social categories. There is also a much larger amount of change in the identities of individuals over their life than is commonly believed. While most people are aware of the phenomena known as "passing"—people raised as one race who change at some point and claim a different race as their identity, there are similar life course changes in ethnicity that happen all the time and are not given the same degree of attention as "racial passing."

White Americans of European ancestry can be described as having a great deal of choice in terms of their ethnic identities. The two major types of options White Americans can exercise are (1) the option of whether to claim any specific ancestry, or to just be "White" or American, [Lieberson (1985) called these people "unhyphenated Whites"] and (2) the choice of which of their European ancestries to choose to include in their description of their own identities. In both cases, the option of choosing how to present yourself on surveys and in everyday social interactions exists for Whites because of social changes and societal conditions that have created a great deal of social mobility, immigrant assimilation, and political and economic power for Whites in the United States. Specifically, the option of being able to not claim any ethnic identity exists for Whites of European background in the United States because they are the majority group—in terms of holding political and social power, as well as being a numerical majority. The option of choosing among different ethnicities in their family backgrounds exists because the degree of discrimination and social distance attached to specific European backgrounds has diminished over time.

The Ethnic Miracle

When European immigration to the United States was sharply curtailed in the late 1920s, a process was set in motion whereby the European ethnic groups already in the United States were for all intents and purposes cut off from any new arrivals. As a result, the composition of the ethnic groups began to age generationally. The proportion of each ethnic group made up of immigrants or the first generation began to gradually decline, and the proportion made up of the children, grandchildren, and eventually great-grandchildren began to increase. Consequently, by 1990 most European-origin ethnic groups in the United States were composed of a very small number of immigrants, and a very large proportion of people whose link to their ethnic origins in Europe was increasingly remote.

This generational change was accompanied by unprecedented social and economic changes. The very success of the assimilation process these

groups experienced makes it difficult to imagine how much the question of the immigrants' eventual assimilation was an open one at the turn of the century. At the peak of immigration from southern and central Europe there was widespread discrimination and hostility against the newcomers by established Americans. Italians, Poles, Greeks, and Jews were called derogatory names, attacked by nativist mobs, and derided in the press. Intermarriage across ethnic lines was very uncommon—castelike in the words of some sociologists (Pagnini and Morgan 1990). The immigrants and their children were residentially segregated, occupationally specialized, and generally poor.

After several generations in the United States, the situation has changed a great deal. The success and social mobility of the grandchildren and great-grandchildren of that massive wave of immigrants from Europe has been called "The Ethnic Miracle" (Greeley 1976). These Whites have moved away from the inner-city ethnic ghettos to White middle-class suburban homes. They are doctors, lawyers, entertainers, academics, governors, and Supreme Court justices. But contrary to what some social science theorists and some politicians predicted or hoped for, these middle-class Americans have not completely given up ethnic identity. Instead, they have maintained some connection with their immigrant ancestors' identities—becoming Irish American doctors, Italian American Supreme Court justices, and Greek American presidential candidates. In the tradition of cultural pluralism, successful middle-class Americans in the late twentieth century maintain some degree of identity with their ethnic backgrounds. They have remained "hyphenated Americans." So while social mobility and declining discrimination have created the option of not identifying with any European ancestry, most White Americans continue to report some ethnic background.

With the growth in intermarriage among people of European ethnic origins, increasingly these people are of mixed ethnic ancestry. This gives them the option of which ethnicity to identify with. The U.S. census has asked a question on ethnic ancestry in the 1980 and 1990 censuses. In 1980, 52 percent of the American public responded with a single ethnic ancestry, 31 percent gave multiple ethnic origins (up to three were coded, but some individuals wrote in more than three), and only 6 percent said they were

American only, while the remaining 11 percent gave no response. In 1990 about 90 percent of the population gave some response to the ancestry question, with only 5 percent giving American as a response and only 1.4 percent reporting an uncodeable response such as "don't know" (McKenney and Cresce 1992; U.S. Bureau of the Census 1992b).

Several researchers have examined the pattern of responses of people to the census ancestry question. These analyses have shown a pattern of flux and inconsistency in ethnic ancestry reporting. For instance, Lieberson and Waters (1986; 1988, p. 93) have found that parents simplify children's ancestries when reporting them to the census. For instance, among the offspring in situations where one parent reports a specific single White ethnic origin and the other parent reports a different single White origin, about 40 percent of the children are not described as the logical combination of the parents ancestries. For example, only about 60 percent of the children of English-German marriages are labeled as English-German or German-English. About 15 percent of the children of these parents are simplified to just English, and another 15 percent are reported as just German. The remainder of the children are either not given an ancestry or are described as American (Lieberson and Waters 1986, 1993).

In addition to these intergenerational changes, researchers have found changes in reporting ancestry that occur at the time of marriage or upon leaving home. At the ages of eighteen to twenty-two, when many young Americans leave home for the first time, the number of people reporting a single as opposed to a multiple ancestry goes up. Thus while parents simplify children's ancestries when they leave home, children themselves tend to report less complexity in their ancestries when they leave their parents homes and begin reporting their ancestries themselves (Lieberson and Waters 1986, 1988; Waters 1990).

These individual changes are reflected in variability over time in the aggregate numbers of groups determined by the census and surveys. Farley (1991) compared the consistency of the overall counts of different ancestry groups in the 1979 Current Population Survey, the 1980 census, and the 1986 National Content Test (a pretest for the 1990 census). He found much less consistency in the numbers for northern European ancestry groups whose immigration peaks were early in the nineteenth century—the English, Dutch, Germans, and other northern European groups. In other words each of these different surveys and the census yeilded a different estimate of the number of people having this ancestry. The 1990 census also showed a great deal of flux and inconsistency in some ancestry groups. The number of people reporting English as an ancestry went down considerably from 1980, while the number reporting German ancestry went up. The number of Cajuns grew dramatically. This has led officials at the Census Bureau to assume that the examples used in the instructions strongly influence the responses people give. (Cajun was one of the examples of an ancestry given in 1990 but not in 1980, and German was the first example given. English was an example in the 1980 instructions, but not in 1990.)

All of these studies point to the socially variable nature of ethnic identity—and the lack of equivalence between ethnic ancestry and identity. If merely adding a category to the instructions to the question increases the number of people claiming that ancestry, what does that mean about the level of importance of that identity for people answering the census? Clearly identity and ancestry for Whites in the United States, who increasingly are from mixed backgrounds, involves some change and choice.

Symbolic Ethnicities for White Americans

What do these ethnic identities mean to people and why do they cling to them rather than just abandoning the tie and calling themselves American? My own field research with suburban Whites in California and Pennsylvania found that later-generation descendants of European origin maintain what are called "symbolic ethnicities." Symbolic ethnicity is a term coined by Herbert Gans (1979) to refer to ethnicity that is individualistic in nature and without real social cost for the individual. These symbolic identifications are essentially leisure time activities, rooted in nuclear family traditions and reinforced by the voluntary enjoyable aspects of being ethnic (Waters 1990). Richard Alba (1990) also found later-generation Whites in Albany, New York, who chose to keep a tie with an ethnic identity because of the enjoyable and voluntary aspects to those identities, along with the feelings of specialness they entailed. An example of symbolic ethnicity is individuals who identify as Irish, for example, on occasions such as Saint Patrick's Day, on family holidays, or for vacations. They do not usually belong to Irish American organizations, live in Irish neighborhoods, work in Irish jobs, or marry other Irish people. The symbolic meaning of being Irish American can be constructed by individuals from mass media images, family traditions, or other intermittent social activities. In other words, for later-generation White ethnics, ethnicity is not something that influences their lives unless they want it to. In the world of work and school and neighborhood, individuals do not have to admit to being ethnic unless they choose to. And for an increasing number of European-origin individuals whose parents and grandparents have intermarried, the ethnicity they claim is largely a matter of personal choice as they sort through all of the possible combinations of groups in their genealogies.

Individuals can choose those aspects of being Italian, for instance, that appeal to them, and discard those that do not. Or a person whose father is Italian, and mother part Polish and part French, might choose among the three ethnicities and present herself as a Polish American. For instance, a nineteen-year-old college student, interviewed in California in 1986, told me he would have answered Irish on the 1980 census form that asked about ethnic ancestry. These are his reasons:

Q: Why would you have answered that?

A: Well my Dad's name is Kerrigan and my mom's name is O'Leary, and I do have some German in me, but if you figure it out, I am about 75% Irish, so I usually say I am Irish.

Q: You usually don't say German when people ask?

A: No, no, I never say I am German. My dad just likes being Irish. . . . I don't know I just never think of myself as being German.

Q: So your dad's father is the one who immigrated?

A: Yes. On his side is Irish for generations. And then my grandmother's name is Dubois, which is French, partly German, partly French, and then the rest of the family is all Irish. So it is only the maternal grandmother who messes up the line. (Waters 1990, p. 10)

Thus in the course of a few questions, this man labeled himself Irish, admitted to being part German but not identifying with it, and then as an afterthought added that he was also part French. This is not an unusual case. With just a little probing, many people will describe a variety of ancestries in their family background, but do not consider these ancestries to be a salient part of their own identities. Thus the 1990 census ancestry question, which estimated that 30 percent of the population is of mixed ancestry, most surely underestimates the degree of mixing among the population. My research, and the research of Richard Alba (1990), shows that many people have already sorted through what they know of their ethnic ancestries and simplified their responses before they ever answer a census or survey question (Waters 1990).

But note that this freedom to include or exclude ancestries in your identification to yourself and others would not be the same for those defined racially in our society. They are constrained to identify with the part of their ancestry that has been socially defined as the "essential" part. African Americans, for example, have been highly socially constrained to identify as Blacks, without other options available to them, even when they know that their forebears included many people of American Indian or European background. Up until the mid-twentieth century, many state governments had specific laws defining one as Black if as little as one-thirty-second of one's ancestors were defined as Black (Davis 1991; Dominguez 1986; Spickard 1989). Even now when the one drop rule has been dropped from our legal codes, there are still strong societal pressures on African Americans to identify in a particular way. Certain ancestries take precedence over others in the societal rules on descent and ancestry reckoning. If one believes one is part English and part German and identifies in a survey as German, one is not in danger of being accused of trying to "pass" as non-English and of being "redefined" English by the interviewer. But if one were part African and part German, one's self identification as German would be highly suspect and probably not accepted if one "looked" Black according to the prevailing social norms.

This is reflected in the ways the census collects race and ethnic identity. While the ethnic ancestry question used in 1980 and 1990 is given to all Americans in the sample regardless of race and allows multiple responses that combine races, the primary source of information on people defined racially in the United States is the census race question or the Hispanic question. Both of these questions require a person to make a choice about an identity. Individuals are not allowed to respond that they are both Black and White, or Japanese and Asian Indian on the race question even if they know that is their background. In fact, people who disobey the instructions to the census race question and check off two races are assigned to the first checked race in the list by the Census Bureau.

In responding to the ancestry question, the comparative latitude that White respondents have does not mean that Whites pick and choose ethnicities out of thin air. For the most part people choose an identity that corresponds with some element of their family tree. However, there are many anecdotal instances of people adopting ethnicities when they marry or move to a strongly identified neighborhood or community. For instance Micaela di Leonardo (1984) reported instances of non-Italian women who married into Italian American families and "became Italian." Karen Leonard (1992) describes a community of Mexican American women who married Punjabi immigrants in California. Some of the Punjabi immigrants and their descendants were said to have "become Mexican" when they joined their wives kin group and social worlds. Alternatively she describes the community acknowledging that Mexican women made the best curry, as they adapted to life with Indian-origin men.

But what do these identities mean to individuals? Surely an identity that is optional in a number of ways—not legally defined on a passport or birth certificate, not socially consequential in terms of societal discrimination in terms of housing or job access, and not economically limiting in terms of blocking opportunities for social mobility—cannot be the same as an identity that results from and is nurtured by societal exclusion and rejection. The choice to have a symbolic ethnicity is an attractive and widespread one despite its lack of demonstrable content, because having a symbolic ethnicity combines individuality with feelings of community. People reported to me that they liked having an ethnic identity because it gave them a uniqueness and a feeling of being special. They often contrasted their own specialness by virtue of their ethnic identities with "bland" Americanness. Being ethnic makes people feel unique and special and not just "vanilla" as one of my respondents put it. For instance, one woman describes the benefits she feels from being Czech American:

> I work in an office and a lot of people in there always talk about their background. It's weird because it is a big office and people are of all different backgrounds. People are this or that. It is interesting I think to find out. Especially when it is something you do not hear a lot about. Something that is not common like Lithuania or something. That's the good part about being Czech. People think it is something different. (Waters 1990, p. 154)

Because "American" is largely understood by Americans to be a political identity and allegiance and not an ethnic one, the idea of being "American" does not give people the same sense of belonging that their hyphenated American identity does. When I asked people about their dual identities—American and Irish or Italian or whatever—they usually responded in a way that showed how they conceived of the relationship between the two identities. Being an American was their primary identity; but it was so primary that they rarely, if ever, thought about it—most commonly only when they left the country. Being Irish American, on the other hand was a way they had of differentiating themselves from others whom they interacted with from day to day—in many cases from spouses or in laws. Certain of their traits—being emotional, having a sense of humor, talking with their hands—were understood as stemming from their ethnicity. Yet when asked about their identity as Americans, that identity was both removed from their day-to-day consciousness and understood in terms of loyalty and patriotism. Although they may not think they behave or think in a certain way because they are American, being American is something they are both proud of and committed to.

Symbolic ethnicity is the best of all worlds for these respondents. These White ethnics can claim to be unique and special, while simultaneously finding the community and conformity with others that they also crave. But that "community" is of a type that will not interfere with a person's individuality. It is not as if these people belong to ethnic voluntary organizations or gather as a group in churches or neighborhoods or union halls. They work and reside within the mainstream of American middle-class life, yet they retain the interesting benefits—the "special-ness"—of ethnic allegiance, without any of its draw-backs.

It has been suggested by several researchers that this positive value attached to ethnic ancestry, which became popular in the ethnic revival of the 1970s, is the result of assimilation having proceeded to an advanced stage for descendants of White Europeans (Alba 1985; Crispino 1980; Steinberg 1981). Ironically, people celebrate and embrace their ethnic backgrounds precisely because assimilation has proceeded to the point where such identification does not have that much influence on their day-to-day life. Rather than choosing the "least ethnic" and most bland ethnicities, Whites desire the "most ethnic" ones, like the once-stigmatized "Italian," because it is perceived as bringing the most psychic benefits. For instance, when an Italian father is married to an English or a Scottish or a German mother, the likelihood is that the child will be reported to the census with the father's Italian ancestry, rather than the northern European ancestries, which would have been predicted to have a higher social status. Italian is a good ancestry to have, people told me, because they have good food and a warm family life. This change in the social meaning of being Italian American is quite dramatic, given that Italians were subject to discrimination, exclusion, and extreme negative stereotyping in the early part of the twentieth century.

Race Relations and Symbolic Ethnicity

However much symbolic ethnicity is without cost for the individual, there is a cost associated with symbolic ethnicity for the society. That is because symbolic ethnicities of the type described here are confined to White Americans of European origin. Black Americans, Hispanic Americans, Asian Americans, and American Indians do not have the option of a symbolic ethnicity at present in the United States. For all of the ways in which ethnicity does not matter for White Americans, it does matter for non-Whites. Who your ancestors are does affect your choice of spouse, where you live, what job you have, who your friends are, and what your chances are for success in American society, if those ancestors happen not to be from Europe. The reality is that White ethnics have a lot more choice and room for maneuver than they themselves think they do. The situation is very different for members of racial minorities, whose lives are strongly influenced by their race or national origin regardless of how much they may choose not to identify themselves in terms of their ancestries.

When White Americans learn the stories of how their grandparents and great-grandparents triumphed in the United States over adversity, they are usually told in terms of their individual efforts and triumphs. The important role of labor unions and other organized political and economic actors in their social and economic successes are left out of the story in favor of a generational story of individual Americans rising up against communitarian, Old World intolerance and New World resistance. As a result, the "individualized" voluntary, cultural view of ethnicity for Whites is what is remembered.

One important implication of these identities is that they tend to be very individualistic. There is a tendency to view valuing diversity in a pluralist environment as equating all groups. The symbolic ethnic tends to think that all groups are equal; everyone has a background that is their right to celebrate and pass on to their children. This leads to the conclusion that all identities are equal and all identities in some sense are interchangeable—"I'm Italian American, you're Polish American. I'm Irish American, you're African American." The important thing is to treat people as individuals and all equally. However, this assumption ignores the very big difference between an individualistic symbolic ethnic identity and a socially enforced and imposed racial identity.

My favorite example of how this type of thinking can lead to some severe misunderstandings between people of different backgrounds is from the *Dear Abby* advice column. A few years back a person wrote in who had asked an acquaintance of Asian background where his family was from. His acquaintance answered that this was a rude question and he would not reply. The bewildered White asked Abby why it was rude, since he thought it was a sign of respect to wonder where people were from, and he certainly would not mind anyone asking HIM about where his family was from. Abby asked her readers to write in to say whether it was rude to ask about a person's ethnic background. She reported that she got a large response, that most non-Whites thought it was a sign of disrespect, and Whites thought it was flattering:

Dear Abby,

I am 100 percent American and because I am of Asian ancestry I am often asked "What are you?" It's not the personal nature of this question that bothers me, it's the question itself. This query seems to question my very humanity. "What am I? Why I am a person like everyone else!"

Signed, A REAL AMERICAN

Dear Abby,

Why do people resent being asked what they are? The Irish are so proud of being Irish, they tell you before you even ask. Tip O'Neill has never tried to hide his Irish ancestry.

Signed, JIMMY.

In this exchange JIMMY cannot understand why Asians are not as happy to be asked about their ethnicity as he is, because he understands his ethnicity and theirs to be separate but equal. Everyone has to come from somewhere—his family from Ireland, another's family from Asia—each has a history and each should be proud of it. But the reason he cannot understand the perspective of the Asian American is that all ethnicities are not equal; all are not symbolic, costless, and voluntary. When White Americans equate their own symbolic ethnicities with the socially enforced identities of non-White Americans, they obscure the fact that the experiences of Whites and non-Whites have been qualitatively different in the United States and that the current identities of individuals partly reflect that unequal history.

In the next section I describe how relations between Black and White students on college campuses reflect some of these asymmetries in the understanding of what a racial or ethnic identity means. While I focus on Black and White students in the following discussion, you should be aware that the myriad other groups in the United States—Mexican Americans, American Indians, Japanese Americans—all have some degree of social and individual influences on their identities, which reflect the group's social and economic history and present circumstance.

Relations on College Campuses

Both Black and White students face the task of developing their race and ethnic identities. Sociologists and psychologists note that at the time people leave home and begin to live independently from their parents, often ages eighteen to twenty-two, they report a heightened sense of racial and ethnic identity as they sort through how much of their beliefs and behaviors are idiosyncratic to their families and how much are shared with other people. It is not until one comes in close contact with many people who are different from oneself that individuals realize the ways in which their backgrounds may influence their individual personality. This involves coming into contact with people who are different in terms of their ethnicity, class, religion, region, and race. For White students, the ethnicity they claim is more often than not a symbolic one—with all of the voluntary, enjoyable, and intermittent characteristics I have described above.

Black students at the university are also developing identities through interactions with others who are different from them. Their identity development is more complicated than that of Whites because of the added element of racial discrimination and racism, along with the "ethnic" developments of finding others who share their background. Thus Black students have the positive attraction of being around other Black students who share some cultural elements, as well as the need to band together with other students in a reactive and oppositional way in the face of racist incidents on campus.

Colleges and universities across the country have been increasing diversity among their student bodies in the last few decades. This has led in many cases to strained relations among students from different racial and ethnic backgrounds. The 1980s and 1990s produced a great number of racial incidents and high racial tensions on campuses. While there were a number of racial incidents that were due to bigotry, unlawful behavior, and violent or vicious attacks, much of what happens among students on campuses involves a low level of tension and awkwardness in social interactions.

Many Black students experience racism personally for the first time on campus. The upper-middle-class students from White suburbs were often isolated enough that their presence was not threatening to racists in their high schools. Also, their class background was known by their residence and this may have prevented attacks being directed at them. Often Black students at the university who begin talking with other students and recognizing racial slights will remember incidents that happened to them earlier that they might not have thought were related to race.

Black college students across the country experience a sizeable number of incidents that are clearly the result of racism. Many of the most blatant ones that occur between students are the result of drinking. Sometimes late at night, drunken groups of White students coming home from parties will yell slurs at single Black students on the street. The other types of incidents that happen include being singled out for special treatment by employees, such as being followed when shopping at the campus bookstore, or going to the art museum with your class and the guard stops you and asks for your I.D. Others involve impersonal encounters on the street—being called a nigger by a truck driver while crossing the street, or seeing old ladies clutch their pocketbooks and shake in terror as you pass them on the street. For the most part these incidents are not specific to the university environment, they are the types of incidents middle-class Blacks face every day throughout American society, and they have been documented by sociologists (Feagin 1991).

In such a climate, however, with students experiencing these types of incidents and talking with each other about them, Black students do experience a tension and a feeling of being singled out. It is unfair that this is part of their college experience and not that of White students. Dealing with incidents like this, or the ever-present threat of such incidents, is an ongoing developmental task for Black students that takes energy, attention, and strength of character. It should be clearly understood that this is an asymmetry in the "college experience" for Black and White students. It is one of the unfair aspects of life that results from living in a society with ongoing racial prejudice and discrimination. It is also very understandable that it makes some students angry at the unfairness of it all, even if there is no one to blame specifically. It is also very troubling because, while most Whites do not create these incidents, some do, and it is never clear until you know someone well whether they are the type of person who

could do something like this. So one of the reactions of Black students to these incidents is to band together.

In some sense then, as Blauner (1992) has argued, you can see Black students coming together on campus as both an "ethnic" pull of wanting to be together to share common experiences and community, and a "racial" push of banding together defensively because of perceived rejection and tension from Whites. In this way the ethnic identities of Black students are in some sense similar to, say, Korean students wanting to be together to share experiences. And it is an ethnicity that is generally much stronger than, say, Italian Americans. But for Koreans who come together there is generally a definition of themselves as "different from" Whites. For Blacks reacting to exclusion, there is a tendency for the coming together to involve both being "different from" but also "opposed to" Whites.

The anthropologist John Ogbu (1990) has documented the tendency of minorities in a variety of societies around the world, who have experienced severe blocked mobility for long periods of time, to develop such oppositional identities. An important component of having such an identity is to describe others of your group who do not join in the group solidarity as devaluing and denying their very core identity. This is why it is not common for successful Asians to be accused by others of "acting White" in the United States, but it is quite common for such a term to be used by Blacks and Latinos. The oppositional component of a Black identity also explains how Black people can question whether others are acting "Black enough." On campus, it explains some of the intense pressures felt by Black students who do not make their racial identity central and who choose to hang out primarily with non-Blacks. This pressure from the group, which is partly defining itself by not being White, is exacerbated by the fact that race is a physical marker in American society. No one immediately notices the Jewish students sitting together in the dining hall, or the one Jewish student sitting surrounded by non-Jews, or the Texan sitting with the Californians, but everyone notices the Black student who is or is not at the "Black table" in the cafeteria.

An example of the kinds of misunderstandings that can arise because of different understandings of the meanings and implications of symbolic versus oppositional identities concerns questions students ask one another in the dorms about personal appearances and customs. A very common type of interaction in the dorm concerns questions Whites ask Blacks about their hair. Because Whites tend to know little about Blacks, and Blacks know a lot about Whites, there is a general asymmetry in the level of curiosity people have about one another. Whites, as the numerical majority, have had little contact with Black culture; Blacks, especially those who are in college, have had to develop bicultural skills—knowledge about the social worlds of both Whites and Blacks. Miscommunication and hurt feelings about White students' questions about Black students' hair illustrate this point. One of the things that happens freshman year is that White students are around Black students as they fix their hair. White students are generally quite curious about Black students' hair—they have basic questions such as how often Blacks wash their hair, how they get it straightened or curled, what products they use on their hair, how they comb it, etc. Whites often wonder to themselves whether they should ask these questions. One thought experiment Whites perform is to ask themselves whether a particular question would upset them. Adopting the "do unto others" rule, they ask themselves, "If a Black person was curious about my hair would I get upset?" The answer usually is "No, I would be happy to tell them." Another example is an Italian American student wondering to herself, "Would I be upset if someone asked me about calamari?" The answer is no, so she asks her Black roommate about collard greens, and the roommate explodes with an angry response such as, "Do you think all Black people eat watermelon too?" Note that if this Italian American knew her friend was Trinidadian American and asked about peas and rice the situation would be more similar and would not necessarily ignite underlying tensions.

Like the debate in *Dear Abby*, these innocent questions are likely to lead to resentment. The issue of stereotypes about Black Americans and the assumption that all Blacks are alike and have the same stereotypical cultural traits has more power to hurt or offend a Black person than vice versa. The innocent questions about Black hair also bring up a number of asymmetries between the Black and White experience. Because Blacks tend to have more knowledge about Whites than vice versa, there is not an even exchange going on, the Black freshman is likely to have fewer basic questions about his White

roommate than his White roommate has about him. Because of the differences historically in the group experiences of Blacks and Whites there are some connotations to Black hair that don't exist about White hair. (For instance, is straightening your hair a form of assimilation, do some people distinguish between women having "good hair" and "bad hair" in terms of beauty and how is that related to looking "White"?). Finally, even a Black freshman who cheerfully disregards or is unaware that there are these asymmetries will soon slam into another asymmetry if she willingly answers every innocent question asked of her. In a situation where Blacks make up only 10 percent of the student body, if every non-Black needs to be educated about hair, she will have to explain it to nine other students. As one Black student explained to me, after you've been asked a couple of times about something so personal you begin to feel like you are an attraction in a zoo, that you are at the university for the education of the White students.

Institutional Responses

Our society asks a lot of young people. We ask young people to do something that no one else does as successfully on such a wide scale—that is to live together with people from very different backgrounds, to respect one another, to appreciate one another, and to enjoy and learn from one another. The successes that occur every day in this endeavor are many, and they are too often overlooked. However, the problems and tensions are also real, and they will not vanish on their own. We tend to see pluralism working in the United States in much the same way some people expect capitalism to work. If you put together people with various interests and abilities and resources, the "invisible hand" of capitalism is supposed to make all the parts work together in an economy for the common good.

There is much to be said for such a model—the invisible hand of the market can solve complicated problems of production and distribution better than any "visible hand" of a state plan. However, we have learned that unequal power relations among the actors in the capitalist marketplace, as well as "externalities" that the market cannot account for, such as long-term pollution, or collusion between corporations, or the exploitation of child labor, means that state regulation is often needed. Pluralism and the

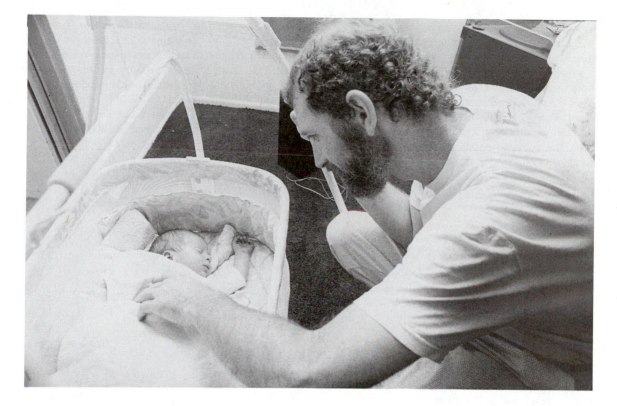

Given their ancestry, as well as the social construction of race and ethnicity in a particular society, some people have many identities available to them. Here the American-born child of an Israeli immigrant may one day have a vast portfolio from which to choose. The father, a recent Israeli immigrant, was born in Hungary and first came to the United States to obtain a Master's degree. Back in Israel, both he and his wife had been involved in the kibbutz movement. (Photo by Steve Gold)

relations between groups are very similar. There is a lot to be said for the idea that bringing people who belong to different ethnic or racial groups together in institutions with no interference will have good consequences. Students from different backgrounds will make friends if they share a dorm room or corridor, and there is no need for the institution to do any more than provide the locale. But like capitalism, the invisible hand of pluralism does not do well when power relations and externalities are ignored. When you bring together individuals from groups that are differentially valued in the wider society and provide no guidance, there will be problems. In these cases the "invisible hand" of pluralist relations does not work, and tensions and disagreements can arise without any particular individual or group of individuals being "to blame." On college campuses in the

1990s some of the tensions between students are of this sort. They arise from honest misunderstandings, lack of a common background, and very different experiences of what race and ethnicity mean to the individual.

The implications of symbolic ethnicities for thinking about race relations are subtle but consequential. If your understanding of your own ethnicity and its relationship to society and politics is one of individual choice, it becomes harder to understand the need for programs like affirmative action, which recognize the ongoing need for group struggle and group recognition, in order to bring about social change. It also is hard for a White college student to understand the need that minority students feel to band together against discrimination. It also is easy, on the individual level, to expect everyone else to be able to turn their ethnicity on and off at will, the way you are able to, without understanding that ongoing discrimination and societal attention to minority status makes that impossible for individuals from minority groups to do. The paradox of symbolic ethnicity is that it depends upon the ultimate goal of a pluralist society, and at the same time makes it more difficult to achieve that ultimate goal. It is dependent upon the concept that all ethnicities mean the same thing, that enjoying the traditions of one's heritage is an option available to a group or an individual, but that such a heritage should not have any social costs associated with it.

As the Asian Americans who wrote to *Dear Abby* make clear, there are many societal issues and involuntary ascriptions associated with non-White identities. The developments necessary for this to change are not individual but societal in nature. Social mobility and declining racial and ethnic sensitivity are closely associated. The legacy and the present reality of discrimination on the basis of race or ethnicity must be overcome before the ideal of a pluralist society, where all heritages are treated equally and are equally available for individuals to choose or discard at will, is realized.

34

The Complexity of Racial Attitudes in America

HOWARD SCHUMAN
CHARLOTTE STEEH

Recent incidents on college campuses and in neighborhoods across the nation have raised new questions about the attitudes of White Americans toward African Americans. The incidents, ranging from verbal harassment to murder, began during a political period—often summarized as the Reagan era—when the nation's commitment to civil rights and to the full incorporation of minorities seemed to lessen substantially—indeed, even to move backward toward a time characterized by racial intolerance and conflict. Thus it is useful to ask how the incidents that create newspaper headlines appear when placed in the context of long-term trends in racial attitudes, measured, as they usually must be, by sample interview surveys of the American population.

We draw here on national survey data gathered over five decades to show trends in the racial attitudes of White Americans since World War II. Reference will also be made to related data on the attitudes of Black Americans, but such evidence is much more limited, partly because it does not go back as far in time and partly because the typical national samples used in surveys include too few Black Americans for detailed analysis. We will also mention data gathered in 1990 on undergraduates at the University of Michigan; as an age and status group, college students are a critically important part of American society. Along the way we will comment on doubts that people often have when they consider evidence from surveys, such as the relation of attitudes expressed in survey interviews to behavior outside the interview setting.

Principles and Their Implementation

The oldest questions for which we have long-term trend data on the national population were asked in the early 1940s, soon after the United States entered World War II. Race as such had not been a salient issue during the depression years—so far as we can tell, the only question explicitly about race that was asked in Gallup and other polls during the 1930s con-

This research was funded in part by grants from the Spencer Foundation and the National Science Foundation (SES-8411371 and 8713633).

cerned the exclusion in 1934 by the Daughters of the American Revolution of the great singer, Marian Anderson, from performing in its auditorium in Washington, D.C., simply because she was Black (Sheatsley 1966). By the early 1940s, however, government concerns about racial tensions during World War II led to a first attempt to use the recently developed method of the sample survey to assess the attitudes of White Americans toward basic issues of discrimination and segregation (Hyman and Sheatsley 1956). National surveys of African Americans were not undertaken until the early 1960s, when Black activism was finally recognized as a potent force; in the immediate postwar years, America's racial issues were treated by survey researchers as, in Gunnar Myrdal's words, "a White man's problem" (Myrdal 1944).

Figure 1 shows trends for seven questions that have been repeated in national surveys over the years since World War II.[1] One of these dealing with the separation of the races in public schools (Same Schools) came from that first national survey in 1942, and the question about job discrimination (Equal Jobs) goes back almost as far.[2] The remaining questions were introduced much later, two of them as late as the mid-1980s. All were asked in surveys of adequately large and fairly representative cross-sections of White Americans, though there is always some error in surveys. Errors are due not only to the 3 or 4 percent "margin of error" referred to in media presentations of polls, but also to those who were never reached, to those who declined to be interviewed, and to those who may have deliberately misstated their attitudes. More will be said about the last kind of error at a later point.

The 1942 Same Schools question that asked simply whether White students and Black students should go to the same schools or to separate schools was formulated at a time when separate schools were legally required throughout the American South, including the U.S. capital city of Washington, D.C. As Figure 1 shows, only about a third of the White population in 1942 favored the same schools for Blacks and Whites, and when the sample is divided into regions, the percentage was a vanishingly small 2 percent in the South and only 42 percent in the North. The year 1942 may seem like ancient history to younger Americans today, but perhaps its relative recentness can be brought home by noting that George Bush turned eighteen years of age in 1942. Thus he, like others of his generation, grew up in a

time when total racial segregation was entirely legal in this country and had the support of the great majority of the White population. Indeed, in those years, a half-century ago, senators, congressmen, and other leaders made public statements in defense of segregation that would not be tolerated today.

The question called Equal Jobs in Figure 1, first asked in 1944, spoke not only of segregation, but of out-and-out job discrimination. The question posed the issue of whether Blacks should have an equal chance with Whites to be hired at any kind of job, or whether Whites should be given preference. In 1944—George Bush was now twenty years of age—the majority of White Americans openly endorsed discrimination against Blacks in employment. And there was, in fact, pervasive and quite open discrimination of this kind, with African Americans barred from most professional, white-collar, and skilled manual jobs in the United States.

Figure 1 shows that since the early 1940s there has been a steady and almost linear change toward rejection of racial discrimination and segregation in schools and jobs. Indeed, these changes in the opinions of White Americans constitute one of the most massive shifts in public attitudes that we know of. By 1972 the question on White preferences in jobs ceased to be asked at all: rejection of job discrimination had reached so close to 100 percent that there was nothing left to measure. Approval of school desegregation also became so widespread that after 1985 the General Social Survey dropped Same Schools from its battery of standard items on racial issues. Thus, since World War II, opinions on these two matters have run the gamut, from majority rejection to almost total acceptance of integration and equal treatment, and today the vast majority of White Americans affirm these broad principles for employment and schooling in their answers to survey questions.

How this change came about is itself an interesting story and one well worth serious investigation. We can be pretty sure that it was not simply a spontaneous shift in individual attitudes, but rather the result of a more fundamental transformation of American society. Some historians believe that in the beginning the transformation had little or nothing to do with relations between Blacks and Whites, but instead evolved predominantly from struggles over the entry of eastern European immigrant groups into the White Anglo-Saxon Protestant mainstream

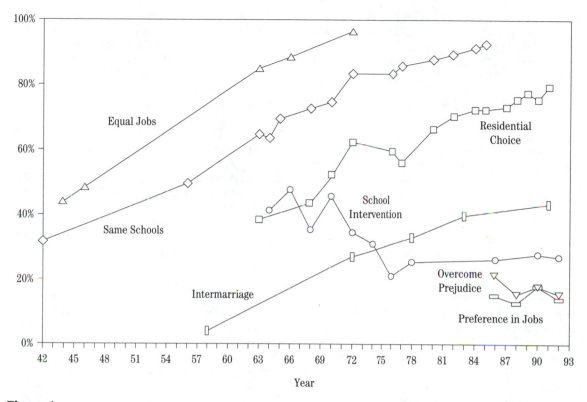

Figure 1
Trends in White Racial Attitudes

(Weiss 1979). Others place primary emphasis on the role of academics and intellectuals, like Franz Boas, in challenging widespread assumptions about innate differences between ethnic and racial groups. Whatever the origins of the initial change, the transformation was probably greatly facilitated by events connected with America's ideological struggle in World War II against Nazi racism, by the postwar decolonization of most non-White areas of the world, and by the glaring contradiction that racial discrimination in the United States posed to a nation claiming leadership of what was then regularly described as "the Free World." Also important was the migration of Blacks out of the rural South, where they were almost entirely disenfranchised, and into northern cities where their political voices could begin to be heard. The change was not limited to relations between Blacks and Whites, however, for other evidence indicates that it extended also to attitudes toward Jews and other minorities that had previously been excluded from major areas of American life

(Stember and others 1966). Indeed, it was not until John Kennedy's election in 1960 that participation by Catholics in national politics became accepted.

Thus, what happened in the years following World War II was a basic change in American *norms*, such that the discrimination and segregation once openly accepted were increasingly defined as un-American and undemocratic. The further changes in individual attitudes since then are really a catching up to these new norms by the general White population. A norm is usually defined by sociologists as a morally binding expectation for how people should behave in particular situations. The extension after World War II of the fundamental norms of nondiscrimination and equality before the law to African Americans and other ethnic groups began to resolve the conflict between American ideals and the actions of its citizens that had plagued the United States since its beginning.

What concerns most readers today, however, is less the origins of the change, or the terminology used to describe it, but its present depth and its future course. For this purpose, it is useful to focus on the other five questions in Figure 1, which tap different dimensions of racial attitudes among adult White Americans. Turning first to the question on intermarriage, we see that in 1958, when these attitudes were first measured, only 4 percent of the White population approved of such a possibility. Yet since that time, responses to the Intermarriage question have shown essentially the same trend as the Equal Jobs and Same Schools questions, namely, a steady movement upward, with the latest reading in 1991 reflecting some 44 percent approval of racial intermarriage.

Of course, this level of approval for intermarriage is still much lower than the level of approval for school integration or job equality. Support for neighborhood integration (Residential Choice in Figure 1), on the other hand, lies in between. In 1963, when this question first appeared in a national survey, only 39 percent of Whites disagreed with the statement that "White people have a right to keep Blacks out of their neighborhoods if they want to and Blacks should respect that right." By 1991, however, disagreement had climbed to 80 percent. Here again, a substantial increase in opposition to discrimination occurred with the passage of time.

The remaining three trends pictured in Figure 1 define the limits of White Americans' support for racial equality. First of all, the racially liberal position on each issue is endorsed by a very small percentage of Whites, much smaller even than the percentage who approve of intermarriage. One of these questions, School Intervention, has been asked for some time and measures whether or not people believe the federal government should "see to it" that Black and White children go to the same schools. By coupling the notion of government coercion to the notions of integration and nondiscrimination, the question, and a few others like it dating from the early 1960s, injected a new element into the national debate about racial justice. As Figure 1 illustrates, overall support for government intervention to integrate schools in the United States was at its highest level from 1964 to 1970, no doubt reflecting the activism that produced such landmark legislation as the 1964 Civil Rights Act, the 1965 Voting Rights Act, and the 1968 Fair Housing Act. After 1970, however, the level of approval declined precipitously as government intervention to desegregate schools became increasingly

associated with the busing of children across districts. Since 1976, when support fell to a low of 21 percent, the trend has not reversed, and only 28 percent of Whites in 1990 favored government efforts to desegregate schools. Of all the racial attitudes that have been measured over time in national surveys, only School Intervention has shown an unmistakable decline.

The final two trends in Figure 1 are too short to draw reliable conclusions about underlying patterns of change. What is significant, however, is that the support of White Americans for the particular policy option mentioned in each question is virtually nonexistent. The trend for the question we have called Overcome Prejudice indicates very little disagreement with the statement that Blacks should overcome prejudice and work their way up without special favors just as the Irish, Italians, Jews, and other ethnic groups have done. The average disagreement from 1986 to 1992 is about 17 percent, although another 11 percent or so say they neither agree nor disagree. Therefore, smaller percentages of White Americans believe that minorities should get special assistance in overcoming prejudice than support government enforcement of school desegregation or approve of intermarriage. Similarly the trend for Preference In Jobs shows that affirmative action is somewhat less popular. On average, only about 15 percent of adult White Americans during the late 1980s felt that employers should give preference in hiring and promotions to Blacks. Thus we find that the White public in the late 1980s and early 1990s endorses norms that can be seen as contradictory. Especially in questions that deal with government intervention and affirmative action, the new norms of racial equality and racial justice appear to clash with the individualism and self-reliance that have been integral parts of American ideology since the days when Benjamin Franklin gave them expression and before.[3]

The evidence in Figure 1 suggests an important point about changes in the attitudes of White Americans. The very fact that they make sharp distinctions in their verbal answers between questions on intermarriage and questions on school integration, for example, or between questions on residential integration and questions on affirmative action in employment, indicates that there is no simple lip-service answering going on. If there were, it would be just as easy to appear racially liberal on the one question as on the other, yet this does not happen. Thus we have in these radically different levels of support for different types of integration and for different policy options some evidence of the validity of the answers. Of course, it is important not to take any percentage too literally, partly because sometimes seemingly small variations in question wording can change responses appreciably, and partly because we are speaking of norms, which influence but do not entirely determine behavior in any area of life. What we must focus on in examining the kinds of changes shown in Figure 1 are broad trends and relative levels of support, not precise numbers.

More generally, we have been able to track some thirty to forty questions on racial attitudes covering many different issues, and although most of them present the same upward movement over time, each also has its own characteristic level. Moreover, the differences in responses make sense in terms of the degree of intimacy described by the question. Experience and intuition indicate that school integration would be much more acceptable to White Americans than intermarriage or neighborhood integration. In fact, this is also true for Black Americans, but to a lesser extent: nearly 100 percent of Blacks in a national probability sample supported the integration of schools in 1985 when the question was last asked, but only 70 percent approved of intermarriage in 1991 (Gallup and Newport 1991).

Not only are there differences according to the degree of intimacy implied by various questions, but also, as we have already seen from Figure 1, by whether or not federal enforcement of the principle of integration is proposed. It has been generally accepted by most scholars studying racial attitudes that a large majority of White Americans today claim to accept racial equality in principle. However, public approval of the implementation of these principles through federal action is clearly a great deal lower and has not increased appreciably over time (see Schuman, Steeh, and Bobo 1985, chap. 3). The example in Figure 1 provides the most striking contrast we have yet discovered. Placing the trend for Same Schools, which gauges support for the principle of school desegregation, next to the trend for School Intervention, exposes a principle-implementation gap of 22 percentage points in 1964 and approximately 60 percentage points by 1976. Although this is the only case in which the gap increased substantially over time, pairings of principle and implementation ques-

tions in other domains show similarly large discrepancies in support. As Figure 1 reveals, Equal Jobs reached 97 percent in favor by 1972, but the change in a question that asks whether the government in Washington "ought to see to it" that Blacks get fair treatment in jobs (not included in Figure 1) has been remarkably stable. The percentage saying the government should see to it that Black Americans get fair treatment in jobs was 38 percent in 1964, 39 percent in 1972 (a gap of 58 percentage points), 30 percent in 1988, and 33 percent in 1992.

Some Explanations

In one sense the gap between principle and implementation can be called hypocritical, since without government action it may be impossible to produce changes. Indeed, the whole history of racial discrimination in this country, going back to the desegregation of public transportation in Montgomery, Alabama, in 1956, and the desegregation of Little Rock High School in 1957, shows the importance of government action in overcoming local resistance to efforts at integration. Yet this resistance by White Americans to government intervention seems not to be based solely on racial content. For one thing, we find a similar though much smaller gap among African American respondents: whereas support for nondiscrimination in employment is essentially 100 percent among them, support for government action to guarantee fair treatment in jobs was about 71 percent in 1992—very high when compared with White Americans, but definitely not 100 percent.

Moreover, we recently carried out a series of experiments in the area of housing in which opposition to government enforcement of nondiscrimination against Black Americans was compared with resistance to enforcement in the case of discrimination against other groups, in particular, Japanese Americans and Jews (Schuman and Bobo 1988). Table 1 presents two of these comparisons, and it turns out that there is not much difference regardless of the group involved—that is, about the same proportions oppose government action no matter which minority is described.[4] Yet we can be pretty sure that actual discrimination against Blacks in the housing market is much greater today than it is for Japanese Americans or for Jews, and in fact other data indicate greater personal resistance to neighborhood integra-

Table 1   Experiments Contrasting Government Enforcement for the Single Black Family
and for Other Ethnic Group Families

1. September 1986

Black Family[a]		Japanese American Family[a]	
Suppose a Black family plans to move into a house in an all-White neighborhood, and some White people in the neighborhood want to stop them from moving in. Do you think the government should enforce the Black family's right to live wherever they can afford to, *or* that it should be left entirely up to the White neighborhood residents to decide?		Suppose a Japanese American family plans to move into a house in an all-White neighborhood, and some White people in the neighborhood want to stop them from moving in. Do you think the government should enforce the Japanese American family's right to live wherever they can afford to, *or* that it should be left entirely up to the White neighborhood residents to decide?	
1. Government enforcement	60.5	1. Government enforcement	56.8
5. Leave it up to White neighbors	11.5	5. Leave it up to White neighbors	18.1
7. Other (volunteered)	26.8	7. Other (volunteered)	24.5
8. Don't know	1.3	8. Don't know	0.6
	100		100
	(157)		(155)

2. August 1986

Black Family[b]		Jewish Family[b]	
The same as above.		The same as above, but substitute "Jewish" for "Japanese American," and "Christian" for "White."	
1. Government enforcement	61.3	1. Government enforcement	69.6
5. Leave it up to White neighbors	20.4	5. Leave it up to Christian neighbors	14.2
7. Other (volunteered)	17.6	7. Other (volunteered)	15.5
8. Don't know	0.7	8. Don't know	0.7
	100		100
	(142)		(148)

Note: Ns are shown in parentheses
[a]White respondents only
[b]White, non-Jewish respondents only.

tion involving Blacks and Whites than to integration involving other minorities. What this combination of findings means is that resistance to federal intervention to integrate neighborhoods is itself a widely accepted principle in America, and one that makes it easy for White Americans to oppose government enforcement of residential integration on principle, whatever the underlying reason. Effective government action in this area of life will be very difficult.

Most of the issues we have considered thus far have been what might be called the classic issues of racial attitudes in the United States in the post–World War II era—even those involving government en-

forcement of equal treatment. In Figure 1 the trend for Preference In Jobs also makes clear that today we face a series of issues, loosely classified under the heading of "affirmative action," that carry government intervention one step further. The idea expressed in these questions is one of compensation for the discrimination and poverty that African Americans, in particular, have historically experienced and for the continuing disadvantage they face in a largely White society. The data in Figure 1 demonstrate strikingly that there is much less support among White Americans for affirmative action policies than there is for the broad principles of integration and equal treatment discussed earlier, or even for some of the forms of government intervention to enforce integration that we have looked at (Lipset and Schneider 1978). On the other hand, Black respondents are overwhelmingly in favor of preferences in hiring and promotion (75 percent in data from the 1992 National Election Study), but they agree with White respondents (60 percent and 84 percent, respectively, in 1991) that getting jobs and being admitted to college should be based on ability, not on preferential treatment (Gallup and Newport 1991).

Moreover, substantial majorities of White Americans disagree with the proposition that Blacks are now being treated badly in the United States, and since such a proposition forms one of the main bases for affirmative action, it is not surprising that affirmative action itself lacks appreciable support. On a somewhat brighter note, there is substantially more approval among White Americans for various training programs intended to redress the disadvantages that Black Americans have endured in the past than for policies that endorse preferential treatment. In 1988, 73 percent of White Americans said they supported affirmative action programs for Blacks in employment, provided there were no rigid quotas (Sigelman and Welch 1991, p. 129). Thus, recent Supreme Court decisions that narrowed affirmative action programs can be seen as largely in accord with the dominant opinion of White Americans.

For the same reason, the success of David Duke in capturing a majority of the White votes in a 1990 senatorial election in Louisiana was not wholly unexpected, since Duke played heavily on opposition to affirmative action and related policies that lack substantial support among White Americans—though the willingness of so many White voters to ignore or discount his past Nazi and Klan associations was more surprising. If policies of affirmative action are to be continued, they will probably have to be done in ways that minimize direct conflict with majority public opinion and with the courts.

Further Complications

By tracing support for a range of different racial issues over time, we have glimpsed some of the tensions that currently characterize the attitudes of White Americans. If anything, these tensions have been underestimated. Even when we stay on the apparently firm ground of abstract principles, we discover that the questions have seldom been phrased in a way that brings out all the nuances. For example, we learned from Figure 1 that most White Americans today claim to accept equal treatment and integration as general principles, but we still do not know exactly what this means in concrete situations, especially when these general principles conflict, as they often do in real life, with other preferences and principles. Let us look now at some of these complications.

First, there is the meaning of the word "integration" itself—for example, what it means to say that White children and Black children should go to the same schools. Figure 2 shows the results from 1972 to the present for three questions that ask White respondents whether they themselves would have any objection to sending their children to a school where different proportions of the other children are Black. Virtually no White respondents today indicate any objection if a few of the children in a school are Black (97 percent in 1991), and the trend prior to the 1970s (data not shown here) is similar to those we saw earlier. But the figure drops to about 80 percent agreement when half the children are said to be Black, and then further down to around 47 percent when the proportion of Black children is more than half. Moreover, in the last case there is little evidence of change over time and no difference by respondent education. Thus when White Americans say they favor the same schools for White and Black children, many are really thinking of situations where White children remain dominant in terms of numbers and no doubt in terms of general influence as well.[5]

Indeed, in companion studies of attitudes toward integrated housing in metropolitan Detroit conducted in 1976 and 1992, the investigators showed diagrams of neighborhoods with different degrees of integration (Farley et al. 1978, 1993). Each diagram was on a separate card, as shown in Figure 3. In 1976, White respondents were only willing to move into neighborhoods that had one Black house out of a total of fifteen (7 percent Black), while Black respondents favored areas that had a substantial number of houses owned by White Americans (anywhere from 27 percent to approximately 50 percent). By 1992, a majority of White respondents indicated that they would be willing to move into neighborhoods that contained three Black households (20 percent

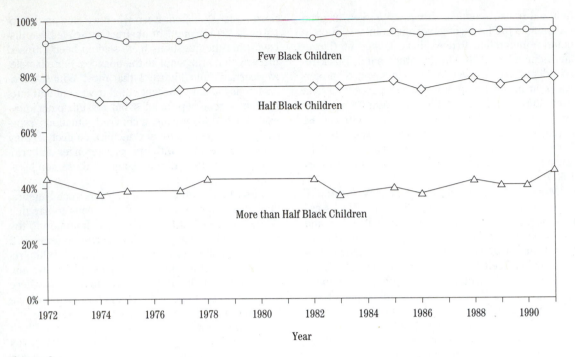

Figure 2

No Objections to Own Children in Integrated School

Black), a significant increase over 1976. Although Blacks maintained their overwhelming preference for a racial mix that was one-quarter to one-half White, they also tended to be somewhat more favorable to all or mostly Black neighborhoods. Thus, substantial numbers of both groups in each year could believe they favored integration, yet the conception of integration each held was quite different. Furthermore, the data from the 1992 survey show that Blacks and Whites today are only a little closer than they were in 1976 to sharing a common definition. This is not really as strange as it may at first sound: most people, Black or White, probably do not feel completely comfortable in a situation where they are part of a tiny minority. African Americans, of course, have had to accept such situations for a great many years, but most White Americans have never been part of a racial minority and reject the idea when it is put to them in concrete terms.

Another complication related to the ideas of integration and racial equality arises when they conflict with other values or preferences that respondents have. As we have already noted, one of the most obvious instances occurs in affirmative action questions

where the norm of racial equality conflicts with the traditional norm of self-reliance. When survey questions attempt to emulate these real-life dilemmas, we have an opportunity to study situational influences on racial attitudes. For example, in one set of questions some years ago, Schuman (1972) asked whether men should be hired for top management without paying attention to race. Perhaps not surprisingly, almost everyone (85 percent) said yes to this question. The next question noted that in this particular situation, hiring a Black engineer as manager was likely to lead to "serious friction" at the top that could ruin the company. On hearing this revised version of the question, about 20 percent of these same respondents switched and said that in that situation it would be better not to hire the Black engineer. Although one might think of this as hypocritical, it is probably more sensible to say that the value of integration is simply not as strong for these people as the value they put on business success. Alternatively, the endorsement of equal treatment may be seen as varying according to the dynamics of particular situations. In this case, attitudes toward hiring the best qualified person regardless of race was conditional

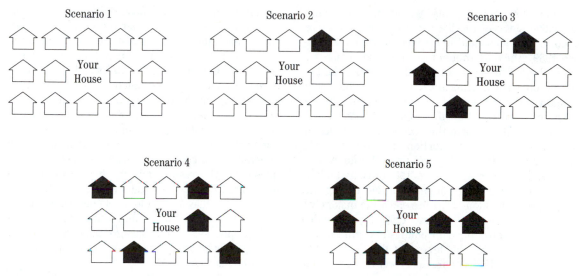

Figure 3

Neighborhood Diagrams Used for White Respondents in the 1976 and 1992 Detroit Area Studies

on the number and status of people in that setting who might object to working with a Black manager. Indeed, at the extreme it is probably possible to think of situations where almost everyone, White or Black, would discriminate—for example, if one genuinely believed that the safety of one's family would be irreparably harmed by a particular attempt at integration. In real life such situations seldom arise— and many purported examples turn out not to hold real danger—and so the serious issue here is the degree to which a person retreats from integration because of a possible threat that is not likely to materialize.

Generational and Other Differences Among White Americans

Having expressed these caveats, we can now note the main social and demographic factors that are generally associated with racially liberal attitudes among Whites. First, as already indicated, southern Whites are regularly less accepting on the average of integration and racial equality than are northern Whites, although the difference diminishes as the trends converge toward 80 or 90 percent. Second, more-educated White Americans are more accepting

of the principles of equal treatment and integration in their responses than are less-educated Whites, though this variation by education tends to disappear with questions that involve implementation by the federal government. Finally, younger generations of White Americans are pretty consistently more liberal in their opinions about racial issues than are older generations. Although White Americans of all ages seemed to have changed their attitudes during the height of the civil rights revolution in the 1960s, the national trends over the last few years have continued to move upward largely because older, less tolerant cohorts of White Americans are gradually being replaced by the younger and more liberal cohorts of the 1950s and 1960s.

This leads to a question about the youth of today, those, for example, of college age who came to maturity after the 1960s Civil Rights movement had ended—that is, from about 1975 to the present. The issue is especially important given the pivotal role that younger generations are assigned in promoting attitude change. Over the life cycle of an issue, public opinion supposedly moves from either total rejection of an idea to total acceptance, or vice versa (Schreiber 1978). We have seen this kind of change depicted in Figure 1 for the questions that measure approval of equal opportunity in jobs and the desegregation of schools. Because young adults are in the formative stage of the life cycle and thus are more open to new ideas and influences (Krosnick and Alwin 1989), they are thought to spearhead change by being the first to adopt unorthodox or unpopular positions on issues. Thus the current distribution of opinion among them should forecast the future shape of opinion within the total population, *ceteris paribus*. For this reason the acceptance or rejection of racial equality by this portion of the population is of particular interest. The cohort of White Americans who came of age in the 1980s reached adulthood, and inclusion in national samples, during the Reagan years and long after the end of the major civil rights actions of the 1950s and 1960s. As a result of having been introduced to politics at a time of national retreat from the tenets of the Civil Rights movement, the 1980s cohort may well be less liberal today on racial issues than the 1970s cohort that immediately precedes them. This hypothesis would certainly fit reports of new outbursts of bigotry on college campuses—instances of flyers containing threats against Black students, crude jokes and stereotypes, scrawlings of Nazi swastikas on dormitory and fraternity houses, and other expressions of racial and ethnic intolerance and hostility.

It is difficult to study college-age students as part of the national population, both because they are a small proportion of the total and because many are not reached by ordinary sample surveys. However, over the past few years we have tried to look at national data to determine whether there are negative trends that fit the picture of a younger generation moving backwards in terms of racial attitudes (Steeh and Schuman 1992). To oversimplify a bit, we have compared three groups: those who reached age eighteen during the 1960s, those who reached the same age during the 1970s, and those who reached that age during the 1980s. So far as we can tell from our analysis (not included here) of the questions on school integration, job equality, intermarriage, and residential integration that are pictured in Figure 1 for the entire adult population, there has been no important drop in support for integration and equal treatment among the young adults in the population. Thus we decided to focus on questions about racial policies and affirmative action, such as government intervention to ensure job equality and preferences in hiring and promotion, which were asked in the most recent period of 1984 to 1990. Since empirical

evidence suggests that these attitudes are not as readily explained by the demographic factors we enumerated above, we felt that they constituted the strongest test of the hypothesis that the cohorts coming of age during the late 1970s and 1980s would have less liberal attitudes than the cohorts coming of age during and immediately after the Civil Rights movement. If we assume that there are no inevitable changes in racial attitudes simply on the basis of aging itself, then we can focus our attention entirely on generational or cohort effects on responses to these questions. The results indicate that on most questions, the youngest generation—those eighteen to twenty-nine today—are at least as liberal in their attitudes about racial policies and affirmative action as their predecessors when background characteristics such as education, income, gender, and region are taken into account.

A Student Sample

A different perspective on the subject of affirmative action comes from a survey of 500 undergraduate students at the University of Michigan conducted in 1990. Their answers to a question on whether they favored or opposed affirmative action with regard to admission to the University of Michigan indicate substantial support among undergraduates, but also some noticeable opposition. About 70 percent of the White students signify some or strong support for this type of affirmative action, while 30 percent indicate some or strong opposition (Schulz and Schuman 1990). Student respondents were also asked what the term "affirmative action" means to them, and those who support affirmative action explain their position mainly by saying that it helps make up for past inequality or contributes to equal opportunity, while those who oppose affirmative action most often see it as an unfair advantage or a form of reverse discrimination. The small sample of Black students, about 6 percent of the total of 500, is almost completely in support of affirmative action, though divided in strength of support between "strong" and "somewhat." Thus, in the university undergraduate body there is probably enough support for now to sustain affirmative action policies, but also enough opposition to create tensions as a result of such policies. Unfortunately we do not have data for these students on any of the questions discussed earlier for the general population, but, as we have seen, it is important to distinguish questions on affirmative action from the classic issues of integration and equal treatment. Of course, the Michigan student body is almost certainly more liberal than the total adult population on most or all of these issues.

Other findings from this survey may shed some light on the difficult problem of determining what is happening today on college campuses. The survey measured sentiment on an issue of salience to the campus concerning "remarks made during classroom discussion [that] may offend members of another racial or ethnic group." Most students are opposed to any attempt by the university administration to prohibit such remarks, with more than 80 percent of the White student body indicating clear opposition and more than two-thirds of the Black students also opposed. As to personal experience with prejudice or discrimination at any point in their university career, approximately half of our Black student sample report having had such negative experiences, with roughly similar percentages of White women reporting their experiences with sexism and of Jewish women reporting their experiences with anti-Semitism.

It is difficult to evaluate these reports without data from earlier points in time, but even such data would leave the interpretation unclear. More generally, the crude remarks and actions that have occurred outside the classroom at Michigan and on other campuses is puzzling in light of the national survey data, but we believe they arise from three factors. First, there are more Black students on campuses and the increasing contact can lead to friction as well as to understanding. About half our sample of White students claim to have at least one Black friend, but the term "friend" is vague, and observation of the campus shows a good deal of de facto racial segregation. Second, there has been an increase nationally in extreme forms of speech, most obviously in the continued pushing against past conventional limits by entertainers of all kinds, both White and Black. This probably has a "modeling effect," making it easier for offensive remarks to be made by ordinary students. Third, student opposition to affirmative action is not trivial in numbers, as noted earlier, and this gives rise to increased criticism by some White Americans and to defensiveness by many African Americans. One must add the preva-

lence of drinking on college campuses, which has been associated with some of the more extreme instances of bigotry reported.

Attitudes and Actions

Despite the complexity of racial attitudes revealed in as brief an account as this report, doubts about the validity of self-reported tolerance abound. Apparent discrepancies between attitudes described as tolerant and behavior assessed as discriminatory support the claim that true attitudes can only be properly gauged when the measurement is unobtrusive to the subject. It should be clear by now that one does not need to draw very much on the distinction between attitudes as revealed in survey interviews and behavior outside the survey interview in order to explain why most White Americans endorse racial integration in principle and yet live in a largely segregated society. The endorsement of integration by White Americans really means the rejection of complete segregation. In that sense it is a very profound change from the 1940s and 1950s. In fact, it is exactly this type of change that has occurred in American public life. Black Americans are no longer barred from public areas such as hotels, restaurants, and the like. They are also represented, though in very small numbers, at the highest levels of government—as a cabinet secretary or military chief of staff. This is a fundamental change from several decades ago when Black and White Americans could not even sit at a table together in government restaurants in Washington, D.C. Moreover, Black political leaders who are viewed as nonthreatening by White Americans can in certain settings win a substantial number of their votes, as happened with Tom Bradley, elected multiple times as mayor of Los Angeles and nearly elected governor of California in 1982, and more recently with the election of Douglas Wilder as governor of Virginia.

However, just as the survey data reveal clear limits to the numbers or power that Blacks are permitted to obtain in America, so we still have very little integration in housing in the United States, and in cities where Blacks constitute about half the population, voting by race becomes the major political division, as it is today in Chicago. Thus, looked at broadly, there is no great discrepancy between the survey trends and the trends in nonsurvey behavior that one would describe for the same time period.

This is not to claim that there is no conscious or unconscious deception in answers to survey questions on race. The Wilder election provided strong evidence that some of the White respondents who claimed they would vote for the Black candidate, in fact, did not do so. Indeed, we know from a number of experiments where the race of the interviewer is varied that there can be substantial differences in the ways that both Black and White respondents answer depending on the race of the person who asks the questions, and a recent study strongly suggests that such interviewer effects caused the discrepancy manifest in the Virginia election (Finkel, Guterbock, and Borg 1991). Important as it is to keep that in mind, however, it is at least equally important to recognize that an estimated 40 percent of the White population in Virginia did vote for the Black gubernatorial candidate in a state that not very long ago enforced legal segregation.

Moreover, it is a serious mistake to assume that people lie about their attitudes in survey interviews, but are totally honest in their other behavior: much behavior is less than completely sincere, whether in relatively innocuous ways, as when we are polite to someone we may dislike, or in larger ways, as when a president appoints an African American to be a cabinet member for purely political purposes. Behavior is controlled to a considerable extent by exactly the same norms that control the expression of attitudes in surveys, and one should not look to either for final evidence as to what goes on in the hearts of men and women, White or Black. As to what really goes on in your heart or in ours, that is a problem no survey or direct observation of behavior can be sure to uncover. Recent theoretical and experimental work in social psychology has emphasized the ambivalence of many White Americans toward Blacks and toward racial issues (Katz 1986). One finds not a simple positive or simple negative stance, but a mixture of positive and negative attitudes. For example, there is some evidence that White Americans are more positive toward highly successful Blacks than toward equally successful Whites, but that the opposite holds when relatively unsuccessful Blacks and unsuccessful Whites are compared. This work is still at an early stage, but it provides a more complex view of White racial attitudes that is probably closer to the truth than arguments over degrees of sincerity.

Conclusion

Standing back from the particular data we have examined, it is possible to look at the future of race relations in America in either an optimistic or a pessimistic light. On the optimistic side, we can point to the transformation of many White racial attitudes and behaviors that have been described here. One can also point to the growth of a substantial Black middle class, far larger, more influential, and better off financially than it was just a few decades ago. African Americans are now in positions of high prestige and authority in the United States, and these are local Black political leaders in a great many cities and in some states. Also of great importance is the continued need in a largely non-European world for the United States—and its national officials—to draw on its unique constellation of races and peoples for its own good name, as well as to make optimum use of its human resources.

However, it is equally possible to paint a deeply pessimistic picture of race in America. Despite the increase in White Americans' support of equal rights over the last forty to fifty years, there is certainly a large amount of discrimination that, among other things, maintains a de facto apartheid in housing and, therefore, also in schools. A large Black underclass seems trapped in inner cities, with massive unemployment, low education, and substantial involvement in crime and drugs. Black leaders and some White liberal leaders call for government economic aid and support for affirmative action, but there is little money going for that purpose, and, as we have seen, little support for affirmative action by either the public or the courts.

Beyond these, great gulfs in trust and communication exist between most Blacks and most Whites, fueled by dramatic incidents, often real, though sometimes exaggerated. Certainly the acquittal of several Los Angeles police officers charged with the beating of Rodney King has left a bitter legacy that cannot be quickly overcome, and the assault on Reginald Denny in the subsequent disturbances in Los Angeles fuels the hostility of White Americans as well. Candor in discussing issues related to race is rare on all sides, and the possibilities of offense through a careless word or act are ever present. Looked at from these standpoints, race relations may be moving in a more and more intractable direction, with the positive changes too slow and too incomplete to prevent the kind of White reaction reflected in the David Duke candidacy.

Which of these pictures of the future is closer to the truth? No one can know for sure, but it seems highly unlikely that the United States will approach the extremes of either genuine integration or large-scale racial war in the foreseeable future. Continued strain, at times extreme, and possibly violent, and accentuating the differences of opinion within the White population, and to some extent within the Black population as well, confronts this country for many years to come.

Appendix: Wording of Racial Attitude Questions

National Election Study (NES)

Overcome Prejudice   Irish, Italians, Jewish and many other minorities overcame prejudice and worked their way up. Blacks should do the same without any special favors. Responses: 1. Agree strongly 2. Agree somewhat 3. Neither agree nor disagree 4. Disagree somewhat 5. Disagree strongly.

Preference in Jobs   Some people say that because of past discrimination, Blacks should be given preference in hiring and promotion. Others say that such preference in hiring and promotion of Blacks is wrong because it gives Blacks advantages they haven't earned. What about your opinion—are you for or against preferential hiring and promotion of Blacks? Responses: 1. Strongly against 2. Not strongly against 3. Not strongly for 4. Strongly for.

Fair Treatment in Jobs   Some people feel that if Black people are not getting fair treatment in jobs, the government in Washington ought to see to it that they do. Others feel that this is not the federal government's business. Have you had enough interest in this question to favor one side over the other? (IF YES) How do you feel? Should the government in Washington see to it that Black people get fair treatment in jobs or is this not the Federal Government's business? Responses: 1. Government should see to it 3. No interest and depends 5. Not the government's business.

School Intervention.   Some people say that the government in Washington should see to it that White and Black children go to the same schools. Others claim that this is not the government's business. Have you been concerned enough about this question to favor one side over the other? (IF YES) Do you think the government in Washington should see to it that White and Black children go to the same schools, or stay out of this matter as it is not its business? Responses: 1. Government should see to it 3. No interest and depends 5. Government should stay out.

General Social Survey

Equal Jobs   Do you think Blacks should have as good a chance as White people to get any kind of job, or do you think White people should have the first chance at any kind of job? Responses: 1. As good a chance 2. White people first.

Same Schools   Do you think White students and Black students should go to the same school or to separate schools? Responses: 1. Same 2. Separate.

Residential Choice   Here are some opinions other people have expressed in connection with Black-White relations. Which statement on the card comes closest to how you yourself feel? White people have a right to keep Blacks out of their neighborhoods if they want to, and Blacks should respect that right. Responses: 1. Agree strongly 2. Agree slightly 3. Disagree slightly 4. Disagree strongly.

Few Black Children   Would you, yourself, have any objection to sending your children to a school where a few of the children are Blacks? Responses: 1. Yes 2. No.

Half Black Children (If No or DK to Few)   Where half of the children are Blacks? Responses: 1. Yes 2. No.

More than Half Black Children (If No or DK to Half)   Where more than half of the children are Blacks? Responses: 1. Yes 2. No.

Gallup

Intermarriage   Do you approve or disapprove of marriage between Whites and non-Whites? Responses: 1. Approve 2. Disapprove.

Preferential Treatment vs. Ability   Some people say that to make up for past discrimination, women and members of minority groups should be given preferential treatment in getting jobs and places in colleges. Others say that their ability, as determined in test scores, should be the main consideration. Which point of view comes closer to how you feel on the subject? Responses: 1. Preferential treatment 2. Test scores only.

Detroit Area Study

Neighborhood Integration   Now, I'd like you to imagine yourself in a different situation. Suppose you have been looking for a house and have found a nice one you can afford. This house could be located in several different types of neighborhoods, as shown on these cards (Diagrams 1–5). Would you consider moving into any of these neighborhoods? (IF YES) Show me *all* the neighborhoods you would move into (CHECK *ALL* MENTIONED).

Hiring a Black Engineer as Manager   Suppose a good Black engineer applied for a job as an engineering executive. The Personnel Director explained to him: "Personally I'd never given your race a thought, but the two men you would have to work with most closely—the plant manager and the chief engineer—both have strong feelings about Blacks. I can offer you a job as a regular engineer, but not at the executive level, because any serious friction at the top could ruin the organization."

   A. Was it all right for the personnel director in this case to refuse to hire the Black engineer as an executive in order to avoid friction with the other employees? Responses: 1. Yes 2. No.

   B. Should the personnel manager have asked the other men how they would feel about working with a Black engineer and then made his decision on the basis of their wishes? Responses: 1. Yes 2. No. C. In general, do you think employers should hire men for top management without paying any attention to whether they are White or Black? Responses: 1. Yes 2. No.

University of Michigan Student Survey

Affirmative Action in Admissions   Do you favor or oppose affirmative action with regard to admission to the University of Michigan? Would you say you: 1. Strongly favor it 2. Somewhat favor it 3. Somewhat oppose it 4. Strongly oppose it.

What Affirmative Action Means   Many people hold different ideas about "Affirmative Action." What does the term "Affirmative Action" mean to you? (IF DK, ASK FOR BEST GUESS) Responses: open.

Speech Codes   Sometimes remarks made during classroom discussions may offend members of another racial or ethnic group. Some people think that the university should try to prohibit such remarks. Others think that the university should not try to control classroom discussion in this way. Which of these two views is closer to your own? Responses: 1. University should prohibit 2. University should not prohibit.

Experience of Discrimination   Have *you ever personally* experienced sexism, racism, anti-Semitism, or other forms of prejudice or discrimination *while attending the University of Michigan*? Responses: 1. Yes 2. No.

Notes

1. The exact wording of all the questions referenced in this study can be found in the Appendix.

2. The data in the text and both figures not otherwise referenced are drawn from tables and figures in Schuman, Steeh, and Bobo (1985) but have been updated to 1992. The original surveys were carried out by three organizations: the National Opinion Research Center at the University of Chicago, the Institute for Social Research at the University of Michigan, and the Gallup Poll.

3. This conflict has led recent scholars to redefine the American dilemma not as a choice between the American creed and the specifics of individual and group living that Myrdal (1944) observed but as a "choice . . . among the very values that have made up the American creed" (Sniderman, Tetlock, and Carmines 1993, p. 10).

4. Somewhat similar results from the 1992 Detroit Area Study indicate that opposition to preferences in hiring and promotion is also just as great when women, rather than Black Americans, are the beneficiaries.

5. It has been argued that the last survey item in this series is a tainted indicator of White Americans' intolerance for settings in which Black students outnumber White. According to this view, the measure is as much an indictment of the quality of public schools in the inner city as a statement of prejudice against associating with large numbers of Black Americans. We have no empirical evidence that would allow us to evaluate the validity of these claims. We simply point out that the desperate plight of public education in the nation's largest urban areas can only be a possible confounding factor in the late 1970s and 1980s. It does not explain the small numbers of respondents who accepted this degree of integration in the 1950s and 1960s when public education in the nation's largest cities had not yet declined.

35

Contesting the Meaning of Race in the Post–Civil Rights Movement Era

MICHAEL OMI
HOWARD WINANT

Anaheim High School in California is a predominantly Latino school about one mile from Disneyland. Recently, the formation of a European American Club there has been at the center of an escalating controversy (Freedberg 1992).

The club was ostensibly created to provide scholarship information and help with college entrance exams for students excluded from "minority-oriented" programs. One student, whose great-grandparents came from Portugal and Scotland, said, "Without this club, I kind of felt out of place." She noted that there was an Asian Club, a Black Student Union, and a Mexican American Engineering Society, but nothing for White students. She felt rejected when college recruiters seemed interested in talking to only minority students—at a school where Whites (who comprise 16 percent of the student body) are the *minority*.

The principal of the school defended the club saying, "It fits on this campus as much as any other club." Critics, however, say such clubs are representative of a White backlash and that such organizations contradict the principles and vision of the Civil Rights movement. In their view, such European American clubs are merely polite organizational expressions of White supremacist sentiments. Those who formed the group deny any racist intentions. They say they are only trying to reclaim a heritage of their own, one based on European ethnicity rather than an affirmation of White identity.

What are we to make of all this? We think that the controversy surrounding the European American Club raises a host of questions regarding the relationship of majority and minority cultures, fairness and representation, and, in the immortal words of Aretha Franklin, "Who's zoomin' who?"; in essence, issues that debate the very meaning of racial integration and equality in American life.

While educational institutions have been formally integrated for decades, we are nonetheless witnessing a growing balkanization—a pattern of separation, segregation, and assertion of "difference"—among students of different racial backgrounds, and a parallel increase in racial conflict and tension. Such problems are by no means confined to schools. Workplaces, neighborhoods, hospitals, the media, and political parties are equally the site of conflicts and dilemmas regarding their racial organization and makeup. Despite legal guarantees of equality and access, race and ethnicity continue to be fundamental organizing principles of individual identity and collective action. Rather than the declining significance of race, as the title of William J. Wilson's book (1978) suggests, we are witnessing the continuing significance of race. This, we argue, is a result of its *contested meaning* in the post–Civil Rights movement period.

Over twenty-five years since the passage of key pieces of federal legislation—such as the Civil Rights Act—outlawing racial discrimination in jobs, public accommodations, immigration policy, and voting rights, we remain consumed by political and cultural attempts to define, and redefine, the meaning of race for institutional life and individual identity. The scope and meaning of discrimination are still debated, legislated, and litigated. Self-segregation in daily practice and institutional life is both denounced and defended. While an overwhelming majority of Whites favor egalitarian principles, according to survey data, only a minority support government attempts to ensure equality (Schuman, Steeh, and Bobo 1985).

Despite the Civil Rights movement's profound impact on racial attitudes and institutional arrangements, in the post–Civil Rights movement era, the issue of race remains more controversial than ever. Why is this so? Because specific patterns of *racialization* have emerged in the contemporary period that create new racial subjects and significantly transform existing ones. Far from declining in significance, the racial dimensions of political and social life have grown.

Racialization

Over the past several decades we have witnessed attempts from across the political spectrum to define the appropriate meaning of race in institutional life, and to establish coherent racial identities based on that meaning. Indeed, the very meaning of political labels such as conservative, liberal, and radical have been transformed by ongoing debates about race and ethnicity in such issues as affirmative action, social welfare policy, and immigration reform. The quest to articulate a clear and consistent definition of race and its meaning has proved elusive. This is because race is preeminently a social construct. Racial categories and the meaning of race are given concrete expression by the specific historical context and social relations in which they are embedded. As

such, race is inherently subject to contestation; its meaning is intrinsically unstable.

From a *racial formation* perspective (Omi and Winant 1986), race is understood as a fluid and unstable complex of social meanings constantly being transformed by political conflict. Race is seen as both a constituent part of the individual psyche and of relationships among individuals, as well as an integral component of collective identities and social structures. As such, it is necessary to interpret the meaning of race, not in terms of definitions, but in terms of the processes of racial formation. Chief among these is the construction of racial identity and meaning, the process which we call *racialization*.

The concept of racialization denotes the extension of racial meaning to a previously racially unclassified relationship, social practice, or group. A historical example is the consolidation of the racial category of Black in the United States from Africans whose specific identity had been Ibo, Yoruba, or Bakaongo, among other West African tribal identities. Parallel to this was, as Winthrop Jordan ([1968] 1977) observes, the emergence of White as a term of self-identity evolving from earlier conceptions of Christian, English, and free.

Stanley Lieberson and Mary Waters (1988) underscore that racial and ethnic groups should not be viewed as static categories, but "as products of labeling and identification processes that change and evolve over time." They suggest that "a continuous process of combining and recombining" takes place in which "groups appear and disappear" (Lieberson and Waters 1988, p. 252). We employ the concept of racialization to capture how race and racial meanings have been significantly transformed by the Civil Rights movement. In the wake of the movement, Americans have witnessed the introduction and institutionalization of policy reforms, such as affirmative action, and demographic changes in the racial composition of the nation, resulting, in large part, from the liberalization of immigration policy. Along with these changes have come dramatic shifts in our own sociocultural understandings of race and racism.

In this post–Civil Rights movement era, new forms and expressions of racialization have unfolded. These include the emergence and consolidation of new racial categories, the appearance of differences and divisions within previously well-defined racial groups, and the phenomenon of groups confronting previously unexamined questions regarding their racial identity and status. Three key examples of these tendencies are the development of new racial subjects as a result of a new panethnic consciousness, the increasing significance of class for African Americans, and the crisis of White identity.

Panethnicity and the Development of New Racial Subjects

In the post–Civil Rights movement era, groups whose previous national or ethnic identities were quite distinct have become consolidated into a single racial category.

Prior to the late 1960s, for example, there were no "Asian Americans." In the wake of the Civil Rights movement, distinct Asian ethnic groups—primarily Chinese, Japanese, Filipino, and Korean Americans—began to frame and assert their common identity as Asian Americans. This political label reflected the similarity of treatment that these groups historically encountered at the hands of state institutions and the dominant culture at large. Different Asian ethnic groups had been subject to exclusionary immigration laws, restrictive naturalization laws, labor market segregation, and patterns of ghettoization by a polity and culture that treated all Asians as alike.

The racialization of Asian Americans involved muting the profound cultural and linguistic differences, and minimizing the significant historical antagonisms, that had existed among the distinct nationalities and ethnic groups of Asian origin. In spite of enormous diversity, Asian American activists found this new political label a crucial rallying point for raising political consciousness about the problems of Asian ethnic communities and for asserting demands on state institutions.

The racialization of Asian ethnic groups that resulted from the emergence of panethnic consciousness was paralleled by the racialization of other groups, notably Latinos and Native Americans. Such panethnic activism was inspired by the Civil Rights movement and anticolonial nationalist movements in Asia, Africa, and Latin America. Somewhat ironically, the very movements which sought to end racial discrimination at home and colonial rule abroad, also fostered an increased political awareness among formerly fragmented ethnic groups that they comprised a larger racially defined entity.

David Lopez and Yen Espiritu define panethnicity as "the development of bridging organizations and solidarities among subgroups of ethnic collectivities that are often seen as homogeneous by outsiders" (1990, p. 198). Such a development, they claim, is a crucial feature of ethnic change—"supplanting both assimilation and ethnic particularism as the direction of change for racial/ethnic minorities" (1990, p. 198). Lopez and Espiritu suggest that panethnic formations are not merely "alliances of convenience" but are shaped by an ensemble of cultural factors, such as common language and religion, and structural factors, such as race, class, generation, and geographical concentration. They conclude, however, that a specific concept of race is fundamental to the construction of panethnicity since, "[t]hose . . . groups that, from an outsider's point of view, are most racially homogeneous are also the groups with the greatest panethnic development" (1990, pp. 219–220).

In our view, the rise of panethnicity is a process of racialization which is driven by a dynamic relationship between the specific group being racialized and the state. The elites representing such groups find it advantageous to make political demands by using the numbers and resources which panethnic formations can mobilize. The state, in turn, can more easily manage claims by recognizing and responding to large blocs as opposed to dealing with the specific claims of a plethora of ethnically defined interest groups. Conflicts often occur over the precise definition and boundaries of various racially defined groups, and their adequate representation in census counts, reapportionment debates, and minority aid programs.

Panethnic consciousness and organization are, to a large extent, contextually and strategically determined. At times it is advantageous to be in a panethnic bloc, and at times it is desirable to mobilize along particular ethnic lines. Therefore both a politics of *inclusion* and *exclusion* are involved in panethnicity, as racial and ethnic definitions and boundaries are contested.

Two examples illustrate the situational nature of this dynamic. In an attempt to boost their political clout and benefits from land trust arrangements, native Hawaiians voted four to one in January 1990 to expand the definition of their people to anyone with a drop of Hawaiian "blood." Previously only those with at least 50 percent Hawaiian "blood" were eligible for certain benefits (Essoyan 1990). By contrast, in June 1991 in San Francisco, Chinese American architects and engineers protested the inclusion of Asian Indians under the city's minority business enterprise ordinance. Citing a Supreme Court ruling that requires cities to narrowly define which groups had suffered discrimination to justify specific affirmative action programs, Chinese Americans contended that Asian Indians should not be considered "Asian" (Chung 1991). At stake were obvious economic benefits accruing to designated "minority" businesses.

The post–Civil Rights movement era has witnessed the rise of panethnicity as a phenomenon of racialization. Groups that were previously self-defined in terms of specific nationalities or ethnic backgrounds, and which were marginalized by the seemingly more central dynamic of Black-White relations, began to confront their own racial identity and status in a political environment of heightened racial consciousness and mobilization.

Panethnicity will continue to be an enduring feature of political life as we enter the next century. The dramatically changing demographic landscape, the transformation of global, regional, and sectoral economies, and the contested nature of political power will all ensure a role for panethnic identity, consciousness, and political organization in the near future.

The Increasing Significance of Class for African Americans

Law professor Roy Brooks states that, "[d]eep class stratification within African American society is without a doubt the most significant development in the 'American dilemma' since the civil rights movement of the 1960s" (1990, p. xi). Few analysts today would take issue with him. When William Julius Wilson argued in 1978, however, that the contemporary life chances of individual African Americans "have more to do with their economic class position than with their day-to-day encounters with Whites" (Wilson 1978, p. 1), he ushered in a raging storm of debate about the relative importance of race and class in America.

Since then, many scholars have emphasized the primary importance of social characteristics other than race, particularly social class, in shaping African

American life chances in this post–Civil Rights movement era. Thomas Sowell (1983), for example, asserts that differences in racial and ethnic group economic performance solely result from a group's human capital (education, training, and the like), and not from societywide discrimination. Wilson himself, in *The Truly Disadvantaged* (1987), argues that the impersonal forces of the market economy explain more about the current impoverishment of inner-city African American poor than outright racial discrimination. While he does not dismiss the effects of historical racial discrimination, he concludes that capital is "color blind," and that the large-scale demographic, economic, and political changes that have negatively affected the ghetto have little to do with race.

Stressing the importance of social class in shaping the African American experience is nothing new. As pioneering studies by African American sociologists W. E. B. DuBois ([1899] 1967) and E. Franklin Frazier (1957) demonstrated, class antagonisms within the African American community have a venerable history. Socioeconomic homogeneity has never existed in any racially defined community. Are social class divisions then any more significant today than they were in the past?

In many respects, the answer is yes. Previous conflicts between the "black bourgeoisie" and the "black masses" took place in a context of nearly complete segregation. However much these two sectors of the community viewed each other with suspicion and mistrust, they were forced to live together, and frequently to ally against the system of White supremacy.

But the Civil Rights movement created a new context for racialization. With the passage of significant pieces of legislation, the patterns of racial segregation in a range of institutions were severely challenged. A generation after the enactment of major civil rights reforms, the African American community is both the beneficiary and victim of its own success. A community once knitted together by survival imperatives in a segregated society and bound up by the thick relationships of labor, commerce, residence, and religion within its own community has now been dispersed.

This dispersion has occurred primarily along social class lines. African Americans who could take advantage of the slow but real lowering of racial barriers in education, employment, and housing have been able to achieve an unprecedented degree of upward mobility. This does not mean that they are shielded from discriminatory acts, or that they have abandoned their identification as "Black." But it does mean that they are living in a far more integrated world, and that racial identity and racism no longer determine their fates or futures as inexorably as before (Landry 1987). At the same time, the decline of manufacturing, capital flight, and the fiscal crisis of the cities have further impoverished low-skilled and undereducated African Americans leading to the much-publicized problem of the "underclass" or "ghetto poor" (Wilson 1991). As a result, African Americans have been racialized differently along the lines of class, with important consequences for individual identity, collective consciousness, and political organization.

For the African American middle class, for example, this generates profound ambivalence about racial identity. As Dr. Alvin Poussaint notes:

> There's a lot of pressure on the black middle class to stay black. . . . It's kind of a contradiction. Your kids are living in an integrated community, and you want them to feel part of the community, participating equally in it. Then you feel very ambivalent about it psychologically, when they do. (quoted in Garreau 1987)

Nor are middle-class African Americans shielded from discriminatory treatment by virtue of their social class position. In a recent study of anti-Black discrimination in public places, Joe Feagin found that African American shoppers were subject to excessive surveillance and frequently received curt and discourteous services from clerks: "No matter how affluent and influential, a black person cannot escape the stigma of being black, even while relaxing or shopping" (1991, p. 107).

This research suggests that racial identity is still conflictual, and racial discrimination is still problematic, for all strata of the African American community. Such analyses also reveal, however, that the effects of these conflicts and problems vary widely by class. The middle-class ambivalence about racial identity described by Poussaint contrasts sharply with the bitter frustration and pervasive violence through which impoverished ghetto youth experience being Black. The sites of discrimination, the types of discrimination experienced, and the range of available

responses to them, obviously differ by where one is located in the class structure. Yet other instances, such as the malevolent attention which police devote to African Americans, reveal a frightening uniformity across all social classes. The Report of the Independent Commission on the Los Angeles Police Department chaired by Warren Christopher, for example, found a pattern of racial harassment against African Americans, including off-duty African American police officers being stopped and detained by White officers (Dunne 1991, p. 68).

The "increasing significance of class," does not, therefore, suggest the "declining significance of race." In the wake of the reforms ushered in by the Civil Rights movement, distinct paths of opportunity opened up which created a modicum of upward mobility for the African American middle-class, but did little or nothing to halt the precipitous decline of the ghetto poor. Despite this dramatic transformation, African Americans remain a preeminently racialized group in politics, cultural representation, and social life. They are not, however, racialized in a uniform manner across social classes.

It is important to realize this since much of the race versus class debate suffers from the imposition of rigid categories and analyses that tend to degenerate into dogmatic assertions regarding the primacy of one category over the other. A more fruitful mode of inquiry would seek to account for the impact of class transformation within a social order still highly structured by race. We need a way, in other words, to grasp the increasing significance of class for African Americans (and other racially defined groups) without denying the centrality of race in the formation of identity and everyday experience.

The Crisis of White Identity

The 1990 U.S. Census revealed that the racial composition of the nation changed more dramatically over the past decade than at any time this century (Barringer 1991). Today, nearly one in every four Americans is of African, Asian, Latino, or Native American ancestry. The stunning increase in what used to be thought of as "minority" populations in the United States renders much of the very language of race relations obsolete and incongruous. By 2003, White Americans are expected make up less than 50 percent of the state of California, for example. And the demographics of the workplace and the campus are changing faster than those of the general population.

The prospect that White Americans may no longer constitute a clear majority nor exercise unquestioned racial domination in various institutions has led to a crisis of White identity. As previous assumptions erode, White identity is losing its transparency—the easy elision with "racelessness" that accompanies racial dominance. Now, the very meaning of "White" has become a matter of anxiety and concern. In this respect, Whites have been racialized in the post–Civil Rights movement era.

During the 1970s, as the influence of the Civil Rights movement waned, there was a backlash against the institutionalization of civil rights reforms and to the political realignments set in motion by the social movements that took hold in the 1960s. Resistance to affirmative action programs grew among White Americans who felt that it was they, and not racially defined minorities, who were discriminated against—reverse discrimination—by government policies ostensibly designed to promote racial equality in the schools, the workplace, and other settings. As one respondent noted in *The Report on Democratic Defection* (Greenberg 1985, p. 70), the "average American White guy" gets a "raw deal" from the government because "blacks get advantages, Hispanics get advantages, Orientals get advantages. Everybody but the White male race gets advantages now."

The idea that a White racial identification could be a handicap was unprecedented. Under conditions of White supremacy, for example, the meaning of being White was relatively unproblematic. In the midst of the Civil Rights movement, though, it became necessary to evaluate more seriously the nature of Whiteness.

One response was to assert the primacy of ethnicity over race. In this account, there was no such thing as a homogeneous White majority. Many White Americans were really minorities in their own right, "unmeltable" ethnic minorities (Novak 1972). More recent research, by contrast, suggests that most White Americans do not experience their ethnicity as a definitive aspect of their social identity. Rather, they perceive it dimly and irregularly, picking and choosing among its varied strands that allows them to exercise, as Mary Waters (1990) suggests, an "ethnic option." The specifically ethnic components of White identity are fast receding with each generation's ad-

ditional distance from the Old Country. White Americans are increasingly unable to speak the language of their immigrant forebears, uncommitted to ethnic endogamy, and unaware of their ancestors' traditions—if in fact they can still identify their ancestors as, say, Polish or Scottish, rather than a combination of four or five European (and non-European) groups. As such, White Americans may also be undergoing a racializing panethnicity as "Euro-Americans" (Alba 1990).

The "twilight of White ethnicity," as Alba calls it, in a racially defined, and increasingly polarized, environment means that a White racial identity will grow in salience. The racialization process for White Americans is very evident on university campuses, where White students encounter a heightened awareness of race that calls their own identity into question. In a recent study on racial diversity (Institute for the Study of Social Change 1991, p. 37) conducted at the University of California at Berkeley, students expressed the new conflictual nature of White identity:

> Many Whites don't feel like they have an ethnic identity at all and I pretty much feel that way too. It's not something that bothers me tremendously, but I think that maybe I could be missing something that other people have, that I am not experiencing.

> Being White means that you're less likely to get financial aid. . . . It means that there are all sorts of tutoring groups and special programs that you can't get into, because you're not a minority.

> If you want to go with the stereotypes, Asians are the smart people, the Blacks are great athletes, what is White? We're just here. We're the oppressors of the nation.

Here we see many of the themes and dilemmas of a White identity in the post–Civil Rights movement era: the "absence" of a clear culture and identity, the "disadvantages" of being White with respect to the distribution of resources, and the stigma of being perceived as the "oppressors of the nation."

Such themes are echoed outside of the campus community as well. Dale Warner, the secretary of the European American Study Group in San Jose, California, said that unless European Americans start speaking out as European Americans, they will not be "engaged in the public discourse of our times" (quoted in Ness, 1992). Warner suggests that European Americans need to organize in order to avoid being increasingly marginalized in political life:

> We've left our chair at the multicultural table empty, and the multicultural table is where the debate is, where the deals are being made—about voting rights, immigrant services, immigration laws, redistricting, health policy. More and more, it's now being built around ethnicity.

We disagree with the last statement and contend that the issue remains one of race. The very term "Euro-American" reinforces the myth of a racially "White" Europe. The term is used in different settings as a substitute for the racial category of White, and little attention is given to how Black or Chinese Britons should be categorized when they immigrate to and settle in the United States. Despite the confusion in categories, the political sentiments remain clear. White Americans are concerned that their interests as Whites are not being articulated, addressed, and advanced.

The big question remains how a White identity will be articulated in the coming period, and the political and ideological elements which will be called into service in its refashioning. Already, far-right political actors, such as David Duke, actively seek to organize White Americans to defend their supposedly threatened racial privileges, all in the name of equality. Such racist populism is not all that distinct from the demagogic use of the code word "quotas" by Jesse Helms and George Bush for the purpose of attracting the votes of White Americans.

On the other hand, the resentments of White Americans cannot be wholly dismissed; they are not solely the result of a racist demagogy or a last-gasp attempt to retain some vestige of racial privilege. Their sentiments also reflect a loyalty to an idealized and seemingly threatened civic culture in which individual equality was enshrined as a core democratic principle. That culture was never guaranteed to most White Americans, much less to others, but it was espoused across the political and cultural spectrum as a central ideal. At present it seems to many Americans to be a receding ideal, obscured by hypocrisy and greed. Avoiding racial polarization in our society may well depend on resuscitating and rearticulating

that very vision so as to go beyond race-specific demands to a society of greater social justice for all.

Here we emphasize that White Americans have not been immune to the process of racialization. Racial mobilization in the post–Civil Rights movement era has not been limited to "people of color," and White Americans have had to consider the racial implications of an order that formally disavows "White skin privilege." The changing demographic landscape, global economic competition, and the perception that America has fallen from grace have provoked a profound crisis of White identity. What direction this takes politically—whether in the form of support for David Duke's National Association for the Advancement of White People or opposition to a "politically correct" multicultural education—remains to be seen.

Conclusion: Racial Integration, Race Consciousness, and Common Sense

The Civil Rights movement challenged racist understandings and practices and ushered in a period of efforts at desegregation, mandates for equal opportunity, and other state reforms. Various forms of racial discrimination—in the labor market, in housing, in public accommodations, and in marriage laws—were declared unconstitutional and illegal. Despite these tremendous accomplishments, patterns of institutional discrimination proved to be quite obstinate, and the precise meaning of race—in politics and law as well as in everyday life—remained unclear and ill-defined.

The ambiguity of race in the post–Civil Rights movement era has now reached the point where any hint of race consciousness is suspiciously viewed as an expression of racism. Mounting evidence in the field of medicine, for example, suggests that race and ethnicity can and should be factors in the diagnosis and treatment of specific illnesses. African Americans with high blood pressure, for example, generally respond better than Whites when treated with diuretics (drugs which lower salt and water levels), while Asian American respond to some antipsychotic drugs at doses one-tenth of the level recommended for Whites. Some medical researchers and practitioners, however, fear that such "race consciousness"

In this original pen and ink drawing, the artist, George Vargas, depicts American society as being transformed by immigration. The Statue of Liberty, of European origin, issues from the column of a Mexican pre-Colombian temple of Indian origin. From its side issue the advertisements and graffiti typical of a *barrio* in an American city. While speaking of a Chicano identity, these also speak of the thwarted aspirations that accompany it.

could result in a return to racist, eugenic notions (Leary 1990).

Ironically, the present confusion and contestation over the meaning of race is part of the legacy of the Civil Rights movement. While the movement cannot be seen as a homogeneous or monolithic entity, from its very inception it contained an irresolvable dilemma regarding racial consciousness and identity, and their meaning for social and political life. On the one hand, the movement sought an end to racial inequality by advancing a vision of an integrated, color-blind society. On the other, it simultaneously sought to increase awareness and pride in racial identification among African Americans and other people of color. This contradictory stance was the result not only of the movement's ideology and strategy, but also of its conflictual and accommodative relationship to the state.

Different currents within the Civil Rights movement sought to synthesize the two movement objectives, arguing that integration and the politics of racial difference could coexist side by side. All such attempts—ranging from cultural pluralism to neo-Marxism to various forms of nationalism—were ultimately unsuccessful. The two horns of the movement dilemma apparently could not be articulated by a single political and cultural outlook.

As a result, the issues of racial equality and identity were later vulnerable to *rearticulation* from projects on the political right (Omi and Winant 1987). Rearticulation involves the discursive reorganization and reinterpretation of ideological themes and interests. Beginning in the early 1970s, notions of racial equality and opportunity were recast, in a context of economic crisis and political instability, as reverse discrimination. Responses have ranged from "White racial nationalism" (Walters 1987) to neoconservative critiques of affirmative action and other "color-conscious" remedies. Despite their clear ability to mobilize sectors of the White American electorate, the arguments of the right have been no more successful than those of the left in establishing a new understanding of the relationship between racial equality and racial identity. The right's failure to resolve the dilemma has led to continuing controversies surrounding the role race plays, or should play, in various institutions and in the United States as a whole.

Neither the Civil Rights movement nor the racial reaction to it have been able to advance and consolidate a new racial common sense—a general conception of the role of race, if any, in a good and just society. But both political forces have succeeded in weakening and even demolishing the system of racial categories, meanings, and institutions which they opposed. The repressive racial order of legally sanctioned segregation has been overthrown, but no clear and consensual racial order has taken its place.

The paradoxical result has been that, far from decreasing, the significance of race in American life has grown, and the racial dimensions of politics and culture have proliferated. The process of racialization continues apace. New racial identities and meanings continue to be created as a result of panethnic linkages. Increasing differentiation within racial communities continues to be a result of the partial yet significant policy reforms wrought by the Civil Rights movement. And as the complexion of America changes, in a climate of affirmative racial policy, White Americans too are experiencing the contradictions and conflicts of racial identification. Debates about the meaning of racial equality, the nature of racial identity, and the role of the state with respect to race will deepen and intensify in the immediate future. From the riots in South Central Los Angeles to the European American Club in the shadows of Disneyland, racialization continues in the United States.

36

American Paradox

SILVIA PEDRAZA

There shall be one law for the native and
for the stranger who sojourns among you.
— Exodus 12:49

Immigration—voluntary and involuntary—is at the origin of all multiracial and multicultural nations. In that, the United States is not unique, but is similar to many other nations. Yet the nearly unceasing nature of immigration, both into and within the United States, enabled race and ethnicity to become so deeply intertwined with American history as to produce a distinct set of contradictions in the American social fabric.

The first of these is the contradiction between a society where vast numbers of its citizens were able to achieve "the American dream"—not the excessive wealth of the very rich, but the middle-class standard of decency implicit in a family's ability to own its own home, live in a decent neighborhood, have access to good transportation, and send its children to good schools—yet at the same time was a society that systematically denied access to that basic level of decency to some of its people. It is the contradiction posed by poverty in a land of plenty. It is a problem of poverty.

The second of these is the contradiction between a society founded on the values of democracy, equality, and liberty—values which it sought to institutionalize as equality of opportunity in its major social, educational, and juridical institutions for all its people—yet at the same time was a society that systematically denied that equality of opportunity to some of its people on the basis of their color or race. It is the contradiction posed by segregation in a land of equal opportunity. It is a problem of race.

The third of these is the contradiction between being a society that was built by immigrants from a multiplicity of nations, yet at the same time was a society that systematically rejected the immigrant—the profound nativism and xenophobia that has been expressed in immigration restriction at various historical moments. It is the contradiction posed by nativism in a land of immigrants. It is a problem of immigration.

These three problems are distinct yet they dovetail, supporting each other well. And occasionally they intertwine, as can be seen in the photograph of the poor Haitian woman aboard the U.S. Coast Guard dinghy that accompanies this essay. Because they are distinct yet dovetail and intertwine, and because they remain the unsolved contradictions of American social life, past and present, they constitute an American paradox.

Poverty

The first problem comes from the persistence of poverty in a land of plenty and the disproportionate number of the poor among certain groups, such as African Americans, Latinos, Native Americans, and women, particularly minority women. Such persistence over the course of generations was contrary to the expectations of social scientists of different political persuasion, such as Park and Burgess (1921), Wirth (1928), and Glazer (1971), who stressed the social disorganization of ghetto life in the short run, but believed in the eventual assimilation of racial and ethnic minorities, over the course of time, into the mainstream of American social life—an essential continuity with the experience of the European immigrants. As Alice O'Connor (1992) pointed out, "subsequent historical research on immigrants and the Black urban experience have shown the inadequacies of the Chicago school assimilationist framework, whether as a description of the migrant experience or as a predictor of how Black migrants would fare in the city."

Moreover, in the halcyon days of post–World War II in America, rapid economic growth generated an affluence that, together with social policy programs such as the GI Bill, went on to create the expansive middle-class sector in American society that came to typify the American experience—a society where vast numbers of people moved to the suburbs in search of tranquility, lawns, good schools, and status. Precisely because of the great success of what came to be called "the affluent society," the poor became invisible. Such was Michael Harrington's ([1962] 1981) message when he pointed to the other America in our midst:

> To be sure, the other America is not impoverished in the same sense as those poor nations where millions cling to hunger as a defense against starvation. This country has escaped such extremes. That does not change the fact that tens of millions of Americans are, at this very moment, maimed in body and spirit, existing at levels beneath those necessary for human decency. If these people are not starving, they are hungry, and sometimes fat with hunger, for that is what cheap foods do. They are without adequate housing and education and medical care. (Harrington [1962] 1981, pp. 1–2)

In 1960, this other America was peopled by unskilled workers, migrant farmworkers, minorities, the aged, and others who lived a precarious existence: about 22 percent of the American population was poor, with incomes below the officially defined poverty line, represented by $3,022 for a nonfarm family of four. That percentage represented approximately 39.9 millions of poor people in the United States. In 1990, thirty years later, the proportion of the population who lived below the poverty line, now represented by $13,354, was around 13.5 percent, which still constituted 33.6 million people (see Table 1).

In the 1970s, however, as Harrington (1984) fully expected, the poor became invisible again. The reasons were various: the success of the War on Poverty declared by Lyndon Johnson in 1964, and the social welfare programs it put in place, together with the domestic growth that resulted from the Vietnam War, did succeed in lowering the rates of poverty in America. From 1969 to 1979, a full ten years, the proportion of the poor fluctuated at about 9 percent of the population. And the very optimism generated by the success of the Civil Rights movement in opening up the society's legal and juridical institutions to those who had been excluded from equal and fair participation in them also conspired to relegate the issue of poverty to the background of social concern.

Despite the controversy that the use of the term "the underclass" gave rise to (Feagin 1989), research on the underclass, or the ghetto poor, succeeded in placing the issue of poverty on the national agenda of policymakers. That research began in earnest just when the overall poverty rate had climbed up to about 14 percent, around which it hovers at present. In his *The Declining Significance of Race*, William J. Wilson (1985) first introduced the term as an analytical concept with which to point to the schism that, he argued, had developed in the Black community after the success of the Civil Rights movement. The schism consisted of the different "life chances" of talented middle-class and working-class Blacks, for whom life had become better because they were able to benefit from the affirmative action programs that spun off from the Civil Rights movement, and the life chances of the underclass, for whom life did not get better but, rather, became worse as a result of the structural shifts in the economy that progressively and consistently left them out.

Table 1 Poverty in the United States, 1959-1991

Year	Number below poverty level (millions)				Percent below poverty level				Average income cutoffs[c]
	All races[a]	White	Black	Hispanic[b]	All races[a]	White	Black	Hispanic[b]	At poverty level
1959	39.5	28.5	9.9	NA	22.4	18.1	55.1	NA	$ 2,973
1960	39.9	28.3	NA	NA	22.2	17.8	NA	NA	3,022
1966	28.5	20.8	8.9	NA	14.7	12.2	41.8	NA	3,317
1969	24.1	16.7	7.1	NA	12.1	9.5	32.2	NA	3,743
1970	25.4	17.5	7.5	NA	12.6	9.9	33.5	NA	3,968
1975	25.9	17.8	7.5	3.0	12.3	9.7	31.3	26.9	5,500
1976	25.0	16.7	7.6	2.8	11.8	9.1	31.1	24.7	5,815
1977	24.7	16.4	7.7	2.7	11.6	8.9	31.3	22.4	6,191
1978	24.5	16.3	7.6	2.6	11.4	8.7	30.6	21.6	6,662
1979	26.1	17.2	8.1	2.9	11.7	9.0	31.0	21.8	7,412
1980	29.3	19.7	8.6	3.5	13.0	10.2	32.5	25.7	8,414
1981	31.8	21.6	9.2	3.7	14.0	11.1	34.2	26.5	9,287
1982	34.4	23.5	9.7	4.3	15.0	12.0	35.6	29.9	9,862
1983	35.3	24.0	9.9	4.6	15.2	12.1	35.7	28.0	10,178
1984	33.7	23.0	9.5	4.8	14.4	11.5	33.8	28.4	10,609
1985	33.1	22.9	8.9	5.2	14.0	11.4	31.3	29.0	10,989
1986	32.4	22.2	9.0	5.1	13.6	11.0	31.1	27.3	11,203
1987	32.2	21.2	9.5	5.4	13.4	10.4	32.4	28.0	11,611
1988	31.7	20.7	9.4	5.4	13.0	10.1	31.3	26.7	12,092
1989	31.5	20.8	9.3	5.4	12.8	10.0	30.7	26.2	12,674
1990	33.6	22.3	9.8	6.0	13.5	10.7	31.9	28.1	13,359
1991	35.7	23.7	10.2	6.3	14.2	11.3	32.7	28.7	13,924

NA—not available

[a] Includes other races not shown separately.

[b] Persons of Hispanic origin may be of any race.

[c] Prior to 1981, for a nonfarm family of 4; beginning in 1981, income cutoffs for nonfarm families are applied to all families, both farm and nonfarm.

Source: U.S. Bureau of the Census, *Current Population Reports*, P60–181. In *Statistical Abstract of the United States: 1993* (Washington, D.C.: U.S. Government Printing Office), Table 735.

Despite the difficulties that existed when trying to define exactly who really fit under the rubric of "the underclass" (cf. Auletta 1982; Mincy, Sawhill, and Wolf 1990), it is clear that a new era of research on and concern with the nature of American poverty was thereby initiated (see Jarrett 1994; Jencks and Peterson 1991; Pfeffer 1994). With time, Wilson's argument became clearer as he shifted the locus of the argument from the declining significance of race

(1985), to identifying the truly disadvantaged (1987), to explaining the existence of a "new urban poverty" in America today (1994).

The new urban poverty consists of poor segregated neighborhoods in which a substantial majority of the adults are not working—either because they are unemployed or because they have dropped out of the labor force. Wilson further argued that the recent growth of these areas of joblessness have in turn aggravated the racial tensions in our cities. For example, among Chicago's seventy-seven community areas, in 1990 twelve had poverty rates of over 40 percent, with only 35 percent of the adults who lived in them employed. Each of these areas was predominantly Black (Wilson 1994, p. 250). But contrary to the optimism for the future that imbued most Americans and social analysts in the 1960s, for a very large proportion of Black Americans the reality has been one of progressive economic deterioration. Wilson stresses that this economic deterioration has been taking place since the 1950s, when many of the same overwhelmingly Black neighborhoods exhibited much less joblessness and the working poor stood out, there was more class integration and stability, and less social disorganization than today. The social disorganization of these neighborhoods is evidenced in the social problems that spin off from joblessness—crime, drug trafficking, and family break-ups. As joblessness and crime increases, the availability of marriageable young men decreases, rendering family formation and stability ever more precarious.

Numerous analysts have criticized the use of the concept of social disorganization as an accurate description of the lives of the poor, which, rather, they see as socially organized—but along other criteria than those of the middle class (for example, Whyte 1943; Gans 1962; Suttles 1968; Tilly 1968; Valentine 1968; Feagin 1974). Yet one can hardly take issue with Wilson's emphasis on the centrality of work in people's lives:

> Regular employment provides the anchor for the temporal and spatial aspects of daily life. In the absence of regular employment, life, including family life, becomes more incoherent. Unemployment and irregular employment preclude the elaboration of a rational planning of life, the necessary condition of adaptation to an industrial economy. (Wilson 1994, p. 252)

Wilson argues that the deterioration of living standards for Blacks is due to two factors: changes in the economy and changes in the racial and social composition of neighborhoods. The first is the structural shift from an industrial, manufacturing, goods-producing economy to a postindustrial services-producing economy. This shift has entailed an increasing polarization of the society into high-wage and low-wage sectors, an income polarization that is eroding the expansive middle-class sector that was the central characteristic of "the affluent society." That structural shift has also entailed the relocation of industry away from the inner city, leaving many young men in inner-city neighborhoods without substantial opportunities, and prone to crime, drugs, and violence. That, in turn, has made them less desirable to employers who have increasingly turned to other alternative labor forces now available—immigrants and women, in particular.

The deterioration of living standards for Blacks is also due to changes in the class and racial composition of such neighborhoods—the creation of new poverty areas as both Black and White families who are not poor have left those neighborhoods, and as poverty has risen among those who remained in them, at the same time that new poor families have moved in. Such compositional changes, Wilson argues, have made the new poverty area more vulnerable to the profound changes taking place in the economy and ever less able to sustain the institutions—stores, banks, credit institutions, restaurants, professional services—that constitute the backbone of the community, rendering the ghetto extremely unstable.

Wilson also underscores that all of these social conditions have been aggravated by federal government policies that have tolerated racial segregation, encouraged separate public housing projects, and, since 1980, have withdrawn support for basic urban programs, such as direct aid to cities, mass transit, public service jobs and job training, compensatory education, economic assistance programs, and the like. This disinvestment in the city is taking place at the same time that the cities have declined in population and become predominantly peopled by Blacks and Latinos, while the suburbs, where now nearly half of the population of the country lives, remain predominantly White. In cities divested of opportunities, racial tensions are heightened as the competition for jobs, government contracts, political power, and public services grows between poor Blacks, working-class Whites, and other immigrants and ethnics, such as Latinos and Asians, who live side by side (cf. Portes and Stepick 1993).

While recognizing the strength of Wilson's argument, Skocpol (1994, p. 277) stresses that to arrive at a full understanding of the problem of poverty in America, we need to bring women into the analysis since "It is not just that young men are unemployed and sometimes turn to crime. It is also that young women are apparently living as non-employed single mothers, getting by on the only meager income supports realistically available to them, the only ones U.S. society offers them, via AFDC." Because the United States lacks universal health insurance, effective enforcement of child support from absentee fathers, and other supports for working parents, welfare programs like AFDC become a way of trapping young mothers in poverty (Luker 1991). Skocpol stresses that this is particularly demeaning at a time, such as now, when social expectations have changed and women are expected to both work and raise children: "We in the United States must find ways to *smooth* transitions from non-work to work" (1994, p. 279) by making low-wage jobs financially adequate, dignified, and covered by health insurance.

The public policy initiatives that issue from both Wilson and Skocpol's analyses are that they favor race-neutral social welfare programs that increase job skills and job opportunities for all Americans, Black or White, universalistic programs that will gather the broadest support from all.

Research on Latino poverty in the United States does not have the same long pedigree as research on Blacks, because until recently most of the large data sources publicly available did not incorporate Latinos in sufficient detail to permit it. Nonetheless, in the 1990s it has finally become part of the intellectual agenda and the search for the most adequate theoretical model to conceptualize it has begun. A central concern has been whether the underclass model is conceptually suitable to describe and understand poverty among Latinos in the United States (cf. Moore and Pinderhughes 1993). Massey (1993) has argued that Hispanics and Blacks differ in such fundamental ways that theories of the underclass with their standard methods are inappropriate for studying Latino poverty. Black Americans, he stresses, share a distinct history in this country, thus a common historical memory; Latinos represent many variegated experiences both because they come from different countries, for very different reasons at varying points in time, and also because their historical processes of incorporation into American society have been vastly different. So much so, that at present, Latinos can be said to constitute a community of culture, but not a community of interest (Pedraza 1994).

Even more, theories of Latino poverty cannot ignore the impact of immigration, which is a central dynamic that increases the incidence of poverty both because of the selectivity of the migration and because new immigrants may lead to competition with and the job displacement of other poor Hispanic Americans (Meléndez 1993). Immigration plays a small part in the development of Black poverty. As such, Massey argues that the underclass model may well be not only inadequate but misleading for the study of poverty among Latinos in the United States.

An exception, however, may be the Puerto Rican case (cf. Tienda 1989), over which there is clear disagreement. Meléndez (1993) argues that the Puerto Rican case resembles that of Black Americans given its high levels of welfare dependency and families headed by single women; their concentration in areas, such as New York, that have experienced profound economic restructuring; the steep decline of industries, such as the garment industry, in which they were overwhelmingly concentrated; and the impact of race and discrimination on their life chances. And even in the case of Puerto Ricans, the selectivity of migration also plays a role. Gurak and Falcón's (1990) research on poverty among Puerto Rican families has suggested that a double selectivity may be operating: the women most likely to migrate from the island to the U.S. mainland are those with less labor force experience, less education, more children, and whose unions are more unstable; while those most likely to migrate from the mainland to the island are the ones whose unions are more stable, have fewer children, and more education. If so, this clearly contributes to the high proportion of female-headed families and poverty among Puerto Ricans in New York.

While to the academic community, the poor are not now invisible, to most Americans they continue to be. The dramatic separation now between life in the suburbs and life in the city means that most middle-class, White Americans no longer see the lives of the poor up close in a daily fashion, as they once did. Today the poor only intrude into the lives of most middle-class Americans when they contemplate the stark reality of the homeless in our cities' downtowns. Such segregated existences can only result in lack of empathy.

Race

The second problem is the problem of race, of segregation—physical and social—in a land of equality of opportunity. The problem, of course, issues from the existence of slavery, an institution that starkly contradicted the principles of liberty and equality embedded in the constitution that founded this nation. Pierre Van den Berghe (1967) underscored that it was that very contradiction that accounted for the particular virulence of the American strain of racism. For "the peculiar institution" (Stampp 1964) and, thereafter, the Jim Crow system of "separate but equal" legal segregation instituted throughout the Deep South could only be justified by stressing the inherent inferiority of its victims—inferiority that justified the denial of both their humanity and their citizenship (cf. Gossett 1965).

Over the issue of race, in the middle of the nineteenth century America fought a civil war—a war so bloody that, as Abraham Lincoln stated in his Second Inaugural Address of 1865, he had come to believe that it was the price God had demanded for having perpetrated the inhumanity of slavery (quoted in Handlin and Handlin 1980). In the middle of the twentieth century, America fought a Civil Rights movement. Led by men and women of enormous personal mettle, the Civil Rights movement did succeed in delivering civil and political rights. Without it, Blacks would still be riding in the back of the bus, attending separate schools, and drinking water in fountains marked "For Colored Only." But the movement was not able to conquer social and economic rights at that time, and the legacy of the racism of the past became mired in the problems of poverty and segregation today.

Roger Wilkins disagrees with the policy recommendation that issues from Wilson and Skocpol's analysis because his own analysis departs from the idea that "the new American poverty has to be viewed as part of the old American racism" (1994, p. 282), without which it cannot be understood. Because of racism, historically as well as today, Blacks have been assigned the role of surplus labor in this society. As unskilled labor, they "have only fared well in America when the country had a desperate need for unskilled labor," such as that generated by World War I and World War II, as well as the expanding post–World War II economy and the Vietnam War, expansion that ended in the early 1970s when "the bottom fell out for unskilled workers, both Black and White" (Wilkins 1994, p. 285).

That the problem of poverty issues from the problem of race in American society is also the conclusion that Massey (1990), Massey and Denton (1989), and Massey, Gross, and Eggars (1991) analyses of segregation in major American cities comes to. The Fair Housing Law of 1968 that outlawed segregation in all aspects of the sale or rental of housing was one of the major achievements of the Civil Rights movement. Yet despite its passage there has been little decline in residential segregation over time, as data from the successive censuses of 1970, 1980, and 1990 show. Moreover, Massey's (1990, p. 330) abstract analysis showed that, in the absence of racial segregation, industrial restructuring and the changing social and racial composition of neighborhoods "would not have produced concentrated poverty or led to the emergence of a socially and spatially isolated underclass."

Massey and Denton (1989) underscore that for Latinos in the United States segregation is more of a variable—one that depends on their level of acculturation, their socioeconomic status in the community, their skin color or phenotype (cf. Arce, Murguia, and Frisbie 1987; Rodriguez 1991), the region of the country, and the rate of immigration. For Black Americans, by contrast, it is more of a constant since it has not declined over time. That constancy indicates that race itself—prejudice and discrimination—is playing a major role in that segregation. Again, the case of Puerto Ricans is the exception among Hispanics, in that their pattern of segregation resembles Blacks', for whom color clearly matters.

Massey and Denton (1989) came to understand segregation as composed of several different measures—evenness, exposure, clustering, centralization, and concentration—and used separate indices to capture each so as to compare the patterns of segregation among Blacks and Hispanics. Evenness, for example, as measured by the classic index of dissimilarity, measures the extent to which two groups are evenly distributed in a residential area (cf. Duncan and Duncan 1955; Taeuber and Taeuber 1965; Farley 1977). But two different cities with the same degree of evenness can have a different level of exposure, which is a measure of neighborhood composition of the extent to which groups are exposed to one another by virtue of living in the same neighborhood side by side. And concentration refers to the density of concentration of an ethnic group. Due to the prevailing racial discrimination of the times, Blacks migrating from the south to the north became extremely concentrated in the cities they arrived to. Indeed, in St. Louis, Missouri, the story was often told that when the train from Mississippi arrived, White policemen with clubs directed the Black newcomers to cross the river—to East St. Louis, Illinois; to this day one of the most blighted ghetto areas in the United States.

Because Massey found that Blacks were highly segregated under all five of these measures in six of the largest cities of the United States—Baltimore, Chicago, Cleveland, Detroit, Milwaukee, and Philadelphia—and in four of these measures in others—Gary, Los Angeles, Newark, and St. Louis—he used the term "hypersegregation" to denote the conditions under which a very substantial part of the Black population still lives. By contrast, Latinos showed low to moderate levels of segregation, even in the cities of large Hispanic populations, such as Los Angeles, San Antonio, Miami, New York, and Chicago. In no metropolitan area of the sixty they studied could Latinos be said to be suffering from hypersegregation. Hence, Massey and Denton (1989) rightly concluded that the social isolation of Blacks in this country is unique.

In the final analysis, more important than whether the driving force behind social inequality is class or race, or even, as Massey (1990, p. 354) so aptly put it, "how race *and* class interact," are the very important social and economic consequences that follow from living in segregation. People who live in highly segregated, poor neighborhoods live in poor stock housing, whose value depreciates over time, rendering them even poorer; they are very often victims of crime, particularly burglary and assault, as well as of police insensitivity and disrespect; they have poor access to public services, including transportation, and, therefore, to good jobs; they live in an environment where not having a job and being on welfare is the norm; their children are often victims of street and drug cultures that thrive in hopeless environments; and, despite the best intentions of good teachers and other mentors, the schools that their children can go to are poor in quality (cf. Williams and Kornblum 1985). In sum, the quality of life in these neighborhoods is low and the life chances of their residents are poor, making it improbable that they can move out—spatially—and up—socially.

Farley and Frey (1994) explored the changes in patterns of racial segregation from 1980 to 1990 across 232 cities—places that differ widely in historical development. They take a historical view of the problem in two ways. First, they trace the development of segregation over time in the United States and the factors that contributed to it, such as the prevalence of discriminatory mortgage lending policies, the use of intimidation and violence against Blacks, and the use of zoning ordinances and housing covenants to keep Blacks out. Second, they examine the recent trends in those cities in which the ghettos first emerged (cf. Spear 1969) between 1900 and World War II (such as Detroit or Chicago) and those cities in which the second ghettos emerged (cf. Hirsch 1983), the "necklace of White suburbs surrounding predominantly Black central cities that appeared after World War II" (Farley and Frey 1994, p. 24) (such as Sacramento or Riverside).

The Civil Rights movement targeted the problem of residential segregation along with the drive to achieve political participation via the vote. From its achievements spun some of the major developments that Farley and Frey (1994) examined: the changes in federal housing policies, the liberalization of Whites' attitudes towards Blacks, and the growth of the Black middle class. Together with new housing construction, these set the stage for reducing segregation.

The general pattern of change from 1980 to 1990 that Farley and Frey (1994, p. 30) found in the 232 cities they examined was one of "modest declines," as measured by slight declines in the index of dissimilarity. But, in addition, they found wide variation by the economic and ecological characteristics of the cities, such as the age of a city (built up before World War II or after), the type of city (university town, manufacturing center), the region the city is in (the Midwest or the South), the age of the housing stock in the city (old versus new construction), the growth of White and other minority groups, such as Latinos and Asians, in the same multiethnic city (see also Frey and Farley 1993), and the changes in the economic status of Blacks vis-à-vis Whites. As a result, they found that segregation between Blacks and Whites changed little in Detroit, or Chicago over this decade, but fell substantially in Los Angeles, and rather dramatically in Fort Worth. This wide variation in outcomes was the result of, first, "the heavy hand of the past" that maintains segregation in the old cities of the Midwest and the Northeast, where the type of city, suburban patterns, and housing stock all discourage integration. By contrast, newness of housing in other cities is linked to declines in segregation. Last, the racial attitudes of Whites continue to matter, as all the largest decreases in segregation occurred in cities in which the presence of Blacks in White neighborhoods was small.

Despite the real gains of the Civil Rights movement in opening up a society that was once closed to all who were not considered to be White, and the liberalization in the attitudes of Whites, clearly we are still a society in which race matters. It matters both in the life chances for social and economic advancement that people of different racial and ethnic groups can expect to have—the kind of head start or handicap that they will start life with; and it matters in the experiences that their race or ethnicity will most likely subject them to throughout their lives—the kind of "public and psychological wage" that W. E. B. Du Bois ([1935] 1992, pp. 700–701) emphasized Whites received because they were White, the "public deference" that even poor Whites received. Even when it had a small effect upon their economic situation, it "had great effect upon their personal treatment," weakening the potential for solidarity between Blacks and Whites (cf. Roediger 1991). By contrast, Blacks were subject to "public insult."

The policy implications that derive from seeing race as the force driving other social problems, such as poverty, of course differ. Wilkins's (1994) policy conclusions, for example, differ from Wilson's and Skocpol's because in his judgment "poor Blacks need to be treated differently because we have been hated uniquely. Sustained disdain and shunning has created sustained and unique damage," the most serious of which is the damage to their family structure. Hence, Wilkins calls for targeted social programs—family policies aimed at supporting and strengthening the poorest Black families in this society.

Terry Williams (1994, p. 292) takes yet another tack when he argues that the new urban poverty is generated, and fueled, by "the new racism"—a society "suffering from compassion fatigue that has made these poor communities into zones of indifference," where the poor folk contained there are seen as human disabilities and have become increasingly invisible once again, as supports for them are aban-

doned. Williams supports public policy programs aimed at youth that would involve, first and foremost, getting them through high school, and would also consist of finding employers who would give them real jobs supported by real training and counseling.

Wilson stresses that, "Given its complex nature, it is not surprising that most people neither understand the forces that have generated the new urban poverty, nor have sympathy for the people who represent or are directly affected by it" (1994, p. 265). Hence, "blaming the victim" type explanations (cf. Ryan 1971) take hold, such as those that thirty years ago became encapsulated in the notion of "the culture of poverty" (Lewis 1966) that portrayed the culture of poverty not as a consequence of poverty but as its main cause. Although Lewis himself expected that the culture of poverty would develop under certain precise structural and historical conditions—as "both an adaptation and a reaction of the poor to their marginal position in a class-stratified, highly individuated, capitalistic society" (Lewis 1966, p. xliv), as an effort to cope with feelings of hopelessness and despair, in the final analysis Lewis thought that, once it came into existence, the culture of poverty would tend to perpetuate itself. As Lewis himself put it (1966, p. li), "it is much more difficult to eliminate the culture of poverty than to eliminate poverty *per se*."

Indeed, in a society where most of its people are not poor (in a land of plenty, such as the United States) rather than in a society where poverty is much more widely shared (as is often the case in the Third World), poverty becomes interpreted as a personal deficiency, as the result of one's own failings, giving rise to the notion of "the undeserving poor." Such lack of sympathy for the poor usually becomes expressed in a lack of commitment to social welfare programs, commitment that is only spurred by the social protest of the poor (cf. Piven and Cloward 1979).

Since race and class dovetail as well as they do, the old urban poverty, the new urban poverty, the old racism, and the new racism all serve to uphold each other—strengthening the social problems which they undergird. The United States has yet to develop realistic and humane poverty policy. Realistic in the sense that it will not encourage excessive dependency on nor trap recipients in welfare systems that become a permanent way of life rather than provide the temporary shelter they were intended to give. But humane also in the sense that such forms of coverage should be available to all, in a universalistic fashion, that due to life's erratic circumstances should suddenly need to be protected—as a social right, not as an act of mercy. Moreover, it is important to recognize that the legacy of the racism and discrimination of the past may well require that particular groups be targeted so that yesterday's wrongs may not be inherited today anew.

Immigration

The last problem comes from the role that immigration has played in American history. Other core developed countries in western Europe—such as France, Germany, Holland, and Spain—have recently become rather reluctant hosts to immigrants from their own periphery—such as Turkey, Spain, Greece, and North Africa. But the phenomenon there is very recent, ushered in by these nations' own post–World War II development and decolonization. Contrary to the American experience, none of these countries were shaped by a nearly constant history of immigration. As Muller and Espenshade expressed it in *The Fourth Wave* (1985), the United States is now experiencing only the most recent of its four major waves of migration. Even in the interim years between the time when the United States closed its door to the vast hordes of migration from Europe in 1924, not to reopen it until the 1965 amendments to the 1952 Immigration and Nationality Act went into effect, the great internal migrations of Blacks, Native Americans, Mexican Americans, and Puerto Ricans from south to north took place. They also entailed social transformations of the same magnitude as those of the great migrations that, yesterday as well as today, crossed the sea. Because immigration has been coterminous with American history, one should not be surprised at the truth of Oscar Handlin's ([1951] 1973, p. 3) encounter when, seeking to write a history of the immigrants in America, he discovered "that the immigrants *were* American history."

Because the history of America is also the history of American immigration, then, immigration is also a part of the American identity—of its definition of self. This also is not true in other countries, such as the western European nations, where the experience of being peopled by immigrants is very recent. In *A Nation of Immigrants* (in itself evidence of the consolidation of that self-definition), then President John F.

Kennedy himself expressed it well: "three large forces—religious persecution, political oppression, and economic hardship—provided the chief motives for the mass migrations to our shores" (1964, p. 6). This American tradition of providing not only an opportunity for a better life but also a refuge from tyranny was nowhere else so well expressed as in that poem every schoolchild learns—Emma Lazarus's poem (1883) written for the Statue of Liberty:

> Give me your tired, your poor,
> Your huddled masses yearning to breathe free,
> The wretched refuse of your teeming shore,
> Send these, the homeless, tempest-tossed, to me:
> I lift my lamp beside the golden door.

Hence, for the United States, a nation where both its history and some of its major values center around immigration, immigration is not only an economic issue or even a political issue but also a moral issue. Yet both the controversies over the massive arrival of southern and eastern European immigrants at the turn of the century as well as the recent controversies over the arrival of illegal aliens from Mexico and Central America and the massive exodus of Cuban and Haitian refugees at this century's end betray a profound nativism. Such nativism consists of the desire to define "the real Americans" (those descended from yesterday's immigrants) as "us" versus "them" (today's immigrants). In Edward Said's (1979) terminology, they constitute "the Other"—within. This echoes Georg Simmel's ([1908] 1971, pp. 143–144) insistence that the stranger is not the man who comes today and goes tomorrow, but rather "the man who comes today and stays tomorrow," the stranger as a member of the group itself—"an element whose membership within the group involves both being outside it and confronting it."

Evidence for the perennial ambivalence toward the immigrant permeates American history. When France had its gift of the Statue of Liberty ready to the shipped to the United States, in the United States its part of the bargain—the pedestal—was only half finished. Likewise, Emma Lazarus's poem was virtually ignored by her contemporaries. It was only in 1903, on the twentieth anniversary of the poem, that it was put on a bronze tablet on an interior wall of the statue's pedestal, where it passed largely unnoticed (in Higham [1975] 1984, pp. 72–74; see also Sollors 1993).

Despite the ambivalence and disregard shown them, both the Statue of Liberty and Emma Lazarus's poem went on to become major symbols of America's identity as a nation of immigrants. So much so that even anti-immigrant sentiments need to first invoke them when, laying all the society's ills at the feet of the most recent immigrants, they call for immigration restriction. For example, right after the first massive arrival of poor immigrants—many of them Black—from Haiti and from Cuba in 1980, then Governor Richard Lamm (1981) called for us to "disenthrall ourselves from our past myths" and to redefine the meaning of the Lady in the Harbor as symbolizing *Liberty*, not *immigration*." Governor Lamm's call for restricting immigration rang loud:

> We hate to say "no" to that worthy individual from the poverty-stricken country who just wants to do a little better. It seems selfish to us to set limits. The Lady in the Harbor would not understand. However, I believe dramatic reform is necessary and inevitable, and that the sooner we recognize this, the better off we will be. (Lamm 1981)

At the turn of the century, when immigration from southern and eastern Europe rose to the highest levels experienced to date, immigrants from Italy, Greece, Spain, Poland, Hungary, and Russia heard themselves described as "inferior races" (Handlin [1951] 1973, pp. 262–263). Amidst the xenophobia generated by World War I, that nativism led to the closing of the door to immigrants from southern and eastern Europe and became embedded in the discriminatory national origins quotas of 1924. When the 1965 amendments abolished them, Edward Kennedy (1966, p. 149) expressed that the national origins quota system "was conceived in a radical period of our history—a period when bigotry and prejudice stalked our streets" and were reaffirmed by the McCarthy era. Most recently, as we turn to yet another century, that nativism has again become expressed in, for example, California's Proposition 187, which states that the people of California "have suffered and are suffering" personal injury and damage, and "economic hardship caused by the presence of illegal aliens in this state." Supported by Governor Wilson, this ballot initiative seeks to deny schooling, health care, and other services to illegal immigrants. As it is, illegal immigrants are most unprotected. Their lives hang by the thread of the goodwill of

strangers and the chance of luck. Such denials would condemn them even further to the shadows of the American labor force, and impede the possibility of their joining the mainstream of American life, eventually achieving a regular status.

The contradiction between the definition of America as a nation of immigrants and particularly its tradition of extending a hand to those who flee from tyranny and oppression has never been more vivid than in the U.S. government's recent treatment of Cuban and Haitian refugees—our interdiction at sea of "the wretched refuse" from the "teeming shore(s)" of Cuba and Haiti, to be indefinitely housed at Guantánamo Bay instead. Looking back at the typology of migration (see Figure 1) developed, we can see that two different axes define the types: on the one hand, the immigrants' own motivation for migration (whether it is predominantly economic or political) and, on the other hand, the government's response (whether it grants or denies) to the immigrants' claim through its legal system.

Of the four types that this typology yields—legal aliens, undocumented aliens, legal refugees, and undocumented refugees—probably the most embarrassing and disputed is the last, because one can deny legitimate claims to be granted the status of legal immigrant on the simple basis that the nation lacks the capacity to absorb any more such immigrants. That is, one can simply say that, regrettably, "the boat is too full." But denying claims to political asylum, to be considered as refugees, involves denying the truth of their motivation. That is, it is not that "the boat is simply too full" but that "they do not rightfully belong on the boat." Ultimately that wounds not only the persons whose claims to fleeing oppression are thereby falsified, but also that nation that has consistently claimed to be a refuge from tyranny. The United States' refusal to accept, first, Haiti's refugees and, more recently, Cuba's refugees who henceforth will be returned to Cuba when found at sea, has eroded the belief of many in what America has always stood for, tarnishing its national image and hurting its credibility. The photograph chosen for the conclusion of this book captures the nature of this moral issue well. The Haitian woman's face betrays her fright and anger as, aboard a U.S. Coast Guard ship at sea, her effort to flee Haiti's violence, poverty, and terror resulted in her interdiction at sea. In the background, against a clear, blue sky, the American flag, out of focus, is unfurled.

Legal and Political Status

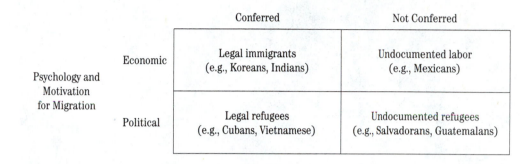

Figure 1
Typology of Migration

The 1980 Refugee Act changed the operating criterion for refugee admissions to the United States from one that, in effect, granted asylum to all immigrants "from communism" and from the repressive governments in the Middle East under the Attorney General's "parole authority" contained in the Immigration and Nationality Act of 1952 that until then was the authority used to admit large numbers of refugees (Rose 1993; Zucker and Zucker 1992). In its place, the 1980 Refugee Act placed the new criterion of considering each claim to asylum on an individual basis by the strict test established by the United Nations Convention (1951) and Protocol (1968) on the status of refugees. As such, a refugee became a person who is unwilling or unable to return to his or her country of nationality or habitual residence because of "a well-founded fear of persecution on account of race, religion, nationality, membership in a particular social group, or political opinion." With this new definition the Refugee Act attempted to assure greater equity in the treatment of refugees. As Edward Kennedy (1981, p. 143) put it, "the new law is intended to end years of *ad hoc* programs and different policies for different refugees by putting the U.S. refugee programs on a firm basis" as well as to institutionalize the procedures and federal programs to assist refugee resettlement.

In practice, however, the evidential burden placed on individuals who claim asylum is rather onerous and the criteria as to what constitutes persecution is unclear. Moreover, the approval rates for asylum cases granted to particular nationalities continued to reflect that country's distance from U.S. foreign pol-

icy concerns. For example, the U.S. Immigration and Naturalization Service published asylum statistics for 1983 to 1986 that show that 60 percent of Iran's claims were granted, as were 45 percent of Czechoslovakia's, 34 percent of Poland's, 32 percent of Hungary's, 26 percent of Vietnam's, 26 percent of Vietnam's, 21 percent of China's, 14 percent of Nicaragua's, 3 percent of El Salvador's, 3 percent of Honduras's, and 1 percent of Guatemala's (U.S. Committee for Refugees 1986a, p. 9). This pattern continued into the 1990s (see Wassem 1992) and led to the *American Baptist Churches v. Thornburg* class action suit in 1990, in which the judge ruled that all Salvadorans and Guatemalans who had been denied asylum since 1980 were entitled to have their cases reopened. For as long as the Cold War between the United States and the Soviet Union raged, refugee admissions continued to be shaped by the symbolic political function that refugees played in the United States. Those who succeeded in the flight to freedom from communist nations exemplified the rightness of our cause, and served to legitimate U.S. foreign policy objectives (Pedraza-Bailey 1985).

With the end of the Cold War and the collapse of communism in the Soviet Union and eastern Europe, refugees from communism lost their symbolic value. That the most recent exodus of *balseros* from Cuba—the most desperate people ever to leave the island—was met by the interdiction of the U.S. Coast Guard that routed them to the camps of Guantánamo Bay, insisting that they were trying to enter the United States illegally, makes the case. Peter Rose (1993, p. 15) emphasized the situational nature of the politics

American society needs to resolve the problems of poverty, race, and immigration that keep it from fulfilling its promise as a nation. Here, a Haitian woman who fled the political turmoil and economic despair in Haiti in 1990 has just been picked up at sea by the U.S. Coast Guard. In the background, the U.S. flag is unfurled. (Photo by C. W. Griffin. *The Miami Herald*)

of rescue: "It is no exaggeration to say that, when it comes to American refugee policy, the quality of mercy has long been strained by overriding economic, social, and political considerations. Our reception and treatment of would-be asylees is—and has long been—a textbook example of situational morality."

Moreover, the criteria for admissions imposed by the 1980 Refugee Act of a "well-founded fear of persecution" should be closely scrutinized. For while it undoubtedly was well intended, seeking to bring greater equity to the process of decision making, it recognizes only one type of refugee as a refugee—namely, those who are *targets* of violence specifically directed at them (cf. Zolberg, Suhrke, and Aguayo 1989). Not all refugees, however, are targets—as are supporters of the *ancien régime* or those who later become activists or dissidents within the new order.

Rather, some are *victims* of the cruel social processes they have lived through but scarcely participated in. As Zolberg, Suhrke, and Aguayo (1989, p. 30) put it, victims are "persons displaced by societal or international violence that is not necessarily directed at them as individuals but makes life in their own country impossible." As victims, they live lives marked by fear, though they themselves may not belong to a social category that is deliberately persecuted, nor would they dare to act in such a way as to incur the attention—much less the wrath—of others. In Cuban slang, those who hide their political disaffection and live as if they were loyal supporters of a regime in which they no longer believe but fear are called *tapaditos*—those who live under a cover, under a daily dissimulation (cf. Pedraza-Bailey 1983). As a nation, we should question whether we have imposed standards that perhaps hark back to the European reality of fascism in midcentury (whose targets were also its victims, such as all Jews), but that do not recognize the distorted nature of victims' lives in the Third World today.

The United States has yet to develop a realistic and humane immigration policy. Realistic in the sense that the numbers of immigrants who can legitimately get on the boat will have to be curtailed and delimited by its capacity as a society to absorb them and by its need to protect other poor Americans—immigrants and native both, Black, White, Hispanic, all—from the labor displacement that excessive immigration can lead to. But humane also in the sense that not only must it consider the plight of people who flee economic and political oppression, but also must consider that it has to speak for and to the self-definition of this nation as a nation of immigrants, for its tradition as a haven for the oppressed. At its most basic, to provide safety for people who flee in fear is an issue in human rights, a duty as old as human history—"You shall not oppress a stranger; you know the heart of a stranger, for you were strangers in the land of Egypt" (Exodus 23:9). It is also an American tradition. A cornerstone of the foundation of this nation, George Washington expressed in 1783 that he had sought to "establish an asylum for the poor and oppressed of all nations and all religions" (quoted in Flexner 1974, p. 176).

Only as the United States strives to solve its paradox—these profound contradictions—can she become true to others and remain true to herself, becoming an America that is not just for some of its people but for all of its people. Perhaps the problems of race, poverty, and immigration in its midst will give rise to the struggles to solve them as we enter the twenty-first century. Following the achievement of the Voting Rights Act and the Civil Rights Act in 1964, Martin Luther King, Jr., turned to highlighting the issue of poverty among all Americans, Black and White. It is well to remember what he underscored then: that equal access to public accommodations did not cost the nation anything, the right to vote did not cost the nation anything, but that now "we are grappling with basic class issues between the privileged and the underprivileged," to solve which will require the restructuring of the architecture of American society. And that "will cost the nation something"—financially and morally (quoted in Schulke 1976). But until, as a nation, we are prepared to pay those costs, children in this country will not be able to grow up with the singular gift that Howell Raines (1991), as a pampered White child, received from Grady Williams, the Black American maid who, while working in his house, helped to raise him: the gift of a free and unhateful heart.

Appendix

Guidelines for the Research Paper Assignment

Think of your family as an immigrant family. Trace back, as far as you can, its history as an immigrant or ethnic history (not just a family chronology) and see to what extent was its members' experience typical or atypical of the group's.

To accomplish this:

1. Read *ethnic histories* of that group to see to what extent your family history was like the history of the ethnic group.

2. Interview your *family members,* thinking of them as members of an immigrant or ethnic group.

3. *Link* the two, using your family members as *examples* to illuminate the *social history* of that group. The family interviews need to be linked to the library research on the history of that group. A paper that centers only on the family interviews is only half a paper and will be regarded as such.

The *central question* in this paper is: Was your family's experience, over the course of generations, typical or atypical of the experience of that group? Keep asking yourself that question all the way through. For example, when you read the ethnic histories you may discover that the occupations that were typical of the group were the "middleman minority" occupations—self-employed entrepreneurs. Were your family members a couple of generations ago employed in those types of occupations? Why or why not?

Both in your interviews with your family and your library research, the following questions should help you pay attention to what is important. A good paper need not answer every single question but should be guided by them.

1. Which was the *immigrant generation?* When did they first come over? Was their experience typical of the first generation?

For example, did they live in immigrant enclaves (Little Italy, Little Haiti, Little Havana)? Were their lives filled with nostalgia for the past?

2. What *"push" and "pull" factors* motivated the migration?

For example, poverty; the transition from an agricultural to an industrial society; revolution; the Irish potato famine; immigration law were all the backdrop to peoples' movements. If there was more than one wave of migration for the ethnic group, what wave did your family come over in?

For example, African Americans migrated north in two distinct waves—"pulled" up by World War I or World War II.

What did coming to the United States (or going north) mean to your family?

3. Did your family migrate within the United States? From what area to what area? What "push" and "pull" factors were there for this *internal migration*?

For example, both the decline of agriculture in the south and northern industrial expansion during the world wars motivated huge waves of internal migration.

4. What particular *state or region of the country* did your family settle in? Did that make a difference?

For example, did they settle in Texas, California, Alabama, New York City, or Chicago?

Was the place of settlement mostly rural or urban? What role did that play in their lives?

5. What opportunities did members of this ethnic group find in the historical period during which they migrated?

For example, the nineteenth century was a period of frontier expansion; in 1924 the Immigration and Nationality Act curtailed immigration; in the 1930s there was the Depression; in the 1940s World War II gave rise to northern industrial expansion; from the 1890s to the Civil Rights Movement, Jim Crow laws were in effect.

6. What were the *occupations* of your family members in the old country? In America, what sorts of jobs were available to them? Which occupations were typical of the group?

For example, Jewish immigrants became concentrated in the garment industry; Norwegians in shipping and farming; Mexicans in agriculture, mining, and the railroads.

7. Was there social mobility in your family from generation to generation?
  a. *Educational mobility:*
    Which was the first generation to go to college? If they did not attend college, why not?

Did some children go and not others? Did the men go but not the women?
  b. *Occupational mobility:*
    For example, for the European immigrants and their descendants, this is an "ideal typical" progression over the generations:
    1st Generation: farming
    2nd Generation: blue-collar working class
    3rd Generation: white-collar labor; self-employed in small business
    4th Generation: professionals; business executives; managers
  c. *Residential mobility:*
    For example, occupational mobility often went hand in hand with residential mobility: rural area → central city → suburbs

8. Were there barriers to your family's social mobility? Were there *legal and social impediments* to their lives, their life chances?

For example, Jim Crow laws, quotas in universities, the Klu Klux Klan or the Texas Rangers, housing discrimination, and job discrimination all blocked the progress of some groups.

Were there any name changes or other forms of "passing"?

9. Were there *stereotypes* about your family's group? Did those stereotypes have an effect on their lives? Did the ethnic group experience prejudice directly? How?

10. Were they strongly identified with a particular political party? Was their ethnicity linked to their *political participation*? Were they active in local or state politics, in a social movement? What goals did the group aim to achieve as a result of their political participation?

11. Was there a relationship between the ethnic group and the American *labor movement*?

For example, were members of the ethnic group often "union men"? Did the women sew in a shop unionized by the International Ladies and Garments Workers Union?

12. What *cultural bonds* were there in the family? What language was spoken in the family? Was the ethnic tradition passed on through the celebration of ethnic holidays, food, customs? Did the family members acculturate?

13. To what *church* did they belong? What role did religion play in their lives? How important was it to them?

For example, did the immigrant generation worship and practice as Orthodox Jews, but their children as Reformed or Conservative? Did religion give substance and meaning to their struggle for equal civil rights? Did their church give them support?

14. Were there *marriages* outside the ethnic group? Did they follow religious lines?

15. On the whole, was the ethnic group successful in this country? Why or why not? Did the group manage to assimilate well? Was there *cultural assimilation*? *Structural assimilation*? What was lost? What was gained? What kind of future would you expect?

References

Abad, R. G., and E. U. Eviota. 1984. "Reproducing Development Inequalities: Some Effects of Filipino Emigration to the United States." Paper prepared for the Conference on Asia-Pacific Immigration to the United States, East-West Center, Honolulu.

Abbott, Carl. 1990. "Dimensions of Regional Change in Washington, D.C." *American Historical Review* 95: 1367–1393.

Abbott, Edith. 1924. *Immigration: Select Documents and Case Record.* Chicago: University of Chicago Press.

Ablon, Joan. 1964. "Relocated American Indians in the San Francisco Bay Area: Social Interactions and Indian Identity." *Human Organization* 23: 296–304.

———. 1965. "American Indian Relocation: Problems of Dependency and Management in the City." *Phylon* 26: 362–371.

———. 1971. "Retention of Cultural Values and Differential Urban Adaptation: Samoans and American Indians in a West Coast City." *Social Forces* 49: 385–393.

Abrams, Israel. 1911. "Jews." Pp. 371–410 in *Encyclopaedia Britannica*, 11th ed., vol. XV.

AC (Archives of the Archdiocese of Chicago). Rev. Francis J. Karabasz File, Archdiocesan Personnel Records.

———. Sacred Heart of Jesus Parish (Honore Street) Correspondence File.

Acuña, Rodolfo. 1988. *Occupied America: A History of Chicanos*, 3rd ed. New York: Harper & Row.

Adams, Henry. 1931. *The Education of Henry Adams.* New York: Random House.

Adams, John. [1780] 1856. *Life and Works.* Boston: Little, Brown.

Agüeros, Jack. 1981. "Halfway to Dick and Jane: A Puerto Rican Pilgrimage." Pp. 85–105 in *The Immigrant Experience: The Anguish of Becoming American*, ed. Thomas C. Wheeler. New York: Penguin.

Aguirre, Benigno E. 1976. "The Differential Migration of Cuban Social Races." *Latin American Research Review* 11: 103–124.

Akwesasne Notes. 1974a. *Trail of Broken Treaties: "B.I.A., I'm Not Your Indian Anymore,"* 2d ed. Mohawk Nation, Rooseveltown: Akwesasne Notes.

———. 1974b. *Voices from Wounded Knee.* Mohawk Nation, Rooseveltown: Akwesasne Notes.

Alba, Richard D. 1985. *Italian Americans: Into the Twilight of Ethnicity.* Englewood Cliffs, N.J.: Prentice-Hall.

———. 1988. "Cohorts and the dynamics of ethnic change." Pp. 211–228 in *Social Structures and Human Lives*, ed. Matilda White Riley, Bettina J. Huber, and Beth B. Hess. Newbury Park, Calif.: Sage.

———. 1990. *Ethnic Identity: The Transformation of White America.* New Haven: Yale University Press.

———. 1992. "'What then is the European American?' Some answers." *Altreitalie* 8: 93–99.

———. 1995. "Assimilation's Quiet Tide." *The Public Interest* 119: 3–18.

Alba, Richard D., and Reid Golden. 1986. "Patterns of interethnic marriage in the United States." *Social Forces* 65: 203–223.

Alba, Richard D., and John R. Logan. 1993. "Minority Proximity to Whites in Suburbs: An Individual-Level Analysis of Segregation." *American Journal of Sociology* 6: 1388–1427.

Aldrich, Nelson W., Jr. 1988. *Old Money: The Mythology of America's Upper Crust*. New York: Knopf.

Alegado, Dean T. 1991. "The Filipino Community in Hawaii: Development and Change," p. 33 in *Social Process in Hawaii*. Department of Sociology, University of Hawaii at Manoa.

Alfonso, Pablo M. 1984. *Cuba, Castro y los Católicos*. Miami: Ediciones Hispamérican.

Alfred, William. 1981. "Pride and Poverty: An Irish Integrity." In *The Immigrant Experience: The Anguish of Becoming American*, ed. Thomas G. Wheeler. New York: Penguin.

Allen, Walter R., and Reynolds Farley. 1986. "The Shifting Social and Economic Tides of Black America, 1950–1980." *Annual Review of Sociology* 12: 277–306.

Allsup, Carl. 1982. *The American G.I. Forum: Origins and Evolution*. Austin: Center for Mexican American Studies, University of Texas at Austin.

Almaraz, Felix D., Jr. 1989. *The San Antonio Missions and Their System of Land Tenure*. Austin: University of Texas Press.

Alsop, Joseph W. 1992. *I've Seen the Best of It: Memoirs*. New York: W. W. Norton.

Alvarez, Julia. 1992. *How the García Girls Lost Their Accents*. New York: Plume.

Alvarez, Robert R. 1987. *Familia: Migration and Adaptation in Baja and Alta California, 1800–1975*. Berkeley and Los Angeles: University of California Press.

Alvarez, Roberto. 1994. "*Un Chileno en La Academia*: Sifting, Shifting and the Recruitment of Minorities in American Anthropology." In *Race*, ed. Roger Sanjek and Steven Gregory. New Brunswick, N.J.: Rutgers University Press.

Alvarez, Rodolfo. 1973. "The Psycho-Historical and Socioeconomic Development of the Chicano Community in the United States." *Social Science Quarterly* 53: 920–942.

Amaro, Nelson. 1977. "Mass and Class in the Origins of the Cuban Revolution." Pp. 221–251 in *Cuban Communism*, ed. Irving Louis Horowitz. New Brunswick, N.J.: Transaction Books.

Amaro, Nelson, and Alejandro Portes. 1972. "Una Sociología del Exilio: Situación de los Grupos Cubanos en los Estados Unidos." *Aportes* 23: 6–24.

Ambler, Marjane. 1990. *Breaking the Iron Bonds: Indian Control of Energy Development*. Lawrence: University Press of Kansas.

American Friends Service Committee. 1970. *Uncommon Controversy: Fishing Rights of the Muckleshoot, Pyyallup, and Nisqually Indians*. Seattle: University of Washington Press.

Amott, Teresa L., and Julie A. Matthaei. 1991. *Race, Gender and Work*. Boston: South End Press.

Andersen, Margaret L. 1993. *Thinking About Women: Sociological Perspectives on Sex and Gender*, 3rd ed. New York: Macmillan.

Aptheker, Herbert. 1943. *American Negro Slave Revolts*. New York: International Publishers.

Aponte, Robert. 1991. "Urban Hispanic Poverty: Disaggregations and Explanations." *Social Problems* 38: 516–528.

Arce, Carlos H., Edward Murguia, and W. Parker Frisbie. 1987. "Phenotype and Life Chances Among Chicanos." *Hispanic Journal of the Behavioral Sciences* 9: 19–32.

Archdeacon, Thomas. 1976. *New York City, 1664–1710: Conquest and Change*. Ithaca, N.Y.: Cornell University Press.

——— . 1982. *Becoming American: An Ethnic History*. New York: Free Press.

Arcinas, Fe. 1987. *Asian Migrant Workers in the Gulf Region: The Philippine Case*. Diliman, Quezon City: University of the Philippines.

Arensberg, Conrad M., and Solon T. Kimball. 1940. *Family and Community in Ireland*. Cambridge: Harvard University Press.

Armstrong, Virginia Irving, comp. 1971. *I Have Spoken: American History through the Voices of the Indians*. Chicago: Swallow Press.

Arndt, Karl J. R., and May E. Olson. 1965. *German-American Newspapers and Periodicals, 1732–1955: History and Bibliography*. New York: Johnson Reprint.

Arnold, Fred, Unmil Minocha, and James T. Fawcett. 1987. "The Changing Face of Asian Immigration to the United States." Pp. 105–152 in *Pacific Bridges: The New Immigraton from Asia and the Pacific Islands*, ed. James T. Fawcett and Benjamin V. Cariño. New York: Center for Migration Studies in association with the East-West Population Institute, East-West Center, University of Hawaii.

Arroyo, Luis L. 1975. "Notes on Past, Present and Future Directions of Chicano Labor Studies." *Aztlán* (summer).

Auletta, Ken. 1982. *The Underclass*. New York: Random House.

Ayres, Ian. 1991. "Fair Driving: Gender and Race Discrimination in Retail Car Negotiations." *Harvard Law Review* 104 (4): 817–872.

Baca Zinn, Maxine, and D. Stanley Eitzen. 1993. *Diversity in Families*, 3rd ed. New York: HarperCollins College.

Bach, Robert L. 1980. "The New Cuban Immigrants: Their Background and Prospects." *Monthly Labor Review* 103: 39–46.

Bach, Robert L., and Rita Argiros. 1991. "Economic Progress Among Southeast Asian Refugees in the United States." Pp. 322–343 in *Refugee Policy: Canada and the United States*, ed. Howard Adelman. Toronto: York Lanes Press.

Bach, Robert L., Jennifer B. Bach, and Timothy Triplett. 1981/1982. "The Flotilla 'Entrants': Latest and Most Controversial." *Cuban Studies* 11/12: 29–48.

Báez Evertsz, Franc, and Frank D'Oleo Ramírez. 1985. *La emigración de Dominicanos a Estados Unidos: Determinantes socio-economicos y consecuencios*. Santo Domingo: Fundación Friedrich Ebert.

Bailey, Guy, Natalie Maynor, and Patricia Cukor-Avila. 1991. *The Emergence of Black English*. Philadelphia: J. Benjamins.

Baker, Ray Stannard. 1908. *Following the Color Line*. Reprint 1964. New York: Harper & Row.

Baker, Reginald P., and David S. North. 1984. *The 1975 Refugees: Their First Five Years in America*. Washington, D.C.: New TransCentury Foundation.

Balderrama, Francisco E. 1982. *In Defense of La Raza: The Los Angeles Mexican Consulate and the Mexican Community, 1919–1936*. Tucson: University of Arizona Press.

Ball, Eve. 1980. *Inde h. An Apache Odyssey*. Provo, UT: Brigham Young University Press.

Balmaseda, Liz. 1994a. "Balserita Violinista Toca pero También Escribe como los Angeles." *El Nuevo Herald*, 19 October, 1B.

——. 1994b. "Cuba Bleeds, and the Drops are Called Rafts." *Miami Herald*, 17 August, 1B.

Baltzell, E. Digby. 1957. *Philadelphia Gentlemen: The Making of a National Upper Class*. New York: Free Press.

——. 1964a. "The Immigrants' Progress and the Theory of the Establishment." Pp. 324–338 in *The Logic of Social Hierarchies*, ed. Edward O. Laumann, Paul M. Siegel, and Robert W. Hodge. Chicago: Markham.

——. 1964b. *The Protestant Establishment: Aristocracy and Caste in America*. New York: Random House.

——. 1970. "The Immigrants' Progress and the Theory of the Establishment." Pp. 324–338 in *The Logic of Social Hierarchies*, ed. Edward O. Laumann, Paul M. Siegel, and Robert W. Hodge. Chicago: Markham.

——. 1991. *The Protestant Establishment Revisited*. Ed. Howard G. Schneiderman. New Brunswick, N.J.: Transaction Books.

Baltzell, E. Digby, and Howard G. Schneiderman. 1988. "Social Class in the Oval Office." *Society* 25 (6).

——. 1991. "From Rags to Robes: The Horatio Alger Myth and the Supreme Court." *Society* 28 (4).

Bannon, John F. 1970. *The Spanish Borderlands Frontier, 1513–1821*. New York: Holt, Rinehart & Winston.

Barrera, Mario. 1979. *Race and Class in the Southwest*. Notre Dame, Ind.: University of Notre Dame Press.

Barrett, James R. 1987. *Work and Community in the Jungle: Chicago's Packinghouse Workers, 1894–1922*. Urbana: University of Illinois Press.

Barringer, Felicity. 1991. "Census Shows Profound Change in Racial Makeup of the Nation." *New York Times*, 11 March, A1, A10.

Barsh, Russell Lawrence, and James Youngblood Henderson. 1980. *The Road: Indian Tribes and Political Liberty*.

Berkeley and Los Angeles: University of California Press.

Barth, Frederik. 1969. *Ethnic Groups and Boundaries*. Boston: Little, Brown.

Bascom, William R. 1941. "Acculturation Among the Gullah Negroes." *American Anthropologist* 63: 43–50.

——. 1969. *The Yoruba of Southwestern Nigeria*. New York: Holt, Rinehart & Winston.

Basler, Roy P. 1953. *The Collected Works of Abraham Lincoln*. 8 vols. New Brunswick, N.J.: Rutgers University Press.

Bayor, Ronald. 1988. *Neighbors in Conflict: The Irish, Germans, Jews, and Italians and New York City, 1929–1941*. Urbana: University of Illinois Press.

Bean, Frank D., Barry Edmonston, and Jeffrey S. Passel. 1990. *Undocumented Migration to the United States*. Washington, D.C.: Urban Institute.

Bean, Frank D., B. Lindsay Lowell, and Lowell J. Taylor. 1988. "Undocumented Mexican Immigrants and the Earnings of Other Workers in the United States." *Demography* 25 (1): 35–52.

Bean, Frank D., E. E. Telles, and B. L. Lowell. 1987. "Undocumented Migration to the United States: Perceptions and Evidence." *Population and Development Review* 13: 671–690.

Bean, Frank D., and Marta Tienda. 1987. *The Hispanic Population of the United States*. New York: Russell Sage.

Bell, C. C., K. Taylor-Crawford, E. J. Jenkins, and D. Chalmers. 1988. "Need for Victimization Screening in a Black Psychiatric Population." *Journal of the National Medical Association* 80(1).

Bender, Eugene. 1969. "Reflections of Negro-Jewish Relations: The Historical Dimensions." *Phylon* 30: 56–65.

Berlin, Ira. 1974. *Slaves Without Masters: The Free Negro in the Ante-Bellum South*. New York: Oxford University Press.

Bernal, Guillermo. 1982. "Cuban Families." Pp. 187–207 in *Ethnicity and Family Therapy*, ed. Monica McGoldrick, John Pearce, and Joseph Giordano. New York: Guilford Press.

Bernard, Jessie. 1966. *Marriage and Family Among Negroes*. Englewood Cliffs, N.J.: Prentice-Hall.

Bernstein, Iver. 1990. *The New York City Draft Riots: Their Significance for American Society and Politics in the Age of the Civil War*. New York: Oxford University Press.

Berrally, AnCita, and T. L. McCarty. 1990. "The Navajo Language Today." Pp. 234–246 in *Perspectives on Official English*, ed. K. L. Adams and D. J. Brink. New York: de Gruyter.

Billington, Ray Allen. 1938. *The Protestant Crusade, 1800–1860: A Study in the Origins of American Nativism*. New York: Macmillan.

Bishop, John. 1977. *Jobs, Cash Transfers, and Marital Instability: A Review of the Evidence*. Madison: Institute for Research on Poverty, University of Wisconsin.

Blalock, Herbert. 1967. *Toward a Theory of Minority Group Relations*. New York: John Wiley & Sons.

Blauner, Robert. 1969. "Internal Colonialism and Ghetto Revolt." *Social Problems* 16: 393–408.

——— . 1992. "Talking Past Each Other: Black and White Languages of Race" *American Prospect* (summer): 55–64.

Bluestone, Barry. 1982. *The Deindustrialization of America: Plant Closings, Community Abandonment, and the Dismantling of Basic Industry*. New York: Basic Books.

Blumberg, Rhoda Lois. 1984. *Civil Rights: The 1960s Freedom Struggle*. Boston: Twayne Publishers.

Blumenfield, Ruth. 1965. "Mohawks: Round Trip to the High Steel." *TransAction* 3: 19–22.

Blumer, Herbert. 1982. "Industrialization and Race Relations." In *Majority and Minority: The Dynamics of Race and Ethnicity in American Life*, ed. Norman R. Yetman and C. Hoy Steele. Boston: Allyn & Bacon.

Bobo, Lawrence. 1983. "Whites' Opposition to School Busing: Symbolic Racism or Realistic Group Conflict?" *Journal of Personality and Social Psychology* 45: 1196–1210.

Bobo, Lawrence, et al. 1992. *Public Opinion Before and After a Spring of Discontent: A Preliminary Report on the 1992 Los Angeles County Social Survey*. Los Angeles: Center for the Study of Urban Poverty, Institute for Social Science Research, University of California, Los Angeles.

Bobo, Lawrence, and Franklin D. Gilliam, Jr. 1990. "Race, Sociopolitical Participation, and Black Empowerment." *American Political Science Review* 84: 377–393.

Bodnar, John. 1985. *The Transplanted: A History of Immigrants in Urban America*. Bloomington: Indiana University Press.

Boe, Eugene. 1981. "Pioneers to Eternity: Norwegians on the Prairie." In *The Immigrant Experience: The Anguish of Becoming American*, ed. Thomas G. Wheeler. New York: Penguin.

Boissevain, Jeremy. 1984. "Small Entrepreneurs in Comparative Europe." Pp. 20–38 in *Ethnic Communities in Business*, ed. Robin Ward and Richard Jenkins. New York: Cambridge University Press.

Bolton, Herbert E. 1963. *Spanish Exploration in the Southwest, 1541–1706*. Reprint. New York: Barnes & Noble.

Bonacich, Edna. 1972. "A Theory of Ethnic Antagonism: The Split Labor Market." *American Sociological Review* 37: 547–559.

——— . 1973. "A Theory of Middleman Minorities." *American Sociological Review* 38 (October): 583–594.

——— . 1976. "Advanced Capitalism and Black/White Relations in the United States: A Split-Labor Market Interpretation." *American Sociological Review* 41: 34–53.

——— . 1984. "Some Basic Facts: Patterns of Asian Immigration and Exclusion." Pp. 60–78 in *Labor Immigration Under Capitalism: Asian Workers in the United States Before World War II*, ed. Lucie Cheng and Edna Bonacich. Berkeley and Los Angeles: University of California Press.

——— . 1993. "Asian and Latino Immigrants in the Los Angeles Garment Industry." Pp. 51–74 in *Immigration and Entrepreneurship: Culture, Capital, and Ethnic Networks*, ed. Ivan Light and Parminder Bhachu. New York: Transaction Publishers.

Bonacich, Edna, and John Modell. 1980. *The Economic Basis of Ethnic Solidarity: Small Business in the Japanese American Community*. Berkeley and Los Angeles: University of California Press.

Bonney, Rachel A. 1977. "The Role of AIM Leaders in Indian Nationalism." *American Indian Quarterly* 3 (3): 209–224.

Boone, Margaret S. 1985. "Social and Cultural Factors in the Etiology of Low Birthweight among Disadvantaged Blacks." *Social Science and Medicine* 20: 1001–1011.

Borjas, George J. 1990a. *Friends or Strangers: The Impact of Immigrants on the U.S. Economy*. New York: Basic Books.

——— . 1990b. "Illegal Aliens: The Black Market for Immigrants." In *Friends or Strangers: The Impact of Immigrants to the U.S. Economy*. New York: Basic Books.

Borjas, George J., and Marta Tienda. 1987. "The Economic Consequences of Immigration." *Science* 235 (February 6): 613–620.

——— . 1993. "The Employment and Wages of Legalized Immigrants." *International Migration Review* 27: 712–747.

Bose, Christine. 1984. "Household resources and U.S. women's work: Factors affecting gainful employment at the turn of the century." *American Sociological Review* 49: 474–490.

Boswell, Thomas D., and James R. Curtis. 1984. *The Cuban-American Experience: Culture, Images and Perspectives*. Totowa, N.J.: Rowman & Allanheld.

Botifol, Luis J. 1985. "How Miami's New Image Was Created." Occasional Paper no. 1985–1, Institute of Interamerican Studies, University of Miami.

Bourne, Edward Gaylord, ed. 1922. *Narratives of the Career of Hernando de Soto*. 2 vols. New York: Allerton.

Bouvier, Leon F., and Anthony J. Agresta. 1987. "The Future Asian Population of the United States." Pp. 285–301 in *Pacific Bridges: The New Immigration from Asia and the Pacific Islands*, ed. James T. Fawcett and Benjamin V. Cariño. Staten Island, N.Y.: Center for Migration Studies.

Bouvier, Leon F., and Robert W. Gardner. 1986. "Immigration to the U.S.: The Unfinished Story." *Population Bulletin* 41 (November).

Bowling, Kenneth R. 1991. *The Creation of Washington D.C.* Fairfax, Va.: George Mason University Press.

Boyd, Monica. 1989. "Family and Personal Networks in International Migration: Recent Developments and New Agendas." *International Migration Review* 23: 638–670.

Bozorgmehr, Mehdi. 1992. "Internal Ethnicity: Armenian, Bahai, Jewish, and Muslim Iranians in Los Angeles." Ph.D. diss., University of California, Los Angeles.

Bozorgmehr, Mehdi, and Georges Sabagh. 1991. "Iranian Exiles and Immigrants in Los Angeles." Pp. 121–144 in *Iranian Refugees and Exiles Since Khomeini*. Costa Mesa, Calif.: Mazda Publishers.

Brain, Jeffrey P. 1971. "The Natchez 'Paradox.'" *Ethnology* 10: 215–222.

Branch, Taylor. 1988. *Parting the Waters*. New York: Simon & Schuster.

Breckinridge, Sophonisba, and Edith Abbott. 1911. "Housing Conditions in Chicago III: Back of the Yards." *American Journal of Sociology* 16: 433–468.

Bridges, Amy. 1984. *A City in the Republic: Antebellum New York and the Origins of Machine Politics*. Cambridge: Cambridge University Press.

Brimelow, Peter. 1992. "Time to Rethink Immigration?" *National Review* (22 June): 30–46.

Brody, David. 1960. *Steelworkers in America: The Nonunion Era*. Cambridge: Harvard University Press.

——— . 1964. *The Butcher Workmen*. Cambridge: Harvard University Press.

——— . 1965. *Labor in Crisis: The Steel Strike of 1919*. Philadelphia: Lippincott.

Brookhiser, Richard. 1991. *The Way of the WASP*. New York: Free Press.

Brooklyn Historical Society. 1989. *Brooklyn's Hispanic Communities*. Brooklyn, N.Y.: Brooklyn Historical Society.

Brooks, Roy L. 1990. *Rethinking America's Race Problem*. Berkeley and Los Angeles: University of California Press.

Brophy, William A., and Sophie D. Aberle. 1966. *The Indian: America's Unfinished Business*. Norman: University of Oklahoma Press.

Brown, Claude. 1965. *Manchild in the Promised Land*. New York: American Library.

Brown, Thomas N. 1966. *Irish-American Nationalism: 1870–1890*. Philadelphia: Lippincott.

Broyles-Gonzales, Yolanda. 1994. *El Teatro Campesino*. Austin: University of Texas Press.

Bryce, James. 1889. *The American Commonwealth*. New York: Macmillan.

——— . 1921. *Modern Democracies*, vol. II. New York: Macmillan.

——— . 1993. *Hindrances to Good Citizenship*. Ed. Howard G. Schneiderman. New Brunswick, N.J.: Transaction Books.

Bryce-Laporte, Roy S., ed. 1980. *Source Book on the New Immigration*. New Brunswick, N.J.: Transaction Books.

Bukowczyk, John. 1987. *And My Children Did Not Know Me: A History of Polish Americans*. Bloomington: Indiana University Press.

Bullard, R., and B. H. Wright. 1987. "Environmentalism and the Politics of Equity: Emergent Trends in the Black Community." *Mid-American Review of Sociology* 12: 21–38.

Burgess, Ernest W. 1923. "The Growth of the City: An Introduction to a Research Project." *Publications of the American Sociological Society* 18: 85–97.

Burt, Larry W. 1982. *Tribalism in Crisis: Federal Indian Policy, 1953–1961*. Albuquerque: University of New Mexico Press.

Bushnell, Charles J. 1902. *The Social Problem in the Chicago Stockyards*. Chicago: University of Chicago Press.

Bustamante, Jorge A. 1977. "Undocumented Immigration From Mexico: Research Report." *International Migration Review* 11: 149–189.

Butcher, Kristin F., and David Card. 1991. "Immigration and Wages: Evidence from the 1980s." *American Economic Review* 81 (2): 292–296.

Butler, Jon. 1983. *The Huguenots in America: A Refugee People in New World Society*. Cambridge: Harvard University Press.

Byron, Christopher. 1992. *Skin Tight: The Bizarre Story of Guess v. Jordache*. New York: Simon & Schuster

Cabeza de Vaca, Alvar Núñez. 1961. *Adventures in the Unknown Interior of America*. Trans. and ed. Cyclone Covey. New York: Collier.

Cabezas, Amado, Larry Shinagawa, and Gary Kawaguchi. 1986–1987. "New Inquiries into the Socio-economic Status of Filipino Americans in California." *Amerasia* 13: 1–22.

Caces, Fe. 1985. "Personal Networks and the Material Adaptation of Recent Immigrants: A Study of Filipinos in Hawaii." Doctoral dissertation, University of Hawaii.

——— . 1986–1987. "Immigrant Recruitment into the Labor Force: Social Networks among Filipinos in Hawaii." *Amerasia* 13: 23–38.

Caldwell, Cleopatra Howard, Angela Dungee Greene, and Andrew Billingsley. 1992. "The Black Church as a Family Support System: Instrumental and Expressive Functions." *National Journal of Sociology* 6: 21–40.

Camarillo, Albert. 1979. *Chicanos in a Changing Society: From Mexican Pueblos to American Barrios in Santa Barbara and Southern California, 1848–1930*. Cambridge, MA: Harvard University Press.

Campa, Arthur L. 1979. *Spanish Culture in the Southwest*. Norman: University of Oklahoma Press.

Campisi, Paul. 1948. "Ethnic family patterns: The Italian family in the United States." *American Journal of Sociology* 53: 443–449.

Canelo, Frank J. 1982. *Dónde, porqué, de qué, y como viven los dominicanos en el extranjero: un informe sociológico sobre la e/inmigración dominicana, 1961–62*. Santo Domingo: Alfa y Omega.

Capeci, Dominic, Jr. 1985. "Black-Jewish Relations in Wartime Detroit: The Marsh, Loving, Wolf Surveys and the Race Riots of 1943." *Jewish Social Studies* 5: 221–242.

Caplan, Nathan, Marcella H. Choy, and John K. Whitmore. 1991. *Children of the Boat People: A Study of Educational Success*. Ann Arbor: University of Michigan Press.

Caplan, Nathan, John K. Whitmore, and Marcella H. Choy. 1989. *The Boat People and Achievement in America: A Study of Family Life, Hard Work and Cultural Values*. Ann Arbor: University of Michigan Press.

Cardoso, Lawrence A. 1980. *Mexican Emigration to the United States 1897–1931*. Tucson: University of Arizona Press.

Cariño, Benjamin V. 1981. "Filipinos on Oahu, Hawaii." Paper no. 72, East-West Population Institute, East-West Center, University of Hawaii.

——— . 1987. "The Philippines and Southeast Asia: Historical Roots and Contemporary Linkages." Pp. 305–325 in *Pacific Bridges: The New Immigraton from Asia and the Pacific Islands*, ed. James T. Fawcett and Benjamin V. Cariño. New York: Center for Migration Studies in association with the East-West Population Institute, East-West Center, University of Hawaii.

——— . 1992. "Migrant Workers from the Philippines." In *Philippine Labor Migration: Impact and Policy,* ed. Graziano Battistella and Anthony Paganoni. Quezon City, Philippines: Scalobrini Migration Center.

Cariño, Benjamin V., James T. Fawcett, Robert W. Gardner, and Fred Arnold. 1990. "The New Filipino Immigrants to the United States: Increasing Diversity and Change." Paper no. 115, East-West Population Institute, East-West Center, University of Hawaii.

Carlson, Leonard A. 1981. *Indians, Bureaucrats, and Land*. Westport, Conn.: Greenwood Press.

Carnegie Foundation for the Advancement of Teaching. 1989. *Tribal Colleges: Shaping the Future of Native America*. Princeton, N.J.: Princeton University Press.

Caroli, Betty Boyd. 1973. *Italian Repatriation from the United States, 1900–1914*. New York: Center for Migration Studies.

Carr, Norma. 1989. "The Puerto Ricans in Hawaii: 1900–1958." Doctoral disseration. University of Hawaii.

Carson, Clayborne. 1981. *In Struggle*. Cambridge: Harvard University Press.

CASAS (Comprehensive Adult Student Assessment System). 1989. *A Survey of Newly Legalized Persons in California*. Prepared for the California Health and Welfare Agency. San Diego: Comprehensive Adult Student Assessment System.

Casal, Lourdes. 1979. "Cubans in the United States: Their Impact on U.S.-Cuban Relations." Pp. 109–136 in *Revolutionary Cuba in the World Arena*, ed. Martin Weinstein. Philadelphia: Ishi.

Castaneda, Carlos. 1936–1958. *Our Catholic Heritage in Texas*. 7 vols. Austin: Von Boeckmann-Jones.

Castro, Max. 1985. "Dominican Journey: Patterns, Context, and Consequences of Migration From the Dominican Republic to the United States." Ph.D. diss., University of North Carolina, Chapel Hill.

Castro, Tony. 1974. *Chicano Power: The Emergency of Mexican America*. New York: Saturday Review Press.

Center for Education Statistics, U.S. Department of Education. 1987. *The American Indian in Higher Education 1975–76 to 1984–85*. Washington, D.C.: U.S. Government Printing Office.

Chadwick, Bruce A., and L. C. White. 1973. "Correlates of Length of Urban Residence Among Spokane Indians." *Human Organization* 34: 359–369.

Chadwick, Bruce A., and Joseph Stauss. 1975. "The Assimilation of American Indians into Urban Society: The Seattle Case." *Human Organization* 34: 359–369.

Chalasinski, Josef. 1935. "Parafja i Szkola Parafijalna Wsrod Emigracji Polskiej w Ameryce. Studium Dzielnicy Polskiej w Poludniowym Chicago" (The Parish and Parish School During the Polish Emigration to America: A Study of the Polish District in South Chicago). *Przegld Socjologiczny* 3: 633–771.

Chan, Sucheng. 1984. "The Chinese in California Agriculture, 1860–1900." Pp. 67–84 in *The Chinese American Experience*, ed. Genny Lim. San Francisco: Chinese Historical Society of America and Chinese Culture Foundation.

——— . 1991. *Asian Americans: An Interpretive History*. Boston: Twayne Publishers.

Chang, Edward. 1990. "New Urban Crisis: Korean–Black Conflicts in Los Angeles." Ph.D. diss., University of California, Berkeley.

Chang, Patricia, David R. Williams, Ezra E. H. Griffith, and John L. Young. 1994. "Church-Agency Relationships in the Black Community." *Nonprofit and Voluntary Sector Quarterly* 23 (2): 91–105.

Chávez, John R. 1984. *The Lost Land: The Chicago Image of the Southwest*. Albuquerque: University of New Mexico Press.

Chavez, Leo R. 1988. "Settlers and Sojourners: The Case of Mexicans in the United States." *Human Organization* 47: 95–108.

——— . 1990. "Coresidence and Resistance: Strategies for Survival among Undocumented Mexicans and Central Americans in the United States." *Urban Anthropology* 19: 31–61.

——— . 1991. "Outside the Imagined Community: Undocumented Settlers and Experiences of Incorporation." *American Ethnologist* 18: 257–278.

——— . 1992. *Shadowed Lives: Undocumented Immigrants in American Society*. Fort Worth, Tex.: Harcourt Brace Jovanovich College.

Chavez, Leo R., Estevan T. Flores, and Marta Lopez-Garza. 1990. "Here Today, Gone Tomorrow? Undocumented Settlers and Immigration Reform." *Human Organization* 49: 193–205.

Chen, Hsiang-Shui. 1992. *Chinatown No More: Taiwan Immigrants in Contemporary New York*. Ithaca, N.Y.: Cornell University Press.

Cheng, Lucie. 1984. "Free, Indentured, Enslaved: Chinese Prostitutes in Nineteenth Century America." Pp. 402–434 in *Labor Immigration Under Capitalism: Asian Immigrant Workers in the United States Before World War II*, ed. Lucie Cheng and Edna Bonacich. Berkeley and Los Angeles: University of California Press.

Cheng, Lucie, and Yen Espiritu. 1989. "Korean Businesses in Black and Hispanic Neighborhoods." *Sociological Perspectives* 32: 521–534.

Chey, Youn-Cha Shin. 1992. "Media Commentary: The Enemy This Time." *San Francisco Chronicle*, 8 May, 5.

Chicago Commission on Race Relations. 1920. *The Negro in Chicago*. Chicago: University of Chicago Press.

Chicago Department of Health. 1894–1900. *Reports, 1894–1900*. Chicago: Chicago Department of Health.

———. 1911. *Report of the Department of Health of the City of Chicago for the Years 1907, 1908, 1909, 1910*. Chicago: Chicago Department of Health.

Chipman, Donald E. 1992. *Spanish Texas, 1519–1821*. Austin: University of Texas Press.

Chiswick, Barry R. 1991. "Jewish Immigrant Skill and Occupational Attainment at the Turn of the Century." *Explorations in Economic History* 28: 64–80.

Chow, Esther Ngan-ling. 1987a. "The Development of Feminist Consciousness Among Asian American Women." *Gender & Society* 1: 284–299.

———. 1987b. "Job Decision, Household Work and Gender Relations in the Asian American Families." Paper presented at the annual meeting of the American Sociological Association, Chicago, Illinois.

———. 1994. "Asian American Women at Work: Survival, Resistance, and Coping." Pp. 203–227 in *Women of Color in U.S. Society*, ed. Maxine Baca Zinn and Bonnie Thornton Dill. Philadelphia: Temple University Press.

———. 1995. "From Pennsylvania Avenue to H Street, N.W.: The Transformation of Chinatown in Washington, D.C." In *Urban Odyssey of Washington, D.C.: Many Voices on a Common Ground,* ed. Francine Cary. Washington, D.C.: Smithsonian Institute Press.

Chow, Esther Ngan-ling, and Catherine White Berheide. 1988. "The Interdependence of Family and Work: A Framework for Family Life Education, Policy, and Practice." *Family Relations* 37: 23–28.

Christopher, Robert C. 1989. *Crashing the Gates: The De-WASPing of America's Power Elite*. New York: Simon & Schuster.

Chung, L. A. 1991. "S.F. Includes Asian Indians in Minority Law." *San Francisco Chronicle*, 25 June, A14.

Clark, Juan M. 1975. "The Exodus from Revolutionary Cuba (1959–1974): A Sociological Analysis." Ph.D. diss., University of Florida.

Clinton, Lawrence, Bruce A. Chadwick, and Howard M. Bahr. 1975. "Urban Relocation Reconsidered: Antecedents of Employment Among Indian Males." *Rural Sociology* 40: 117–133.

Cohen, Abner. 1969. *Custom and Politics in Urban Africa*. Berkeley and Los Angeles: University of California Press.

Cohen, Fay G. 1973. "The Indian Patrol in Minneapolis: Social Control and Social Change in an Urban Context." Ph.D. diss., University of Minnesota.

———. 1986. *Treaties on Trial: The Continuing Controversy over Northwest Indian Fishing Rights*. Seattle: University of Washington Press.

Cohen, Miriam. 1993. *Workshop to Office: Two Generations of Italian Women in New York City, 1900–1950*. Ithaca, N.Y.: Cornell University Press.

Cohen, Nathan. 1970. *The Los Angeles Riots: A Sociological Study*. New York: Praeger.

Cohen, Steven. 1988. *American Assimilation or Jewish Revival?* Bloomington: University of Indiana Press.

Cohn, D'Vera, and Evelyn Hsu. 1991. "Hispanics, Asians Change Face of N. Virginia." *Washington Post*, 3 February, B1.

Comer, James. 1988. *Maggie's American Dream: The Life and Times of a Black Family*. New York: New American Library.

Commission for Racial Justice. 1987. *Toxic Wastes and Race in the United States: A National Report on the Racial and Socioeconomic Characteristics of Communities with Hazardous Waste Sites*. New York: United Church of Christ.

Commons, John R. 1904. "Labor Conditions in Meat Packing and the Recent Strike." *Quarterly Journal of Economics* 19: 1–32.

Connell, Kenneth H. 1968. *Irish Peasant Society: Four Historical Essays*. Oxford: Clarendon Press.

Conzen, Kathleen Neils. 1980. "Germans." In *Harvard Encyclopedia of American Ethnic Groups*. Cambridge: Harvard University Press.

———. 1985. "Peasant Pioneers: Generational Succession Among German Farmers in Frontier Minnesota." Pp. 259–292 in *The Countryside in the Age of Capitalist Transformation*, ed. Steven Hahn and Jonathan Prude. Chapel Hill: University of North Carolina Press.

Conzen, Kathleen Neils, David Gerber, Ewa Morawska, George Pozzetta, and Rudolph Vecoli. 1992. "The invention of ethnicity: A perspective from the U.S.A." *Journal of American Ethnic History* 12: 3–41.

Coolidge, Mary. 1909. *Chinese Immigration*. New York: Henry Holt.

Cooper, Richard S., and Richard David. 1986. "The Biological Concept of Race and its Application to Public Health and Epidemiology." *Journal of Health Politics, Policy and Law* 11: 97–116.

Cooper, Richard, and Brian E. Simmons. 1985. "Cigarette Smoking and Ill Health Among Black Americans." *New York State Journal of Medicine* 85: 344–349.

Cornelius, Wayne A. 1978. *Mexican Migration to the United States: Causes, Consequences, and U.S. Responses*. Cambridge: Massachusetts Institute of Technology, Center for International Studies.

———. 1980. "America in the Era of Limits." Working paper No. 3. La Jolla, Calif.: Center for U.S.-Mexican Studies, University of California, San Diego.

———. 1988. "The Role of Mexican Labor in the North American Economy of the 1990s." Paper presented at the Fourth Annual Emerging Issues Program for State Legislative Leaders, San Diego, Calif.

Cornelius, Wayne A., Leo R. Chavez, and Jorge A. Castro. 1982. "Mexican Immigrants in Southern California: A

Summary of Current Knowledge." Research report no. 36. LaJolla, Calif.: Center for U.S.-Mexican Studies, University of California, San Diego.

Cornell, Stephen. 1988. *The Return of the Native: American Indian Political Resurgence*. New York: Oxford University Press.

Cornell, Stephen, and Joseph P. Kalt. 1990. "Pathways from Poverty: Economic Development and Institution Building on American Indian Reservations." *American Indian Culture and Research Journal* 14: 89–125.

——— . 1992. "Reloading the Dice: Improving the Chances for Economic Development on American Indian Reservations." Pp. 1–59 in *What Can Tribes Do? Strategies and Institutions in American Indian Economic Development*, ed. Stephen Cornell and Joseph P. Kalt. Los Angeles: American Indian Studies Center, University of California, Los Angeles.

——— . 1993. "The Redefinition of Property Rights in American Indian Reservations: A Comparative Analysis of Native American Economic Development." In *Self-Governance and Economic Development for American Indians*, ed. Lyman H. Legters and Fremont J. Lyden. Westport, Conn.: Greenwood Press.

Council on Ethical and Judicial Affairs. 1990. "Black-White Disparities in Health Care." *Journal of the American Medical Association* 263: 2344–2346.

Covello, Leonard. 1972. *The Social Background of the Italo-American School Child*. Totowa, N.J.: Rowman & Littlefield.

Cowan, Paul. 1982. *An Orphan in History: Retrieving a Jewish Legacy*. New York: Doubleday.

Craig, Richard B. 1971. *The Bracero Program: Interest Groups and Foreign Policy*. Austin: University of Texas Press.

Crassweller, Robert D. 1966. *Trujillo: The Life and Times of a Caribbean Dictator*. New York: Macmillan.

Crawford, James. 1986. "Conservative Groups Take Aim at Federal Bilingual Programs." *Education Week* (March 19): 1.

Crispino, James A. 1980. *The Assimilation of Ethnic Groups: The Italian Case*. Staten Island, N.Y.: Center for Migration Studies.

Crosby, Alfred W., Jr. 1972. *The Columbian Exchange*. Westport, Conn.: Greenwood Press.

——— . 1976. "Virgin Soil Epidemics as a Factor in the Aboriginal Depopulation in America." *William and Mary Quarterly* 33: 289–299.

Cruz, Gilbert R. 1988. *Let There Be Towns: Spanish Municipal Origins in the American Southwest, 1610–1810*. College Station: Texas A&M Press.

Cruz, Robert David. 1990. "The Industry Composition of Production and the Distribution of Income by Race and Ethnicity in Miami." Unpublished paper, Department of Economics, Florida International University.

Cullen, Louis M. 1981. *The Emergence of Modern Ireland, 1600–1900*. New York: Holmes and Meier.

Cummings, Scott, ed. 1980. *Self-Help in Urban America: Patterns of Minority Business Enterprise*. Port Washington, N.Y.: Kennikat Press.

Cummins, James. 1981. "Empirical and Theoretical Underpinnings of Bilingual Education." *Journal of Education* 163 (winter): 16–29.

Dahrendorf, Ralf. 1967. *Society and Democracy in Germany*. Garden City, N.Y.: Doubleday.

Daniels, Roger. 1973. *The Politics of Prejudice*. New York: Atheneum.

——— . 1990. *Coming to America: A History of Immigration and Ethnicity in American Life*. New York: HarperCollins.

Daniels, Roger, and Harry L. Kitano. 1990. *American Racism*. Englewood-Cliffs, N.J.: Prentice-Hall.

David, F. James. 1991. *Who Is Black: One Nation's Definition*. University Park: Pennsylvania State University Press.

Davidman, Lynn. 1991. *Tradition in a Rootless World: Women Turn to Orthodox Judaism*. Berkeley and Los Angeles: University of California Press.

Davis, Allison, Burleigh Gardner, and Mary Gardner. [1941] 1965. *Deep South: A Social Anthropological Study of Caste and Class*. Chicago: University of Chicago Press.

Davis, Mike. 1990. *City of Quartz: Excavating the Future in Los Angeles*. London: Verso.

Davis, Ronald M. 1987. "Current Trends in Cigarette Advertising and Marketing." *New England Journal of Medicine* 316: 725–732.

Day, Robert C. 1972. "The Emergence of Activism as a Social Movement." Pp. 506–532 in *Native Americans Today: Sociological Perspectives*, ed. Howard M. Bahr, Bruce A. Chadwick, and Robert C. Day. New York: Harper & Row.

de la Garza, Rodolfo, A. Kruszewski, and Tomas Arciniega, eds. 1973. *Chicanos and Native Americans*. Englewood Cliffs, N.J.: Prentice-Hall.

De León, Arnoldo. 1982. *The Tejano Community, 1836–1900*. Albuquerque: University of New Mexico Press.

——— . 1983. *They Called Them Greasers: Anglo Attitudes toward Mexicans in Texas, 1821–1900*. Austin: University of Texas Press.

Deloria, Philip S. 1986. "The Era of Indian Self-Determination: An Overview." Pp. 191–207 in *Indian Self-Rule: First-Hand Accounts of Indian-White Relations from Roosevelt to Reagan*, ed. Kenneth R. Philp. Salt Lake City: Howe Brothers.

Deloria, Vine, Jr. 1974. *Behind the Trail of Broken Treaties: An Indian Declaration of Independence*. New York: Dell.

Deloria, Vine, Jr., and Clifford Lytle. 1984. *The Nations Within: The Past and Future of American Indian Sovereignty*. New York: Pantheon.

Del Rio Independent School District v. Salvatierra, Tex Civ App, 4th Dist. (1930).

Del Rio Independent School District v. Salvatierra, 33 S.W. 2d 790, Cert. denied, 284 US 580 (1931).

Denevan, William M., ed. 1992. *The Native Population of the Americans in 1492*, 2d ed. Madison: University of Wisconsin Press.

Dent, David J. 1992. "The New Black Suburbs." *New York Times*, 14 June, sec. 6, 22.

Department of City Planning, New York City. 1992. *The Newest New Yorkers: Analysis of Immigration into New York City During the 1980s*. New York: City of New York.

———. 1993. *Socioeconomic Profiles: A Portrait of New York City's Community Districts from the 1980 and 1990 Censuses of Population and Housing*. New York: City of New York.

Dershowitz, Alan. 1991. *Chutzpa*. New York: Touchstone.

Dewing, Rolland. 1985. *Wounded Knee: The Meaning and Significance of the Second Incident*. New York: Irvington.

Díaz, Guarioné M., ed. 1980. *Evaluation and Identification of Policy Issues in the Cuban Community*. Miami: Cuban National Planning Council.

Díaz-Briquets, Sergio. 1983. "Demographic and Related Determinants of Recent Cuban Emigration." *International Migration Review* 17: 95–119.

———. 1984. "Cuban-Owned Businesses in the United States." *Cuban Studies* 14 (2): 57–68.

———. 1989. "The Central America Demographic Situation: Trends and Implications." In *Mexican and Central American Population and U.S. Immigration Policy*, ed. Frank D. Bean, Jurgen Schmandt, and Sidney Weintraub. Austin: Center for Mexican American Studies, University of Texas.

di Leonardo, Micaela. 1984. *The Varieties of Ethnic Experience: Kinship, Class and Gender among Italian Americans*. Ithaca, N.Y.: Cornell University Press.

Dill, Bonnie Thornton. 1988. "Our Mothers' Grief: Racial Ethnic Women and the Maintenance of Families." *Journal of Family History* 13: 415–431.

Dillard, J. L. 1985. *Toward a Social History of American English*. New York: Mouton.

Diner, Hasia R. 1984. *Erin's Daughters in America: Irish Immigrant Women in the Nineteenth Century*. Baltimore: Johns Hopkins University Press.

Dinnerstein, Leonard, Roger L. Nichols, and David M. Reimers. 1990. *Natives and Strangers: Blacks, Indians and Immigrants in America*, 2d ed. New York: Oxford University Press.

Dinnerstein, Leonard, and David Reimers. 1988. *Ethnic Americans: A History of Immigration*. New York: Harper & Row.

Divine, Robert A. 1972. *American Immigration Policy, 1924–1952*. New York: Da Capo Press.

Dixon, Heriberto. 1988. "Black Cubans in the United States: A Case of Conflicts between Race and Ethnicity." Paper presented at the annual meeting of the American Studies Association, Miami, Florida, October 27–30.

Dobrowolski, Kazimierz. 1971. "Peasant Traditional Culture." In *Peasants and Peasant Society*, ed. Teodor Shanin. Harmondsworth, U.K.: Penguin.

Dobyns, Henry F. 1966. "Estimating Aboriginal American Population: An Appraisal of Techniques with a New Hemispheric Estimate." *Current Anthropology* 7: 395–416.

———. 1983. *Their Number Become Thinned*. Knoxville: University of Tennessee Press.

Dolan, Jay P. 1975. *The Immigrant Church: New York's Irish and German Catholics, 1815–1865*. Baltimore: Johns Hopkins University Press.

———. 1983. *The Immigrant Church: New York's Irish and German Catholics, 1815–1865*. Notre Dame, Ind.: Notre Dame University Press.

Domínguez, Jorge I. 1975. "La Tradición Liberal y la Emigración Cubana. *Areíto* 1: 4–5.

———. 1991. *Cooperating with the Enemy? U.S. Immigration Policies toward Cuba*. Cambridge: Harvard University, Center for International Affairs.

———. 1992. "Cooperating with the Enemy? U.S. Immigration Policies Toward Cuba." Pp. 31–88 in *Western Hemisphere Immigration and United States Foreign Policy*, ed. Christopher Mitchell. University Park: Pennsylvania State University Press.

Domínguez, Virginia. 1986. *White by Definition: Social Classification in Creole Louisiana*. New Brunswick, N.J.: Rutgers University Press.

Douglass, Frederick. [1845] 1963. *Narrative of the Life of Frederick Douglass, An American Slave, Written By Himself*. Garden City, N.Y.: Doubleday.

Dowd, Gregory Evans. 1992. *A Spirited Resistance: The North American Indian Struggle for Unity, 1745–1815*. Baltimore: Johns Hopkins University Press.

Doyle, David N. 1981. *Ireland, Irishmen, and Revolutionary America, 1760–1820*. Dublin: Mercier Press.

Dozier, E. P. 1966. "Problem Drinking Among American Indians: The Role of Socio-Cultural Deprivation." *Quarterly Journal of Alcohol Studies* 27: 72–87.

Du Bois, W. E. B. [1899] 1967. *The Philadelphia Negro: A Social Study*. New York: Schocken.

———. [1903] 1989. *The Souls of Black Folk*. New York: Bantam.

———. [1935] 1992. *Black Reconstruction in the United States, 1860–1880*. New York: Atheneum.

———. 1964. *Black Reconstruction in America: An Essay Toward a History of the Part Which Black Folk Played in the Attempt to Reconstruct Democracy in America, 1860–1880*. Cleveland: World Publishing.

Dudden, Arthur Power. 1992. *The American Pacific: From the Old China Trade to the Present*. New York: Oxford University Press.

Duis, Perry. 1976. *Chicago: Creating New Traditions*. Chicago: Chicago Historical Society.

Duncan, Otis Dudley, and Beverly Duncan. 1955. "A Methodological Analysis of Segregation Indices." *American Sociological Review* 20: 210–217.

Dunne, John G. 1967. *Delano*. New York: Ambassador Books.

———. 1991. "Law & Disorder in Los Angeles," pt. 2. *New York Review of Books* (24 October): 62–70.

East-West Population Institute. 1990. *Recent Filipino Immigration to the United States: A Profile*. Honolulu: East-West Center, University of Hawaii.

Edwards, Robert Dudley, and T. Desmond Williams, eds. 1957. *The Great Famine: Studies in Irish History, 1845–1852*. New York: New York University Press.

EEOC (Equal Employment Opportunity Commission). 1968. "Hearings Held Before the U.S. EEOC on Discrimination in White Collar Employment, New York City, January 15–18, 1968." Washington D.C.: U.S. Government Printing Office.

El Nuevo Herald. 1993. "Cifra Anual de Balseros Cubanos Marcó Récord." 26 December, 4A.

"El Plan Espiritual de Aztlan." 1970. Pp. 20–23 in *Educating the Mexican American*, ed. H. S. Johnson and W. J. Hernández. Valley Forge, Pa.: Judson Press.

Elkins, Stanley. 1958. *Slavery*. Chicago: University of Chicago Press.

Ellis, Frances. 1954. "German Instruction in the Public Schools of Indianapolis, 1869–1919." *Indiana Magazine of History* 50: 119–138, 251–276, 357–380.

Embree, John F. 1939. *Suye Mura: A Japanese Village*. Chicago: University of Chicago Press.

Enchautegui, María E. 1992. "Immigration and County Employment Growth." PRIP-UI-23. Washington, D.C.: Urban Institute.

Encyclopaedia Britannica. 1993. *Britannica Book of the Year*. Chicago: Encyclopedia Britannica.

Ericksen, Chuck. 1970. "Uprising in the Barrio." In *Educating the Mexican American*, ed. H. S. Johnson and W. J. Hernandez. Valley Forge, Pa.: Judson Press.

Erie, Steven. 1988. *Rainbow's End: Irish-Americans and the Dilemmas of Urban Machines Politics, 1840–1985*. Berkeley and Los Angeles: University of California Press.

Ernst, Robert. 1949. *Immigrant Life in New York City, 1825–1863*. New York: King's Crown Press.

———. 1965. *Immigrant Life in New York City, 1825–1863*. Port Washington, N.Y.: Ira J. Friedman.

Espenshade, Thomas J., and Tracy Ann Goodis. 1985. *Recent Immigrants to Los Angeles: Characteristics and Labor Market Impacts*. Washington, D.C.: Urban Institute.

Espinal Jacobo, Flavio Dario. 1993. "Return Migration and Public Policies in the Dominican Republic." Mimeo. Report prepared for the Hemispheric Migration Project, Georgetown University.

Essoyan, Susan. 1990. "Native Hawaiians Vote for Clout." *San Francisco Chronicle*, 31 January, A6.

Evans, Peter B., Dietrich Rueschemeyer, and Theda Skocpol, eds. 1985. *Bringing the State Back In*. London: Cambridge University Press.

Fagen, Richard R., Richard A. Brody, and Thomas J. O'Leary. 1968. *Cubans in Exile: Disaffection and the Revolution*. Stanford, Calif.: Stanford University Press.

Fairclough, Adam. 1987. "A Study of the Southern Christian Leadership Conference and the Rise and Fall of the Nonviolent Civil Rights Movement." Ph.D. diss., University of Keele.

Falcon, Angelo. 1993. "The Puerto Rican Community: A Status Report." In *Di logo*. New York: Puerto Rican Policy Group.

Falcón, Luis M. 1991. "Migration and Development: The Case of Puerto Rico." Pp. 146–181 in *Determinants of Emigration from Mexico, Central America, and the Caribbean*, ed. Sergio Díaz-Briquets and Sidney Weintraub. Boulder, Colo.: Westview Press.

Fallows, James. 1983. "Immigration: How It's Affecting Us." *Atlantic Monthly* 252: 45–106.

Farber, Bernard, Charles H. Mindel, and Bernard Lazerwitz. 1988. "The Jewish American Family." Pp. 400–437 in *Ethnic Families in America*, 3rd ed., ed. Charles H. Mindel, Robert Habenstein, and Roosevelt Wright, Jr. New York: Elsevier.

Farley, Reynolds. 1977. "Residential Segregation in Urbanized Areas of the United States in 1970: An Analysis of Social Class and Racial Differences." *Demography* 14: 497–529.

———. 1984. *Blacks and Whites: Narrowing the Gap?* Cambridge: Harvard University Press.

———. 1991. "The New Census Question about Ancestry: What Did it Tell Us?" *Demography* 28: 411–429.

Farley, Reynolds, and Walter R. Allen. 1989. *The Color Line and the Quality of Life in America*. New York: Oxford University Press.

Farley, Reynolds, and William H. Frey. 1994. "Changes in the Segregation of Whites from Blacks during the 1980s: Small Steps toward a More Integrated Society." *American Sociological Review* 59: 23–45.

Farley, Reynolds, Howard Schuman, Suzanne Bianchi, Diane Colasanto, and Shirley Hatchett. 1978. "Chocolate City, Vanilla Suburbs: Will the Trend toward Racially Separate Communities Continue?" *Social Science Research* 7: 319–344.

Farley, Reynolds, Charlotte Steeh, Tara Jackson, Maria Krysan, and Keith Reeves. 1993. "The Causes of Continued Racial Residential Segregation: Chocolate City, Vanilla Suburbs Revisited." *Journal of Housing Research* 4: 1–38.

Fawcett, James T., and Robert W. Gardner. 1992. "Asian Immigrant Entrepreneurs and Non-Entreprenuers: A Comparative Study of Recent Korean and Filipino Immigrants." *Population and Environment*.

Feagin, Joe R. 1974. "Community Disorganization: Some Critical Notes." Pp. 123–146 in *The Community: Approaches and Implications*, ed. Marcia Pelly Effrat. New York: Free Press.

———. 1978. *Racial and Ethnic Relations*. Englewood Cliffs, N.J.: Prentice-Hall.

———. 1989. "The Black 'Underclass' Ideology in Race Relations Analysis." *Social Justice* 16: 13–34.

———. 1991. "The Continuing Significance of Race: Antiblack Discrimination in Public Places." *American Sociological Review* 56: 101–117.

Feagin, Joe R., and Douglas Lee Eckberg. 1980. "Discrimination: Motivation, Action, Effects, and Context." *Annual Review of Sociology* 6: 1–20.

Femminella, Francis, and Jill Quadagno. 1976. "The Italian American Family." In *Ethnic Families in America: Patterns and Variations*, ed. Charles Mindell and Robert Haberstein. New York: Elsevier.

Ferguson, F. N. 1968. "Navajo Drinking: Some Tentative Hypotheses." *Human Organization* 27: 159–167.

Fernández, Gastón. 1982. "The Freedom Flotilla: A Legitimacy Crisis of Cuban Socialism?" *Journal of Interamerican Studies and World Affairs* 24: 183–209.

Fernández, Roberto M., and Francois Nielsen. 1986. "Bilingualism and Hispanic Scholastic Achievement: Some Baseline Results." *Social Science Research* 15: 43–70.

Ferree, Myra Max. 1979. "Employment Without Liberation: Cuban Women in the United States." *Social Science Quarterly* 60: 35–50.

Finkel, Stephen E., Thomas M. Guterbock, and Marian J. Borg. 1991. "Race-of-Interviewer Effects in a Preelection Poll." *Public Opinion Quarterly* 55: 313–330.

Fishman, Joshua A. 1969. "A Sociolinguistic Census of a Bilingual Neighborhood." *American Journal of Sociology* 75: 323–339.

———. 1988. "'English Only': Its Ghosts, Its Myths, and Dangers." *International Journal of the Sociology of Language* 74: 125–140.

Fishman, Joshua A., and Robert L. Cooper. 1969. "Alternative Measures of Bilingualism." *Journal of Verbal Learning and Behavior* 8: 276–282.

Fishman, Joshua A., and Frank R. Solano. 1989. "Cross-Polity Linguistic Homogeneity/Heterogeneity and Per-capita Gross National Product: An Empirical Exploration." *Language Problems and Language Planning* 2: 103–118.

———. 1990. "Cross-Polity Perspective on the Importance of Linguistic Heterogeneity as a 'Contributory Factor' in Civil Strife." *Canadian Review of Studies in Nationalism* 1 (2): 131–146.

Fishman, Joshua A., and Charles Terry. 1969. "The Validity of Census Data on Bilingualism in a Puerto Rican Neighborhood." *American Sociological Review* 34: 636–650.

Fitzgerald, Francis. 1980. *America Revised*. New York: Random House.

Fitzhugh, George. [1857] 1960. *Cannibals All or Slaves Without Masters*. Cambridge: Harvard University Press.

Fitzpatrick, Joseph P. 1987. *Puerto Rican Americans: The Meaning of Migration to the Mainland*, 2d ed. Englewood Cliffs, N.J.: Prentice-Hall.

Fixico, Donald L. 1986. *Termination and Relocation: Federal Indian Policy, 1945–1960*. Albuquerque: University of New Mexico Press.

Flexner, James Thomas. 1974. *Washington: The Indispensable Man*. Boston: Little, Brown.

Fligstein, Neil. 1981. *Going North*. New York: Academic Press.

Florant, Lyonel C. 1942. "Negro Internal Migration." *American Sociological Review* 7: 782–791.

Foley, Douglas E. 1988. *From Peones to Politicos: Class Ethnicity in a South Texas Town, 1900–1987*. Austin: University of Texas Press.

Foner, Nancy E. 1978. *Jamaica Farewell: Jamaican Migrants in London*. Berkeley and Los Angeles: University of California Press.

Forman, Sheila M. 1991. "Filipino Participation in Civil Rights Policies and Practices in Hawaii." *Social Process in Hawaii* 33: 1–11.

Fortunate Eagle, Adam. 1992. *Alcatraz! Alcatraz! The Indian Occupation of 1969–1971*. Berkeley, Calif.: Heyday Books.

Fox, Geoffrey E. 1971. "Cuban Workers in Exile." *Trans-Action* 8: 21–30.

Franklin, Benjamin. 1959. *The Papers of Benjamin Franklin*. Ed. Leonard W. Labarre. New Haven: Yale University Press.

Franklin, John Hope. 1967. *From Slavery to Freedom: A History of Negro Americans*, 3rd ed. New York: Knopf.

———. 1993. *The Color Line: Legacy for the Twenty-First Century*. Columbia: University of Missouri Press.

Frazier, E. Franklin. 1957. *Black Bourgeoisie*. New York: Free Press.

———. [1957] 1971. *The Negro in the United States*. New York: Macmillan.

———. 1966. *The Negro Church in America*. New York: Schocken.

———. 1969. *The Negro in the United States*. New York: Macmillan.

Freedberg, Louis. 1992. "School Club of the New Outsiders." *San Francisco Chronicle*, 15 January, A1, A6.

Freeman, James M. 1989. *Hearts of Sorrow: Vietnamese-American Lives*. Stanford, Calif.: Stanford University Press.

Freeman, Thomas W. 1957. *Pre-Famine Ireland: A Study in Historical Geography*. Manchester, U.K.: Manchester University Press.

Frey, William H., and Reynolds Farley. 1993. "Latino, Asian, and Black Segregation in Multi-Ethnic Metro Areas: Findings from the 1990 Census." Research report, Population Studies Center, University of Michigan.

Fuentes, Carlos. 1992. *The Buried Mirror: Reflections on Spain and the New World*. New York: Houghton Mifflin.

Gabaccia, Donna. 1984. *From Sicily to Elizabeth Street: Housing and Social Change Among Italian Immigrants, 1880–1930*. Albany: State University of New York Press.

Gabarino, James, Nancy Dubrow, Kathleen Kostelny, and Carole Pardo. 1992. *Children in Danger: Coping with the Consequences of Community Violence*. San Francisco: Jossey-Bass.

Galarza, Ernesto. 1964. *Merchants of Labor: The Mexican Bracero Story*. Santa Barbara, CA: McNally and Loffin.

Gale, Dennis E. 1987. *Washington, D.C.: Inner-City Revitalization and Minority Suburbanization*. Philadelphia: Temple University Press.

Galenson, David W. 1981. *White Servitude in Colonial America: An Economic Analysis*. Cambridge: Cambridge University Press.

Gallup, George, Jr., and Frank Newport. 1991. "Blacks and Whites Differ on Civil Rights Progress." *Gallup Poll Monthly* (August): 54–62.

Gallup Report. 1985. *Religion in America—50 Years*. Princeton, N.J.: Princeton Religious Research Center.

Gans, Herbert. 1962. *The Urban Villagers: Group and Class in the Life of Italian Americans*. New York: Free Press.

——— . 1979. "Symbolic ethnicity: The future of ethnic groups and cultures in America." *Ethnic and Racial Studies* 2: 1–20.

——— . 1988. *Middle American Individualism: The Future of Liberal Democracy*. New York: Free Press.

García, Ignacio. 1989. *United We Win: The Rise and Fall of La Raza Unida Party*. Tucson: University of Arizona Press.

García, Manny, and Joanne Cavanaugh. 1993. "One Hope, One Boat, Two Fates?" *Miami Herald*, December 19, 1A.

García, Mario T. 1994. *Memories of Chicano History: Bert Corona*. Berkeley and Los Angeles: University of California Press.

García, Richard A. 1991. *Rise of the Mexican American Middle Class: San Antonio, 1929–1941*. College Station: Texas A&M Press.

García y Griego, Manuel, and Leo R. Chavez. 1993. "The Auditor General's Estimates of the Fiscal Impact of Undocumented Immigrants Residing in San Diego County: A Critique." Mimeo. Presented to the Assembly Select Committee on California–Mexico Affairs, California State Legislature, Sacramento, California, February 23.

Garcilaso de la Vega, The Inca. 1951. *The Florida of the Inca*. Trans. and ed. John G. Varner and Jeannette J. Varner. Austin: University of Texas Press.

Gardner, R. 1970. *Grito: Reis Tijerina and the New Mexico Land Grant War of 1967*. New York: Harper & Row.

Garfinkle, Irwin, and Sara McLanahan. 1986. *Single Mothers and Their Children*. Washington, D.C.: Urban Institute.

Garland, Hamlin. [1893] 1956. "The Creamery Man." Pp. 141–158 in *Main-Travelled Roads*. New York: Harper & Row.

Garreau, Joel. 1987. "Competing Bonds of Race and Class." *Washington Post*, 30 November, A1, A8, A9.

——— . 1991. *The Edge City, Life on the New Frontier*. New York: Doubleday.

Garrow, David. 1986. *Bearing the Cross*. New York: William Morrow.

Genovese, Eugene. 1966. *The Political Economy of Slavery: Studies in the Economy and Society of the Slave South*. New York: Pantheon.

——— . 1969. *The World the Slaveholders Made: Two Essays in Interpretation*. New York: Pantheon.

——— . 1979. *From Rebellion to Revolution: Afro-American Slave Revolts in the Making of the Modern World*. Baton Rouge: Louisiana State University Press.

Georges, Eugenia. 1984. *New Immigrants and the Political Process: Dominicans in New York*. Occasional papers no. 45, Center for Latin American and Caribbean Studies, New York University.

——— . 1990. *The Making of a Transnational Community: Migration, Development, and Cultural Change in the Dominican Republic*. New York: Columbia University Press.

——— . 1993. "Dominicans." In *Encyclopedia of New York City*, ed. Kenneth T. Jackson. New Haven: Yale University Press.

Gillette, Howard, and Alan Kraut. 1986. "The Evolution of Washington, D.C.'s Italian-American Community, 1890 to World War Two." *Journal of American Ethnic History* 6: 7–27.

Gilkes, Cheryl. 1990. "The Black Church as a Therapeutic Community: Suggested Areas for Research into the Black Religious Experience." *Journal of the Interdenominational Theological Center* 8: 29–44.

Glazer, Nathan. 1954. "Ethnic Groups in America." Pp. 158–173 in *Freedom and Control in Modern Society*, ed. M. Berger, T. Abel, and C. Page. New York: Van Nostrand.

——— . 1955. "The Social Characteristics of American Jews." *American Jewish Year Book* 56: 3–42.

——— . 1971. "Blacks and Ethnic Groups: The Difference, and the Political Difference it Makes." *Social Problems* 18: 444–461.

——— . 1975. *Affirmative Discrimination: Ethnic Inequality and Public Policy*. New York: Basic Books.

——— . 1990. "American Jewry or American Judaism." Pp. 31–41 in *American Pluralism and the Jewish Community*, ed. Seymour Martin Lipset. New Brunswick, N.J.: Transaction Books.

Glazer, Nathan, and Daniel Patrick Moynihan. 1963. *Beyond the Melting Pot: The Negroes, Puerto Ricans, Jews, Italians, and Irish of New York City*. Cambridge: MIT Press.

——— . 1970. *Beyond the Melting Pot: The Negroes, Puerto Ricans, Jews, Italians, and Irish of New York City*, 2d ed. Cambridge: MIT Press.

Glenn, Evelyn Nakano. 1983. "Split Household, Small Producer, and Dual Wage Earner: An Analysis of Chinese-American Families Strategies." *Journal of Marriage and the Family* 45: 35–46.

——— . 1986. *Issei, Nisei, Warbride: Three Generations of Japanese American Women in Domestic Service*. Philadelphia: Temple University Press.

Glenn, Susan. 1991. *Daughters of the Shtetl*. Ithaca, N.Y.: Cornell University Press.

Gliwicowna, Marja. 1936. "Drogi Emigracji" (The Immigrant Road). *Przeglad Socjologiczny* 4: 502–520.

Gold, Steven J. 1992. *Refugee Communities: A Comparative Field Study*. Newbury Park, Calif.: Sage.

Goldscheider, Calvin. 1986. *Jewish Continuity and Change.* Bloomington: Indiana University Press.

Goldscheider, Calvin, and Frances E. Kobrin. 1978. *The Ethnic Factor in Family Structure and Mobility.* Cambridge, Mass.: Ballinger.

———. 1980. "Ethnic Continuity and the Process of Self-Employment." *Ethnicity* 7: 256–278.

Goldscheider, Calvin, and Alan S. Zuckerman. 1984. *The Transformation of the Jews.* Chicago: University of Chicago Press.

Gómez-Quiñones, Juan. 1978. *Mexican Students por La Raza.* Santa Barbara, Calif.: Editorial la Causa.

———. 1991. *Chicano Politics.* Albuquerque: University of New Mexico Press.

González. 1992. "Dominican Immigration Alters Hispanic New York." *New York Times,* 1 September, A1.

———. 1993. "Unmasking Roots of Washington Heights Violence." *New York Times,* 21 October, B1.

González, Gilbert G. 1985. "Segregation of Mexican Children in a Southern California City: The Legacy of Expansion and the American Southwest." *Western Historical Quarterly* 16: 58–76.

Gonzáles, Rodolfo. 1967. *I Am Joaquin.* Denver, Colo.: Crusade for Justice.

Goodfriend, Joyce D. 1992. *Before the Melting Pot: Society and Culture in Colonial New York City, 1664D1730.* Princeton, N.J.: Princeton University Press.

Gordon, Milton M. 1964. *Assimilation in American Life.* New York: Oxford University Press.

Gorelick, Sherry. 1981. *City College and the Jewish Poor.* New York: Schocken.

Goren, Arthur A. 1970. *New York Jews and the Quest for Community.* New York: Columbia University Press.

Gossett, Thomas F. 1965. *Race: The History of an Idea in America.* New York: Schocken.

Gottlieb, Peter. 1987. *Making Their Own Way: Southern Blacks' Migration to Pittsburgh, 1916–38.* Urbana: University of Illinois Press.

Graham, Otis L., Jr., and Roy Beck. 1992. "To Help City, Cut Flow of Immigrants." *Los Angeles Times,* 19 May, B11.

Grasmuck, Sherri. 1992. "Una gota de sangre: el significado del color en dos lados de la frontera." *Quehaceres* (Santo Domingo) 12 (September): 9.

Grasmuck, Sherri, and Patricia R. Pessar. 1991. *Between Two Islands: Dominican International Migration.* Berkeley and Los Angeles: University of California Press.

Graves, Theodore D. 1971. "Drinking and Drunkeness Among Urban Indians." Pp. 274–311 in *The American Indian in Urban Society,* ed. Jack O. Waddell and O. Michael Watson. Boston: Little, Brown.

Greeley, Andrew M. 1971. *Why Can't They Be Like Us?* New York: Dutton.

———. 1972. *That Most Distressful Nation: The Taming of the American Irish.* Chicago: Quadrangle Books.

———. 1974. *Ethnicity in the U.S.* New York: John Wiley & Sons.

———. 1976. "The Ethnic Miracle." *Public Interest* 45 (fall): 20–36.

Green, Constance McLaughlin. 1962. *Washington: Village and Capital, Washington: Capital City, 1878D1950.* Princeton, N.J.: Princeton University Press.

———. 1967. *The Secret City, A History of Race Relations in the Nation's Capital.* Princeton, N.J.: Princeton University Press.

Greenberg, Cheryl. 1991. *Or Does It Explode?: Black Harlem in the Great Depression.* New York: Oxford University Press.

Greenberg, Stanley B. 1985. "Report on Democratic Defection." Unpublished report prepared for the Michigan House Democratic Campaign Committee.

Greenblatt, Stephen J. 1976. "Learning to Curse: Aspects of Linguistic Colonialism in the Sixteenth Century." Pp. 561–580 in *First Images of America: The Impact of the New World on the Old,* vol. 2, ed. F. Chiapelli, M. J. B. Allen, and R. L. Benson. Berkeley and Los Angeles: University of California Press.

Greene, Victor. 1968. *The Slavic Community on Strike: Immigrant Labor in Pennsylvania Anthracite.* Notre Dame, Ind.: Notre Dame University Press.

Grier, George. 1992. *Poverty in Washington: Its Dimensions and its Human Impact.* Washington, D.C.: Greater Washington Research Center.

Griffith, Ezra, Thelouizs English, and Violet Mayfield. 1980. "Possession, Prayer and Testimony: Therapeutic Aspects of the Wednesday Night Meeting in a Black Church." *Psychiatry* 43: 12–128.

Griffith, Ezra, John Young, and Dorothy Smith. 1984. "An Analysis of the Therapeutic Elements in a Black Church Service." *Hospital and Community Psychiatry* 35: 464–469.

Griswold del Castillo, Richard. 1990. *The Treaty of Guadalupe Hidalgo.* Norman: University of Oklahoma Press.

Gromada, Thaddeus V. 1982. "Goral Regionalism and Polish Immigration to America." In *Pastor of the Poles: Polish American Essays Presented to Right Reverend Monsignor John P. Wodarski,* ed. Stanislaus A. Blejwas and Mieczylaw B. Biskupski. Polish Studies Program Monographs. New Britain: Central Connecticut State College.

Groneman, Carole. 1977. "'She Earns as a Child—She Pays as a Man': Women Workers in a Mid-Nineteenth Century New York Community." Pp. 33–46 in *Immigrants in Industrial America, 1850–1920,* ed. Richard L. Ehrlich. Charlottesville: University of Virginia Press.

Gross, Bert Crawford. 1929. "Factors Affecting the Origin of Livestock Receipts at Chicago, 1923–27." Master's thesis, University of Chicago.

Grossman, James. 1989. *Land of Hope.* Chicago: University of Chicago Press.

———. 1991. "The White Man's Union: The Great Migration and the Resonance of Race and Class in Chicago, 1916–1922." Pp. 83–105 in *The Great Migration in Historical Perspective: New Dimensions of Race, Class and Gender*, ed. Joe W. Trotter, Jr. Bloomington: Indiana University Press.

Guarnizo, Luis Eduardo. 1992. "One Country in Two: Dominican-Owned Firms in New York and in the Dominican Republic." Ph.D. diss., Johns Hopkins University.

———. 1993. "Going Home: Class, Gender, and Household Transformation Among Dominican Returned Migrants." Mimeo. Report prepared for the Commission for the Hemispheric Migration and Refugee Policy, Georgetown University.

Guest, Avery, and James Weed. 1976. "Ethnic residential segregation: Patterns of change." *American Journal of Sociology* 81: 1088–1111.

Guillemin, Jeanne. 1975. *Urban Renegades: The Cultural Strategy of American Indians*. New York: Columbia University Press.

Gulick, Sydney L. n.d. *New Factors in Japanese-American Relations and a Constructive Proposal*. New York: National Committee on Japanese American Relations.

Gundlach, James H., and Alden E. Roberts. 1978. "Native American Indian Migration and Relocation: Success or Failure." *Pacific Sociological Review* 12: 117–128.

Gurak, Douglas T., and Luis M. Falcón. 1990. "The Puerto Rican Family and Poverty: Complex Paths to Poor Outcomes." Pp. 39–83 in *Puerto Ricans: Breaking Out of the Cycle of Poverty*. Washington, D.C.: National Puerto Rican Coalition.

Gurak, Douglas, and Luis Falcón-Rodríguez. 1987. "The Social and Economic Situation of Hispanics in the United States and New York City in the 1980's." In *Hispanics in New York: Religious, Cultural and Social Experiences*, vol. 2, ed. Office of Pastoral Research.

Gurin, Patricia, Shirley Hatchett, and James S. Jackson. 1989. *Hope and Independence: Blacks' Response to Electoral and Party Politics*. New York: Russell Sage.

Gutiérrez, Ramón A. 1991. *When Jesus Came, the Corn Mothers Went Away: Marriage, Sexuality, and Power in New Mexico, 1500–1846*. Stanford, Calif.: Stanford University Press.

Guzmán, Ralph C. 1976. *The Political Socialization of the Mexican American People*. New York: Arno Press.

Ha, Julie. 1992. "Reactions from KA Students at UCLA." *Korean Times Los Angeles* (English ed.), 11 May, 7.

Haan, Mary N., and A. George Kaplan. 1986. "The Contribution of Socioecomic Position to Minority Health." Pp. 69–103 in *Report of the Secretary's Task Force on Black and Minority Health*, vol. 2. Washington, D.C.: U.S. Department of Health and Human Services.

Hacker, Andrew. 1957. "Liberal Democracy and Social Control." *American Political Science Review* 51 (4).

Hacker, A., R. Collins, and M. Jacobson. 1987. *Marketing Booze to Blacks*. Washington, D.C.: Center for Science in the Public Interest.

Hagan, William T. 1979. *American Indians*. Chicago, IL: University of Chicago Press.

Hainer, P., C. Hines, E. Martin, and G. Shapiro. 1988. *Research on Improving Coverage in Household Surveys*. Fourth annual research conference proceedings. Washington, D.C. U.S. Bureau of the Census.

Haines, David W., ed. 1985. *Refugees in the United States: A Reference Handbook*. Westport, Conn.: Greenwood Press.

———, ed. 1989. *Refugees as Immigrants: Cambodians, Laotians and Vietnamese in America*. Totowa, N.J.: Rowman & Littlefield.

Hall, Thomas D. 1989. *Social Change in the Southwest, 1350–1880*. Lawrence, KS: University Press of Kansas.

Hamilton, Gary. 1978. "Pariah Capitalism: Paradox of Power and Dependence." *Ethnic Groups* 2: 1–15.

Handlin, Oscar. 1941. *Boston's Immigrants, 1790–1860: A Study in Acculturation*. Cambridge: Harvard University Press.

———. [1951] 1973. *The Uprooted: The Epic Story of the Great Migration that Made the American People*. Boston: Little, Brown.

———. 1959. *The Newcomers: Negroes and Puerto Ricans in a Changing Metropolis*. Cambridge: Harvard University Press.

Handlin, Oscar, and Lilian Handlin. 1980. *Abraham Lincoln and the Union*. Boston: Little, Brown.

Handman, Max S. 1926. "The Mexican Immigrant in Texas." *Southwestern Political and Social Science Quarterly* 7: 33–40.

Hansen, Marcus Lee. 1940. *The Atlantic Migration, 1607–1860*. Cambridge: Harvard University Press.

Harburg, Ernest, John Erfurt, Catherine Chape, Louise Havenstein, William Scholl, and M. A. Schork. 1973. "Socioecological Stressor Areas and Black-White Blood Pressure: Detroit." *Journal of Chronic Disease* 26: 595–611.

Hareven, Tamara K. 1976. "Modernization and Family History: Perspectives on Social Change." *Signs: Journal of Women in Culture and Society* 2: 190–206.

Harrington, Michael. [1962] 1981. *The Other America: Poverty in the United States*. New York: Penguin.

———. 1984. *The New American Poverty*. New York: Holt, Rinehart & Winston.

Harrison, Paul. 1981. *Inside the Third World: The Anatomy of Poverty*. New York: Penguin.

Hastings, Lansford. [1845] 1932. *The Emigrants' Guide to Oregon and California*. Reproduced in facsimile. Princeton, NJ: Princeton University Press.

Hatch, Orrin G. 1980. "Senator Hatch Wrestles with Affirmative Action." *Personnel Administrator* 25: 78.

Hayashi, Yoshie. 1989. "Prostitutes for Foreigners: Women's Contribution to the Japanese Economy." Unpublished paper.

Hayes-Bautista, David E., Werner O. Schink, and Jorge Chapa. 1988. *The Burden of Support: Young Latinos in an Aging Society*. Stanford, Calif.: Stanford University Press.

Hechter, Michael. 1975. *Internal Colonialism: The Celtic Fringe in British National Development, 1536–1966*. London: Routledge and Kegan Paul.

Hegel, G. W. F. 1970. *The Phenomenology of Mind*. 2 vols. Trans. J. B. Baillie. London: S. Sonnenshein.

Hein, Jeremy. 1993. *States and International Migrants: The Incorporation of Indochinese Refugees in the United States and France*. Boulder, Colo.: Westview Press.

Heizer, Robert F., and Theodora Kroeber, eds. 1979. *Ishi, the Last Yahi: A Documentary History*. Berkeley and Los Angeles: University of California Press.

Helbich, Wolfgang. 1988. "*Alle Menschen sind dort gleich . . .*": *Die deutsche Amerika-Auswanderung im 19. und 20. Jahrhundert*. Düsseldorf: Schwann.

Henderson, Vivian W. 1967. "Regions, Race, and Jobs." In *Employment, Race, and Poverty*, ed. Arthur M. Ross and Herbert Hill. New York: Harcourt, Brace, and World.

Henige, David. 1986. "Primary Source by Primary Source? On the Role of Epidemics in New World Depopulation." *Ethnohistory* 33: 293–312.

Henri, Florette. 1975. *Black Migration: Movement North 1900–1920*. Garden City, N.Y.: Anchor Press/Doubleday.

Herbst, Alma. 1930. "The Negro in the Slaughtering and Meat Packing Industry in Chicago." Ph.D. diss., University of Chicago.

Herd, Denise. 1985. "Migration, Cultural Transformation and the Rise of Black Liver Cirrhosis Mortality." *British Journal of Addiction* 80 (4): 397–410.

———. 1990. "Subgroup Differences in Drinking Patterns among Black and White Men: Results from a National Survey." *Journal of Studies on Alcohol* 51 (3): 221–232.

Herman, Pini, and David LaFontaine. 1983. "In Our Footsteps: Israeli Migration to the U.S. and Los Angeles." Master's thesis, Hebrew Union College.

Herring, Cedric. 1989. "Convergence, Polarization, or What?: Racially Based Changes in Attitudes and Outlooks, 1964–1984." *Sociological Quarterly* 30: 267–281.

Herring, Cedric, and Sharon Collins. 1995. "Retreat From Equal Opportunity?: The Case of Affirmative Action." In *The Bubbling Cauldron*, ed. M. P. Smith and J. R. Feagin. Minneapolis: University of Minnesota Press.

Herring, Cedric, James S. House, and Richard P. Mero. 1991. "Racially Based Changes in Political Alienation in America." *Social Science Quarterly* 72: 123–134.

Herskovits, Melville J. 1990. *The Myth of the Negro Past*. Boston: Beacon Press.

Hertzberg, Hazel W. 1971. *The Search for an American Indian Identity*. Syracuse, N.Y.: Syracuse University Press.

Hidalgo, Ariel. 1994. *Disidencia*. Miami: Ediciones Universal.

Higham, John. 1955. *Strangers in the Land: Patterns of American Nativism, 1860–1925*. New Brunswick, N.J.: Rutgers University Press.

———. 1963. *Strangers in the Land: Patterns of American Nativism, 1860–1925*. New York: Atheneum.

———. 1970. *Strangers in the Land: Patterns of American Nativism, 1860–1925*. New York: Atheneum.

———. [1975] 1984. *Send These to Me: Immigrants in Urban America*. Baltimore: Johns Hopkins University Press.

———. 1978. *Strangers in the Land: Patterns of American Nativism, 1860–1925*. New Brunswick, N.J.: Rutgers University Press.

Hill, Robert. 1981. *Economic Policies and Black Progress*. Washington, D.C.: National Urban League.

Hirata, Cheng Hirata. 1979. "Free, Indentured, Enslaved: Chinese Prostitutes in Nineteenth-Century America." *Signs: Journal of Women in Culture and Society* 5: 3–29.

Hirsch, Arnold R. 1983. *Making the Second Ghetto: Race and Housing in Chicago: 1940–1960*. Chicago: University of Chicago Press.

Hirsch, E. D., Jr. 1987. *Cultural Literacy*. Boston: Houghton Mifflin.

———. 1988. *The Dictionary of Cultural Literacy*. Boston: Houghton Mifflin.

Hitchens, Christopher. 1990. *Blood, Class, and Nostalgia: Anglo-American Ironies*. New York: Farrar, Straus & Giroux.

Hodge, Robert W. 1973. "Toward a Theory of Racial Differences in Employment." *Social Forces* 52 (1): 16–31.

Hoffman, Abraham. 1974. *Unwanted Americans: Mexican Americans in the Great Depression, Repatriation Pressures, 1929–1939*. Tucson: University of Arizona Press.

Hoffman, Michael. 1993. "More Jews in Congress: Does it Make a Difference?" *Moment* (February): 32–39.

Hofstadter, Richard. 1955. *The Age of Reform*. New York: Knopf.

Hollinger, David A. 1995. *Postethnic America: Beyond Multiculturalism*. New York: Basic Books.

Holmes, Samuel J. 1926. "An Argument Against Mexican Immigration." *Commonwealth Club of California Transaction* 21: 1–34.

Holzer, Harry J. 1991. "The Spatial Mismatch Hypothesis: What has the Evidence Shown?" *Urban Studies* 28 (1): 105–122.

———. 1993. "Black Employment Problems: New Evidence, Old Questions." Unpublished paper, Department of Economics, Michigan State University.

Horsman, Reginald. 1981. *Race and Manifest Destiny: The Origins of American Racial Anglo-Saxonism*. Cambridge: Harvard University Press.

Horton, John. 1989. "The Politics of Ethnic Change: Grass-Roots Responses to Economic and Demographic Restructuring in Monterey Park." *Urban Geography* 10: 578–592.

Hourwich, Isaac A. 1912. Immigration and Labor: The Economic Aspects of European Immigration to the United States. New York: Putnam's Sons.

House, James S., Karl R. Landis, and Debra Umberson. 1988. "Social Relationships and Health." *Science* 241: 540–545.

Houstoun, M. F., R. G. Kramer, and J. M. Barrett. 1984. "Female Predominance of Immigration to the United States since 1930: A First Look." *International Migration Review* 18: 908–963.

Howe, Irving. 1976. *World of Our Fathers*. New York: Harcourt Brace Jovanovich.

Hoxie, Frederick E. 1984. *A Final Promise: The Campaign to Assimilate the Indians, 1880–1920*. Lincoln, Neb: University of Nebraska Press.

Huckfeldt, Robert, and Carol Weitzel Kohfeld. 1989. *Race and the Decline of Class in American Politics*. Urbana: University of Illinois Press.

Hudson, Charles, Marvin T. Smith, and Chester B. DePratter. 1984. "The Hernando De Soto Expedition: From Apalachee to Chiaha." *Southeastern Archaeology* 3: 65–77.

Hudson, Charles, Marvin Smith, David Hally, Richard Polhemus, and Chester DePratter. 1985. "Coosa: A Chiefdom in the Sixteenth-Century Southeastern United States." *American Antiquity* 50: 723–737.

Hudson, William T., and Walter D. Broadnax. 1982. "Equal Employment Opportunity as Public Policy." *Public Personnel Management* 11: 268–276.

Hunt, Chester, and Lewis Walker. 1974. *Ethnic Dynamics*. Homewood, Ill.: Dorsey.

Hurh, Won Moo, and Kwang Chung Kim. 1988. "Uprooting and Adjustment: A Sociological Study of Korean Immigrants' Mental Health." Final report submitted to the National Institute of Mental Health. Department of Sociology and Anthropology, Western Illinois University.

——— . 1990. "Religious Participation of Korean Immigrants in the United States." *Journal for the Scientific Study of Religion* 29: 19–34.

Hutchinson, Edward P. 1953. "Immigration Policy Since World War One." Pp. 116–121 in *Immigration: American Dilemma*, ed. Benjamin Munn Ziegler. Boston: Little, Brown.

——— . 1956. *Immigrants and Their Children, 1850–1950*. New York: John Wiley & Sons.

Hyman, Herbert H., and Paul B. Sheatsley. 1956. "Attitudes Toward Desegregation." *Scientific American* 195: 35–39.

Ichihashi, Yamato. 1932. *Japanese in the United States*. Stanford, Calif.: Stanford University Press.

Ichioka, Yuji. 1977. "Ameyuki-san: Japanese Prostitutes in Nineteenth-Century America." *Amerasia* 4: 1–15.

——— . 1980. "Japanese Immigrant Women in the United States, 1900–1924." *Pacific Historical Review* 49: 339–357.

Indians of All Tribes. 1972. "Proclamation to the Great White Father and All His People." Pp. 40–42 in *Alcatraz is Not an Island*, ed. Peter blue cloud. Berkeley, Calif.: Wingbow Press.

Institute for the Study of Social Change. 1991. "The Diversity Project: The Final Report." University of California, Berkeley.

Institute of Cuban Studies. 1982. *Itinerario Ideológico: Antología de Lourdes Casal*. Miami: Ediciones Di spora.

Inter-American Development Bank. 1992. *Economic and Social Progress in Latin America: 1992 Report*. Washington, D.C.: Inter-American Development Bank.

Ito, Kazuo. 1973. *Issei: A History of Japanese Immigrants in North America*. Trans. Shinichiro Nakamura and Jean S. Gerard. Seattle: Japanese Community Service.

Ivy, A. C., and Irwin Ross. 1957. "Discrimination in College Admissions." Pp. 133–144 in *American Minorities*, ed. Milton L. Barron. New York: Knopf.

Jackson, Frank D. 1885. *Census of Iowa for the Year 1885*. Des Moines.

Jackson, Kenneth T. 1985. *Crabgrass Frontier, The Suburbanization of the United States*. New York: Oxford University Press.

Jackson, Thomas F. 1993. "Reframing the 'Underclass' Debate." Pp. 266–283 in *The Underclass Debate: Views From History*, ed. Michael Katz. Princeton, N.J.: Princeton University Press.

Jarrett, Robin L. 1994. "Living Poor: Family Life among Single Parent, African-American Women." *Social Problems* 41: 30–49.

Jasso, Guillermina, and Mark R. Rosenzweig. 1990. *The New Chosen People: Immigrants in the United States*. New York: Russell Sage.

Jaynes, Gerald D. 1990. "Labor Market Status of Black Americans: 1939–1985." *Journal of Economic Perspectives* 4: 9–24.

Jaynes, Gerald D., and Robin M. Williams, eds.. 1989. *A Common Destiny: Blacks and American Society*. Washington, D.C.: National Academy Press.

Jehl, Douglas. 1992. "Buchanan Raises Specter of Intolerance, Critics Say." *Los Angeles Times*, 17 March, A1.

Jencks, Christopher, and Paul E. Peterson, eds. 1991. *The Urban Underclass*. Washington, D.C.: Brookings Institution.

Jennings, Veronica T. 1993. "Calling Racial Attack 'Savagery,' Judge Gives Man 60 Years." *Washington Post*, 4 February, B1.

Johansen, Bruce, and Roberto Maestas. 1979. *Wasi'chu: The Continuing Indian Wars*. New York: Monthly Review Press.

Johnson, Daniel M., and Rex R. Campbell. 1981. *Black Migration in America: A Social Demographic History*. Durham, N.C.: Duke University Press.

Johnson, James, Colyzelle Jones, Walter Farrell, and Melvin Oliver. 1992. "The Los Angeles Rebellion: A Retrospective View." *Economic Development Quarterly* 6: 356–372.

Jones, Jacqueline. 1985. *Labor of Love, Labor of Sorrow: Black Women, Work and the Family from Slavery to the Present*. New York: Basic Books.

———. 1992. *The Dispossessed*. New York: Basic Books.

Jones, Maldwyn Allen 1960. *American Immigration*. Chicago: University of Chicago Press.

Jordan, Terry G. 1966. *German Seed in Texas Soil: Immigrant Farmers in Nineteenth Century Texas*. Austin: University of Texas Press.

Jordan, Winthrop. [1968] 1977. *White Over Black: American Attitudes Toward the Negro, 1550–1812*. New York: W. W. Norton.

Jorgensen, Joseph G., Richard O. Clemmer, Ronald L. Little, Nancy J. Owens, and Lynn A. Robbins. 1978. *Native Americans and Energy Development*. Cambridge, Mass.: Anthropology Resource Center.

Jorgenson, Lloyd P. 1987. *The State and the Non-Public School, 1825–1925*. Columbia: University of Missouri Press.

Joselit, Jenna Weissman. 1983. *Our Gang: Jewish Crime and the New York Jewish Community, 1900–1940*. Bloomington: Indiana University Press.

Joseph, Samuel. 1914. *Jewish Immigration to the United States from 1881 to 1910*. New York: Arno Press and *New York Times*.

Kahan, Arcadius. 1978. "Economic Opportunities and Some Pilgrims' Progress: Jewish Immigrants from Eastern Europe in the U.S., 1890–1914." *Journal of Economic History* 38 (1): 235–251.

———. 1986. *Essays in Jewish Social and Economic History*. Chicago: University of Chicago Press.

Kalmijn, Matthijs. 1991. "Shifting Boundaries: Trends in Religious and Educational Homogamy." *American Sociological Review* 56: 786–800.

Kamphoefner, Walter D. 1984. "The German Agricultural Frontier: Crucible or Cocoon?" *Ethnic Forum* 4: 21–35.

———. 1987. *The Westfalians: From Germany to Missouri*. Princeton, N.J.: Princeton University Press.

———. 1992. "Implications of German-American Bilingualism: Evidence from the 1940 Census." Paper presented at the 85th annual meeting of the Organization of American Historians, Chicago, April 2–5.

Kamphoefner, Walter D., Wolfgang Helbich, and Ulrike Sommer, eds. 1991. *News From the Land of Freedom: German Immigrants Write Home*. Ithaca, N.Y.: Cornell University Press.

Karnow, Stanley. 1991. *Vietnam: A History*, rev. ed. New York: Penguin Press.

Karst, K. L. 1989. *Belonging to America: Equal Citizenship and the Constitution*. New Haven: Yale University Press.

Kasarda, John P. 1985. "Urban Change and Minority Opportunities." In *The New Urban Reality*, ed. Paul E. Peterson. Washington D.C.: Brookings Institution.

Kasinitz, Philip. 1992. *Caribbean New York: Black Immigrants and the Politics of Race*. Ithaca, N.Y.: Cornell University Press.

Kass, Drora, and Semour Martin Lipset. 1982. "Jewish Immigration to the United States from 1967 to the Present: Israelis and Others." Pp. 272–294 in *Understanding American Jewry*, ed. Marshall Sklare. New Brunswick, N.J.: Transaction Books.

Katz, Irwin. 1986. "Racial Ambivalence, Value Duality, and Behavior." Pp 35–60 in *Prejudice, Discrimination, and Racism*, ed. John F. Dovidio and Samuel L. Gaertner. San Diego: Academic Press.

Katz, Jacob. 1971. *Tradition and Crisis: Jewish Society at the End of the Middle Ages*. New York: Schocken.

Katzman, David M. 1981. *Seven Days a Week: Women and Domestic Service in Industrializing America*. Urbana: University of Illinois Press.

Katznelson, Ira. 1930. *Black Men, White Cities*. Chicago: University of Chicago Press.

Keith, Verna M., and Cedric Herring. 1991. "Skin Tone and Stratification in the Black Community." *American Journal of Sociology* 97: 760–778.

Kelley, Bruce. 1990. "El Mosco." *Los Angeles Times Magazine*, 18 March, 10–43.

Kennedy, Edward M. 1966. "The Immigration Act of 1965." *Annals of the American Academy of Political and Social Sciences* 367: 137–149.

———. 1981. "Refugee Act of 1980." *International Migration Review* 15: 141–156.

Kennedy, J. C., et al. 1914. *Wages and Family Budgets in the Chicago Stockyards District*. Chicago: University of Chicago Press.

Kennedy, John F. 1964. *A Nation of Immigrants*. New York: Harper & Row.

Kennedy, Louise. 1930. *The Negro Peasant Turns Cityward*. New York: Columbia University Press.

Kennedy, Robert E., Jr. 1962. "The Protestant Ethic and the Parsis." *American Journal of Sociology* 68: 11–20.

———. 1973. *The Irish: Emigration, Marriage, and Fertility*. Berkeley and Los Angeles: University of California Press.

Kessler, Ronald C., Katherine A. McGonagle, Shanyang Zhao, Christopher B. Nelson, Michael Hughes, Suzann Eshleman, Hans-Ulrich Wittchen, and Kenneth S. Kendler. 1994. "Lifetime and 12-Month Prevalence of DSM-III-R Psychiatric Disorders in the United States." *Archives of General Psychiatry* 51: 8–19.

Kessner, Thomas. 1977. *The Golden Door: Italian and Jewish Immigrant Mobility in New York City, 1880–1915*. New York: Oxford University Press.

Kessner, Thomas, and Betty Boyd Caroli. 1978. "New Immigrant Women at Work: Italians and Jews in New York City, 1880–1905." *Journal of Ethnic Studies* 5: 19–32.

Key, V. O. 1949. *Southern Politics in State and Nation*. New York: Knopf.

———. 1987b. "The Koreans: Small Business in an Urban Frontier." Pp. 219–242 in *New Immigrants in New York*, ed. Nancy Foner. New York: Columbia University Press.

King, Gary, and David R. Williams. In press. "Race and Health: A Multi-Dimensional Approach to African American Health." In *Society and Health: Foundation for a Nation*, ed. Sol Levine, Diana Chapman Walsh, Benjamin C. Amick, and Alvin R. Tarlov. New York: Oxford University Press.

Kitano, Harry L. 1960. "Housing of Japanese-Americans in the San Francisco Bay Area." Pp. 178–197 in *Studies in Housing and Minority Groups*, ed. Nathan Glazer and Davis McEntire. Berkeley and Los Angeles: University of California Press.

Kloss, Heinz. 1977. *The American Bilingual Tradition*. Rowley, Mass.: Newbury House.

Knobel, Dale T. 1986. *Paddy and the Republic: Ethnicity and Nationality in Antebellum America*. Middletown, Conn.: Wesleyan University Press.

Knowles, John. 1977. "The Responsibility of the Individual." Pp. 57–80 in *Doing Better and Feeling Worse*, ed. John Knowles. New York: W. W. Norton.

Knowles-Robbins, Katherine. 1918. "The Work of the American Speech Committee of the Chicago Woman's Club, and Notes upon its School Survey." *English Journal* 7: 163–176.

Koenig, Rev. Msgr. Harry C., ed. 1980. *A History of the Parishes of the Archdiocese of Chicago*, 2 vols.

Korea Times Los Angeles. 1992a. "Korean Stores Selectively Targeted for Arson." 11 June, 1.

———. 1992b. "A Black Radio Instigated Black Gangs to Target Korean Stores for Arson and Looting." 22 June, 1.

———. 1992c. "$4,752,571 Was Raised in the U.S. Korean Community." 23 July, 23.

———. 1992d. "2,400 Koreans in Orange County Registered to Vote in 1992 Alone." 31 October, 3.

Korea Times New York. 1985. "A Proposal for a Law for Rent Regulation Gains Support." 14 May, 1.

———. 1992. "A Hearing on Need for a Rent Regulation Law." 24 July, 1.

Korean National Bureau of Statistics. 1984. *Korean Statistical Yearbook, 1984*, vol. 31. Seoul: Economic Planning Board, Republic of Korea.

———. 1987. *1985 Population and Housing Census*. Seoul: Economic Planning Board, Republic of Korea.

Korn, Peter. 1991. "Agent Orange in Vietnam: The Persisting Poison." *The Nation* 252 (13): 440–446.

Kornweibel, Theodore. 1975. *No Crystal Stair*. Westport, Conn.: Greenwood Press.

Kosmin, Barry. 1991. "Exploring and Understanding the Findings of the 1990 National Jewish Population Survey." Paper presented at the Hollander Colloquium, Los Angeles, California, July.

Kotkin, Joel. 1994. "An Emerging Asian Tiger: The Vietnamese Connection." *Los Angeles Times*, 24 April, M1.

Krieger, Nancy. 1987. "Shades of Difference: Theoretical Underpinnings of the Medical Controversy on Black/White Differences in the United States, 1830–1870." *International Journal of Health Services* 17: 259–278.

———. 1990. "Racial and Gender Discrimination: Risk Factors for High Blood Pressure?" *Social Science and Medicine* 30: 1273–1281.

Krieger, N., and M. Bassett. 1986. "The Health of Black Folk: Disease, Class, and Ideology in Science." *Monthly Review* 38: 74–85.

Kritz, Mary M., and Fe Caces. 1992. "Science and Technology Transfers and Migration Flows." Pp. 221–242 in *International Migration Systems: A Global Approach*, ed. Mary M. Kritz, Lin Lean Lim, and Hania Zlotnik. New York: Oxford University Press.

Krosnick, Jon A., and Duane F. Alwin. 1989. "Aging and Susceptibility to Attitude Change." *Journal of Personality and Social Psychology* 57: 416–425.

Kula, Witold. 1973. *Listy Emigranto'w Z Brazylii i Stano'w Ziednoczonych* (Emigrant Letters From Brazil and the United States). Warsaw: Ludowa Spo'ldzielnia Wydawnicza.

Kunz, E. F. 1973. "The Refugee in Flight: Kinetic Models and Forms of Displacement." *International Migration Review* 7: 125–146.

———. 1981. "Exile and Resettlement: Refugee Theory." *International Migration Review* 15: 42–51.

Kurtines, William M., and Luke Miranda. 1980. "Differences in Self and Family Role Perception among Acculturating Cuban-American College Students: Implications for the Etiology of Family Disruption among Migrant Groups." *International Journal of Intercultural Relations* 4: 167–184.

Kuznets, Simon. 1975. "Immigration of Russian Jews to the United States: Background and Structure." *Perspectives in American History* 9: 35–124.

Kwong, Peter. 1987. *The New Chinatown*. New York: Noonday Press.

La Course, Richard. 1974. "FBI, BIA Probing Indian Movement." *Race Relations Reporter* 5 (September): 1–4.

LaFeber, Walter. 1983. *Inevitable Revolutions: The United States in Central America*. New York: W. W. Norton.

Laguerre, Michel S. 1984. *American Odyssey: Haitians in New York City*. Ithaca, N.Y.: Cornell University Press.

Laird, Charlton G. 1970. *Language in America*. New York: World Publishing.

Lambert, Wallace E., and G. Richard Tucker. 1972. *Bilingual Education of Children: The St. Lambert Experiment*. Rowley, Mass.: Newbury House.

Lamm, Richard D. 1981. "America Needs Fewer Immigrants." *New York Times*, 12 July, E21.

Lamm, Richard D., and Gary Imhoff. 1985. *The Immigration Time Bomb*. New York: Truman Talley Books.

Landesman, Alter F. 1969. *Brownsville*. New York: Block Press.

Laird, Charlton G. 1970. *Language in America*. New York: World Publishing.

Lambert, Wallace E., and G. Richard Tucker. 1972. *Bilingual Education of Children: The St. Lambert Experiment*. Rowley, Mass.: Newbury House.

Lamm, Richard D. 1981. "America Needs Fewer Immigrants." *New York Times*, 12 July, E21.

Lamm, Richard D., and Gary Imhoff. 1985. *The Immigration Time Bomb*. New York: Truman Talley Books.

Landesman, Alter F. 1969. *Brownsville*. New York: Block Press.

Landis, Kenesaw M. 1948. *Segregation in Washington, A Report of the National Committee on Segregation in the Nation's Capital*. Chicago: National Committee on Segregation in the Nation's Capital.

Landry, Bart. 1987. *The New Black Middle Class*. Berkeley and Los Angeles: University of California Press.

Larson, Eric M., and Wolfgang Opitz. 1988. "Sex Ratio and Vital Statistics–based Estimates of Emigration from the Dominican Republic." Paper presented at the Conference on Dominican Migration to the United States. Santo Domingo: Fundación Friedrich Ebert.

LaRuffa, Anthony. 1988. *Monte Carmelo: An Italian-American Community in the Bronx*. New York: Gordon & Breach.

Lasker, Bruno. 1969. *Filipino Immigration*. New York: Arno Press.

La Sorte, Michael. 1985. *La Merica: Images of Italian Greenhorn Experience*. Philadelphia: Temple University Press.

Laviera, Tato. 1992. *La Carreta Made a U-Turn*. Houston, Tex.: Arte Público Press.

LaViest, Thomas A. 1992. "The Political Empowerment and Health Status of African-Americans: Mapping a New Territory." *American Journal of Sociology* 97: 1080–1095.

Leap, W. L. 1981. "American Indian Languages." Pp. 116–144 in *Language in the USA*, ed. C. A. Ferguson and S. B. Heath. Cambridge: Cambridge University Press.

Leary, Warren E. 1990. "Uneasy Doctors Add Race-Consciousness to Diagnostic Tools." *New York Times*, 25 September, B5, B7.

Leavell, R. H. 1983. "Negro Migration from Mississippi."

Lee, Everett S. 1966. "A Theory of Migration." *Demography* 3: 47–57.

Lemann, Nicholas. 1992. *The Promised Land: The Great Black Migration and How It Changed America*. New York: Vintage Books.

Leonard, Karen. 1992. *Making Ethnic Choices: California's Punjabi Mexican Americans*. Philadelphia: Temple University Press.

Lesko, Kathleen M., Valerie Babb, and Carrol R. Gibbs. 1991. *Black Georgetown Remembered*. Washington, D.C.: Georgetown University Press.

Levin, Myron. 1988. "The Tobacco Industry's Strange Bedfellows." *Business and Society Review* 65 (spring): 11–17.

Levitan, Sar A., and William B. Johnston. 1975. *Indian Giving: Federal Programs for Native Americans*. Baltimore: Johns Hopkins University Press.

Lewis, Oscar. 1966. "The Culture of Poverty." Pp. xlii-lii in *La Vida*. New York: Vintage Books.

Lewontin, R. 1974. *The Genetic Basis of Evolutionary Change*. New York: Columbia University Press.

——— . 1982. *Human Diversity*. New York: Scientific American Books.

Li, Peter S. 1977. "Fictive Kinship, Conjugal Tie and Kinship Claim Among Chinese Immigrants in the United States." *Journal of Comparative Family Studies* 8: 47–64.

Lieberson, Stanley. 1963. *Ethnic Patterns in American Cities*. New York: Free Press.

——— . 1980. *A Piece of the Pie: Blacks and White Immigrants Since 1880*. Berkeley and Los Angeles: University of California Press.

——— . 1985. "Unhyphenated Whites in the United States." *Ethnic and Racial Studies* 8: 159–180.

Lieberson, Stanley, Guy Dalto, and Mary Ellen Johnston. 1975. "The Course of Mother Tongue Diversity in Nations." *American Journal of Sociology* 81 (July): 34–61.

Lieberson, Stanley, and Mary Waters. 1986. "Ethnic Groups in Flux: The Changing Ethnic Responses of American Whites." *Annals of the American Academy of Political and Social Science* 487: 79–91.

——— . 1988. *From Many Strands: Ethnic and Racial Groups in Contemporary America*. New York: Russell Sage.

——— . 1993. "The Ethnic Responses of Whites: What Causes their Instability, Simplification and Inconsistency?" *Social Forces* 72 (2): 421–450.

Light, Ivan. 1972. *Ethnic Enterprise in America*. Berkeley and Los Angeles: University of California Press.

——— . 1983. *Cities in World Perspective*. New York: Macmillan.

——— . 1988. "Los Angeles." Pp. 56–96 in *The Metropolis Era: Megacities*, vol. 2, ed. Mattei Dogan and John Kasarda. Newbury Park, Calif.: Sage.

Light, Ivan, and Edna Bonacich. 1988. *Immigrant Entrepreneurs: Koreans in Los Angeles, 1965–1982*. Berkeley and Los Angeles: University of California Press.

Light, Ivan, Hadas Har-Chvi, and Kenneth Kan. 1994. "Black/Korean Conflicts in Los Angeles." Pp. 72–87 in *Managing Divided Cities*, ed. Seamus Dunn. London: Keele University Press.

Light, Ivan, Georges Sabagh, Mehdi Bozorgmehr, and Claudia Der-Martirosian. 1993. "Internal Ethnicity in the Ethnic Economy." *Ethnic and Racial Studies* 16: 581–597.

Light, Ivan, and Angel Sanchez. 1987. "Immigrant Entrepreneurship in 272 SMSAs." *Sociological Perspectives* 30: 373–399.

Limerick, Patricia Nelson. 1987. *Legacy of Conquest: The Unbroken Past of the American West*. New York: W. W. Norton.

Lincoln, C. Eric, and Lawrence H. Mamiya. 1990. *The Black Church in the African American Experience*. Durham, N.C.: Duke University Press .

Link, Bruce, and Bruce P. Dohrenwend. 1980. "Formulation of Hypotheses about the True Prevalence of Demoralization." Pp. 114–132 in *Mental Illness in the United States: Epidemiological Estimates*, ed. Bruce P. Dohrenwend. New York: Praeger.

Lipset, Seymour Martin. 1990. "A Unique People in an Exceptional Country." Pp. 3–29 in *American Pluralism and the Jewish Community*, ed. Seymour Martin Lipset. New Brunswick, N.J.: Transaction Books.

Lipset, Seymour Martin, and William Schneider. 1978. "The Bakke Case: How Would It be Decided at the Bar of Public Opinion?" *Public Opinion* (March/April): 38–44.

Liu, John M., Paul M. Ong, and Carolyn Rosenstein. 1992. "Rural Chain Migration: Post-1965 Filipino Immigration to the United States." *International Migration Review*.

Llanes, José. 1982. *Cuban-Americans: Masters of Survival*. Cambridge, Mass.: ABT.

Loewen, James W. 1988. *The Mississippi Chinese: Between Black and White*, 2d ed. Prospect Heights, Ill.: Waveland Press.

Longbrake, David B., and Woodrow W. Nichols, Jr. 1976. *Sunshine and Shadows in Metropolitan Miami*. Cambridge, Mass.: Ballinger.

Loo, Chalsa, and Paul Ong. 1982. "Slaying Demons with a Sewing Needle: Feminist Issues for Chinatown Women." *Berkeley Journal of Sociology* 27: 77–88.

López, Adalberto. 1980. "Birth of a Nation: Puerto Rico in the Nineteenth Century." In *The Puerto Rican: Their History, Culture, and Society*. Cambridge, Mass.: Schenkman.

Lopez, David, and Yen Espiritu. 1990. "Panethnicity in the United States: A Theoretical Framework." *Ethnic and Racial Studies* 13: 198–224.

Lopreato, Joseph. 1970. *Italian Americans*. New York: Random House.

Lozano, Wilfredo. 1992. "La urbanización de la pobreza: economia informal, familia e identidades colectivas." Unpublished paper presented at the conference Ciudades en el Caribe, Santo Domingo, Facultad Latinoamericana de Ciencias Sociales (FLASCO), August.

Luebke, Frederick C. 1974. *Bonds of Loyalty: German Americans and World War I*. DeKalb: Northern Illinois University Press.

——— . 1990. "Legal Restrictions on Foreign Languages in the Great Plains States, 1917–23." Pp. 31–50 in *Germans in the New World: Essays in the History of Immigration*. Chicago/Urbana: University of Illinois Press.

Luker, Kristin. 1991. "Dubious Conceptions: The Controversy over Teen Pregnancy." *American Prospect* 4: 73–83.

Lurie, Nancy Oestreich. 1961. "The Voice of the American Indian: Report on the American Indian Chicago Conference." *Current Anthropology* 2 (5): 478–500.

——— . 1971. "The Contemporary American Indian Scene." Pp. 418–480 in *North American Indians in Historical Perspective*, ed. Eleanor Burke Leacock and Nancy Oestreich Lurie. New York: Random House.

Lyman, Stanford M. 1971. "Strangers in the City: The Chinese in the Urban Frontier." Pp. 159–187 in *Roots: An Asian American Reader*, ed. Amy Tachiki, Eddie Wong, and Franklin Odo with Buck Wong. Los Angeles: Asian American Studies Center, University of California.

——— . 1974. *Chinese Americans*. New York: Random House.

——— . 1977. *The Asian in North America*. Santa Barbara, Calif.: ABC Clio.

Lyman, Stanley David. 1991. *Wounded Knee 1973: A Personal Account*. Lincoln: University of Nebraska Press.

Lynch, Hollis R. 1973. *The Black Urban Condition, A Documentary History 1866–1971*. New York: Thomas Y. Crowell.

Mann, Arthur. 1959. *La Guardia: A Fighter Against His Times, 1882–1933*. Philadelphia: Lippincott.

Mannheim, Karl. 1952. *Essays in the Sociology of Knowledge*. New York: Oxford University Press.

Manning, Robert D. 1989. "Industrial Restructuring, Immigrant Workers, and the American State: The Political Economy of Mexican Migration." Ph.D. diss., Johns Hopkins University.

——— . 1995. "The Post-Industrial Crisis of Urban America: Clintonomics and The Politics of Multiculturalism." In *The First Hundred Days of the Clinton Administration*, ed. Barbara Driscoll. Mexico City: National University of Mexico (UNAM) Press.

——— . 1996. "Multicultural Washington, D.C.: The Changing Social and Economic Landscape of a Post-Industrial Metropolis." *Ethnic and Racial Studies*.

Manning, Robert D., and Anita Butera. 1994. "From Ellis Island to the Golden Arches: Immigration, Native Minorities, and the Post-Industrial Metropolis." *Proteus* 2: 10–16.

Manning, Robert D., and David E. Pedersen. 1992. "Post-Industrial America: Minority Rights in a Multicultural Society?" *People and Rights: Nordic Journal of Human Rights*, Special Issue, 4: 339–356.

Marable, Manning. 1984. *Race, Reform and Rebellion: The Second Reconstruction in Black America, 1945–1982*. Jackson: University Press of Mississippi.

Marckwardt, Albert H. 1980. *American English*. New York: Oxford University Press.

Marcus, George E., and Michael M. Fischer. 1986. *Anthropology as Cultural Critique*. Chicago: University of Chicago Press.

Marin, C. 1975. "Rodolfo 'Corky' Gonzales: The Mexican American Movement Spokesman, 1966–1972." *Journal of the West* 14: 4.

Marks, Carole. 1989. *Farewell—We're Good and Gone: The Great Black Migration*. Bloomington: University of Indiana Press.

Marmora, Lelio. 1988. "Social Integration and Employment of Central American Refugees." Pp. 142–155 in *When Borders Don't Divide*, ed. Patricia Pessar. New York: Center for Migration Studies.

Marmot, M. G., M. Kogevinas, and M. A. Elston. 1987. "Social/Economic Status and Disease." *Annual Review of Public Health* 8: 111–135.

Marmot, M. G., and M. E. McDowall. 1986. "Mortality Decline and Widening Social Inequalities." *Lancet* 2: 274–276.

Márquez, Benjamin. 1993. *LULAC: The Evolution of a Texas Mexican American Political Organization*. Austin: University of Texas Press.

Marshall, Thurgood. 1978. "A History of Discrimination Requires Compensation." Dissenting opinion, *Regents of California vs. Allan Bakke*, U.S. Supreme Court.

Martí, José. 1891. "Nuestra América." *Revista Ilustrada de Nueva York*, 1ro. de Enero.

Martin, Elizabeth, T. J. DeMaio, and P. C. Campanelli. 1990. "Context Effects for Census Measures of Race and Hispanic Origin." *Public Opinion Quarterly* 54: 551–566.

Martin, Philip, and Elizabeth Midgley. 1994. "Immigration to the United States: Journey to an Uncertain Destination." *Population Bulletin* 49 (September).

Massey, Douglas S. 1990. "American Apartheid: Segregation and the Making of the Underclass." *American Journal of Sociology* 96: 329–357.

——— . 1993. "Latino Poverty Research: An Agenda for the 1990s." *Social Science Research Council Items* 47: 7–11.

Massey, Douglas S., Rafael Alarcón, Jorge Durand, and Humberto González. 1987. *Return to Aztlán: The Social Process of International Migration from Western Mexico*. Berkeley and Los Angeles: University of California Press.

Massey, Douglas S., Joaquín Arango, G. Hugo, A. Kouaouci, A. Pellegrino, and J. E. Taylor. 1993. "Theories of International Migration: A Review and Appraisal." *Population and Development Review* 19 (3): 431–466.

Massey, Douglas S., and Nancy A. Denton. 1988. "Suburbanization and Segregation in U.S. Metropolitan Areas." *American Journal of Sociology* 94: 592–626.

——— . 1989. "Hypersegregation in U.S. Metropolitan Areas: Black and Hispanic Segregation Along Five Dimensions." *Demography* 26: 373–391.

——— . 1993. *American Apartheid: Segregation and the Making of the Underclass*. Cambridge: Harvard University Press.

Massey, Douglas S., and Mitchell L. Eggers. 1990. "The Ecology of Inequality: Minorities and the Concentration of Poverty, 1970–1980." *American Journal of Sociology* 95 (5): 1153–1188.

Massey, Douglas, Andrew B. Gross, and Mitchell L. Eggers. 1991. "Segregation, the Concentration of Poverty, and the Life Chances of Individuals." *Social Science Research* 20: 397–420.

Masud-Piloto, Felix Roberto. 1988. *With Open Arms: Cuban Migration to the United States*. Totowa, N.J.: Rowman & Littlefield.

Mayer, Egon. 1991. "Jewishness Among the Intermarried." Paper presented at the Hollander Colloquium, Los Angeles, California, July.

McAdam, Doug. 1982. *Political Process and the Development of Black Insurgency, 1930–1970*. Chicago: University of Chicago Press.

McCarthy, Kevin F., and R. Burciaga Valdez. 1986. *Current and Future Effects of Mexican Immigration in California*. Santa Monica, Calif.: RAND Corp.

McComb, David G. 1989. *Texas: A Modern History*. Austin: University of Texas Press.

McCord, Colin, and Harold Freeman. 1990. "Excess Mortality in Harlem." *New England Journal of Medicine* 322: 173–177.

McDowell, Mary. n.d. "City Waste." In *Mary McDowell and Municipal Housekeeping: A Symposium*, ed. Caroline M. Hill. Chicago: Millar Publishing.

——— . 1970. "A Quarter of a Century in the Stockyards District." In *An Illinois Reader*, ed. Clyde C. Walton. De Kalb: Northern Illinois University Press.

McKay, Reynolds R. 1982. "Texas-Mexican Repatriation During the Great Depression." Ph.D. diss., University of Oklahoma.

McKenney, Nampeo R., and Arthur R. Cresce. 1992. "Measurement of Ethnicity in the United States: Experiences of the U.S. Census Bureau." Paper presented at the Joint Canada-United States Conference on the Measurement of Ethnicity, Ottawa, Canada, April 1–3, 1992.

McManus, Edgar. 1966. *A History of Negro Slavery in New York*. Syracuse, N.Y.: Syracuse University Press.

McNeill, William H. 1976. *Plagues and Peoples*. Garden City, N.Y.: Anchor/Doubleday.

McNickle, D'Arcy. 1973. *Native American Tribalism: Indian Survivals and Renewals*. New York: Oxford University Press.

McPhearson, James M. 1965. *The Negro's Civil War: How American Negroes Felt and Acted During the War for the Union*. New York: Pantheon.

McWilliams, Carey. 1933. "Getting Rid of the Mexican." *American Mercury* 28: 323.

——— . 1939. *Factories in the Field: The Story of Migratory Farm Labor in California*. Boston: Little, Brown.

——— . 1968. *North From Mexico*. Westport, Conn.: Greenwood Press.

Mei, June. 1984. "Socioeconomic Development Among the Chinese in San Francisco, 1848–1906." Pp. 370–401 in *Labor Immigration Under Capitalism: Asian Workers in the United States Before World War II*, ed. Lucie Cheng and Edna Bonacich. Berkeley and Los Angeles: University of California Press.

Meinig, D. W. 1969. *Imperial Texas: An Interpretive Essay in Cultural Geography*. Austin: University of Texas Press.

Meléndez, Edwin. 1993. "Understanding Latino Poverty." *Sage Race Relations Abstracts* 18: 3–42.

Mendez et al. v. Westminster School District of Orange County et al., 64 F. Supp. 544 (1946).

Merton, Robert K. 1949. "Discrimination and the American Creed." In *Discrimination and the National Welfare*, ed. Robert M. MacIver. New York: Harper & Row.

Mesa-Lago, Carmelo. 1978. *Cuba in the 1970s: Pragmatism and Institutionalization*. Albuquerque: University of New Mexico Press.

———. 1994. "Will Cuba's Economic Reforms Work?" *Miami Herald*, 2 January, 1M.

Metro-Dade. 1988. "Changes in Employment and Occupations: Dade County, Florida: 1980–1987." Unpublished paper, Metro-Dade County Planning Department, Miami, Florida, 14 September.

Meyer, Melissa L., and Russell Thornton. 1988. "Indians and the Numbers Game: Quantitative Methods in Native American History." Pp. 5–29 in *New Directions in American Indian History*, ed. Colin G. Calloway. Norman: University of Oklahoma Press.

———. 1990. "American Indian Tribal Enrollment: The Blood Quantum Quandary." Paper presented at the annual meetings of the American Society of Ethnohistory, Toronto, Canada.

Meyers, Howard Barton. 1929. "The Policing of Labor Disputes in Chicago: A Case Study." Ph.D. diss., University of Chicago.

Milanich, Jerald T. 1990. "The European Entrada into La Florida." Pp. 3–26 in *Columbian Consequences*. Vol. 2 of *Archaeological and Historical Perspectives on the Spanish Borderlands East*, ed. David Hurst Thomas. Washington, D.C.: Smithsonian Institution Press.

Miller, Hunter, ed. 1937. *Treaties and Other International Acts of the United States of America*, vol. V. Washington, D.C.: U.S. Government Printing Office.

Miller, Kerby A. 1985. *Emigrants and Exiles: Ireland and the Irish Exodus to North America*. New York: Oxford University Press.

Miller, Randall M. 1977. "Introduction." Pp. xi–xxii in *Immigrants and Religion in Urban America*, ed. Randall M. Miller and Thomas D. Marzik. Philadelphia: Temple University Press.

Millis, H. A. 1915. *The Japanese Problem in the United States*. New York: Macmillan.

Mills, C. Wright. 1961. *The Sociological Imagination*. New York: Grove Press.

Min, Pyong Gap. 1984. "From White-Collar Occupations to Small Business: Koreans' Occupational Adjustment." *Sociological Quarterly* 25: 333–352.

———. 1987. "Factors Contributing to Small Business: A Comprehensive Synthesis." *International Journal of Comparative Sociology* 28: 173–193.

———. 1988a. *Ethnic Business Enterprise: Korean Small Business in Atlanta*. New York: Center for Migration Studies.

———. 1988b. "Korean Immigrant Entrepreneurship: A Multivariate Model." *Journal of Urban Affairs* 10: 197–212.

———. 1989. "Some Positive Functions of Ethnic Business for an Immigrant Community: Koreans in Los Angeles." Final report submitted to the National Science Foundation. Department of Sociology, Queens College of the City University of New York.

———. 1990. "Problems of Korean Immigrant Entrepreneurship." *International Migration Review* 24: 436–455.

———. 1991. "Cultural and Economic Boundaries of Korean Ethnicity: A Comparative Analysis." *Ethnic and Racial Studies* 14: 225–241.

———. 1992. "Korean Immigrant Wives' Overwork." *Korea Journal of Population and Development* 21: 23–26.

———. 1993a. "The Prevalence and Causes of Black Hostility Toward Korean Merchants." Unpublished manuscript, Department of Sociology, Queens College of the City University of New York.

———. 1993b. "Ethnic Business and Ethnic Solidarity: Koreans in New York." Unpublished manuscript, Department of Sociology, Queens College of the City University of New York.

Mincy, Ronald B., Isabel V. Sawhill, and Douglas A. Wolf. 1990. "The Underclass: Definition and Measurement." *Science* 248: 450–453.

Minero, Alberto. 1993. "Cine dominicano para el mundo." *El Diario*, 1 October, 26.

Mintz, Sidney W. 1970. "Foreword." Pp. 1–16 in *Afro-American Anthropology*, ed. N. E. Whitten, Jr., and J. F. Szwed. New York: Free Press.

———. 1971. "Toward an Afro-American History." *Journal of World History* 13: 317–332.

Mitchell, Christopher. 1992. "U.S. Foreign Policy and Dominican Migration to the United States." Pp. 89–124 in *Western Hemisphere Immigration and United States Foreign Policy*, ed. Christopher Mitchell. University Park: Pennsylvania State University Press.

Miyamoto, Frank. 1939. "Social Solidarity among the Japanese in Seattle." *University of Washington Publications in Social Sciences* 11: 57–130.

Mohl, Raymond A. 1983. "Miami: The Ethnic Cauldron." Pp. 67–72 in *Sunbelt Cities: Politics and Growth Since World War II*, ed. R. M. Bernard and B. R. Rice. Austin: University of Texas Press.

———. 1986. "The Politics of Ethnicity in Contemporary Miami." *Migration World* 14 (3).

Mollenkopf, John Hull. 1992. *A Phoenix in the Ashes: The Rise and Fall of the Koch Coalition in New York City Politics*. Princeton, N.J.: Princeton University Press.

Montagu, Ashley. 1965. *The Idea of Race*. Lincoln: University of Nebraska Press.

Montejano, David. 1987. *Anglos and Mexicans in the Making of Texas, 1836–1986*. Austin: University of Texas Press.

Montes Mozo, S., and J. J. García Vásquez. 1988. *Salvadoran Migration to the United States: An Exploratory Study*. Washington, D.C.: Center for Immigration Policy and Refugee Assistance, Georgetown University.

Montgomery, Paul. 1981. "For Cuban Refugees, Promise of U.S. Fades." *New York Times*, 19 April, 1; 1,32.

Mooney, James. 1910. "Population." Pp. 286–287 in *Handbook of American Indians North of Mexico*, vol. 2. Ed. Frederick W. Hodge. Smithsonian Institution, Bureau of American Ethnology Bulletin no. 30. Washington, D.C.: U.S. Government Printing Office.

——. 1928. "The Aboriginal Population North of Mexico." Pp. 1–40 in Smithsonian Miscellaneous Collections, vol. 80. Ed. John R. Swanton. Washington, D.C.: Smithsonian Institution.

Moore, Joan. 1989. "Is there a Hispanic Underclass?" *Social Science Quarterly* 70 (2): 265–284.

——. 1970. "Colonialism: The Case of the Mexican Americans." *Social Problems* 17: 463–473.

Moore, Joan, and Raquel Pinderhughes, eds. 1993. *Latinos and the Underclass Debate*. New York: Russell Sage.

Moore, John H. n.d. "Structured Dispersion: Cheyenne Response to Epidemic Disease." Department of Anthropology, University of Oklahoma.

Morales, Armando. 1972. *Ando Sangrando*. La Puente, Calif.: Perspectiva Publishing.

Morris, Aldon. 1984. *Origins of the Civil Rights Movement*. New York: Free Press.

——. 1992. "The Future of Black Politics: Substance versus Process and Formality." *National Political Science Review* 3: 168–174.

——. 1993a. "Birmingham Confrontation Reconsidered: An Analysis of the Dynamics and Tactics of Mobilization." *American Sociological Review* 58: 621–636.

——. 1993b. "Centuries of Black Protest: Its Significance for America and the World." In *Race in America*, ed. H. Hill and J. Jones, Jr. Madison: University of Wisconsin Press.

Moynihan, Daniel P. 1965. *The Negro Family: The Case for National Action*. Washington, D.C.: U.S. Government Printing Office.

——. 1972. "The Schism in Black America." *Public Interest* 27: 3–24.

Muller, Thomas, and Thomas J. Espenshade. 1985. *The Fourth Wave: California's Newest Immigrants*. Washington, D.C.: Urban Institute.

Munoz, Carlos, Jr. 1989. *Youth, Identity, and Power: The Chicano Movement*. New York: Verso.

Murray, Charles. 1984. *Losing Ground*. New York: Basic Books.

Myerson, A. R. 1992. "Dominicans Thrive Where Big Chains Won't Go." *New York Times*, 7 January, C1, C3.

Myrdal, Gunnar. 1944. *An American Dilemma: The Negro Problem and Modern Democracy*. New York: Random House.

Myrdal, Gunnar. 1964. *An American Dilemma*. New York: McGraw-Hill.

Nabokov, Peter. 1969. *Tijerina and the Courthouse Raid*. Albuquerque: University of New Mexico Press.

——. 1991. *Native American Testimony: A Chronicle of Indian-White Relations from Prophecy to the Present, 1492–1992*. New York: Viking.

Nadel, Stanley. 1990. *Little Germany: Ethnicity, Religion and Class in New York City, 1845–1880*. Urbana: University of Illinois Press.

Nagel, Joane. 1986. "The Political Construction of Ethnicity." Pp. 93–112 in *Competitive Ethnic Relations,* Susan Olzak and Joane Nagel (eds.). Orlando, FL: Academic Press.

——. 1989. "American Indian Repertoires of Contention." Paper presented at the annual meeting of the American Sociological Association, San Francisco, California.

——. 1995. *American Indian Ethnic Renewal: Red Power and the Resurgence of Identity and Culture*. New York: Oxford University Press.

Nagel, Joane, Carol Ward, and Timothy Knapp. 1988. "The Politics of American Indian Economic Development: The Reservation/Urban Nexus." Pp. 39–76 in *Public Policy Impacts on American Indian Economic Development*, ed. C. Matthew Snipp. Albuquerque: University of New Mexico Press.

National Center for Health Statistics. 1991. *Vital Statistics of the United States, 1988*, vol. II, *Mortality*, pt. A. Washington, D.C.: Public Health Service.

——. 1993. *Health, United States, 1992 and Healthy People 2000 Review*. Washington, D.C.: Public Health Service.

National Congress of American Indians. 1944. *Proceedings of the First Convention of the National Congress of American Indians*. Records of the National Congress of American Indians, National Anthropological Archives, Smithsonian Institution, Washington, D.C.

National Park Service. 1991. *Ellis Island & the Statue of Liberty*. San Francisco, Calif.: Pali Arts Communications.

NEDA (National Economic and Development Authority). 1992. *A Physical Planning Strategy in the Philippines*. Manila: National Economic and Development Authority.

Nee, Victor G., and B. Nee. 1974. *Longtime Californ'*. Boston: Houghton Mifflin.

Nee, Victor, and Herbert Y. Wong. 1985. "Asian American Socioeconomic Achievement: The Strength of the Family Bond." *Sociological Perspectives* 28: 281–306.

Neidert, Lisa, and Reynolds Farley. 1985. "Assimilation in the United States: An Analysis of Ethnic and Generation Differences in Status and Achievement." *American Sociological Review* 50: 840–850.

Nelli, Humbert. 1980 "Italians." Pp. 545–560 in *The Encyclopedia of American Ethnic Groups*, ed. Stephan Thernstrom, Ann Orlov, and Oscar Handlin. Cambridge: Harvard University Press.

——. 1983. *From Immigrants to Ethnics: The Italian Americans*. Oxford: Oxford University Press.

Nelson, E. 1966. *Huelga!* Delano, Calif.: Farm Workers Press.

Ness, Carol. 1992. "Eurocentrism Gains a Voice." *San Francisco Examiner*, 2 February, B1, B4.

Newsweek. 1992. "What They [Primary Candidates] Think." *Newsweek* (16 March): 33.

Nichols, Charles H. 1963. *Many Thousands Gone: The Ex-Slaves' Account of Their Bondage and Freedom*. Leiden, The Netherlands: E. J. Brill.

Niebuhr, Reinhold. 1952. *The Irony of American History*. New York: Scribners.

Nielsen, Francois. 1985. "Towards a Theory of Ethnic Solidarity in Modern Societies." *American Sociological Review* 50 (2): 133–149.

Nieman, Donald G. 1991. *Promises to Keep: African-Americans and the Constitutional Order, 1776 to the Present*. New York: Oxford University Press.

Niess, Frank. 1990. *A Hemisphere to Itself: A History of U.S.-Latin American Relations*. London: Zed Books.

Nixon, Richard. 1970. "Message to Congress on Indian Affairs." Pp. 213–230 in *Red Power: The American Indians' Fight for Freedom*, ed. Alvin M. Josephy, Jr. New York: McGraw-Hill.

Noel, Peter. 1993. "Word to my Brother: A Crisis of Conscience in Crown Heights." *Voice* (July 6): 11.

Northrup, Solomon. [1857] 1968. *Twelve Years a Slave*. Ed. Sue Eakin and Joseph Logsdon. Baton Rouge: Louisiana State University Press.

Novak, Michael. 1972. *The Rise of the Unmeltable Ethnics*. New York: Macmillan.

Noyelle, Thierry J., and Thomas M. Stanback, Jr. 1984. *The Economic Transformation of American Cities*. Totowa, N.J.: Rowman & Allanheld.

Obidinski, Eugene, and Helen Stankiewicz Zand. 1987. *Polish Folkways in America: Community and Family*. Lanham, Md.: University Press of America.

O'Brien, Sharon. 1990. *American Indian Tribal Governments*. Norman: University of Oklahoma Press.

O'Connor, Alice. 1992. "Race and Class in Chicago Sociology, 1920–1990." Paper presented at the meetings of the Social Science History Association, Chicago, Illinois.

Ogbu, John. 1990. "Minority Status and Literacy in Comparative Perspective." *Daedalus* 119: 141–169.

O'Hare, William. 1987. "Best Metros for Hispanic Businesses." *American Demographics* (November): 31–33.

Okumura, Jonathan Y., and Richard Coller. n.d. *The Social Adjustments of Filipino Non-Migrants and Emigrants to Hawaii*. Collaborative report of the Clinical Research Branch, National Institute of Mental Health, Rockville, Md., and the Institute of Behavioral Sciences, Honolulu, Hawaii.

Olshansky, S. J. 1985. "Pursuing Longevity: Delay vs. Elimination of Degenerative Diseases." *American Journal of Public Health* 75: 754–757.

Olson, Mary. 1988. "The Legal Road to Economic Development: Fishing Rights in Western Washington." Pp. 77–112 in *Public Policy Impacts on American Indian Economic Development*, ed. C. Matthew Snipp. Albuquerque: Institute for Native American Development, University of New Mexico.

Olzak, Susan. 1983. "Contemporary Ethnic Mobilization." *Annual Review of Sociology* 9: 355–374.

Omi, Michael, and Howard Winant. 1986. *Racial Formation in the United States: From the 1960s to the 1980s*. New York: Routledge.

——— . 1987. "Race and the Right: The Politics of Reaction." Pp. 145–177 in *Research in Social Policy: Historical and Contemporary Perspectives*, vol. 1, ed. John H. Stanfield II. Greenwich, Conn.: JAI Press.

Ong, Paul, and Suzanne Hee. 1992. *Losses in the Los Angeles Civil Unrest*. Los Angeles: Center for Pacific Rim Studies, University of California, Los Angeles.

Orfalea, Gregory. 1988. *Before the Flames: A Quest for the History of Arab Americans*. Austin: University of Texas Press.

Orsi, Robert Anthony. 1985. *The Madonna of 115th Street: Faith and Community in Italian Harlem, 1880–1950*. New Haven: Yale University Press.

Osofsky, Gilbert. 1966. *Harlem: The Making of a Ghetto, Negro New York, 1890–1930*. New York: Harper & Row.

Pacyga, Dominic. 1987. "Polish America in Transition: Social Change and the Chicago Polonia, 1945 to 1980." *Polish American Studies* 44: 38–55.

——— . 1989. "The Russell Square Community Committee: An Ethnic Response to Urban Problems." *Journal of Urban History* 15: 159–184.

——— . 1991. *Polish Immigrants and Industrial Chicago: Workers on the South Side, 1880–1922*. Columbus: Ohio State University Press.

Pagnini, Deanna L., and S. Philip Morgan. 1990. "Intermarriage and Social Distance among U. S. Immigrants at the Turn of the Century." *American Journal of Sociology* 96 (2): 405–432.

Palmer, David S. 1992. "Central America: Starting Over." *Hemisfile: Perspectives on Political and Economic Trends in the Americas* 3: 1–2.

Palmer, Dewey H. 1967. "Moving North: Migration of Negroes During World War I." *Phylon* 27: 29–42.

Park, In-Sook Han, James Fawcett, and Robert Gardner. 1990. "Korean Immigrants to the United States: A Pre-Departure Analysis." Paper no. 114, East-West Population Institute, East-West Center, University of Hawaii.

Park, Robert E. 1914. "Racial Assimilation in Secondary Groups." *American Journal of Sociology* 19: 606–623.

——— . 1922. *The Immigrant Press and its Control*. New York: .

——— . [1926] 1950. *Race and Culture*. Glencoe, Ill.: Free Press.

——— . 1928. "Human Migration and the Marginal Man." *American Journal of Sociology* 33: 881–893.

Park, Robert E., and Ernest W. Burgess. 1921. *Introduction to the Science of Sociology*. Chicago: University of Chicago Press.

Parsons, Talcott. 1955. "Social Strains in America." In *The New American Right*, ed. Daniel Bell. New York: Criterion.

Passell, Jeffrey S. 1994. "Immigrants and Taxes: A Reappraisal of Huddle's 'The Cost of Immigrants.'" PRIP-UI-29. Washington, D.C.: Urban Institute.

Passel, J. S., and K. A. Woodrow. 1984. "Geographic Distribution of Undocumented Aliens Counted in the 1980 Census by State." *International Migration Review* 18: 642–671.

Patterson, Orlando. 1982. *Slavery and Social Death: A Comparative Study*. Cambridge: Harvard University Press.

Paul, Rodman W. 1988. *The Far West and the Great Plains in Transition, 1859–1900*. New York: Harper & Row.

Peal, Elizabeth, and Wallace E. Lambert. 1962. "The Relation of Bilingualism to Intelligence." *Psychological Monographs* 76 (27): 1–23.

Pedraza, Silvia. 1991. "Women and Migration: The Social Consequences of Gender." *Annual Review of Sociology* 17: 303–325.

———. 1992. "Cubans in Exile, 1959–1989: The State of the Research." In *Cuban Studies Since the Revolution*, ed. Dami n J. Fernandez. Gainesville: University Presses of Florida.

———. 1994. "Ethnic Identity: Developing a Hispanic-American Identity." Paper presented at the meetings of the American Sociological Association, Los Angeles, California, August.

Pedraza, Silvia, and Lisa J. Neidert. 1990. "Discrimination vs. Exclusion: The Experience of Immigrants vs. Racial Minorities in America." Paper presented at the 85th annual meeting of the American Sociological Association, Washington, D.C., August 11–15, 1991.

Pedraza-Bailey, Silvia. 1983. "Reflections on a Revolution." *Washington University Magazine* 52: 12–16.

———. 1985. *Political and Economic Migrants in America: Cubans and Mexicans*. Austin: University of Texas Press.

———. 1990. "Immigration Research: A Conceptual Map." *Social Science History* 14 (spring): 43–67.

Pelling, Henry. 1960. *American Labor*. Chicago: University of Chicago Press.

———. 1968. *American Labor*. Chicago: University of Chicago Press.

People v. Zammora, 66 CA2d 166; 152 P2d 180 (1944).

Pérez, Glauco. 1981. "Dominican Illegals in New York: Selected Preliminary Findings." Paper presented to the Center for Inter-American Affairs, New York University.

Pérez, Lisandro. 1985. "The Cuban Population of the United States: The Results of the 1980 U.S. Census of Population." *Cuban Studies/Estudios Cubanos* 15: 1–18.

———. 1986. "Immigrant Economic Adjustment and Family Organization: The Cuban Success Story Reexamined." *International Migration Review* 20: 4–20.

———. 1988. "Cuban Women in the U.S. Labor Force: A Comment." *Cuban Studies* 18: 159–164.

Pérez Firmat, Gustavo. 1994. *Life on the Hyphen: The Cuban-American Way*. Austin: University of Texas Press.

Pernia, Ernesto M. 1976. "The Question of the Brain-drain from the Philippines." *International Migration Review* 10: 63–72.

Pernicone, Carol Groneman. 1973. "The 'Bloody Ould Sixth:' A Social Analysis of a New York City Working-Class Community in the Mid-Nineteenth Century." Ph.D. diss., University of Rochester.

Perrigo, Lynn I. 1971. *The American Southwest: Its People and Cultures*. Albuquerque: University of New Mexico.

Pessar, Patricia R. 1984. "The Linkage Between the Household and Workplace of Dominican Women in the U.S." *International Migration Review* 18: 1188–1211.

Petras, E. M. 1981. "The Global Labor Market in the Modern World." In *Global Trends in Migration: Theory and Research on International Population Movements*, ed. Mary M. Kritz et al. New York: Center for Migration Studies.

Pettigrew, Thomas F. 1985. "New Black-White Patterns: How Best to Conceptualize Them?" *Annual Review of Sociology* 11: 329–346.

Pfeffer, Max. 1994. "Low-Wage Employment and Ghetto Poverty: A Comparison of African-American and Cambodian Day-Haul Farm Workers in Philadelphia." *Social Problems* 41: 9–29.

Philippine Overseas Employment Administration. 1988. *Statistical Report*. Manila: Philippine Overseas Employment Administration.

Phillips, Bruce. 1991. "Sociological Analysis of Jewish Identity." Pp. 3–25 in *Jewish Identity in America*, ed. David M. Gordis and Yoav Ben-Horin. Los Angeles: Wilstein Institute.

Pido, Antonio J. A. 1986. *The Filipinos in America: Macro/Micro Dimensions of Immigration and Integration*. New York: Center for Migration Studies.

Piore, Michael J. 1979. *Birds of Passage: Migrant Labor and Industrial Societies*. Cambridge: Cambridge University Press.

Pitt, Leonard. 1966. *The Decline of the Californios: A Social History of the Spanish-Speaking Californians, 1846–1900*. Berkeley and Los Angeles: University of California Press.

Piven, Frances Fox, and Richard A. Cloward. 1979. *Poor People's Movements: Why They Succeed, How They Fail*. New York: Random.

———. 1993. *Regulating the Poor*. New York: Vintage Books.

Plyler v. Doe, 457 US 202 (1982).

Polednak, A. P. 1989. *Racial and Ethnic Differences in Disease*. New York: Oxford University Press.

Polenberg, Richard. 1980. *One Nation Divisible: Class, Race and Ethnicity in the U.S. Since 1938*. New York: Viking Penguin.

Portes, Alejandro. 1978. "Immigrant Aspirations." *Sociology of Education* 51: 241–260.

——— . 1981. "Modes of Structural Incorporation and Present Theories of Labor Immigration." Pp. 279–297 in *Global Trends in Migration: Theory and Research on International Population Movements*, ed. M. M. Kritz, C. B. Keely, and S. M. Tomasi. New York: Center for Migration Studies.

——— . 1982. "Immigrants' Attainment: An Analysis of Occupation and Earnings among Cuban Exiles in the United States." Pp. 91–111 in *Social Structure and Behavior: Essays in Honor of William Hamilton Sewell*, ed. R. M. Hauser et al. New York: Academic Press.

——— . 1984. "The Rise of Ethnicity: Determinants of Ethnic Perceptions among Cuban Exiles in the United States." *American Sociological Review* 49: 383–397.

——— . 1987 "The Social Origins of the Cuban Enclave Economy of Miami." *Sociological Perspectives* 30: 476–485.

——— , ed. 1994. "The New Second Generation." *International Migration Review* 28: 4.

Portes, Alejandro, and Robert L. Bach. 1985. *Latin Journey: Cuban and Mexican Immigrants in the United States*. Berkeley and Los Angeles: University of California Press.

Portes, Alejandro, Juan M. Clark, and Robert L. Bach. 1977. "The New Wave: A Statistical Profile of Recent Cuban Exiles to the United States." *Cuban Studies* 7: 1–32.

Portes, Alejandro, and Luis Guarnizo. 1991. "Tropical Capitalists: U.S. Bound Immigration and Small Enterprise Development in the Dominican Republic." In *Migration, Remittances, and Small Business Development, Mexico and Caribbean Basin Countries*, ed. Sergio Díaz-Briquets and Sydney Weintraub. Boulder, Colo.: Westview Press.

Portes, Alejandro, and Robert D. Manning. 1986. "The Immigrant Enclave: Theory and Empirical Examples." Pp. 47–68 in *Competitive Ethnic Relations*, ed. Susan Olzak and Joane Nagel. Orlando, Fla.: Academic Press.

Portes, Alejandro, and Rubén G. Rumbaut. 1990. *Immigrant America: A Portrait*. Berkeley and Los Angeles: University of California Press.

Portes, Alejandro, and Alex Stepick. 1993. *City on the Edge: The Transformation of Miami*. Berkeley and Los Angeles: University of California Press.

Portes, Alejandro, and John Walton. 1981. *Labor, Class and the International System*. New York: Academic Press.

Portes, Alejandro, and Min Zhou. 1993. "The New Second Generation: Segmented Assimilation and its Variants among Post-1965 Immigrant Youth." *Annals of the American Academy of Political and Social Sciences* 530 (November): 74–96.

Posadas, Barbara M. 1986–1987. "Editorial Forum: A View from the Interior." *Amerasia* 13: xi–xiii.

Pratt, John W. 1961. "Governor Seward and the New York City School Controversy, 1840–1842: A Milestone in the Advance of Nonsectarian Public Education." *New York History* 42: 351–364.

Price, James A. 1968. "The Migration and Adaptation of American Indians to Los Angeles." *Human Organization* 27: 168–175.

Prieto, Yolanda. 1987. "Cuban Women in the U.S. Labor Force: Perspectives on the Nature of the Change." *Cuban Studies* 17: 73–94.

Prucha, Francis Paul. 1984. *The Great Father.* Lincoln: University of Nebraska Press.

Pula, James S., and Eugene E. Dziedzic. 1990. *United We Stand: The Role of Polish Workers in the New York Mills Textile Strikes, 1912 and 1916*. Boulder, Colo.: East European Monographs.

Puzo, Mario. 1981. "Choosing a Dream: Italians in Hell's Kitchen." In *The Immigrant Experience: The Anguish of Becoming American*, ed. Thomas G. Wheeler. New York: Penguin.

Quinn, David B. 1991. *Ireland and America: Their Early Associations: 1500–1640*. Liverpool, U.K.: Liverpool University Press.

Rabinowitz, Howard. 1978. *Race Relations in the Urban South 1865–1890*. New York: Oxford University Press.

Rabow, J., and R. Watt. 1982. "Alcohol Availability, Alcohol Beverage Sales and Alcohol-Related Problems." *Journal of Studies on Alcohol* 43: 767–801.

Raines, Howell. 1977. *My Soul Is Rested*. New York: Bantam Books.

——— . 1991. "Grady's Gift." *New York Times Magazine*, 1 December.

Ramenofsky, Ann F. 1987. *Vectors of Death: The Archaeology of European Contact*. Albuquerque: University of New Mexico Press.

Ramírez, Nelson, Pablotactuk, and Minerva Breton. 1977. *La migración interna en la República Dominica*. Santo Domingo: Alfa y Omega.

Ransom, Roger, and Richard Sutch. 1977. *One Kind of Freedom*. New York: Cambridge University Press.

Rawick, George P., ed. 1972a. *The American Slave: A Composite Autobiography*. 19 vols. Westport, Conn.: Greenwood Press.

——— . 1972b. *From Sundown to Sunup: The Making of the Black Community*. Vol. 1 of *The American Slave: A Composite Autobiography*, ed. George P. Rawick. Westport, Conn.: Greenwood Press.

Rea, Louis M., and Richard A. Parker. 1992. "A Fiscal Impact Analysis of Undocumented Immigrants Residing in San Diego County." Mimeo. San Diego: Rea & Parker, Inc.

Reed, Wornie L. 1992. "Lead Poisoning: A Modern Plague Among African American Childrren." In *Health Issues in the Black Community*, ed. Ronald L. Braithwaite and Sandra E. Taylor. San Francisco: Jossey-Bass.

Reimers, David M. 1985. *Still the Golden Door: The Third World Comes to America*. New York: Columbia University Press.

————. Forthcoming. "New York City's Irish: 1945-present." In *A History of the Irish in New York City*, ed. Ronald Bayor and Tim Meager. Baltimore: Johns Hopkins University Press.

Reinhold, Robert. 1991. "In California, New Discussion on Whether to Bar the Door." *New York Times*, 3 December, A1.

Rendon, Armando. 1971. *Chicano Manifesto*. New York: Macmillan.

Reskin, Barbara, and Patricia A. Roos. 1990. *Job Queues, Gender Queues: Explaining Women's Inroads into Male Occupations*. Philadelphia: Temple University Press.

Reynolds, William Bradford. 1983. "The Justice Department's Enforcement of Title VII." *Labor Law Journal* 34: 259–265.

Richardson, James and J. A. Farrell. 1983. "The New Indian Wars." *Denver Post*. Special reprint November 20–27, Denver, CO.

Rieder, Jonathan. 1985. *Canarsie: The Jews and Italians of Brooklyn Against Liberalism*. Cambridge: Harvard University Press.

Rieff, David. 1987a. *Going to Miami: Exiles, Tourists, and Refugees in the New America*. Boston: Little, Brown.

————. 1987b. "A Reporter at Large: The Second Havana." *New Yorker* (May 18): 65–83.

————. 1991. *Los Angeles: Capital of the Third World*. New York: Simon & Schuster.

————. 1993. *The Exile: Cuba in the Heart of Miami*. New York: Simon & Schuster.

Ries, Peter W. 1990. *Health of Black and White Americans, 1985–87*. Data from the National Health Interview Survey 10(171), PHS 90–1599. Washington, D.C.: U.S. Department of Health and Human Services.

Rinder, Irwin. 1959. "Strangers in the Land: Social Relations in the Status Gap." *Social Problems* 8: 253–261.

Rischin, Moses. 1962. *The Promised City: New York's Jews, 1870–1914*. Cambridge: Harvard University Press.

Rivas-Porta, Guillermo. 1994. "El Pueblo Cubano: Protagonista, Víctima, y Espectador." *Desafíos* 1 (Agosto/Septiembre): 4–5.

Robins, L. N., and D. A. Regier, eds. 1991. *Psychiatric Disorders in America: The Epidemiologic Catchment Area Study*. New York: Free Press.

Robinson, Alfred. 1846. *Life in California*. New York: Peregrine Press.

Robinson, James. 1984. "Racial Inequality and the Probability of Occupation-related Injury or Illness." *Milbank Memorial Fund Quarterly* 62: 567–590.

Roca, Sergio. 1977. "Cuban Economic Policy in the 1970s: The Trodden Paths." Pp. 83–118 in *Cuban Communism*, ed. Irving Louis Horowitz. New Brunswick, N.J.: Transaction Books.

Rodgers, Mary Beth. 1990. *Cold Anger: A Story of Faith and Power Politics*. Denton: University of North Texas Press.

Rodríguez, Clara E. 1989. *Puerto Ricans: Born in the U.S.A.* Boston: Unwin Hyman.

Rodríguez, Clara E. 1991. *Puerto Ricans: Born in the U.S.A.* Boulder, Colo.: Westview Press.

Rodríguez-Chavez, Ernesto. 1993. "Tendencias Actuales del Flujo Migratorio Cubano." *Cuadernos de Nuestra América* 10: 114–137.

Roediger, David R. 1991. *The Wages of Whiteness: Race and the Making of the American Working Class*. New York: Verso.

Rogler, Lloyd H., and Rosemary Santana-Cooney. 1984. *Puerto Rican Families in New York: Intergenerational Processes*. Maplewood, N.J.: Waterfront Press.

Romano-V, Ocatavio I. 1968. "The Anthropology and Sociology of the Mexican Americans: The Distortion of Mexican American History." In *El Grito*. Berkeley, Calif.: Quinto Sol Publishing

Romo, Harriet. 1990. *Latinos and Blacks in the Cities: Policies for the 1990s*. Austin: LBJ Library and the LBJ School of Public Affairs.

Romo, Ricardo. 1977. "The Urbanization of Southwestern Chicanos in the Early Twentieth Century." In *New Directions in Chicano Scholarship*, ed. Ricardo Romo and Raymund Paredes. San Diego: University of California, San Diego.

————. l983. *East Los Angeles: History of a Barrio*. Austin: University of Texas Press.

————. 1986. "George I. Sánchez and the Civil Rights Movement." *La Raza Law Journal* 1: 342–362.

————. 1990. "Southern California and the Origins of Latino Civil-Rights Activism." *Western Legal History* 3 (2): 379–406.

Rose, Peter I. 1981. "Some Thoughts about Refugees and the Descendants of Theseus." *International Migration Review* 15: 8–15.

————. 1993. "Tempest-Tost: Exile, Ethnicity, and the Politics of Rescue." *Sociological Forum* 8: 5–24.

Rosenbaum, Robert J. 1981. *Mexicano Resistance in the Southwest: The Sacred Right of Self-Preservation*. Austin: University of Texas Press.

Rosenblum, Gerald. 1973. *Immigrant Workers*. New York: Basic Books.

Rosenwaike, Ira. 1972. *Population History of New York City*. Syracuse, N.Y.: Syracuse University Press.

Roskolenko, Harry. 1981. "America, the Thief: A Jewish Search for Freedom." In *The Immigrant Experience: The Anguish of Becoming American*, ed. Thomas G. Wheeler. New York: Penguin.

Roszak, Theodore. 1969. *The Making of a Counter Culture*. New York: Anchor Books.

Rotondaro, Alfred. 1991. "Ethnicity at work." *Altreitalie* 6: 119–123.

Roy, Prodipto. 1962. "The Measurement of Assimilation: The Spokane Indians." *American Journal of Sociology* 67: 641–651.

Royce, Josiah. [1887] 1970. *California: From the Conquest in 1846 to the Second Vigilance Committee in San Francisco*. Santa Barbara, Calif.: Peregrine.

Rumbaut, Rubén G. 1985. "Mental Health and the Refugee Experience: A Comparative Study of Southeast Asian Refugees." Pp. 433–486 in *Southeast Asian Mental Health*, ed. Tom C. Owan. Rockville, Md.: National Institute of Mental Health.

——. 1989a. "Portraits, Patterns and Predictors of the Refugee Adaptation Process." Pp. 138–182 in *Refugees as Immigrants: Cambodians, Laotians and Vietnamese in America*, ed. David W. Haines. Totowa, N.J.: Rowman & Littlefield.

——. 1989b. "The Structure of Refuge: Southeast Asian Refugees in the United States, 1975–1985." *International Review of Comparative Public Policy* 1: 97–129.

——. 1991a. "The Agony of Exile: A Study of the Migration and Adaptation of Indochinese Refugee Adults and Children." Pp. 53–91 in *Refugee Children: Theory, Research, and Practice*, ed. Frederick L. Ahearn, Jr., and Jean L. Athey. Baltimore: Johns Hopkins University Press.

——. 1991b. "Migration, Adaptation, and Mental Health." Pp. 383–427 in *Refugee Policy: Canada and the United States*, ed. Howard Adelman. Toronto: York Lanes Press.

——. 1991c. "Passages to America: Perspectives on the New Immigration." Pp. 208–244 in *America at Century's End*, ed. Alan Wolfe. Berkeley and Los Angeles: University of California Press.

——. 1992. "The Americans: Latin American and Caribbean Peoples in the United States." Pp. 275–307 in *Americas: New Interpretive Essays*, ed. Alfred Stefan. New York: Oxford University Press.

Rumbaut, Rubén G., Leo R. Chavez, Robert J. Moser, Sheila M. Pickwell, and Samuel M. Wishik. 1988. "The Politics of Migrant Health Care: A Comparative Study of Mexican Immigrants and Indochinese Refugees." *Research in the Sociology of Health Care* 7: 143–202.

Rumbaut, Rubén G., and Wayne A. Cornelius, eds. 1995. *Children of Immigrants: Theory, Research, and Implications for Educational Policy*. La Jolla, Calif.: Center for U.S.DMexican Studies, University of California, San Diego.

Rumbaut, Rubén G., and Kenji Ima. 1988a. *The Adaptation of Southeast Asian Refugee Youth: A Comparative Study*. Washington, D.C.: U.S. Office of Refugee Resettlement.

——. 1988b. "Determinants of Educational Attainment Among Indochinese Refugees and Other Immigrant Students." Paper presented at the annual meeting of the American Sociological Association, Atlanta, Georgia, August.

Rumbaut, Rubén G., and John R. Weeks. 1986. "Fertility and Adaptation: Indochinese Refugees in the United States." *International Migration Review* 20 (2): 428–466.

——. 1989. "Infant Health Among Indochinese Refugees: Patterns of Infant Mortality, Birthweight, and Prenatal Care in Comparative Perspective." *Research in the Sociology of Health Care* 8: 137–196.

Ryan, William. 1971. *Blaming the Victim*. New York: Vintage Books.

Sabagh, Georges. 1991. "Los Angeles, A World of New Immigrants: An Image of Things to Come?" Pp. 79–105 in *Migration Policies in Europe and the United States*, ed. Giacomo Luciani. Dordrecht, The Netherlands: Kluwer Academic.

Sachar, Howard. 1992. *A History of the Jews in America*. New York: Knopf.

Sacred Heart Parish. 1960. *Sacred Heart of Jesus Parish Golden Jubilee Book, 1910–1960*. Chicago: Sacred Heart Parish.

Sae Gae Times. 1989. "Koreans Support Dinkins." 23 August, 1.

Said, Edward W. 1979. *Orientalism*. New York: Vintage Books.

Samora, Julian. 1971. *Los Mojados: The Wetback Story*. Notre Dame: University of Notre Dame Press.

Sánchez, David. 1978. *Expeditions Through Aztlan*. La Puente, Calif.: Perspectiva Press.

Sánchez-Korrol, Virginia 1981. *From Colonia to Community: The History of Puerto Ricans in New York City, 1917–1948*. Westport, Conn.: Greenwood Press.

Sando, Joe S. 1976. *The Pueblo Indians*. San Francisco: Indian Historian Press.

San Miguel, Guadalupe, Jr. 1987. *Let All of Them Take Heed: Mexican Americans and the Campaign for Educational Equality in Texas, 1910–1981*. Austin: University of Texas Press.

Santana, Ray, and Mario Esparza. 1974. "East Los Angeles Blowouts." Pp. 1–9 in *Parameters of Institutional Change*, ed. Southwest Network. Hayward, Calif.: Southwest Network.

Sarachek, Bernard. 1980. "Jewish American Entrepreneurs." *Journal of Economic History* XL (2): 359–372.

Sassen, Saskia. 1988. *The Mobility of Labor and Capital: A Study in International Investment and Labor Flow*. Cambridge: Cambridge University Press.

——. 1990. "Finance and Business Services in New York City: International Linkages and Domestic Effects." *International Social Science Journal* 42: 287–306.

Sassen-Koob, Saskia. 1979. "Formal and Informal Associations: Dominicans and Colombians in New York." *International Migration Review* 13: 319–332.

——. 1984. "Notes on the Incorporation of Third World Women into Wage-Labor Through Immigration and Off-Shore Production." *International Migration Review* 18: 1144–1167.

Satterfield, David. 1987. "Growth Is on Rise; Focus Is in Dade." Special Report: Outlook '87, *Miami Herald*, 19 January, 29–31.

Schecter, Arnold, et al. 1995. "Agent Orange and the Vietnamese: The Persistence of Elevated Dioxin Levels in Human Tissues." *American Journal of Public Health* 85 (4): 516–522.

Scheiner, Seth M. 1965. *Negro Mecca: A History of the Negro in New York City, 1865–1920*. New York: New York University Press.

Schneiderman, Howard G. 1989. "Presidents, Privilege, and Performance." *Society* 26 (3).

———, ed. 1991. *The Protestant Establishment Revisited*. New Brunswick, N.J.: Transaction Books.

———. 1992a. "Out of the Golden Ghetto." *Society* March/April 29 (3).

———. 1992b. "Religious Regimes and Elite Cultures." Pp. 131–153 in *Faith and Polity*, ed. Mart Bax et al. Amsterdam: VU University Press.

———. 1993. "Hindrances to Good Citizenship." Introduction to James Bryce, *Hindrances to Good Citizenship*. New Brunswick, N.J.: Transaction Books.

Schoenberger, Karl. 1993. "Breathing Life Into Southland: Chinese Immigrants Boost Economy." *Los Angeles Times*, 4 October, A1.

Schoultz, Lars. 1992. "Central America and the Politicization of U.S. Immigration Policy." Pp. 157–220 in *Western Hemisphere Immigration and United States Foreign Policy*, ed. Christopher Mitchell. University Park: Pennsylvania State University Press.

Schrag, Peter. 1970. *The Decline of the WASP*. New York: Simon & Schuster.

Schreiber, Anna P. 1973. "Economic Coercion as an Instrument of Foreign Policy: U.S. Economic Measures against Cuba and the Dominican Republic." *World Politics* 25: 387–413.

Schreiber, E. M. 1978. "Education and Change in American Opinions on a Woman for President." *Public Opinion Quarterly* 42: 171–182.

Schulke, Flip, ed. 1976. *Martin Luther King, Jr.: A Documentary . . . Montgomery to Memphis*. New York: W. W. Norton.

Schulweis, Harold M. 1991. "The Role of the Synagogue in Jewish Identity." Pp. 159–165 in *Jewish Identity in America*, ed. David M. Gordis and Yoav Ben-Horin. Los Angeles: Wilstein Institute.

Schulz, Amy, and Howard Schuman. 1990. "A Survey of Racial Attitudes and Experience of LS&A Students at the University of Michigan." Unpublished report.

Schuman, Howard. 1972. "Attitudes vs. Actions versus Attitudes vs. Attitudes." *Public Opinion Quarterly* 36: 347–354.

Schuman, Howard, and Lawrence Bobo. 1988. "Survey-based Experiments on White Racial Attitudes toward Residential Integration." *American Journal of Sociology* 94: 273–299.

Schuman, Howard, Charlotte Steeh, and Lawrence Bobo. 1985. *Racial Attitudes in America: Trends and Interpretations*. Cambridge: Harvard University Press.

Schwartz, Anita. 1988. "The Secular Seder: Continuity and Change among Left-Wing Jews." Pp. 105–127 in *Between Two Worlds: Ethnographic Essays on American Jewry*, ed. Jack Kugelmass. Ithaca, N.Y.: Cornell University Press.

Schwartz, Michael. 1976. *Radical Protest and Social Structure: The Southern Farmers' Alliance and Cotton Tenancy, 1880–1890*. New York: Academic Press.

Scott, Allen J. 1990. "The Technopoles of Southern California." *Environment and Planning* 22: 1575–1605.

Scott, Emmett J. 1919a. "Additional letters of Negro Migrants of 1916–1918." *Journal of Negro History* 4: 412–465.

———. 1919b. "Letters of Negro Migrants of 1916–1918." *Journal of Negro History* 4: 290–390.

———. 1920. *Negro Migration During the War*. New York: Oxford University Press.

Scott, Kenneth. 1961. "The Slave Insurrection in New York City, 1712." *New York Historical Quarterly* 45: 43–74.

Seavoy, Ronald E. 1986. *Famine in Peasant Societies*. Westport, Conn.: Greenwood Press.

Seeman, Melvin, and Carolyn S. Anderson. 1983. "Alienation and Alcohol: The Role of Work, Mastery, and Community in Drinking Behavior." *American Sociological Review* 48: 60–77.

Segura, Denise A. 1989. "Chicana and Mexican Immigrant Women at Work: The Impact of Class, Race, and Gender on Occupational Mobility." *Gender and Society* 3: 37–52.

Seidler-Feller, Chaim. 1991. "Response to Perry London and Allissa Hirchfeld." Pp. 61–65 in *Jewish Identity in America*, ed. David M. Gordis and Yoav Ben-Horin. Los Angeles: Wilstein Institute.

Shea, John Gilmary. 1886–1892. *The History of the Catholic Church in the United States*. 4 vols. New York: J. G. Shea.

Sheatsley, Paul B. 1966. "White Attitudes Toward the Negro." *Daedalus* 95: 117–238.

Shockley, John Staples. 1973. "Crystal City: Los Cinco Mexicanos." Pp. 303–313 in *Chicano: The Evolution of a People*, ed. R. Rosaldo et al. Minnesota: Winston Press.

———. 1974. *Chicano Revolt in a Texas Town*. Notre Dame, Ind.: University of Notre Dame Press.

Sibley-Butler, John. 1991. *Entrepreneurship and Self-Help Among Black Americans*. Albany: State University of New York Press.

Sigelman, Lee, and Susan Welch. 1991. *Black Americans Views' of Racial Inequality: The Dream Deferred*. New York: Cambridge University Press.

Silvestri, George, and John Lukasiewicz. 1991. "Outlook: 1990–2005, Occupational Employment Projections." *Monthly Labor Review* (November): 64–94.

Simkins, Francis Butler. 1947. *The South Old and New 1820–1947*. New York: Knopf.

Simmel, Georg. [1908] 1971. "The Stranger." Pp. 143–149 in *Georg Simmel: On Individuality and Social Forms*, ed. Donald N. Levine. Chicago: University of Chicago Press.

Simon, Julian L. 1989. *The Economic Consequences of Immigration*. Cambridge, Mass.: Basil Blackwell.

Simon, Rita J. 1985. *Public Opinion and the Immigrant*. Lexington, Mass.: Lexington Books.

Singer, Merril. 1986. "Toward a Political Economy of Alcoholism." *Social Science and Medicine* 23: 113–130.

Sitkoff, Harvard. 1981. *The Struggle for Black Equality.* New York: Hill and Wang.

Siu, Peter C. T. 1952. "The Sojourners." *American Journal of Sociology* 8: 32–44.

Skocpol, Theda. 1979. *States and Social Revolution.* London: Cambridge University Press.

———. 1994. "The New Urban Poverty and U.S. Social Policy." *Michigan Quarterly Review* 33: 274–281.

Slayton, Robert. 1986. *Back of the Yards: The Making of a Local Democracy.* Chicago: University of Chicago Press.

Sleeper, Jim. 1990. *The Closest of Strangers: Liberalism and the Politics and Race in New York.* New York: W. W. Norton.

Smith, A. Wade. 1981. "Racial Tolerance as a Function of Group Position." *American Sociological Review* 46: 558–573.

Smith, Abbot E. 1947. *Colonists in Bondage: White Servitude and Convict Labor in America, 1607–1776.* Chapel Hill: University of North Carolina Press.

Smith, Marvin T. 1984. "Depopulation and Culture Change in the Early Historic Period of the Interior Southeast." Ph.D. diss., University of Florida.

Smith, Peter C. 1976. "The Social Demography of Filipinos' Migrations Abroad." *International Migration Review* 10: 307–351.

Sniderman, Paul M., Philip E. Tetlock, and Edward G. Carmines, eds. 1993. *Prejudice, Politics, and the American Dilemma.* Stanford, Calif.: Stanford University Press.

Snipp, C. Matthew. 1988. "Public Policy Impacts on American Indian Economic Development." Pp. 1–22 in *Public Policy Impacts on American Indian Economic Development,* ed. C. Matthew Snipp. Albuquerque, NM: Insitute for Native American Development, University of New Mexico.

———. 1989. *American Indians: The First of this Land.* New York: Russell Sage.

Snipp, C. Matthew, and Gary Sandefur. 1988. "Earnings of American Indians and Alaska Natives: The Effects of Residence and Migration." *Social Forces* 66: 994–1008.

Snipp, C. Matthew, and Gene F. Summers. 1991. "American Indian Development Policies." Pp. 166–180 in *Rural Policies for the 1990s,* ed. Cornelia B. Flora and James A. Christenson. Boulder, Colo.: Westview Press.

Snow, Dean, and Kim M. Lanphear. 1988. "European Contact and Indian Depopulation in the Northeast: The Timing of the First Epidemics." *Ethnohistory* 35: 15–33.

Soja, Edward, Rebecca Morales, and Goetz Wolff. 1983. "Urban Restructuring: An Analysis of Social and Spatial Change in Los Angeles." *Economic Geography* 59: 195–230.

Sollors, Werner. 1992. "'Of Plymouth Rock and Jamestown and Ellis Island' or, Ethnic Literature and Some Redefinitions of 'America'." Pp. 272–311 in *Multiculturalism and the Canon of American Culture,* ed. Hans Bak. Amsterdam: VU Press.

Sori, Ercole. 1979. *L'Emigrazione Italiana dall' Unità alla Seconda Guerra Mondiale.* Bologna: Il Mulino.

Sorin, Gerald. 1990. *The Nurturing Neighborhood: The Brownsville Boys Club and Jewish Community in Urban America, 1940–1990.* New York: New York University Press.

Sorkin, Alan L. 1978. *The Urban American Indian.* Lexington, Mass.: Lexington Books.

Southwest Voter Research Institute. 1991. *1988–1989 Biennial Report.* San Antonio, Tex.: Southwest Voter Research Institute.

Sowell, Thomas P. 1981. *Ethnic America.* New York: Basic Books.

———. 1983. *The Economics and Politics of Race: An International Perspective.* New York: Quill.

Spann, Edward K. 1981. *The New Metropolis: New York City, 1840–1857.* New York: Columbia University Press.

Spear, Allan H. 1969. *Black Chicago: The Making of a Negro Ghetto, 1890–1920.* Chicago: University of Chicago Press.

Specter, Michael. 1991. "Illegal Child Labor Resurging in U.S." *Washington Post,* 14 April.

Spero, Sterling D., and Abram L. Harris. 1931. *The Black Worker.* New York: Columbia University Press.

Spicer, Edward H. 1969. *A Short History of the Indians of the United States.* New York: Van Nostrand.

Spickard, Paul R. 1989. *Mixed Blood.* Madison: University of Wisconsin Press.

Stack, Carol B. 1974. *All Our Kin: Strategies for Survival in a Black Community.* New York: Harper & Row.

Stampp, Kenneth M. 1956. *The Peculiar Institution: Slavery in the Ante-Bellum South.* New York: Knopf.

———. 1964. *The Peculiar Institution: Slavery in the Ante-Bellum South.* New York: Vintage Books.

Staples, Robert. 1993. "The Illusion of Racial Equality: The Black American Dilemma." In Gerald Early, ed., *Lure and Loathing: Essays on Race, Identity, and the Ambivalence of Assimilation.* New York: Penguin Books, pp. 227–244.

Stauss, Joseph A., and Bruce A. Chadwick. 1979. "Urban Indian Adjustment." *American Indian Culture and Research Journal* 3: 23–38.

Steeh, Charlotte, and Howard Schuman. 1992. "Young White Adults: Did Racial Attitudes Change in the 1980s?" *American Journal of Sociology* 98: 340–367.

Stein, Barry N. 1981. "The Refugee Experience: Defining the Parameters of a Field of Study." *International Migration Review* 15: 320–330.

Steinberg, Stephen. 1981. *The Ethnic Myth: Race, Ethnicity, and Class in America.* Boston: Beacon Press.

Steiner, Stan. 1968. *The New Indians.* New York: Dell.

———. 1970. *La Raza: The Mexican Americans.* New York: Harper & Row.

Stember, Charles Herbert, and others. 1966. *Jews in the Mind of America*. New York: Basic Books.

Stevens, Alden. 1952. "Voice of the Native: Arizona's and New Mexico's redskins could swing the election in those two states." *New York Times*, 2 November, 9.

Stevens, Evelyn P. 1973. "*Marianismo*: The Other Face of *Machismo* in Latin America." Pp. 89–102 in *Female and Male in Latin America*, ed. Ann Pescatello. Pittsburgh: University of Pittsburgh Press.

Stevens, Gillian. 1992. "The Social and Demographic Context of Language Use in the U.S." *American Sociological Review* 57: 171–185.

Stewart, Sally Ann. 1993. "New Page in English-Only Debate." *USA Today*, 25 June, 8A.

Stinchcombe, Arthur L. 1984. "The Origins of Sociology as a Discipline." *Acta Sociologica* 27: 51–61.

Stockton, Ronald R. 1994. "Recognize the Benefits from Our Arab Neighbors," *Detroit News*, 3 April, 12A.

Stone, Russell H. 1974. "Religious Ethic and the Spirit of Capitalism in Tunisia." *International Journal of Middle East Studies* 5: 260–273.

Stouffer, Samuel. 1955. *Communism, Conformity, and Civil Liberties*. New York: Doubleday.

Strong, Edward K., Jr. 1933. *Japanese in California*. Stanford, Calif.: Stanford University Press.

Sturtevant, William C., ed. 1978. *Handbook of North American Indians*. 20 vols. Washington, D.C.: Smithsonian Institution Press.

Sullivan, Louis W. 1991. "Effects of Discrimination and Racism on Access to Health Care." *Journal of the American Medical Association* 266: 2674.

Summers, Gene F. n.d. "Social Characteristics of Reservations." Mimeo. Department of Rural Sociology, University of Wisconsin, Madison.

Sunoo, Paik Brenda. 1992. "Out of Ashes, Solidarity." *Korean Times Los Angeles* (English ed.), 11 May, 1.

Suro, Roberto. 1994. *Remembering the American Dream: Hispanic Immigration and National Policy*. New York: Twentieth Century Fund.

Suttles, Gerald D. 1968. *The Social Order of the Slum*. Chicago: University of Chicago Press.

Symmes, Patrick. 1992. "Under the Bridge . . . Francisco, Umberto, and the Nguyens Longing for Home and Waiting for the Fish to Bite." *Washington City Paper* 12: 22–32.

Szapocznik, José, and Roberto Hernández. 1988. "The Cuban American Family." Pp. 160–172 in *Ethnic Families in America*, 3rd ed., ed. Charles H. Mindel, Robert W. Habenstein, and Roosevelt Wright, Jr. New York: Elsevier.

Szapocznik, José, Mercedes A. Scopetta, and Wayne Tillman. 1978. "What Changes, What Remains the Same, and What Affects Acculturative Change in Cuban Immigrant Families." Pp. 35–49 in *Cuban Americans: Acculturation, Adjustment and the Family*, ed. José Szapocznik and María Cristina Herrera. Washington, D.C.: National Coalition of Hispanic Mental Health and Human Services Organization.

Szasz, Margaret C. 1988. *Indian Education in the American Colonies, 1607–1783*. Albuquerque, NM: University of New Mexico Press.

Tabori, Paul. 1972. *The Anatomy of Exile*. London: Harrap.

Taeuber, Karl E., and Alma F. Taeuber. 1965. *Negroes in Cities: Residential Segregation and Neighborhood Change*. Chicago: Aldine.

Takaki, Ronald. 1989. *Strangers From a Different Shore: A History of Asian Americans*. Boston: Little, Brown.

———. 1993. *A Different Mirror: A History of Multicultural America*. Boston: Little, Brown.

Tate, Katherine. 1992. "The Impact of Jesse Jackson's Presidential Bids on Blacks' Relationship with the Democratic Party." *National Political Science Review* 3: 184–197.

Taylor, Graham D. 1980. *The New Deal and American Indian Tribalism: The Administration of the Indian Reorganization Act, 1934–1945*. Lincoln: University of Nebraska Press.

Taylor, Jerome, and Beryl Jackson. 1990. "Factors Affecting Alcohol Consumption in Black Women. Part II." *International Journal of Addictions* 25 (12): 1415–1427.

Taylor, Paul S. [1928] 1971. *An American-Mexican Frontier: Nueces County, Texas*. New York: Russell & Russell.

Taylor, Phillip. 1971. *The Distant Magnet: European Emigration to the U.S.A.* New York: Harper & Row.

Taylor, R. J., and L. M. Chatters. 1988. "Church Members as a Source of Informal Social Support." *Review of Religious Research* 30: 193–203.

Taylor, Ronald B. l975. *Chavez and the Farm Workers*. Boston: Beacon Press.

Thernstrom, Stephan. 1964. *Poverty and Progress: Social Mobility in a Nineteenth Century City*. Cambridge: Harvard University Press.

———. 1973. *The Other Bostonians: Poverty and Progress in the American Metropolis, 1880–1970*. Cambridge: Harvard University Press.

Thomas, Hugh. 1977. *The Cuban Revolution*. New York: Harper & Row.

Thomas, John F. 1967. "Cuban Refugees in the United States." *International Migration Review* 2: 46–57.

Thomas, Melvin, and Hayward Derrick Horton. 1992. "Race, Class and Family Structure: The Case of Family Income, 1968–1988." *Sociological Perspectives* 35: 433–450.

Thomas, Melvin, and Michael Hughes. 1986. "The Continuing Significance of Race: A Study of Race, Class, and Quality of Life in America, 1972–1985." *American Sociological Review* 51: 830–841.

Thomas, Piri. 1967. *Down These Mean Streets*. New York: Knopf.

Thomas, Robert K., 1965. "Pan-Indianism." *Mid-Continent America Studies Journal* 6: 75–83.

Thomas, William I., and Florian Znaniecki. 1919. *The Polish Peasant in Europe and America*. 5 vols. Chicago: University of Chicago Press.

Thomas, W. I., and Florian Znaniecki. 1920. *The Polish Peasant in Europe and America*. Boston: Gorham Press.

Thomas, W. I., and Florian Znaniecki. 1927. *The Polish Peasant in Europe and America*. 2 vols. New York: Knopf.

Thornton, Russell. 1981. "Demographic Antecedents of a Revitalization Movement: Population Change, Population Size and the 1890 Ghost Dance." *American Sociological Review* 46: 88–96.

———. 1984a. "Cherokee Population Losses During the 'Trail of Tears:' A New Perspective and a New Estimate." *Ethnohistory* 31: 289–300.

———. 1984b. "Social Organization and the Demographic Survival of the Tolowa." *Ethnohistory* 31: 187–196.

———. 1986a. "History, Structure and Survival: A Comparison of the Yuki (Unkomno'n) and Tolowa (Hush) Indians of Northern California." *Ethnology* 25: 119–130.

———. 1986b. *We Shall Live Again: The 1870 and 1890 Ghost Dance Movements as Demographic Revitalization*. New York: Cambridge University Press.

———. 1987. *American Indian Holocaust and Survival: A Population History Since 1492*. Norman: University of Oklahoma Press.

———. 1990. *The Cherokees: A Population History*. Lincoln: University of Nebraska Press.

———. 1994a. "Population." Pp. 461–464 in *Encyclopedia of Native Americans in the 20th Century*. New York: Garland.

———. 1994b. "Urbanization." Pp. 670–672 in *Encyclopedia of Native Americans in the 20th Century*. New York: Garland.

Thornton, Russell, and Joan Marsh-Thornton. 1981. "Estimating Prehistoric American Indian Population Size for United States Area: Implications of the Nineteenth Century Population Decline and Nadir." *American Journal of Physical Anthropology* 55: 47–53.

Thornton, Russell, Tim Miller, and Jonathan Warren. 1991. "American Indian Population Recovery following Smallpox Epidemics." *American Anthropologist* 93: 28–45.

Thornton, Russell, Gary D. Sandefur, and Harold Grasmick. 1982. *The Urbanization of American Indians*. Bloomington, IN: Indiana University Press.

Thornton, Russell, Jonathan Warren, and Tim Miller. 1992. "Depopulation in the Southeast after 1492." Pp. 187–195 in *Disease and Demographics in the Americas*, ed. John W. Verano and Douglas H. Ubelaker. Washington, D.C.: Smithsonian Institution Press.

Tienda, Marta. 1980. "Familial and Structural Assimilation of Mexican Immigrants in the United States." *International Migration Review* 14: 383–408.

———. 1989. "Puerto Ricans and the Underclass Debate." *Annals of the American Academy of Political and Social Science* 501 (January): 105–119.

Tienda, Marta, and Katharine Donato. 1993. "Changes in the Employability of Minority Men, 1960–1980." Paper presented at the annual meetings of the American Association for the Advancement of Science, Boston, February.

Tienda, Marta, Katharine Donato, and Héctor Cordero-Guzmán. 1992. "Schooling, Color and Labor Force Activity of Women." *Social Forces* 71 (2): 365–395.

Tienda, Marta, and Jennifer Glass. 1985. "Household Structure and Labor Force Participation of Black, Hispanic and White Mothers." *Demography* 22 (3): 381–394.

Tienda, Marta, and Leif Jensen. 1988. "Poverty and Minorities: A Quarter Century Profile of Color and Socioeconomic Disadvantage." Pp. 23–61 in *Divided Opportunities*, ed. Gary D. Sandefur and Marta Tienda. New York: Plenum Publishers.

Tienda, Marta, Leif Jensen, and Robert Bach. 1984. "Immigration, Gender and the Process of Occupational Change in the United States, 1970–80." *International Migration Review* 18: 1021–1044.

Tienda, Marta, and Zai Liang. 1993. "Poverty and Immigration in Policy Perspective." In *Poverty and Public Policy*, ed. Sheldon H. Danziger, Gary D. Sandefur, and Daniel H. Weinberg. Cambridge: Harvard University Press.

Tienda, Marta, and Haya Stier. 1991. "Joblessness or Shiftlessness: Labor Force Activity in Chicago's Inner City." In *The Urban Underclass*, ed. Christopher Jencks and Paul Peterson. Washington D.C.: Brookings Institution.

Tilly, Charles. 1968. "Race and Migration to the American City." Pp. 136–157 in *The Metropolitan Enigma*, ed. James Q. Wilson. Cambridge: Harvard University Press.

Tilly, Louise A., and Joan W. Scott. 1978. *Women, Work and Family*. New York: Holt, Rinehart & Winston.

Tocqueville, Alexis de. 1945a. *Democracy in America*, vol. I. New York: Knopf.

———. 1945b. *Democracy in America*, vol. II. New York: Knopf.

———. 1955. *The Old Regime and the French Revolution*. New York: Doubleday.

Torres-Saillant, Silvio. 1989. "Dominicans as a New York Community: A Social Appraisal." *Punto 7 Review* 2: 7–25.

Tsai, Shih-Shan Henry. 1986. *The Chinese Experience in America*. Bloomington: Indiana University Press.

Tuttle, William. 1974. *Race Riot*. New York: Atheneum.

Ubelaker, Douglas H. 1976. "Prehistoric New World Population Size: Historical Review and Current Appraisal of North American Estimates." *American Journal of Physical Anthropology* 45: 661–666.

———. 1988. "North American Indian Population Size, A.D. 1500 to 1985." *American Journal of Physical Anthropology* 77: 289–294.

Ubelaker, Douglas H., and P. Wiley. 1978. "Complexity in Arikara Mortuary Practice." *Plains Anthropologist* 23: 69–74.

Upham, Steadman. 1986. "Smallpox and Climate in the American Southwest." *American Anthropologist* 88: 115–128.

——— . 1987. "Understanding the Disease History of the Southwest: A Reply to Reff." *American Anthropologist* 89: 708–710.

U.S. Bureau of the Census. 1872. *Ninth Census*, vol. I, *The Statistics of the Population of the United States*. Washington, D.C.: U.S. Government Printing Office.

——— . 1883. *Tenth Census: Statistics of the Population of the United States*. Washington, D.C.: U.S. Government Printing Office.

——— . 1895. *Eleventh Census: Report on Population of the United States*, pt. I. Washington, D.C.: U.S. Government Printing Office.

——— . 1902. *Twelfth Census of the United States, Census Reports*, vol. II, *Population*, pt. II. Washington, D.C.: U.S. Government Printing Office.

——— . 1904. *Special Reports, Occupations of the Twelfth Census*. Washington, D.C.: U.S. Government Printing Office.

——— . 1913. *Thirteenth Census of the United States*, vol. I, *Population, General Report and Analysis*. Washington, D.C.: U.S. Government Printing Office.

——— . 1922. *Fourteenth Census*, vol. II, *Population, General Report and Analytic Tables*. Washington, D.C.: U.S. Government Printing Office.

——— . 1923. *Fourteenth Census of the United States Taken in the Year 1920*, vol. IV, *Population*. Washington, D.C.: U.S. Government Printing Office.

——— . 1932. *Fifteenth Census of the United States: 1930 Population*, vol. II, p. 65. Washington, D.C.: U.S. Government Printing Office.

——— . 1933. *Fifteenth Census of the United States*, vol. II, *Population, General Report, Statistics by Subject* and vol. V, *Occupation: General Report*. Washington, D.C.: U.S. Government Printing Office.

——— . 1943a. *Sixteenth Census of the Population: 1940. Population Characteristics of the Nonwhite Population by Race*. Washington, D.C.: U.S. Government Printing Office.

——— . 1943b. *Sixteenth Census, Mother Tongue of the White Population*, Series P-15, no. 10, p. 4. Washington, D.C.: U.S. Government Printing Office.

——— . 1953. *U.S. Census of the Population: 1950*, vol. IV, *Special Reports*, pt. 3, chap. B, *Nonwhite Population by Race*. Washington, D.C.: U.S. Government Printing Office.

——— . 1960. *Historical Statistics of the United States: Colonial Times to 1957*. Washington, D.C.: U.S. Government Printing Office.

——— . 1963. *U.S. Census of Population: 1960. Subject Reports. Nonwhite Population by Race*. Final Report PC (2)-1C. Washington, D.C.: U.S. Government Printing Office.

——— . 1973a. *Census of the Population: 1970. Persons of Spanish Origin*. Washington, D.C.: U.S. Government Printing Office.

——— . 1973b. *Census of Population: 1970. Subject Reports: Japanese, Chinese, and Filipinos in the United States*. Final Report PC (2)-1G. Washington, D.C.: U.S. Government Printing Office.

——— . 1975a. *Historical Statistics of the United States, Colonial Times to 1970*, Bicentennial ed., pt. 1. Washington, D.C.: U.S. Government Printing Office.

——— . 1975b. *Historical Statistics of the United States, Colonial Times to 1970*, Bicentennial ed., pt. 2. Washington, D.C.: U.S. Government Printing Office.

——— . 1978. *Social and Economic Conditions of Negroes in the United States*. Washington, D.C.: U.S. Government Printing Office.

——— . 1982. *1980 Census of Population, General Social and Economic Characteristics*. Washington, D.C.: U.S. Government Printing Office.

——— . 1983a. *1980 Census of Population: Ancestry of the Population by State, Supplementary Report*. Washington, D.C.: U.S. Government Printing Office.

——— . 1983b. *1980 Census of Population*, vol. 1, *Characteristics of the Population*, chap. C, *General Social and Economic Characteristics*, pt. 1, *United States Summary*, PC80–1-C1. Washington, D.C.: U.S. Government Printing Office.

——— . 1983c. *1980 Census of Population, General Population Characteristics, United States Summary*. Washington, D.C.: U.S. Government Printing Office.

——— . 1983d. *1980 Census of Population, General Social and Economic Characteristics*. Washington, D.C.: U.S. Government Printing Office.

——— . 1984. *1980 Census of the Population: Detailed Population Characteristics*. Washington, D.C.: U.S. Government Printing Office.

——— . 1988a. *The Hispanic Population in the United States: March 1988 (Advance Report)*. Washington, D.C.: U.S. Government Printing Office.

——— . 1988b. *Statistical Abstracts, 1988*. Washington, D.C.: U.S. Government Printing Office.

——— . 1990a. *1990 Census Special Tabulations, Persons of Hispanic Origin*. Washington, D.C.: U.S. Government Printing Office.

——— . 1990b. *1990 Census of Population*. Summary Tape File 3-A. Washington, D.C.: U.S. Government Printing Office.

——— . 1990c. *Summary Tape File, New York City*. Washington, D.C.:U.S. Government Printing Office.

——— . 1991a. "Census Bureau Releases 1990 Census Counts on Specific Racial Groups." *U.S. Department of Commerce News*, 12 June, CB91–215.

——— . 1991b. *Current Population Reports, The Hispanic Population in the United States: March 1991*. Series P-20, no. 455. Washington, D.C.: U.S. Government Printing Office.

———. 1992a. *Census of Population: 1990. Characteristics of the Population.* Washington, D.C.: U.S. Government Printing Office.

———. 1992b. *Census of Population and Housing, 1990: Detailed Ancestry Groups for States.* Supplementary Reports CP-S-1–2. Washington, D.C.: U.S. Government Printing Office.

———. 1992c. *1990 Census of Population. General Population Characteristics: United States.* 1990 CP-1–1. Washington, D.C.: U.S. Government Printing Office.

———. 1992d. *1990 Census of Population, General Social and Economic Characteristics.* Washington, D.C.: U.S. Government Printing Office.

———. 1993a. *1990 Census of Population and Housing, 1990 PUMS.* Washington, D.C.: U.S. Government Printing Office.

———. 1993b. "1990 Census of Population and Housing." Summary Tape File 3C (District of Columbia and Washington, D.C.—MD—VA), unpublished statistics.

———. 1993c. "Population Projections of the United States, by Age, Sex, Race, and Hispanic Origin: 1993 to 2050." *Current Population Reports* P25–1104. Washington, D.C.: U.S. Government Printing Office.

———. 1993d. *We, the American Asians.* Washington, D.C.: U.S. Government Printing Office.

———. 1993e. *Asians and Pacific Islanders in the United States.* Washington, D.C.: U.S. Government Printing Office.

U.S. Commission on Civil Rights. 1971. *Mexican American Education Study, Report 1: Ethnic Isolation of Mexican Americans in the Public Schools of the Southwest.* Washington, D.C.: U.S. Government Printing Office.

———. 1992. *Civil Rights Issues Facing Asian Americans in the 1990s.* Washington, D.C.: U.S. Government Printing Office.

———. 1993. *Racial and Ethnic Tensions in American Communities: Poverty, Inequality, and Discrimination,* vol. I, *The Mount Pleasant Report.* Washington, D.C.: U.S. Government Printing Office.

U.S. Committee for Refugees. 1985. *Cambodians in Thailand: People on the Edge.* Washington, D.C.: U.S. Committee for Refugees.

———. 1986a. "Despite a Generous Spirit: Denying Asylum in the United States." American Council for Nationalities Service.

———. 1986b. *Refugees from Laos: In Harm's Way.* Washington, D.C.: U.S. Committee for Refugees.

———. 1987. *Uncertain Harbors: The Plight of Vietnamese Boat People.* Washington, D.C.: U.S. Committee for Refugees.

———. 1993. *World Refugee Survey: 1993 in Review.* Washington, D.C.: U.S. Committee for Refugees.

U.S. Congress. 1976. "Public Law 93–638: The Indian Self-Determination and Education Assistance Act." *United States Statutes at Large,* vol. 88, pt. 2.

———. 1986. *Indian Health Care.* Office of Technology Assessment. Washington, D.C.: U.S. Government Printing Office.

U.S. Department of Health, Education, and Welfare. 1978. *Cuban Refugee Program Fact Sheet.* Washington, D.C.: U.S. Government Printing Office.

———. 1979. *Healthy People: The Surgeon General's Report on Health Promotion and Disease Prevention.* Washington, D.C.: U.S. Government Printing Office.

U.S. Department of Health and Human Services. 1985. *Report of the Secretary's Task Force on Black and Minority Health.* Washington, D.C.: U.S. Government Printing Office.

U.S. GAO (General Accounting Office). 1994. *Vietnamese Amerasian Resettlement: Education, Employment, and Family Outcomes in the United States.* GAO/PEMD-94–15. Washington, D.C.: U.S. General Accounting Office.

U.S. House. 1986. "Immigration Reform and Control Act of 1986. Conference Report." 99th Cong., 2d sess., rep. 99–1000. Washington, D.C.: U.S. Government Printing Office.

U.S. INS (Immigration and Naturalization Service). 1948. *Annual Report.* Washington, D.C.: U.S. Government Printing Office.

———. 1990–1994. *Statistical Yearbook, 1990–1994.* Washington, D.C.: U.S. Government Printing Office.

———. 1991. *1990 Statistical Yearbook of the Immigration and Naturalization Service.* Washington, D.C.: U.S. Government Printing Office.

———. 1992a. *Advance Report.* Washington D.C.: U.S. Government Printing Office.

———. 1992b. *Immigration Reform and Control Act: Report on the Legalized Alien Population.* Washington, D.C.: U.S. Government Printing Office.

———. 1992c. *1991 Statistical Yearbook of the Immigration and Naturalization Service.* Washington, D.C.: U.S. Government Printing Office.

———. 1993. *1992 Statistical Yearbook.* Washington, D.C.: U.S. Government Printing Office.

U.S. ORR (Office of Refugee Resettlement). 1985. *Annual Report.* Washington, D.C.: U.S. Department of Health and Human Services.

———. 1991. *Annual Report.* Washington, D.C.: U.S. Department of Health and Human Services.

———. 1992. *Annual Report.* Washington, D.C.: U.S. Department of Health and Human Services.

U.S. Senate. 1911. *Steerage Conditions, Importation and Harboring of Women for Immoral Purposes, Immigrant Homes and Aid Societies, Immigrant Banks.* 61st Cong., 3rd sess., S. Doc. 753.

———. 1916. *Final Report and Testimony Submitted to Congress by the Commission on Industrial Relations.* 64th Cong., 2d sess., S. Doc. 415, p. 4.

———. 1976. *Revolutionary Activities Within the United States: The American Indian Movement.* Committee on the Judiciary. Washington, D.C.: U.S. Government Printing Office.

USP (University Settlement House Papers), Chicago Historical Society/Hedger, Caroline. 1908. "Health—Summer of 1908." Typed manuscript in USP Box 2, Folder 13.

USP/McDowell, Mary E. n.d. "Standard of Living." Typed manuscript in USP Box 2, Folder 13.

——. 1914a. "Beginnings." Typed manuscript in USP Box 1, Folder 3.

——. 1914b. "Civic Experience." Typed manuscript in USP Box 3, Folder 14.

Utley, Robert M. 1963. *The Last Days of the Sioux Nation*. New Haven: Yale University Press.

——. 1984. *The Indian Frontier of the American West, 1846–1890*. Albuquerque: Universtiy of New Mexico Press.

Valentine, Charles A. 1968. *Culture and Poverty*. Chicago: University of Chicago Press.

Van den Berghe, Pierre L. 1967. *Race and Racism: A Comparative Perspective*. New York: John Wiley & Sons.

Van Haitsma, Martha. 1989. "A Contextual Definition of the Underclass." *Focus* 12 (spring/summer): 27–31.

Vargas, Zaragosa. 1993. *Proletarians of the North: A History of Mexican Industrial Workers in Detroit and the Midwest, 1917–1933*. Berkeley and Los Angeles: University of California Press.

Vecoli, Rudolph. 1978. "The Coming of Age of the Italian Americans: 1945–1974." *Ethnicity* 5: 119–147.

Vega, William A., and Rubén G. Rumbaut. 1991. "Ethnic Minorities and Mental Health." *Annual Review of Sociology* 17: 351–383.

Veltman, Calvin. 1983. *Language Shift in the United States*. Berlin: Mouton.

Vigil, James Diego. n.d. "Chicano Highschoolers in a Changing Los Angeles: Educational Performance Through a Prison of Culture and Ethnic Identity." Unpublished manuscript.

——. 1974. *Early Chicano Guerrilla Fighters*. La Mirada, Calif.: Advanced Graphics.

——. 1980. *From Indians to Chicanos: The Dynamics of Mexican American Culture*. Prospect Heights, Ill: Waveland Press.

——. 1984. *From Indians to Chicanos*. Prospect Heights, Ill.: Waveland Press.

——. 1991. "History and Place in the Formation of Chicano Ethnicity." Paper presented at the Hispanic History and Culture Conference, University of Wisconsin, Milwaukee, April 25–27.

——. 1994. "Chicano and Latino Activism and Political Change." Pp. 309–327 in *Handbook of Hispanic Cultures in the United States: Anthropology*, ed. Thomas Weaver. Houston, Tex.: Arte Publico Press.

Viglucci, Andrés. 1986. "Hispanics Growing in Numbers, Clout." *Miami Herald*, 26 March 26, 1B.

Vinje, David L. 1985. "Cultural Values and Economic Development on Reservations." Pp. 155–75 in *American Indian Policy in the Twentieth Century*, ed. Vine Deloria, Jr. Norman: University of Oklahoma Press.

Vizenor, Gerald. 1976. *Tribal Scenes and Ceremonies*. Minneapolis: Nodin Press.

Vogt, Ezra Z. 1957. "The Acculturation of American Indians." *Annals* 311: 137–146.

Wade, Louise. 1987. *Chicago's Pride: The Stockyards, Packingtown and Environs in the Nineteenth Century*. Urbana: University of Illinois Press.

Wagley, Charles. 1968. "The Concept of Social Race in the Americas." Pp. 155–174 in *The Latin American Tradition*, ed. Charles Wagley. New York: Columbia University Press.

Waldinger, Roger. 1985. "Immigrant Enterprise and the Structure of the Labor Market." Pp. 213–227 in *New Approaches to Economic Life*, ed. Bryan Robert, Ruth Finnegan, and Duncan Gallie. London: Manchester University Press.

——. 1986. *Through the Eye of the Needle: Immigrants and Enterprise in the New York Garment Trade*. New York: New York University Press.

——. 1992. "When the Melting Pot Boils Over: The Irish, Jews, Blacks and Koreans of New York." Mimeo, Department of Sociology, University of California, Los Angeles.

Waldinger, Roger, and Mehdi Bozorgmehr, eds. Forthcoming. *Ethnic Los Angeles*. New York: Russell Sage.

Waldinger, Roger, Robin Ward, and Howard Aldrich. 1985. "Ethnic Business and Occupational Mobility in Advanced Societies." *Sociology* 19: 586–597.

Waldman, Lois. 1956. "Employment Discrimination Against Jews in the United States—1955." *Jewish Social Studies* 18: 208–216.

Walkosz Family. n.d. A collection of various documents and notes concerning the family's history in the possession of the author.

Walsh, Bryan O. 1971. "Cuban Refugee Children." *Journal of Inter-American Studies and World Affairs* 13: 378–415.

Walters, Ronald V. 1987. "White Racial Nationalism in the United States." *Without Prejudice* 1: 7–29.

Ward, David. 1971. *Cities and Immigrants: A Geography of Change in Nineteenth Century America*. New York: Oxford University Press.

Ware, Caroline. 1935. *Greenwich Village, 1920–1930*. Boston: Houghton Mifflin.

Warren, Robert. 1988. "Legalization Data and Other Statistical Information About Dominican Migration to the United States." Paper presented at the Conference on Dominican Migration, Fundacion Friedrich Ebert, Santo Domingo.

Warren, Robert, and Ellen Percy Kraly. 1985. *The Elusive Exodus: Emigration from the United States*. Washington, D.C.: Population Reference Bureau.

Warren, Robert, and Jeffery S. Passel. 1987. "A Count of the Uncountable: Estimates of Undocumented Aliens Counted in the 1980 United States Census." *Demography* 24: 375–393.

Wassem, Ruth Ellen. 1992. "U.S. Response to Asylum Seekers, 1946 to Present." Washington, D.C.: Library of Congress, Congressional Research Service.

Washington, Booker T. [1899] 1969. *The Future of the American Negro*. New York: New American Library.

Watanabe, Theresa. 1977. "A Report from the Japanese American Community Study." Department of Anthropology, University of Washington.

Waters, Mary C. 1990. *Ethnic Options: Choosing Identities in America*. Berkeley and Los Angeles: University of California Press.

Wattenberg, Ben J., and Richard M. Scammon. 1973. "Black Progress and Liberal Rhetoric." *Commentary* 55: 35–44.

Waxman, Chaim. 1990. "Is the Cup Half-Full or Half-Empty?: Perspectives on the Future of the American Jewish Community." Pp. 71–85 in *American Pluralism and the Jewish Community*, ed. Seymour Martin Lipset. New Brunswick, N.J.: Transaction Books.

Weatherspoon, Floyd D. 1985. *Equal Employment Opportunity and Affirmative Action*. New York: Garland Publishing.

Weaver, Robert C. 1948. *The Negro Ghetto*. New York: Russell & Russell.

Webber, Melvin. 1944. "Order in Diversity: Community without Propinquity." Pp. 23–54 in *Cities and Space: The Future Use of Urban Land*, ed. Lowdon Wingo, Jr. Baltimore: Johns Hopkins University Press.

Weber, David, ed. 1973. *Foreigners in Their Own Land*. Albuquerque: University of New Mexico Press.

Weber, Max. 1921. *Economy and Society: An Outline of Interpretive Sociology*. Ed. Guenther Roth and Claus Wittich, trans. Ephraim Fischoff. New York: Bedminster Press.

——. [1922] 1946. "The Social Psychology of the World Religions." Pp. 267–301 in *From Max Weber: Essays in Sociology,* ed. H. H. Gerth and C. Wright Mills. New York: Oxford University Press.

——. 1947. *The Theory of Social and Economic Organizations*. Ed. Talcott Parsons. New York: Free Press.

——. 1952. *The Protestant Ethic and the Spirit of Capitalism*. Trans. Talcott Parsons. New York: Scribners.

Weeks, O. Douglas. 1929. "The League of United Latin-American Citizens: A Texas-Mexican Civic Organization." *Southwestern Political and Social Science Quarterly* 10 (3): 257–278.

Weeks, John R., and Rubén G. Rumbaut. 1991. "Infant Mortality Among Ethnic Immigrant Groups." *Social Science and Medicine* 33 (3): 327–334.

Weibel-Orlando, Joan. 1991. *Indian Country, L.A.* Urbana, IL: University of Illinois Press.

Weinberg, Sydney Stahl. 1988. *The World of Our Mothers*. Chapel Hill: University of North Carolina Press.

Weisberger, Bernard A. 1994. "A Nation of Immigrants." *American Heritage* (March): 75–91.

Weiss, Richard. 1979. "Ethnicity and Reform: Minorities and the Ambience of the Depression Years." *Journal of American History* 66: 566–585.

Weisser, Michael R. 1985. *A Brotherhood of Memory: Jewish Landsmanshaften*. Ithaca, N.Y.: Cornell University Press.

Wells, Robert V. 1991. "Demographic Change and Family Life in American History: Some Reflections." Pp. 43–62 in *The Family Experience*, ed. Mark Hutter. New York: Macmillan.

Weyler, Rex. 1982. *Blood of the Land: The Government and Corporate War Against the American Indian Movement*. New York: Everest House.

Wheeler, Thomas G. 1981. *The Immigrant Experience: The Anguish of Becoming American*. New York: Penguin.

White, Richard. 1978. "The Winning of the West: The Expansion of the Western Sioux in the Eighteenth and Nineteenth Centuries." *Journal of American History* 65: 319–346.

White, Robert H. 1990. *Tribal Assets: The Rebirth of Native America*. New York: Henry Holt.

White, Shane. 1991. *Somewhat More Independent: The End of Slavery in New York City, 1770–1810*. Athens: University of Georgia Press.

Whyte, William F. 1943. "Social Organization in the Slums." *American Sociological Review* 8: 34–39.

——. 1955. *Street Corner Society*, 2d ed. Chicago: University of Chicago Press.

Wilkins, Roger. 1994. "Progress and Policy." *Michigan Quarterly Review* 33: 282–288.

Wilkinson, Charles F., and Eric R. Biggs. 1977. "The Evolution of the Termination Policy." *American Indian Law Review* 5 (1): 139–184.

Wilkinson, R. G. 1986. "Socioeconomic Differences in Mortality: Interpreting the Data on Their Size and Trends. Pp. 1–20 in *Class and Health*, ed. R. G. Wilkinson. London: Tavistock.

Williams, Brett. 1988. *Upscaling Downtown: Stalled Gentrification in Washington, D.C*. Ithaca, N.Y.: Cornell University Press.

Williams, David R. 1990. "Socioeconomic Differentials in Health: A Review and Redirection." *Social Psychology Quarterly* 52 (2): 81–99.

——. 1991. "Social Structure and the Health Behavior of Blacks." Pp. 59–64 in *Aging, Health Behaviors and Health Outcomes*, ed. K. W. Schaie, James S. House, and D. Blazer. Hillsdale, N.J.: Erlbaum.

Williams, David R., and An-Me Chung. In press. "Racism and Health." In *Health in Black America*, ed. Rose Gibson and James Jackson. New York: Russell Sage.

Williams, David R., and James S. House. 1991. "Stress, Social Support, Control and Coping: A Social Epidemiologic View." Pp. 147–172 in *Health Promotion Research: Towards a New Social Epidemiology*, ed. B.

Badura and I. Kickbusch. Copenhagen: World Health Organization.

Williams, David R., Risa Lavizzo-Mourey, and Rueben C. Warren. 1994. "The Concept of Race and Health Status in America." *Public Health Reports* 109 (1): 26–41.

Williams, David R., Leon Wilson, and An-Me Chung. 1992. "Socioeconomic Status, Psychosocial Factors and Health in Urban Guyana." *Sociological Focus* 25 (4): 279–294.

Williams, John A. 1981. "Time and Tide: Roots of Black Awareness." In *The Immigrant Experience: The Anguish of Becoming American*, ed. Thomas G. Wheeler. New York: Penguin.

Williams, Linda F. 1990. "White/Black Perceptions of the Electability of Black Political Candidates." *National Political Science Review* 2: 45–64.

Williams, Terry. 1994. "Moving Beyond the Academy." *Michigan Quarterly Review* 33: 289–294.

Williams, Terry, and William Kornblum. 1985. *Growing Up Poor*. Lexington, Mass.: D. C. Heath.

Wilson, Cynthia A., James H. Lewis, and Cedric Herring. 1991. *The 1991 Civil Rights Act: Restoring Basic Protections*. Chicago: Urban League.

Wilson, Frank H. 1992. "The Changing Distribution of the African American Population in the United States, 1980–1990." *Urban League Review* 15: 53–74.

Wilson, Kenneth L., and Alejandro Portes. 1980. "Immigrant Enclaves: An Analysis of the Labor Market Experiences of Cubans in Miami." *American Journal of Sociology* 86: 259–319.

Wilson, William Julius. 1978. *The Declining Significance of Race: Blacks and Changing American Institutions*. Chicago: University of Chicago Press.

——— . 1985. *The Declining Significance of Race*. Chicago: University of Chicago Press.

——— . 1987. *The Truly Disadvantaged: The Inner City, the Underclass, and Public Policy*. Chicago: University of Chicago Press.

——— . 1991. "Studying Inner-City Social Dislocations: The Challenge of Public Agenda Research." *American Sociological Review* 56 (February): 1–14.

——— . 1994. "The New Urban Poverty and the Problem of Race." *Michigan Quarterly Review* 33: 247–273.

Winnick, Lou. 1990. *New People in Old Neighborhoods: The Role of New Immigrants in Rejuvenating New York's Communities*. New York: Russell Sage.

Wirth, Louis. 1928. *The Ghetto*. Chicago: University of Chicago Press.

——— . 1938. "Urbanism as a Way of Life." *American Journal of Sociology* 44 (1): 3–24.

Witt, Shirley Hill. 1968. "Nationalistic Trends among American Indians." Pp. 93–127 in *The American Indian Today*, ed. Stuart Levine and Nancy O. Lurie. New York: Penguin.

Wittke, Carl. 1957. *The German-Language Press in America*. Lexington: University of Kentucky Press.

Wolf, Eric. 1959. *Sons of the Shaking Earth*. Chicago: University of Chicago Press.

Wollenberg, C. 1974. "Mendez v. Westminster: Race, Nationality and Segregation in California Schools." *California Historical Society Quarterly* 53: 317–322.

——— . 1978. *All Deliberate Speed: Segregation and Exclusion in California Schools, 1855–1975*. Berkeley and Los Angeles: University of California Press.

Wong, Bernard. 1987. "The Chinese: New Immigrants in New York's Chinatown." Pp. 243–271 in *New Immigrants in New York*, ed. Nancy Foner. New York: Columbia University Press.

Woo, Debarah. 1995. "The Gay Between Striving and Achieving: The Case of Asian American Women." Pp. 218–227 in *Race, Class, and Gender: An Anthology*, ed. Margaret L. Andersen and Patricia Hill Collins. Belmont, Calif.: Wadsworth Publishing.

Woodham-Smith, Cecil. 1962. *The Great Hunger: Ireland, 1845–1849*. New York: Harper & Row.

Woodrow, K. A., and J. S. Passel. 1990. "Post-IRCA Undocumented Immigration to the United States: An Assessment Based on the June 1988 CPS." Pp. 33–75 in *Undocumented Migration to the United States*, ed. Frank D. Bean, Barry Edmonston, and Jeffrey S. Passel. Washington, D.C.: Urban Institute.

Woodward, C. Vann. 1951. *The Origins of the New South: 1877–1913*. Baton Rouge: Louisiana State University Press.

——— . [1955] 1974. *The Strange Career of Jim Crow*. 3rd. ed. New York: Oxford University Press.

Worthman, Paul, and James Green. 1971. "Black Workers in the New South." In *Key Issues in the Afro-American Experience,* eds. Nathan I. Huggins, Martin Kelson, and Daniel Fox. New York: Harcourt Brace Jovanovich.

Woofter, Thomas. 1919. "Migration of Negroes from Georgia." Pp. 75–91 in *Negro Migration in 1916–1917*, ed. George Haynes. Washington, D.C.: U.S. Government Printing Office.

Wright, Gavin. 1986. *Old South, New South: Revolutions in the Southern Economy Since the Civil War*. New York: Basic Books.

Wright, Lawrence. 1994. "One Drop of Blood." *The New Yorker,* July 15, pp. 46–55.

Wright, Richard. 1969. *12 Million Black Voices*. New York: Amo Press.

Wrigley, E. A. 1969. *Population and History*. London: Weidenfeld and Nicolson, World University Library.

Wu, Yuan-li, ed. 1980. *The Economic Condition of Chinese Americans*. Chicago, Ill.: Pacific/Asian American Mental Health Research Center.

Yancey, William L., Eugene P. Ericksen, and Richard N. Juliani. 1976. "Emergent Ethnicity: A Review and Reformulation." *American Sociological Review* 41: 391–403.

——— . 1977. "Reply to Levine and Bergesen." *American Sociological Review* 42: 825–826.

Yanagisako, Sylvia Junko. 1975. "Two Processes of Change in Japanese-American Kinship." *Journal of Anthropological Research* 31: 196–224.

Yans-McLaughlin, Virginia. 1977. *Family and Community: Italian Immigrants in Buffalo, 1880–1930*. Ithaca, N.Y.: Cornell University Press.

Yinger, J. Milton. 1985. "Assimilation in the United States: The Mexican Americans." In *Mexican Americans in Comparative Perspective*, ed. Walker Connor. Washington D.C.: Urban Institute.

Yoon, In-Jin. 1991. "The Changing Significance of Ethnic and Class Resources in Immigrant Businesses: The Case of Korean Immigrant Businesses in Chicago." *International Migration Review* 25: 303–332.

——. 1993. "The Social Origin of Korean Immigration to the United States from 1965 to the Present." Paper no. 121, East-West Population Institute, East-West Center, University of Hawaii.

Yuan, D. Y. 1969. "Division of Labor Between Native-Born and Foreign-Born Chinese in the United States: A Study of Their Traditional Employments." *Phylon* 30: 160–169.

Yung, Judy. 1986. *Chinese Women in American: A Pictorial History*. Seattle: University of Washington Press.

Zeitlin, Maurice. 1966. "Political Generations in the Cuban Working Class." *American Journal of Sociology* 71: 493–508.

Zenner, Walter. 1991. *Minorities in the Middle: A Cross-Cultural Analysis*. Albany: New York State University Press.

Zhou, Min. 1992. *Chinatown: The Socioeconomic Potential of an Urban Enclave*. Philadelphia: Temple University Press.

Zolberg, Aristide R., Astri Suhrke, and Sergio Aguayo. 1989. *Escape from Violence: Conflict and the Refugee Crisis in the Developing World*. New York: Oxford University Press.

Zucker, Norman L., and Naomi Flink Zucker. 1987. *The Guarded Gate: The Reality of American Refugee Policy*. New York: Harcourt Brace Jovanovich.

——. 1992. "From Immigration to Refugee Redefinition: A History of Refugee and Asylum Policy in the United States." Pp. 54–70 in *Refugees and the Asylum Dilemma in the West*, ed. Gil Loescher. University Park: Pennsylvania State University Press.

About the Authors

Richard D. Alba was born and raised in the Bronx and attended Columbia University in New York City, where he received his Ph.D. in 1974. He is currently professor of sociology and public policy at the State University of New York at Albany. He is the author of *Ethnic Identity: The Transformation of White America* and *Italian Americans: Into the Twilight of Ethnicity*. He has written numerous articles on ethnicity, especially among the descendants of immigrants from Europe, and on methodology. He has twice visited the Federal Republic of Germany as a Fulbright scholar.

Frederick M. Binder is professor of history and chair of the history department at the College of Staten Island, City University of New York. He is a graduate of Boston University and earned his doctorate from Columbia University. He is the author of *The Color Problem in Early National America as Viewed by John Adams, Jefferson and Jackson; The Age of the Common School, 1830–1865*; and coauthor with David Reimers of *All the Nations Under Heaven: An Ethnic and Racial History of New York City*.

Mehdi Bozorgmehr was born and raised in Tehran, Iran. After completing high school, he went to England and then came to the United States in pursuit of higher education. He has received all of his degrees in California, including a Ph.D. in sociology from UCLA. He has taught at San Diego State University and at UCLA, and is currently an assistant professor of sociology at the City College, City University of New York. Over the last decade he has conducted three major research projects on immigration, ethnicity, and economy in Los Angeles, based on which he has published several articles and book chapters. His most relevant publication to his chapter in this volume is *Ethnic Los Angeles* (coedited with Roger Waldinger), published by the Russell Sage Foundation in 1996.

Benjamin V. Cariño is currently dean and professor, School of Urban and Regional Planning, University of the Philippines. He received his Ph.D. from Indiana University. He has authored and coauthored numerous monographs, journal articles, and book chapters on such subjects as urbanization and development planning, internal and international migration, education for planners, and other related topics. He is the coeditor (with James T. Fawcett) of *Pacific Bridges: The New Immigrants from Asia and the Pacific Islands*, published by the Center for Migration Studies in New York.

Héctor Carrasquillo was born in Puerto Rico and is assistant professor in the department of Puerto Rican studies and director of the Center for Latino Studies at Brooklyn College, City University of New York. He earned his doctorate from Syracuse University. His current work focuses on the family and Latino aged. Among his most recent publications is a chapter on "The Puerto Rican Family" in *Minority*

Families in the United States: A Multicultural Perspective, edited by Ronald L. Taylor.

Leo R. Chavez is associate professor of anthropology at the University of California, Irvine. He received his Ph.D. from Stanford University. Born in Alamogordo, New Mexico, he is a thirteenth-generation New Mexican on his father's side, and fourth-generation on his mother's side. He was a research associate for the Center for U.S.–Mexican Studies at the University of California, San Diego, and at El Colegio de la Frontera Norte in Tijuana, Mexico. He has conducted research on many aspects of the immigrant experience, particularly of undocumented immigrants from Mexico and Central America. He is the author of *Shadowed Lives: Undocumented Immigrants in American Society* (1992), and has also published numerous articles on immigrant health issues, families and households, and formation of community consciousness.

Esther Ngan-ling Chow is professor of sociology at the American University in Washington, D.C. She received her doctorate in sociology from UCLA. She is a feminist scholar, researcher, and community activist who has conducted studies on the acculturation of Asian American women, the work and family life of Chinese American women, labor stratification in the high-tech industries of Taiwan, and ethnic transformation of Washington, D.C.'s Chinatown. She was one of the guest editors of a special issue of *Gender & Society* (1992) on race, class, and gender. She is also coeditor and coauthor of *Women, the Family, and Policy: A Global Perspective* (1994) and *Common Bonds, Different Voices: Race, Class, and Gender* (forthcoming).

Stephen Cornell is associate professor of sociology at the University of California, San Diego. He received his Ph.D. from the University of Chicago. He is the author of *The Return of the Native: American Indian Political Resurgence*, and coeditor of *What Can Tribes Do? Strategies and Institutions in American Indian Economic Development*. He is engaged in research on the nature of ethnic ties and the dynamics of ethnic group formation, and also serves as codirector of the Harvard Project on American Indian Economic Development, an extended study of poverty and development on American Indian reservations.

Hasia Diner is professor of American studies at the University of Maryland at College Park. She is the author of *In the Almost Promised Land: American Jews and Blacks, 1915–1935*, *Erin's Daughters in America: Irish Immigrant Women in the Nineteenth Century*, and *A Time for Gathering: The Second Migration, 1820–1880*, which is the second volume in the Johns Hopkins University series, *The Jewish People in America*. She is currently involved in a comparative study of immigrant adaptation in America through the medium of foodways and diet.

Evelynn Nakano Glenn is chair of women's studies and professor of ethnic studies and women's studies at the University of California, Berkeley. The granddaughter of Japanese immigrants, she spent her first three years in an American concentration camp, along with members of her extended family. The impacts of this experience fueled her interest in immigration, racism, and labor exploitation. These issues are explored in depth in her book *Issei, Nisei, War Bride: Three Generations of Japanese American Women in Domestic Service*. Her other publications include *Mothering: Ideology, Experience and Agency*, and numerous articles on women, work, and technology; racial ethnic women's labor; and the political economy of families and households.

Steven J. Gold is associate professor of sociology at Michigan State University and Senior Fellow at the Wilstein Institute. He received his Ph.D. from the University of California, Berkeley in 1985, and taught at Whittier College. He has published articles on immigrant adaptation, ethnic self-employment, and community development in *Ethnic and Racial Studies*, *International Migration Review*, *Research in Community Sociology*, *The American Jewish Yearbook*, and the *Western Journal of Medicine*, as well as several edited volumes. The former president of the International Visual Sociology Association, his book *Refugee Communities: A Comparative Field Study* was published by Sage Press in 1992. He is currently involved in studies of ethnic philanthropy and Israeli immigrants in Los Angeles.

Sherri Grasmuck, a native Texan, of German, Scots-Irish-Welsh, and a wee-bit of Cherokee-Choctaw stock, received her Ph.D. from the University of Texas at Austin. She has served as director of women's studies at Temple University and is currently an associate professor of sociology there. She is the coauthor of *Between Two Islands: Dominican International Migration*, and more recently has conducted research on household structure, gender, and microentrepreneurs in the Dominican Republic.

Guillermo J. Grenier is the director of the Florida Center for Labor Research and Studies and associate professor of sociology at Florida International University. Born in Havana, Cuba, he received his undergraduate education at Emory University and Georgia State University in Atlanta, and his Masters and Ph.D. from the University of New Mexico in Albuquerque. He is the author of *Inhuman Relations: Quality Circles and Anti-Unionism in American Industry*, and coauthor of *Employee Participation and Labor Law in the American Workplace* (with Ray Hogler). He coedited *Miami Now! Immigration, Ethnicity and Social Change* (with Alex Stepick), and *Newcomers in the Workplace: Immigrants and the Restructuring of the American Workplace* (with Louise Lamphere and Alex Stepick). He is currently completing work on his upcoming volume *This Land is Our Land: Newcomers and Established Residents in Miami*.

Cedric Herring is professor of sociology at the University of Illinois at Chicago, and faculty associate at the University of Chicago. He received his Ph.D. from the University of Michigan in 1985. He is the author of *Splitting the Middle: Political Alienation, Acquiescence, and Activism,* and several articles on such issues as discrimination over the life course (published in *Social Problems*), skin tone and stratification in the Black community (*American Journal of Sociology*), ethnicity and entrepreneurship (*Sociological Perspectives*), and political responses to underemployment (*National Political Science Review*). He is a former Summer Fellow at the Center for Advanced Studies in the Behavioral Sciences, and was the president of the Association of Black Sociologists in 1994–1995.

Walter D. Kamphoefner, a Missouri-German farm boy, earned his Ph.D. in 1978 at the University of Missouri and presently teaches immigration history at Texas A&M University. He has authored *The Westfalians: From Germany to Missouri* (1985), and coedited *News from the Land of Freedom: German Immigrants Write Home* (1991). His current research is exemplified by "German-American Bilingualism: *Cui Malo*? Mother Tongue and Socioeconomic Status among the Second Generation in 1940," in *International Migration Review* (Winter 1994).

Joanne Leonard is professor of art at the School of Art, University of Michigan. Her work has been exhibited and reproduced widely. Exhibitions include museums and galleries including the George Eastman House (Rochester, N.Y.), San Francisco Museum of Modern Art, Whitney Museum Downtown (NYC), Museum of Contemporary Crafts (NYC), and the Detroit Institute of Art. Publications include, most notably, Time Life's *Library of Photography* (1970), Lucy Lippard's *From the Center* (1976), H. W. Janson's *History of Art* (1986), and Helen Gardner's *Art Through the Ages* (1991).

Ivan Light is professor of sociology at the UCLA. He has long-standing research interests in internal and international immigration, the ethnic economy, entrepreneurship of ethnic and racial minorities, ethnic conflict and accommodation, urban sociology, and organized crime. His books include *Ethnic Enterprise in America* (1972), *Cities in World Perspective* (1983), *Immigrant Entrepreneurs: Koreans in Los Angeles* (with Edna Bonacich, 1988), *Immigration and Entrepreneurship* (with Parminder Bhachu, 1993), and *Race, Ethnicity, and Entrepreneurship in Urban America* (with Carolyn Rosenstein, forthcoming). His current research concerns Iranians in Los Angeles. The Italian Mafia Commission recently translated his organized crime articles into Italian for publication in the commission archives.

Robert D. Manning is assistant professor of sociology at American University. A past Fulbright lecturer to Mexico, he received his Ph.D. from The Johns Hopkins University.

He is coauthor of *What's Debt Got To Do With It?* (Basic Books, 1996), and is preparing a manuscript on the social and economic consequences of U.S. industrial restructuring. He has published articles on Mexican and Caribbean migration to the United States as well as the role of immigrant/minority entrepreneurship in ethnic community development. Presently he is conducting research on the changing patterns of multicultural relations in the postindustrial society.

Carole C. Marks is director of the Black American Studies Program at the University of Delaware, where she also holds a joint appointment in the sociology department. She received her Ph.D. from New York University, has held a National Institute of Mental Health postdoctoral fellowship at Duke University, and a research fellowship at the W. E. B. DuBois Institute at Harvard University. The author of numerous publications and professional papers, she is well known for her 1989 book, *Farewell, We're Good and Gone: The Great Black Migration*. She has been a member of the Board of Managers of Haverford College since 1983, is currently serving as the secretary of the National Association of Black Sociologists, and was recently appointed to the State of Delaware Human Relations Commission.

Pyong Gap Min is professor of sociology at Queens College and the graduate school of the City University of New York. His research interests include Korean small business and ethnic solidarity, immigrant families, and ethnic attachment among Korean and other Asian American groups. He is the author of *Ethnic Enterprise: Korean Small Business in Atlanta* (1988) and *Middlemen in Contemporary America: Koreans in New York and Los Angeles, 1970–1994* (forthcoming). He edited *Asian Americans: Contemporary Trends and Issues* (1995), to which he contributed several chapters. He received a bachelor's degree from Seoul National University (in Korea) and a master's degree from Georgia State University, both in history, and two Ph.D. degrees from Georgia State University, one in educational philosophy and the other in sociology.

Aldon Morris is professor of sociology and chair of the department of sociology at Northwestern University. He received his Ph.D. from the State University of New York at Stony Brook. His first academic appointment was at the University of Michigan. He is the author of *The Origins of the Civil Rights Movement: Black Communities Organizing for Change*. He is the coeditor of *Frontiers in Social Movement Theory*, and has authored numerous publications and professional papers. He believes that scholarship and efforts to achieve positive social change are enriched when they remain in close dialogue.

Michael Omi is associate professor of ethnic studies and chair of the Asian American studies program at the University of California, Berkeley. He received an A.B. in sociology from the University of California, Berkeley, and a Ph.D. in sociology from the University of California, Santa Cruz. Along with Howard Winant, he is the author of *Racial Formation in the United States* (2nd edition, 1994) and numerous articles on racial theory and politics. He has also written about right-wing political movements, Asian Americans and race relations, and race and popular culture. In 1990 he was the recipient of Berkeley's Distinguished Teaching Award.

Dominic A. Pacyga received his Ph.D. in history from the University of Illinois at Chicago in 1981. He has coauthored two books concerning Chicago's neighborhoods and has written a study of two Polish working-class community's on the city's South Side during the period 1880 to 1922 titled *Polish Immigrants and Industrial Chicago* (1991). Pacyga is currently a faculty member in the Liberal Education Department at Columbia College in Chicago.

Rhacel Parreñas immigrated in the mid-1980s to the United States from the Philippines at the age of thirteen. A doctoral candidate in ethnic studies at the University of California at Berkeley, she is currently coediting the second volume of the anthology by and about Asian American women, *Making Waves*. Her research interests include women of color, feminism, and women and migration.

Silvia Pedraza was born in Havana, Cuba, is a graduate of the University of Michigan, and earned her doctorate from the University of Chicago. She taught at Washington University in St. Louis, and at the University of Michigan where at present she is associate professor of sociology. She is the author of *Political and Economic Migrants in America: Cubans and Mexicans* (University of Texas Press, 1985), and recently was guest editor of a special issue of *Social Problems* on "Immigration, Race, and Ethnicity in America" (February 1994). She is currently working on a study of Cuba's revolution and exodus.

Lisandro Pérez is associate professor and director of the Cuban Research Institute at Florida International University. He received his B.A. from the University of Miami and his M.A. and Ph.D. from the University of Florida. He taught at Louisiana State University before joining FIU in 1985 as chair of the sociology and anthropology department. His research has been in the areas of demographics and social change in Cuba, and on the dynamics of the Cuban community in the United States. His writings have appeared in several academic journals as well as in edited volumes, and his columns on Cuba and Cuban Americans have been published in the *Miami Herald*, the *New York Times*, the *London Observer*, and other newspapers. He was born in Havana, Cuba, and migrated to the United States in 1960 at the age of eleven.

Patricia R. Pessar is associate professor of American studies and anthropology at Yale University. She is the coauthor of *Between Two Islands: Dominican International Migration* (1991), author of *A Visa for a Dream: Dominicans in New York* (forthcoming), and editor of *When Borders Don't Divide: Labor Migration and Refugee Movements in the Americas* (1988).

Bruce Phillips is professor of Jewish communal studies at Hebrew Union College in Los Angeles. His published research on American Jews spans social history, demography, and ethnography. He is a member of the National Technical Advisory Committee for the National Jewish Population Survey of 1990 and for the year 2000. He is currently doing research on Jewish intermarriage.

Alejandro Portes was born in Havana, Cuba, and educated in universities in Havana, Buenos Aires, and the Midwest. He received his Ph.D. from the University of Wisconsin-Madison in 1970, and has taught at the University of Illinois, the University of Texas at Austin, Duke, and Johns Hopkins. He is currently John Dewey professor of sociology and international relations and chair of the sociology department at Johns Hopkins. His recent books include *City on the Edge: The Transformation of Miami* (with Alex Stepick), *En Torno a la Informalidad* (Editorial Porrua, Mexico City), and *The Economic Sociology of Immigration*.

The late **George P. Rawick** was professor of sociology at Washington University, St. Louis, and of history at the University of Missouri-St. Louis. He wrote *From Sundown to Sunup: The Making of the Black Community* (1972) as the lead volume to accompany 18 other volumes of oral histories of 2,000 former slaves, conducted largely by the Federal Writer's Project of the Work Progress Administration in the 1930s. Over a 20-year period, Rawick single-handedly collected and readied these narratives for publication, preserving their authenticity. As teacher, activist, and scholar, Rawick influenced and inspired many scholars today, such as George Lipsitz and Dave Roediger, whose focus on race, like Rawick's, is part of their interest in the development of the American working class.

David M. Reimers is professor of history at New York University. A specialist in social history and American ethnic history, he has also taught at the City University of New York. He is the author of *Still the Golden Door: The Third World Comes to America* (2nd edition, 1992), and the coauthor of *Ethnic Americans, Natives and Strangers*, and most recently (with Frederick Binder) *All the Nations Under Heaven: An Ethnic and Racial History of New York City* (1995). He is currently working on a study of American nativism since 1960.

Ricardo Romo, a native of San Antonio, Texas, completed his undergraduate studies at the University of Texas at Austin and earned his doctorate in history at UCLA. He has taught at the University of California, San Diego, and the University of Texas at Austin, where he currently serves as vice provost. He has also held appointments as the chancellor's distinguished lecturer at the University of California, Berkeley, and as a fellow at the Center for Advanced Studies in the Behavioral Sciences at Stanford University. His principal teaching and research areas concern twentieth-century American social and legal history, with a current focus on major Latino civil rights legislation. His book, *East Los Angeles: History of a Barrio*, a University of Texas Press publication, recently went into its seventh printing.

Rubén G. Rumbaut is professor of sociology at Michigan State University. He received his Ph.D. from Brandeis University, and has taught at the University of California, San Diego, and San Diego State University. He is the coauthor of *Immigrant America: A Portrait* (with Alejandro Portes), and coeditor of *California's Immigrant Children: Theory, Research, and Implications for Educational Policy* (with Wayne A. Cornelius). He has published widely on Asian and Latin American immigrants in the United States, particularly on refugees from Vietnam, Laos, and Cambodia. A native of Havana, Cuba, he is the founding chair of the Section on International Migration of the American Sociological Association, and serves on research committees on immigration issues of the National Academy of Sciences and the Social Science Research Council.

Georges Sabagh is professor of sociology at UCLA. He recently authored a book chapter on "Los Angeles, A World of New Immigrants: An Image of Things to Come?" He has conducted several research projects in Los Angeles, including a survey of the adaptation of Iranian immigrants and another survey of the growth of Mexican American families. Based on these studies, he has published a wide range of papers on the following topics: Iranian exiles and immigrants, ethno-religiosity of Iranian Muslims, the Iranian ethnic economy, Armenian ethnicity, Chicanos' attitudes toward abortions, and the effects of religiosity and ethnicity on Chicano fertility.

Virginia Sánchez Korrol is professor and chairperson of the department of Puerto Rican studies at Brooklyn College. She authored from *Colonia to Community: The History of Puerto Ricans in New York City*, and coedited *The Puerto Rican Struggle: Essays on Survival in the U.S.*, and *The Way It Was and Other Writings*. In addition, she researches the experiences of women in Puerto Rican and Latin American societies. She is the founding president of the Puerto Rican Studies Association, and was appointed by the New York State regents and the commissioner of education to the Social Studies Syllabus Review and Development Commission, charged with laying the foundations for the revision of the state's social studies framework.

Richard Schauffler is a doctoral student in the Department of Sociology at The Johns Hopkins University who has conducted extensive fieldwork among immigrant families in Miami and San Diego. After eight years of community and trade union organizing in San Francisco, from 1984 to 1990 he organized and chaired an annual academic exchange program with sociologists in Bulgaria focused on workers' self-management and economic reform. He is the coauthor of *The Iron Fist and the Velvet Glove: An Analysis of the U.S. Police*, and of various articles on immigration in the United States as well as on the informal economy in Latin America. He is currently on leave and working as a private investigator in San Francisco.

Howard G. Schneiderman is professor of sociology at Lafayette College. He has written widely about politics and social class, and most recently about the Protestant Establishment in three volumes that he has edited: *The Protestant Establishment Revisited, Hindrances to Good Citizenship*, and *Judgment and Sensibility*. He has also been active in social science publishing, and is currently executive editor of the interdisciplinary social science journal *SOCIETY*, and serves on the editorial board and board of directors of Transaction Publishers.

Howard Schuman is a research scientist in the Institute for Social Research and a professor of sociology, both at the University of Michigan. Together with Charlotte Steeh and others, he is completing a revision of the book *Racial Attitudes in America: Trends and Interpretations*, first published in 1985 (Harvard University Press). He recently wrote a chapter on "Attitudes, Beliefs, and Behavior" for the edited volume *Sociological Perspectives on Social Psychology*, and he continues to write in two other areas: "generations and collective memory" and the "question-answer process in surveys."

C. Matthew Snipp is professor of rural sociology and sociology, and director of the American Indian studies program at the University of Wisconsin-Madison. He has been a research fellow of the U.S. Bureau of the Census and a fellow at the Center for Advanced Studies in the Behavioral Sciences at Stanford University. Snipp has published numerous works on American Indian demography, economic development, poverty, and unemployment. He is the author of *American Indians: The First of This Land* and *Public Policy Impacts on American Indian Economic Development*. His current research and writing deals with poverty and unemployment on American Indian reservations, and American Indian ethnic identity. His tribal heritage is Oklahoma Cherokee and Choctaw.

Charlotte Steeh is director of the Detroit Area Study in the Department of Sociology at the University of Michigan. After receiving her Ph.D. from the University of Michigan, she conducted research for the Survey Research Center, University of Michigan, that traced changes in response rates to sample surveys over approximately 30 years. She coauthored *Racial Attitudes in America: Trends and Inter-*

pretations and has also published articles on survey methodology and white attitudes toward racial policies. She is currently a consulting editor of the *American Journal of Sociology* and on the editorial board of *Public Opinion Quarterly*.

Haya Stier is a lecturer in sociology and labor studies at Tel Aviv University in Israel. She received her Ph.D. in sociology from the University of Chicago. Her work deals with the incorporation of women and minority groups into the labor force, with special attention to institutional constraints on their market activity and structural barriers to socioeconomic equality.

Russell Thornton is professor of anthropology at UCLA. He has taught previously at the University of Pennsylvania, the University of Minnesota, and the University of California, Berkeley. He received his Ph.D. in sociology from Florida State University. Born in Oklahoma, he is a registered member of the Cherokee Nation of Oklahoma. Among his publications are *We Shall Live Again* (1986), *American Indian Holocaust and Survival* (1987), and *The Cherokees: A Population History* (1990).

Marta Tienda is Ralph Lewis professor of sociology and chair of the sociology department at the University of Chicago. She received her Ph.D. from the University of Texas at Austin. She is the coauthor of *The Hispanic Population of the United States* (with Frank D. Bean), coeditor of *Divided Opportunities* (with Gary D. Sandefur) and *Hispanics in the U.S. Economy* (with George J. Borjas), and former editor of the *American Journal of Sociology*. She has published extensively in areas of race and gender inequality, persistent poverty and welfare participation, the economic and social consequences of immigration, and various aspects of the sociology of economic life. Currently she is researching ethnic variation in the school to work transition (with V. Joseph Hotz) and the contextual determinants of minority educational achievement and entrepreneurship in Chicago's ethnic neighborhoods.

George Vargas was born in 1948 in Carrizo Springs, Texas, of Mexican American parents who, much like many other Mexicans, migrated north in search of better economic opportunity. After following the migrant stream to the Midwest, his family settled down in Michigan to begin a new life as sharecroppers on a large farm. Eventually his mother found steady work as a laundress and his father as a factory worker. After high school, he joined his father working in factories. He later attended the University of Michigan in Ann Arbor, where he earned his B.F.A. and Ph.D. Presently he is assistant professor of Latin American art history at the University of Texas at El Paso, where he is researching Chicano murals and exploring borderland art. He is also a visual artist, producing assemblages and collages.

James Diego Vigil is professor of anthropology and research fellow at the Center for Multiethnic and Transnational Studies, University of Southern California. A native of Los Angeles, where he grew up in a multicultural community near the coliseum and was an active participant in the Chicano movement of the 1960s, he received his doctorate from UCLA in 1976. He has taught at a variety of levels, from elementary, junior and senior high school to community college and university, and was chair of Chicano studies at the University of Wisconsin-Madison. His interest in youth issues and his research on street gangs and the role of street socialization are reflected in numerous publications, especially in his books *Barrio Gangs*, *Chicano Highschoolers in a Changing Los Angeles: Educational Performance Through a Prism of Culture and Ethnic Identity*, and *A Rainbow of Gangs* (forthcoming).

Mary C. Waters is associate professor of sociology at Harvard University. She received a B.A. in philosophy from The Johns Hopkins University, an M.A. in demography, and an M.A. and Ph.D. in sociology from the University of California, Berkeley. She has written two books on white ethnics in the United States, *From Many Strands: Ethnic and Racial Groups in Contemporary America* (with Stanley Lieberson) and *Ethnic Options: Choosing Identities in America*. She is the author of numerous articles on immigration and ethnicity in the United States. She has been a Guggenheim fellow, a visiting scholar at the Russell Sage Foundation, and she is a member of the International Migration Committee of the Social Science Research Council.

David R. Williams is associate professor of sociology, and associate research scientist in the Institute for Social Research, at the University of Michigan. He holds masters' degrees in public health and religion, and received his doctorate from the University of Michigan. His previous academic appointment was at Yale University. His research has focused on socioeconomic status differences in health and the health of the African American population. He is an associate editor of *Ethnicity and Disease*, and is the secretary-treasurer of the Medical Sociology Section of the American Sociological Association. He is a member of the National Committee on Vital and Health Statistics and chairs its subcommittee on minority and other special populations. He is also a member of the National Academy of Sciences panel and the National Science Foundation's Board of Overseers for the General Social Survey.

Howard Winant teaches sociology and Latin American studies at Temple University in Philadelphia. He is the coauthor (with Michael Omi) of *Racial Formation in the United States: From the 1960s to the 1990s* (2nd edition, 1994). He has also written *Racial Conditions: Politics, Theory, Comparisons* (1994) and *Stalemate: Political Economic Origins of Supply-Side Policy* (1988).

Index